BMVC92

Proceedings of the British Machine Vision
Conference, organised by the British
Machine Vision Association

22-24 September 1992
Leeds

Edited by
David Hogg and Roger Boyle

Springer-Verlag
London Berlin Heidelberg New York
Paris Tokyo Hong Kong
Barcelona Budapest

David Hogg, BSc, MSc, DPhil
and
Roger Boyle, BA, PhD
School of Computer Studies, University of Leeds, Leeds LS2 9JT, UK

British Library Cataloguing in Publication Data
BMVC92: Proceedings of the British Machine Vision Conference,
Organised by the British Machine Vision Association, 22-24 September
1992, University of Leeds
  I. Hogg, David   II. Boyle, Roger
  006.37
  ISBN-13: 978-3-540-19777-5

Library of Congress Cataloging-in-Publication Data
British Machine Vision Conference (1992: University of Leeds)
BMVC92: proceedings of the British Machine Vision Conference, 22-24
September, University of Leeds/organised by the British Machine Vision
Association; edited by David Hogg and Roger Boyle.
    p.   cm.
Includes bibliographical references and index.
ISBN-13: 978-3-540-19777-5   e-ISBN-13: 978-1-4471-3201-1
DOI: 10.1007/ 978-1-4471-3201-1
1. Computer vision–Congresses. I. Hogg, David. II. Boyle, Roger.
III. British Machine Vision Association   IV. Title.
TA 1632.B75 1992   92-25630
006.3'7–dc20   CIP

Typesetting: Camera ready by author
34/3830-543210 Printed on acid-free paper

Foreword

This book contains the 61 papers that were accepted for presentation at the 1992 British Machine Vision Conference. Together they provide a snapshot of current machine vision research throughout the UK in 24 different institutions. There are also several papers from vision groups in the rest of Europe, North America and Australia. At the start of the book is an invited paper from the first keynote speaker, Robert Haralick.

The quality of papers submitted to the conference was very high and the programme committee had a hard task selecting around half for presentation at the meeting and inclusion in these proceedings. It is a positive feature of the annual BMVA conference that the entire process from the submission deadline through to the conference itself and publication of the proceedings is completed in under 5 months.

My thanks to members of the programme committee for their essential contribution to the success of the conference and to Roger Boyle, Charlie Brown, Nick Efford and Sue Nemes for their excellent local organisation and administration of the conference at the University of Leeds.

David Hogg
University of Leeds
July 1992

Acknowledgement

Cover illustration: The photograph used is a radar image of volcanoes and fractures on Venus, obtained by the NASA spaceprobe "Magellan" (NASA Jet Propulsion Laboratory), courtesy of University of London Observatory Planetary Image Centre.

BMVC92

Programme Committee

Bill Adaway
Computer Recognition Systems Ltd

Roger Boyle
University of Leeds

Bernard Buxton
GEC Research Centre

Adrian Clark
University of Essex

Tim Ellis
City University

Edwin Hancock
University of York

David Hogg
University of Leeds

John Illingworth
University of Surrey

Josef Kittler
University of Surrey

Andrew Sleigh
Defence Research Agency

Geoff Sullivan
University of Reading

Chris Taylor
University of Manchester

Margaret Varga
Defence Research Agency

Mick Brown
British Aerospace

Conference Chair:	David Hogg
Local Organisation:	Roger Boyle
Administration:	Charlie Brown
Exhibition:	Nick Efford

Contents

Contents

Performance Characterization in Computer Vision

Robert M. Haralick
University of Washington
Seattle WA 98195

Abstract

Computer vision algorithms are composed of different sub-algorithms often applied in sequence. Determination of the performance of a total computer vision algorithm is possible if the performance of each of the sub-algorithm constituents is given. The problem, however, is that for most published algorithms, there is no performance characterization which has been established in the research literature. This is an awful state of affairs for the engineers whose job it is to design and build image analysis or machine vision systems.

This suggests that there has been a cultural deficiency in the computer vision community: computer vision algorithms have been published more on the merit of an experimental or theoretical demonstration suggesting that some task can be done, rather than on an engineering basis. Such a situation was tolerated because the interesting question was whether it was possible at all to accomplish a computer vision task. Performance was a secondary issue.

Now, however, a major interesting question is how to quickly design machine vision systems which work efficiently and which meet requirements. To do this requires an engineering basis which describes precisely what is the task to be done, how this task can be done, what is the error criterion, and what is the performance of the algorithm under various kinds of random degradations of the input data.

In this paper, we discuss the meaning of performance characterization in general, and then discuss the details of an experimental protocol under which an algorithm performance can be characterized.

1 Introduction

A major interesting question is how to quickly design machine vision systems which work efficiently and which meet requirements. To do this requires an engineering basis which describes precisely what is the task to be done, how this task can be done, what is the error criterion, and what is the performance of the algorithm under various kinds of random degradations of the input data. To accomplish this in the general case means propagating random perturbations through each algorithm stage in an open loop systems manner. To accomplish this for adaptive algorithms requires being able to do a closed loop engineering analysis. To perform a closed loop engineering analysis requires first doing an open loop engineering analysis and closing the loop by adding a constraint relation and solving for the output and the output random perturbation parameters.

The purpose of this discussion is to raise our sensitivity to these issues so that our field can more rapidly transfer the research technology to a factory floor technology. To initiate this dialogue, we will first expand on the meaning of performance characterization in general, and then discuss the experimental protocol under which an algorithm performance can be characterized.

2 Performance Characterization

What does performance characterization mean for an algorithm which might be used in a machine vision system? The algorithm is designed to accomplish a specific task. If the input data is perfect and has no noise and no random variation, the output produced by the algorithm ought also to be perfect. Otherwise, there is something wrong with the algorithm.

So measuring how well an algorithm does on perfect input data is not interesting. Performance characterization has to do with establishing the correspondence of the random variations and imperfections which the algorithm produces on the output data caused by the random variations and the imperfections on the input data. This means that to do performance characterization, we must first specify a model for the ideal world in which only perfect data exist. Then we must give a random perturbation model which specifies how the imperfect perturbed data arises from the perfect data. Finally, we need a criterion function which quantitatively measures the difference between the ideal output arising from the perfect ideal input and the calculated output arising from the corresponding randomly perturbed input.

Now we are faced with an immediate problem relative to the criterion function. It is typically the case that an algorithm changes the data unit. For example, an edge-linking process changes the data from the unit of pixel to the unit of a group of pixels. An arc segmentation/extraction process applied to the groups of pixels produced by an edge linking process produces fitted curve segments. This data unit change means that the representation used for the random variation of the output data set may have to be entirely different than the representation used for the random variation of the input data set. In our edge-linking/arc extraction example, the input data might be described by the false alarm/misdetection characteristics produced by the preceding edge operation, as well as the standard deviation in the position and orientation of the correctly detected edge pixels. The random variation in the output data from the extraction process, on the other hand, must be described in terms of fitting errors (random variation in the fitted coefficients) and segmentation errors. Hence, the random perturbation model may change from stage to stage in the analysis process.

Consider the case for segmentation errors. The representation of the segmentation errors must be natural and suitable for the input of the next process in high-level vision which might be a model-matching process, for example. What should this representation be to make it possible to characterize the identification accuracy of the model matching as a function of the input segmentation errors and fitting errors? Questions like these, have typically not been addressed in the research literature. Until they are, analyzing the performance of a machine vision algorithm will be in the dark ages of an expensive experimental trial-and-error process. And if the performance of the different pieces of a total algorithm cannot be used to determine the performance of the total algorithm, then there cannot be an engineering design methodology for machine vision systems.

This problem is complicated by the fact that there are many instances of algorithms which compute the same sort of information but in forms which are actually non-equivalent. For example, there are arc extraction algorithms which operate directly on the original image along with an intermediate vector file obtained in a previous step and which output fitted curve segments. There are other arc extraction algorithms which operate on groups of pixels and which output arc parameters such as center, radius, and endpoints in addition to the width of the original arc.

What we need is the machine vision analog of a system's engineering methodology. This methodology can be encapsulated in a protocol which has a modeling component, an experimental component, and a data analysis component. The next section describes in greater detail these components of an image analysis engineering protocol.

3   Protocol

The modeling component of the protocol consists of a description of the world of ideal images, a description of a random perturbation model by which non-ideal images arise, a description of a random perturbation process which characterizes the output random perturbation as a function of the parameters of the input random perturbation and a specification of the criterion function by which the difference between the ideal output and the computed output arising from the imperfect input can be quantified. The experimental component describes the experiments performed under which the data relative to the performance characterization can be gathered. The analysis component describes what analysis must be done on the experimentally observed data to determine the performance characterization.

3.1   Input Image Population

This part of the protocol describes how, in accordance with the specified model, a suitably random, independent, and representative set of images from the population of ideals is to be acquired or generated to constitute the sampled set of images. This acquisition can be done by taking real images under the specified conditions or by generating synthetic images. If the population includes, for example, a range of sizes of the object of interest or if the object of interest can appear in a variety of situations, or if the object shape can have a range of variations, then the sampling mechanism must assure that a reasonable number of images are sampled with the object appearing in sizes, orientations, and shape variations throughout its permissible range. Similarly, if the object to be recognized or measured can appear in a variety of different lighting conditions which create a similar variety in shadowing, then the sampling must assure that images are acquired with the lighting and shadowing varying throughout its permissible range.

Some of the variables used in the image generation process are ones whose values will be estimated by the computer vision algorithm. We denote these variables by $z_1, \ldots, z_K$. Other of these variables are nuisance variables. Their values provide for variation. The performance characterization is averaged over their values. We denote these variables by $w_1, \ldots, w_M$. Other of the variables specify the parameters of the random perturbation and noise process against which the performance is to be characterized. We denote these variables by $y_1, \ldots, y_J$. The generation of the images in the population can then be described by $N = J + K + M$ variables. If these N variables having to do with the kind of lighting, light position, object position, object orientation, permissible object shape variations, undesired object occlusion, environmental clutter, distortion, noise etc., have respective range sets $R_1, \ldots, R_N$, then the sampling design must assure that images are selected from the domain $R_1 \times R_2 \times \ldots \times R_N$ in a representative way. Since the number of images sampled is likely to be a relatively small fraction of the number of possibilities in $R_1 \times R_2 \times \ldots \times R_N$, the experimental design may have to make judicious use of a Latin square layout.

3.2 Random Perturbation and Noise

Specification of random perturbation and noise is not easy because the more complex the data unit, the more complex the specification of the random perturbation and noise. Each specification of randomness has two potential components. One component is a small perturbation component which affects all data units. It is often reasonable to model this by an additive Gaussian noise process on the ideal values of the data units. This can be considered to be the small variation of the ideal data values combined with observation or measurement noise. The other component is a large perturbation component which affects only a small fraction of the data units. For simple data units it is reasonable to model this by replacing its value by a value having nothing to do with its true value. Large perturbation noise on more complex data units can be modeled by fractionating the unit into pieces and giving values to most of the pieces which would follow from the values the parent data unit had and giving values to the remaining pieces which have nothing to do with the values the original data unit had.

This kind of large random perturbation affecting a small fraction of units is replacement noise. It can be considered to be due to random occlusion, linking, grouping, or segmenting errors. Algorithms which work near perfectly on small amounts of random perturbation on all data units, often fall apart with large random perturbation on a small fraction of the data units. Much of the performance characterization of a complete algorithm will be specified in terms of how much of this replacement kind of random perturbation the algorithm can tolerate and still give reasonable results. Algorithms which have good performance even with large random perturbation on a small fraction of data units can be said to be robust.

3.3 Performance Characterization

Some of the variables used in the image generation are those whose values are to be estimated by the machine vision algorithm. Object kind, location, and orientation are prime examples. The values of such variables do not make the recognition and estimation much easier or harder, although they may have some minor effect. For example, an estimate of the surface normal of a planar object viewed at a high slant angle will tend to have higher variance than an estimate produced by the planar object viewed at a near normal angle. The performance characterization of an image analysis algorithm is not with respect to this set of variables. From the point of view of what is to be calculated, this set of variables is crucial. From the point of view of performance characterization, the values for the variables in this set as well as the values in the nuisance set are the ones over which the performance is averaged.

Another set of variables characterize the extent of random perturbations which distort the ideal input data to produce the imperfect input data. These variables represent variations which degrade the information in the image, thereby increasing the uncertainty of the estimates produced by the algorithm. Such variables may characterize object contrast, noise, extent of occlusion, complexity of background clutter, and a multitude of other factors which instead of being modeled explicitly are modeled implicitly by the inclusion of random shape perturbations applied to the set of ideal model shapes.

Finally, there may be other variables governing parameter constants that must be set in the image analysis algorithm. The values of these variables may to a large or small extent change the performance of the algorithm.

The variables characterizing the input random perturbation process and the variables which are the algorithm tuning constants constitute the set of variables in terms of which the performance characterization must be measured. Suppose there are I algorithm parameters $x_1, \ldots, x_I$, which can be set, J different variables $y_1, \ldots, y_J$ characterizing the random perturbation process, and K different measurements $\hat{z}_1, \ldots, \hat{z}_K$ to be made on each image. There will be a difference between the true ideal values $z_1, \ldots, z_K$ of the measured quantities and the measured values $\hat{z}_1, \ldots, \hat{z}_K$ themselves. The nature of this difference can be characterized by the parameters $q_1, \ldots, q_L$ of the output random perturbation process: $(q_1, \ldots, q_L) = f(x_1, \ldots, x_I, y_1, \ldots, y_J, z_1, \ldots, z_K)$.

The last step of a total algorithm not only has a characterization of the output random perturbation parameters, but also an error criterion e which is application and domain specific. The error criterion, $e(z_1, \ldots, z_K, \hat{z}_1, \ldots, \hat{z}_K)$, must state how the comparison between the ideal values and the measured values will be evaluated. Its value will be a function of the I algorithm parameters and the J random perturbation parameters.

An algorithm can have two different dimensions to the error criterion. To explain these dimensions, consider algorithms which estimate some parameter such as position and orientation of an object. One dimension the error criterion can have is reliability. An estimate can be said to be reliable if the algorithm is operating on data that meets certain requirements and if the difference between the estimated quantity and the true but known value is below a user specified tolerance. An algorithm can estimate whether the results it produces are reliable by making a decision on estimated quantities which relate to input data noise variance, output data covariance, and structural stability of calculation. Output quantity covariance can be estimated by estimating the input data noise variance and propagating the error introduced by the noise variance into the calculation of the estimated quantity. Hence the algorithm itself can provide an indication of whether the estimates it produces have an uncertainty below a given value. High uncertainties would occur if the algorithm can determine that the assumptions about the environment producing the data or the assumptions required by the method are not being met by the data on which it is operating or if the random perturbation in the quantities estimated is too high to make the estimates useful.

Characterizing this dimension can be done by two means. The first is by the probability that the algorithm claims reliability as a function of algorithm parameters and parameters describing input data random perturbations. The second is by misdetection false alarm operating curves. A misdetection occurs when the algorithm indicates it has produced a reliable enough result when in fact it has not produced a reliable enough result. A false alarm occurs when the algorithm indicates that it has not produced a reliable enough result when in fact it has produced a reliable enough result. A misdetection false alarm rate operating curve results for each different noise and random perturbation specification. The curve itself can be obtained by varying the algorithm tuning constants, one of which is the threshold by which the algorithm determines whether it claims the estimate it produces is reliable or not.

The second dimension of the error criterion would be related to the difference between the true value of the quantity of interest and the estimated value. This criterion would be evaluated only for those cases where the algorithm indicates that it produces a reliable enough result. A scalar error criterion would weight both of these dimensions in an appropriate manner.

Each estimated quantity $\hat{z}_k$ is a random variable which is a function of the ideal input data, the values of the algorithm tuning parameters $x_1, \ldots, x_I$ and the random perturbation parameters $y_1, \ldots, y_J$ characterizing the random perturbation process distorting the ideal input.

Each ideal quantity z_k is a function only of the algorithm constants $x_1, \ldots, x_I$. The expected value E of $e(z_1, \ldots, z_K, \hat{z}_1, \ldots, \hat{z}_K)$ is taken over the input data set subpopulation consistent with $z_1, \ldots, z_K$ and the random perturbation process. It is, therefore, a function of $x_1, \ldots, x_I$ and $y_1, \ldots, y_J$. Performance characterization of the estimated quantity with respect to the error criterion function then amounts to expressing in graph, table or analytic form $E[e(z_1, \ldots, z_K, \hat{z}_1, \ldots, \hat{z}_K)]$ for each $z_1, \ldots, z_K$ as a function of $x_1, \ldots, x_I$ and $y_1, \ldots, y_J$.

3.4 Experiments

In a complete design, the values for the algorithm constants $x_1, \ldots, x_I$ and the values for the random perturbation parameters $y_1, \ldots, y_J$ will be selected in a systematic and regular way. The values for $z_1, \ldots, z_K$ and the values for the nuisance variables $w_1, \ldots, w_M$ will be sampled from a uniform distribution over the range of their permissible values.

The values for $z_1, \ldots, z_K$ specify the equivalence class of ideal images. The values for $y_1, \ldots, y_J$ characterize the random perturbations and noise which are randomly introduced into the ideal image and/or object(s) in the ideal image. In this manner, each noisy trial image is generated. The values for $x_1, \ldots, x_I$ specify how to set the tuning constants required by the algorithm. The algorithm is then run over the trial image producing estimated values $\hat{z}_1, \ldots, \hat{z}_K$ for $z_1, \ldots, z_K$.

The data analysis plan for the characterization of the output random perturbation process generates records of the form

$$x_1, \ldots, x_I, y_1, \ldots, y_J, z_1, \ldots, z_K, \hat{z}_1, \ldots, \hat{z}_K$$

From an assumed model of the output random perturbation process, the data analysis plan will have a way of estimating the parameters $q_1, \ldots, q_L$ of the output random perturbation process from the $z_1, \ldots, z_K, \hat{z}_1, \ldots, \hat{z}_K$ part of the records. Thus $q_1, \ldots, q_L$ will be a function of $x_1, \ldots, x_I, y_1, \ldots, y_J, z_1, \ldots, z_K$. The data analysis plan must also specify how this dependence will be determined by an estimating or fitting procedure.

If we apply the error criterion to each record, we then produce the values $e(z_1, \ldots, z_K, \hat{z}_1, \ldots, \hat{z}_K)$. The data produced by each trial then consists of a record

$$x_1, \ldots, x_I, y_1, \ldots, y_J, e(z_1, \ldots, z_K, \hat{z}_1, \ldots, \hat{z}_K)$$

The data analysis plan for the error criterion describes how the set of records produced by the experimental trials will be processed or analyzed to compactly express the performance characterization. For example, an equivalence relation on the range space for $y_1, \ldots, y_J$ may be defined and an hypothesis may be specified stating that all combinations of values of $y_1, \ldots, y_J$ in the same equivalence class have the same expected error. The data analysis plan would specify the equivalence relation and give the statistical procedure by which the hypothesis could be tested. Performing such tests are important because they can reduce the number of variable combinations which have to be used to express the performance characterization. For example, the hypothesis that all other variables being equal, whenever y_{J-1}/y_J has a ratio of k, then the expected performance is identical. In this case, the performance characterization can be compactly given in terms of k and $y_1, \ldots, y_{J-2}$.

Once all equivalence tests are complete, the data analysis plan would specify the kinds of graphs or tables employed to present the experimental data. It might specify the form of a simple regression equation by which the expected error, the probability of claimed reliability, the probability of misdetection, the probability of false alarm, and the computational complexity or execution time can be expressed in terms of the independent variables $x_1, \ldots, x_I, y_1, \ldots, y_J$. As well it would specify how the coefficients of the regression equation could be calculated from the observed data. Finally, when error propagation can be done analytically using the parameters associated with input data noise variance and the ideal noiseless input data, the data analysis plan can discuss how to make the comparison between the expected error computed analytically and the observed experimental error.

Finally, if the computer vision algorithm must meet certain performance requirements, the data analysis plan must state how the hypothesis that the algorithm meets the specified requirement will be tested. The plan must be supported by a theoretically developed statistical analysis which shows that an experiment carried out according to the experimental design and analyzed according to the data analysis plan will produce a statistical test itself having a given accuracy. That is, since the entire population of images is only sampled, the sampling variation will introduce a random fluctuation in the test results. For some fraction of experiments carried out according to the protocol, the hypothesis to be tested will be accepted but the algorithm, in fact, if it were tried on the complete population of image variations, would not meet the specified requirements; and for some fraction of experiments carried out according to the protocol, the hypothesis to be tested will be rejected but if the algorithm were tried on the complete population of image variation, it would meet the specified requirements. The specified size of these errors of false acceptance and missed acceptance will dictate the number of images to be in the sample for the test. This relation between sample size and false acceptance rate and missed acceptance rate of the test for the hypothesis must be determined on the basis of statistical theory. One would certainly expect that the sample size would be large enough so that the uncertainty caused by the sampling would be below 20%.

For example, suppose the error rate of a quantity estimated by a machine vision algorithm is defined to be the fraction of time that the estimate is further than ϵ_0 from the true value. If this error rate is to be less than $\frac{1}{1,000}$, then in order to be about 85% sure that the performance meets specification, 10,000 tests will have to be run. If the image analysis algorithm performs incorrectly 9 or fewer times, then we can assert that with 85% probability, the machine vision algorithm meets specification [1].

4 Conclusion

We have discussed the problem of the lack of performance evaluation in the published literature on computer vision algorithms. This situation is causing great difficulties to researchers who are trying to build up on existing algorithms and to engineers who are designing operational systems. To remedy the situation, we suggested the establishment of a well-defined protocol for determining the performance characterization of an algorithm. Use of this kind of protocol will make using engineering system methodology possible as well as making possible well-founded comparisons between machine vision algorithms that perform the same tasks. We hope that our discussion will encourage a thorough and overdue dialogue in the field so that a complete engineering methodology for performance evaluation of machine vision algorithms can finally result.

References

[1] Haralick, R.M., "Performance Assessment of Near Perfect Machines," *Machine Vision and Applications*, Vol. 2, No. 1, 1989, pp. 1-16.

Training Models of Shape from Sets of Examples

T.F.Cootes, C.J.Taylor, D.H.Cooper and J.Graham
Department of Medical Biophysics
University of Manchester
Oxford Road
Manchester M13 9PT
email: bim@wiau.mb.man.ac.uk

Abstract

A method for building flexible shape models is presented in which a shape is represented by a set of labelled points. The technique determines the statistics of the points over a collection of example shapes. The mean positions of the points give an average shape and a number of modes of variation are determined describing the main ways in which the example shapes tend to deform from the average. In this way allowed variation in shape can be included in the model. The method produces a compact flexible 'Point Distribution Model' with a small number of linearly independent parameters, which can be used during image search. We demonstrate the application of the Point Distribution Model in describing two classes of shapes.

1 Introduction

We have previously described a method for modelling two dimensional shape, based on the statistics of chord lengths over a set of examples [12]. Although this provided a means of automatically parameterising shape variability, the method was difficult to use, requiring an iterative procedure to reconstruct a shape given a set of parameters. The method has computational complexity $O[n^2]$ where n is the number of points used to describe the shape. In this paper we present a new method which produces a more compact representation, allows direct reconstruction of a shape from a set of parameters and offers $O[n]$ computational complexity.

Image interpretation using rigid models is well established [1,2]. However, in many practical situations objects of the same class are not identical and rigid models are inappropriate. This is particularly true in medical applications, but also many industrial applications involve assemblies with moving parts, or components whose appearance can vary. In such cases flexible models, or deformable templates, can be used to allow for some degree of variability in the shape of the imaged object.

Yuille, Cohen and Hallinan [3] and Lipson et al [4] use deformable templates for image interpretation. Unfortunately their templates are hand–crafted with modes of variation which have to be individually tailored for each application. Kass, Witkin and Terzopoulos [5] described 'Active Contour Models', flexible snakes which can stretch and deform to image features. These have been extended to apply constraints to their deformation by adjusting the elasticity and stiffness of the model [6,7]. Pentland and Sclaroff [8] model objects as lumps of elastic clay, generating different shapes using combinations of the modes of vibration of the clay. However this does not always lead to a very compact description of the variability within a particular class of objects. Bookstein [9] has studied the statistics of shape deformation by representing objects as sets of 'landmark points', but has not applied this to the problem of shape modelling. Mardia, Kent and Walder [10] represent the boundary of a shape as a sequence of points with distributions related by a covariance matrix. To fit a model to an image they cycle through the points to find the most likely position given the image and the current shape. The examples given seem to be local models, in that deforming one part of the boundary does not affect the rest of it until the change has been propagated round the boundary by the updating method.

In this paper we describe a new method of shape modelling based on the statistics of labelled points placed on a set of training examples. The sets of points are automatically aligned so that their mean positions and main modes of variation can be calculated. Aligning the shapes allows the positions of equivalent points in different examples to be compared simply by examining their co-ordinates. A model consists of the mean positions of the points and a number of vectors describing the modes of variation.

2 Point Distribution Models

Suppose we wish to derive a model to represent the shape of resistors as they appear on a printed circuit board, such as those shown in Figure 1. Different examples of resistor have sufficiently different shapes that a rigid model would not be appropriate. Figure 2 shows some examples of resistor boundaries which were obtained from backlit images of individual resistors. Our aim is to build a model which describes both typical shape and allowed variability, using the examples in Figure 2 as a training set.

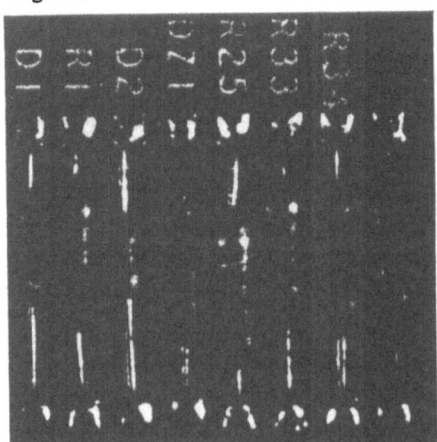
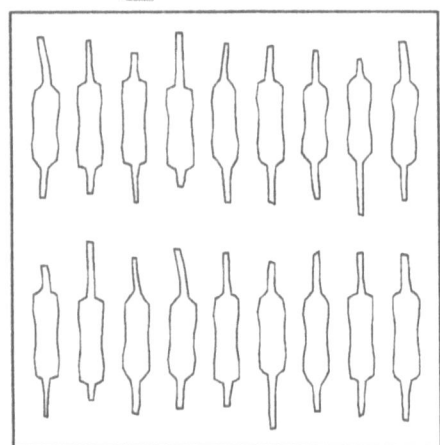

Figure 1 : Image of printed circuit board showing examples of resistors.

Figure 2 : Examples of resistor shapes from a training set.

2.1 Labelling The Training Set

In order to model a shape, we represent it by a set of points. For the resistors we have chosen to place points around the boundary, as shown in Figure 3. This must be done for each shape in the training set. The labelling of the points is important, each labelled point represents a particular part of the object or its boundary. For instance, in the resistor model, points 0 and 31 always represent the ends of a wire, points 3, 4 and 5 represent one end of the body of the resistor and so on. The method works by modelling how different labelled points tend to move together as the shape varies. If the labelling is incorrect, with a particular point placed at different sites on each training shape, the method will fail to capture shape variability.

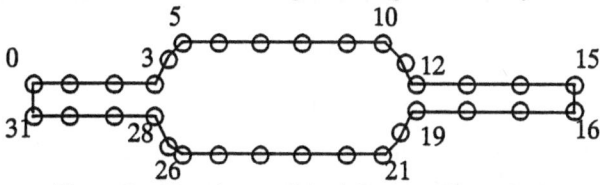

Figure 3 : 32 point model of the boundary of a resistor.

It is important that the points are placed correctly on each example image. This will usually require someone familiar with the application to choose the most appropriate set of points and to be able to reproducably place them on different examples. This procedure can be time consuming, though we are developing tools to speed up the process. It should be noted that though the labelling of the training set is done manually, finding the mean shape and main modes of variation is automatic. Deducing such a set of modes would be very difficult by hand, particularly for more complex biological shapes.

2.2 Aligning The Training Set

Our modelling method works by examining the statistics of the co-ordinates of the labelled points over the training set. In order to be able to compare equivalent points from different shapes, they must be aligned in the same way with respect to a set of axes. If they are not, we would not be comparing like with like and any statistics derived would be meaningless. We achieve the required alignment by scaling, rotating and translating the training shapes so that they correspond as closely as possible. We aim to minimise a weighted sum of squares of distances between equivalent points on different shapes. This is a form of Generalised Procrustes Analysis [11].

We will first consider aligning a pair of shapes. Let x_i be a vector describing the n points of the ith shape in the set;

$$x_i = (x_{i0}, y_{i0}, x_{i1}, y_{i1}, \ldots, x_{ik}, y_{ik}, \ldots, x_{in-1}, y_{in-1})^T$$

Let $M_j[x_j]$ be a rotation by θ_j and a scaling by s_j. Given two similar shapes, x_i and x_j we can choose θ_j, s_j and a translation $(t_x, t_y)_j$ mapping x_i onto $M_j[x_j]$ so as to minimise the weighted sum

$$E_j = (x_i - M_j(x_j))^T W(x_i - M_j(x_j)) \tag{1}$$

where
$$M_j \begin{pmatrix} x_{jk} \\ y_{jk} \end{pmatrix} = \begin{pmatrix} (s_j \cos \theta)x_{jk} - (s_j \sin \theta)y_{jk} + t_{jx} \\ (s_j \sin \theta)x_{jk} + (s_j \cos \theta)y_{jk} + t_{jy} \end{pmatrix} \tag{2}$$

and W is a diagonal matrix of weights for each point.

Details are given in Appendix A.

The weights can be chosen to give more significance to those points which tend to be most 'stable' over the set – the ones which move about least with respect to the other points in a shape. We have used a weight matrix defined as follows: let R_{kl} be the distance between points k and l in a shape; let $V_{R_{kl}}$ be the variance in this distance over the set of shapes; we can choose a weight, w_k, for the kth point using

$$ w_k = \left(\sum_{l=0}^{n-1} V_{R_{kl}} \right)^{-1} \tag{3} $$

If a point tends to move around a lot with respect to the other points in the shape, the sum of variances will be large, and a low weight will be given. If, however, a point tends to remain fixed with respect to the others, the sum of variances will be small, a large weight will be given and matching such points in different shapes will be a priority.

In order to align all the shapes in a set we use the following algorithm.

    1) Rotate, scale and translate each of the shapes in the set to align to the first shape.

  Repeat

    2) Calculate the mean of the transformed shapes

    3) Either

        a) Adjust the mean to a default scale, orientation and origin,

        b) Rotate, scale and translate the mean to align to the first shape

    4) Rotate, scale and translate each of the shapes again to match to the adjusted mean.

  Until convergence.

Stage 3 inside the iteration loop is required to renormalise the mean. Without this the algorithm is ill-conditioned – there are in effect $4(N_s-1)$ constraints on $4N_s$ variables (θ, s, t_x, t_y for each shape) – and will not converge – the mean will shrink, rotate or slide off to infinity. Constraints on the pose and scale of the mean allow the equations to have a unique solution. Either the mean is scaled, rotated and translated so it matches the first shape, or an arbitrary default setting can be used, such as choosing an origin at its centre of gravity, an orientation so that a particular part of the shape is at the top and a scale so that the distance between two points is one unit.

The convergence condition can be tested by examining the average difference between the transformations required to align each shape to the recalculated mean and the identity transformation. Experiments suggest that the method converges to the same result independent of which shape is aligned to in the first stage, though a formal proof of convergence has yet to be devised.

2.3 Capturing the Statistics of a Set of Aligned Shapes

Once a set of aligned shapes is available the mean shape and variability can be found. The mean shape, $\bar{x}$, is calculated using

$$\bar{x} = \frac{1}{N_s} \sum_{i=1}^{N_s} x_i \tag{4}$$

The modes of variation, the ways in which the points of the shape tend to move together, can be found by applying principal component analysis to the deviations from the mean as follows.

For each shape in the training set we calculate its deviation from the mean, dx_i, where

$$dx_i = x_i - \bar{x} \tag{5}$$

We can then calculate the $2n$ x $2n$ covariance matrix, S, using

$$S = \frac{1}{N_s} \sum_{i=1}^{N_s} dx_i dx_i^T \tag{6}$$

The modes of variation of the points of the shape are described by the unit eigenvectors of S, p_i *(i = 1 to 2n)* such that

$$Sp_i = \lambda_i p_i \tag{7}$$

(where λ_i is the i'th eigenvalue of S, $\lambda_i \geq \lambda_{i+1}$)

$$p_i^T p_i = 1 \tag{8}$$

It can be shown that the eigenvectors of the covariance matrix corresponding to the largest eigenvalues describe the most significant modes of variation in the variables used to derive the covariance matrix, and that the proportion of the total variance explained by each eigenvector is equal to the corresponding eigenvalue [13]. Most of the variation can usually be explained by a small number, t, modes. One method for calculating t would be to chose the smallest number of modes such that the sum of variance explained was a sufficiently large proportion of λ_T, the total variance of all the variables, where

$$\lambda_T = \sum_{i=1}^{2n} \lambda_i \tag{9}$$

The i'th eigenvector affects point k in the model by moving it along a vector parallel to (dx_{ik}, dy_{ik}), which is obtained from the k'th pair of elements in p_i,

$$(dx_{i0}, dy_{i0}, \ldots, dx_{ik}, dy_{ik}, \ldots, dx_{in-1}, dy_{in-1}) \tag{10}$$

Any shape in the training set can be approximated using the mean shape and a weighted sum of these deviations obtained from the first t modes

$$x = \bar{x} + Pb \tag{11}$$

where $P = (p_1\ p_2\ \ldots\ p_t)$ is the matrix of the first t eigenvectors,

$$\mathbf{b} = (b_1 \ b_2 \ ... \ b_t)^T \quad \text{is a vector of weights for each eigenvector}$$

the eigenvectors are orthogonal, $\quad \mathbf{P}^T\mathbf{P} = \mathbf{I} \quad$ so

$$\mathbf{b} = \mathbf{P}^T(\mathbf{x} - \bar{\mathbf{x}}) \tag{12}$$

The above equations allow us to generate new examples of the shapes be varying the parameters (b_i) within suitable limits. The parameters are linearly independent, though there may be non–linear dependencies still present. The limits for b_i are derived by examining the distributions of the parameter values required to generate the training set. Since the variance of b_i over the training set can be shown to be λ_i, suitable limits are likely to be of the order of

$$-3\sqrt{\lambda_i} \leq b_i \leq 3\sqrt{\lambda_i} \tag{13}$$

since most of the population lies within three standard deviations of the mean.

3 Practical Examples

The techniques described above have been used to generate shape models for both manufactured and biological objects. We present results for the set of resistor shapes shown in Figure 2 and a set of hand shapes.

3.1 Resistor Example

The resistor shapes were aligned using the method described above, arranging the mean shape to be horizontal and scaling so the average distance of each point of the mean from its centre of gravity is one unit. The most significant eigenvalues of the covariance matrix derived are shown in Table 1.

Table 1 : Eigenvalues of the covariance matrix derived from a set of resistor shapes.

Eigenvalue	λ_i	$\dfrac{\lambda_i}{\lambda_T}$ x 100%	$\sqrt{\lambda_i}$
λ_1	0.207	66%	0.46
λ_2	0.026	8%	0.16
λ_3	0.017	5%	0.13
λ_4	0.013	4%	0.11
λ_5	0.010	3%	0.10
λ_6	0.008	3%	0.09

Figure 4 shows the plot of b_1 against b_2 for the training set. The lack of structure in the scatter plot suggests that the parameters can be treated as independent. We are currently working on deriving more formal tests of independence. Any dependencies between the parameters would imply non–linear relationships between the original point positions and would results in some combinations of parameters generating 'illegal' shapes. By varying the first three parameters separately we can generate examples of the shape as shown in Figures 5–7. Each parameter 'represents' a mode of variation of the shape which can frequently be associated with an intuitive description of the deformation. Compare Figures 5–7 with Figure 2. Varying the first parameter (b_1) adjusts the position of the body of the resistor up and down the wire. The second parameter varies the shape of the ends of the main body of the resistor, between tapered and square. The third parameter affects the curvature of the wires at either end. Subsequent parameters have smaller effects, including the wires bending in opposite directions. These modes of variation effectively capture the variability which was present in the training set.

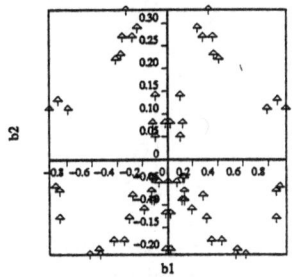

Figure 4 : Plot of b_1 vs b_2 for a training set of resistor shapes.

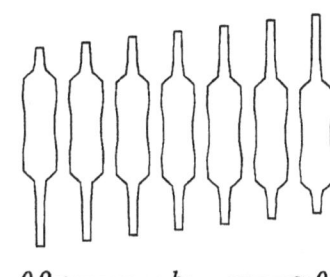

$-0.9 \longleftarrow \quad b_1 \quad \longrightarrow 0.9$

Figure 5 : Effects of varying the first parameter of the resistor model.

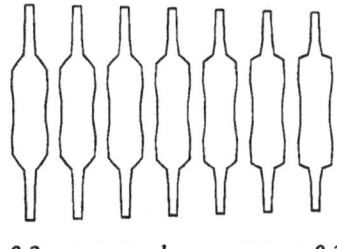

$-0.3 \longleftarrow \quad b_2 \quad \longrightarrow 0.3$

Figure 6 : Effects of varying the second parameter of the resistor model.

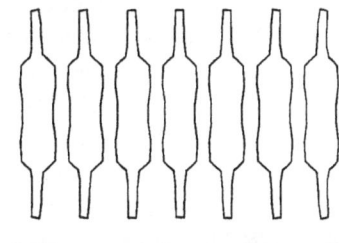

$-0.25 \longleftarrow \quad b_3 \quad \longrightarrow 0.25$

Figure 7 : Effects of varying the third parameter of the resistor model.

3.2 Hand Example

A set of 18 hand shapes was generated from images of the right hand of one of the authors (Figure 8). Each was represented by 72 points around the boundary. These were planted on the examples by locating 12 control points at the ends and joints of the fingers and filling in the rest equally along the connecting boundaries. A model was trained on the data, and it was found that 96% of the variance could be explained by the first 6 modes of variation. The first three modes are shown in Figure 9, and consist of combinations of movements of the fingers. Again, a compact parameterised model has been generated.

4 Discussion and Conclusions

The method outlined above allows a compact, flexible shape model to be built, representing a class of shapes by the mean positions of a set of labelled points and a small number of modes of variation about the mean. The model points do not have to lie only on the boundary of objects, they can represent internal features, and even sub–components of a complex assembly. In the latter case the model describes both

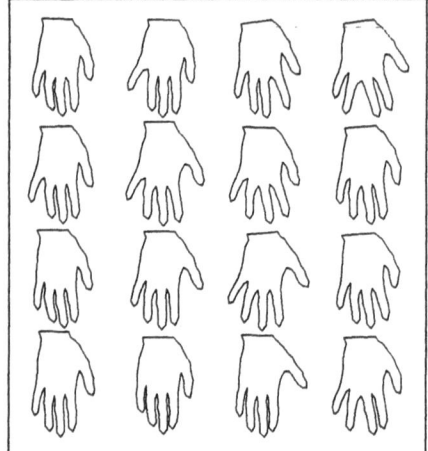

Figure 8 : Training set of hand shapes, each defined by 72 points.

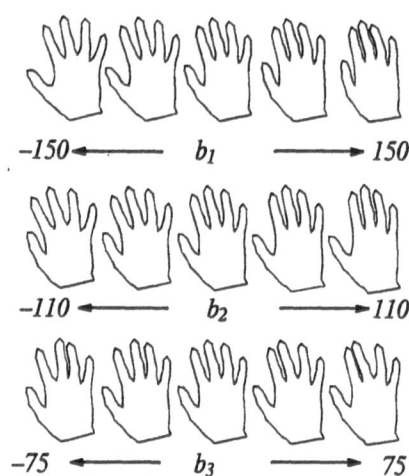

Figure 9 : Effects of varying each of the first three parameters of the hand model individually.

the variations in the shapes of the sub–components and the geometric relationships between components. Such a model, representing a section through the ventricles in the brain in MR scans is described by Hill *et al* in [16].

It is important to arrange that all the examples used to train the model are similarly aligned with respect to a set of axes, to ensure that the labelled points in different shapes are being compared correctly. In some cases an obvious alignment is apparent, but in others, particularly medical cases where the shapes of organs are very flexible, the automatic least squares alignment method is essential. The method has been used successfully to model a variety of objects from both industrial and biological domains.

The models we build are linear. Varying each parameter individually moves the points along straight lines. The method is inefficient at modelling non–linear effects such as bending or rotation of one sub–component about another. To deal with such cases a non–linear model of the modes of variation would be required. We have begun experimenting with a system which represents each mode using a polynomial curve rather than a straight line [14]. Some promising results have been produced which will be the subject of a further paper.

Point Distribution Models have been used in image search. A local optimiser called the Active Shape Model has been developed [15] which provides a way of iteratively improving an initial estimate of the position, pose and shape parameters of a model fitted to image data. The model has also been used in conjunction with a generate and test strategy based around Genetic Algorithms [16]. The hand and resistor models described above have been successfully used to find examples in images with both techniques.

The model can also be used in a classifier. Given an example of a shape, an estimate can be made of how likely that example is to be a member of the class of shapes described by a model. If labelled points are placed on the example and the point set aligned with the mean shape, Equation 12 can be used to calculate the model parameters required to generate the example. The distributions of the parameters can be estimated from the training set, allowing probabilities to be assigned. This technique has been successfully used in a simple handwritten character recognition application [17].

The models are compact and easy to use. Given a set of parameters an example of the model can be calculated rapidly. The models are well suited to generate-and-test image search strategies in many domains.

Acknowledgements

This work is funded by SERC under the IEATP Initiative (Project Number 3/2114). The authors would like to thank the other members of the Wolfson Image Analysis Unit for their help and advice, particularly D.Bailes and A.Hill.

Appendix : Aligning A Pair of Shapes

Given two similar shapes, x_1 and x_2 we would like to choose a rotation, θ, a scale s and a translation (t_x, t_y) mapping x_2 onto $M(x)$ so as to minimise the weighted sum

$$E = (x_1 - M(x_2))^T W(x_1 - M(x_2)) \tag{1}$$

where
$$M\begin{pmatrix} x_{jk} \\ y_{jk} \end{pmatrix} = \begin{pmatrix} (s \cos \theta)x_{jk} - (s \sin \theta)x_{jk} + t_x \\ (s \sin \theta)x_{jk} + (s \cos \theta)x_{jk} + t_y \end{pmatrix} \tag{2}$$

and W is a diagonal matrix of weights for each point.

If we write

$$a_x = s \cos \theta \qquad a_y = s \cos \theta$$

then least squares approach (differentiating with respect to each of the variables a_x, a_y, t_x, t_y) leads to a set of four linear equations;

$$\begin{pmatrix} X_2 & -Y_2 & W & 0 \\ Y_2 & X_2 & 0 & W \\ Z & 0 & X_2 & Y_2 \\ 0 & Z & -Y_2 & X_2 \end{pmatrix} \begin{pmatrix} a_x \\ a_y \\ t_x \\ t_y \end{pmatrix} = \begin{pmatrix} X_1 \\ Y_1 \\ C_1 \\ C_2 \end{pmatrix} \tag{14}$$

where

$$X_i = \sum_{k=0}^{n-1} w_k x_{ik} \qquad Y_i = \sum_{k=0}^{n-1} w_k y_{ik} \tag{15}$$

$$Z = \sum_{k=0}^{n-1} w_k(x_{2k}^2 + y_{2k}^2) \qquad W = \sum_{k=0}^{n-1} w_k \tag{16}$$

$$C_1 = \sum_{k=0}^{n-1} w_k(x_{1k}x_{2k} + y_{1k}y_{2k}) \tag{17}$$

$$C_2 = \sum_{k=0}^{n-1} w_k(y_{1k}x_{2k} - x_{1k}y_{2k}) \tag{18}$$

These can be solved for a_x, a_y, t_x, and t_y using standard matrix methods.

References

[1]   R. Chin and C.R. Dyer, Model–Based Recognition in Robot Vision. Computing Surveys 1986; Vol 18, No 1

[2]   W.E.L. Grimson, Object Recognition by Computer : The Role of Geometric Constraints, The MIT Press, Cambridge, MA, USA, 1990.

[3]   A.L. Yuille, D.S. Cohen and P. Hallinan, Feature extraction from faces using deformable templates, Proc. Computer Vision and Pattern Recognition (1989) pp104–109.

[4]   P. Lipson, A.L. Yuille, D. O'Keeffe, J. Cavanaugh, J. Taaffe and D. Rosenthal, Deformable Templates for Feature Extraction from Medical Images, Proceedings of the First European Conference on Computer Vision (Lecture Notes in Computer Science, ed. O. Faugeras, pub. Springer–Verlag) 1990 pp413–417.

[5]   M. Kass, A. Witkin and D. Terzopoulos, Snakes: Active Contour Models. First International Conference on Computer Vision, pub. IEEE Computer Society Press, 1987, pp 259–268.

[6]   L.H. Staib and J.S. Duncan, Parametrically Deformable Contour Models. IEEE Computer Society conference on Computer Vision and Pattern Recognition, San Diego, 1989

[7]   D. Terzopoulos and D. Metaxas, Dynamic 3D Models with Local and Global Deformations : Deformable Superquadrics. IEEE Trans. on Pattern Analysis and Machine Intelligence 1991; Vol.13 No.7 pp703–714.

[8]   A. Pentland and S, Sclaroff, Closed–Form Solutions for Physically Based Modelling and Recognition. IEEE Trans. on Pattern Analysis and Machine Intelligence 1991; Vol.13 No.7 pp703–714.

[9]   F.L. Bookstein, Morphometric Tools for Landmark Data. Cambridge University Press, 1991.

[10] K.V. Mardia, J.T. Kent and A.N. Walder, Statistical Shape Models in Image Analysis. Proceedings of the 23rd Symposium on the Interface, Seattle 1991, pp 550–557.

[11] J.C. Gower, Generalized Procrustes Analysis. Psychometrika. 40, 1975, 33–51.

[12] T.F. Cootes, D. Cooper, C.J. Taylor and J. Graham, A Trainable Method of Parametric Shape Description. Proc. BMVC 1991 pub. Springer–Verlag, pp54–61.

[13] K. Fukunaga and W.L.G. Koontz, Application of the Karhunen–Loeve Expansion to Feature Selection and Ordering. IEEE Trans. on Computers 1970; 4.

[14] J. Graham, T.F. Cootes, D.Cooper and C.J. Taylor, VISAGE Progress Report – Deliverable D4, Wolfson Image Analysis Unit, Manchester University 1992.

[15] T.F. Cootes and C.J. Taylor, Active Shape Models – 'Smart Snakes'. This Volume.

[16] A. Hill, T.F. Cootes and C.J. Taylor, A Generic System for Image Interpretation Using Flexible Templates. This Volume.

[17] A. Lanitis, Optical Character Recognition of Hand–written Characters using Flexible Templates. Internal Report, Wolfson Image Analysis Unit, Manchester University 1992.

The Delaunay/Voronoi Selection Graph: a Method for Extracting Shape Information from 2-D Dot-Patterns with an Extension to 3-D.

Glynn Robinson, Lewis Griffin & Alan Colchester.
Department of Neurology,
Guy's Hospital, London, England, SE1 9RT.

Abstract

In this paper we present the *Delaunay/Voronoi selection graph (DVSG)*, an approach to the representation of the shape of 2-D objects which does not require a complete segmentation, merely a pattern of dots which are believed to lie on the edges of objects. The technique produces both a skeleton and boundary representation and, with subsequent processing, generates a hierarchical description of the topology of individual objects. We compare the DVSG to other methods used for obtaining information from dot-patterns, and show how this technique can be extended to 3-D.

1   Introduction

We require a method of representing the shapes of objects following an initial segmentation of an image. Since segmentation techniques often fail to produce a complete segmentation, we chose to treat candidate edge points as individual dots and compute both a skeleton and boundary representation for the perceived objects within the dot pattern. The method of shape representation we have developed is based on the Delaunay triangulation and its dual the Voronoi diagram of a set of points [1]. We refer to the method as the Delaunay/Voronoi selection graph (DVSG). Subsequent processing produces a hierarchical representation both of the boundary and the skeleton of objects. The hierarchical structure reduces its sensitivity to small changes along the object boundary and also facilitates coarse to fine matching of image features to model entities.

Many approaches to shape representation have been proposed, and more extensive reviews can be found in references [2,3]. Boundary representations of the shape of objects such as those described in references [4,5] tend to be sensitive to small changes along the object boundaries, and hierarchical representation is often difficult, as is the sub-division of objects into their sub-parts. The hierarchical approach of the curvature primal sketch [6] overcomes some of the problems of boundary representations, but the use of multiple Gaussian scales may cause problems especially when primitives are close together.

Skeleton representations such as proposed in [5,7] allow objects to be represented in terms of the relationships between their sub-parts. Spurious skeleton branches can be generated by small protrusions on the object boundary.

Nackman & Pizer [8] propose an approach which overcomes the problems of these spurious branches by generating the skeleton of an object at multiple Gaussian scales, and Arcelli [9] proposes a hierarchy of skeletons in terms of the object's boundary curvature.

The problem of grouping together dots to form perceptual groups has been attempted by a number of authors. Fairfield [10,11], like ourselves, is concerned with both the detection of the boundary of objects from dots, and also the segmenting of these objects into their sub-parts. He uses the Voronoi diagram to detect areas of internal concavity and replacing Voronoi diagram sides with the corresponding Delaunay triangulation side to produce both the object boundary and sub-parts. This work is dependent on a user defined threshold and does not differentiate between object boundaries and the sub-part boundaries. Ogniewicz et. al. [12] use the Voronoi diagram of a set of points to produce a medial axis description of objects. This method requires that the points making up the boundary have known connectivity, and a threshold is used to prune the skeleton description. Ahuja et. al. [13,14] propose the use of the Voronoi diagram and the properties of the individual Voronoi cells to classify points as boundary points, interior points, isolated points, or points on a curve.

Perceptual grouping algorithms based on the Delaunay Triangulation (DT) have been proposed by several other authors [15,16]. Three subgraphs of the DT which are of particular interest are: the Gabriel Graph (GG) [17]; the Relative Neighbourhood Graph (RNG) [18]; the Minimum Spanning Tree (MST). These graphs are shown in Figure 3 and defined below.

The **Gabriel Graph** is defined as follows: any edge $<u,v>$ of the DT is an edge of the GG iff the circle with $<u,v>$ as diameter contains no points in its interior. The **Relative Neighbourhood Graph** is defined as follows: any edge $<u,v>$ of the DT is an edge of the RNG iff the lune formed by the intersection of the circles centred at u and v with radius $|<u,v>|$ contains no points in its interior. The **Minimum Spanning Tree** is the tree of minimal total length which visits every point. Section 3 compares these three graphs to the graph generated by our technique.

Our method, the DVSG, does not require connected boundaries as input, merely a set of points (dots) which are believed to be edge points of objects. It produces distinct objects from these potential edge points, and concurrently generates both a skeleton and boundary representation of the shape of these objects.

2 Methods

2.1 Selecting skeleton and boundary sections.

Input to our technique comes in the form of disconnected dots. Each dot is assumed to lie on the edge of an unspecified object(s). We assume that the sides in the Delaunay triangulation provide a superset of the boundaries of the perceived objects within the dot-pattern. Likewise, we assume that the sides of the Voronoi diagram form a superset of the skeletons of the perceived objects.

In our approach, the problem of boundary and skeleton definition is simplified to selecting from the Voronoi diagram those sides that form the object skeleton, and selecting from the Delaunay triangulation those sides that form the object boundary. An example is shown in figure 1. We observe that when a Delaunay side is perceived to lie on the boundary (e.g. AB in figure 1b), the Voronoi side corresponding to this Delaunay side is NOT perceived to be on the skeleton (e.g. CD in figure 1b). Further, when a Voronoi side is perceived to lie on the skeleton, the corresponding Delaunay side is NOT perceived to lie on the boundary. The problem is thus further refined into a simple choice between a Delaunay side and its corresponding Voronoi side. We make this choice by keeping the shorter of the Delaunay side or its associated Voronoi side. If the Delaunay side is shorter then this is added to the list of boundary sections; if on the other hand the Voronoi side is shorter, this is added to the list of skeleton sections. This can be viewed as deciding whether any two points connected by the Delaunay triangulation are to be regarded as adjacent points on the same boundary, in which case we choose to keep the Delaunay side connecting them (e.g. AB in figure 1b); or whether they are lying opposite each other and separated by a skeleton, in which case we choose to keep the associated Voronoi section as part of the skeleton (e.g. CE in figure 1b). We call the graph resulting from this selection process the *Delaunay/Voronoi Selection Graph (DVSG)*. Figure 2a shows an example image, and figure 2b shows a series of points outlining the major features within the image. Figure 2c shows the result of the above selection criterion for the points in figure 2b (skeleton sections are shown as dashed lines; boundary sections as solid lines).

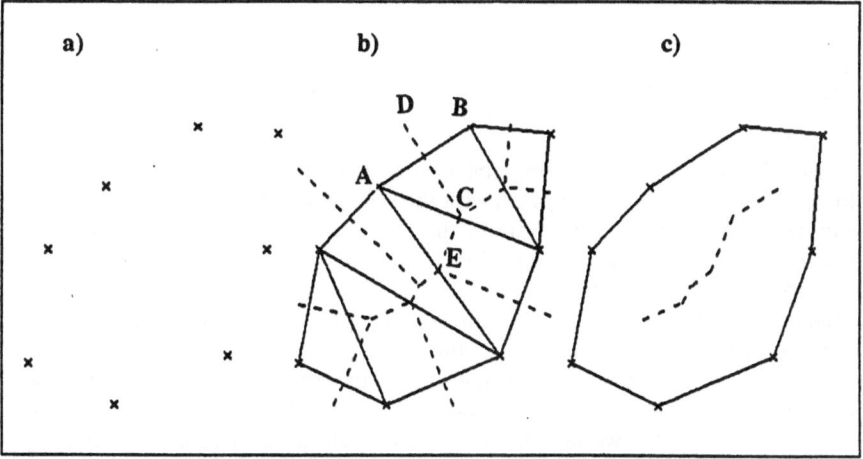

Figure 1 a) sample input points; **b)** Delaunay triangulation (solid) and Voronoi Diagram (dashed) **c)** result of the selection process.

2.2 Defining objects

Since no prior knowledge about the connectivity between edge points is necessarily provided, and there is no information about the number of objects present, it is necessary to define the objects that have been created following the creation of the DVSG. Following the initial selection of boundary and skeleton sections, sequences of connected skeleton sections (which may posses multiple branches) exist. These connected skeleton sections are surrounded by boundary sections. We define an object as a sequence of connected skeleton branches and the surrounding boundary sections. Each pair of points which was connected in the original Delaunay triangulation is now classed as either defining a boundary side or a skeleton section, and is also associated with one object (if the pair defines a skeleton section) or one or two objects (if the pair defines a boundary side). Figure 2d shows the computed boundary for the largest object in figure 2c.

2.3   Splitting objects into sub-parts

Each branching object is next segmented into its sub-parts. This is accomplished in the following way. All skeleton branches are three way and occur where all three Voronoi sides corresponding to one Delaunay triangle were shorter than their respective triangle sides. Each branch point in an object's skeleton is considered as either: *1) the meeting point of a small part of the object with the larger, main body of the object.* The smaller part is considered as less important than the main body and is broken off as a sub-part; *2) the meeting point of two smaller sections with the main body of the object.* Both of the smaller sections are then broken off as sub-parts.

Associated with each branch we define an area and a direction. The area is the sum of the areas of all of the triangles along that branch. The direction is the direction of the line from the branch point through the centre of gravity of the mid-points of the skeleton sections making up that branch. The centre of gravity is calculated from the mid-points of each Voronoi section in the skeleton of that branch, and each mid-point is weighted by the length of its skeleton section. The allows longer skeleton sections to have more effect on the direction of the branch than smaller ones.

For 1) above to be chosen, the branch with the smallest area must be significantly different from the branch with the second smallest area. This measure of significance is defined by the ratio of the two areas. Branch "a" is significantly smaller than branch "b" if area(a)/area(b) < 0.2. The branch with the smallest area is then broken off. Failing this, the sine of the difference between the direction of the smallest and the largest branch is compared to the sine of difference between the direction of the second smallest branch and the largest branch. The sine is taken as this gives us a large value if the directions are similar and a small value if the directions are dissimilar. If the ratio of the smaller of these two values over the larger falls below 0.2 we assume that one branch has a direction significantly different to that of the largest branch, and we break off the branch with a direction most different to that of the largest branch.

If neither of the above two conditions are met we assume that 2) above is true, and we break off both of the smaller branches.

When a branch is classified as belonging to a sub-part, its Voronoi section emanating from the branch is flagged as a *cut stem*. All of the skeleton and boundary sections of that sub-part are then labelled accordingly. The Delaunay sides corresponding to the skeleton cut stems form the *virtual boundaries* between sub-parts. This is an example of how the duality between the Voronoi diagram and the Delaunay triangulation makes it very easy to add/remove sections of an object by simply toggling a triangle side. Figure 2e shows the part boundaries (dotted lines) for the object of figure 2d.

2.4   Generating the intra object hierarchy

The sub-parts of each object are next represented hierarchically as a coarse-to-fine description. For each object, the sub-parts are ordered in terms of decreasing area. A level in the hierarchy is placed wherever there is a large change in area between successive sub-parts in this ordered list. Currently this is accomplished by looking for maxima in the derivative of the list. By successively adding the different levels of the hierarchy to the object description a more detailed version of the object can be obtained. Unlike many other approaches to hierarchial, coarse-to-fine descriptions, the position of the boundaries does not change as we go from coarse to fine; it is simply the case that some virtual boundaries are replaced either by virtual boundaries at a more detailed level of description, or by the final boundary of the object. Figure 2f-h show the different levels of the hierarchy for the object of figure 2d. The finest detail is shown in figure 2h.

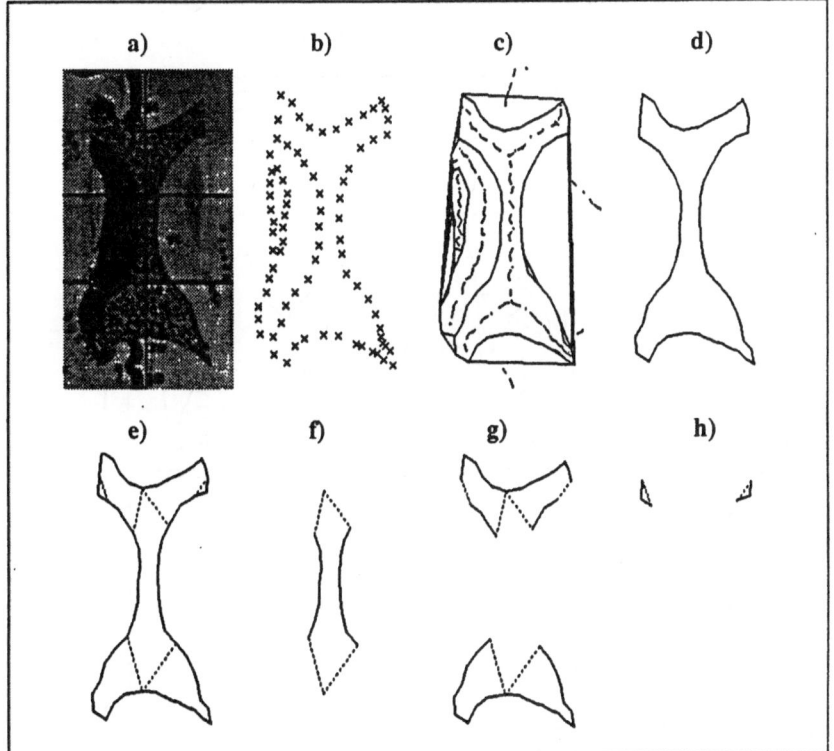

Figure 2 a) Original image; **b)** input points; **c)** Result of selection; **d)** Major object; **e)** parts of major object; **f-h)** hierarchy for major object.

3   Comparison with other subgraphs of the Delaunay triangulation

This section compares the DVSG with the GG, RNG and MST (defined in section 1). Figure 3 shows an example of these graphs. Several points can be made from comparing these graphs:

1.   It is well known that the relation DT ⊃ GG ⊃ RNG ⊃ MST holds [19]. Inspection of figure 3 shows that the DVSG does not fit into this sequence and in fact no inclusion relationship exists between the DVSG and the GG, RNG or MST.
2.   The GG, RNG and MST are necessarily connected whilst the DVSG is not.
3.   The DVSG always contains the convex hull whilst the GG, RNG and MST do not.

This lack of relation between the DVSG and the GG, RNG and MST can be understood by appreciating that the DVSG was developed with the aim of connecting points into coherent boundaries, whereas the other graphs have had there main application in clustering points into dense groups.

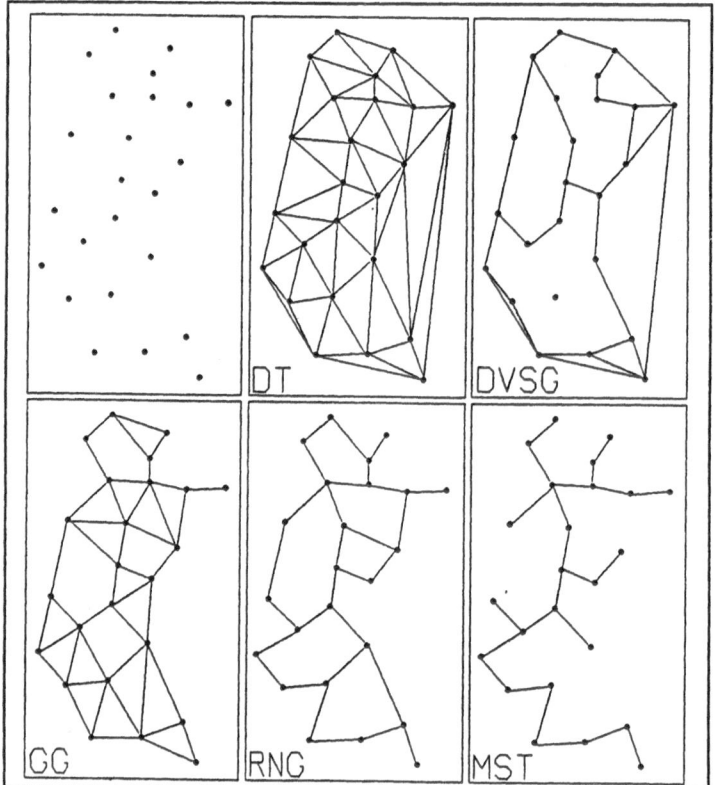

Figure 3 Comparison of the DVSG with subgraphs of the DT.

4   Extension to 3-D

The 2-D Delaunay triangulation has a 3-D equivalent. In 3-D points are connected into tetrahedra (c.f. triangles in 2-D). Each tetrahedron has no points in the interior of its circumsphere. The Voronoi diagram is also extendable to 3-D. Lines connecting the circumsphere centres of adjacent tetrahedra form Voronoi sides. Each Voronoi cell is the volume around a point closer to that point than any other.

Each face of a Delaunay tetrahedron has associated with it one <u>side</u> from the 3-D Voronoi diagram. Likewise, each face of the 3-D Voronoi diagram has associated with it one <u>side</u> of a Delaunay tetrahedron.

As in the 2-D case, input points to the 3-D extension are assumed to be potential edge points of 3-D objects.

Each face of a Delaunay tetrahedron is assumed to be a potential surface facet of an object. The decision that needs to be made for every face is whether it should be kept as a surface facet (when its points lie close to each other) or whether it should be ignored and the corresponding Voronoi side kept as a 3-D skeleton section. The method used here compares the absolute length of the Voronoi side to the mean length of the three sides making up the Delaunay face to which it corresponds. If the mean length of the Delaunay sides is less than the Voronoi side then the face is retained as a surface facet. Otherwise the Voronoi side is retained as a skeleton section.

The surface facets are individual Delaunay faces. Had the selection process been based around individual Delaunay sides then it would have been possible to generate a single one dimensional line as an individual surface. This does not seem desirable, especially when assuming that the tetrahedron faces form a superset of the object surface.

Skeletons can take a number of forms. At their simplest they can be single points corresponding to the circumsphere centre of a tetrahedra which has had all four of its faces retained as surface. More complex skeleton sections can be created. These more complex sections comprise *1) skeleton lines* which are stretches of one dimensional polylines connecting the centres of circumspheres of adjacent tetrahedra; *2) skeleton sheets* which arise when all of the faces that a particular Delaunay side is an edge of are retained as skeleton sections. Sheets occur when the surface generated cannot be described by a one dimensional axis. Figure 4a shows two tetrahedra formed from five points. Figure 4b shows the creation of a skeleton line which replaces a single tetrahedron face. Figure 5a shows four tetrahedra formed from six points. Figure 5b shows a skeleton sheet formed by replacing all of the tetrahedron faces of which the Delaunay triangle side AB is a part with Voronoi sections.

Objects are defined in the same way as for the 2-D case. They are sequences of unbroken (and possibly branching) skeleton sections (normal and sheets), and surrounding Delaunay faces.

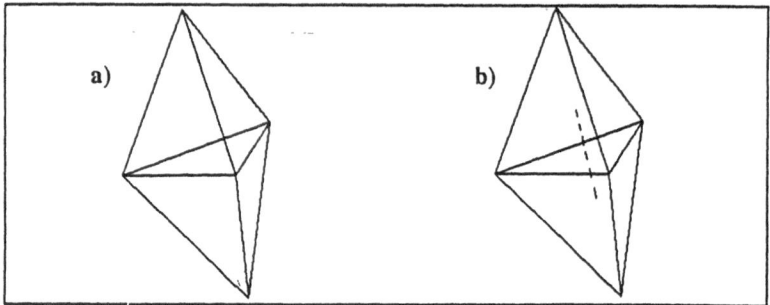

Figure 4 a) Two tetrahedra; **b)** tetrahedra and skeleton section (dashed).

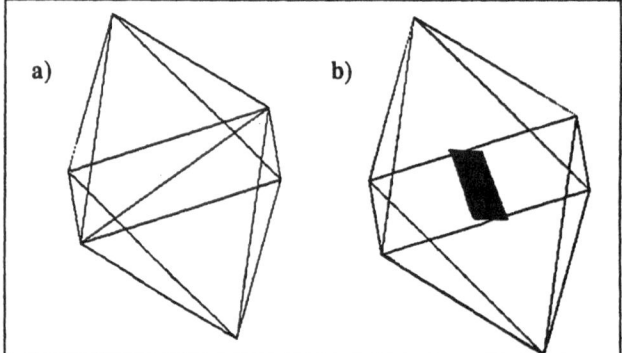

Figure 5 a) four tetrahedra; **b)** tetrahedra and skeleton sheet (solid)

5 Discussion

We construct a shape description of each object by decomposing the object into a hierarchy of sub-parts. Our technique simultaneously extracts boundary and skeleton representations of multiple objects from incomplete edge data. The input candidate edge points require no connectivity information. Objects are defined simultaneously in terms of their intact skeleton and boundary.

The connections between the points are generated through the algorithm using local geometry only, but a simple modification allows the user to specify partial connectivity or complete connectivity between points. Using information about connectivity means that the technique can be used as a simple edge connection algorithm [20] The purely local decision could be improved by incorporating information about continuity into the algorithm.

The creation of the skeleton and boundary sections for objects is a separate process from that of segmenting the objects into sub-parts and generating the hierarchical representation. It should be noted that the technique proposed here for generation of the skeleton and boundary sections can be used in conjunction with any other algorithm for segmenting the skeleton; and also that the technique proposed here for segmenting the skeleton could be used with any other method for generating sub-parts and hierarchies.

The current method of segmenting individual objects into their sub-parts usually appears to correspond well to subdivisions perceived by observers. This implies that the criteria of size and direction are sensible ones to use. However there are occasions where a different segmentation would be desirable. This is usually where a skeleton has no branch but the boundary that it separates has a significant narrowing. Use of the rate of change of the boundary-to-skeleton distance could be included as an extra criterion when deciding where to segment objects into sub-parts. We currently use an empirically determined threshold for deciding whether to split an object into sub-parts. The threshold is a ratio of local values (in the feature space) and is relatively insensitive to small changes in these values. However, a more robust approach is required.

The use of minima in the derivative of the ordered list of sub-parts provides a generally good hierarchical description. The major problem with the current approach is that there are superfluous levels in the hierarchy towards the fine end description, since the levels only use local criteria, and small perturbations in the size of the sub-parts means there are a large number of minima in the first differential. Alternative criteria for the hierarchical ordering of sub-parts could be used such as large percentage change in the size of sub-parts.

Our technique provides an easy method for splitting one object into two or combining two objects into one by simply changing the labelling of the Voronoi section and its corresponding Delaunay section [21]. This splitting and merging could be controlled by high-level knowledge in an attempt to match features in a model to objects generated by the proposed shape description. We are currently undertaking work in defining efficient metrics with which to match object descriptions extracted from the input data to those held within the high level model. In conjunction with this, we are extending our high level model[22] such that all features have a shape representation in the same form as the one described in this paper.

The technique can also be extended to 3-D. We have shown preliminary results of an extension to 3-D using a proximity measure computed from the mean length of the sides of a Delaunay tetrahedron face and the length of the associated Voronoi side.

6   References

[1]   D. T. Lee and B. J. Schacter, Two algorithms for constructing a Delaunay triangulation. Int J Comp Info Sci 1980; 9 No. 3: 219-242.

[2]   D. R. Bailes and D. H. Cooper, A Literature Review of Representation of Two Dimensional Shape, Uni. Manchester Tech. Report Mobprim/Mu/Lr1/881209, 1988. (UnPub)

[3]   S. Marshall, Review of Shape Coding Techniques. Image and Vision Computing 1989; 7 No. 4: 281-294.

[4]   H. Freeman, On the encoding of arbitrary geometric configurations. IRE Trans Electronic Computers 1961; June: 260-268.

[5]   N. J. Ayache, A model-based vision system to identify and locate partially visible industrial parts. Proceedings of the IEEE Computer Society Conference on Computer Vision and Pattern Recognition. IEEE, New York, 1983, pp 492-494.

[6]   H. Asada and M. Brady, The Curvature Primal Sketch. IEEE Trans Pat

Anal Machine Intel 1986; PAMI-8 N0. 1: 2-14.

[7] H. Blum, Biological Shape and Visual Science. J Theor Biol 1973; 38: 205-287.

[8] L. R. Nackman and S. M. Pizer, Three dimensional shape description using the symmetric axis transform. IEEE Trans Pat Anal Machine Intel 1985; PAMI-9 No. 4: 505-511.

[9] C. Arcelli, Pattern thinning by contour tracing. Comput Graphics Image Process 1981; 17: 130-144.

[10] J. R. C. Fairfield, Segmenting blobs into subregions. IEEE Trans Sys Man Cyb 1983; SMC-13 No. 3: 363-384.

[11] J. R. C. Fairfield, Segmenting dot patterns by Voronoi diagram concavity. IEEE Trans Pat Anal Machine Intel 1983; PAMI-5 No. 1: 104-110.

[12] R. Ogniewicz and M. Ilg, Skeletons with euclidean metric and correct topology and their application in object recognition and document analysis. Proceedings 4th International Symposium on Spatial Data Handling. Zurich, Switzerland, 1990, pp 15-24.

[13] N. Ahuja, Dot Pattern Processing Using Voronoi Neighbourhoods. IEEE Trans Pat Anal Machine Intel 1982; PAMI-4 No. 3.: 336-343.

[14] N. Ahuja and M. Tuceryan, Extraction of Early Perceptual Structure in Dot Patterns: Integrating region, boundary & component Gestalt. Comp Vis Graph Image Proc 1989; 48: 304-346.

[15] F. Meyer, The perceptual graph: a new algorithm. Proc. Int. Congr. Acoust. Speech Signal Process. Paris, 1982,

[16] G. T. Toussaint, The relative neighbour graph of a finite planar set. Patt Recog 1980; 12(4): 1324-1347.

[17] K. R. Gabriel and R. R. Sokal, A new statistical approach to geographic variations analysis. Systematic Zoology 1969; 18: 259-278.

[18] P. M. Lankford, Regionalisation: theory and alternative algorithms. Geograph Anal 1992; 1: 196-212.

[19] L. Vincent, Graphs and mathematical morphology. Signal Processing 1989; 16: 365-388.

[20] G. P. Robinson, A. C. F. Colchester and L. D. Griffin, A hierarchical shape representation for use in anatomical object recognition. In: Achcrya, R. S., Cogswell, C. J. and Goldof, D. B. (ed) Proc. SPIE Biomedical Image processing III and 3-D Microscopy. San Jose, 1992, pp 594-605.

[21] G. P. Robinson, A. C. F. Colchester, L. D. Griffin, et al., Integrated skeleton and boundary shape representation for medical image interpretation. In: Sandini, G. (ed) Proc. European Conference Computer Vision. LNCS-Series Vol 588. Springer-Verlag, 1992, pp 725-729.

[22] C. I. Attwood, G. D. Sullivan, G. P. Robinson, et al., Model-based interpretation of anatomical structures in cranial MR images. Proc. British Machine Vision Conference. Sheffield University Press, 1990, pp 145-150.

Range Recovery using Virtual Multi-camera Stereo

David W. Murray and Paul A. Beardsley

Robotics Research Group,
Department of Engineering Science,
Parks Road, Oxford University, OX1 3PJ, U.K.

Abstract

We describe the principles of a device comprised of a static camera and rotating plane mirror that enables the passive recovery of 3D range information. The range recovery can be regarded as either structure from known motion or as virtual multi-camera stereo. Two advantages of the arrangement are that the camera-mirror system is compact and that range information can be recovered over a wide, near panoramic, field of view.

1   Introduction

The most powerful methods of depth recovery using passive vision, shape from stereo and shape from motion, both involve obtaining images of a scene from different viewpoints, where the change of viewpoint must involve a translation of the optic centre of the camera. At the level of mechanism, this requirement detracts somewhat from the elegance of vision as a simple sensor. Either one has to have two (or more) cameras able to provide the different viewpoints simultaneously, or one has to have a device to move the camera to its new viewpoint. Both routes result in relatively complex and bulky apparatus.

We were concerned to find whether there were ways of making viewpoint displacements more simply to produce a compact, mechanically neat, sensor. The solution we present here exploits the near perfect specular reflection of broadband visible radiation from smooth planar metallic surfaces. In short, we do it with mirrors.

By pointing the camera at a mirror and rotating the mirror we provide a means of changing the viewpoint in a known way. Physically, the scene points are reflected by the moving mirror, and the resulting moving virtual scene is imaged in a stationary real camera, yielding image data which can be used to drive range recovery using "structure from known motion". Because the mirror's rotation axis is offset from the optic centre, rotation induces a translation of the scene relative to the optic centre. An equivalent but informative way of regarding the system draws on Fermat's principle, which indicates that one can consider the scene to be unreflected, but viewed by a virtual camera created by reflection in the mirror. As the mirror is moved, so this virtual camera moves, providing the multiple views. The mirror's rotation axis becomes the fixation point for stereo using multiple "virtual cameras".

As well as providing a compact mechanism, the device described recovers range over a wide field of view: apart from two blind directions, covering say 30°, this is panoramic. A further feature is that it is possible to recover range in a plane using only 1D image measurements; indeed this possibility is the one we explore experimentally here. Such a device may be useful for navigation.

The use of mirrors in not new in vision. They are of course used routinely in devices using active illumination (e.g. [5] describes an active IR rangefinder using a rotating mirror). In passive vision, plane mirrors were used by Cornog [2] in an early mechanism for redirecting gaze, and are now used for the same purpose in commercially available tele-operated surveillance systems. A conical mirror [7] has been used to obtain a panoramic view of a scene, and by establishing correspondence with a second similar panoramic image taken from a different viewpoint, panoramic stereo recovery is possible. (A similar approach has been made using spherical projection with a fish-eye lens [6].)

We have, however, been unable to find a similar application of moving mirrors to range recovery. The closest analogue to our system is that of Ishiguro *et al* [3, 4], in which *real* cameras are moved in the same way that our *virtual* camera moves (see Figure 2b). These authors however analyse their results quite differently. They establish the range of a scene point from correspondence between just two points one from each of a panoramic stereo pair. In this work we recover each range value from several tens of image measurements, using a form of the spatio-temporal (or, in our case, spatio-angular) epipolar analysis of Bolles *et al* [1].

We describe and analyse the rotating mirror system in the following section. In Section 3, we show experiments using a implementation of the device using 1D image measurements to recover 2D scene information. The results are discussed in Section 4, along with a modification which removes blind directions, though at the expense of the 1D image solution.

2 Imaging using a mirror system

Figure 1 shows the arrangement of camera and mirror for the first device to be experimented with. The camera's optic axis defines $\hat{z}$ and the optic centre is at the origin of

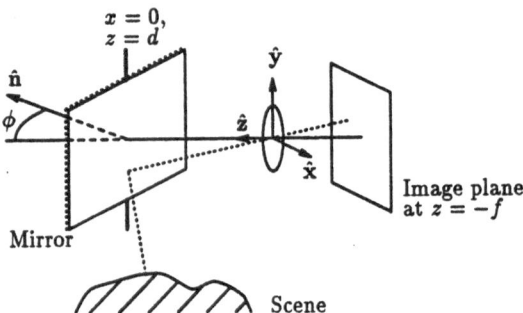

Figure 1: The rotating mirror device. The camera is placed in front of mirror which rotates about an axis parallel to $\hat{y}$. Images are formed under perspective on the image plane at $z = -f$.

the physical camera's right-handed coordinate system $(\hat{x}, \hat{y}, \hat{z})$. Images are formed under perspective projection on the image plane at $z = -f$. The mirror is placed at a distance d along the optic axis, and rotates about the axis $(x = 0, z = d)$, parallel to $\hat{y}$. The normal to the mirror is $\hat{n}$, pointing into the mirror surface, so that when the mirror is rotated by angle ϕ as shown

$$\hat{n} = (-\sin\phi, 0, \cos\phi)^T . \qquad (1)$$

Figure 2 sketches the two ways of modelling the system outlined in the introductory section. Either (a) we consider the virtual scene imaged by the real camera, or (b) image the real scene in a virtual camera (b), where we note that in the latter case the coordinate system attached to the virtual camera has reversed parity due to reflection. The latter model makes clear that the virtual camera rotates about the mirror axis. Because this

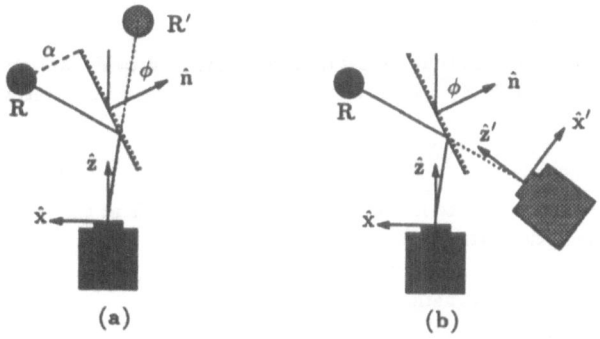

Figure 2: Two different ways of modelling the system. In (a) the virtual scene point (half filled) is imaged by the real camera (filled) and in (b) the real scene point is imaged in a "virtual camera".

axis does not pass through the real camera's optic centre, the optic centre of the virtual camera rotates *and* translates, the latter being necessary for depth recovery.

It is slightly simpler to derive the imaging conditions using model (a). Consider a scene point $\mathbf{R} = (X, Y, Z)^T$, its virtual reflection at $\mathbf{R}'$, and the image formed of it in the camera $\mathbf{r} = (x, y)^T$. Suppose that when the mirror is rotated by ϕ, the distance from $\mathbf{R}$ to the mirror is α, so that

$$\mathbf{R}_m = \mathbf{R} + \alpha \hat{\mathbf{n}} \tag{2}$$

is a point in the plane of the mirror. The equation of the mirror plane is

$$\mathbf{R}_m \cdot \hat{\mathbf{n}} = d \cos \phi , \tag{3}$$

whence

$$\alpha = d \cos \phi - \mathbf{R} \cdot \hat{\mathbf{n}} . \tag{4}$$

The virtual scene point is formed equidistant behind the mirror at

$$\mathbf{R}' = \mathbf{R} + 2\alpha \hat{\mathbf{n}} \tag{5}$$

$$= \begin{bmatrix} X \cos 2\phi + (Z - d)\sin 2\phi \\ Y \\ X \sin 2\phi - (Z - d)\cos 2\phi + d \end{bmatrix} . \tag{6}$$

Under perspective projection the virtual scene point is imaged at

$$\mathbf{r} = (x, y)^T = -f\mathbf{R}'/\mathbf{R}' \cdot \hat{\mathbf{z}} , \tag{7}$$

where f, the focal length, is assumed known from calibration. The image measurements are then of the form

$$m_x = \frac{x}{f} = -\left(\frac{X \cos 2\phi + (Z - d)\sin 2\phi}{X \sin 2\phi - (Z - d)\cos 2\phi + d} \right) \tag{8}$$

$$m_y = \frac{y}{f} = -\left(\frac{Y}{X \sin 2\phi - (Z - d)\cos 2\phi + d} \right) .$$

Now consider what happens as the angle of rotation of the mirror ϕ is changed. The image of a particular scene point moves across the image, eventually leaving the field of view. By establishing correspondence, we construct the image locus $\mathbf{r}(\phi)$ of this point, from which we obtain a set of measurements $\{\ldots, m_{xi}, m_{yi}, \ldots\}$ where $m_{xi} = x(\phi_i)/f$ and $m_{yi} = y(\phi_i)/f$, all arising from the same scene point.

2.1 Recovering range

It is evident from equation (8) that the set of measurements provides an over constrained linear system for $\mathbf{R} = (X, Y, Z)^T$. In fact, it is most straightforward to recover

$$\mathbf{R}^* = (X^*, Y^*, Z^*)^T = (X, Y, Z - d)^T . \qquad (9)$$

In other words, the natural place for the origin of coordinates is at the axis of rotation of the mirror, not the optic centre of the camera.

At a particular angle ϕ_i, equation 8 can be rewritten as

$$\begin{bmatrix} m_{xi}\sin 2\phi_i + \cos 2\phi_i & 0 & \sin 2\phi_i - m_{xi}\cos 2\phi_i \\ m_{yi}\sin 2\phi_i & 1 & -m_{yi}\cos 2\phi_i \end{bmatrix} \mathbf{R}^* = -d \begin{bmatrix} m_{xi} \\ m_{yi} \end{bmatrix} . \qquad (10)$$

For k samples at different angles ϕ_i, k such matrices are blocked together to form

$$[\mathbf{A}]\mathbf{R}^* = \mathbf{b} \qquad (11)$$

where $[\mathbf{A}]$ is an $2k \times 3$ matrix and $\mathbf{b}$ has length $2k$. Assuming independent measurements, with measurement i having weight W_{ii}, a least squares solution for the over-constrained system can be found by solving

$$[\mathbf{A}^T\mathbf{W}\mathbf{A}]\mathbf{R}^* = [\mathbf{A}^T\mathbf{W}]\mathbf{b} , \qquad (12)$$

where $[\mathbf{A}^T\mathbf{W}\mathbf{A}]$ is a real symmetric 3×3 matrix, and $[\mathbf{W}]$ is the diagonal weight matrix.

Although this solution is straightforward enough, equation 8 indicates that we can recover depth without the m_y measurements using

$$\begin{bmatrix} m_{xi}\sin 2\phi_i + \cos 2\phi_i \\ \sin 2\phi_i - m_{xi}\cos 2\phi_i \end{bmatrix}^T \begin{bmatrix} X^* \\ Z^* \end{bmatrix} = -dm_{xi} . \qquad (13)$$

This solution is especially useful provided we track features along the central horizontal raster, $y = 0$, from which which can recover range in the $Y = 0$ plane. For k measurements, equation (13) can be rewritten in terms of an $k \times 2$ matrix $[\mathbf{A}_{1D}]$ and length k vector $\mathbf{b}_{1D}$ analogous to $[\mathbf{A}]$ and $\mathbf{b}$, and the least squares solution is found by solving

$$[\mathbf{A}^T{}_{1D}\mathbf{W}\mathbf{A}_{1D}] \begin{bmatrix} X^* \\ Z^* \end{bmatrix} = [\mathbf{A}^T{}_{1D}\mathbf{W}]\mathbf{b}_{1D} . \qquad (14)$$

Our experimental implementation of the rotating mirror system has pursued this recovery of 2D scene data from 1D image measurements in both simulation and using imagery. It is convenient to define 2D range, ρ, in a particular direction, γ, measured from the mirror's rotation axis as

$$\rho = \sqrt{X^{*2} + Z^{*2}} \quad \text{and} \quad \gamma = \tan^{-1}\left(\frac{+X^*}{-Z^*}\right) \qquad (15)$$

respectively. This definition of γ relates simply to the definition of ϕ: if the mirror is set at angle ϕ, then the $x = 0$ vertical strip in the image in viewing in direction $\gamma = 2\phi$.

3   Experimental Results

As a preliminary test of noise sensitivity, range recovery was tested from artificially generated image contours. A typical trial is shown in Figure 3. One can imagine the device placed so that the mirror rotation axis is at the centre of a cross shaped room. At a number of directions γ around the device the actual (X^*, Z^*) values of the "walls" were used to simulate the locus that would be obtained on the image under mirror rotation from $\phi = -90°$ to $\phi = 90°$. Because the reflected ray rotates at twice the rate of the mirror rotation (the 2ϕ dependence of equations (8), this range covers the entire 360° field of view around the device. The field of view of the camera itself was limited to 40°. The image positions of the scene points were synthesized and then corrupted with Gauss random noise, and the least squares technique method described in Section 2 used to recover the values of X^* and Z^*, and hence the range and direction, ρ and γ.

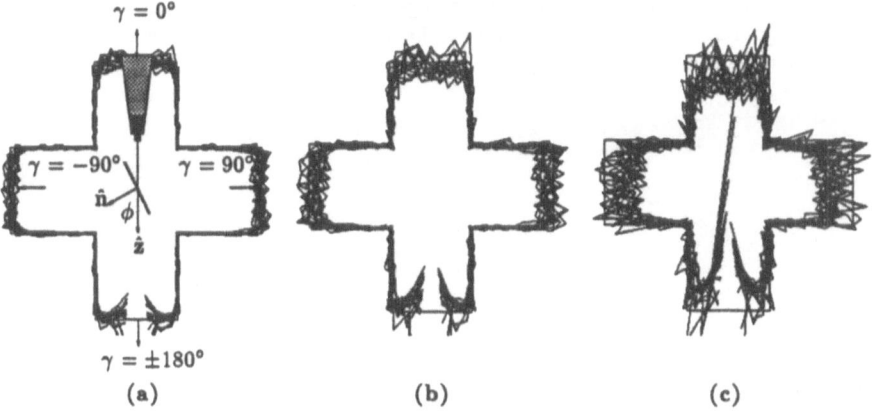

(a)   (b)   (c)

Figure 3: Simulations of range recovery using the rotating mirror under increasing error in image positions: (a) recovery with 2%, (b) with 4%, (c) with 8% noise added to locus positions.

Figure 3 shows the original outline of the "room" overlaid on the recovered range from several scans of the mirror for increasing values of image noise. The results show that range recovery is nearly panoramic. The shaded region indicates the invisible region around $\gamma = 0$ blocked out by the camera. The gaps ahead of the camera at $\gamma \approx \pm180°$ occur when the mirror is edge on — that is, $\phi \approx \pm90°$. In these positions, the m_{xi} are all close to zero and equation (13) degenerates to $X^* = 0$, leaving insufficient information to recover Z^*.

Another interesting feature is that as noise is introduced the range recovered tends to reduce, a useful conservative feature for navigation. The plots (especially (c)) suggest that this reduction in ρ is uniform in all directions (notwithstanding the impossibility of recovering depth when $\phi = 0$), and it is possible to show that the degradation is graceful. Suppose we assume that because of noise the measurements m_{xi} are actually independent of X^* and Z^*. Then equation 13 must be split into two parts, each independent of the m_{xi}. The parts are

$$m_{xi}\sin 2\phi_i X^* - m_{xi}\cos 2\phi_i Z^* = -dm_{xi} \tag{16}$$
$$\cos 2\phi_i X^* + \sin 2\phi_i Z^* = 0 . \tag{17}$$

Dividing out the first of these by m_{xi} and solving:

$$X^* = -d\sin 2\phi_i \quad Z^* = d\cos 2\phi_i , \tag{18}$$

whence

$$\rho = d \ , \tag{19}$$

independent of ϕ, as observed.

3.1 Experiments with imagery

As a prelude to building a dedicated device, we have exploited a robot arm to provide rotation.

The device is housed in a perspex safety cage in the corner of a visually cluttered laboratory, whose approximate plan view is shown in Figure 4. To establish scene points with known range, most of the inside of the perspex cage (A1 and A2) was "wallpapered" with black on white patterns, as shown in the Figure, although at (B) the device could see out. Objects were scattered about the walls (C1 and C2). For operational reasons, the wallpaper was place on the outside of the cage door at (D). These features are also marked on Figure 5, a panoramic view around the device. This image is for explanation only, and plays no part in the analysis.

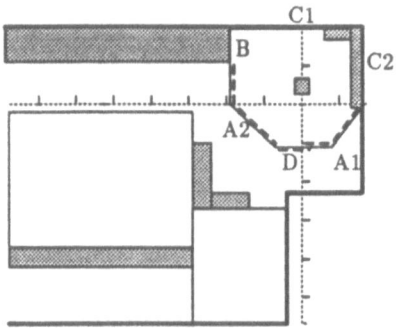

Figure 4: A plan of the workspace inside the cage. Thick lines are walls, the half shaded regions are populated with visually interesting equipment, and the lines are perspex safety cages. The dashed line represents the "wallpaper". The scale is in metres.

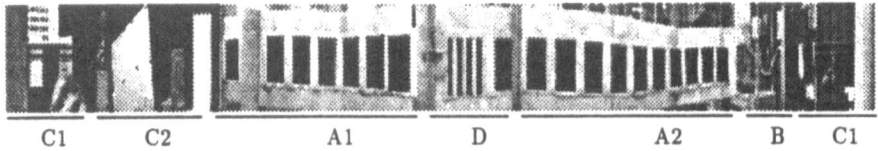

| C1 | C2 | A1 | D | A2 | B | C1 |

Figure 5: A panoramic view inside the cage. The labels refer to the text and Fig.4.

The camera and image capture electronics were calibrated to determine the optic centre relative to the top-left of the framestore ($x_c = 256.6(4)$ pixel, $y_c = 255.4(4)$ pixel) and focal length ($f_x = 1302(8)$pixel) and aspect ratio $s = 1.54(1)$, allowing framestore coordinates to be converted to world coordinates. The distance d between rotation centre and optic axis was determined as $0.176(1)m$. A single rotation scan was performed, moving the mirror angle ϕ between $-90°$ and $90°$ in $0.25°$ steps. Image rasters $-15 < y < +15$ centred about $y = 0$ (in world coordinates) were captured and edges derived using the Canny operator followed by hysteresis linking and thresholding, and the resulting edge information from the central $y = 0$ raster, consisting of $x-$position, orientation and contrast, stored. Near-horizontal edges were discarded as their intersection with the horizontal raster is likely to be uncertain.

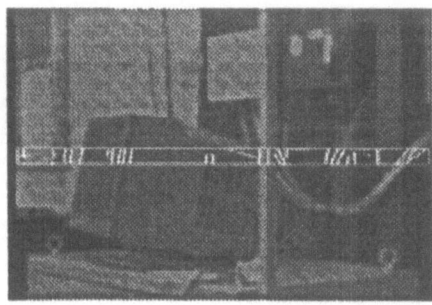

Figure 6: Edgels were computed in a central image block, and those of sufficient contrast linked into extended contours. The x−position, contrast and orientation of those in the central raster $y = 0$ were stored.

A simple matcher was used to link corresponding edgels up to form extended contours $x(\phi)$, using expected position, contrast and orientation as matching attributes. The change is x−position is bounded by $\Delta x = 0$ if the scene point is at range $\rho = 0$, and $\Delta x \approx f(2\Delta\phi)$ when the range is infinite. In our case this maximum was about 14 pixel. Of course, as the slope $dx/d\phi$ of a contour does not change markedly, the search range can be reduced after the first contour measurements are made. The angle of view of the camera itself was around 23°, and the maximum number of matches in a contour was thus 46. We retained (somewhat arbitrarily) all those with greater than 30 matches for further analysis. The contours $x(\phi)$ are shown in Figure 7.

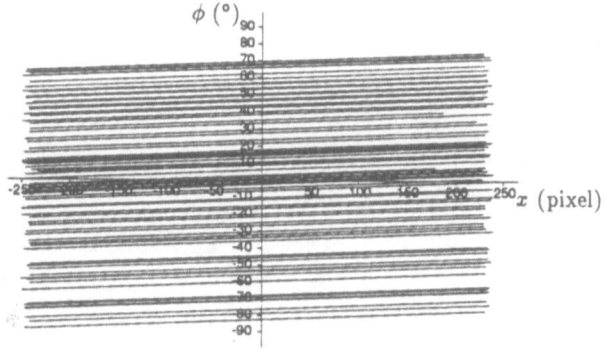

Figure 7: Contours $x(\phi)$ created by matching edgels in successive frames. The x-axis is in pixels, the ϕ-axis in degrees.

The set of discrete measurements from each contour was analysed using the least squares method of Section 2, and the recover scene points plotted in Figure 8. The outline of the cage is recovered well, as are objects from the side walls. Note that the depths where the device could see out of the cage are indeed greater. The two points recovered at $Z^* \approx -5m$ caused some surprise. However, as we noted earlier, the wallpaper on the cage door was on the *outside* and these points correspond to some strong reflections of the equipment at A in the perspex cage.

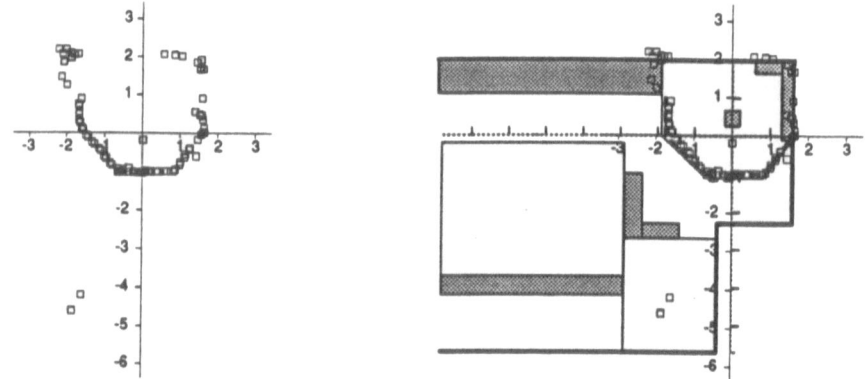

Figure 8: The recovered scene points (left) and overlaid on the room plan (right). Distances are in metres.

In a further experiment, the "wallpaper" was removed, and three objects (polystyrene mannequins) introduced closer to the device. The panorama (Figure 9) shows that much of the scene is visually cluttered, and we again expect to suffer problems of reflections in the cage walls, making it difficult to interpret the results in terms of specific objects. Nonetheless, the back wall is now interpretable, and all the depths are broadly as expected (Figure 10).

Figure 9: A panoramic view of the entire laboratory.

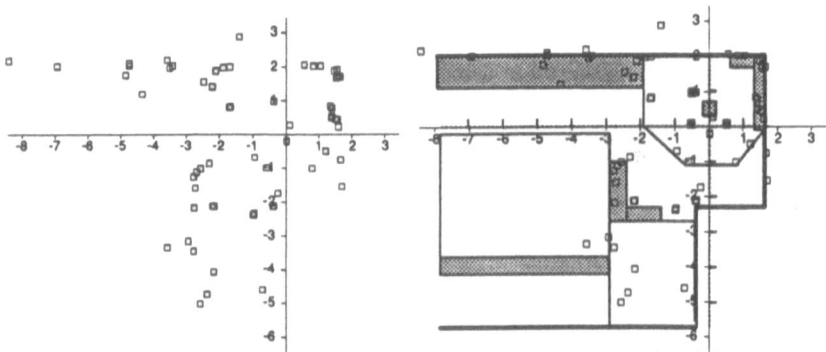

Figure 10: On the left we show the recovered scene points, and on the right we overlay them on the room plan for comparison. The scale is in metres.

4   Discussion

The device described shows considerable promise as a compact, wide angle of view passive vision sensor, and seems most applicable to autonomous navigation. We now discuss some of the perceived drawbacks and merits of the system.

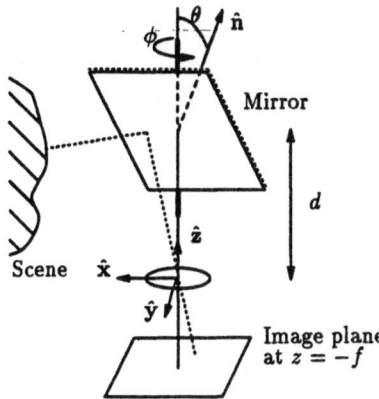

Figure 11: The second system. The camera is placed with its optic axis vertical and the mirror tilted by θ ($\theta = 45°$ is the likely choice). The mirror rotates about the camera's optic axis $\hat{\mathbf{z}}$.

First it is appropriate to compare the merits of our analysis with that in Ishiguro *et al* [3, 4] in their analagous real camera system. As their camera rotates, by taking two *vertical* strips of one pixel width from each image they build up a panoramic stereo pair of images. By establishing correspondence between pairs of features in the two images they are able to compute range. Essentially then they must ensure that a feature observed in one pixel-wide strip is captured in the other, and so they must rotate the camera in angular increments of approximately $1/f$ radian, where f is the focal length in pixels. As $f \approx 1000$ this requires high angular resolution. Our approach is to establish correspondence and track features through each successive image at much lower angular resolution. In the experimental system that recovers range in the 2D plane, this involves taking a single *horizontal* raster from each image. By doing this we can rotate by larger angles, but still recover several tens of image measurements from which to recover the range of single scene point, rather than just two measurements.

One disadvantage of the mirror system compared with rotating a real camera with the configuration described is that a panoramic view is marred by blind spots when the mirror is edge-on and when the camera looks at itself. A different configuration which remedies this problem is shown in Figure 11. The fixed camera is positioned with its optic axis $\hat{\mathbf{z}}$ vertical and the mirror is tilted at some fixed angle θ (likely to be 45°) and rotates through angles ϕ about the optic axis. The analysis closely follows that of Section 2, but the normal to the mirror is now

$$\hat{\mathbf{n}} = (\sin\theta\cos\phi, \ \sin\theta\sin\phi, \ \cos\theta)^T \ . \tag{20}$$

The perpendicular distance to the mirror plane is now fixed at $d\cos\theta$, and the virtual scene point is therefore at

$$\mathbf{R}' = \mathbf{R} + 2(d\cos\theta - \mathbf{R}\cdot\hat{\mathbf{n}})\hat{\mathbf{n}} \tag{21}$$

The measurements are

$$m_{xi} = \frac{x(\phi_i)}{f} = -\frac{X'(\phi_i)}{Z'(\phi_i)} \tag{22}$$

$$m_{yi} = \frac{y(\phi_i)}{f} = -\frac{Y'(\phi_i)}{Z'(\phi_i)} \ , \tag{23}$$

and as in Section 2 these can be used as a set of linear equations for $\mathbf{R}^*$:

$$[A]\mathbf{R}^* = \mathbf{b} \ . \tag{24}$$

where for k measurements $[A]$ is an $2k \times 3$ matrix and $\mathbf{b}$ has length $2k$. The contribution from the ith measurement to $[A]$ is

$$[A] = \begin{bmatrix} \vdots & \vdots & \vdots \\ (2S_\theta^2 C_{\phi_i}^2 - 1 + m_{xi}S_{2\theta}C_{\phi_i}) & (S_\theta^2 S_{2\phi_i} + m_{xi}S_{2\theta}S_{\phi_i}) & (S_{2\theta}C_{\phi_i} + m_{xi}C_{2\theta}) \\ (S_\theta^2 S_{2\phi_i} + m_{yi}S_{2\theta}C_{\phi_i}) & (2S_\theta^2 S_{\phi_i}^2 - 1 + m_{yi}S_{2\theta}S_{\phi_i}) & (S_{2\theta}S_{\phi_i} + m_{yi}C_{2\theta}) \\ \vdots & \vdots & \vdots \end{bmatrix} \tag{25}$$

and to $\mathbf{b}$ is

$$\mathbf{b} = d \begin{bmatrix} \vdots \\ m_{xi} \\ m_{yi} \\ \vdots \end{bmatrix} , \tag{26}$$

where $S_\theta^2 = \sin^2 \theta$, $S_{2\theta} = \sin 2\theta$ and so on. The obvious disadvantage with this modified system is that there is no 1D version available using matching within a single raster. The image will rotate about the optic centre, and tracking would have to be performed in 2D, using say a corner detector.

Another difficulty with the present treatment of the data is that a sparse depth map is recovered. Of course this is inevitable if features invariant to raw intensity changes are sought. However, as the viewpoint changes are small, it may be that a matching process using attributes closer to the raw intensity would suffice. The technique that immediately suggests itself is dynamic time warping. This technique might also address a further and most pressing problem with the present arrangement, that of speed. For the device to be practical, we must complete a complete scan in say a couple of seconds. Using $\Delta\phi = 0.25°$ then requires an acquisition and processing time per frame of order 3 ms. This suggests that we should consider only the 1D device as feasible at present and use fast linear sensor arrays with dedicated processing hardware.

References

[1] R C Bolles, H H Baker, and D H Marimont. Epipolar-plane image anlysis: an approach to determining structure from motion. *International Journal of Computer Vision*, 1:7–55, 1987.

[2] K H Cornog. Smooth pursuit and fixation for robot vision. Master's thesis, Department of Electrical Engineering and Computer Science, MIT, 1985.

[3] H Ishiguro, M Yamamoto, and S Tsuji. Omni-direction stereo for making global map. In *3rd International Conference on Computer Vision, Osaka, 1990*, pages 540–547, Washington DC, 1990. IEEE Computer Society Press.

[4] H Ishiguro, M Yamamoto, and S Tsuji. Omni-directional stereo. *IEEE Transactions on PAMI*, 14(2):257–262, 1992.

[5] G L Miller and E R Wagner. An optical rangefinder for autonomous cart navigation. In *Proceedings SPIE Mobile Robots II*, volume 852, pages 132–144, Bellingham, Washington, 1987. SPIE.

[6] T Morita. Measurement in three dimensions by motion stereo and spherical mapping. In *Proceedings of the IEEE Conference on Computer Vision and Pattern Recognition, 1989*, pages 422–428, Washington DC, 1989. IEEE Computer Society Press.

[7] Y Yagi and S Kawato. Panoramic scene analysis with conic projection. In *IEEE International Workshop on Intelligent Robots and Systems*, pages 181–190, Washington DC, 1990. IEEE Computer Society Press.

Robust Recovery of 3D Ellipse Data

Stephen Pollard and John Porrill

AI Vision Research Unit, University of Sheffield
Sheffield, England.

Abstract

This paper is concerned with robust, accurate and computationally tractable methods for the automatic recovery of 3D ellipse data from edge based stereo. The processing paradigm relies heavily on the 2D image as a rich and robust source of scene feature hypotheses (in this case ellipses). Rather than attempt to recover 3D scene descriptions by grouping unstructured estimates of disparity and/or depth, a processes of automatic 2D feature hypothesis is used in conjunction with an appropriate disparity grouping constraint (in the case of ellipse hypotheses we use an affine disparity plane constraint) to recover more accurate 3D scene descriptors.

1 Introduction

Exploratory work in [1] and [2] has shown the recovery of accurate ellipse data in 2D from pure edge data and in 3D from edge based stereo respectively. Here we concentrate attention upon the difficult problem of resolving the scene segmentation problem. That is the task of decomposing the continuous noisy edge contours recovered from the scene in terms of their constituent elliptical conic sections and the subsequent combination of ellipse data recovered from different contours into more complete and useful descriptions.

2 Conic fitting

We are given a string of edge points in 2D with some given accuracy of measurement and we wish to find the best estimate of the conic fitting these points. We represent a conic by its quadratic form

$$ax^2 + 2bxy + cy^2 + 2dx + 2ey + f = 0$$

the 6 coefficients are determined only up to a constant factor. Since we are not interested in fitting hyperbolae we normalise the coefficients using the additional linear constraint

$$a + c = 1$$

so that the conic is completely described by the measurement equation

$$z = \mathbf{h} \cdot \mathbf{x}$$

where

$$\mathbf{x} = (a \quad b \quad d \quad e \quad f)'$$
$$\mathbf{h} = (x^2 - y^2 \quad 2xy \quad 2x \quad 2y \quad 1)'$$
$$z = -y^2$$

Assuming that a given measurement point lies on the conic gives us a single linear equation for the conic coefficients. Given five points on the conic we can determine it uniquely (given the above constraint). Given a string of error prone measurements we can determine the least squares solution by standard methods. Though computationally cheap, the result is often highly unsatisfactory, showing a marked bias to high curvature conics. This problem is resolved by the use of the bias corrected linearised Kalman filter [1]. (Using the usual formulation of the linearised Kalman filter does not lead to an improvement, for most implicit equations the linearisation involved takes place about the *wrong point*).

Another advantage of the Kalman filter approach is that it supplies estimates of the variance and covariances of the estimated conic coefficients, given the known accuracy of the data, in the form of a 5 by 5 covariance matrix S. This allows us to test the likelihood that the conic fitted to a small section of data extends to pass through a given point (x, y) by treating the weighted square error

$$\frac{(z - \mathbf{h} \cdot \mathbf{x})^2}{\mathbf{h}' S \mathbf{h}}$$

as a chi squared variable on one degree of freedom. Using this test we can find all areas into which a search for extensions to a fitted conic is reasonable at some confidence level.

3   Edge contour segmentation

The task of decomposing extended edge contours into meaningful subsections has received considerable attention from both the vision and graphics communities. In essence two methods exist. First those that perform an analysis of the local curvature along an edge with a view to identifying points of rapidly changing curvature to act as segmentation or knot points[3][4]. The sections between knot points are then fitted with an appropriate approximating curve. As the curvature computation relies upon the ratio of discrete first and second difference operators it has the effect of amplifying noise. Hence the segmentation is inherently unstable and are unable to overcome errors that arise in the imaging process and are added as a result of edge detection .

We prefer to adopt the alternative strategy in which we fit primitives directly to those parts of the edge contours that lie between knot points without first making the knot points explicit. As we are fitting directly to the edge data it is straightforward to ensure that the method is robust to local deviations of the underlying edge contour from the fitted primitive. Our method has been used previously to obtain robust polygonal approximations [2], it has now been extended to allow generic decomposition of edge contours into any algebraic form.

Before describing our algorithm we shall outline 4 important criteria that affect the segmentation process. *Robustness*: the effect of local noise upon the decomposition process should remain local; preferably the method should be immune to a large class of normally expected errors. *Stability*: the partitioning should be stable and invariant.

Compactness: the decomposition should be concise. The minimal number of approximating primitives should be used. *Efficiency:* the method should be computationally efficient. Preferably the computational complexity should reflect the physical complexity of the edge contour rather than some other implementation dependent metric, such as its length in pixels.

In common with [4] and [5] we believe that an important goal in the design of a segmentation algorithm is that it should recover as close as possible a description of a contour at its most 'natural' scale or scales. Only then can the requirements for robustness, stability, compactness and efficiency be addressed satisfactorily. We differ from previous work in (1) that we try to describe contours by primitives at the largest scale that gives a sufficiently accurate description, and (2) in the nature of the algorithm chosen to perform the segmentation.

3.1 Finding ellipses at a single scale

Imagine that we have sampled an edge contour, that includes at least one elliptical section (e.g. look ahead to figure 1), at a scale 'natural' for segmenting out a particular ellipse. If we consider each set of 5 successive sample points along the contour as the 'starting point' of a potential ellipse then a 'natural' sampling frequency is one for which at least one 5 point span along the contour is guaranteed to almost cover the entire elliptical section.

The algorithm to identify the ellipse is straight-forward. Start by "testing" each 5 point span along the contour to see which if any are appropriate for description as an ellipse. The "test" is performed between the fitted ellipse and the underlying edge contour. The 5 point span sections of the contour that pass the initial ellipse decomposition phase then undergo an iterative refinement process during which the fitted ellipse is 'grown' in each direction along the contour. Each iteration of the growing procedure has three stages:

i. Apply the bias corrected Kalman filter to all the edge data covered by the current ellipse fit. This results in a more accurate ellipse description.

ii. Test that the fitted conic is still consistent with the underlying edge data. If not go back to the previous estimate.

iii. Extend the fitted conic in each direction along the edge contour up to the point where it no longer fits.

The ellipse recovery process completes when no further extension is possible.

We employ a form of hysteresis thresholding in both the testing and growing phases of our segmentation scheme. Three parameters control this process, an upper threshold than must not be exceeded, a lower threshold that can be exceeded transiently and a deviation count threshold that determines the maximum number of steps for which the lower threshold can be exceeded. Through the use of a large upper threshold and deviation count the segmentation scheme can be made relatively insensitive to moderate local deviations from the true ellipse and at the same time employing a small lower threshold ensures that a tight rein is kept upon slow but systematic deviation from the fitted data.

3.2 Segmenting edge contours at all scales

Edge contours are recursively decomposed into geometric primitives starting from the largest and leading to the smallest using a sophisticated form of split and merge algorithm (although merging is replaced by growing). The algorithm is first applied at a coarse scale of description and then recursively re-applied at ever finer sub-scales until all sections of the edge contour have either been described or the remaining fragments fall below a threshold for that type of primitive.

The recursive procedure to describe an edge contour has the following form:

```
contour_segment(start, end, sample, step)
ContourPoint      start
ContourPoint      end
IntegerNumber     sample
IntegerNumber     step
{
    [1] consider in turn each span, of length "sample", along
        the contour until a suitable fit is found between the
        chosen primitive and the underlying edge contour (this
        could be the best such fit).

    [2] if the fitting process was successful then grow the
        primitive along the edge contour keeping track of its
        start and end (pstart and pend respectively).
        Now recursively call
        contour_segment(start, pstart, sample, step)
        contour_segment(pend, end, sample, step)
        to describe the remaining sections of the contour and
        EXIT this function call.

    [3] reduce the sample and step lengths by a fraction.

    [4] if the sample is now below a sample length threshold
        then no fit is possible for this contour section so EXIT
        this function call.

    [5] recursively call
        contour_segment(start, end, sample, step)
        with the reduced sample length and step size.
}
```

In this recursive procedure the arguments **start** and **end** define the two end points of the edge contour or a contiguous sub-contour extracted from it. The argument **sample** is the span length parameter that defines the current sample frequency and **step** determines the degree of overlap of neighbouring samples. The effect of a call to this procedure is to keep increasing the initial sampling frequency by reducing the sample length until it is possible to get a satisfactory fit between the primitive and a contour sub-section. This is the largest one (or the first equal largest one) and is described at approximately its natural scale. In practice it is admissible to impose an upper threshold on the sample length without unduly affecting the behaviour of the algorithm. The recovered primitive is then extended along the contour within the bounds defined by **start** and **end**. The procedure then deals in turn with the remaining sub-sections of the contour.

Despite the apparent algorithmic complexity of this approach the method is in fact computationally effective. The complexity of the algorithm is independent of the length of the edge contour *per se*; for example two single span ellipse sections of different lengths will be recovered in the same time. It depends instead upon the physical complexity of the contour. As geometrical primitives are identified at approximately their largest natural scale repetitive work at finer scales is avoided. Hence the algorithm has a favourable computational workload when compared to other algorithms used for contour decomposition. Furthermore, fitting and growing large sections first also improves the robustness of segmentation. Short but convoluted contour sections do not effect this algorithm.

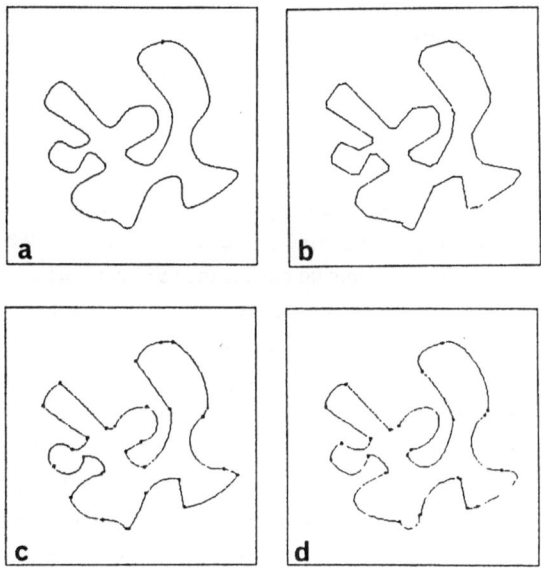

Figure 1. *Example of segmenting a complex edge string. The edge contour shown in (a) was recovered from a real image using a local implementation of the Canny edge detector and a nearest neighbour edge linking algorithm. The decompositions in (b), (c) and (d) are obtained with straight line, circle and ellipse based segmentation schemes respectively. The knot points are used to illustrate the less obvious segmentations of (c) and (d). All 3 segmentations used the same standard set of parameter and thresholding values.*

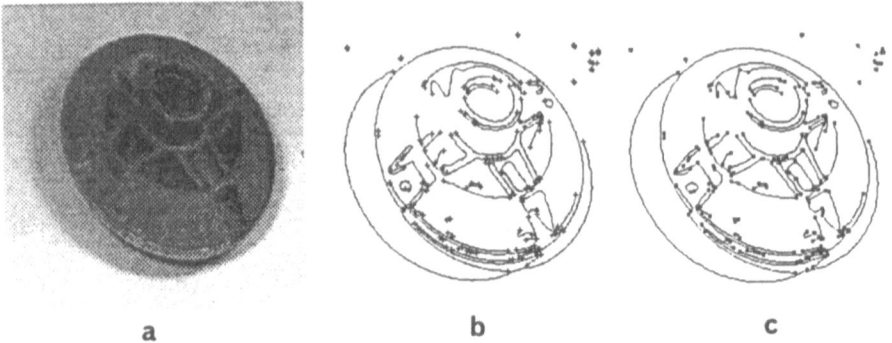

a   b   c

Figure 2. *Example of an industrial part. The theme object used throughout the remainder of the paper is shown on (a), with Canny edge contours shown in (b), and segmented ellipses in (c).*

4   Feature combination

We now address the issue of feature combination. That is, which elliptical curves can be combined with other ellipses or straight lines to produce larger and more robust descriptions. Larger primitives form a better basis for subsequent integration with disparity data and result in more accurate 3D primitives. However, improper grouping at this stage may result in the creation of accidental 2D features that have no 3D counterparts. Hence we adopt a fairly conservative grouping strategy.

The principal criteria for allowable grouping is the statistical test provided by the confidence interval of the Kalman filter. As a hard and fast rule we reject for combination those primitives whose mid-points lie outside the 99% confidence limit with respect to the ellipse that is under consideration for extension. Based upon the this very liberal initial criteria we have developed a rapid index into the set of possible primitives that could be used to extend the current ellipse. First we produce a spatial index of the visual array on the basis of a regular grid of bins into each of which we store the list of geometrical primitives whose mid-points lie within it. Hence each feature is represented exactly once in the indexing array. Clearly it is now possible to use the chi-square test with respect to the grid points of the index to determine which lists of image features are candidates for inclusion in the extension. If either the extrapolated ellipse (and not the physical part of the ellipse already recovered) passes through a bin of the index grid or one or more vertices of that bin lies within the 99% confidence limit then the list of primitives whose mid-points are stored within it are chosen as potential candidates. What we have in effect is a quantised form of the confidence envelope of the original ellipse.

Rather than exhaustively searching the whole spatial index for just those few grid elements that fall within the confidence envelope, we exploit the monotonic nature with which confidence falls of with respect to distance away from the extrapolated ellipse. In other words the confidence envelope is fully connected and once we have identified a single grid element that lies within the confidence envelope we know it must be indirectly connected in the eight neighbour sense to all other such elements. Points along the extrapolated curve (especially its midpoint) provide prime and easily computable candidates for a simple bush-fire technique to identify all elements inside the envelope (see figure 3).

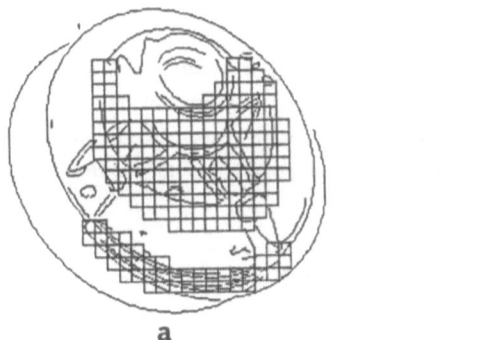 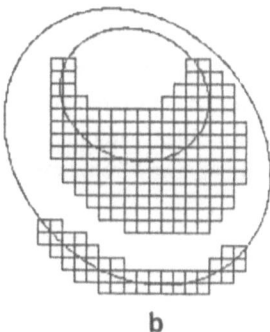

a   b

Figure 3. *Confidence envelope indexing. The results of using the quantised confidence envelope indexing scheme are illustrated for 2 partial ellipses identified in the theme figure. In (a) the indexed members of the spatial array are shown superimposed over the original ellipse data. In (b) the ellipses that gave rise to them are shown in isolation, extended along the loci of highest confidence recovered by the Kalman filter. Notice how the longer more complete ellipse provides a much tighter constraint envelope.*

In addition to high values of chi-square confidence (90% for example) we have identified a number of additional heuristics that help select good candidates for possible conic extension. They are:

Proximity. That the closest end-points of the current ellipse and the candidate feature fall within a small threshold.

Co-tangency. That the tangent directions of the closest end-points of the current ellipse and the candidate feature are almost collinear.

Exclusivity. Curve segments must not overlap along their lengths.

Maximum Gap. Do not grow a conic over a gap that is larger than either one of the constituent parts.

Invariant Curvature Ratio. Any two points on the same ellipse should have an invariant curvature ratio of almost unity. This ratio is defined as:

$$\frac{\kappa_1 \sin^3 \theta_2}{\kappa_2 \sin^3 \theta_1}$$

where κ is curvature and θ is the angle between the chord (between the 2 points) and the tangent.

Consistency. Is the combined ellipse consistent with each of the constituent edge contours.

Note that in order to deal with slight over extension of the original ellipses that may have occurred during their growing phase (especially if they are not bounded by tangent discontinuities) we only insist upon consistency with respect to the combined conic description away from the endpoints of the original edge contours (see figure 4).

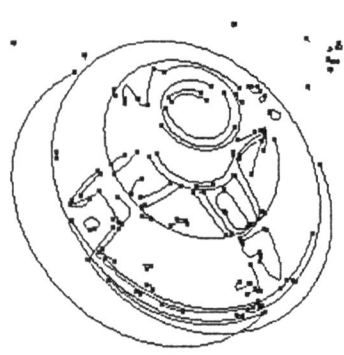

Figure 4. *Combined conics.*

5 Plane fitting

As we perform stereo matching and conic fitting in a 'rectified' parallel camera coordinate frame that is equivalent to the original camera geometry (which is itself recovered by accurate calibration procedure) then the transform of a plane curve between the images will be a y preserving affine as all epipolars are horizontal. That is, the transform from left to right image points is given by:

$$X_l = aX_r + bY + c \quad (1)$$

where X_l and X_r are the left and right horizontal components of position and Y is the common vertical component.

Rearranging we get

$$X_r = \frac{X_l}{a} - \frac{bY}{a} - \frac{c}{a} \quad (2)$$

Using (1) we can solve for the affine parameters using the weighted least squares form of:

$$\begin{pmatrix} X_r & Y & 1 \\ & \cdot & \\ & \cdot & \\ & \cdot & \end{pmatrix} \begin{pmatrix} a \\ b \\ c \end{pmatrix} = \begin{pmatrix} X_l \\ \cdot \\ \cdot \\ \cdot \end{pmatrix}$$

Disparity D is defined as $X_r - X_l$, hence from (2):

$$D = X_l \left(\frac{1}{a} - 1 \right) + Y \left(\frac{-b}{a} \right) - \frac{c}{a}$$

Which defines a plane in a disparity space based upon the left image.

$$D = X_l A + YB + C$$

From which it is straight forward to recover the plane in world coordinates based upon the optical centre of the virtual 'rectified' left camera.

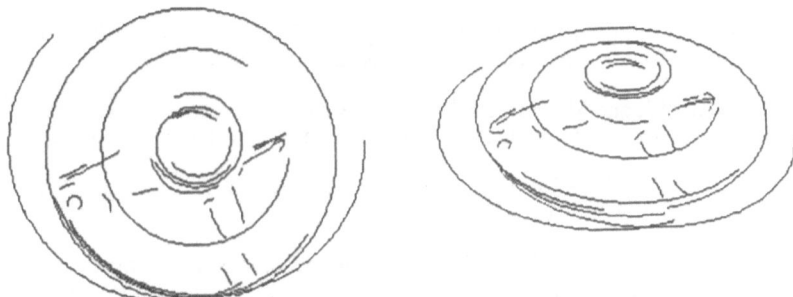

Figure 5. *Two very different views of the reconstructed 3D ellipse data recovered from our theme image are shown in (a) and (b).*

48

6   Conclusions

We have shown that is possible to exploit the rich source of structural information that exists in the 2D image in order to drive the recovery of 3D information. The resulting 2D and 3D geometrical primitives are both robust and accurate. Furthermore the algorithms that we have exploited are computational feasible on modern day computer hardware. Our emerging competence at representing and from there recognising complex industrial parts goes some way to reducing the gap between that exist between state of the art computer vision systems and the requirements of complex visual automata for the industrial market-place. However, extensions to the work presented here could and probably should be directed along the following 3 directions.

1. Introduction of higher level grouping constraints in both 2D and 3D. Examples include co-centricity, co-linearity, co-planarity, orthogonality, cylindricality, etc.

2. Exploitation of more data from the 2D image. Currently we restrict our image based grouping and description operators to work upon edge data alone.

3. Two eyes are better than one. Ideally the stereo description should tend towards the union rather than the intersection of the information provided by each image.

To summarise, the act of forcing a unique description for a particular section of an edge string is not always a reliable strategy. Only in the context of higher level descriptions (e.g. full conics, cylinders, etc.) which are verified against the underlying image intensity data can some lower level ambiguities be resolved. The descriptions obtained in this paper can be used to provide a powerful initial hypothesis for such a strategy.

References

[1] Porrill J (1990) Fitting ellipses and predicting confidence envelopes using a bias corrected Kalman filter, Image and Vision Computing, vol. 8, no 1, 37-41.

[2] Pollard S B, Porrill J, and Mayhew J E W (1991) Recovering partial 3D wire frames descriptions from stereo data, Image and Vision Computing, vol 9, no 1, 58-65.

[3] Asada H, and Brady M (1986) The curvature primal sketch, IEEE Trans. Pattern Anal. Machine Intell., vol. PAMI-8, no 1, 2-14.

[4] Wuecher D M and Boyer K L (1991) Robust contour decomposition using a constant curvature criterion, IEEE Trans. Pattern Anal. Machine Intell., vol. PAMI-13, no 1, 41-51.

[5] Hoffman D D (1983) Representing shapes for visual recognition, PhD Thesis Massachusetts Institute of Technology.

[6] Lowe D G (1988) Organisation of smooth image curves at multiple scales, in Proc. 2nd Int. Con. Computer Vision, 558-567.

Affine and Projective Structure from Motion

Sabine Demey[*]

Katholieke Universiteit Leuven

Department of Mechanical Engineering, Division PMA

Heverlee, Belgium

Andrew Zisserman and Paul Beardsley[†]

Robotics Research Group, Department of Engineering Science

Oxford University, OX1 3PJ.

Abstract

We demonstrate the recovery of 3D structure from multiple images, without attempting to determine the motion between views. The structure is recovered up to a transformation by a 3D linear group - the affine and projective group. The recovery does not require knowledge of camera intrinsic parameters or camera motion.

Three methods for recovering such structure based on point correspondences are described and evaluated. The accuracy of recovered structure is assessed by measuring its invariants to the linear transformation, and by predicting image projections.

1 Introduction

A number of recent papers have discussed the advantages of recovering structure alone, rather than structure and motion simultaneously, from image sequences [4, 5, 6]. Briefly, structure can be recovered up to a 3D global linear transformation (affine or projective) without the numerical instabilities and ambiguities which normally plague SFM algorithms. In this paper we compare and evaluate three methods for obtaining such structure. The novelty of this approach is that camera calibration, extrinsic or intrinsic, is not required at any stage. The absence of camera calibration facilitates simple and general acquisition: Structure can be recovered from two images taken with different and unknown cameras. All that is required for unique recovery is point correspondences between images. Here the points are polyhedra vertices.

The three methods are labelled by the minimum number of points required to compute the epipolar geometry (see below). If more points are available a least squares minimisation can be used, though which error measure should be minimised for noise in non-Euclidean structure recovery is an unresolved question.

1. 4 point - affine structure

This assumes affine projection[1]. It requires the least number of points

---

[*]SD acknowledges the support of ERASMUS

[†]AZ and PAB acknowledge the support of the SERC

[1]a generalisation of weak perspective or scaled orthography, valid when the object depth is small compared to its distance from the camera, see the Appendix in [13].

of the three methods. The structure is recovered *modulo* a 3D affine transformation: Lengths and angles are not recovered. However, affine invariants are determined. For example: parallelism, length ratios on parallel lines, ratio of areas.

2. **6 point, 4 coplanar - projective structure**
   The camera model is a perspective pin-hole and structure is recovered *modulo* a 3D projective transformation (i.e. multiplication of the homogeneous 4-vectors representing the 3D points by a 4×4 matrix). Affine structure is not recovered, so parallelism cannot be determined. However, projective invariants, such as intersections, coplanarity and cross ratios can be computed. The invariants are described in more detail below.

   Only two more points (than the affine case) are required to cover perspective projection rather than its affine approximation. However, the planarity requirement is a limitation on the type of object to which the method is applicable.

3. **8 point - projective structure**
   Again perspective pin hole projection is assumed, and structure is recovered *modulo* a 3D projective transformation.

   This method makes no assumption about object structure, but requires more points than the other two methods.

The methods are described in sections 3 and 4.

   Structure known only up to a 3D linear transformation, is sufficient to compute images from arbitrary novel viewpoints. The process of rendering new images given only *image(s)* of the original structure is known as *transfer*. This is described in section 2 and evaluated in section 5.2. Transfer has several significant visual applications:

1. **Verification in model based recognition**
   Model based recognition generally proceeds in two stages: first, a recognition hypothesis is generated based on a small number of image features; second, this hypothesis is *verified* by projecting the 3D structure into the image and examining the overlap of the projected structure with image features (edges) not used in generating the hypothesis. This is used routinely for planar objects [15] where the projections can be sourced from an image of the object - it is not necessary to measure the actual objects. The transfer method described here generalises this to 3D objects with similar ease of model acquisition - the model is extracted directly from images, and no camera calibration is required at any stage. In contrast, for 3D structures, previous verification methods (e.g. [9]) have required full Euclidean structure for the model and known camera calibration.

2. **Tracking**
   The performance of trackers, such as snakes or deformable templates, in efficiently tracking 3D structure, is markedly improved if the image position of the tracked features can be estimated from previous motion. This reduces the search region, which facilitates faster tracking, and gives greater immunity to the tracker incorrectly attaching itself to background clutter. The work here demonstrates that by tracking a small number of

features on an object it is possible to predict the image projection of the entire structure.

Since structure is recovered only up to a transformation, invariants to the transformation contain all the available (coordinate free) information. We assess the quality of the recovered structure by measuring these invariants. The invariants are described in sections 3 and 4 and evaluated in section 5.3.

2   Point Transfer

Transfer is most simply understood in terms of epipolar geometry. This is described here for the eight point case. In the other cases the principle is exactly the same, but less reference points are required.

It is assumed that two acquisition images, $imA1$ and $imA2$, have been stored with n known point correspondences ($n > 8$). Consider transfer to a third image, imT. The epipolar geometry between $imA1$ and imT is determined from eight reference point correspondences. A ninth point (and any other point) then generates an epipolar line in imT. Similarly, between $imA2$ and imT the epipolar geometry is determined, and each extra point defines an epipolar line in imT. The transferred point in imT lies at the intersection of these two epipolar lines. (Note, it is not necessary to explicitly compute correspondences between $imA2$ and imT. Once the correspondences between $imA1$ and imT are known, the correspondences required between $imA2$ and imT are determined from the acquisition correspondences between $imA1$ and $imA2$).

3   Affine invariants and point transfer

This approach is similar to that adopted by [6, 14, 17] and Barrett in [13].

3.1   Affine invariants

The 3D affine group is 12 dimensional, so for N general points we would expect[2] $3N - 12$ affine invariants - i.e. 3 invariants for each point over the fourth. The four (non-coplanar) reference points[3] $\mathbf{X}_i, i \in \{0, .., 3\}$ may be considered as defining a 3D affine basis (one for the origin $\mathbf{X}_0$, the other three specifying the axes $\mathbf{E}_i = \mathbf{X}_i - \mathbf{X}_0$ $i \in \{1, .., 3\}$, and unit point) and the invariants, α, β, γ, thought of as affine coordinates of the point, i.e. $\mathbf{X}_4 = \mathbf{X}_0 + \alpha\mathbf{E}_1 + \beta\mathbf{E}_2 + \gamma\mathbf{E}_3$.

Under a 3D affine transformation (with $\mathbf{A}$ a general 3×3 matrix and $\mathbf{T}$ a 3-vector), $\mathbf{X}' = \mathbf{A}\mathbf{X} + \mathbf{T}$ the transformed vectors are $\mathbf{X}_4' - \mathbf{X}_0' = \alpha\mathbf{E}_1' + \beta\mathbf{E}_2' + \gamma\mathbf{E}_3'$, which demonstrates that α, β, γ are affine invariant coordinates. Following the basis vectors in this manner (they can be identified after the transformation) allows the retrieval of α, β, γ after the transformation by simple linear methods.

Projection with an affine camera may be represented by $\mathbf{x} = \mathbf{M}\mathbf{X} + \mathbf{t}$, where $\mathbf{x}$ is the two-vector of image coordinates, $\mathbf{M}$ is a general 2×3 matrix, $\mathbf{t}$ a general

---

[2]By the counting argument in the introduction of [13].

[3]We adopt the notation that corresponding points in the world and image are distinguished by large and small letters. Vectors are written in bold font, e.g. $\mathbf{x}$ and $\mathbf{X}$. $\mathbf{x}$ and $\bar{\mathbf{x}}$ are corresponding image points in two views.

2-vector, and $\mathbf{X}$ a three vector for world coordinates. Differences of vectors eliminate $\mathbf{t}$. For example the basis vectors project as $\mathbf{e}_i = \mathbf{ME}_i \ \ i \in \{1, .., 3\}$. Consequently,

$$\mathbf{x}_4 - \mathbf{x}_0 \ = \ \alpha \mathbf{e}_1 + \beta \mathbf{e}_2 + \gamma \mathbf{e}_3 \tag{1}$$

A second view gives

$$\bar{\mathbf{x}}_4 - \bar{\mathbf{x}}_0 \ = \ \alpha \bar{\mathbf{e}}_1 + \beta \bar{\mathbf{e}}_2 + \gamma \bar{\mathbf{e}}_3 \tag{2}$$

Each equation (1) and (2) imposes two linear constraints on the unknown α, β, γ. All the other terms in the equations are known from image measurements (for example the basis vectors can be constructed from the projection of reference points $\mathbf{X}_i, i = \{0, .., 3\}$). Thus, there are four linear simultaneous equations in the three unknown invariants α, β, γ, and the solution is straightforward.

3.2 Epipolar line construction

Equation (1) gives two linear equations in three unknowns, which determines β and γ in terms of α, namely:

$$\beta \ = \ [v(\mathbf{x}_4 - \mathbf{x}_0, \mathbf{e}_3) - \alpha v(\mathbf{e}_1, \mathbf{e}_3)] / v(\mathbf{e}_2, \mathbf{e}_3)$$
$$\gamma \ = \ [-v(\mathbf{x}_4 - \mathbf{x}_0, \mathbf{e}_2) + \alpha v(\mathbf{e}_1, \mathbf{e}_2)] / v(\mathbf{e}_2, \mathbf{e}_3)$$

where the notation $v(\mathbf{a}, \mathbf{b}) = a_x b_y - a_y b_x$.

These are used to generate the epipolar line in another view. From (2) $\bar{\mathbf{x}}_4$ lies on the line

$$\bar{\mathbf{x}} \ = \ \bar{\mathbf{x}}_0 + [v(\mathbf{x}_4 - \mathbf{x}_0, \mathbf{e}_3)\bar{\mathbf{e}}_2 - v(\mathbf{x}_4 - \mathbf{x}_0, \mathbf{e}_2)\bar{\mathbf{e}}_3] / v(\mathbf{e}_2, \mathbf{e}_3)$$
$$+ \alpha \left(\bar{\mathbf{e}}_1 + [-v(\mathbf{e}_1, \mathbf{e}_3)\bar{\mathbf{e}}_2 + v(\mathbf{e}_1, \mathbf{e}_2)\bar{\mathbf{e}}_3] / v(\mathbf{e}_2, \mathbf{e}_3) \right)$$

which is the equation of a line parameterised by α. Note, all epipolar lines are parallel with a direction independent of $\mathbf{x}_4$.

4 Projective invariants and point transfer

4.1 6 point, 4 coplanar, projective transfer

The 3D projective group is 15 dimensional so for N general points we would expect $3N - 15$ independent invariants. However, the coplanarity constraint loses one degree of freedom leaving only two invariants for the 6 points.

The meaning of the 3D projective invariants can most readily be appreciated from figure 1. The line formed from the two off plane points intersects the plane of the four coplanar points in a unique point. This construction is unaffected by projective transformations. There are then 5 coplanar points and consequently two plane projective invariants - which are also invariants of the 3D transformation.

4.1.1 Epipolar geometry

The algorithm for calculating the epipolar geometry is described briefly below, more details are given in [1, 12]

We have 6 corresponding points $\mathbf{x}_i, \bar{\mathbf{x}}_i, i \in \{0, .., 5\}$ in two views, with the first 4 $i \in \{0, .., 3\}$ the projection of coplanar world points.

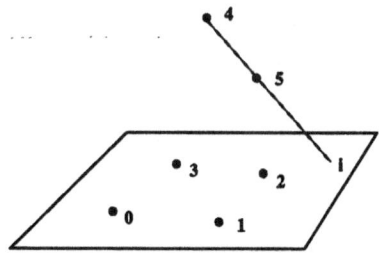

Figure 1: The projective invariant of 6 points, 4 coplanar (points 0-3), can be computed by intersecting the line through the non-planar points (4 and 5) with the common plane. There are then 5 coplanar points, for which two invariants to the plane projective group can be calculated.

1. Calculate plane projective transformation matrix $\mathbf{T}$, such that $\bar{\mathbf{x}}_i = \mathbf{T}\mathbf{x}_i, i \in \{0,..,3\}$.

2. Determine the epipole, $\bar{\mathbf{p}}$, in the $\bar{\mathbf{x}}$ image as the intersection of the lines $\mathbf{T}\mathbf{x}_i \times \bar{\mathbf{x}}_i$, $i \in \{4,5\}$.

3. The epipolar line in the $\bar{\mathbf{x}}$ image of any other point $\mathbf{x}$ is given by $\mathbf{T}\mathbf{x} \times \bar{\mathbf{p}}$.

4.1.2 Projective invariants

1. Determine the $\bar{\mathbf{x}}$ image of the intersection, $\bar{\mathbf{x}}_I$, of the plane and the line as the intersection of the lines $\mathbf{T}\mathbf{x}_4 \times \mathbf{T}\mathbf{x}_5$ and $\bar{\mathbf{x}}_4 \times \bar{\mathbf{x}}_5$ [14].

2. Calculate the two plane projective invariants of five points (in this case the four coplanar points and $\bar{\mathbf{x}}_I$) by

$$I_1 = \frac{|m_{320}||m_{I10}|}{|m_{310}||m_{I20}|} \qquad I_2 = \frac{|m_{310}||m_{I21}|}{|m_{321}||m_{I10}|}$$

where m_{jkl} is the matrix $[\bar{\mathbf{x}}_j \bar{\mathbf{x}}_k \bar{\mathbf{x}}_l]$ and $|m|$ its determinant.

4.2   8 point projective transfer

The construction described is a projective version [4, 5] of Longuet-Higgins' 8 point algorithm [7]. As is well known [8, 10] if points lie on a critical surface the epipolar geometry cannot be recovered. The method will clearly fail in these cases.

We have 8 corresponding points $\mathbf{x}_i, \bar{\mathbf{x}}_i, i \in \{0,..,7\}$ in two views.

1. Calculate essential matrix $\mathbf{Q}$, such that $\bar{\mathbf{x}}_i^t \mathbf{Q}\mathbf{x}_i = 0, i \in \{0,..,7\}$.

2. The epipolar line in the $\bar{\mathbf{x}}$ image of any other point $\mathbf{x}$ is given by $\mathbf{Q}\mathbf{x}$.

5   Experimental results and discussion

The images used for acquisition and assessment are shown in figure 2.

Figure 2: Images of a hole punch captured with different lenses and viewpoints. These are used for structure acquisition and transfer evaluation.

5.1 Segmentation and tracking

For the acquisition images the aim is to obtain a line drawing of the polyhedron. A local implementation of Canny's edge detector [2] is used to find edges to sub-pixel accuracy. These edge chains are linked, extrapolating over any small gaps. A piecewise linear graph is obtained by incremental straight line fitting. Edgels in the vicinity of tangent discontinuities ("corners") are excised before fitting as the edge operator localisation degrades with curvature. Vertices are obtained by extrapolating and intersecting the fitted lines. Figure 3 shows a typical line drawing.

Correspondence between views is achieved by tracking corners with a snake. This stage is currently being developed and some hand matching is necessary at present.

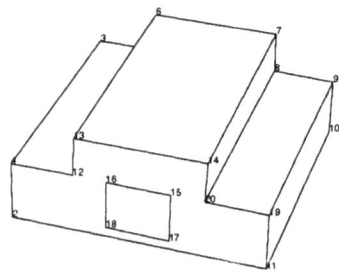

Figure 3: Line drawing of the hole punch extracted from image A in figure 2. Points 1 and 5 are occluded in this view.

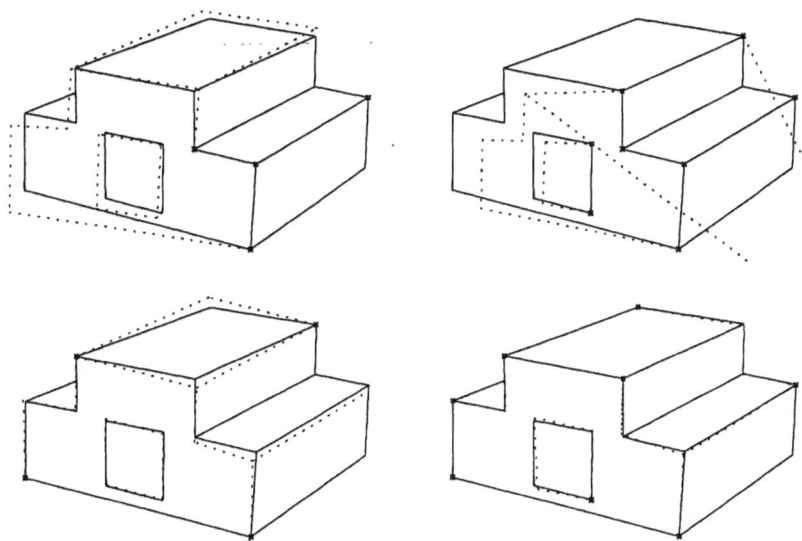

Figure 4: The effect of not spreading base points across the entire object. The transfer is computed from view A and B to view C. The "correct" graph structure of view C is shown by a solid line and the transferred view by a dashed line. "Correct" refers to the corners extracted from the real image. Base points are indicated by a cross. The left figures are affine transfer, the right projective 8 point transfer. Note, the graph structure is for visualization only, points not lines are transferred.

5.2 Transfer Results

The method is evaluated using two acquisition images plus a third image. Correspondence is established between the reference points in the third and acquisition images (e.g. 4 points in the case of affine transfer). Other points in the acquisition image are then transferred, the difference between the transferred points and the actual position of the corresponding points giving the transfer error. This error is taken as the Euclidean distance $d(\mathbf{p}_t, \mathbf{p}_c)$ between the transferred point $\mathbf{p}_t$, and its actual image position $\mathbf{p}_c$ (i.e. the position extracted from the actual image). Two measures are used for the evaluation:

1. mean error $E_{mean} = \frac{1}{n} \sum_{i=1}^{n} d(\mathbf{p}_t^i, \mathbf{p}_c^i)$

2. maximum error $E_{max} = \max_i \ d(\mathbf{p}_t^i, \mathbf{p}_c^i) \quad i \in \{1, .., n\}$

Method	Spread out points		Not spread out	
	mean error	max. error	mean error	max. error
affine	4.09	8.72	9.56	23.72
8 points	1.60	4.64	51.84	421.11

Table 1: Mean and maximum errors for the transfers shown in figure 4.

Figure 5: Typical transfers using the 6 point method with different base points. Left figure: using off plane points are 6 and 9, measured mean and max. errors are 3.88 and 15.88. Right figure: using off plane points 7 and 10, mean and max. errors are 1.76 and 4.91. See figure 3 for point numbering.

Method	mean error	max. error
affine	4.09	8.72
8 points	1.60	4.64
6 points (4 coplanar)	3.58	12.09
6+ points (5 coplanar)	2.93	5.58

Table 2: Comparison of typical results of the various transfer methods from view A and B to C. Note, the improvement in the 6 point method obtained by taking an additional coplanar point. The points used for the transfer are 11,13,14,17,(4),6,9. The point in brackets is the additional fifth coplanar point.

In both cases n is the number of points transferred. This varies between transfer methods as differing number of reference points are involved.

We have found that all methods perform poorly when the base points only partially cover the figure, see table 1 and figure 4. A similar result was noted in [11] in the case of planar transfer.

The affine transfer methods is very stable and does not suffer from the dramatic errors shown in the projective case (see figure 4). However, as would be expected, its performance degrades as perspective effects become more significant.

The six point transfer method can produce very good results, but success is very dependent on the "correctness" of the four point planarity. Of course, four real world points are never coplanar, but here there are additional errors from measured image positions. Some typical examples are given in figure 5. Stability can be improved by using more than four coplanar points to estimate the plane to plane projection matrix. This least squares estimate tends to cancel out some of the localisation errors in the image measurements (a total of 7 points for projective transfer, is still an advantage over the eight point method). This improvement is demonstrated in table 2 which also compares the three methods for a typical example. The eight point method does achieve the best performance, but the price paid is additional complexity of finding and matching extra points.

Images	α	β	γ
A,B	0.610	-0.221	-0.009
A,C	0.664	-0.268	-0.023
A,D	0.597	-0.213	-0.002
B,D	0.594	-0.214	-0.003
B,C	0.636	-0.242	-0.012
C,D	0.681	-0.285	-0.031
A,B,D	0.601	-0.217	-0.004
B,C,D	0.637	-0.244	-0.014
A,C,D	0.670	-0.274	-0.027
A,B,C	0.637	-0.243	-0.013

Images	I_1	I_2
D,A	0.440	-0.968
D,B	0.378	-1.117
B,A	0.371	-1.170
C,E	0.370	-1.150
F,A	0.333	-1.314
D,A,B	0.372	-1.151
C,E,D	0.369	-1.148
F,A,C	0.370	-1.196
C,A,B,D,E	0.375	-1.140
F,A,B,C,D,E	0.369	-1.170

Table 3: Invariants for varying sets of images. **Left**: affine coordinates (α, β, γ) of point 20 with respect to the base points 11,13,2,7. Note, measurements are spread over a smaller range when more images are used. **Right**: 6 point invariants using points 2,4,14,17 and the line between points 6 and 13. See figure 3 for point numbering.

5.3 Invariance Results

We find in general that the invariant values are more stable than transfer would suggest. This is probably because extra errors are incurred in measuring reference points in the transfer image.

5.3.1 Affine invariants

Equation (1) and (2) are four linear constraints on the three unknown affine invariants. Least-squares solution (by using singular value decomposition) immediately confers some immunity to noise. Further improvement is obtained by including corresponding equations from additional views. The stability and benefit of additional views is illustrated in table 3. In a tracked sequence robust estimates can be built in real-time using a recursive filter.

5.3.2 Projective invariants

Although invariants obtained from two views are fairly stable, improvements in stability are again achieved by augmenting with measurements from other views. See table 3. In this case by providing a least squares estimate of the line plane intersection.

6 Conclusion

In this paper, we have presented three methods to recover structure from two or more images taken with unknown cameras. All that is required is point correspondences between the images. Structure is recovered up to a linear transformation, but this is sufficient for transfer and computation of invariants of the 3D point set.

Experimental results show the methods perform well except for some sensitivity to corner detection errors. Future work will be based on automatically tracking corners using snakes.

Acknowledgements

We are grateful for helpful discussions with Roberto Cipolla, Richard Hartley, Joe Mundy and David Murray. Rupert Curwen provided the snake tracker and Charlie Rothwell the segmentation software.

References

[1] Beardsley, P., Sinclair, D., Zisserman, A., Ego-motion from Six Points, Insight meeting, Catholic University Leuven, Feb. 1992.

[2] Canny J.F. "A Computational Approach to Edge Detection," *PAMI-6*, No. 6. p.679-698, 1986.

[3] Curwen, R.M., Blake, A. and Cipolla, R. Parallel Implementation of Lagrangian Dynamics for real-time snakes. Proc. BMVC91, Springer Verlag, 29-35, 1991.

[4] Faugeras, O., What can be seen in 3D with an uncalibrated stereo rig?, *ECCV*, 1992.

[5] R. Hartley, R. Gupta and Tom Chang, "Stereo from Uncalibrated Cameras" Proceedings of CVPR92.

[6] Koenderink, J.J. and Van Doorn, A.J., Affine Structure from Motion, *J. Opt. Soc. Am. A*, Vol. 8, No. 2, p.377-385, 1991.

[7] Longuet-Higgins, H.C., A Computer Algorithm for Reconstructing a Scene from Two Projections, *Nature*, Vol. 293, p.133-135, 1981.

[8] Longuet-Higgins, H.C., The Reconstruction of a Scene from two Projections - Configurations that Defeat the 8-point Algorithm, *Proc. 1st IEEE Conference on Artificial Intelligence Applications*, p.395-397, December 1984.

[9] Lowe, D.G., *Perceptual Organization and Visual Recognition*, Kluwer, 1985.

[10] Maybank, S.J., Longuet-Higgins, H.C., The Reconstruction of a Scene from two Projections - Configurations that Defeat the 8-point Algorithm, *Proc. 1st IEEE Conference on Artificial Intelligence Applications*, p.395-397, December 1984.

[11] Mohr, R. and Morin, L., Relative Positioning from Geometric Invariants, *Proc. CVPR*, p.139-144, 1991.

[12] Mohr, R., Projective geometry and computer vision, To appear in *Handbook of Pattern Recognition and Computer Vision*, Chen, Pau and Wang editors, 1992.

[13] Mundy, J.L. and Zisserman, A., editors, *Geometric Invariance in Computer Vision*, MIT Press, 1992.

[14] Quan, L. and Mohr, R., Towards Structure from Motion for Linear Features through Reference Points, *Proc. IEEE Workshop on Visual Motion*, 1991.

[15] Rothwell C.A., Zisserman A., Forsyth D.A., and Mundy J.L., "Using Projective Invariants for Constant Time Library Indexing in Model Based Vision", *Proc. BMVC91, Springer Verlag*, 62-70, 1991.

[16] Semple, J.G. and Kneebone, G.T. *Algebraic Projective Geometry*, Oxford University Press, 1952.

[17] Ullman, S. and Basri, R., Recognition by Linear Combination of Models, *PAMI-13*, No. 10, p.992-1006, October, 1991.

Planar Region Detection and Motion Recovery

D. Sinclair, A. Blake, S. Smith and C. Rothwell

Robotics Research Group
Department of Engineering Science
Oxford University OX1 3PJ

Abstract

This paper presents a means of segmenting planar regions from two views of a scene using point correspondences. The initial selection of groups of coplanar points is performed on the basis of conservation of two five point projective invariants (groups for which this invariant is conserved are assumed to be coplanar). The correspondences for four of the five points are used to define a projectivity which is used to predict the change in position of other points assuming they lie on the same plane as the original four. A distance threshold between actual and predicted position is used to find extended planar regions. If two distinct planar regions can be found then a novel motion direction estimator suggests itself.

1 Introduction

Classically the structure from motion problem has been seen as obtaining the distance from the camera optical center to points in the world from their motions in a sequence of images. This approach stems from the fact that if the position of a point within the field of view is known then the only piece of information left to recover is the point's 'depth'. Individual depth estimates are noisy and sparse. This makes looking for qualitative scene structure difficult or impossible.

The problems of recovering camera motion and scene structure are inextricably linked [6], [4], [7] . If the motion of the camera (or equivalently stereo geometry) is accurately known then it is straight forward to recover scene depth, once the correspondence problem has been solved. Alternatively if the depths of scene points are known then motion direction recovery is easy. Longuet-Higgins [5] derived a scheme for recovering camera motion if the scene was planar or points within it were coplanar. The solution suffered from a two fold ambiguity and was not demonstrated on real data. No a priori method was presented for determining whether or not points were coplanar. A mathematically equivalent scheme was independently derived by Tsai [14] with equivalence being shown by Faugeras in [2]. Most general methods, though, return the depths of a series of points. By themselves these depths convey little useful qualitative information about the scene. Authors have performed Delaurnay triangulation on the points to provide a kind of surface. The surface does not reflect the underlying surface of the scene and for the most part any depth estimate derived from it will be wrong.

It is well known that the mapping between two projected views of a plane is completely specified by a 3 by 3 transformation matrix [11], [12]. The group of these matrices is called the projective group PGL(2). Members of this group have 8 rather than 9 degrees of freedom. This means that a projectivity is completely specified by four point correspondences if no three of the projected points are co-linear. If a fifth point correspondence is available then 2 projectively invariant quantities are defined. If these two quantities are not conserved between the two views then the five points do not lie on a plane. Conic invariants have been used in a similar manner in [3].

In this paper a simple test for planarity of sets of five points is derived. Implicit in the test is an estimate of the variance in the position of points in an image. This variance is used to provide an estimate of the variance in the values of two 'projective invariants'. Conservation of the values of the two invariants is taken to mean that a set of five points lies in a plane. The accepted difference in the values of the invariant between views is the linearised variance of the invariants themselves.

The above process provides groups of five points lying on possibly different planes. Four of the five points are used to generate the projectivity associated with the transformation of any point on the same plane as the set of five points between views. This transformation allows the new position of any point on the plane to be predicted. The predicted position and the actual new position are compared. A simple distance threshold, based on an assumed variance in the position of the point undergoing prediction, is used to decide whether or not a given point lies on the plane of the five points. This allows planar regions to be grown.

If two or more distinct planar regions have been found in an image then the two projectivities associated with them may be used to recover unambiguously the motion direction and the line of intersection of the planes in the image. Once the motion direction has been recovered it is then possible to solve for the camera's rotation and the normal of the planes. This does however require a calibrated camera.

Section 2 defines the notation to be used in this paper. Section 3 details the projective invariants used and the method of selecting groups of five points. Section 5 shows how the projectivity associated with the motion of one of the planes may be derived and section 6 how it may be used to look for other points on the same plane. Section 7 covers the recovery of camera motion.

2 Notation

In this chapter the following notation is adopted, x a vector in the projective plane, x_i the ith vector of a set of vectors, x' a transformed vector, $\mathcal{P}$ any invertible 3 by 3 projective transformation matrix. In projective space the following identification is made;

$$x = \lambda x, \tag{1}$$

where λ is any constant, hence 3 dimensional vectors in the projective plane only have 2 degrees of freedom. x'_i is therefore given by

$$x'_i = \lambda_i \mathcal{P} x_i. \tag{2}$$

The vectors x are then the homogeneous co-ordinates of the image positions of corners. The matrix whose columns are the vectors x_i, x_j and x_k is wrritten

$$M_{ijk} = (x_i, x_j, x_k). \tag{3}$$

3 The Two Five Point Planar Invariants

Projective invariants are quantities which do not change under projective transformations. A full review of the uses of invariants is given in [10]. There are two convenient invariants that may be defined for groups of five points. They correspond to the two degrees of freedom of the projective position of the fifth point with respect to the first four. The two invariants may conveniently be written as the ratios of determinants of matrices of the form M_{ijk}.

$$I_1 = \frac{|M_{124}| \, |M_{135}|}{|M_{134}| \, |M_{125}|} \tag{4}$$

and

$$I_2 = \frac{|M_{241}| \, |M_{235}|}{|M_{234}| \, |M_{215}|}. \tag{5}$$

these two quantities may be seen to be conserved under a projective transformation if x' is substituted for x,

$$\frac{|M'_{124}| \, |M'_{135}|}{|M'_{134}| \, |M'_{125}|} = \frac{|\lambda_1 \mathcal{P} x_1, \lambda_2 \mathcal{P} x_2, \lambda_4 \mathcal{P} x_4| \, |\lambda_1 \mathcal{P} x_1, \lambda_3 \mathcal{P} x_3, \lambda_5 \mathcal{P} x_5|}{|\lambda_1 \mathcal{P} x_1, \lambda_3 \mathcal{P} x_3, \lambda_4 \mathcal{P} x_4| \, |\lambda_1 \mathcal{P} x_1, \lambda_2 \mathcal{P} x_2, \lambda_5 \mathcal{P} x_5|}, \tag{6}$$

which gives,

$$\frac{|M'_{124}| \, |M'_{135}|}{|M'_{134}| \, |M'_{125}|} = \frac{\lambda_1^2 \lambda_2 \lambda_3 \lambda_4 \lambda_5 \, |\mathcal{P}| \, |M_{124}| \, |M_{135}|}{\lambda_1^2 \lambda_2 \lambda_3 \lambda_4 \lambda_5 \, |\mathcal{P}| \, |M_{134}| \, |M_{125}|}. \tag{7}$$

From a combinatorial point of view it might be thought that there were 10 independent invariants. The determinant $|M_{ijk}|$ does not change under cyclic permutations of the indices and only changes sign under acyclic permutations. However there are only two. Generally four points have eight degrees of freedom all of which are required to define the transformation into the canonical frame. The fifth point has two degrees of freedom which form the values of the two invariants [10].

In selecting the groups of five points it is important that they be sufficiently far apart that image measurement noise does not swamp the invariant. To this end the nearest four points outside a circle of radius 25 pixels are selected for the invariant. Other selection strategies are under investigation as the invariants degenerate when any three of the five points are collinear.

4 Covariance Matrices of the 2 Invariants

There is a measurement uncertainty associated with the estimated position of corner features in any image. These errors are assumed to be normally distributed with variances σ_x and σ_y which are additionally assumed to be isotropic and equal to σ. This value will vary with the type of corner detector used. The linearised variance of the invariants may be computed as follows,

$$\delta I_1^2 = \sum_{i=1}^{5} \left(\frac{\partial I_1}{\partial x_i} \cdot dx_i \right) \left(\frac{\partial I_1}{\partial x_i} \cdot dx_i \right) \tag{8}$$

or as noise is uncorrelated [1] (by the law of propagation of error),

$$\mathrm{Var}(I_1) = \sum_{i=1}^{5} \frac{\partial I_1^T}{\partial x_i} \begin{pmatrix} \sigma^2 & 0 \\ 0 & \sigma^2 \end{pmatrix} \frac{\partial I_1}{\partial x_i}. \tag{9}$$

The vectors $\mathbf{x}_i$ have components,

$$\mathbf{x}_i = \begin{pmatrix} x_i \\ y_i \\ f \end{pmatrix}, \tag{10}$$

where f is the focal length of the camera. The derivatives of I_1 are then given by,

$$\frac{\partial I_1}{\partial x_1} = I_1 \left(\frac{|(\mathbf{e}_1, \mathbf{x}_2, \mathbf{x}_4)|}{|(\mathbf{x}_1, \mathbf{x}_2, \mathbf{x}_4)|} + \frac{|(\mathbf{e}_1, \mathbf{x}_3, \mathbf{x}_5)|}{|(\mathbf{x}_1, \mathbf{x}_3, \mathbf{x}_5)|} - \frac{|(\mathbf{e}_1, \mathbf{x}_3, \mathbf{x}_4)|}{|(\mathbf{x}_1, \mathbf{x}_3, \mathbf{x}_4)|} - \frac{|(\mathbf{e}_1, \mathbf{x}_2, \mathbf{x}_5)|}{|(\mathbf{x}_1, \mathbf{x}_2, \mathbf{x}_5)|} \right) \tag{11}$$

where,

$$\mathbf{e}_1 = \begin{pmatrix} 1 \\ 0 \\ 0 \end{pmatrix}, \tag{12}$$

$$\frac{\partial I_1}{\partial x_2} = I_1 \left(\frac{|(\mathbf{x}_1, \mathbf{e}_1, \mathbf{x}_4)|}{|(\mathbf{x}_1, \mathbf{x}_2, \mathbf{x}_4)|} - \frac{|(\mathbf{x}_1, \mathbf{e}_1, \mathbf{x}_5)|}{|(\mathbf{x}_1, \mathbf{x}_2, \mathbf{x}_5)|} \right) \tag{13}$$

etc.

$$\frac{\partial I_1}{\partial y_1} = I_1 \left(\frac{|(\mathbf{e}_2, \mathbf{x}_2, \mathbf{x}_4)|}{|(\mathbf{x}_1, \mathbf{x}_2, \mathbf{x}_4)|} + \frac{|(\mathbf{e}_2, \mathbf{x}_3, \mathbf{x}_5)|}{|(\mathbf{x}_1, \mathbf{x}_3, \mathbf{x}_5)|} - \frac{|(\mathbf{e}_2, \mathbf{x}_3, \mathbf{x}_4)|}{|(\mathbf{x}_1, \mathbf{x}_3, \mathbf{x}_4)|} - \frac{|(\mathbf{e}_2, \mathbf{x}_2, \mathbf{x}_5)|}{|(\mathbf{x}_1, \mathbf{x}_2, \mathbf{x}_5)|} \right) \tag{14}$$

where,

$$\mathbf{e}_2 = \begin{pmatrix} 0 \\ 1 \\ 0 \end{pmatrix}, \tag{15}$$

$$\frac{\partial I_1}{\partial y_2} = I_1 \left(\frac{|(\mathbf{x}_1, \mathbf{e}_2, \mathbf{x}_4)|}{|(\mathbf{x}_1, \mathbf{x}_2, \mathbf{x}_4)|} - \frac{|(\mathbf{x}_1, \mathbf{e}_2, \mathbf{x}_5)|}{|(\mathbf{x}_1, \mathbf{x}_2, \mathbf{x}_5)|} \right), \tag{16}$$

and so forth. The same analysis may be performed in order to derive a similar expression for the derivatives of I_2. These expressions permit the computation of the variance of the two invariants. To test whether it is possible that five points visible in two views lie on a plane the difference in each of the two invariants must obey the condition

$$|\mathbf{I}' - \mathbf{I}| < 2\sqrt{\text{Var}(I)}. \tag{17}$$

Figure 1 shows an image from a motion sequence with motion vectors from tracked corners superimposed on it. Figure 2 shows all the starting positions of the flow vectors as square boxes. Two of the groups of five points found using the invariant planarity test are marked by darker symbols. From this sequence a total of 13 groups of five planar points were found. In this case the value of σ was taken to be 0.2. Points at extreme distance will tend to behave as if they were on a plane as well as those on flat surfaces.

5 Finding the Projectivity between four Points in two Views

If four coplanar points are available in two views then a projectivity may be defined which will transform the first set of points into the second (provided no three of the

Figure 1: *An image from a motion sequence with flow vectors of tracked corners superimposed on it.*

points are collinear). The same projectivity will predict the position of any point on the plane in the first image in the second image [11].

If x_i are the initial four points and x'_i the transformed four points then,

$$x'_i = \lambda_i \mathcal{P} x_i. \tag{18}$$

The easiest way of finding $\mathcal{P}$ is first to transform both x_i and x'_i to the canonical frame e_i where e_1 and e_2 are as before and

$$e_3 = \begin{pmatrix} 0 \\ 0 \\ 1 \end{pmatrix}, \tag{19}$$

$$e_4 = \begin{pmatrix} 1 \\ 1 \\ 1 \end{pmatrix}. \tag{20}$$

$$x_i = \lambda_i \mathcal{M}_1 e_i, \tag{21}$$

where using i from 1 to 3 gives,

$$\mathcal{M}_1 = (\alpha x_1, \beta x_2, \gamma x_3) \tag{22}$$

and using i = 4 gives

$$x_4 = (x_1, \ x_2, \ x_3) \begin{pmatrix} \alpha \\ \beta \\ \gamma \end{pmatrix} \tag{23}$$

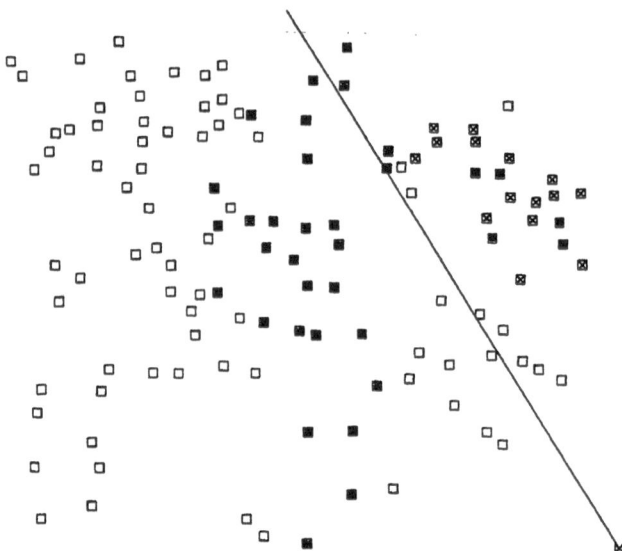

Figure 2: *The starting positions of the flow vectors in figure 1 are marked. Two of the groups of five planar points are marked by darker boxes. The boxes marked with stars were found to be on the same plane as the left hand group of five. The boxes marked with crosses were found to be on the same plane as the right hand group of five. The line represents the line of intersection of the two planes.*

$$\begin{pmatrix} \alpha \\ \beta \\ \gamma \end{pmatrix} = (\mathbf{x}_1, \; \mathbf{x}_2, \; \mathbf{x}_3)^{-1} \mathbf{x}_4, \qquad (24)$$

this may be solved for α, β, γ by inverting the matrix whose columns are $\mathbf{x}_1$, $\mathbf{x}_2$, $\mathbf{x}_3$. The corresponding matrix $\mathcal{M}_2$ that maps $\mathbf{x}'_i$ to $\mathbf{e}_i$ must then be found and then,

$$\mathcal{P} = \mathcal{M}_2 \mathcal{M}_1^{-1}. \qquad (25)$$

Three matrix inversions must hence be performed to find $\mathcal{P}$. Analytical expressions are available for the inverse of 3 by 3 matrices therefore, with the aid of Mathematica the variance of $\mathcal{P}$ with respect to either set of four points can be determined.

Alternative methods of calculating this projectivity have been explored. A pseudo inverse technique [9] [13] using all five of the points was tried. A minimum eigenvalue technique similar to that used by Kanatani [8] was also used. The most 'useful' technique in terms of predicting the motion of planar points was the four point method detailed above.

6 Finding Additional Coplanar Points

Once it has been established that four points lie on a plane then the projectivity that maps the four points to their images may be used to predict the new image position of any point on the same plane [11]. If the projectivity between the two frames is $\mathcal{P}$ then,

$$\hat{\mathbf{x}} = \lambda \mathcal{P} \mathbf{x}, \tag{26}$$

with λ constrained so that $\hat{\mathbf{x}} \cdot \mathbf{e}_3 = f$. If a rapid answer is required a simple distance threshold between predicted and actual image position may be used. We derive here an expression for the x and y variances of the predicted position $\hat{\mathbf{x}}$ of a point $\mathbf{x}$. The variance of the x component of the predicted position is given by,

$$\delta \hat{x}^2 = \left(\frac{\partial \hat{x}}{\partial x}\right)^2 dx^2 + \frac{\partial \hat{x}}{\partial y} dy \frac{\partial \hat{x}}{\partial x} dx + \left(\frac{\partial \hat{x}}{\partial y}\right)^2 dy^2 \tag{27}$$

or as dx and dy are uncorrelated,

$$\mathrm{Var}(\hat{x}) = \left(\frac{\partial \hat{x}}{\partial x}, \frac{\partial \hat{x}}{\partial y}\right) \begin{pmatrix} \sigma^2 & 0 \\ 0 & \sigma^2 \end{pmatrix} \begin{pmatrix} \frac{\partial \hat{x}}{\partial x} \\ \frac{\partial \hat{x}}{\partial y} \end{pmatrix} \tag{28}$$

where

$$\frac{\partial \hat{x}}{\partial x} = \frac{f}{\mathcal{P}_{31}x + \mathcal{P}_{32}y + \mathcal{P}_{33}f} \mathcal{P}_{11} - (\mathcal{P}_{11}x + \mathcal{P}_{12}y + \mathcal{P}_{13}f) \frac{f\mathcal{P}_{31}}{(\mathcal{P}_{31}x + \mathcal{P}_{32}y + \mathcal{P}_{33}f)^2} \tag{29}$$

$$\frac{\partial \hat{x}}{\partial y} = \frac{f}{\mathcal{P}_{31}x + \mathcal{P}_{32}y + \mathcal{P}_{33}f} \mathcal{P}_{12} - (\mathcal{P}_{11}x + \mathcal{P}_{12}y + \mathcal{P}_{13}f) \frac{f\mathcal{P}_{32}}{(\mathcal{P}_{31}x + \mathcal{P}_{32}y + \mathcal{P}_{33}f)^2} \tag{30}$$

and σ is as before. The corresponding variance in $\hat{y}$ is derived analogously. Points are accepted as being on the plane associated with the projectivity if the difference between actual and predicted position in the second frame is less than two standard deviations in the x or y directions. That is,

$$|\hat{x} - x'| \leq 2\sqrt{\mathrm{Var}(\hat{x})} \tag{31}$$

and

$$|\hat{y} - y'| \leq 2\sqrt{\mathrm{Var}(\hat{y})}. \tag{32}$$

The permissible difference in position may be anisotropic. Figure 2 shows the additional groups of points found to be on the same plane as the initial two groups of five.

It would have been possible to insert successive points into the five point planar invariant keeping four of them constant and using the analysis of section 4. This however would take more time and as will be seen in the following sections the projectivity associated with a plane in motion has additional uses. For simply detecting planar regions no camera calibration at all is required. It is only if information like plane normal or motion direction are required that calibration is necessary.

7 Finding the Motion Direction

As is well known if four planar point correspondences are available then the motion direction may be recovered up to a two fold ambiguity [5] [14]. If two non-parallel planes are visible then this ambiguity may be resolved. A more direct method of recovering an observer's motion direction is presented here. Let the two projectivities associated with the motion of points on the two planes be $\mathcal{P}_1$ and $\mathcal{P}_2$. Projectivity $\mathcal{P}_1$ may be used to predict the new position of any point, $\mathbf{x}$,

$$\hat{\mathbf{x}} = \lambda \mathcal{P}_1 \mathbf{x} \tag{33}$$

and likewise $\mathcal{P}_2$ may also be used to predict a new position of any point,

$$\hat{\mathbf{x}}' = \lambda \mathcal{P}_2 \mathbf{x}. \tag{34}$$

The two predicted positions will only coincide for certain points. The predictions will agree for the points on the line of intersection of the two planes and the motion direction, (the two predictions will agree on the epipole). Why this is so is shown in figure 3. The line of intersection (in the image) of the two planes and the motion

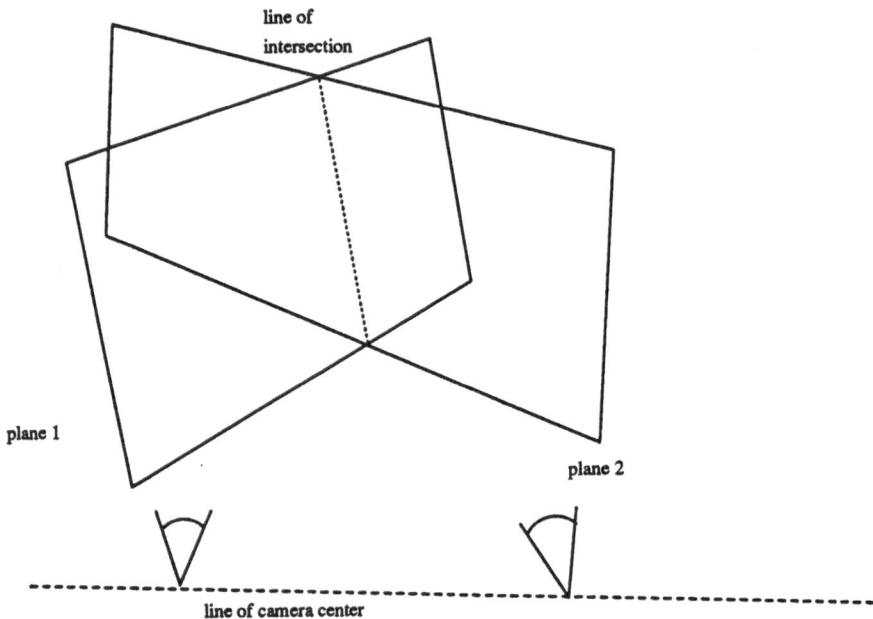

Figure 3: *This figure show two camera positions relative to two planes. The line of intersection is preserved between views of the planes. The only other image direction which both projectivities predict go to the same position is the motion direction of the camera.*

direction may be found by solving the following eigenvector equation,

$$\mathcal{P}_1 \mathbf{x} = \lambda \mathcal{P}_2 \mathbf{x}. \tag{35}$$

If the camera undergoes any rotation at all the real eigenvalue will have eigen direction parallel to the motion direction and the degenerate pair will have eigen directions that span the line of intersection of the two planes. The method will fail when the system has three degenerate roots. This corresponds to the case when the motion direction points towards a point on the line of intersection of the two planes. Figure 2 shows the line of intersection of the two planes in figure 1.

8 Possible Extension

The central idea in the paper is that the transformation, associated with a change in viewing position, of a group of planar points may be written as 3 by 3 matrix. This fact permits the prediction of the new position of any point on the plane. The following section suggests a way of making use of this fact.

8.1 Using Lines in the Invariant

In a man made world planar regions are often bounded by lines (books, roads, sides of buses, houses, etc). If a line has been found in an image from a motion sequence then is it possible to use points lying on one side of the line to construct invariants, conservation of which will imply that the lines and points are coplanar. Line are the dual of points in P(2) and hence a minimum of one line and four points are required to construct two invariants.

A line l through a set of points x is defined by the equation,

$$l \cdot x = 0 \tag{36}$$

this means that l transforms according as,

$$l' = \lambda_l \mathcal{P}^{-T} l. \tag{37}$$

Noting that,

$$l' \cdot x'_i = \lambda_l \lambda_i l \cdot x_i, \tag{38}$$

where x_i are points not on the line l, invariants may be constructed of the form,

$$I = \frac{|M_{123}| \, l \cdot x_4}{|M_{124}| \, l \cdot x_3}. \tag{39}$$

The covariance of these invariants may be derived in a similar manner to section 4. These may be used to look for groups of four points coplanar with the line. This remains to be implemented.

9 Conclusions

The mechanism proposed provides a robust and rapid means of isolating sets of planar points from two views of a scene. If two planar regions are available then it is straightforward to recover the camera's motion direction. The direction of the plane normal may be isolated from the projectivity associated with a particular plane and the motion direction used to resolve the two fold ambiguity. If estimates of plane normal and motion direction have been obtained then the camera's rotation may be estimated.

10 Acknowledgements

We acknowledge discussions with Professor J. M. Brady, Dr A. Zisserman, Dr. D. Murray, P. Beardsley and R. Cipolla of the Robotics Research Group. We are grateful for support from the SERC, the EEC Esprit program and DRA RARDE Chertsey.

68

References

[1] Coehlo C., Heller A., Mundy J., and Forsyth D. An experimental evaluation of projective invariants. In *DARPA Image Understanding Workshop*, 273–294, 1991.

[2] O.D. Faugeras and Maybank S.J. Motion from point matches: Multiplicity of solutions. *Int. Journal of Computer Vision*, 4:225–246, 1988.

[3] D.A. Forsyth, J.L. Mundy, A.P. Zisserman, and C.M. Brown. Projectively invariant representations using implicit algebraic curves. In *Proc. 1st European Conf. on Computer Vision*, 427–436. Springer-Verlag, 1990.

[4] C.G. Harris. Determination of ego - motion from matched points. In *3rd Alvey Vision Conference*, 189–192, 1987.

[5] Longuet-Higgins H.C. The reconstruction of a plane surface from two projections. In *Proc. R. Soc. Lond.*, 399–410, 1986.

[6] B.K.P. Horn and B.G. Schunk. Determining optical flow. *Artificial Intelligence*, vol.17:185–203, 1981.

[7] B.K.P. Horn and E.J. Weldon. Direct methods for recovering motion. *Int. Journal of Computer Vision*, vol.2:51–76, 1988.

[8] K. Kanatani. *Geometric Computation for Machine Vision*, volume 1st Edition. MIT Press, 1991.

[9] Carlsson S. Projectively invariant decomposition of planar shapes. In J.L. Mundy and A. Zisserman, editors, *Geometric Invariance in Computer Vision*, 267–276. MIT Press, 1992.

[10] J.L. S. Mundy and A. Zisserman. *Geometric Invariance in Computer Vision*, volume First Edition. MIT Press, 1992.

[11] J.G. Semple and G.T. Kneebone. *Algebraic projective geometry*. Oxford University Press, 1952.

[12] C.E. Springer. *Geometry and Analysis of Projective Spaces*, volume 1. Freeman, 1964.

[13] G. Strang. *Linear Algebra and its Applications*, volume I. Academic Press, 1980.

[14] R.Y. Tsai and T.S. Huang. Estimating three-dimensional motion parameters of a rigid planar patch. *IEEE Trans. on Acoustics, Speech and Signal Processing*, vol.ASSP-29,no.6:1147–1152, 1981.

3D Structure and Motion Estimation from 2D Image Sequences[†]

T. N. Tan, K. D. Baker and G. D. Sullivan

Intelligent Systems Group
Department of Computer Science
University of Reading, ENGLAND

Abstract

Two novel algorithms are presented in this paper for depth estimation using point correspondences and the ground plane constraint. One is a direct non-iterative method, and the other a simple well-behaved iterative technique where the choice of initial value is straightforward. The algorithms are capable of handling any number of points and frames as well as points which become occluded. Once the point depths are determined, motion parameters can be obtained by a linear least squares technique. Extensive test results are included which show that the proposed algorithms are robust to noise, and perform satisfactorily using real outdoor image sequences.

1 Introduction

In previous work [16-17], we have shown that the ground plane constraint (the fact that objects, such as road vehicles, are often confined to move on the ground surface) can be used to develop simple and robust structure from motion (SFM) algorithms using point correspondences from pairs of image frames. In this paper, we discuss the use of multiple (more than two) image frames. We call SFM algorithms that use multiple frames the Multiple Frame SFM or simply MFSFM algorithms. A MFSFM algorithm can be either recursive or batch in nature depending on whether it processes one frame at a time or all frames simultaneously. In general, batch approaches have been shown to be both more accurate and stable [7]. The algorithms presented in this paper belong to the batch group.

Many MFSFM algorithms have been reported [1, 2-12]. These algorithms, however, have a number of limitations: unrealistic assumptions about object and/or camera motion [3-5, 7, 9-11]; requirement of good initial guesses to initialise the iteration process; high computational complexity [2, 7, 9, 11-12]; failure to handle feature occlusion; and unknown performance in real image data [3-4, 6, 8, 11]. These difficulties are mostly due to the scale of the task of solving the six degrees of freedom non-linear problem allowed by the existing MFSFM algorithms.

Many practical tasks in vision need be concerned with fewer degrees of freedom, and object motion is often subjected to physical constraints, such as the commonly occurring ground plane constraint. We show in this paper that the ground plane constraint can be used to develop simple and robust MFSFM algorithms which avoid the

---

†. This work was carried out as part of the ESPRIT project P2152 (VIEWS).

above problems. The work presented in this paper has mainly been motivated by the desire to apply machine vision in automatic monitoring and surveillance in airport and road traffic, but is also applicable to a wide range of potential industrial applications. With autonomous vehicles, for example, the ground plane constraint is equivalent to assuming that the camera is at a known fixed height, tilt and roll. This is frequently the case, at least for brief periods. The algorithms therefore provide robust and efficient methods for the recovery of unknown obstacles for robots moving on a flat surface.

The ground plane constraint is defined in the next section. Section 3 describes two different techniques for recovering point depths using multiple frames and the ground plane constraint. Section 4 outlines an algorithm for optimal 3D motion parameter estimation. Experimental results are presented in Section 5.

2   The ground plane constraint

The scenes considered in this paper concern airport or road traffic, where objects (e.g., aeroplanes, vehicles, etc.) are confined to move on the ground surface, which is, at least in the local region of our interest, approximately planar. We represent the ground surface by the X-Y plane of a WCS whose Z-axis points upwards. The movement of an object only has three degrees of freedom: translations (T_x and T_y) along the X and Y axes on the ground plane, and rotation (θ) about the vertical Z axis. The other three motion parameters, i.e., the rotations (α and β) about the X and Y axes, and the vertical translation (T_z), are all zero:

$$\alpha, \beta, T_z = 0 \tag{1}$$

We call this the *ground plane constraint* (GPC). We observe that when object motion is expressed in the camera-centred frame (as is usually the case in the existing SFM algorithms), then the number of unknown motion parameters under the GPC cannot, in general, be reduced to less than four (although the unknowns have to satisfy one or more equation computable from the GPC). This simply means that the GPC can be used most effectively only by SFM algorithms (such as those presented in this paper) that are defined in the WCS.

The GPC ensures that points on the object are constrained to move in planes parallel to the ground plane. With known camera parameters, there is a one to one correspondence between any such plane and the image. Hence if we know the depth from the camera of a point in one frame, then the plane on which the point is confined to move is uniquely determined, and the depth of the same point in any other frame can easily be computed. In fact it can be shown that under the GPC, the depth λ_j of a point in the jth frame is related to its depth λ_i in the ith frame by [18]

$$\lambda_j = (W_i / W_j) \lambda_i \tag{2}$$

where W_i and W_j are terms computable from known camera parameters and image coordinates.

In the subsequent discussions, we assume that the motion of an image sequence of an object is described by the motions of the object w.r.t. its pose in an arbitrarily chosen frame (we call it the reference frame). Further discussions on the GPC and its use in model-based object pose recovery are described in a companion paper [20].

3   Depth estimation

We now discuss the estimation of point depths (structure parameters) from given point correspondences. We define the following symbols:

$S_F = \{F_0, F_1, F_2, ..., F_{M-1}, F_M\}$: the set of $M + 1$ frames in which points have been detected and matched, and F_0 is used as the reference frame;

$S_P = \{P_1, P_2, ..., P_{N-1}, P_N\}$: the set of points appearing in S_F;

$S_{Pm} = \{P_{m1}, P_{m2}, ..., P_{mN_m}\}$: the set of points present in frame F_m, i.e., $S_{Pm} \in S_P$.

S_{Fi}: the set of frames in which point P_i is present;

$S_{Fij} = S_{Fi} \cap S_{Fj}$: the set of frames in which both P_i and P_j appear.

We do not require $S_{Pm} = S_{Pn}$, $m \neq n$, thus point occlusions are allowed. For convenience, we assume all points are present in the reference frame, i.e., $S_{P0} = S_P$. Let the 3D structure of an object be defined by the depths $\lambda_1, \lambda_2, ..., \lambda_N$ of N points in the reference frame. The problem to be solved is: Given S_F and $S_{Pm}, m \in \{0, 1, 2, ..., M\}$, determine $\lambda_1, \lambda_2, ..., \lambda_N$. Two solutions to this problem are given.

3.1   The direct non-iterative solution

We first consider two points P_1 and P_2 in two frames F_0 and F_m ($F_m \in S_{F12}$). According to the distance invariance property [14] of the rigidity constraint [15], the distance between P_1 and P_2 in F_0 is the same as the distance between the two points in F_m. From (2), this gives the following second-order polynomial equation on the depths λ_1 and λ_2 (both associated with F_0) of P_1 and P_2 [16-17]:

$$A_{m1}\lambda_1^2 + B_{m12}\lambda_1\lambda_2 + A_{m2}\lambda_2^2 = 0 \tag{3}$$

where subscript m signifies F_m, and A_{m1}, B_{m12} and A_{m2} are terms computable from known parameters such as image coordinates and extrinsic camera parameters. Their expressions can be found in [16-17]. By considering the two points in F_0 and each of the other frames in S_{F12}, one at a time, a set of second-order polynomial equations on λ_1 and λ_2 can be obtained:

$$A_{m1}\lambda_1^2 + B_{m12}\lambda_1\lambda_2 + A_{m2}\lambda_2^2 = 0, \quad \forall m, F_m \in S_{F12} \tag{4}$$

The number of equations in (4) equals to the number of frames in S_{F12} or $\#S_{F12}$. Since all constraint equations in (4) are homogeneous in λ_1 and λ_2, depth can only be solved from (4) up to a global scale. We therefore arbitrarily choose $\lambda_1 = 1$, and (4) becomes a set of quadratic equations in λ_2:

$$A_{m2}\lambda_2^2 + B_{m12}\lambda_2 + A_{m1} = 0, \quad \forall m, F_m \in S_{F12} \qquad (5)$$

which can easily be solved for each equation separately using the standard formula. Let L_2 denote the set of all positive roots obtained from (5) (note: according to definition, λ_2 must be positive). Then the task is to derive a suitable solution for λ_2 from L_2. Each equation in (5) produces up to two positive depth solutions. If an equation in (5) does have two distinct positive roots, then one is valid, and the other is due to the reflection caused by the use of the distance invariance property in deriving the depth constraint equations. Therefore L_2 can be divided into two subsets L_{2T} and L_{2F}, with L_{2T} representing the set of physically valid solutions, and L_{2F} the false solutions. We thus first detect L_{2T} from L_2 (for a simple technique, see [18]), and then define the median of L_{2T} as the final solution for λ_2:

$$\lambda_2^{(1)} = \text{median}\,(L_{2T}) \qquad (6)$$

where superscript (1) indicates that the depth solution was obtained by using P_1 as the reference point (i.e., the point whose depth was initially set to 1). By maintaining $\lambda_1 = 1$, we can compute depths of all other points in S_P in a similar way. We write all these solutions collectively as $(\lambda_1^{(1)}, \lambda_2^{(1)}, ..., \lambda_i^{(1)}, ..., \lambda_{N-1}^{(1)}, \lambda_N^{(1)})$ with $\lambda_1^{(1)} = \lambda_1 = 1$.

These solutions have been obtained by treating the point P_1 as a reference point. If this point is disturbed by noise, then the resulting depths of points $P_2, P_3, ..., P_N$ will be in error. To avoid this bias towards P_1, we repeat the above process using each P_i as the reference point independently. This generates N sets of depths for the given N points in S_P:

$$\{ (\lambda_1^{(n)}, \lambda_2^{(n)}, ..., \lambda_i^{(n)}, ..., \lambda_{N-1}^{(n)}, \lambda_N^{(n)}) : n = 1, 2, ..., N \} \qquad (7)$$

where $\lambda_n^{(n)} = 1, n = 1, 2, ..., N$, and the superscript n indicates the depths computed under reference point P_n. The depths of each set in (7) may be normalised with respect to (say) the first depth of the set to give

$$\{ (\tilde{\lambda}_1^{(n)}, \tilde{\lambda}_2^{(n)}, ..., \tilde{\lambda}_i^{(n)}, ..., \tilde{\lambda}_{N-1}^{n}, \tilde{\lambda}_N^{(n)}) : n = 1, 2, ..., N \} \qquad (8)$$

where $\tilde{\lambda}_1^{(n)} = 1, n = 1, 2, ..., N$. Then the final solution for the depths of the N points in the reference frame F_0 is defined as

$$\lambda_i = \text{median}\,\{ \tilde{\lambda}_i^{(n)}, n = 1, 2, ..., N \}, i = 1, 2, ..., N \qquad (9)$$

(9) is justified by the fact that all sets of normalized depths in (8) describe the same relative structure of the given N points.

3.2 The non-linear minimization solution

Given an initial value for λ_2, the depth constraint equations in (5) may also be solved simultaneously using the standard non-linear least squares technique. Then the steps described in (7)-(9) can be followed to get the final depth solutions. For detailed descriptions, the reader is referred to [18].

Several remarks can be made at this point. Under normal viewing conditions, the depth range within an object (i.e., the maximum depth difference of points on the object) is much smaller than the nominal depth of the object. Therefore, the depth value assigned to the reference point provides a good initial guess for the depths of all other points. This makes the choice of initial guesses in this non-linear minimization approach a trivial matter indeed. Since the iteration process involves only one unknown and is provided with a good initial guess, its convergence to the correct solution is extremely fast. The total number of iterations required is typically three and rarely exceeds five.

Once the point depths in the reference frame are determined, those in other frames can easily be obtained using (2). If required, the 3D world coordinates may be computed from known image coordinates and the determined depths [18].

4   Estimation of 3D motion parameters

The motion parameters to be determined consist of the translational and rotational parameters of all frames w.r.t. the reference frame. Under the GPC, the motion between the reference frame F_0 and frame F_m is characterized by three independent motion parameters (expressed in the WCS): the translations T_{xm} and T_{ym} on the ground plane, and the rotation angle θ_m about the Z-axis. It can be shown that using the 3D world coordinates of the points in the reference frame computed in the preceding section and the given 2D image coordinates in frame F_m of the N_m points in F_m, a set of $2N_m$ constraint equations on T_{xm}, T_{ym} and θ_m can be derived [18]:

$$\begin{cases} D_{mi1}\cos\theta_m + E_{mi1}\sin\theta_m + F_{mi1}T_{xm} + G_{mi1}T_{ym} = H_{mi1} \\ D_{mi2}\cos\theta_m + E_{mi2}\sin\theta_m + F_{mi2}T_{xm} + G_{mi2}T_{ym} = H_{mi2} \end{cases}, \forall i, P_i \in S_{Pm} \qquad (10)$$

where D, E, F, G and H are terms computable from known image and world coordinates. By regarding $\cos\theta_m$ and $\sin\theta_m$ as two independent unknowns, (10) can be solved using the standard linear least squares technique to get $\cos\theta_m$, $\sin\theta_m$, T_{xm} and T_{ym}. θ_m is then computed as $\theta_m = \tan^{-1}(\sin\theta_m/\cos\theta_m)$. The correct quadrant of θ_m is determined from the senses of $\cos\theta_m$ and $\sin\theta_m$. Motion parameters of other frames in S_F can be obtained similarly.

5   Experimental results

The two proposed algorithms have been tested using both synthetic and real outdoor image sequences.With the synthetic data, Monte Carlo simulations were conducted as follows. An object was specified by N points randomly chosen from within a cuboid. A sequence of frames was then generated by moving the object on the ground plane. The ideal image coordinates of the points in each frame were perturbed by noise. Relative estimation errors were recorded during simulation. The relative error in a motion parameter was obtained by computing the average absolute relative error in the parameter over all frames in a trial, and then calculating the mean of this over all trials. The accuracy of the recovered 3D structure was measured by the *standard scene error*

(SSE) defined as the average Euclidean distance in the reference frame between the original and the reconstructed points. The SSE was computed at each trial, and its mean over all trials was divided by the diameter of the synthetic cuboid model to yield the relative SSE measure.

5.1 Robustness against image data noise

Noise was simulated by adding zero-mean, uniformly distributed random values to the ideal image coordinates of all points in all frames, the level of noise given by ΔE (in pixels) defining a uniform distribution interval $[-\Delta E, +\Delta E]$. Monte Carlo simulations were performed to investigate the noise robustness of the proposed algorithms using a fixed number of points ($=10$) in a fixed number of frames ($=10$). The results are summarized in Fig.1. It can be seen [18] that the overall performances of the two

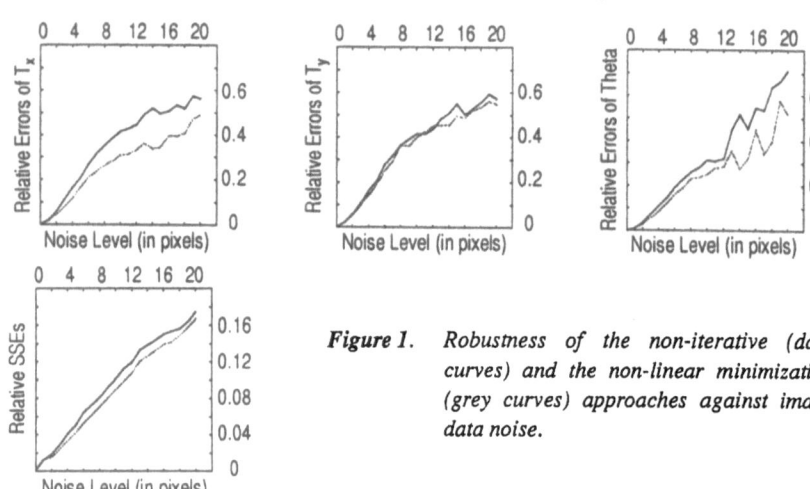

Figure 1. Robustness of the non-iterative (dark curves) and the non-linear minimization (grey curves) approaches against image data noise.

approaches are very similar, with the non-linear minimization approach performing slightly better than the direct technique. Both algorithms are very robust against image data noise. The relative errors in the motion parameters rarely exceed 60%, and the relative SSE is always less than 18% even using unrealistically high noise levels of ±20 pixels.

5.2 Effectiveness of using more frames in noise reduction

Monte Carlo simulations were also carried out to study the benefits of using longer image sequences (i.e., more frames) in noise reduction. The number of points used was fixed at 10, and the noise level was maintained at $\Delta E = 5$ pixels. The results are given in Fig.2. The robustness of the two algorithms is consistently improved by using longer image sequences, with most improvement when the number of frames increases from 3 to 6. Further increase in the number of frames beyond 15 results in barely noticeable improvement.

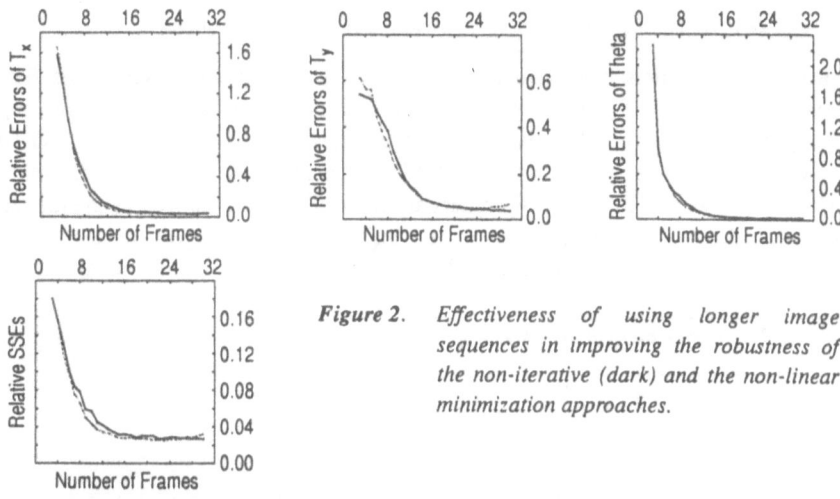

Figure 2. *Effectiveness of using longer image sequences in improving the robustness of the non-iterative (dark) and the non-linear minimization approaches.*

5.3 Effectiveness of using more points in noise reduction

In a further experiment, we kept unchanged the number of frames used (=10) and the level of noise involved (ΔE = 5.0 pixels), and varied the number of points used to examine the effect of the number of points in combating noise. Monte Carlo simulation results are summarized in the plots in Fig.3. The results show that the robustness of the

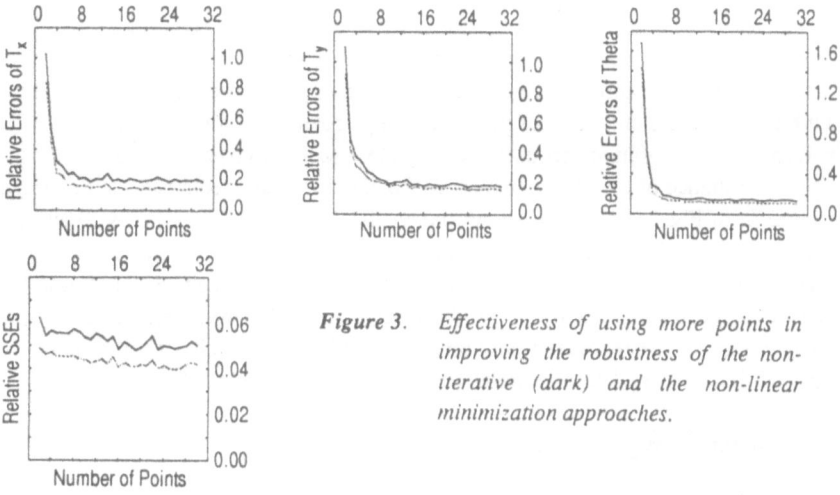

Figure 3. *Effectiveness of using more points in improving the robustness of the non-iterative (dark) and the non-linear minimization approaches.*

two algorithms is consistently improved by using more points. The improvement is most dramatic when the number of points is increased from 2 to 4, and is only marginal for additional points.

5.4 Performance using real outdoor image sequences

The proposed algorithms have also been tested using outdoor image sequences. The Plessey corner finder [19] was applied to detect corner points on a slowly moving (reversing) van (see Fig.4). The image plane trajectories of 14 detected corners over 10 frames were then used as the input to the MFSFM algorithms. Because of the smoothly curved body of the van, the physical corners are not well defined, as well as the limited accuracy of the corner finder, the corner trajectories are subject to significant measurement errors. Table 1 lists the heights (in meters) of the 14 points recovered by

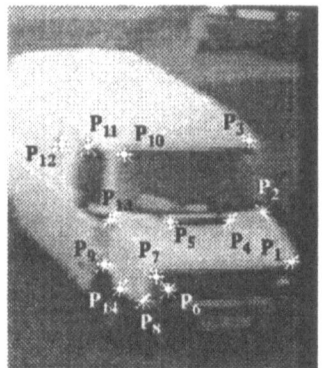

Table 1. Recovered heights of 14 van points

Point	P_1	P_2	P_3	P_4	P_5
Height	0.74	1.08	1.85	1.10	1.10
Point	P_6	P_7	P_8	P_9	P_{10}
Height	0.45	0.57	0.40	0.59	1.73
Point	P_{11}	P_{12}	P_{13}	P_{14}	
Height	1.67	1.72	1.03	0.42	

the non-linear minimization algorithm, where the global scale has been set by assuming P_8 to be 0.4m high. Since the ground truth is not available, the quantitative measures used for synthetic data cannot be calculated. Qualitatively, however, the figures in Table 1 as a whole are consistent with our perception (e.g., relative heights) of the object. The results from the non-iterative approach are very close to those given in Table 1.

The performance of the algorithms using the outdoor image sequence can be further appreciated from Fig.4. The figure shows the originally detected (marked by x) and the reconstructed (marked by +) corner points overlaid on four consecutive frames of the van sequence. If no x is shown near a +, then the corner point marked by the + was not detected in the corresponding frame by the corner finder, but was "predicted" by the algorithm based on the recovered structure and motion. It can be seen that both the detected and the "missing" corners were reconstructed fairly accurately.

6 Conclusions

Novel algorithms have been presented in this paper for 3D structure and motion estimation from 2D image sequences using the ground plane constraint. It has been shown that the depth parameters can be computed using either a non-iterative direct approach or a simple well-behaved non-linear minimization approach, and that the motion parameters can be estimated using the standard linear least squares technique.

The algorithms possess a number of desirable characteristics. They are very robust and perform satisfactorily with real outdoor image sequences. They do not require excessively large numbers of points and/or frames for satisfactory performance and are

Figure 4. Originally detected (marked by x) and reconstructed (marked by +) corners overlaid on four frames of a van sequence.

capable of handling any number of points and/or frames as well as point occlusions. The algorithms are computationally very simple and highly parallel in nature. All these make the algorithms very desirable where applicable. They may be used for a wide range of potential industrial applications.

References

[1] J. K. Aggarwal and N. Nandhakumar, On the Computation of Motion from Sequences of Images - A Review, Proc. of IEEE, vol.76, 1988, pp.917-935.

[2] L. Dreschler and H. -H. Nagel, Volumetric Model and 3D Trajectory of a Moving Car Derived from Monocular TV Frame Sequences of a Street Scene, CGIP, vol.20, 1982, pp.199-228.

[3] Y. Yasumoto and G. Medioni, Robust Estimation of Three-Dimensional Motion Parameters from a Sequence of Image Frames Using Regularization, IEEE Trans. PAMI, vol.8, 1986, pp.464-471.

[4] J. Y. Weng, T. S. Huang and N. Ahuja, 3-D Motion Estimation, Understanding, and Prediction from Noisy Image Sequences, IEEE Trans. PAMI, vol.9, 1987, pp.370-389.

[5] H. Shariat and K. E. Price, Motion Estimation with More Than Two Frames, IEEE Trans. PAMI, vol.12, 1990, pp.417-434.

[6] C. Jerian and R. Jain, Polynomial Methods for Structure from Motion, IEEE Trans. PAMI, vol.12, 1990, pp.1150-1166.

[7] T. J. Broida and R. Chellappa, Estimating the Kinematics and Structure of a Rigid Object from a Sequence of Monocular Images, IEEE Trans. PAMI, vol.13, 1991, pp.497-513.

[8] M. Spetsakis and J. Aloimonos, A Multi-frame Approach to Visual Motion Perception, Inter. J. Comput. Vision, vol.6, 1991, pp.245-255.

[9] R. Kumar, A. Tirumalai and R. C. Jain, A Non-linear Optimization Algorithm for the Estimation of Structure and Motion Parameters, Proc. of CVPR'89, June 4-8 1989, San Diego, USA, pp.136-143.

[10] H. S. Sawhney, J. Oliensis and A. R. Hanson, Description and Reconstruction from Image Trajectories of Rotational Motion, Proc. of ICCV'90, December 1990, Osaka, Japan, pp.494-498.

[11] G. S. Young, R. Chellappa and T. H. Wu, Monocular Motion Estimation Using a Long Sequence of Noisy Images, Proc. of IEEE Inter. Conf. on ASSP, 1991, pp.2437-2440.

[12] N. Cui, J. Y. Weng and P. Cohen, Extended Structure and Motion Analysis from Monocular Image Sequences, Proc. of ICCV90, December 1990, Osaka, Japan, pp.222-229.

[13] J. K. Aggarwal and A. Mitiche, Structure and Motion from Images: Fact and Fiction, The 3rd IEEE Workshop on Vision: Representation and Control, October 1985, pp.127-128.

[14] A. Mitiche and J. K. Aggarwal, A Computational Analysis of Time-Varying Images, in Handbook of Pattern Recognition and Image Processing, T. Y. Young and K. S. Fu, Eds., New York: Academic Press, 1986.

[15] S. Ullman, The Interpretation of Visual Motion, Cambridge, MA: MIT Press, 1979.

[16] T. N. Tan, G. D. Sullivan, and K. D. Baker, Structure from Constrained Motion Using Point Correspondences, British Machine Vision Conference 1991, P. Mowforth Ed., Springer-Verlag, 1991, pp.301-309.

[17] T. N. Tan, G. D. Sullivan, and K. D. Baker, Structure from Motion Using the Ground Plane Constraint, Proc. of ECCV-92, G. Sandini Ed., LNCS-Series Vol.588, Springer-Verlag, 1992.

[18] T. N. Tan, 3D Structure and Motion Estimation from 2D Image Sequences Using the Ground Plane Constraint, ESPRIT II project (P.2152) research report, RU-03-WP•T2122-TNT-03, University of Reading, April 1992.

[19] J. A. Noble, Finding Corners, Proc. of 3rd Alvey Vision Conf., University of Cambridge, England, 15-17 September 1987, pp.267-274.

[20] T. N. Tan, G. D. Sullivan and K. D. Baker, Linear Algorithms for Object Pose Estimation, Proc. of British Machine Vision Conference 1992, Springer-Verlag, 1992.

Statistical Detection of Independent Movement from a Moving Camera

P H S Torr and D W Murray

Robotics Research Group

Department of Engineering Science

Oxford University, Parks Road, Oxford, OX1 3PJ, UK

Abstract

This paper describes the use of a low level, computationally inexpensive closed form motion detector to define regions of interest within an image, based upon statistical measures. The algorithm requires only the first order properties of the image intensities and does not require known camera motion. It has been tested on a variety of real imagery. A b-spline snake is initialised on the occluding contours of this region of interest.

1 Introduction

Amongst the important tasks which rely on motion data is that of motion segmentation. This paper addresses the problem of how to detect a set of moving objects in the two dimensional projection of an otherwise rigid scene, given that the camera is moving in an arbitrary and unpredetermined manner. A closed form analytic solution is supplied for the detection of non-rigid motions given the spatio-temporal gradients.

The problem of motion segmentation has received considerable attention over the years. For example, Nelson [10] has described an algorithm designed to solve the object segmentation problem for the case of known camera translation and rotation. Burt and his co-workers [3] have developed a multi-scale pyramidal motion segmentation algorithm designed for use in conjunction with control of the sensor and the parameters of the algorithm. François and Bouthemy [7] have designed an algorithm that uses qualitative information about the camera motion to aid motion segmentation, using a Markow Random Field (MRF) approach to segment the scene into regions with common affine flows. Adiv [1] and Waxman and Duncan [15] make second order approximations to the flow field over small regions. They then use various methods to test the compatibility of these regions, in order to decide whether or not to merge them. Thompson, Mutch and Berzins [13] and Schunck [12] present algorithms for motion boundary detection based upon early edge detection algorithms using the Laplacian of the Gaussian.

In this paper we propose a global solution to the problem of motion segmentation, in order to overcome the problems posed by regions contributing a paucity of data (e.g. aperture problem), by considering the whole image. Many methods assume dense and accurate velocity fields as input and impose a continuity constraint upon the projected motion field identifying motion boundaries along lines where this continuity is violated. Without highly accurate estimates

of the projected motion, small parts of the image are unlikely to contain sufficient information to reconstruct full flow and its deformation parameters, thus
numerical differentiation of the projected vector field or the merging of small
areas will be very ill-conditioned. As in [4] we assume an affine background
flow, and successive quantitative estimates are made about the affine deformation parameters of the background motion over the image. Parts of the image
that do not accord with this estimate are identified as regions of interest. We
use the first order intensity properties of the image as input to our algorithm,
so that we do not throw any information away.

Our aim then is to partition the five dimensional space of discrete image
points (pixels) and the spatio-temporal intensity gradients calculated at those
pixels $\{(E_x, E_y, E_t, x, y)\}$ into disjoint sets corresponding to either the designated background or to foreground objects undergoing independent motions.
To achieve this we

1. Fit a hyper-plane through the points in a given region using least-squares,
   assuming that the points undergo an affine transform (Sections 2.1 and
   2.2);

2. Check for collinearity and appropriateness of the assumption (Section
   2.3);

3. Identify the outliers to the fit (Section 2.1); and

4. Cluster the outliers to form regions of interest (Section 3).

2 Least squares fit to intensity gradients

Within this section we shall outline how to discover outliers from the flow
predicted by the affine scene model. These outliers will usually arise from
either noise or occlusions. The interested reader is referred to [2, 6] for a more
thorough coverage of the theory and methods of diagnostic techniques.

2.1 Testing for Outliers within Least Squares

Given a set of equations

$$y_\alpha = \vec{d}_\alpha \vec{b}^T \qquad\qquad \alpha = 1 \ldots n \qquad\qquad (1)$$

where y_α is a known variable, $\vec{d}_\alpha$ is a known p dimensional vector and $\vec{b}$ is an
unknown p dimensional vector, termed the vector of coefficients We shall term
y_α the *dependent* variable and $\vec{d}_\alpha$ the vector of *independent* variables. Let $[\mathbf{D}]$
be a matrix whose rows are $\vec{d}_\alpha$ then from equation 1 we can see that:

$$\vec{y}^T = [\mathbf{D}]\vec{b}^T \qquad\qquad (2)$$

we can then use the pseudo inverse to solve for $\vec{b}$:

$$\vec{\beta}^T = [[\mathbf{D}]^T[\mathbf{D}]]^{-1}[\mathbf{D}]^T\vec{b}^T \qquad\qquad (3)$$

where $\vec{\beta}$ is our estimate of the coefficients $\vec{b}$ which is unknown.

The general procedure for assessing the influence of a given point in a regression analysis is to determine the changes that occur when the point is omitted. Several measures of influence exist in the literature. They differ in the particular regression result on which the effect of the deletion is measured, and the standardization used to make them comparable over observations. All the influence measures discussed can be computed from the results of a single regression. Below we discuss three influence measures, each of which takes account of the deletion of the ith observation or equation (e.g. what would be the solution to the set of equations 1 if we delete the equation $y_i = \vec{d_i}\vec{b}^T$) on some regression variable. Cook's D measures the effect on $\vec{\beta}$ our estimator of $\mathbf{b}$. DFFITS–F measures the effects on our prediction of the dependent variable $\vec{y_i'}$ given $\vec{\beta}$. COVRATIO–C which measures the effect on the variance-covariance matrix of the parameter of estimates.

Potentially influential points are data points that are far from the centre of the $[\mathbf{D}]$-space. A measure of the distance of the ith data point from the centroid of all the points in $[\mathbf{D}]$-space $\overline{\mathbf{d}}$ is provided by h_{ii}, the ith diagonal element of the hat matrix $[\mathbf{H}]$. The hat matrix is derived as follows, from equation 3 we can see that:

$$\vec{y'}^T = [\mathbf{D}] \left([\mathbf{D}]^T [\mathbf{D}] \right)^{-1} [\mathbf{D}]^T \vec{y}^T \tag{4}$$

where $\vec{y'}$ is our prediction of $\vec{y}$ given $\vec{\beta}$. $[\mathbf{H}] = [\mathbf{D}] \left([\mathbf{D}]^T [\mathbf{D}] \right)^{-1} [\mathbf{D}]^T$ is termed the hat or the orthogonal projection matrix on the column space of $[\mathbf{D}]$. h_{ii} is termed the *leverage* or *potential* of the ith case in that is gives an indication of the effect of y_i on $\hat{y}_i$, the closer h_{ii} is to 1 the smaller the residual e_i. The estimate of the variance of the dependent variable is

$$\hat{\sigma_t}^2 = \frac{\sum e_i^2}{n-p} \tag{5}$$

Where e_i is the ith element of the residual vector $\vec{e}$. Note that $\vec{e} = ([\mathbf{I}] - [\mathbf{H}])\vec{y}$, $\mathbf{Var}(e) = \mathbf{Var}(([\mathbf{I}] - [\mathbf{H}])\vec{y})$. Thus the residuals do not have common variance. The heterogeneous variances in the residuals are corrected by dividing each residual by an estimate of its standard deviation given by the square root of the diagonal elements of $([\mathbf{I}] - [\mathbf{H}])\hat{\sigma_t}^2$. Standardized (or internally Studentized) residuals s are given by [11]:

$$s_i \stackrel{\text{def}}{=} \frac{\hat{e}_i}{\hat{\sigma_t}\sqrt{1 - h_{ii}}} \tag{6}$$

Belsley, Kuh and Welsch [2] suggest standardizing the residuals with an estimate of its standard deviation independent of the residual. This is accomplished using $\sigma_{(i)}$, the estimate of the standard deviation without the ith observation which can be obtained by:

$$(n - p - 1)\sigma_{(i)}^2 = (n - p)\sigma_i^2 - \frac{e_i^2}{1 - h_{ii}} \tag{7}$$

Let t_i be the ith Studentized residual such that [2]

$$t_i \overset{\text{def}}{=} \frac{e_i}{\sigma_{(i)}(1 - h_{ii})^{\frac{1}{2}}} = s_i \left(\frac{n - p - 1}{n - p - s_i^2} \right)^{\frac{1}{2}} \qquad (8)$$

t_i will follow a t-distribution with $n - p - 1$ degrees of freedom if the errors in y_i are normally distributed. Cook's D [5] test is designed to measure the shift in $\vec{\beta}$ when a particular observation is omitted. Cook's D_i is defined as

$$D_i \overset{\text{def}}{=} \frac{(\vec{\beta}_{(i)} - \vec{\beta})^T ([\mathbf{D}]^T [\mathbf{D}])(\vec{\beta}_{(i)} - \vec{\beta})}{p \sigma_i^2} = \frac{s_i^2}{p} \left(\frac{h_{ii}}{1 - h_{ii}} \right) \qquad (9)$$

Where $\vec{\beta}_{(i)}$ is the set of parameters fitted to the data without the ith observation included. D_i has approximately an F-distribution thus $D_i \approx F_{(\alpha, p, n-p)}$. Cook [5] suggests that if $\alpha = .50$ from omitting a single data point then this is significant. The 50th percentile for F is 1.0 when the numerator and denominator are large thus a value of D_i near 1.0 is significant. This is extreme and the literature suggest a more modest threshold of $\frac{4}{n}$, where we recall n is the number of observations. The **DFFITS** [2] statistic F can be computed from the Studentized residual. F_i gives a measure in the change of $\vec{y}'$ when the ith observation is not included in the estimation of $\vec{\beta}$.

$$F_i \overset{\text{def}}{=} \frac{y_i' - y_{(i)i}'}{\sigma_{(i)}\sqrt{h_{ii}}} = \left(\frac{h_{ii}}{1 - h_{ii}} \right)^{\frac{1}{2}} t_i \qquad (10)$$

where $y_{(i)i}' = x_i \vec{\beta}_{(i)}$ i.e. the estimated y_i' for the ith observation where the ith observation was not used to estimate $\vec{\beta}$. After tests it was found that the significant areas identified by Cook's D and F_i are very nearly identical. An approximation to the impact of the ith observation on the variance of the estimated coefficients is measured by the ratio of the determinants of the two variance-covariance matrices **COVRATIO** $= C$.

$$C \overset{\text{def}}{=} \frac{\left| \sigma_{(i)}^2 \left[[\mathbf{D}]_{(i)}^T [\mathbf{D}]_{(i)} \right]^{-1} \right|}{\left| \sigma_i^2 [[\mathbf{D}]^T [\mathbf{D}]]^{-1} \right|} = \left[\left(\frac{n - p - 1 + t_i^2}{n - p} \right)^p (1 - h_{ii}) \right]^{-1} \qquad (11)$$

The determinant of the variance-covariance matrix is a generalised measure of variance. Thus C reflects the impact of the ith observation on the precision of the estimates of the regression coefficients. Values near 1 indicate that the ith observation has little effect, greater than 1 indicates that the presence of the ith observation increases the precision of the estimates, the converse is true. A range of $1 \pm 3p/n$ is suggested to be considered the extremes for identifying influential points. Thus in this section we have presented a set of computationally simple methods for determining outliers to a set of linear equations.

2.2 Affine Flow

In this paper we assume that the spatial structure of the projected flow, with the exception of independently moving foreground objects, is coherent and may be approximated by a linear vector field. According to the proposed method these foreground objects may be detected as inconsistent with the affine background motion. The affine assumption is approximately correct when the distance to background objects is large when compared to the variations in these distances, this occurs in many outdoor scenes. It is also approximately correct for rotations seen over a small field of view.

Rather than first computing the flow, then fitting, we fit directly to the spatiotemporal image surface. Let the image intensity at inhomogeneous pixel coordinate $\vec{x} = (x, y, -f)$ be $E(x, y)$. The motion constraint equation [9] is:

$$\frac{\partial E}{\partial x}\frac{dx}{dt} + \frac{\partial E}{\partial y}\frac{dy}{dt} + \frac{\partial E}{\partial t} = 0 \tag{12}$$

Verri and Poggio [14] have shown that equation 12 does not hold in general, but increases in accuracy as the spatial gradient increases. Thus in general points are only used if $|\nabla E|$ exceeds a certain threshold. Let $\vec{u} = (u, v)$ be the projected velocity at $\vec{x}$ then $\nabla E \cdot \vec{u} + E_t = 0$ given the flow varies linearly:

$$\begin{pmatrix} E_x & E_y \end{pmatrix} \begin{bmatrix} u & \frac{\partial u}{\partial x}\Delta x & \frac{\partial u}{\partial y}\Delta y \\ v & \frac{\partial v}{\partial x}\Delta x & \frac{\partial v}{\partial y}\Delta y \end{bmatrix} = -E_t \tag{13}$$

We can rewrite equation 13 separating the observables and unobservables into two vectors of the form given by equation 1 where

$$\vec{d} = \begin{pmatrix} E_x & E_y & \Delta x E_x & \Delta y E_x & \Delta x E_y & \Delta y E_y \end{pmatrix}^T$$
$$\vec{b} = \begin{pmatrix} u & v & \frac{\partial u}{\partial x} & \frac{\partial u}{\partial y} & \frac{\partial v}{\partial x} & \frac{\partial v}{\partial y} \end{pmatrix}^T$$
$$\vec{y} = \begin{pmatrix} E_{t_1} & E_{t_2} & \dots & E_{t_N} \end{pmatrix}^T \tag{14}$$

We may solve the set of equations presented in 14 by a least squares method, given N image points for which we know the spatio-temporal derivatives. Outliers to this system of equations are then be deemed to be independently moving objects.

2.3 Collinearity

If there are near-singularities among the columns of [D] then we have insufficient data within this region to reconstruct the fit. This could have arisen from a highly structured image e.g. a series of vertical bars. Alternatively sparse or unbalanced data could give rise to collinearity. When the data is collinear then we must rely on past estimates for information. Waxman [16] referred to this as the *aperture problem in the large* in which insufficient contour structure leaves the set of deformation parameters undetermined, even over large regions of the image. For instance, if the data had arisen from a single conic section then there would be at least one affine dependency in [D] and the Taylor coefficients β would be undetermined.

Geometrically, this means there is poor dispersion in one of the dimensions of [D]-space. The presence of collinearity can be detected by an eigen analysis of $[D]^T[D]$. The six eigenvalues λ_i^2 provide measures of the amount of dispersion for each of the principal component axes in [D]-space [11]. The condition number is defined as the ratio of the largest to the smallest singular value λ_i. This gives a measure of sensitivity of β to small changes in [D]. The condition number concept is extended to the condition index for each (principal component) dimension of the [D]-space. The condition index K_i for the ith principal component dimension in [D]-space is the ratio of the maximum singular value to the ith singular value. We shall take values of the condition index around 10 to indicate moderate dependencies, values from $30 - 100$ to indicate strong dependencies and values in excess of 100 to indicate severe collinearity problems [11]. The number of condition numbers in the critical range indicating the number of near-dependencies. Given a large number of dependencies warning must be given that the result of the regression is suspect.

The size of the eigenvalues depends on the scale of the columns of [D], thus we shall scale [D] so that the length of each column vector is one (i.e the sum of squares of the elements is unity) to prevent the eigen-analysis being dominated by one or two independent variables e.g the term xE_x will be always be larger than the E_x column but we do not wish to give it any greater weight when testing for collinearity.

Thus we have presented a statistically well founded technique for determining whether the image is indeed sufficiently structured enough to allow us to recover β.

3 Clustering

Once we have identified a set of outliers we then need to form a hypothesis about whether they are consistent with one or more rigid three dimensional objects moving independently of the background. We utilise a method of spatial clustering by merging nearby outliers into groups and defining the region of interest as the convex hull of the group, following an algorithm presented in [8].

A problem with this is that a lone outlier (a result of noise for instance) might seriously distort the convex hull. Thus we utilise further robust statistical methods to differentiate between outliers caused by noise and outliers caused by objects moving differently to the background. The image is tessellated into equal sized overlapping regions. Given an estimate of the noise (by observation of some static scenes) we calculate a 99% confidence interval that the number of outliers within the region must exceed to determine that the outliers within that region are not due to noise. From the set of points delineating the convex hull we initialise a b-spline snake onto the occluding contour of the object.

4 Results

We have implemented a movement detector based on the above principals using the COVRATIO test−C and Cook's D measures for outliers on a linear fit. Empirically it was found that the results of DFFITS was indistinguishable from

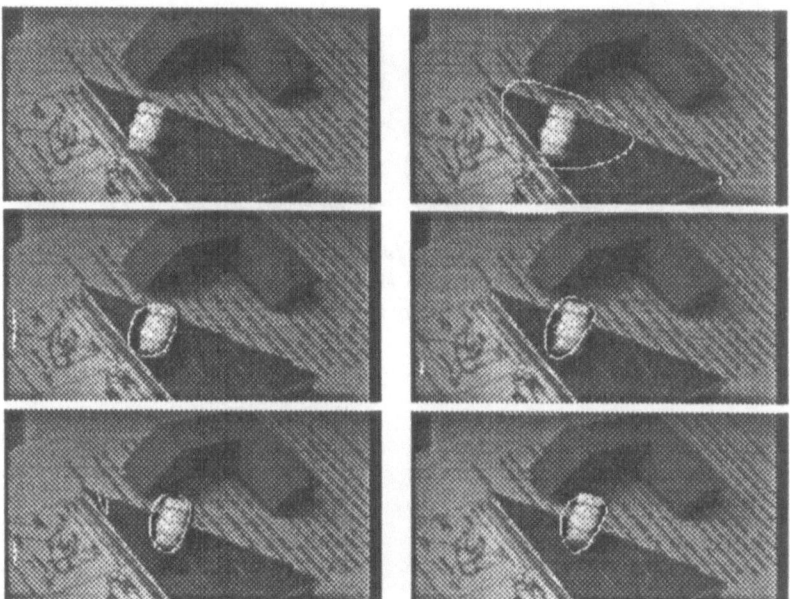

Figure 1: *Showing a sequence of images of a white object translating 4 pixels to the right as the background translates 0 − 2 pixels down and 0 − 1 pixels right depending on the depth.*

Cook's D. The first is of a moving object taken from a camera in motion in an indoor scene. The second is of several moving objects as the camera translates. No heuristics or "magic numbers" were used to derive the thresholds. Instead they were derived from the statistical theory underpinning the work. Furthermore, these thresholds are self scaling to the type of image concerned.

The first sequence of images shown in figure 1 is of a white object translating 4 pixels to the right as the background translates 2 pixels down and $0 - 1$ pixels right depending on the depth. The image is 256×128. To reduce the amount of redundant information in the least squares points with low gradient were excluded, this reduced the number of points under consideration from 32768 to 29039. The reason for the exclusion of these points is that they are clustered about the origin in observation space through which any fit must pass and thus provide redundant information, their exclusion reduces the amount of calculation. The thresholds for points to be considered outlying were $C = 1 - \frac{3p}{n}$ and $D = \frac{4}{n}$ taken from [11], e.g. $C < 0.997864$, $D > 0.000285$. Figure 2 shows the result of the variance test superimposed on the motion shown in in figure 1, areas in black are outliers, white areas are background and the grey areas where points excluded from the fit and are shown in their original intensities. Note that some of the edge of the lower triangle has been indicated as outlying. Care must be taken when handling the output of the outlier tests. As we are making a linear flow assumption depth and velocity discontinuities are both shown. It is hoped that the inclusion of a matching strategy over time might reduce the number of false outliers. The second pair of images, in figure 3, depicts several

Figure 2: *Showing the result of the C-test for outliers superimposed on the images shown in figure 1, white areas are the background and black areas are outliers. Grey areas are the intensities of the original regions with low (E_x, E_y, E_t) that are excluded from the regression.*

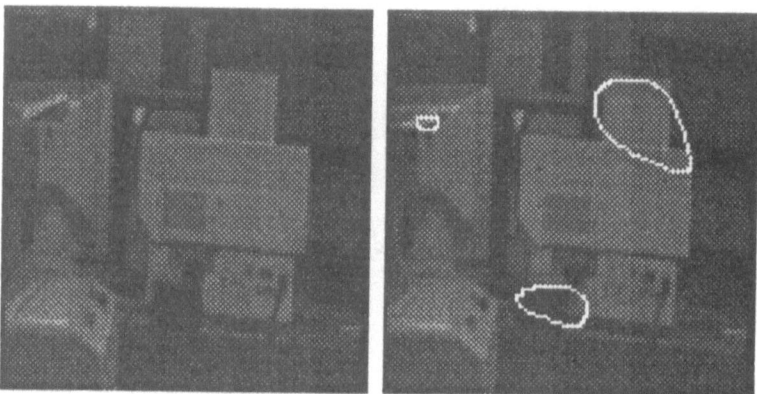

Figure 3: *Showing several lab objects moving in different directions as the background moves $0 - 3$ pixels left. In the bottom centre of the image a black box moves quickly 10 pixels to the left and 3 pixels up, on the top left a piece of paper contracts as it slips down the back of the monitor $3 - 5$ pixels and moves 3 pixels to the left (i.e 0 left pixels relative to the background). In the centre a large white box moves to the right by 6 pixels and down by 1 pixel*

lab objects moving in different directions as the background moves $0 - 3$ pixels left. In the bottom centre of the image a black box moves quickly 10 pixels to the left and 3 pixels up, on the top left a piece of paper contracts as it slips down the back of the monitor $3-5$ pixels and moves 3 pixels to the left (i.e 0 left pixels relative to the background). In the centre a large white box moves to the right by 6 pixels and down by 1 pixel. The image is 256×256, points with low gradient were excluded, this reduced the number of points under consideration from 65536 to 21216. The thresholds for points to be considered outlying were $C < 0.998586$, $D > 0.000189$

Overall given accurate estimates of the first order properties of the image the algorithm successfully localises the independently moving objects, providing that there motion is sufficiently different to the background motion.

5 Conclusions

In this paper we have presented a method for detection of non-rigid motion given information derived from time varying imagery. The method is founded upon the examination of the differences between the observed temporal difference and a predicted form given an affine transformation. Thus we make a global estimate of the background motion using a scene constraint, as a heuristic. The algorithm does not attempt to establish point correspondences, estimate the optic flow, or make a three dimensional reconstruction. It does not require knowledge of camera motion or calibration. Instead successive quantitative estimates are made about the deformation parameters of background motion and parts of the image that do not accord with this estimate are identified as regions of interest. This identification is done using recent statistical work on the analysis of regressions. The thresholds are determined in a principled manner and are self scaling to the variances of the image intensities.

6 Future Work

There are two avenues of current research. The first is to improve the method of grouping the outliers into cohesive groups. The second is generalisation to a more realistic set of motions. An inherent problem with an affine approximation is that it is only valid in a limited number of situations. Current work addresses the problem of how to detect a set of moving objects in the two dimensional projection of an otherwise rigid scene, given that the camera is moving in an arbitrary and unpredetermined manner. We utilise the fact that point correspondences having arisen from a projective $3D$ transformation can be described by a 3×3 *Essential Matrix* [**E**] linking the coordinates of the points before and after the transformation. The Essential Matrix is derived by an analytic $O(N^3)$ least squares method, assuming at least half of the image is undergoing a coherent projective transformation. Thus points with non-rigid motion (modulo a projectivity) are deemed to be those statically inconsistent from the calculated value of [**E**].

Acknowledgements

This work was supported by SERC grant GR/G30003. Thanks are due to Andrew Zisserman and Paul Bearsdley for helpful suggestions and to Charlie Rothwell for the convex hull software.

References

[1] G. Adiv. Inherent ambiguities in recovering 3-d motion and structure from a noisy flow field. In *Proceedings, CVPR '85 (IEEE Computer Society Conference on Computer Vision and Pattern Recognition, San Francisco, CA, June 10-13, 1985)*, IEEE Publ. 85CH2145-1., pages 70–77. IEEE, IEEE, 1985.

[2] E. Belsley, D.A. Kuh and Welsch R. E. *Regression Diagnostics: Identifying Influential Data and Sources of Collinearity*. Wiley, 1980.

[3] P. J. Burt. Image motion analysis made simple and fast, one component at a time. In *Proc. BMVC*, pages 1–8, 1991.

[4] M. Campani and A. Verri. Computing optical flow from an overconstrained system of linear algebraic equations. In *Proceedings of the Third International Conference on Computer Vision*, pages 22–25, 1990.

[5] R.D. Cook and S. Weisberg. Characterizations of an empirical influence function for detecting influential cases in regression. *Technometrics*, 22:337–344, 1980.

[6] R.D. Cook and S. Weisberg. *Residuals and Influence in Regression*. Chapman Hall; London, 1982.

[7] E. François and Bouthemy P. Multiframe based identification of mobile components of a scene with a moving camera. In *Proc. CVPR.*, 1991.

[8] Green and Silverman. Constructing the convex hull of a set of points in the plane. *Computer Journal*, vol.22:262–266, 1979.

[9] B.K.P. Horn and B.G. Schunck. Determining optical flow. *Artificial Intelligence*, 17:185–203, 1981.

[10] R.C. Nelson. Qualitative detection of motion by a moving observer. *IJCV*, pages 33–46, 1991.

[11] J.O. Rawlings. *Applied Regression Analysis*. Wadsworth and Brooks, California, 1988.

[12] B.G. Schunck. Image flow segmentation and estimation by constraint line clustering. *IEEE Transactions on Pattern Analysis and Machine Intelligence*, 11:1010–1027, 1989.

[13] W.B. Thompson, K.M. Mutch, and V.A. Berzins. Dynamic occlusion analysis in optical flow fields. *IEEE Transactions on Pattern Analysis and Machine Intelligence*, 7:374–383, 1985.

[14] A. Verri and T. Poggio. Against quantitative optical flow. In *First International Conference on Computer Vision, (London, England, June 8–11, 1987)*, pages 171–180, Washington, DC., 1987. IEEE Computer Society Press.

[15] A.M. Waxman and J.H. Duncan. Binocular image flows: Steps toward stereo-motion fusion. *IEEE Transactions on Pattern Analysis and Machine Intelligence*, 8:715–729, 1986.

[16] A.M. Waxman and K. Wohn. Contour evolution, neighborhood deformation, and global image flow: Planar surfaces in motion. *International Journal of Robotics Research*, 4(3):95–108, 1985.

Accurate Boundary Location from Motion

J.A. Marchant
Agricultural and Food Research Council,
Silsoe Research Institute,
Wrest Park, Silsoe, Beds. MK45 4HS.

1   Introduction

The ability to monitor a visual scene containing animals and to draw intelligent conclusions automatically would have a significant impact on agricultural practice. For example, if the gait of an animal could be objectively measured, early detection of lameness would be possible. If the motion of a sow and piglets could be analysed, a stockman could be alerted if the piglets were in danger of being crushed or were not feeding properly.

This work forms part of a programme to estimate the weight and hence growth rate of animals from images. In this case, accurate boundaries are required. Animals are often found in visual situations where the background is cluttered and cannot easily be controlled. Also their own surface is often marked either naturally or by contamination from their environment. Segmentation techniques based on thresholding are usually not successful but it may be possible to exploit the fact that animals move whereas the background is stationary.

2   Related work

Methods which seek to segment objects using their motion usually rely on an estimate of the object motion itself. Motion estimates for parts of the image can be gained by correlating between small windows in a pair from an image sequence [1]. Correlation can be done in the spatial domain where a window is fixed in position in one image and moved in the other until some measure of correspondence is maximised [2]. Alternatively, the process can be done in the frequency domain using the Fourier transform which is generally faster and suited to modern special purpose hardware. There is a basic problem in using any technique in which a finite sized window is used -estimates near the boundary of the moving object (the very places which require accurate location in this work) will be poor.

In principle this problem could be overcome by using differential techniques for motion measurement (e.g. [3]). However, these techniques are affected greatly by noise problems.   Also it can be shown that no information can be obtained on the motion component parallel to an edge feature unless extra assumptions are made concerning the form of motion.

Murray et al. [4] exploit the fact that functions of the image intensity and its changes can be chosen which vary rapidly across object boundaries. However, the method still depends on using a finite sized operator to detect peaks in these functions. Rivero and Bouthemy [5] also use differential methods for motion estimation and collect together regions having similar motions. The size of these regions are smaller near to object boundaries but still of a sufficient size to give a very "blocky" appearance to the edge.

In the work reported here a correlation technique is used to avoid noisy motion estimates. Poor estimates near to boundaries are avoided by using a motion model derived with a robust estimator. An accurate and reasonably complete boundary is then built up by integration of successive estimates over a motion sequence.

3 Outline of method

The method starts by finding the area of significant change in an image pair by differencing and thresholding. Such an image pair and the changed area is shown in Fig. 3.1. The changed area contains components from:

a) where background has been uncovered by the object,
b) where background has been covered up,
c) where object pixels have been replaced by other object pixels at a significantly different grey level,
d) noise giving rise to isolated pixels.

These changes occur over an area generally larger than the moving object and the changed area contains many missing pixels where a grey level change is within the threshold. Some of these missing pixels can be filled in by a number of dilation operations followed by an equal number of erosions.

As pointed out by Ostermann [6] the boundary of the moving object can be found by combining the changed area with a knowledge of the object movement as follows:

Calculate the motion vector for each point in the changed area (see below). Place the tail of the vector on each pixel in the changed area. If the head is within the changed area (i.e. the head is also on a changed pixel), the head point is on the moving object. Note that if the motion vector is other than zero, isolated noise points will be removed provided there is no second noise point at the head of the vector.

As the changed area is an incomplete representation of the areas where motion has occurred, this procedure gives an equally incomplete version of the moving object (Fig. 1).

In order to give a more complete rendition of the object the method is applied to a sequence of images. As the method proceeds, an "object mask" is maintained which is a binary image, grey level 255 signifies that particular pixel is on the moving object, level 0 signifies background. As each new image pair is analysed, the existing object mask is "warped" by moving each pixel by the calculated motion vector. Then new points are added to form the new object mask. Thus the object mask tracks the object through the image sequence and becomes more complete in the process. The underlying assumption is that object pixels which do not change significantly at any one iteration of the method (and thus do not form part of the changed area) will change significantly at some other stage in the sequence. To recover the moving object from any image in the sequence the object mask is simply combined with the image by a logical AND.

4 Calculation of motion vectors

The early stages of motion measurement follow the work reported by Burt et al. [1]. Motion vectors are measured by correlation between small windows in the image pair. A fast Fourier transform is used to perform the correlation using a window size of eight pixels square. A coarse to fine procedure limits the motion at each level of the procedure to a few pixels. A pyramid of images is formed , each image being half the resolution of its parent. Some care must be taken when reducing resolution in order to avoid aliaising of frequencies which are present in the finer resolution image but above the Nyquist frequency at the lower resolution. The author convolved each image in the pyramid with a filter having three zeros at frequencies at and above the Nyquist frequency before sampling the filtered image (Appendix).

The sequence analysis depends on an accurate knowledge of the motion of each part of the object. In order to locate boundaries accurately (a major objective of this work) the information must be available at the boundaries of the object. However, these areas are also the points where correlations will be poor and motion estimates inaccurate. To avoid this problem, a motion model is used.

Following Burt et al. [1] the variation of motion across the object is explained by assuming the object to be a rigid body moving with six degrees of freedom in three dimensions. Thus a model for the coherent motion of the object can be obtained by fitting two functions, one each to the x and y components of the motion, of the form:

$$v_x = ax + by + c \qquad\qquad ..... (1)$$
$$v_y = dx + ey + f \qquad\qquad ..... (2)$$

Because of boundary effects, the raw motion data will contain a significant number of outliers. Three methods have been used to combat this problem.

1 After dilating and eroding the changed area (previous section) the area is further eroded to remove from the boundary a width equal to approximately half the window size used for correlation. This new area becomes the basis for raw motion estimates although the original area is retained to calculate the object mask.

2 The cross correlation for each motion estimate is normalised by dividing by

$$\left(\sum g^2 \sum f^2 \right)^{0.5}$$

where g is the grey level in one window, f is the grey level in the other window at maximum correspondence and the sums are over the window areas [2]. This gives a value between 0.0 and 1.0. An average value is calculated over all the points in the changed area and only motions for those points above the average are passed on to the next stage.

3 A robust estimator is used to identify the parameters in Eqns. 1 and 2. In a normal least squares estimator it can be shown that each point is weighted according to its distance from the fitted plane. In the estimator used here [7],

a sinusoidal weighting function is used which peaks at a difference of 1 unit and returns to zero at 2 units. Thus points greater than 2 units from the fitted plane are ignored completely. The procedure results in a non-linear minimisation problem which has been solved here using the Simplex method [7]. The starting values for the estimated parameters are gained from a least squares fit.

5 Results

The method was used on two types of images. Firstly a random pattern of grey levels between 0 and 255 in a window 64 pixels by 48 which was moved against a second random pattern as background. A random number generator was used to produce displacements between ±9 pixels horizontally and ± 6 pixels vertically. For this test the window motion was confined to a translation in the image plane of a whole number of pixels in each direction.

Table 1 shows the number of edge pixels, object pixels, and background pixels found in two cases; firstly where the changed area was not modified by dilation and erosion and secondly where two stages of each were used. The images were numbered from 0 to 7 and sequential pairs were used for analysis.

Table I. Performance of algorithm on a random pattern

image pair	displacement x,y	dilation/erosion = 0			dilation/erosion = 2		
		edge pixels	object pixels	back-ground pixels	edge pixels	object pixels	back-ground pixels
0/1	-7,-2	172	2375	0	220	3072	0
1/2	9,-2	202	2875	0	220	3072	0
2/3	5, 5	215	3021	0	220	3072	0
3/4	8, 0	219	3056	0	220	3072	0
4/5	4,-6	220	3070	0	220	3072	0
5/6	1,-6	220	3071	0	220	3072	0
6/7	7, 6	220	3072	0	220	3072	0
Total no. of edge pixels = 220; total no. of object pixels = 3072							

It should be noted that the problem is made easier by the fact that the warping of the object mask is constrained to give an integer result. As the window was moved by integer amounts, the rounding process removes errors providing they are less than 0.5 pixels.

Table I shows that the algorithm yields the moving window exactly, correctly finding all object pixels (including those on the edge) and no background pixels. With no dilation/erosion of the changed region seven iterations of the algorithm are required. With two stages of dilation and erosion the algorithm finds the window on the first iteration.

In the second set of tests the trunk and head of a person was used as the target object. Two situations were chosen - where the person was wearing patterned clothing against a cluttered background, and where relatively plain clothing was worn against a plain background.

Eight images were used in each sequence numbered zero to seven. Figs. 2 and 3 show image number 7 along with results from image pairs 0/1, 3/4, and 6/7 from each of the two situations. In each case, the movements were a combination of rotations and translations in the dimensions. Some care was taken to ensure that the head/body combination moved as a rigid body i.e. articulation at the neck was kept to a minimum. This restriction was imposed, for this stage of the work, to avoid contravening of the basic assumptions of the method. Note that a second assumption, that the body is planar, was regularly violated. For dilations of the changed area were used followed by four erosions when finding object pixels from motion vectors.

Fig. 2 shows the gradual improvement of the object segmentation throughout a sequence. With the exception of a few isolated background areas, the only significant addition to the object is a small area to the left hand side if the neck. A small part of the boundary on the left shoulder has become slightly ragged. Some areas of the body, notably the forehead and below the neck have been missed. This is due to insufficient texture in these areas giving regions with no significant change in grey level with movement. These areas could be filled in by noting the fact that there are no holes in the real object. As the cause of the problem is lack of texture, a reasonable estimate of the grey level of the holes could be made by averaging the grey levels over the corresponding areas in the whole sequence. However, the performance on the task in hand, finding an accurate boundary, is good. Note that the cracks in the objects are caused by using integer arithmetic in the warping process to produce the object mask.

Where there is less texture in the image (Fig. 3) the performance is worse, as expected. However, with the exception of the area to the right of the neck, the segmentation of the head is good. The ragged boundary to the left and right of the body could possibly be improved by smoothing the boundary direction. As with Fig. 2 the holes (this time much larger) on the body could possibly be filled in with an average of grey levels over the sequence.

6 Conclusions

A method has been proposed which can derive an accurate boundary of an object from an image sequence. The method depends on a number of assumptions, in particular that the motion is due to a planar object which has sufficient texture on its surface.

Tests on random dot patterns which translate an integral number of pixels give perfect results when there is sufficient grey level texture but, as expected, poorer results on more uniformly shaded objects.

In order to use the method on more general animal images, a technique will need to be developed to handle objects which cannot be represented as rigid bodies. For instance, those that deform or articulate. Future work will address this problem.

References

[1] P.J. Burt, J.R. Bergen, R. Hingorani, R. Kolczynski, W.A. Lee, A. Leung, J. Lubin and H. Shvaytser. Object tracking with a moving camera. Proc. IEEE Workshop on Visual Motion, Irvine CA, 1989, pp 2-12.

[2] A. Rosenfeld and A. Kak. Digital picture processing (2nd ed.). Academic Press, Orlando, 1982.

[3] A. Verri, F. Girosi and V. Torre. Differential techniques for optical flow. J. Optical Society of America 1990; 7:912-922.

[4] D.W. Murray and N.S. Williams. Detecting the boundaries between optical flow fields from several moving planar facets. Pattern Recognition Letters 1986; 4:87-92.

[5] J.S. Rivero and P. Bouthemy. A hierarchical likelihood framework for motion based segmentation from image sequences. Proc 5th Scandinavian Conf. on Image Analysis. Stockholm, 1987, pp 623-631.

[6] J. Osterman. Modelling of 3D moving objects for an analysis - synthesis coder. Proc. SPIE Conf. Sensing and reconstruction of 3D objects and scenes, Santa Clara CA, 1990, pp 240-249.

[7] W.H. Press, B.P. Flannery, S.A. Teukolsky and W.T. Vetterling. Numerical recipes in C. Cambridge University Press, Cambridge, 1988.

APPENDIX

Anti-alias filter for image sampling.

If $g(m,n)$ is the grey level of an image at point m,n then the image can be represented in the frequency domain as $G(u,v)$ where G is the two dimensional discrete Fourier transform:

$$G(u, v) = \sum_{m,n} g(m, n) \exp(-2\pi i (um+vn)/N)$$

N is the image size and u and v complex frequency components.

Consider an image at level ℓ in the pyramid of decreasing resolution. This is derived by sampling an image at level ℓ-1 at every other pixel. The Nyquist frequency at level ℓ-1 is π radians/pixel and so that at level ℓ is $\pi/2$ radians pixel.

Note that the pixel dimension for both levels in this explanation is that for level $l-$ 1. Hence, in order to avoid aliasing, all frequencies above $u = v = \pi/2$ should be removed in the image at level $l-1$ before it is sampled. This cannot be achieved exactly but an approximation can be made by convolving three masks

$$
\begin{array}{l}
1\ 1\ ^* \quad 1\ 1\ 1\ ^* \quad 1\ 1\ 1\ 1 \\
1\ 1 \quad\ \ \ 1\ 1\ 1 \quad\ \ \ 1\ 1\ 1\ 1 \\
\quad\ \ \ 1\ 1\ 1 \quad\ \ \ 1\ 1\ 1\ 1 \\
\quad\ \ \ \quad\ \ \ 1\ 1\ 1\ 1
\end{array}
$$

to give a filter mask

1	3	5	6	5	3	1
3	9	15	18	15	9	3
5	15	25	30	25	15	5
6	18	30	36	30	18	6
5	15	25	30	25	15	5
3	9	15	18	15	9	3
1	3	5	6	5	3	1

The 4 x 4 filter has a zero at $\pi/2$, the 3 x 3 at $2\pi/3$ and the 2 x 2 at π, and so the composite filter has zeros at these three frequencies.

Note that the 7 x 7 filter can be implemented by convolving the image firstly with the 1 x 7 filter formed by the first row, then with the 7x1 filter formed by the first column. This procedure speeds up the implementation significantly.

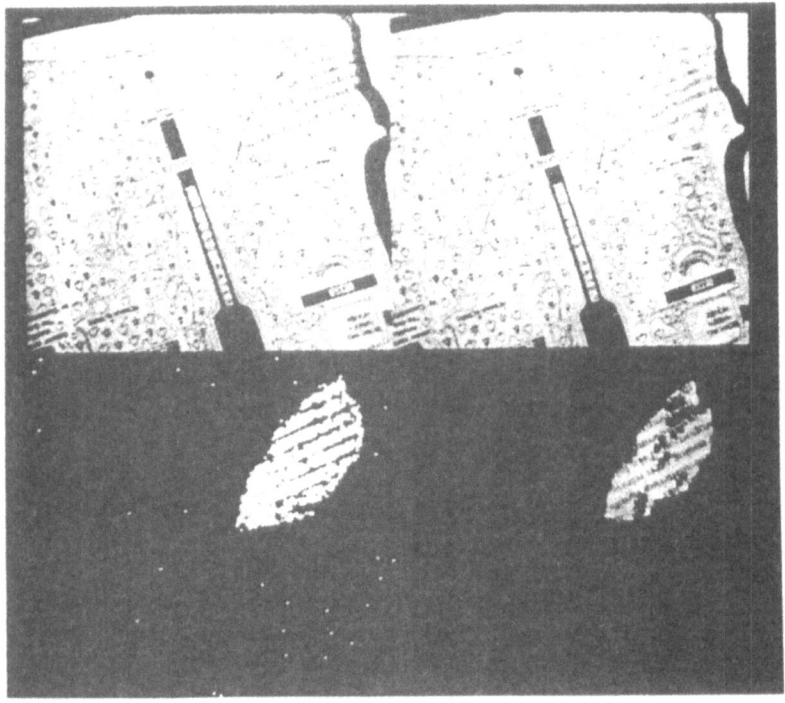

Fig. 1

Top, image pair consisting of an area torn from a page of text moving on a similarly textured background. Bottom left, changed region. Bottom right, incomplete rendition of object.

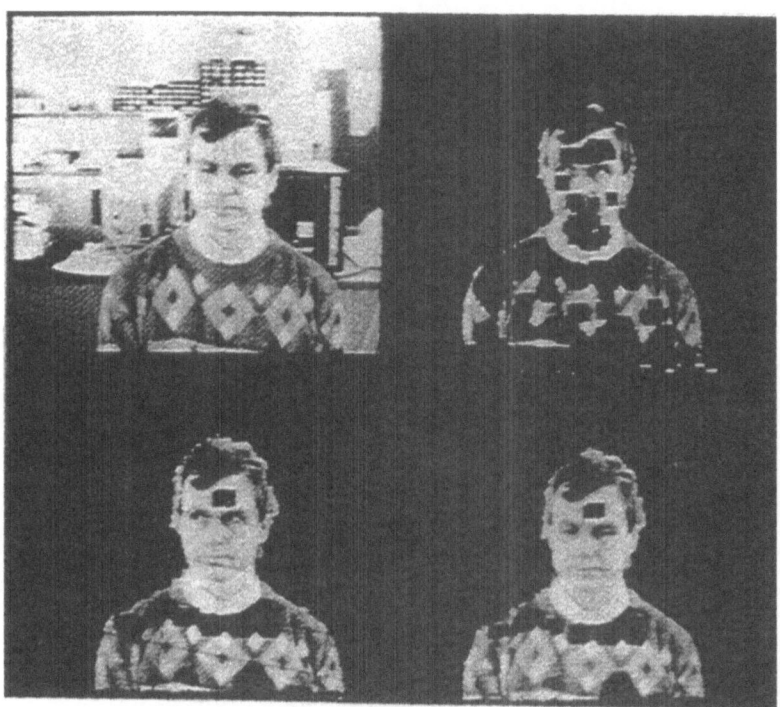

Fig. 2

Results for relatively patterned object on a cluttered background. Top left, image sequence No.7. Top right, bottom left, bottom right, improvements of object rendition.

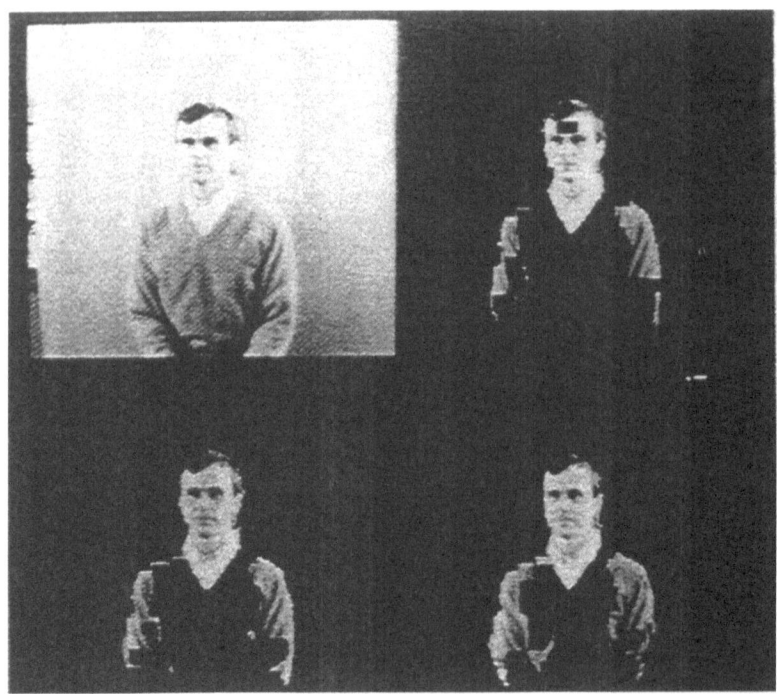

Fig. 3

Results for relatively plain object on plain background. Top left, image sequence No.7. Top right, bottom left, bottom right, improvement of object rendition.

From Features to Perceptual Categories

W. Richards*

J. Feldman   A. Jepson

First we review an analysis of conditions that should be met if features are to provide robust inferences about world properties. Features meeting these conditions provide indices into especially useful categories of visual properties. Then we show that for a given set of elemental concepts the categories associated with these properties have a natural hierarchical (specialization) structure. We argue that this structure provides constraints on the form and type of categories that are inferred when visual objects are classified.

1 Introduction

Perception, or "seeing", involves the assignment of world properties to image elements. Both machine and biological vision systems proceed in this task by recasting the image pixels into meaningful "features", from which object properties are inferred [1]. The features suitable for this task are not arbitrary, but are highly constrained, and have a natural hierarchical structure. This structure mirrors that of natural processes, thereby providing a basis for inferring natural categories. Hence we begin with a review of conditions that constrain the choice of those special features that provide robust inferences about world properties.

2 What Makes a Good Feature?

Let the world consist of various properties P that are associated with various contexts, C. Then $p(P|C)$ denotes the conditional probability of a property, P, such as "has 4 corners" in the context C, which could be sitting "on a plane", "in this region", etc. Similarly the collection of measurements of a property and their conditional probabilities will be specified by F and $p(F|C)$. Note that $p(P|C)$ and $p(F|C)$ are simply objective facts about the world and are *not* statements about the perceiver's model of the world. Our first task is to place conditions on F, P and C that ensure the measurements F constitute a reliable indicator that P occurs in the world.

*Authors contributed equally and their order is arbitrarily permuted. WR and JF are at Mass. Inst. Tech., Dept. Brain & Cog. Sci., Cambridge, MA 02139; AJ is at the Univ. Toronto, Dept. Computer Science, Toronto M5S 1A4. This work was supported by AFOSR 89-504 and NSERC Canada. Correspondence should be addressed to WR.

2.1 Reliable Inferences

The posterior probability of inferring property P given the feature F in context C is $p(P|F\&C)$. A reliable inference makes this probability nearly one, and keeps the probability of an "error", i.e. $p(notP|F\&C)$ near zero. Hence a reliable feature F, in context C, will keep the following ratio, namely R_{post} much larger than one:

$$R_{post} = p(P|F\&C) \, / \, p(notP|F\&C). \tag{1}$$

Using Bayes Rule, R_{post} can be broken down into the product of two components, a likelihood ratio L that relates to the "imaging" of P onto F, and the prior probability R_{prior}, that relates to the genericity of the world property P in context C. Specifically, $R_{post} = L \cdot R_{prior}$, where

$$R_{prior} = p(P|C) \, / \, p(notP|C) \quad \text{and} \quad L = p(F|P\&C) \, / \, p(F|notP\&C). \tag{2}$$

Note that the likelihood ratio captures the intuition that a feature should arise reliably from a given world property, i.e. $L >> 1$. As will be seen in the next section, however, this condition does not insure a reliable inference, because if R_{prior} becomes too small, then R_{post} can become insignificant even in the presence of a high likelihood ratio. (Also see [2].)

2.2 An Example

Consider a world of line segments on a plane seen under orthographic view. Of interest is the special property "two line segments are parallel". Let the threshold for discriminating the orientation difference between two (adjacent lines) be θ, and let $\delta << \theta$ be the limiting resolution of the process that governs straight and parallel. Now let the collective distribution of the orientation ϕ of all line segments be rather flat (Figure 1A). Given this context, we are now presented with two lines that fall within the crosshatched sample for $\phi < \theta$. Hence the two lines appear parallel; should we conclude that these lines indeed arise from a parallel process?

First note that the likelihood ratio, L, is very high, because (i) whenever parallel lines occur in the world, they always will appear parallel in the image, and (ii) our chance of error is vanishingly small – when two lines are not parallel, they will not be seen as such except in the rare case when they lie within our limit of resolution θ. Hence $p(F|P\&C) = 1$ and $p(F|notP\&C)$ is, say 0.01 if θ is 1 part in 100. It appears therefore that we should infer that the lines are indeed parallel in the world. However,

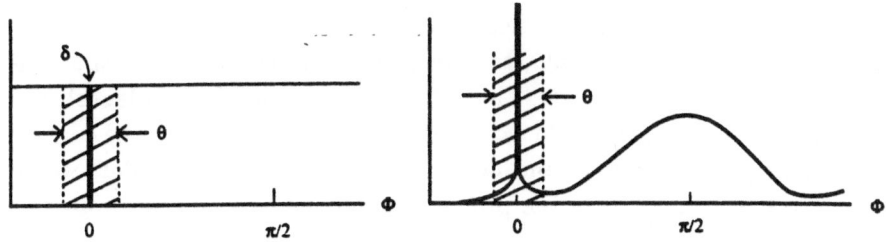

Figure 1 A: Flat distribution function. B: "Modal" distribution.

given our chosen random world context, such an inference is almost always guaranteed to be *wrong*.

Consider the prior probability ratio R_{prior}. Because the prior probability δ of the parallel process occurring is much less than the resolution limit θ, the area occupied by δ in Figure 1A is much less than the area set by θ. Thus $R_{prior} = \delta/(1-\delta) << \theta$, and its product with the likelihood $L \cong 1/\theta$ will give an a posteriori probability ratio $R_{post} < 1$. Hence the odds really favor the conclusion "not parallel". (See [2] and [3] for further details and examples.) In order to raise R_{post} to a significant level, we need significant priors, say a δ in this case such that $\delta/\theta >> 1$. In terms of Figure 1, this is equivalent to requiring that the ϕ distribution function for pairs of lines be biased, such as indicated in Figure 1B where the process "parallel" appears as a mode in the probability distribution function.

3.0 Model Class

The important message of the previous example is that "good" features arise from some modal regularity in the distribution function of world properties. However, not all regularities satisfying the likelihood and prior conditions will be useful. For example, the property "two skewed lines" satisfies these two conditions, but clearly this property is not very informative. Hence what we seek are properties that are not just arbitrary configurations, but rather ones that are in some sense special.

3.1 Two Kinds of Regularities

Structural regularities within a given model class can be divided into two classes: transverse and non-transverse [4]. Transverse relations arise when the elements of the model are postioned arbitrarily such as the above two skewed lines; non-transverse arrangements require careful positioning, as implied by the term "non-accidental" of Binford [5] and Lowe [6]. Unlike the notion of "non-accidental", however, the usage of transverse and non-transverse requires a context. Thus, "two parallel lines" (or planes) in a random stick (or planar) world would be non-transverse, but in the context of a building with windows and doors, etc., the concept "parallel" would become transverse. Within the proper context, non-transverse properties are thus very special. But as we showed earlier, in order to be recoverable from image features, the non-transversality must be an isolated spike in the distribution function as in Figure 1B, with sufficient mass to be "visible". This is what previous researchers meant by "modal" properties [7, 8, 9]. Features that satisfy (2) and which arise from non-transverse regularities provide especially reliable and useful inferences about world properties and are called Key Features (see [2] for "natural" examples taken from motion and color). Loosely speaking, F will be a Key Feature for property P if P is a generic non-transverse mode in the space of world models, and F occurs in the presence of P but never in its absence. Hence the set of properties that image onto the Key Features are an especially useful set of properties, because they are reliably inferable.

3.2 An Example

To illustrate a set of properties that image onto key features in our simplified world of line segments in a plane, assume there are two processes that generate two types of relations between two lines. One is the process "parallel"; the other is a process "coincident", where the lines just touch one another. We take these regularities as generic – i.e. we stipulate that both occur with significantly non-zero probabilities in the given context. First we enumerate those regularities that image to key features. Then in the following section, we will place an ordering on this special set of properties.

The enumeration is equivalent to identifying all the non-transverse configurations between line segments in a plane, given the chosen context. We assume the measurement is the orientation of one line to the other, ϕ, and the position x, y of the end-point of one line with respect to the other. Hence the relative positioning has three degrees of freedom (DOF). Referring to Figure 2, the uninformative, transverse regularity chooses x,y, and ϕ arbitrarily, producing two skewed lines. (Intersection or not was not specified in our model class and an "X" will be treated as equivalent to skew without crossing.) First, with care the end of one line can be placed

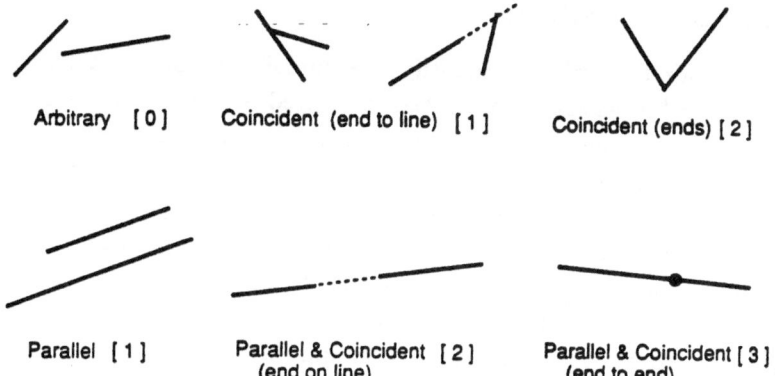

Figure 2 Line-to-line non-transversalities.

on the other (or its extension), eliminating one degree of freedom. These configurations are assigned a codimension of one. Next, with still more care, we can place the end of one line exactly on the end of the other, allowing only the angle ϕ to vary. This arrangement has a codimension of two.

Similarly, if the lines are parallel, then the orientation is fixed and the cost, or codimension of the arrangement is again one. However parallel and coincident lines, with one end allowed to slide along the other, increase the codimension further to two. Finally, we have the last, most special case of positioning of codimension 3 where the two lines merge into one when placed end to end in a parallel arrangement.

4.0 From Features to Categories

Our main point will be that the "interesting" structural regularities in a model class – namely those that satisfy the key feature conditions – can be used as a basis for partitioning the model class into categories. In the previous example, the property space would be built from the end-point position measurements x, y, and the relative pose ϕ. Within this x, y, ϕ space, our proposal is that the partioning should make explicit the line-to-line non-transversalities illustrated in Figure 2. If this scheme is adopted, then the subspaces will preserve the character of the nontransversal modes, thus distinguishing among the interesting properties. Note that the context sensitivity is critical to our set-up, because it permits legitimate reconfigurations of the property space depending upon the observer's goals, etc.

4.1 "Two Stick" Categories

Continuing our example, we identify the modal subspace as that associated with our "two-stick" model class presented earlier in Figure 2. Each of these configurations has a codimension, which allows us to place each of these non-transverse modes in a lattice, where each node depicts a proper subspace in the particular context (Figure 3A). The top node shows the arbitrary two-stick configuration. As we move down the lattice, the nodes below differ at each successive level by the removal of exactly one degree-of-freedom from the configuration. Upward transitions, then, are the elemental ones that locally "break" or "unfold" a non-transversal property but which do not add any additional non-transverse properties. An important example is the missing link between the "V" and "parallel line" nodes (or similarly, the "T" and "collinear" nodes). There is no direct route from one node to the other. The explanation is that the concepts "coincident" impose a constraint on the endpoint position x, y of one line with respect to the other, whereas the concept "parallel" is expressed by an angular relation, ϕ between the two lines. Because position (x, y) is not defined by angle (ϕ) or vice versa in this context, there is no intersection other than the excluded degenerate case of two coincident lines. A similar explanation applies to the missing path from the two "collinear" lines and the "T" node.

At the bottom of the lattice, two nodes have the two sticks collapsed to one. These two nodes have broken outlines to indicate that they are not part of the lattice for the perceptual context because they suggest a "one-stick" configuration. (If the two sticks were each identified in some manner, say by coloring, then the dashed paths and nodes would become part of this "two-stick" category lattice.)

4.2 "Natural Example": Beetle Lattice

In the biological realm, growth processes exhibit regularities [10]. To illustrate how such regularities can be used for a taxonomic classification, we will use a simple modification of the "two-stick" mode lattice of Figure 3A. Let the context be the backs of beetle-like bugs that are marked by two distinctive lines oriented with respect to the symmetry axis of the beetle. As is typical for biological shapes, we assume the markings are generated symmetrically about this axis. Hence, with respect to our "two-stick" mode lattice, one stick – the "reference stick" – will simply be the symmetrical bisector of the beetle's back. The other, namely the "second stick", will thus

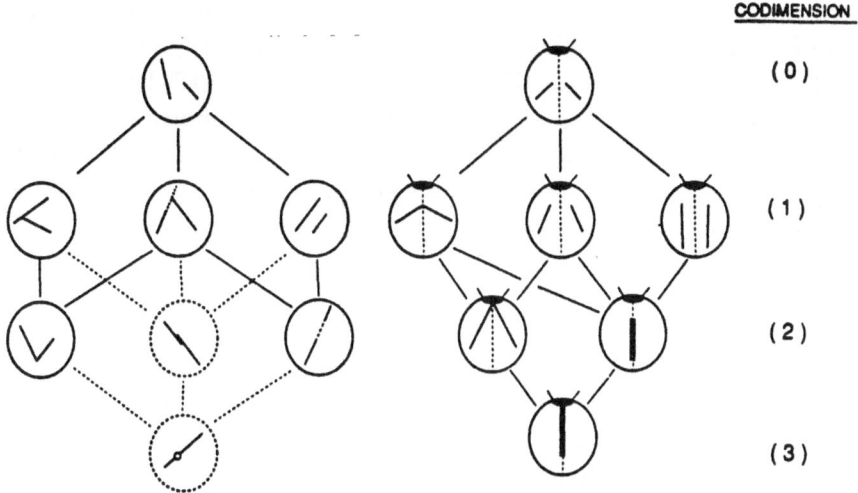

Figure 3 A: "Dot-on-line" categories. "Two-stick" categories given concepts coincidental andf parallel. See text for explanation of dashed paths and nodes. B: Beetle taxonomy, based on a version of a "two-stick" mode lattice.

appear twice in mirror image positions about this symmetric axis. (The situation is equivalent to symmetric markings appearing on a left and right wing.) As before, we assume two possible marking processes, one laying the mark down parallel to the bisecting axis, the other positioning the end point (of either the reference stick axis or the additional marking stick) to be coincident with one of the two lines. All of this sets the context.

Because the two-stick modes in a similar context have already been enumerated, we simply need to recast the previous lattice of Figure 3A in a symmetric form compatible with this revised "biological" context. This has been done in Figure 3B, where now each node depicts the markings on the beetle's back. At the top, the two symmetric marking lines are set arbitrarily with respect to the bisecting axis (dotted). This is the codimension 0 case for this species. At the next level either the coincident or parallel process applies, giving us three codimension 1 subspecies. Next, we have two codimension 2 cases: in one the marking lines form a V, coming together at the "head" of the reference line, or the other where the two marking lines collapse onto the reference line (but do not reach the head of the beetle). Finally we have a single codimension 3 case in this context where the "V" collapses onto the reference bisecting line. Given these generating processes and this context, these are all the types of beetles expected. These types, with the exception of the "generic" beetle at the top, represent the beetle modes or subspecies, each exhibiting a slightly different, but related regularization of the ontology of beetles. Thus the beetle lattice is a convenient hypothesis generator for an observer who is seeking to assign any particular beetle to its "natural" category [11, 12].

5.0 Category Induction

It is easy to imagine that pairs of sticks or beetle markings in a world might obey the regularities depicted in the nodes of the lattice of Figure 3, having all originated from some common underlying processes. Then it would seem plausible that object categories corresponding to the regularities found on the lattice are more prone to occur in the world than completely arbitrary collections of stick-pairs that are not on the lattice (This is the Natural Mode assumption [7, 9, 13].) Putatively, then, when an observer examines a collection of stick-pairs, or beetles, the conclusions about what category processes are responsible for the observed collection will be drawn exclusively from the conceived lattice. Consider the induction problem in Figure 4A. What is the common description of the collection on the left as opposed to the collection on the right? An infinity of different answers are possible, many making entirely different predictions about what "more of the same" would mean on each side. However, the simple answer "V's versus parallel lines" is most intuitively compelling. Note that such predictions about what new examples of either side will look like constitute an enormous inductive leap, since so many alternative solutions are also possible given the small number of exemplars. The problem in Figure 4B can similarly be solved almost instantly, namely T's and V's versus "parallel and collinear lines". Note that our conclusion was not T's versus parallel lines, although the V's and collinear lines are respectively degenerate cases of T's and parallel. Thus each node in the lattice is taken as a separate category in its own right.

The previous example shows that the mixture of two nodes is never on the lattice. This property allows categories to be inferred correctly by an observer even when they are intermingled, because the collapsed compositional category that all objects are of the same kind is not a sanctioned hypothesis (see [14]). Again this is seen in panel C, where now we have mixtures of different beetle subspaces. Hence as long as the world presents

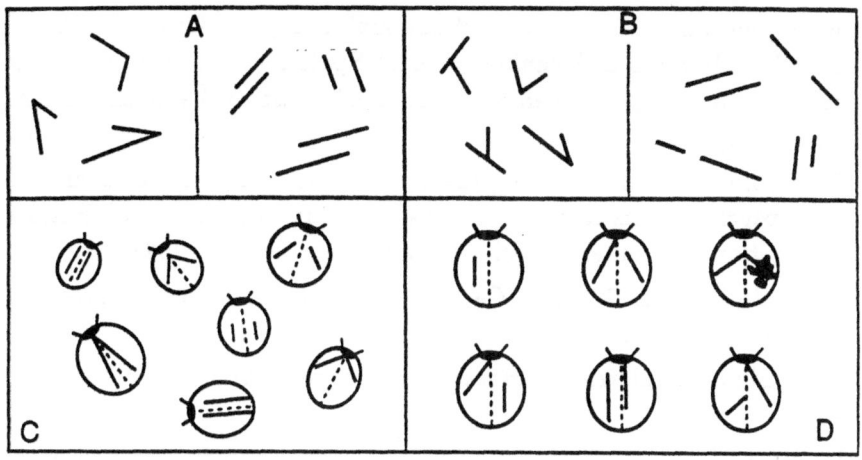

Figure 4 Category induction.

only lattice regularities, an observer can report correctly the number of categories present – a capacity unaccounted for by conventional categorization techniques such as clustering.

Finally, in panel D we have presented a set of beetles that clearly appear defective. Note that these beetles fall outside the normal beetle definition because of their asymmetry. Rather than concluding these forms don't fit, however, instead we attempt to hypothesize that they are altered defective forms of recognizable categories. On the top half of Figure 4D we are successful, in the the bottom half we are not – these latter beetles seem "weird" mutants. Hence whenever possible the lattice is used to "regularize" the defective beetles back to the true form [15]. When such regularization is impossible, then the beetle is seen as inconsistent with our models for the generative process of natural forms. Hence the category lattice built from easily recognizable non-transverse structural modes appears to play a major role in object classification and recognition.

References

[1] D. Marr, Vision: a computational investigation into the human representation and processing of visual information. Freeman, San Francisco, 1982.

[2] A. Jepson and W. Richards, What makes a good feature? To appear in: Harris L and Jenkin M (eds) Spatial vision in humans and robots. Cambridge University Press, 1992. Also available as MIT AI Lab Memo 1356, 1992.

[3] D.C. Knill and D.K. Kersten, Ideal perceptual observers for computation, psychophysics and neural networks. In: Watt R J (ed) Pattern recognition by man and machine. MacMillan London, 1991.

[4] T. Poston and I. Stewart, Catastrophe theory and its applications. Pitman, London, 1981.

[5] T.O. Binford, Inferring surfaces from images. Artif Intell 1981; 17: 205-244.

[6] D. Lowe, Perceptual organization and visual recognition. Klewer, Boston, 1985.

[7] A. Bobick, Natural categorization. MIT Artif Intell Lab Tech Report 1001, 1987.

[8] D. Marr, A theory of cerebral neocortex. Proc R Soc Lond B 1970; 176: 161-234.

[9] W. Richards and A. Bobick, Playing twenty questions with nature. In: Pylyshyn Z (ed) Computational processes in human vision: an interdisciplinary perspective. Ablex Norwood NJ, 1988.

[10] D. Thompson, On growth and form. The University Press, Cambridge, 1952.

[11] J. Feldman, Perceptual simplicity and modes of structural generation. Proc 13th Ann Conf Cog Sci, Chicago, IL, August, 1991, pp 299-304.

[12] M. Leyton, Perceptual organization as nested control. Biol Cybernetics 1984; 51: 141-153.

[13] A. Witkin and J.M. Tenenbaum, On the role of structure in vision. In: Beck J, Hope B and Rosenfeld A (eds) Human and machine vision. Academic New York, 1983.

[14] J. Feldman, Constructing perceptual categories. Proc Comp Vis and Pat Recog, Champaign, IL, June, 1992, pp 244-250.

[15] J. Feldman, A. Jepson and W. Richards, Is perception for real? Proc Conference on Cognition and Representation, SUNY Buffalo, April 1992, in press.

Vanishing Point Detection

A Tai, J Kittler, M Petrou and T Windeatt

Dept. of Electronic and Electrical Engineering,

University of Surrey,

Guildford, Surrey GU2 5XH, United Kingdom

Abstract

Commencing with a review of methods for vanishing point (VP) detection, a new approach is suggested. The proposed approach estimates the location of candidate vanishing points and provides probability measures which reflect the likelihood of those points being the VPs. This new approach allows VPs to be identified in less structured environments compared with its conventional counterparts.

1   Introduction

Geometrical cues and constraints provide valuable information as to how certain image features should be interpreted. For instance, in many man made scenes there exist a number of straight lines which are mutually parallel in 3D. Under perspective projection these lines will meet at a common point known as the vanishing point (VP). Once this point is identified, one can infer 3D structures from 2D features and this constrains the search for other structures. Also, under known camera geometry the orientation of the lines that are grouped together can be determined from the corresponding VP. Furthermore, two or more vanishing points arising from lines which lie on a certain 3D plane give a vanishing line. This property provides an additional constraint which is particularly relevant when analysing, for instance, aerial imagery where one can often assume that structures of interest lie in a common plane – the ground plane.

The relationships amongst camera parameters, structures in 3D scenes and VPs had been established by Haralick [1]. The applications of VP analysis ranges from extracting 3D structures to the calibration of camera parameters.

An obvious approach to locating VPs is to exploit directly the property that all lines with the same orientation in 3D converge to a VP under perspective transformation. Thus the task of VP detection can be treated as locating peaks in a two dimensional array where the intersections of all line pairs in an image plane accumulate. However, the line pairs can intersect anywhere from points within an image to infinity and this poses problem on implementation.

In order to avoid analysing an open space Barnard proposed the projection of image lines onto a Gaussian sphere [2] [3] [4] which neatly represents any 3D orientations. The plane which contains the lens centre and the line segment in the image intersects with the Gaussian sphere centred at the origin to form a great circle. That is a line segment on the image plane is mapped to a great circle. Hence, VPs can be detected as elements on the surface of the Gaussian sphere which have relatively high votings. Obviously, the Gaussian surface has to be partitioned in order to accumulate votes. One popular parameterisation is in terms of the azimuth and elevation angles of the unit vector in the sphere. Note that a uniform partitioning in the Hough space maps to non-uniform area in the image plane. For example, the elementary areas at the poles are small compared with those at the equator. This implies that some lines may intersect within a larger area and still be grouped together to hypothesise a VP whereas others may not. A crude way of ensuring accuracy is to partition the Hough plane into finer bins. This improvement in accuracy is paid for by an increase in memory requirement and computational load.

Magee *et al* [3] compute the vectors pointing towards the intersection of line segments in the image plane using a series of cross-product operations. Instead of incrementing a discrete parameterisation, the actual values of the azimuth and elevation angles are maintained for comparison using an arc distance as a metric. This circumvents the problem of the non-uniform elemental surface area of the Gaussian sphere. Although this allows one to locate VPs to a higher accuracy, the computation of the vectors pointing at the intersection points has transformed the $O(n)$ order problem to an $O(n^2)$ one.

Quan and Mohr [4] propose an efficient way both in terms of the amount of operations and memory requirement for computing VPs. They employ a pyramidal data structure instead of a straight forward two dimensional array. This algorithm is similar to a Fast Hough Transform method. It consists of subdividing recursively a patch on the sphere into 4 sub-patches from a coarse to fine resolution. By doing so, a coarse to fine algorithm can be used to improve the efficiency of the Hough Transform (HT). However, detailed experimental studies of hierarchical approaches to the vote accumulation in the HT suggest that the steps that need to be taken to ensure the detection of all level features may render the technique computationally inferior to standard HT implementation.

The methods outlined above do not handle the issue of noise directly, instead the locations of the vanishing points are assumed to be the mid-point of the bin. Thus, the error of the estimated locations of VPs is a function of bin dimensions. Collins and Weiss [5] treat the task of VP detection as a statistical estimation problem. They note that a vector pointing towards the VP lies in the projection plane of the line, and is thus perpendicular to the projection plane normal. In other words, the projection plane normals of 3D parallel lines lie in a plane through the origin, perpendicular to the orientation of the 3D lines in a noiseless environment. In reality, these normals cluster around a great circle forming an equatorial distribution and this distribution is then modelled using Bingham's distribution [6]. It turns out that this approach gives the same result as fitting the least squares perpendicular error planes corresponding to the line pairs.

VPs are also widely used for camera calibrations and recovery of rotational component of motion. Camera calibration involves the determination of camera rotation and translation matrix, focal length etc. These rely on the relationships between the VP coordinates and the camera parameters both intrinsic and extrinsic. The usefulness of VPs in motion analysis stems from the fact that they represent 3D orientations and are therefore invariant to 3D translations between the camera and the scene.

Liou and Jain [7] devised a scheme for road tracking in image sequences.

Assuming that the locations of VP remains unchanged in contiguous frames, they then fit a template around this VP. The dimensions of this template are determined by a constant and a tilt angle. Since the true VP must fall inside this area, one can locate the road boundaries by considering a pair of convergent lines which maximises an *ad hoc* measure based upon the lengths of lines, directions of edge points supporting the lines etc.

Shigag *et al* [8] compute camera rotation by first classifying lines on the image into horizontal and non-horizontal groups and then estimating the camera rotation as the difference of the angle between the optical axis and the 3D orientation of a certain horizontal line. This method takes advantage of the property of vanishing line (vanishing line is the locus of VPs for all lines lie on the same plane). To identify whether a line is horizontal or not, two constraints are applied: (1) horizontal lines should not intersect the vanishing line, (2) locus of VPs of horizontal lines are different from that of non-horizontal ones.

Wang and Tsai [9] proposed an approach to camera calibration based upon the use of vanishing lines. This technique requires only a single view of a cube. The three principle vanishing points (i.e. VPs that correspond to the orthogonal directions of the world coordinates) are detected and the orthocentre of the triangle thus formed gives the image plane centre. This provides a neat way for calibration. The camera orientation parameters are determined from the slopes of the lines forming the triangle. In addition to these they also establish the relationship between the area of the vanishing triangle for calibration of focal length.

In summary, all existing methods for extraction of VPs perform some form of accumulation of line pair junctions. With the Gaussian sphere parameterisation being most popular, while this is a valued approach, there are several shortcomings associated with it.

As bins in the Hough plane map to non-uniform area on the image plane, some intersection points are grouped together under a more stringent condition than others depending on the locations of the VPs. Problem also arises when votes fall into neighbouring bins which might cause a significant peak to diminish in strength. However, most important of all is that accuracies of detected lines are ignored. As far as VPs are concerned the positional and orientational errors cause incorrect intersection points to be formed which reduce the strength of the 'true' peaks and give rise to spurious intersection points which might in turn group with other points to produce 'false' vanishing points. Additionally, this would also disperse intersection points (as the bin size has an impact on Hough Transform) which inherently belong to the same VP. The above points show the sensitivity of this approach to noise. Furthermore, due to the nature of the algorithm any convergent group which consists of a relatively small number of 3D parallel lines would be left undetected.

Most papers on the topic, analyse images of scenes such as offices and corridors which are highly structured and have strong perspective. Consequently, there are less potential VPs and the strength of true VPs are significantly higher than the background and are therefore distinguishable from random intersections.

More recently, Brillault-O'Mahony [10] took into consideration the uncertainty in the detection of line segments and designed an isotropic accumulator space where the probability of erroneous VP detection is uniformly distributed

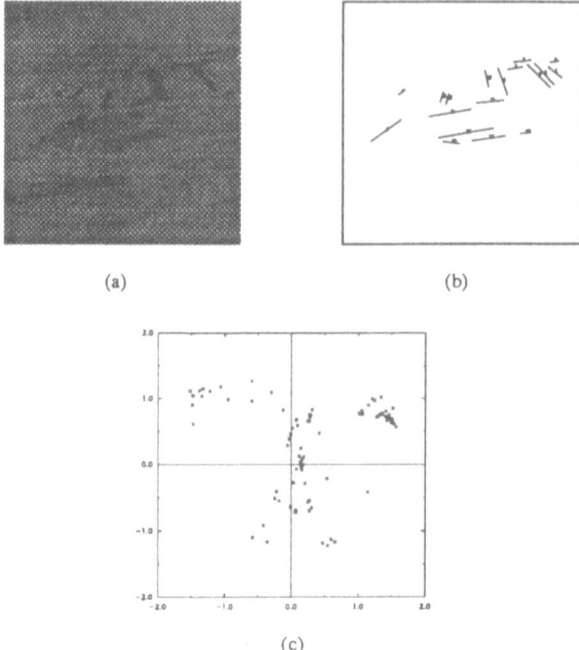

(a)   (b)

(c)

Figure 1: *(a) Original image (b) Hough Transform output (c) Hough Plane accumulation of the azimuth and elevation angles.*

throughout all the cells. In any case, the approach is still based on the accumulator array idea and thus is inappropriate for images with sparse parallel lines.

However, there are many domains of applications where the scene contains only a few lines which are parallel in 3D and on such imagery the existing techniques gave results as illustrated in fig.1. Fig.1(c) shows that there is no dominant peak for the set of lines shown in fig.1(b).

This paper proposes a method which takes a different perspective to detecting vanishing points. Instead of accumulating intersection points, we compute the probability of a group of lines passing the same point. This approach provides a probability measure for discriminating between competing hypotheses irrespective of the size of the vanishing group. In addition, its performance also degrades gracefully in noisy environments. The paper is organised as follows, Section 2 of this paper introduces a novel vanishing point detection method. In Section 3, a probabilistic line representation which is a prerequisite of the method developed. In section 4 we present the experimental results and finally, section 5 offers some conclusions and discussion.

2   Vanishing Point Detection

Let us consider an image with line segments represented by $\rho - \theta$ parameterisation. Due to the geometrical constraints dictated by the image formation process, all perspectively projected line segments having the same orientation in three space converge to a single point – the vanishing point – in the image under a noise free condition. However, both the imaging and low level edge and straight line extraction processes are inherently noisy resulting in uncertainties in the ρ and θ parameters of the detected lines. Errors in ρ and θ will result in a considerable scatter of the intersection points of the pairs of lines segments which makes it difficult to identify true vanishing points. As pointed out earlier, this problem is particularly pertinent when the scene structure contains only a small number of parallel lines.

In this paper the search for vanishing points makes an explicit use of distribution models of the parameters of the detected lines. With such a probabilistic description for each line we can pose the question of how likely a given point is the common intersection point of a group of lines. In this manner for any selected group of lines we can determine the probability distribution $p(x, y)$ for their mutual intersection point (x, y). A vanishing point is then identified as the point of maximum of this probability distribution function which exceeds some pre-specified threshold.

Let us start by considering a single line with parameters (ρ_i, θ_i) and let the distribution of errors $\delta\rho, \delta\theta$ in ρ_i and θ_i be $p_i(\delta\rho, \delta\theta)$ respectively. Now the probability of the line passing through a point (x, y) in the image will be given by compounding all the combinations of errors $\delta\rho$ and $\delta\theta$ such that the true line with parameters

$$\rho = \rho_i + \delta\rho \tag{1}$$
$$\theta = \theta_i + \delta\theta \tag{2}$$

satisfies the constraint equation

$$\rho = x\cos\theta + y\sin\theta \tag{3}$$

the compound probability $P_i(x, y)$ is thus given by

$$P_i(x, y) = \frac{1}{z_i} \int_s p_i(\delta\rho, \delta\theta)\,ds \tag{4}$$

where the integration is performed in the parameter space along the sinusoidal line defined by (3) and z_i is the normalising constant to ensure that $p_i(x, y)$ is a probability density function. In terms of parameter errors the compound probability can be expressed as

$$P_i(x, y) = \frac{1}{z_i} \int_{-\pi}^{\pi} p_i(\delta\rho, \delta\theta)\sqrt{1 + (\frac{d\rho}{d\theta})^2}\,d\theta \tag{5}$$

The compounding process is illustrated in fig.2.

Now let $X = \{e_i | i = 1, 2, ...k\}$ be a group of lines selected from the best of lines output by an image description process, with the measured parameters for each line denoted by vector $\underline{\omega}_i = [\rho_i, \theta_i]^T$ and the associated error distribution by $p_i(\delta\rho, \delta\theta)$. By analogy the probability that the lines jointly pass through a point (x, y) in the image plane (which extends beyond the physical imaging area of the sensor) is given by

$$P(x, y) = \prod_{i=1}^{k} \frac{1}{z_i} \int_{-\pi}^{\pi} p_i(\delta\rho, \delta\theta)\sqrt{1 + (\frac{d\rho}{d\theta})^2}\,d\theta \tag{6}$$

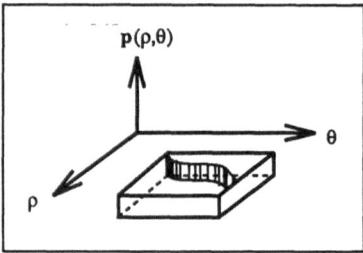

Figure 2: *Rectangular probability density function.*

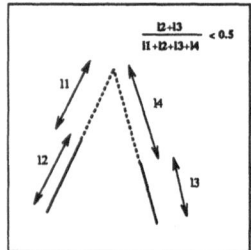

Figure 3: *Evidential Support.*

From the knowledge of $p_i(\delta\rho, \delta\theta)$ the probability of $P(x, y)$ can easily be evaluated. Its mode (x_v, y_v) then defines a vanishing point provided $P(x_v, y_v)$ is above the threshold.

In order to develop a practical procedure based on the above idea we first need to select a suitable groups of lines. Regarding these lines, the method is intended for finding vanishing points of small sets of 3D parallel lines, hence the cardinality of the group should be quite small. Moreover, the computational complexity of the problem could potentially grow combinatorially with the number of lines in the group. In the present approach the initial analysis is performed for line triplets. Any larger group of lines is formed after this first analysis stage by considering the proximity of detected vanishing points and the overlap of the two participating line sets.

To prune the set of all possible triplets each candidate group of lines must satisfy a number of criteria. These include

1. angular constraints (similarity of θ_i values)

2. distance constraint (the perpendicular distance of line pair intersection point from the third line)

3. junction quality constraint [11](the lines should intersect at a point which is remote from all participating line endpoints as illustrated in fig.3.)

4. imaging geometry constraints (if known)

3 Probabilistic Line Representation

In the light of the discussion in the previous section we require an appropriate line representation which can associate uncertainties with its parameters. It is important that this representation is easy to convert to and from the standard Hough Transform(HT) $\rho - \theta$ space representation, since we use this method for the extraction of straight lines.

The parametric representations that we adopted for a perfect line are either $v_1 = [\rho, l, \theta, L]^T$ or $v_2 = [x_m, y_m, \theta, L]^T$, where ρ is the distance between the foot of the normal and the origin, l is the distance from the foot of the normal to the line midpoint; θ and L are the line orientation and length respectively. x_m, y_m are the coordinates of the line midpoint.

Deriche and Faugeras [12] also address the issue of finding an appropriate line representation. Their conclusion is that vector v_2 is a more favourable choice than vector v_1 simply because representation v_1 leads to a covariance matrix that strongly depends upon the position of the associated line segment in the image through the effect of ρ and l. Hence, from this standpoint v_1 does not allow different Kalman filters to be applied on each parameter. However, as far as our application is concerned it does not matter what the interactions between various parameters are. We only need the necessary statistical parameters to build the error models which we can utilise for the development of a formal approach to VP detection.

A simple analysis involving Taylor series expansion leads to the following approximate relationship between the errors in line orientation, line midpoint and the distance from the origin to the foot of the normal:

$$\delta\rho = \frac{-\rho\delta\theta^2}{2} - (x_m \sin\theta - y_m \cos\theta)\delta\theta + \delta x \cos\theta + \delta y \sin\theta \qquad (7)$$

Note that in deriving the above equation, we assume that any terms involving the cross product of positional and orientational errors can be neglected. For lines four pixels long or more the quadratic terms become negligible. Equation (7) then gives a linear relationship between the errors in orientation θ and line segment midpoint position and the errors in ρ. Thus if $\delta\theta, \delta x$ and δy are normally distributed, so will the errors ρ.

A Monte Carlo experiment was performed to check the validity of the approximate model and its dependence on line length. From table 1 it is apparent that provided the line length $L >= 4$ the linear model yields a distribution of errors $\delta\rho$ with negligible skew and curtosis which can be taken to imply that it closely approximates a Gaussian. Thus if $\delta\theta, \delta x$ and δy are Gaussian, the joint distribution of $\delta\rho, \delta\theta$ and δl will be Gaussian with covariance matrix

$$\begin{pmatrix} l^2\sigma_\theta^2 + \sigma^2 & -l\sigma_\theta^2 & (sin2\theta)\sigma^2 - l\rho\sigma_\theta^2 \\ -l\sigma_\theta^2 & \sigma_\theta^2 & \rho\sigma_\theta^2 \\ (sin2\theta)\sigma^2 - l\rho\sigma_\theta^2 & \rho\sigma_\theta^2 & \rho^2\sigma_\theta^2 + \sigma^2 \end{pmatrix} \qquad (8)$$

Note that $\sigma_x^2 = \sigma_y^2 = \sigma^2$. From (8) the distribution of interest $p(\delta\rho, \delta\theta)$ is normal with zero mean and covariance matrix

$$\Sigma = \frac{1}{N} \begin{pmatrix} l^2\sigma_\theta^2 + \sigma^2 & -l\sigma_\theta^2 \\ -l\sigma_\theta^2 & \sigma_\theta^2 \end{pmatrix} \qquad (9)$$

Line		mean ($\bar{\rho}$)	stdev (σ_ρ)	skew (σ_3)	curtosis (σ_4)
$x_m = -36.6, y_m = 136.6, \theta = 150°$					
First	L = 1.0	0.06199	8.77310	-0.01172	0.10832
order	L = 4.0	0.01441	2.40076	0.00618	-0.07306
approx.	L = 10.0	0.00489	1.33047	0.02300	-0.03726

Table 1: *Statistical results of error in ρ ($\delta\rho$) obtained from Monte Carlo experiment.*

where N is the number of points providing evidential support for the line. Note that N may differ form L, the line length as some pixels inside the line segment may be undetected. The covariance matrix in (9) applies if the Hough Transform (HT) line detection scheme has an optimisation facility to estimate the most likely values of ρ and δ. When the standard HT is used with ρ and δ parameter quantisation, the probability distribution $p(\delta\rho, \delta\theta)$ is no longer Gaussian. Instead its shape may vary from the rect to roof function depending on the coarseness of the quantisation process. A study of the relative merits of the various detection schemes and the associated probabilistic line representations is beyond the scope of this paper.

4 Experimental Results

The method was applied to aerial imagery of resolution $256 X 256$. The rectangular line parameter probability distribution was used in the experiment.
   The implementation involved the following steps:

1. Compute the intersection points of all possible pairs and discard those that fall outside the virtual image (this is an imaginary image of size 512X512 centred at the origin).

2. Combine a line with a pair to create a triplet to see if it satisfies the pursuing constraints of section 2.

3. Set up a window around the line pair intersection point and compute the probability $P(x, y)$ for all (x, y) in the window. Find the mode of $P(x, y) - (x_v, y_v)$.

4. Repeat steps (2) and 3 for all other triplets.

   The result shown in fig.4 demonstrates that the vanishing points which grouped together the lines of interest (i.e. the runways and taxiways) are ranked highly and their corresponding VPs are in close proximity to each other (these VPs can be grouped to extract larger vanishing groups). Fig.5 shows the probability envelope of a VP in image space.
   When the conventional VPs finder was applied on the image shown in fig.4, the runways and taxiways were again detected but there is no way of discriminating between the various hypotheses. If the size of the peak were used as a discriminating criterion, false vanishing groups would be extracted.

Figure 4: *New vanishing points finder result on areial imagery.*

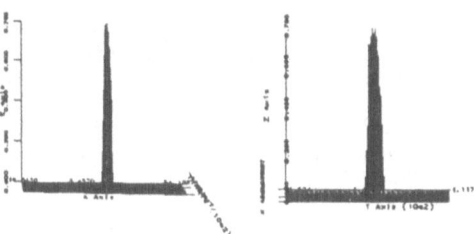

Figure 5: *Probability profile of a certain VP.*

5   Discussion and Conclusions

The parameter space in which the algorithm proposed here operates is an open one, since the intersection point of a line pair can lie somewhere between the image and infinity. This, however, should not present much difficulty as lines whose 3D orientations are the same only cease to converge to a point under a very restricted viewpoint. This also demonstrates the relationship between convergent and parallel groups. 3D parallel features when projected onto the image plane can only transform into convergent or parallel groups.

The algorithm proposed in this report provides an estimate of the VP location as well as producing a performance measure. Thus VP hypotheses can be compared and discarded on the basis of their quality measure values rather than the number of lines which converge to a certain point. This offers a common ground for comparison between convergent groups (i.e small convergent groups can also be detected). However, there is a tradeoff between small convergent groups and the probability of accidental coincidence. As the size of the convergent groups reduces, the chance of accidental coincidence increases, which renders the location of true vanishing points more difficult. Therefore some kind of geometrical cue is needed to recover true vanishing points. Vanishing lines are powerful cues to be exploited for the purpose. They are defined as the locus of VPs formed by 3D parallel groups which lie on the same plane. This property means that the hypothesised points must lie on the vanishing line and thereby provide a constraint for discarding false VPs.

References

[1] R.M. Haralick. Using perspective transformation in scene analysis. *Computer Graphics Image Processing*, 13:191–221, 1980.

[2] S.T. Barnard. Methods for interpreting perspective images. In *Proc. Image Understanding Workshop*, pages 193–203, Palo Alto, California, Sept 1982.

[3] M.J. Magee and J.K. Aggarwal. Determining vanishing points from perspective images. *Computer Vision, Graphics and Image Processing*, 26:256–267, 1984.

[4] L. Quan and R. Mohr. Determining perspective structures using hierachical hough transform. *Pattern Recognition Letters*, 9(4):279–286, 1989.

[5] R.T. Collins and R.S. Weiss. Deriving line and surface orientation by statistical methods. In *Proc. Image Understanding Workshop*, pages 433–438, 1990.

[6] R.T. Collins and R.S. Weiss. Vanishing point calculation as a statistical inference on the unit sphere. *Computer Vision, Graphics and Image Processing*, pages 400–403, 1990.

[7] S. Liou and R. Jain. Road following using vanishing points. *Proc. IEEE Computer Society Conference on Computer Vision and Pattern Recognition*, pages 41–46, 1986.

[8] L. Shigag, S. Tsuji, and M. Imai. Determining of camera rotation from vanishing points of lines on horizontal planes. *Computer Vision, Graphics and Image Processing*, pages 499–502, 1990.

[9] L. Wang and W. Tsai. Computing camera parameters using vanishing-line information from a rectangular parallelepiped. *Machine Vision and Applications*, (3):129–141, 1990.

[10] B. O'Mahony. *A Probabilistic Approach to 3D Interpretation of Monocular Images*. PhD thesis, City University, London, 1992.

[11] A. Etemadi, J-P. Schmidt, G. Matas, J. Illingworth, and J. Kittler. Low-level grouping of straight line segments. In *Proc. British Machine Vision Conference*, pages 118–126, 1991.

[12] R. Deriche and O. Faugeras. Tracking line segments. *ECCV*, pages 259–268, 1990.

Acknowledgements

The authors wish to acknowledge support by the Procurement Executive, Ministry of Defence for this work.

Contextual Junction Finder

J. Matas and J. Kittler

Dept. of Electronic and Electrical Engineering,

University of Surrey,

Guildford, Surrey GU2 5XH, United Kingdom

Abstract

A novel approach to junction detection using an explicit line finder model and contextual rules is presented. Contextual rules expressing properties of 3D-edges (surface orientation discontinuities) limit the number of line intersections interpreted as junctions. Probabilistic relaxation labelling scheme is used to combine the a priori world knowledge represented by contextual rules and the information contained in observed lines.

Junctions corresponding to a vertex (V-junctions) and an occlusion (T-junctions) of a 3D object are detected and stored in a *junction graph*. The information in the junction graph is used to extract higher level features. Results of the most promising method, *the polyhedral object face recovery*, are briefly discussed. The performance of the junction detection process is demonstrated on images from indoor, outdoor, and industrial environments.

1 Introduction

Perceptual groupings of image features have been widely used in computer vision systems to guide scene interpretation and 3D model matching [1, 2, 5, 9, 10]. Of all perceptual groupings studied by psychologists [13, 4] and computer vision researchers we focus our attention on *junctions of line segments* - points of co-termination of lines. As co-termination is a projection-invariant property, the task of junction detection would be relatively simple in an ideal noise-free world. A set of lines terminating at the same point could be interpreted as a projection of edges meeting at a vertex. However, due to the inherent inaccuracy of line (and edge) detection, endpoints of lines can be widely separated even if the lines emanate from a common vertex.

In a novel approach, an explicit error model for line detection in conjunction with *contextual rules* is used to recover junctions. The contextual rules express physical properties of 3D-edges (discontinuities in surface orientation): 1. projections of 3D-edges never cross and 2. visible parts of 3D-edges terminate at either a vertex or a point of occlusion. Both the use of context and an empirically tested explicit error model is a distinguishing feature of the work presented in the paper.

The problem we are facing can be stated as follows: For every line A, find line B which is most likely the line that occluded/had a common vertex in 3D with A. Of all possible assignments for A and B that don't violate rule 1. select the most probable one given the line detector error model. A probabilistic relaxation scheme developed in [6] is applied to the junction detection problem (Section 3). Section 2 specifies the line detection model. Implementation issues are addressed in Section 4. Section 5 presents intermediate level groupings built on top of the junction finder The results are summarised in Section 6.

2 Line detector model

Conceptually, the first stage of the junction finder can be viewed as an attempt to recover *projected lines*. A *projected line* is, by definition, a projection of a 3D-edge and therefore must terminate at an intersection of 2 or more projected lines. A set of all projected lines would be very close to an 'ideal line drawing' (see Fig. 1(c)).

Any endpoint detected by a line finder can be treated as a noisy measurement of a projected endpoint. A statistical model of the noise affecting the line finder is a necessary prerequisite for any attempt to recover projected lines. The junction detection becomes trivial if the filtering process is successful - any point where two lines touch is a junction. See Figs. 1(a)-(c) to compare the line finder output, junction finder output (filtered lines) and a set of projected lines.

The uncertainty in line parameters is a function of the particular edge and line finder used. After extensive (but subjectively evaluated) tests we selected Horaud's implementation [5] of the Deriche filter [3] for edge detection and a line detector based on Hough transform [12, 14] for line detection. In agreement with [9, page 367] we observed:

- very precise estimation of the line angle and of the transversal position

- large uncertainty in the localisation of the endpoint

Simplifying the characteristics of the line finder the following model was adopted: 1. the projected endpoint lies on the straight line defined by the line segment. 2. if d denotes the (oriented) distance from detected endpoint E with positive values of d for point outside the line segment, then the endpoint error distribution is

$$
P(d) = \begin{cases} \frac{k_{in}k_{out}}{k_{in}+k_{out}}e^{k_{in}d} & \text{if } d < 0 \\ \\ \frac{k_{in}k_{out}}{k_{in}+k_{out}}e^{-k_{out}d} & \text{otherwise} \end{cases} \tag{1}
$$

The two constants, k_{in} and k_{out} control the shape of the exponential. At present the values are set to 0.5 and 0.1 pixels respectively (Fig. 2). The choice of exponential is somewhat arbitrary, but Fig. 3 shows that it is in good agreement with empirical data. Test runs with different k_{in} (range 1-0.2) and k_{out} (0.2-0.03) produced similar results suggesting that the performance of the junction finder is not critically sensitive to the shape of the distribution.

3 Junction detection using probabilistic relaxation labelling

Visible parts of 3D edges terminate at either a vertex or a point of occlusion. A *junction* is a projection of such 3D point. The line finder model (section 2) guarantees that all junctions lie at an intersection of straight lines passing through a detected 2D line. Consider the example of Fig. 4. If line A is a

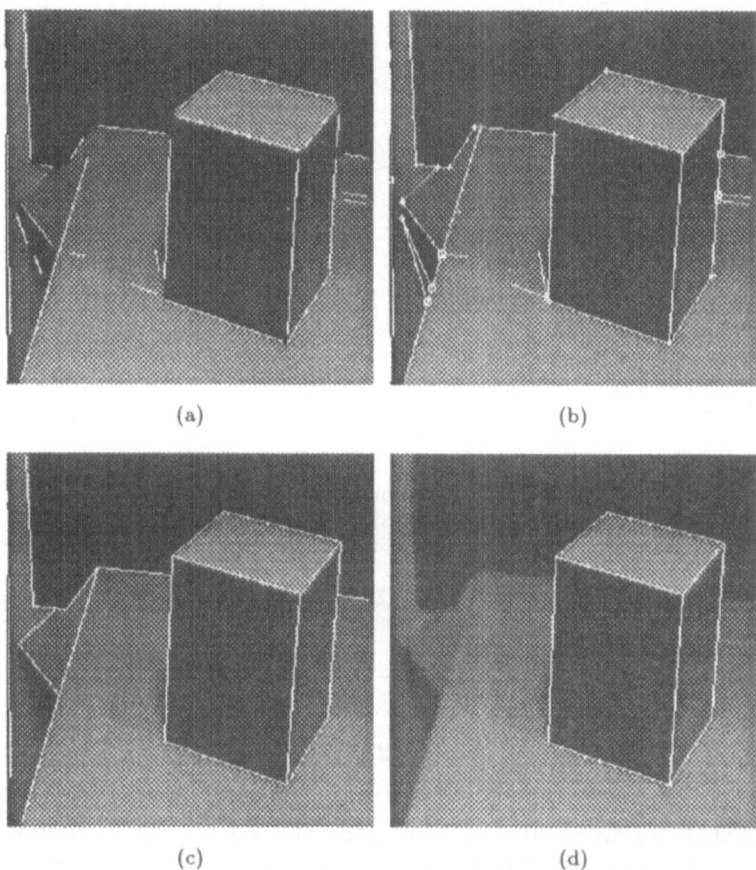

(a)　　　　　　　　　　(b)

(c)　　　　　　　　　　(d)

Figure 1: CUBE. A simple scene with a test object used for stereo camera head calibration in the VAP project [15]. In Fig. (a), lines detected by Hough transform are superimposed over the original image. Note that errors in orientation and transversal position of line segments are negligible. Fig. (c) gives an example of an 'ideal line drawing' of the CUBE scene. The image was prepared by editing the result of the junction finder shown in (b). Fig. (b) illustrates junction finder results. Endpoint errors are filtered out by 'stretching' line segments to junctions. The only significant structural error occurred at the top-left vertex of the cube. The left vertical edge of the cube is associated with the rear edge of the table. The result is explainable; the line corresponding to the vertical edge was terminated very close to the projected rear table edge because there is no gray level gradient between the front face of the cube and the background. Results of the *face recovery* postprocessing (Section 5) are shown in Fig. (d). Using geometric information enabled the recovery of the front face despite the top-left vertex problem described above. *Gap bridging* postprocessing (Fig. 6) joint the two lines on the rightmost vertical edge of the cube.

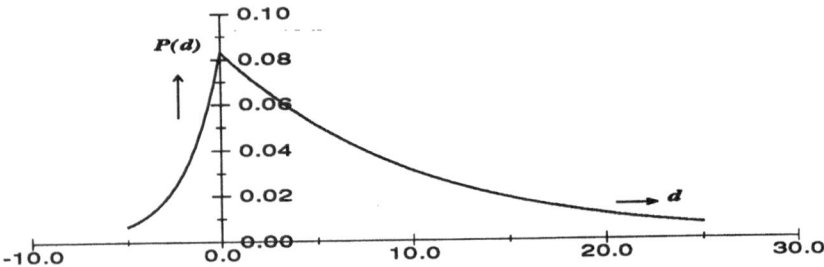

Figure 2: Assumed endpoint error distribution of the line finder. d denotes the distance from an enpoint along the line segment. Negative values refer to points inside the line segment.

projection of a 3D edge it must terminate at a junction that is identical to one of the intersection I_{ab}, I_{ac}, I_{ad}. We are concerned with the problem of computing the probabilities of all events $\{A_e = I_{ai}, \forall i\}$. To exploit the contextual information conveyed by these interacting events we compute the probabilities using a dictionary based relaxation scheme whereby the probabilities at stage $n + 1$ are obtained from probabilities $P^n(A_e = I_{ai})$ at the previous iteration, ie.

$$P^{n+1}(A_e = I_{ai}) = \frac{P^n(A_e = I_{ai})\sum_{\Lambda^k} P(\Lambda^k) \prod_l \frac{P^n(l_e = I_{em})}{P(l_e = I_{em})}}{\sum_j P^n(A_e = I_{aj}) \sum_{\Lambda^k | A_e = I_{aj}} P(\Lambda^k) \prod_l \frac{P^n(l_e = I_{em})}{P(l_e = I_{em})}} \quad (2)$$

The initial probabilities $P^0(A_e = I_{ai})$ are computed using the Bayes formula from the probability distribution of endpoint errors (Eq. 1, Fig. 2) and prior probabilities. All prior probabilities in Eq. 2 are assumed to be equal. The context is introduced using a dictionary Λ that contains all permissible configurations of junction assignments. The dictionary is constructed using to rules obeyed by projections of 3D edges: 1. projected lines must not cross and 2. every projected line must terminate at a single junction.

4   Implementation of the Junction finder

The junction detection process is performed in three stages - initialisation and preprocessing, relaxation, and postprocessing. In the first stage, all intersections of line pairs are considered as possible junctions. Any intersection with endpoint distance outside the margin of error of the line detector is immediately discarded. Each intersection is assigned an initial, non-contextual probability according to Bayes formula. The information associated with every intersection is stored in a node of an *intersection network*. Each node in the network is linked to four other nodes representing its predecessor and successor (with respect to distance) in the list of intersections of one line (Fig. 5(a)).

In the relaxation stage, repeated sweeps through the intersection network are made. At each endpoint, the probability distribution is updated according to context-conveying formula 2. Generally, the probabilities of an intersection being a junction gradually shift either towards 0 or 1. At the end of the sweep, intersections with 0 probabilities are deleted (it follows from formula 2 that

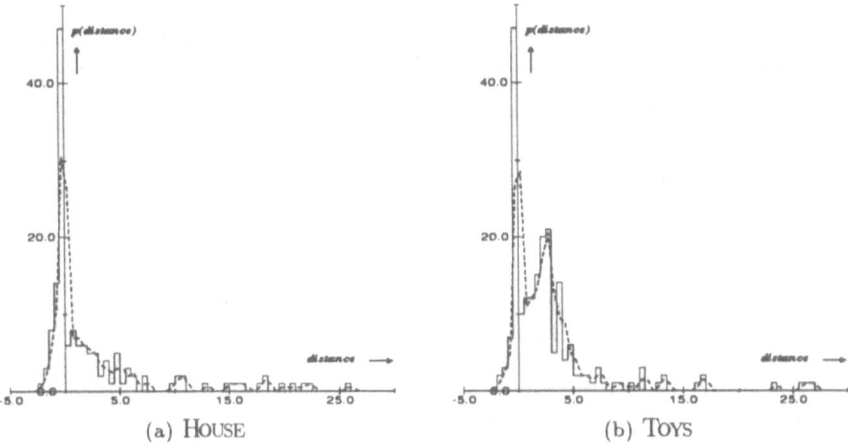

(a) HOUSE   (b) TOYS

Figure 3: Histogram of enpoint distance errors. The output of the junction finder, the *junction graph* is assumed to represent the ground truth (see Fig. 8 to check the validity of this assumptions). Histogram of line end to corresponding junction distances (solid line) shows good agreement with the line detection model (Fig. 2). The dashed line graph shows the running average of two consecutive histogram bins (of 0.5 pixel size).

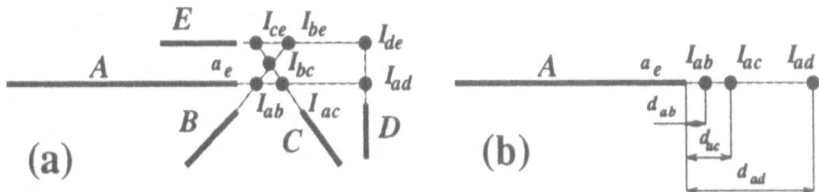

Figure 4: A set of interacting lines

the probability would remain 0). The originally dense network (Fig. 5(a)) is gradually transformed into a structure similar to Fig. 5(b). The iteration loop is exited when the average relative change of intersection probability falls under a preset threshold (default: 2%).

Finally, only intersections with maximum probability at an endpoint are retained and labelled as either V or T junctions. An intersection having a maximum probability with respect to both participating lines is assumed to be a projection of a vertex; the intersection is labelled a V-junction. If the intersection probability is a maximum with respect to one line only then the other line must 'pass through' (other possibilities are suppressed by the relaxation process). The situation indicates that the intersection is a projection of a point where a 3D-edge was occluded; the intersection is labelled as T-junction. The network of intersections is transformed into an attributed graph structure called a *junction graph*, (Fig. 5(c)). Every node of the junction graph represents a junction relation (either V or T) between a pair of lines. The junction

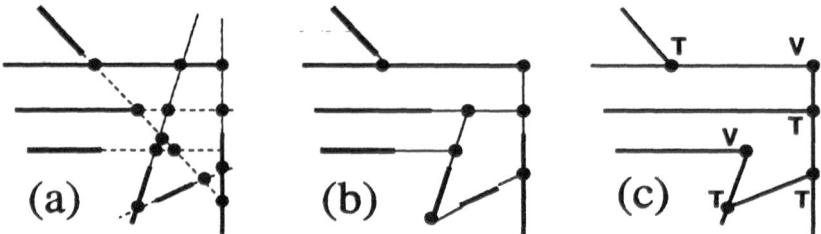

Figure 5: From *intersection network* to *junction graph*. The large margin of error of the line detector allows for multiple interpretations of an endpoint. An *intersection network* structure is created to facilitate the probabilistic updating of intersection probabilities. During the relaxation process, the network becomes sparser as probabilities of some intersections drop to 0 (see Tab. 1). Finally, all non-maximum junctions are discarded - a *junction graph* is created (c).

graph is a semi-symbolic structure; numerical information is attached to both junctions (position, probability) and lines (position). The junction graph can be used for both symbolic and geometric reasoning about the scene (see section 5).

image	HOUSE		WIDGET		TOYS		CUBE	
time	1.6s		0.7s		2.1s		0.4s	
lines	78		44		90		28	
iteration	inters.	change	inters.	change	inters.	change	inters.	change
1	563	84.16	247	84.01	800	86.60	107	83.72
2	368	75.08	164	67.00	489	73.34	64	47.00
3	142	24.03	85	50.29	204	25.65	43	7.87
4	142	2.09	85	7.15	204	2.80	43	1.26
5	142	5.73	85	3.20	204	5.65	43	6.49
6	142	0.38	85	20.41	204	1.04	43	0.99
7			85	3.36				

Table 1: Junction finder performance. The processing time was measured on a SPARC 2 machine. The 'inters.' column shows the number of intersections processed in the $n - th$ iteration. The 'change' column contains information about an average change (in %) of the intersection probabilities

The efficiency-minded readers may express doubt about the speed of the process. The computational complexity of the implementation of preprocessing is $O(N^2)$ (where N is the number of input lines) as all line pairs are examined (an $O(NlogN)$ algorithm can be found in [16]). The theoretical worst-case complexity of the iterative relaxation is even worse. Fortunately, the worst-case complexity is not of practical importance (representing a situation where all lines terminate in a tight cluster); the *average* complexity is extremely hard to derive analytically, but empirical results suggest $O(NlogN)$. Table 1 summarises the junction finder performance. For run-times in the order of seconds 1. optimising performance of the junction finder is not of particular importance and 2. the performance depends more on system specific constants (i/o speed etc.) than on the computational complexity. The good performance is a consequence of the *focus of attention* property of the relaxation labelling. After two or three iterations, app. $2N$ (not the theoretical N^2) intersection probabilities are updated. Most of the processing in later iterations revolves around a set of ambiguous intersections (Tab. 1, column 'inters.'). Empirical data (column 'relative change' in Tab. 1) suggest good stability and fast, although not monotonic, convergence.

5 Beyond the Junction Graph

The *junction graph* can be viewed as a final result of the junction finder. The richness of the information represented by the junction graph invites further exploitation. Three methods, *gap bridging, V3-junction detection and polyhedral object face recovery*, are presented in this section.

Gap bridging (see fig. 6 for full description) corrects edge detector failures at T-junctions. The V3-junction detector finds subgraphs of the junction graph that are likely to be projection of a *3-vertex*, a vertex where 3 visible 3D-edges meet. In Section 2 we made an assumption that transversal positions of lines are error-free. This implies that all 3 (or, more generally n) lines terminating at a common 3(or n)-vertex should intersect at a single point. In practise, the line intersections are clustered in a small region. The implementation is straightforward; for every V-junction: check for V-junctions in a small (default: 1 pixel) neighbourhood. The size of the neighbourhood makes false positives virtually impossible. Examples of V3-junctions can be found in figs. 8 and 1.

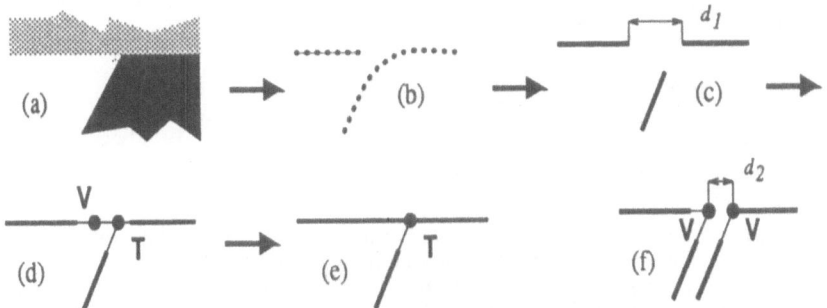

Figure 6: Gap bridging. Conventional edge detectors perform poorly (subfig. (b)) in areas where 3 regions (subfig. (a)) meet (see [11, 8]). This causes a projection of one 3D-edge to be broken into two line segments (c). Spatial arrangement (c) of line segments give rise to subgraph (d) of the junction graph. Situation (d) can be interpreted as either a view of a vertex from a highly improbable, accidental viewpoint or as a manifestation of the above mentioned edge detector problem. The latter interpretation is assumed (the former having a negligible probability) and the subgraph is transformed into the form depicted in subfig. (e). The gap bridging process is *context dependent*; gap of subfig. (f) of length d_2, although significantly smaller then d_1, is left intact. Examples of the gap bridging can be found in the HOUSE (see e.g. roof), WIDGET (upper vertical edge of the from face) and CUBE images (Figs. 8, 1)

Recovery of polyhedral object faces is more complex. If 2D line segments are projections of 3D-edges and junctions projections of vertices then a completely visible face of a polyhedral object must project into a closed loop of V-junctions in the junction graph. Fig. 7 illustrates results of the *face recovery* procedure. Application of *face recovery* for fast 3D pose estimations is described in [7].

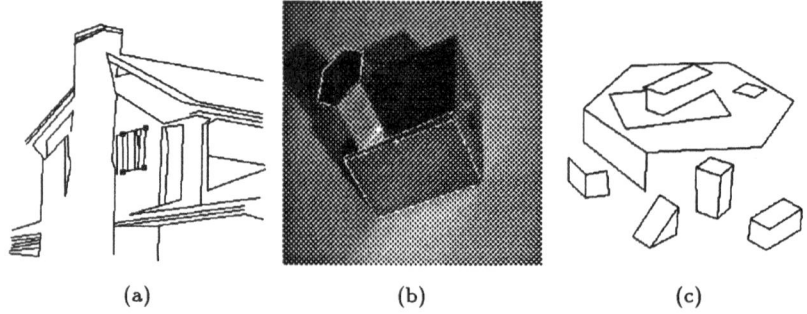

(a)   (b)   (c)

Figure 7: Results of *polyhedral face recovery*. Closed loops in the *junction graph* are assumed to correspond to 3D-edges and vertices of a polyhedral object face. Only one incorrect face is recovered in the TOYS scene (the non-convex polygon inside the large hexagon of Fig. (c); a careful scrutiny of (c) and Fig. 8(f) reveals an accidental alignment of the block and pad leading to junction graph distortion). Fig. (a) shows *corrected lines* - lines stretched to terminate at a junction (compare with Fig. 8(b)). The single face recovered in the HOUSE scene is a projection of the window frame (corners marked by □).

6   Conclusion

We have presented an algorithm for junction detection with two new features: exploitation of spatial context and use of an explicit line detector model. The combination of contextual evidence is not based on an ad hoc method; a well established method, *probabilistic relaxation*, is employed to accomplish the task. The tests performed to validate the line detector model could prove valuable in its own right as an evaluation tool for line detectors.

The junction finder performance has been tested on hundreds of images, mostly running as a part of a continuously operating vision system [7]. Four scenes, HOUSE, TOYS, WIDGET, and CUBE, were selected as representative of different environments (indoor, outdoor, industrial). Results (Fig. 8) show that our main objective has been achieved - vast majority of V and T junctions indeed correspond to vertices and occlusions in the 3D world. Success of the polyhedral face recovery strongly supports this claim. The junction finder possesses two key features vital for continuous operation - it is fast and it doesn't require any user-defined thresholds even in changing conditions. Experiments have shown [7] that the junction finder output provides salient intermediate level features for model invocation and 3D pose estimation.

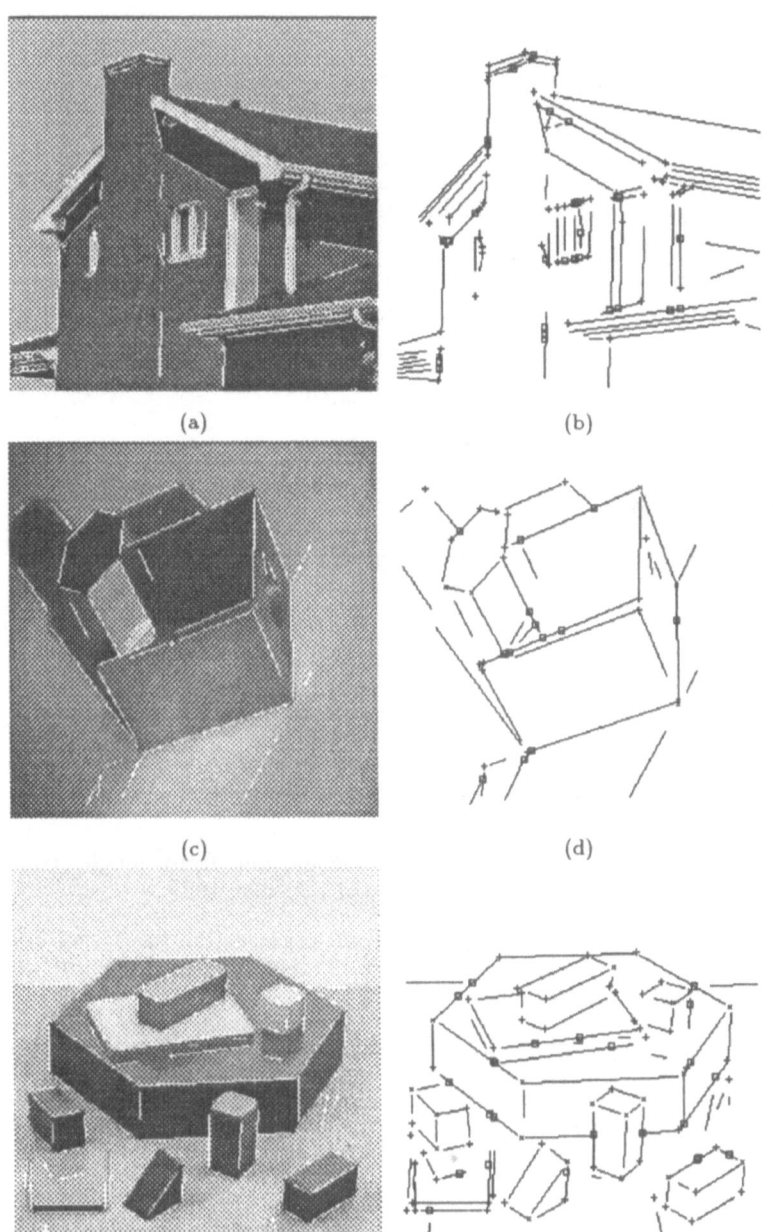

(a)

(b)

(c)

(d)

(e)

(f)

Figure 8: HOUSE, WIDGET and TOYS scenes. Figures (a), (b) and (c) show lines detected by Hough transform superimposed over the original image . Figures (b), (c) and (d) depict lines corrected by contextual gap filling (described in section 6) and the results of the junction finder. *T-junctions* (indicating an occlusion) are marked by □; *V-junctions* (indicating a vertex of two 3D edges) by + and *V3-junctions* (indicating a vertex of three 3D-edges) by ×.

References

[1] R. Bergevin and M. D. Levine. Extraction of line drawing features for object recognition. In *Proceedings of IEEE International Conference Pattern Recognition*, pages 496–501, 1990.

[2] C. Coelho, M. Straforini, and M. Campani. Using geometrical rules and a priori knowledge for the understanding of indoor scenes. In *Proc. British Machine Vision Conference*, pages 229–234, 1990.

[3] R. Deriche. Using Canny's criteria to derive a recursively implemented optimal edge detector. *International Journal of Computer Vision*, 1(2):167, 1987.

[4] R.N. Haber and M. Hershenson. *The Psychology of Visual Perception*. Holt, Rinehart and Winston Inc., U.S.A, 1973.

[5] R. Horaud, F. Veillon, and T. Skordas. Finding geometric and relational structures in an image. In *European Conference on Computer Vision*, pages 373–384, 1990.

[6] J. Kittler and E.R. Hancock. Combining evidence in probabilistic relaxation. *International Journal of Pattern Recognition and Artificial Intelligence*, 3:29–51, 1989.

[7] J. Kittler, J. Illingworth, J. Matas, P. Remagnino, K. C. Wong, H. Christensen, J-O. Eklundh, G. Olofsonn, and M. Li. Symbolic scene interpretation and control of perception. Technical report, ESPRIT BRA Project 3038, March 1992.

[8] Du Li, G.D. Sullivan, and K.D. Baker. Edge detection at junctions. In *AVC*, pages 121–125, 1989.

[9] D. Lowe. Three-dimensional object recognition from single two-dimensional images. *Artifical Intelligence*, 31:355–395, 1987.

[10] R. Mohan and R. Nevatia. Using perceptual organisation to extract 3-d structures. *IEEE Transactions on Pattern Analysis and Machine Intelligence*, 11(11):1121–1139, 1989.

[11] J. A. Nobel. Finding corners. In *AVC*, pages 267–274, 1988.

[12] P. L. Palmer, J. Kittler, and M. Petrou. A Hough transform algorithm with a 2D hypothesis testing kernel. In *Proceedings of IEEE International Conference Pattern Recognition*, September 1992.

[13] J.R Pomerantz. Perceptual organization and information processing. In *Perceptual Organization*, pages 141–180, 365 Broadway, Hillsdale, New Jersey, 1981. Lawrence ERLBRAUM associates.

[14] J. Princen. *Hough Transform Methods for Curve Detection and Parameter Estimation*. PhD thesis, University of Surrey, June 1990.

[15] VAP project. Vision as process. Technical annex, EEC, March 1989.

[16] R. Sedgewick. *Algorithms*. Addison-Wesley, Reading, 1983.

Generation of 3D Dense Depth Maps by Dynamic Vision

An Underwater Application[*]

José Santos–Victor
João Sentieiro

E-mail : d2760@beta.ist.rccn.pt

CAPS/ISR – Instituto Superior Técnico

Av. Rovisco Pais 1, 1096 Lisboa Codex, PORTUGAL

Abstract

This paper presents a dynamic 3D Vision system that is able to estimate dense depth maps from an image sequence. The depth maps computed at each time instant are used in an Extended Kalman filtering structure, that integrates all depth measurements over time, reducing uncertainty. Results with images acquired by an underwater camera, are presented.

1 Introduction

During the last decade, an increasing interest has been devoted to research and development activities in the area of autonomous systems, capable of performing complex tasks in unknown environments, without human intervention. These systems must be able to build and maintain internal models of the observed world. Fields of application range from mobile robots, to autonomous vehicles, space robots, AUVs (autonomous underwater vehicles), etc...

In this paper we address the problem of depth extraction by using an image sequence acquired by a moving camera (with known motion), in an unknown environment. A stochastic approach is adopted to model the existence of uncertainty in every depth estimate, and well established techniques, like Kalman filtering, are used to reduce the uncertainty of the estimated depth maps, over time.

This approach differs from the one presented in [1], both in the matching stage, where geometric constraints arising from the camera motion are included, and in the filtering stage, where a different formulation of the state space model for the motion/observation equations, leading to a clearer formulation of the kalman filtering structure.

The 3D Vision System described in this paper, comprises three major processing modules: the matching process, the regularization procedure and the Kalman filtering stage.

---

[*]This work has been supported in the context of the MOBIUS project, of the EEC MArine Science Technology (MAST) programme. The authors wish to thank Thomson CSF-LER and Thomson Sintra ASM, for providing the images for the underwater application, and Prof. Takeo Kanade for the valuable comments made on this work.

In the matching process, a correlation-like method, based on the *Sum of Squared Differences* method (SSD) [1, 2], is used. To improve the disparity estimate, prior knowledge of the camera motion is considered, as a geometric constraint (epipolar line), and sub-pixel resolution is achieved by interpolating the SSD error surface.

Noting the ill-posed nature of the matching process, a regularization stage is formulated, to constrain the desired disparity field to smooth solutions. In this way, the noise levels of the disparity estimates can be reduced, and new estimates can be calculated to fill in the areas, wherever the matcher was unable to produce estimates.

Kalman filtering is used in the final stage of this work, to integrate multiple disparity measurements over time, in order to produce a more reliable depth estimate [1, 3].

By integrating, over time, several measurements, one can benefit from the advantages of having closely spaced image view points (which simplifies the matching problem, but usually degrades the depth estimates precision) without jeopardizing the precision of the depth estimates computation.

In the following sections, the models involved in the depth from motion algorithms, are described. Results obtained with a sequence of synthesized images and with real underwater images are presented. Finally, we draw some conclusions and some future directions of research are pointed out.

2 Model

This section is devoted to the study and description of all the models involved in the 3D dynamic vision system. First, the dynamics associated to the position of a 3D point relatively to a moving camera referential is derived. Then, the camera model is introduced, and used to obtain the equations that describe the apparent motion induced in the image plane. Finally a model for the uncertainty affecting the system, is established.

Consider a camera moving with relation to a fixed point in the space. Let $\{C\}$ be a cartesian coordinate system (frame) attached to the camera. Let $\vec{\omega}$ and $\vec{T}$, be the angular and translational velocities of the camera with relation to a fixed referential, and let $\vec{P}$ be the position vector of a fixed point in the space. The velocity of P with relation to $\{C\}$ (rigid body motion) is described by the following differential equation [4]:

$$\frac{d\vec{P}}{dt} = -\vec{T} - \vec{\omega} \times \vec{P} \qquad (1)$$

To determine how this motion is perceived in the image plane, we have used a *pinhole* camera model [4, 6]. According to this model, the location of a given image pixel, (x, y), is determined by the perspective projection in the image plane, of the corresponding 3D point, $[X\ Y\ Z]^T$:

$$x = \frac{X}{Z} \qquad\qquad y = \frac{Y}{Z} \qquad (2)$$

By using the camera model in equation (1), and eliminating Z, we finally obtain a new equation set, that expresses the apparent motion (velocity field) induced in the image plane by the actual camera motion [3, 5, 6][1]:

$$\begin{bmatrix} \dot{x} \\ \dot{y} \end{bmatrix} = \frac{1}{Z} \begin{bmatrix} -1 & 0 & x \\ 0 & -1 & y \end{bmatrix} \vec{T} + \begin{bmatrix} xy & -(1+x^2) & y \\ (1+y^2) & -xy & -x \end{bmatrix} \vec{\omega} \qquad (3)$$

$$\dot{Z}_{(t)} = (-\omega_y x + \omega_x y) Z_{(t)} - T_z \qquad (4)$$

Uncertainty (the simplified camera model, errors associated to the camera motion and parameters, the image acquisition process, etc) is modeled by additive white gaussian noise in equations (3,4), and the final model is obtained by approximating derivatives in equation (3, 4) with forward differences (with sampling period τ): [2]

State equation :

$$Z_{[t+\tau]} = a_{[t]} Z_{[t]} + b_{[t]} + \eta \qquad (5)$$

Observer equation (measurements):

$$\vec{d} = \begin{bmatrix} x_{[t+\tau]} - x_{[t]} \\ y_{[t+\tau]} - y_{[t]} \end{bmatrix} = C_{[t]} \frac{1}{Z_{[t]}} + D_{[t]} + \mu \qquad (6)$$

where η is a zero mean gaussian random variable with variance r, μ is a zero mean gaussian random vector with covariance matrix Q and $\{\mu, \eta\}$ are independent. The terms $a_{[t]}$, $b_{[t]}$, $C_{[t]}$ and $D_{[t]}$ depend on the motion parameters and the image plane coordinates of a given pixel [7].

3 System Description

The complete system structure is shown in figure 1. At every sample instant, a new image is acquired by the moving camera and a new depth map is computed. A matching algorithm is applied to each pair of successive images, to compute the disparity vector field and uncertainty estimates. This vector field is the input to a regularization stage, used to decrease the noise levels and fill in unavailable estimates. Every new observation (regularized disparity vector) is finally used in a Kalman filtering module to update an estimated depth map resultant from previous measurements, thus reducing the uncertainty over time.

3.1 Matcher

To determine the displacement of each image pixel, induced by the camera motion in the static environment, we have used a correlation based technique, derived from the Sum of Squared Differences (SSD) method [2, 3].

To evaluate potential match candidates, p'_t and $p'_{t+\tau}$, from images acquired at time t and $t + \tau$, it is assumed that homologous points have similar gray-levels. Hence, the sum of the squared gray level differences, for pixels inside

---

[1]For simplicity, we have not explicitly written the different variables as time functions (e.g. $x(t)$ instead of x), as it should be clear from the context.

[2]Note that the image coordinates in equation (6) correspond to a virtual camera and the actual image pixel coordinates are obtained using the *intrinsic* camera model parameters [4]. See [7] for details.

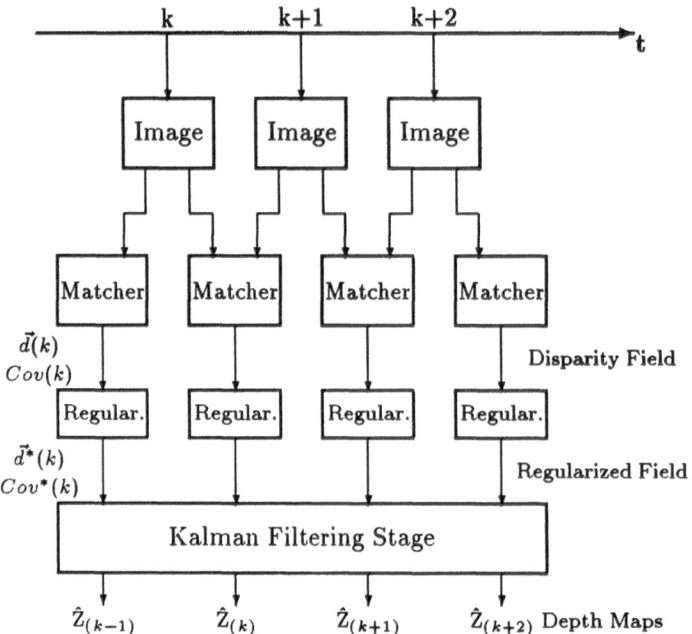

Figure 1: Block Diagram of the Vision System

windows centered in p'_t and $p'_{t+\tau}$, can be used to quantify the "likelihood" of a matching decision [1].

Moreover, the location of a pixel $p'_{t+\tau}$, homologous to p'_t, is constrained to a line, *the epipolar line*, in the image at $t + \tau$, determined by the camera motion and camera parameters [7]. By introducing the prior knowledge of the epipolar line, we define the *Extended Sum of Squared Differences* (ESSD) criterion that measures gray level compatibility and simultaneously penalizes deviations from the epipolar line:

$$ESSD(u, v, x, y) = \sum_{\beta} \sum_{\alpha} \phi_{(\alpha,\beta)} \left[I_{(t,\alpha,\beta)} - I_{(t+\tau,\alpha+u,\beta+v)} \right]^2 + \lambda_{ep} \, d^2_{ep}(x, y, u, v) \quad (7)$$

where $I_{(t,x,y)}$ designates the pixel (x, y) of the image acquired at time t, $d_{ep}(x, y, u, v)$ is the distance of the matching candidate pixel, $I_{(t+\tau,x+u,y+v)}$, to the epipolar line, and $\phi_{(\alpha,\beta)}$ is a weighting function. The contribution of the prior knowledge to the final cost functional[3] is quantified λ_{ep}.

In [1] the disparity vector is estimated by fitting a quadratic surface to a neighbourhood of the minimum SSD point, while in [3] two one dimensional quadratic curves are fit, both in the x and y directions. Due to the increased robustness to noise and simplicity, we adjust two one dimensional quadratic curves in each direction, in a neighbourhood of the minimum ESSD point:

$$q(u) = au^2 + bu + c \quad (8)$$

---

[3] Since the computation of the epipolar line is based on the camera and motion parameters, the value of λ_{ep} should be related to the uncertainty associated to these parameters.

where a, b and c are estimated from the data. The minimum point can then be determined, with sub pixel resolution, yielding the optimal disparity estimate:

$$\hat{u}_{opt} = -\frac{b}{2\,a} \tag{9}$$

whenever the coefficient a is different from 0.

The uncertainty of the estimate is related to the shape of the ESSD error surface [1, 2, 3]. In [2], the uncertainty is a function of the SSD surface curvature along the principal axis, and in [1] error propagation techniques in the SSD cost functional, were used. In each direction, we will approximate the estimate variance by [3]:

$$\sigma_u^2 = \left(\frac{d^2 q(u_{opt})}{du^2}\right)^{-2} q(u_{opt}) \tag{10}$$

The first term of (10) expresses the decrease of the uncertainty with the increase of the error surface curvature, while the second term is a normalizing factor dependent of the minimum ESSD surface value.

3.2 Regularization

Many visual reconstruction processes, aimed at recovering 3D information by processing 2D image data, are inverse, ill-posed problems (eg. the estimation of disparity fields between successive images) [8, 9].

Using the framework of regularization, ill-posed problems can be reformulated as variational principles, by introducing *a priori* knowledge about the solution. Standard Tikhonov regularization, uses stabilizing functionals to restrict the space of admissible solutions to smooth functions.

To determine a regularized function $\mathcal{U}$, using a set of data $\mathcal{D}$, we define an error function $\Psi_d(\mathcal{D},\mathcal{U})$, to measure the compatibility of the proposed solution and the data, a stabilizing functional $\Psi_p(\mathcal{U})$ that quantifies the smoothness conditions on the desired solution, and search for the $\mathcal{U}^*$ that minimizes the following cost criterion [10, 11].

$$\Psi(\mathcal{U},\mathcal{D}) = \Psi_d(\mathcal{U},\mathcal{D}) + \lambda\Psi_p(\mathcal{U}) \tag{11}$$

The choice for both functionals Ψ_p and Ψ_d guarantees that, under certain weak conditions, a solution for the optimization problem exists [2]. For the regularization of the displacement/disparity vector field, a *thin membrane* [2, 8, 10] stabilizing functional was used:

$$\Psi(\mathcal{U},\mathcal{D}) = \lambda\sum_{x,y}(\vec{u}-\vec{d})^T Q^{-1}(\vec{u}-\vec{d}) + \int\int trace\,\{(\nabla\vec{u})(\nabla\vec{u}^T)\}dxdy \tag{12}$$

$$Q = \begin{bmatrix} \sigma_u^2 & 0 \\ 0 & \sigma_v^2 \end{bmatrix} \tag{13}$$

where $\vec{u}(x,y) = [u(x,y)\ \ v(x,y)]^T$ is the regularized solution, σ_u^2 and σ_v^2 are the variances of the x and y components of the measured disparity vectors, ∇

is the gradient operator and λ quantifies the relative weight of the fitness to data term, in the global cost functional.[4]

The domain of the surface $\vec{u}(x, y)$ is usually discretized using either the finite differences method or the finite element method [2, 4, 8]. Applying finite element analysis, as proposed by Terzopoulos [8], to the functionals involved in the minimization problem we obtain:

$$\Upsilon(\mathcal{U}, \mathcal{D}) = \sum_{x,y} \lambda(\vec{u} - \vec{d})^T Q^{-1}(\vec{u} - \vec{d}) + \| \vec{u}_{(x+1,y)} - \vec{u}_{(x,y)} \|^2 + \| \vec{u}_{(x,y+1)} - \vec{u}_{(x,y)} \|^2 \quad (14)$$

To minimize (14), the Gauss-Seidel relaxation algorithm is used, and $\vec{u}$ is obtained iteratively at each point as a function of the values of its neighbours (similar equations are obtained for u and v):

$$u_{(x,y)}^{n+1} = \bar{u}_{(x,y)}^n + \frac{\lambda \sigma_u^{-2}}{1 + \lambda \sigma_u^{-2}} (u_{(x,y)}^0 - \bar{u}_{(x,y)}^n) \quad (15)$$

where $u_{(x,y)}^0$ is the measured x disparity at pixel (x, y) and $\bar{u}_{(x,y)}$ is the local mean value, given by:

$$\bar{u}_{(x,y)} = (u_{(x+1,y)} + u_{(x-1,y)} + u_{(x,y+1)} + u_{(x,y-1)})/4 \quad (16)$$

To determine the uncertainty associated to the regularized disparity field, the uncertainty values must be updated, as the regularization procedure imposes changes in the original field. Unfortunately, as the regularization iterations proceed, $\bar{u}^n$ will depend on an increasing number of samples, thus hardening the calculation of the uncertainty. For computation efficiency, we have adopted the following simpler expression:

$$var[u^{n+1}] = \frac{\alpha_m^2 var[\bar{u}^n] + \alpha_0^2 var[u^0]}{\alpha_m^2 + \alpha_0^2} \quad (17)$$

3.3 Kalman Filtering - Recursive Estimation of Depth Maps

We will now see how to use Kalman filtering theory, to combine different disparity measurements over time, yielding a more reliable single depth estimate.

Using the state space model framework, we can formulate an estimation problem, to determine the value of $Z_{[t]}$, based on the noisy observations $\vec{d}_{[t]}$. Depth is estimated independently at each pixel, while spatial dependencies are embodied in the regularization process.

This state estimation problem can be conveniently dealt with, using Kalman filtering techniques. Since the observation equation is non-linear in the state variable, the discrete-time Extended Kalman Filter (EKF) [12] must be used. The filtering equations comprise a prediction step and an update/filtering step. In the prediction stage, the future values of depth and associated uncertainty are predicted, based on past information and on the dynamic model:

---

[4] Whenever the measured variable $\vec{d}$ is unavailable, λ will be set to zero.

Prediction :

$$\hat{Z}(t/t-1) = a_{[t-1]}\hat{Z}(t-1/t-1) + b_{[t-1]} \tag{18}$$

$$\sigma^2_{Z(t/t-1)} = a^2_{[t-1]}\sigma^2_{Z(t-1/t-1)} + r \tag{19}$$

where $\hat{Z}(t/t-1)$ is the predicted depth value for time t, based on the data available up to time $t-1$, $\sigma^2_{Z(t/t-1)}$ is the corresponding variance and r is the variance of the motion equation noise (see Section 2).

At time t, a new disparity measurement is available, and the predicted depth can be updated. This is the filtering step procedure and the EKF uses a linearized version of the observation equation in a neighbourhood of the predicted value.

$$\vec{d}_{[t]} \approx \vec{d}^L_{[t]} = C^L_{[t]}Z_{[t]} + D^L_{[t]} + \mu \tag{20}$$

where $C^L_{[t]}$, $D^L_{[t]}$ are the coefficients of the linearized model [7]. The filtering step is given by

Filtering :

$$K_t = \sigma^2_{Z(t/t-1)}(C^L_{[t]})^T \, [\, C^L_{[t]}\sigma^2_{Z(t/t-1)}(C^L_{[t]})^T + Q_t \,]^{-1} \tag{21}$$

$$\sigma^2_{Z(t/t)} = (1 - K_t C^L_{[t]})\sigma^2_{Z(t/t-1)} \tag{22}$$

$$\hat{Z}(t/t) = \hat{Z}(t/t-1) + K_t \, (\, \vec{d}_{[t]} - C^L_{[t]}\hat{Z}(t/t-1) - D^L_{[t]} \,) \tag{23}$$

with $\hat{Z}(0)$ and $\sigma^2_{Z(0)}$ being the initial depth and depth uncertainty estimates, and K_t being the Kalman gain.

3.3.1 Warping the Depth Map

To complete the EKF analysis, there is still an additional problem to be solved, concerning the prediction step.

The predicted depth value $\hat{Z}(t/t-1)$ is obtained by applying the motion equation. However, this predicted depth value, no longer corresponds to the original pixel (x, y), since the x and y coordinates have also changed. As the predicted depth values does not coincide with image pixel positions, the depth at the grid point (x, y) must be inferred from these values. This problem can be approached in several ways like bi-linear or bi-cubic interpolation [1, 3].

We compute the depth estimate at (x, y) as a weighted average of the estimates falling within a 3x3 window centered in (x, y). The warped depth uncertainty is estimated using error propagation techniques:

$$Z_{(x,y)} = \frac{\sum_{i=1}^{n} d_i^{-2} Z_{(x',y')}}{\sum_{i=1}^{n} d_i^{-2}} \qquad \sigma^2_{Z(x,y)} = \frac{\sum_{i=1}^{n} d_i^{-4} \sigma^2_{Z(x',y')}}{(\sum_{i=1}^{n} d_i^{-2})^2} \tag{24}$$

where $Z_{(x',y')}$ stands for the predicted depth values, and d_i is the euclidean distance from (x'_i, y'_i) to (x, y).

4 Results

The 3D vision system was tested under a set of different synthetic and real conditions. The next section describes the results obtained using a sequence of synthetic images. The final results consist of tests with real underwater images.

4.1 Results with Synthetic Images

We started by assuming a scene structure composed by several rectangular patches placed at different distances from the camera, and we defined a brightness pattern that allows the generation of an image sequence. Figure 2 shows the last image of the sequence. The background is placed at 10m from the camera, the left rectangle at 5m and the right rectangular patch at 3.33m.

A 5x5 match window size was used, with $\hat{Z}(0) = 5m$, $\sigma^2_{Z(0)} = 25.0m^2$, $r = 1$ and $\lambda_{epip} = 5$. The final depth map is shown in figure 2, where depth is coded in gray level, darker points being closer to the camera. Figures 3 shows the first and final depth profiles along the x direction.

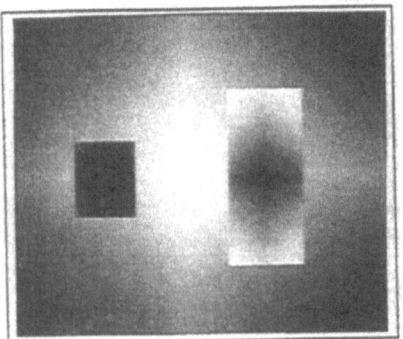 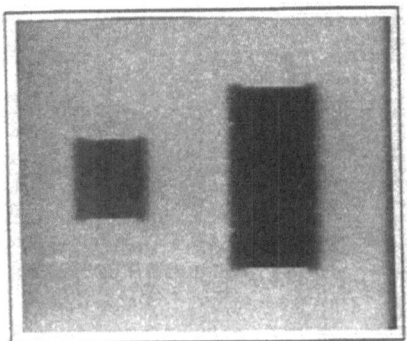

Figure 2: Last image of the synthetic sequence. Gray level coded depth map after the 10th iteration.

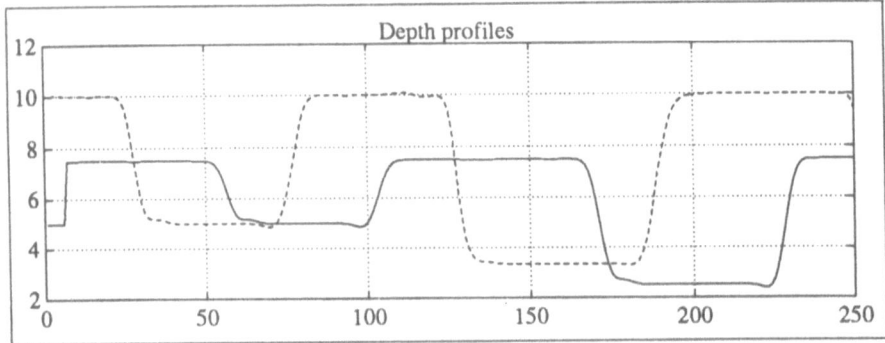

Figure 3: Depth estimate profile along the x direction (y = 128)

Although the synthesized images, are not corrupted with noise, they exhibit occlusion. The results obtained show a high level of precision. Notice the improvement over time, as successive depth estimates are being combined, and the blurring effect in discontinuities. The blurring effect, due to the regularization procedure is more noticeable in the occluded areas, where the matcher fails, and the disparity field is filled in, by the regularization process.

4.2 Results with Real Underwater Images

In this section we present the results obtained using real underwater images acquired with a camera in a special tank. The camera is moving with downwards vertical speed of 4.4 cm/s.

The images were acquired at the video rate of 25 images/s. In order to use a suitable sampling period, we have assumed that all the objects in the scene were located at a distance from the camera in the range [1,6] meters, and to keep the disparity values within 1 and 12 pixels, the sampling period was chosen to be 0.4 seconds.

The *a priori* depth map estimate is 6 meters for every image pixel, with 0.01 m^2 variance. A 7x7 match window was used with $\lambda = 0.1$ and $\lambda_{epip} = 0$. Figure 4 shows the first and final image in the sequence, together with the perspective view of the corresponding depth maps.

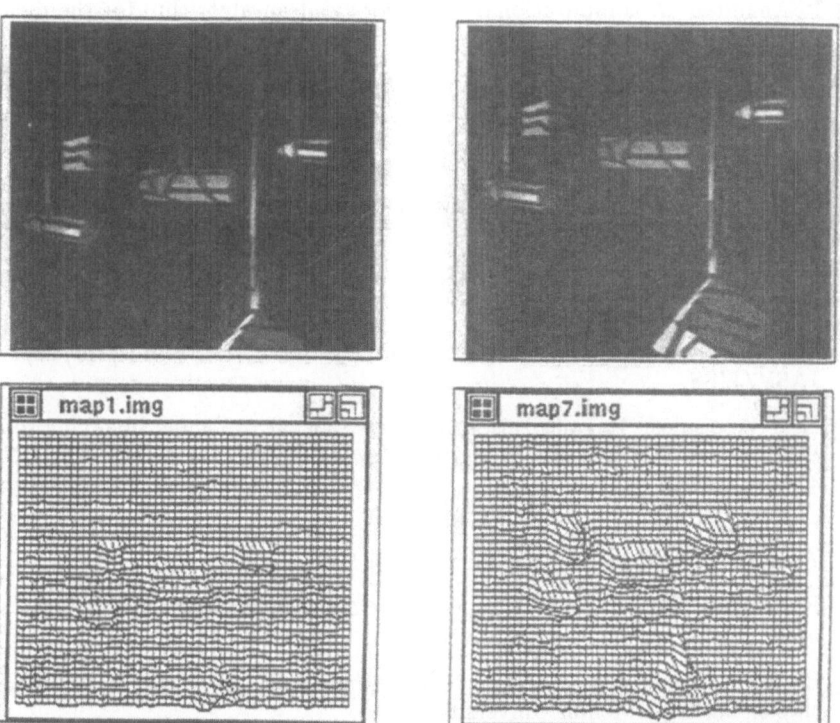

Figure 4: Initial and last (7th) images of the underwater sequence. Perspective of the first and final 3D depth maps

5 Conclusions

We presented a depth from motion vision system designed to compute dense depth maps from an image sequence, acquired by a moving camera. The system is based on a state space description of the depth from motion problem and models the existence of several uncertainty sources. A matching procedure including the epipolar constraint was defined, and regularization is used to filter the disparity vector field. Finally, depth measurements are combined over time, using Kalman filtering, to reduce uncertainty.

The system can be applied to a large number of problems in robotics, where the estimation of the scene structure may be found useful. A particular application related to underwater robotics was presented. Results obtained with synthetic images were presented, together with results with real images acquired in an underwater environment. In both cases the scene structure was recovered with remarkable accuracy.

References

[1] L. Mathies, T. Kanade, and Szelisky R. Kalman filter-based algorithms for estimating depth from image sequences. *Int. J. of Computer Vision*, 4(3):209–238, 1989.

[2] P. Anandan. A computational framework and an algorithm for the measurement of visual motion. *Int. J. of Computer Vision*, 4(2):283–310, 1989.

[3] J. Heel. Dynamic motion vision. In *Proc. of the DARPA Image Understanding Workshop*. Morgan-Kaufman Publishers, May 1989.

[4] B.K. Horn. *Robot Vision*. M.I.T.Press, 1986.

[5] B. Horn and B. Shunck. Determining optical flow. *Artif. Intell.*, 17:185–203, 1981.

[6] D. Ballard and C. Brown. *Computer Vision*. Prentice-Hall, London, 1982.

[7] J. Santos-Victor and J. Sentieiro. A Dynamic 3D Vision System. Technical report, Instituto Superior Técnico, February 1992. Ref. Mobius/rpt/02/92 JASV/JJSS.

[8] D. Terzopoulos. Regularization of inverse visual problems involving discontinuities. *IEEE Trans. on Pattern Anal. and Machine Intell.*, 8(4):413–424, July 1986.

[9] M. Bertero, T. Poggio, and V. Torre. Ill-posed problems in early vision. *Proceedings of the IEEE*, 76(8):869–889, 1988.

[10] R. Szeliski. Bayesian modeling of uncertainty in low-level vision. *Int. J. of Computer Vision*, 5(3):271–301, 1990.

[11] R. Szeliski. *Bayesian Modeling of Uncertainty in Low-Level Vision*. PhD thesis, Carnegie Mellon University, 1988.

[12] Jazwinski. *Stochastic Processes and Filtering Theory*. Academic Press, 1970.

A New Class of Corner Finder

Stephen Smith

Robotics Research Group, Department of Engineering Science,
University of Oxford, Oxford, England,

and

DRA (RARDE Chertsey),
Surrey, England

January 31, 1992

Abstract

An accurate, stable and very fast corner finder (for feature based vision)
has been developed, based on a novel definition of corners, using no image
derivatives. This note describes the algorithm and the results obtained
from its use.

1   Introduction

A lot of vision research lies in the area of 'early vision'. This has as its goal
the reduction of image data so that information becomes more manageable and
more immediately useful. This is commonly achieved by reducing a greyscale
image to a list of edges or a list of 'corners'. (See [1] for a discussion of the merits
of such an approach.) The mathematical description of a greyscale edge has
been fairly well defined, but there are many different mathematical descriptions
of the 2–D image structure which defines a corner. This is not surprising as
the term 'corner' is quite vague. Indeed, an accurate definition of a 'corner'
cannot go much further than *a position in the 2-D array of brightness pixels
which humans would associate with the word 'corner'*.

As a result of this, many different corner finding algorithms have appeared
in vision literature. For a good review of most of the different methods of
corner finding see [2]. Some of the more interesting work which has been done
is covered in [3], [4], [5], [6], [7], [8] and [9]. Differing methods each have an
idea of a definition of what a 'corner' is, and this idea is translated into a
mathematical way of finding corners. Reference [2] in fact shows that most
of the different methods end up with very similar mathematics. This note
describes a completely new type of corner finder, in both its definition and the
outworking of the mathematics. It is called the 'Smallest Univalue Segment
Assimilating Nucleus' ('SUSAN') corner finder and is described below[1]. The
idea behind it is relatively straightforward and is explained in section 2. In
section 3 the algorithm that has been developed is described in detail. Section 4
gives various results of the SUSAN algorithm.

---

[1] Patents have been applied for (by DRA) completely covering the principles of the SUSAN
corner and edge finders.

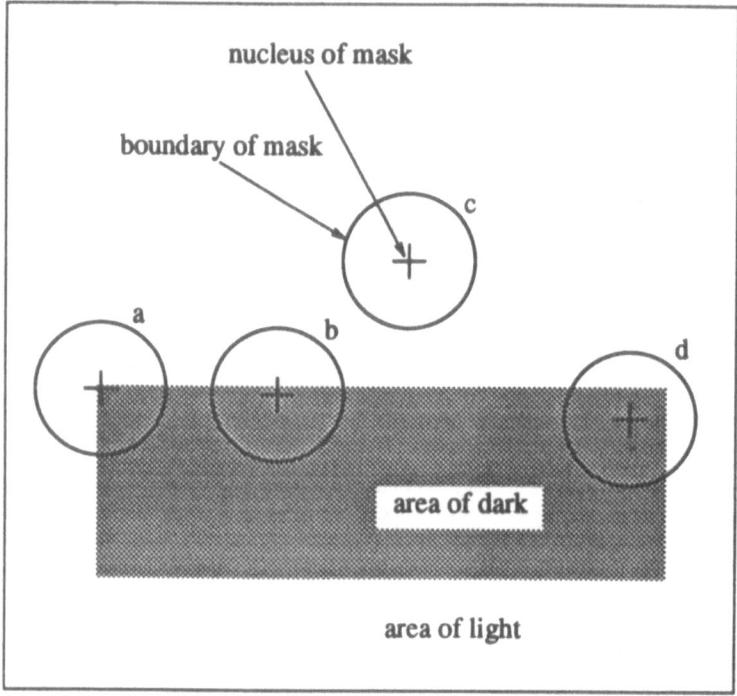

Figure 1: Four 'corner finding' masks at different places on a simple image.

2 The Principle Behind the SUSAN Algorithm

This section describes the SUSAN principle from an algorithmic point of view, as this is the simplest approach.

Each image pixel is used as the central point of a small mask (in this case circular) which contains the few pixels which are closest to the central one. This idea is shown for just four positions in an image in Figure 1. The mask is delimited by a circle and the central pixel (which shall be known as the 'nucleus') is marked by a cross.

The first step is to look at the greyscale (brightness) of the nucleus and compare this with the greyscales of the other pixels within the circular mask. The pixels with similar brightness to the nucleus are then assumed to be part of the same surface in the image (for example the light or dark surface in Figure 1). Figure 2 shows the masks with pixels of similar brightness to the nucleus coloured white, whilst pixels with different brightness are coloured black. The white portion of the mask is now used to determine whether the nucleus is positioned over a 'corner' or not.

The white portion of the mask shall be known as the 'USAN' or 'univalue segment assimilating nucleus'. (Here 'univalue segment' means a segment of the small mask which locally has constant or nearly constant brightness. It is of interest when it contains the nucleus.) Although a detailed study of the

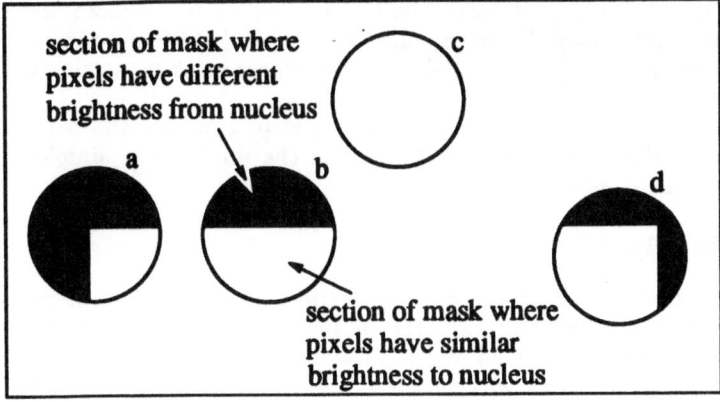

section of mask where
pixels have different
brightness from nucleus

section of mask where
pixels have similar
brightness to nucleus

Figure 2: Four 'corner finding' masks with similarity colouring; USANs are coloured white.

shape of the USAN would enable corners to be detected, this is not necessary. Instead a much simpler rule suffices. It can be observed from Figure 2 that the USAN corresponding to the corner – case (a) – has relatively small area. In fact, if the USAN has an area of less than half the mask area, then the nucleus is placed on or near a convex edge of a univalue surface. Cases (b) and (c) clearly fail this condition. If an upper limit for the USAN's area is set at some fraction less than 1/2 (for example 2/5) of the mask area, then the curvature of the convex edge must be above some minimum level.

It is also clear that a local minimum in USAN area will find the exact point of the corner. For example, the USAN in case (d) of figure 2 will not have as small an area as it will when the nucleus is placed exactly on the corner – compare with case (a). Hence the term 'smallest USAN' gives rise to the algorithm's name.

One of the novel aspects of this corner finder is that it does not use any image brightness spatial derivatives in the algorithm. It has been common until now to use first and second derivatives of the brightness, amplifying the problem of noise in the image. Most former methods have attempted to reduce the effects of noise by smoothing the image or the derivatives, but this inevitably reduces the quality of the localisation (accuracy in 2-D location) of the corners. The SUSAN algorithm is the first corner finder to use no spatial derivatives; it therefore does not need any smoothing process, and so there is no degradation in localisation.

3 The SUSAN Algorithm in Detail

In this section the SUSAN algorithm is described in detail.

The current implementation of SUSAN uses for its mask a 5 pixel by 5 pixel square with 3 pixels added on to the centre of each edge. This shape forms a fairly good digital approximation to a circle. The size was carefully chosen. A smaller mask gave results that were not very stable, that is, a corner was not reliably detected at the same place in several images taken in succession (even if the camera was static). A larger mask (based on a 7 by 7 square) gave stable results but was too large for the smallest structures (which were identifiable in the digitised image) to be detected. Also the larger mask was twice as slow as the one finally used.

The next step in the SUSAN algorithm is to compare the brightness of each pixel in the defined mask with that of the nucleus. A simple function determines this comparison;

$$c(\vec{r}, \vec{r_0}) = 100e^{-(\frac{I(\vec{r})-I(\vec{r_0})}{t})^6},$$ (1)

where $\vec{r_0}$ is the position of the nucleus in the 2–D image, $\vec{r}$ is the position of any other point within the mask, $I(\vec{r})$ is the brightness of any pixel, t is the brightness difference threshold and c is the output of the comparison.

The form of equation 1 was chosen to give a similar response to a box function but with slightly rounded corners. This allows a pixel's brightness to vary slightly without having a large effect on c, even if it is near the threshold position. The exact form for equation 1 was carefully chosen (empirically) to give a balance between good stability about the threshold and the function originally required (namely to count pixels that have similar brightness to the nucleus as "in" the univalue surface and to count pixels with dissimilar brightness as "out" of the surface). The equation is implemented as a look up table for speed.

This comparison is done for each pixel within the mask, and a running total, n, of the outputs (c) is made;

$$n = \sum_{\vec{r}} c(\vec{r}, \vec{r_0}).$$ (2)

This total n is just 100 times the number of pixels in the USAN, i.e. it gives the USAN's area. As described earlier, this total is eventually minimised.

Next, n is compared with a fixed threshold g (the 'geometric threshold') which should be set to just under half of the maximum possible value for n. This prevents straight boundaries from giving false positives.[2]

The two thresholds so far introduced are good examples of two different types of threshold. The geometric threshold clearly affects the 'quality' of the output. Although it affects the number of corners found, much more importantly, it affects the shape of the corners detected.[3] For example, if it were reduced, the allowed corners would be sharper. Thus this threshold can be fixed (to the value previously explained) and will need no further tuning. Therefore no weakness is introduced into the algorithm by the use of the geometric threshold. The brightness difference threshold is very different. It does not affect the quality of the output as such, but does affect the number of corners reported. Because it determines the allowed variation in brightness within the USAN, a reduction in this threshold picks up more subtle variations in the image and gives a correspondingly greater number of reported corners. This threshold

---

[2]In practice, the threshold is set to exactly half of the maximum of 3700 (this number comes from the 37 pixels in the mask; the nucleus automatically contributes a count of 100). This threshold was chosen because a straight edge including the nucleus will always be greater than half of 3700 (1850) by at least 50, due to quantisation of the pixels, and the threshold should be set as large as possible to allow the maximum variety of corners.

[3]This assumes that the USAN is a contiguous region. The refinements described in [10] are designed to enforce this contiguity.

can therefore be used to control the 'quantity' of the output without affecting the 'quality'. This can be seen as a negative or a positive point. On the one hand, it means that this threshold must be set according to the contrast etc. in the image, and on the other, this threshold can be easily tuned to give the required density of corners. In practice, there is no problem. A fixed value of 25 is suitable for almost all real images, and if low contrast images need to be catered for, the threshold can be set automatically, simply by varying it to give the required number of reported corners. When SUSAN was tested on an extremely low contrast image this threshold was reduced to 7. This gave a 'normal' number of corners. Even at this low value, the distribution was still good (i.e. not over-clustered) and the corners found were still quite stable.

Next an intermediate image is created from the value of n found at each image position. If $n(x, y)$ is less than the geometric threshold then $(g - n(x, y))$ is put into the new image at (x, y). If it does not pass this test, then zero is put into the image at this place. This stage is done so that local minima in the values of n which pass the geometric test may be found, to locate corners exactly. In practice therefore, the intermediate image is searched over a square 5 by 5 region for local maxima (above zero). Finally, these local maxima are reported as corners. The entire process, run on a Sun Sparc 2 processor, using a 256 by 256 pixel image takes about 1/3 of a second for an average scene.

This simple algorithm is the basis for a successful corner finder. It is a completely new way of approaching the problem.

A rigorous mathematical analysis of the SUSAN algorithm would be very complicated as its validity, as explained here, though fairly obvious, is more 'intuitively picturesque' than analytic. However, a mathematical analysis has been performed, and can be found in [10]. This analysis gives as the final mathematical interpretation of the SUSAN principle,

$$\frac{dI}{d\vec{r_0}}(\vec{r_0}) = < \frac{dI}{d\vec{r_0}}(\vec{r}) >_{\text{over edges}} . \tag{3}$$

This equation expresses the fact that the intensity differential at the nucleus must be the same as the intensity differential averaged over all of the edges lying within the mask. This must be the case for both the x and y differentials. For this to hold, the nucleus must be placed on a line of reflective symmetry of the boundary pattern within the mask area. This is so that the contributions to the average differential on either side of the nucleus may balance out. This is equivalent to saying that the nucleus must lie on a local maximum in the edge curvature; consider a Taylor's expansion of the edge curvature. At the smallest local level, the nucleus will only lie on a centre of symmetry of the curvature if it lies at a maximum (or minimum) of curvature.

The conditions derived ensure that the nucleus lies not only on an edge, but also that it is placed on a sharp point on the edge. It will as it stands give a false positive result on a straight edge; this is countered by not just finding local minima in n, but by forcing n to be below the geometric threshold g.

4 Results

The results of testing this corner finder are shown and discussed in this section. Firstly, the output of SUSAN given a test image[4] is shown in figure 3 (top). Compare this with figure 3 (bottom), the output from the Plessey corner finder, which looks for large and similar principal curvatures in the local autocorrelation function. The accurate localisation and reliability of the SUSAN algorithm is apparent; it does all it is expected to do. Algorithms based on image derivatives often have problems at some of these junctions, for example where more than two regions touch at a point. There is no problem for SUSAN. This is expected, as the corner finder will look at the point of each region individually; the presence of more than two regions near the nucleus will not cause any confusion. The local non-maximum suppression will simply choose the pixel at the point of the region having the sharpest corner for the exact location of the final marker. The inaccuracies of the Plessey corner finder, even at simple two region corners, are visible. This is discussed and explained in [11]. With respect to speed, SUSAN took 0.3 seconds to process this picture on a single RISC processor; the Plessey corner finder took 3.5 seconds.

The SUSAN algorithm has also been tested with respect to its sensitivity to noise. The results are very good; the quality of its output (both the reliability and localisation) degrades far less quickly than other algorithms tested as noise in the image is increased. In the following example, the original test image had a considerable amount ($\sigma = 3$) of gaussian noise added. The outputs of SUSAN and the Plessey corner finder are shown in figure 4 (top and bottom respectively).

SUSAN has also been tested with very many individual real images; unfortunately they must be seen in [10] due to space constraints.

The temporal stability of SUSAN has also been analysed. The output from several consecutive frames from a moving camera was used as the first stage of the DROID 3–D vision system developed by Plessey (see [1], [12] and [13]). This program tracks corners through time in order to reconstruct a 3–D description of the world. The results obtained when the Plessey corner finder was used were compared with those obtained when SUSAN was used. Several different sequences were tested. Some sequences gave slightly better 3–D output data when using the SUSAN algorithm, and the rest gave similar results with both algorithms. The results were compared by observing the quality of a least squares plane fit through the tracked 3–D points which were on a plane in the world, and also by using this to detect small obstacles in the vehicle's path. It was found that the size of the smallest objects which could be detected by DROID was less when the SUSAN corner detector was used than when the Plessey corner finder was used. This suggests that in this example the quality of the data was slightly better when using the SUSAN corner finder.

In these tests, SUSAN ran on average 10 times faster than the Plessey algorithm.

---

[4]This test image has been developed by the author to include two dimensional structures of many different types, and ones which existing corner finding algorithms often cannot correctly interpret. Note simple 90° corners, corners with angles close to 0° and 180°, various junctions with more than 2 regions meeting at a point, two corners close together, and corners created by taking a single brightness ramp and raising a rectangle in its centre by a uniform brightness.

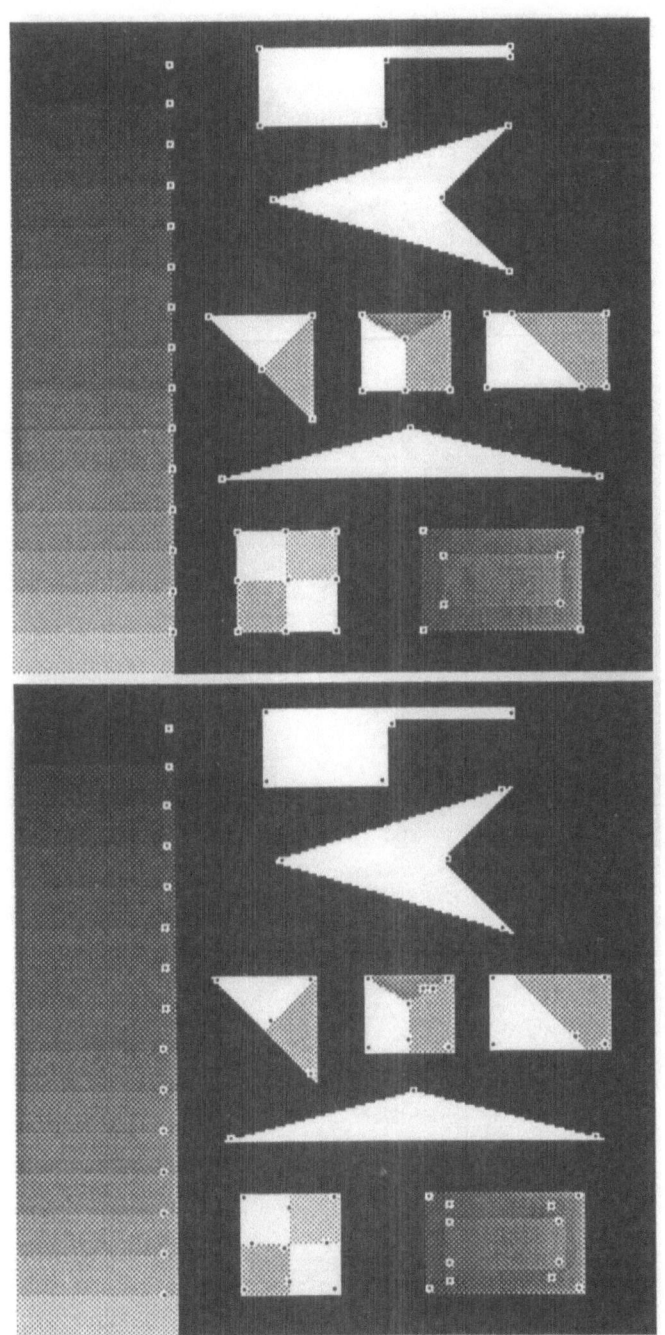

Figure 3: Result of SUSAN corner finder on test image (top), result of Plessey corner finder on test image (bottom).

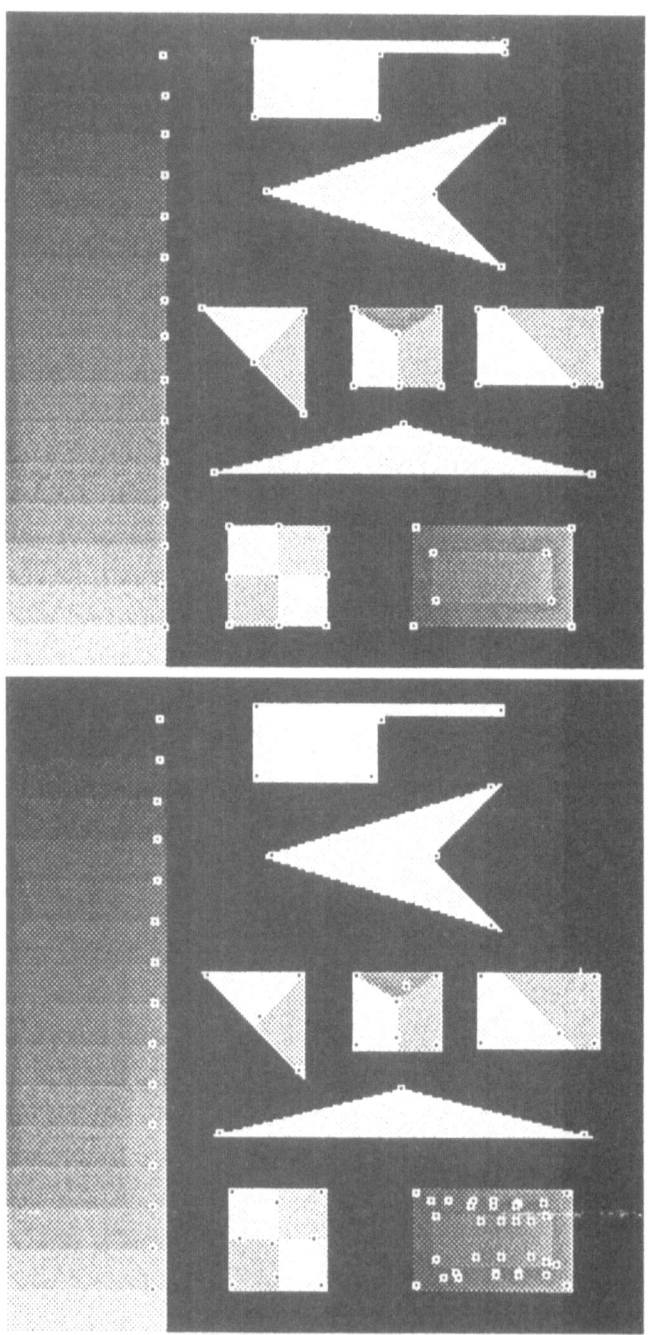

Figure 4: Result of SUSAN corner finder on test image with gaussian noise ($\sigma = 3$) added (top), result of Plessey corner finder on same test image (bottom).

As another quantitative test of its stability and general suitability for real applications, SUSAN has been used to provide corners for a program (at present being developed by the author) which segments a stream of images into independently moving objects. This program uses corner motion to segment the images into parts which have different motion from each other. The results are extremely good, with successful segmentation of two vehicles travelling in front of a third carrying a video camera.

5   Conclusions and Future Work

A completely new approach to corner finding has been developed and tested. It is accurate, stable and very fast. It has been tested for accuracy on both test and real images, and additionally for speed and stability on image sequences. It has given very good results in all the tests and has been used with complete success as a front end for both a 3–D structure-from-motion program and a motion segmentation program.

The SUSAN algorithm has been shown to be related to finding local maxima in edge curvature. However, because the algorithm uses regions to find corners and not first or second image derivatives, it is very good at ignoring noise in the image, and also very good at producing well localised corners. This is the first approach which does not use any image derivatives.

The author has recently extended the principle of minimising the number of similar brightness neighbours to give a combined corner and edge detector. Using the same initial response map as described above, the geometric threshold can be eliminated so that edge enhancement is achieved. Edge direction can be found from USAN centre of gravity and direction of symmetry, and the edges can be thinned and localised to sub-pixel accuracy.

The resulting algorithm has the advantage over most current edge finders that there is continuity at junctions, as the response rises rather than falls at junctions of more than two edges. It is also very fast. As with the SUSAN corner finder, the setting of the remaining threshold is very simple and not sensitive to finding an exact "right" value. The results are extremely good, and the author is preparing a paper giving more details of the algorithm and its results.

6   Acknowledgements

The author thanks his wife, J. M. Brady, J. Savage and R. Taylor for their input and encouragement.

References

[1] D. Charnley, C. Harris, M. Pike, E. Sparks, and M. Stephens. The DROID 3D vision system – algorithms for geometric integration. Technical Report 72/88/N488U, Plessey Research Roke Manor, December 1988.

[2] J.A. Noble. *Descriptions of Image Surfaces.* D.Phil. thesis, Robotics Research Group, Department of Engineering Science, Oxford University, 1989.

[3] L. Dreschler and H.H. Nagel. Volumetric model and 3D trajectory of a moving car derived from monocular TV-frame sequence of a street scene. In *IJCAI*, pages 692–697, 1981.

[4] H.-H. Nagel. Displacement vectors derived from second order intensity variations in image sequences. *Computer Vision, Graphics and Image Processing*, 21(1):85–117, 1983.

[5] C.G. Harris and M. Stephens. A combined corner and edge detector. In *4th Alvey Vision Conference*, pages 147–151, 1988.

[6] L. Kitchen and A. Rosenfeld. Grey-level corner detection. *Pattern Recognition Letters*, 1:95–102, 1982.

[7] G. Medioni and Y. Yasumoto. Corner detection and curve representation using curve b-splines. In *Proc. CVPR*, pages 764–769, 1986.

[8] A. Singh and M. Shneier. Grey level corner detection: A generalisation and a robust real time implementation. *Computer Vision, Graphics and Image Processing*, 51:54–69, 1990.

[9] H. Wang. Corner detection for 3D vision using array processors. In *Proc. BARNAIMAGE 91*, Barcelona, 1991.

[10] S.M. Smith. *Feature Based Image Sequence Understanding.* D.Phil. thesis, Robotics Research Group, Department of Engineering Science, Oxford University, 1992.

[11] S.M. Smith. Extracting information from images. First year D.Phil. report, Robotics Research Group, Department of Engineering Science, Oxford University, June 1990.

[12] D. Charnley and R.J. Blissett. Surface reconstruction from outdoor image sequences. *Image and Vision Computing*, 7(1):10–16, 1989.

[13] C.G. Harris and J.M. Pike. 3D positional integration from image sequences. *Image and Vision Computing*, 6(2):87–90, 1988.

On Evidence Assessment for Model-Based Recognition

L Du[1], G D Sullivan and K D Baker

Department of Computer Science, Reading University
Reading, RG6 2AY

Abstract

Evidence assessment for model based recognition is concerned with determining if a set of correspondences between image features and model features gives sufficient evidence for recognition. While many previous studies have addressed strategies to establish the correspondences, post-model-matching evidence assessment has remained largely primitive and *ad hoc*.

This paper presents a novel two-stage scheme of evidence assessment for model-based vision based on: (i) evidence against coincidental configuration of random image features and (ii) evidence against mis-recognition of other objects. We demonstrate this scheme for model based 3D recognition from 2D image features.

1 Introduction

Model-based vision usually comprises two phases: (i) *model matching* in which a search is carried out for evidence consistent with a pose, and (ii) *evidence assessment* to arrive at a recognition verdict. The second issue has received far less attention than model matching, and the majority of existing systems have employed *ad hoc* thresholds based on some form of goodness measure to give the final verdict.

An example drawn from car recognition [5] is illustrated in Figure 1. Figure1 (a) shows the original image and an hypothesised initial pose; (b) shows the candidate image features; (c) is the final 3D clique. The question addressed is: does this clique provide

s

 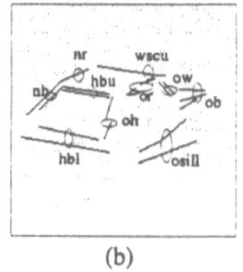 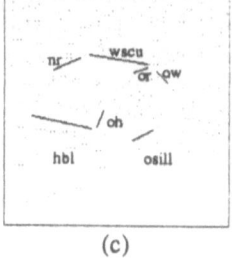

(a)	(b)	(c)

Figure 1   Example of Object Recognition

---

1. Currently a SERC research fellow at the Department of Electrical and Electronic Engineering, University of Surrey, Guildford, GU2 5XH, (email: L.Du@ee.surrey.ac.uk)

Bolles et al [2] treated a detected feature as providing one of 3 types of evidence, (i) positive evidence (a feature found close to the predicted feature); (ii) neutral evidence (a feature between the sensor and the predicted feature); (iii) negative evidence (a feature further from the sensor than the predicted feature). Their decision algorithm was based on an *ad hoc* threshold for a simple sum of these pieces of evidence. Fan, et al [8] presented a measure of the goodness of their graph matching based on 3 factors, (i) the ratio of the number of matched model nodes to the total number of model nodes; (ii) the 3D surface area of matched model nodes comparing to that of all model nodes; (iii) the ratio of the area of matched scene nodes to that of all the scene nodes. This measure is quantitative but the choice of threshold still remains *ad hoc*. Lowe [13] used an arbitrary threshold on the minimum number of matches needed for recognition, which is typical of the majority of model based recognition systems [9] [14].

Brisdon's iconic verification [5, 6] for 2D-3D recognition and Grimson's symbolic verification for 2D-2D recognition [12] have given systematic treatments to the problem of evidence assessment. They both defined evidence assessment as a process intended to eliminate accidental configuration among random image features. Implementation of this definition for two domains has allowed informed selection of verification thresholds, on the basis of a relation between a probability of an accidental configuration (called conspiracy by Grimson) and the quality measure of a set of correspondences.

This paper proposes a novel two-stage evidence assessment for model based vision, which not only assesses accidental configurations caused by random image features but also takes object confusability into consideration.

2 A two stage approach to evidence assessment

Several definitions of evidence assessment have been put forward as a process to assess the significance of a set of matches (against accidental configuration caused by random image features), but this only captures one aspect of the problem. Mistaken matches are also very likely to happen to occur non-randomly, where other structures may be confusable with the target object. For example, a model-based recognition process for a car may well find sufficient evidence from an image of a house.

Therefore, we treat evidence assessment in model-based vision as a two-stage task. Firstly, we assess evidence against the possibility that the set of feature correspondences arises by accidental configuration of random data features; This is the task addressed by traditional evidence assessment. Secondly, we assess the evidence against the possibility of matching the model of an object by mistake to data features caused by other structures; this has not previously been reported. We use the SDT [11] approach to investigate the second issue, by focusing on discrimination between a target object and a highly confusable non-target object (as defined by the specific application). This puts a lower bound on the discrimination ability of a system, and thus gives a measure of total performance.

The above discussion applies to model-based vision using edge, region or surface features. In the remainder of this paper we concentrate on 2D-3D vision using edge features, forming cliques of 2D-3D feature correspondences (referred to as a 3D clique).

3 Significance of a set of feature correspondence

The significance of a 3D clique may be assessed by comparison with the null hypothesis that a clique is due to an accidental configuration among essentially random image features.

3.1 Significance according to cardinality and VCE

Given a 3D clique, we consider two attributes which reflect its significance: (i) the size of the set (the cardinality of the clique), (ii) a measure of viewpoint consistency error, *VCE* (defined as the disagreement between the clique and the 2D model template for the pose derived from clique [5]).

3.2 The VCE distribution of a single random clique

Probability distributions of VCE for random cliques of different cardinalities form the basis for the first stage of evidence assessment with 3D cliques. A probability distribution function is defined as:

$$p_{ca}(v) = \frac{d}{dv} P_{ca}(v) \tag{1}$$

where the subscript *ca* denotes the particular cardinality and $P_{ca}(v)$ is the probability that a random clique of cardinality *ca* scores a VCE $\leq v$.

Grimson analytically established a similar probability for the 2D-2D recognition situation, as a function of the fraction of model feature, and sensor noise. However, in the current situation, these probability functions are extremely difficult to obtain analytically due to the inherent complexity of the 2D-3D problem. An experimental approach has been adopted, in which random 3D cliques are simulated by assigning random image features to features of a cube model. We assume that the VCE distribution obtained on the basis of those random 3D cliques is representative of random 3D cliques formed using general polyhedra.

Each experiment comprised trials of 10,000 random cliques of a fixed cardinality (*ca*). Each random 3D clique was created by assigning *ca* random image features to the same number of model features. A cube has at most 9 visible line features, so that it allows 7 experiments (from *ca* =3 to *ca*= 9), giving 7 histograms of VCE (Figure 2). These histograms are used as empirical approximations to the *VCE* distribution functions for significance testing.

3.3 Significance test for a 3D grouping problem

In order to carry out the significance test, we formulate the null hypothesis that the 3D clique is the result of matching an object to random image features. The test of significance becomes an attempt to contradict the null hypothesis.

Images containing several random features give rise to many cliques of different cardinality. Let *k* be the number of possible cliques of a given cardinality. The probability that at least *one* of the cliques happens to score a *VCE* as low as *v*, is

$$Q_{ca}(v) = 1 - [1 - P_{ca}(v)]^k \tag{2}$$

We assume that a perfect model matching method has been used, which always produces the clique with the lowest VCE among the *k* possible cliques. Therefore, refutation of the null hypothesis depends on a sufficiently small *Q*.

The significance test can be arranged as the following steps:

- Establish look-up tables for $P_{ca}(v)$, off-line.

- Calculate *k* for different cardinalities according to the set of candidate features fed into the model-matching process (those features, which fall into the focus neighbourhood defined by the initial pose estimate before model-matching). Record the cardinality and the VCE.

- Calculate $P_{ca}(v)$.

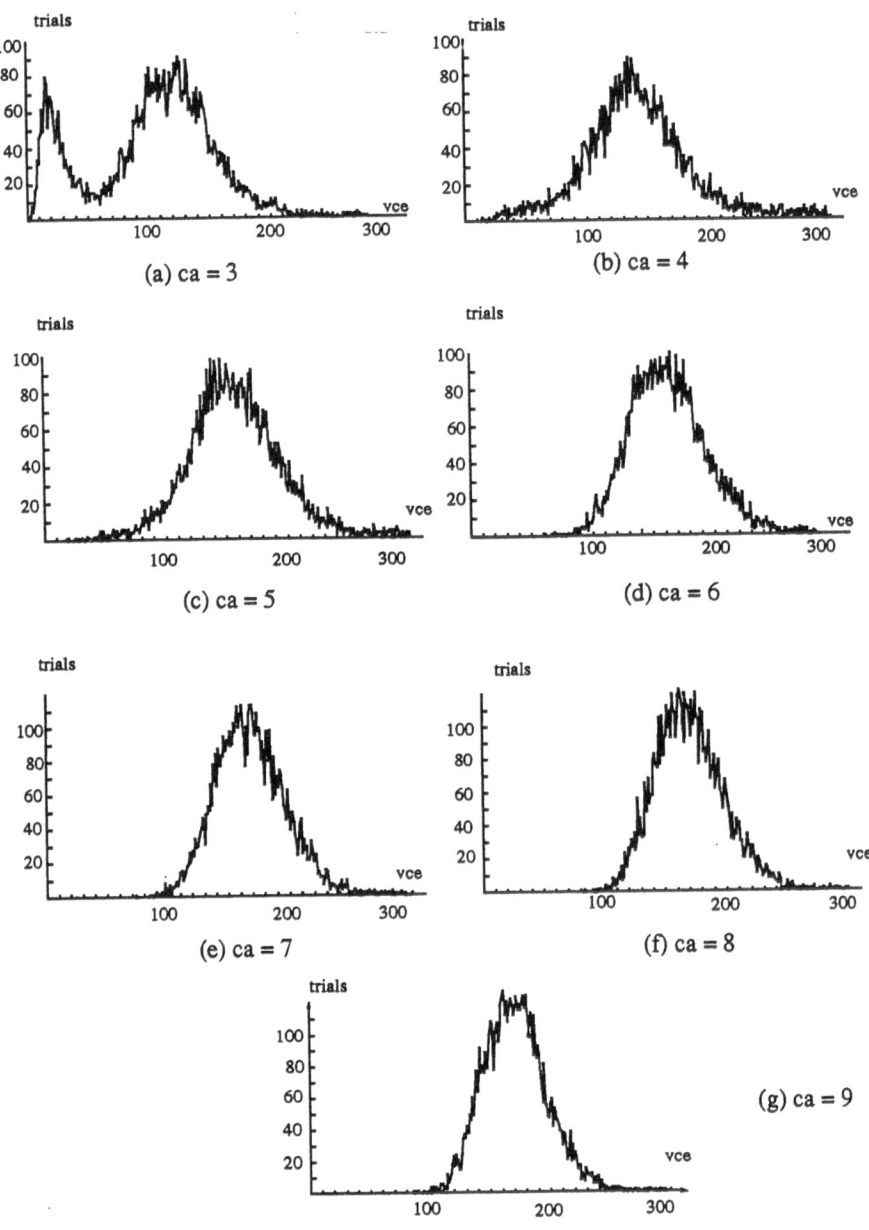

Figure 2   VCE distribution for random cliques

- Select a confidence threshold T_{cfd} to fit the particular application. If $T_{cfd} > Q$, reject the null hypothesis and accept a significant configuration. Otherwise reject the 3D clique.

The process can be illustrated for the example shown in Figure 1. Under the null hypothesis (that the candidate image features are random), the total number of possible cliques of cardinality=6 is k=18080 (obtained by combination and permutation of all potential matches). The VCE of the clique shown in Fig. 1(c) is 4.6. Referring to Fig.2 (d), it is found that the probability for a single random clique of ca=6 to score 4.6 is $Q_6(4.6)=0$. We also find $Q = 0$, by using (2). Therefore, the clique can be taken as highly unlikely to have arisen by chance.

4 Object discrimination

A 3D clique which is significantly different from an accidental configuration of random image features, as described above, may yet be the result of incorrectly matching the target model to features due to a similar object. A second stage is needed to assess the possibility.

4.1 Signal Detection Theory (SDT)

SDT [11] considers classification as a signal detection problem, where a series of events E_i are presented consecutively to a decision process. Each event E_i causes a response in the receiver V_i. The event E_i may be of two types, (i) a signal with noise (E_s), (ii) pure noise (E_n). A rational decision process uses a threshold (Thd) to classify the event according to V_i. The rule is that If $V_i > Thd$ then, classify E_i as arising from pure noise, otherwise classify E_i as arising from signal plus noise.

The Receiver Operating Characteristic (ROC) [16] is defined as the function relating the rate of hits (correct detection of E_s) versus the rate of false alarms (false detection of E_n) over all possible choices of threshold. It provides an accepted method to assess the performance of a detector, irrespective of the selected threshold.

4.2 Object discrimination by using the viewpoint consistency measure

We use the SDT approach to investigate object discrimination using the VCE as a receiver function. Each application of the model-based vision process has one target object. The task in the second evidence assessment stage is to discriminated between the target object and other non-target objects. We try to estimate a lower bound on discrimination performance by considering the target object and a non-target object which is highly confusable with it. The choice of this non-target object depends on the intended application, and should be chosen to provide a lower bound on performance.

Once a non-target object has been chosen, the problem can be fit into the SDT framework. There are two types of cliques: (i) the cliques of features arising from the target object being matched to the target model (called G_c - analogous to E_s) (ii) the cliques with features arising from the other object being matched to the target model (called G_w - analogous to E_n). Monte-Carlo experiments were conducted to establish the two probability distributions of the VCE response to two types of cliques $(p_{G_c}(v)$ and $p_{G_w}(v)$, with random noise added to all features.

Following SDT practice, we use the Receiver Operating Characteristic (ROC) to measure discrimination power (D), which is defined here as twice the area between the positive diagonal line and the ROC curve. If D=0 then the two distributions completely overlap and no discrimination is possible, if D=1 then they are distinct and discrimination is perfect.

To study the performance of the VCE, we measured discrimination the distribution for a number of noise levels, using random perturbation to extracted image features, defined in [7].

154

4.3 Discrimination between a Hatch-back car and Saloon car as function of noise

To illustrate the approach we selected two highly confusable models from the library available at Reading University (Fig. 3). There are small differences in the dimensions of the lines, but the conspicuous difference is in the shape of the back of the vehicle (though this structure is not explicit in the test). The following analysis answer the question: how good is the VCE for discriminating this pair of confusable objects.

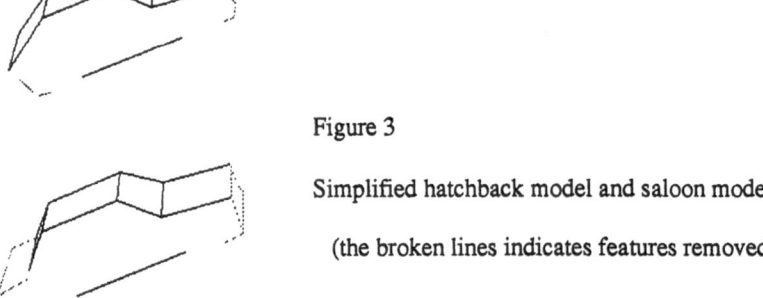

Figure 3

Simplified hatchback model and saloon model

(the broken lines indicates features removed)

Random G_w cliques were simulated by assigning instantiated features from the saloon model to the corresponding features of the hatchback model. In this case of non-identical objects, the wire-frame models for a hatchback and a saloon car were simplified (Figure 4) so that each feature in one model has exactly one corresponding feature of the same name in the other model. Therefore, the creation of G_w cliques meets the assumption of an ideal model-matching process.

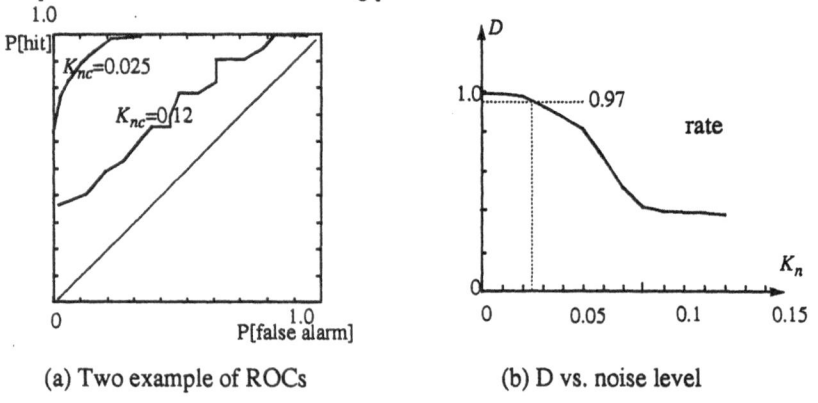

(a) Two example of ROCs

(b) D vs. noise level

Figure 4   Discrimination of saloon and hatchback

Random G_c cliques were simulated by assigning instantiated model features from a hatchback model to features on the same model. Random noise was added to all template features by random perturbation.

The VCE responses to both types of cliques were collected from a large number of G_c and G_w cliques at a range of noise levels. At each level of noise, the histograms for the VCE response to each type of cliques was used to approximate the two distributions $(p_{G_c}(v)$ and $p_{G_w}(v)$. It is found that at all noise levels (0.01~0.12) two distributions overlapped significantly. Therefore, it is impossible to specify a threshold which perfectly discriminates two types of cliques, and SDT applies.

Figure 4 (a) illustrates the ROC curves at two noise levels concerning a hatchback and a saloon car. Figure 4 (b) plots the resulting D as a function of noise level against 12 noise levels (0.0~0.12)

4.4 Using D-curves

Experiments with the two types of cliques produce distributions, which lead to estimates of D for a particular pair of objects. The experiment with G_c cliques can also be plotted to produce an approximate relation between the VCE and the noise level (see Fig. 5).

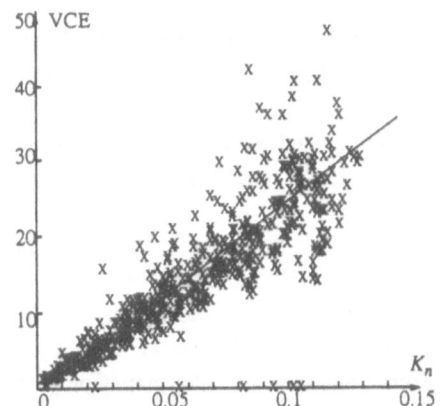

Figure 5   Corelation between noise and VCE value
for correct cliques

An important point is that once the D-curve has been established for a target object (e.g. a hatchback car), D can be estimated for any future image provided we have an estimate of the image noise level. To obtain this estimate, we run the feature extraction process and manually pick out the correct features to form a number of correct cliques. We then run the VCE evaluator, and find their average VCE value. The noise level is acquired by finding an equivalent noise level on the VCE and noise relation.

For example, for the image shown in Fig. 1(a) the average VCE of manually-selected correct cliques is 9.5, which gives an equivalent noise level of 0.025 from Fig. 5. This further gives a D = 0.97 (see Fig. 4(b)).

5   Conclusion

The two-stage evidence assessment process provides an approach to answering the question raised at the beginning of this paper. The evidence assessment as a two-stage statistical process applies to model-based object recognition in general. It allows two important decisions: (1) a significance test to eliminate possible accidental configuration of random features and (2) discrimination of confusable objects by using a threshold on VCE of the set. When a clear choice of the discrimination threshold (100% hit and no false alarm) is not possible, we can still determine the best possible compromise, as a function of the noise level.

The paper also described, in particular, a realisation of this process for 2D-3D model-based vision using edge-based features.

6 References

[1]   Bodington, Sullivan & Baker, Experiments on the use of the ATMS to label features for object recognition, Computer Vision-ECCV'90, Spring-Verlag, 1990.

[2]   Bolles, R, Horaud, P., 3DPO: A Three-Dimensional Part Orientation System, Int. J. of Robotics Research, Vol. 5. No. 3, 1986, pp3-26

[3]   Bray, A., Recognising and Tracking Polyhedral Objects, Ph.D Thesis, Sussex University, UK, 1991.

[4]   Brisdon, K, Evaluation and Verification of Model instances, Proceeding of Alvey Vision Conference'87, Cambridge, 1987, pp33-37

[5]   Brisdon, K Hypothesis Verification using Iconic Matching, Ph.D. Thesis, Reading University.

[6]   Du, L, G D Sullivan and K D Baker, 3D grouping by viewpoint consistency ascent, Image and Vision Computing, Special Issue on BMVC'91, 1992

[7]   Du, L, G D Sullivan and K D Baker, Modelling data complexity for model-based vision, submitted to BMVC'92

[8]   Fan, T, Medioni, G and Nevatia, R Recogising 3D Object Using Surface Descriptions, IEEE PAMI, Vol. 11, No. 11, 1989, pp1140-1157

[9]   Fisher, R, From Surfaces to Objects: Computer Vision and 3 Dimensional Scene Analysis, John Wiley & Sons

[10]   Goad, C., Special Purpose Automatic Programming for 3D model-based vision, Proceeding of the ARPA image understanding Workshop, Arlington, Virginia, 1983.

[11]   Green, D and Swets, J, Signal Detection Theory and Psychophysics, Robert E Krieger Publishing Co. 1974 (Reprint of Wiley & Sons 1966)

[12]   Grimson, L & Huttenlocher, D, On the Verification of Hypothesized Matches in Model-Based Recognition, Proceeding of the European Computer Vision Conference, France, 1990, pp489-498, Spring Verlag

[13]   Jain, A and Hoffman, R, Evidence based Recognition of 3-D Objects IEEE PAMI, Vol. 16, No. 6, 1988, pp783-800

[14]   Ikeuchi, K and Takeo, K, Automatic Generation of Object recognition Programs, IEEE Proceeding, Aug. 1988

[15]   Lowe, D., The viewpoint consistency constraint, International Journal of Computer Vision, 1987

[16]   Schiffmen, H, Sensation and Perception: An Integrated Approach, John Wiley and Sons, Inc, 1976

[17]   Worrall, A., et al, Model Based Tracking, BMVC'91, Glasgow, 1991

Lane Boundary Tracking for an Autonomous Road Vehicle

N. W. Campbell and B.T. Thomas

Advanced Computer Research Centre

University of Bristol

Bristol, UK

Abstract

We describe an algorithm by which an autonomous land vehicle is able to navigate along roads utilising lane boundary markings. This is achieved by defining a six-parameter model of the lane markings and fitting this to the processed monochrome image using non-linear least squares techniques. By qualitative as well as quantitative analysis over many hundreds of images the model used is shown to be justified, resulting in a very robust method, finding lane markings to sub-pixel accuracy.

1 Introduction

There has been a great deal of interest in autonomous road vehicles especially in the last decade, during which real-time image processing has become a reality. Road following is better defined than the more difficult problem of general terrain navigation, but is still extremely complex. There are several approaches to the problem, including:

- Detecting and tracking road-edges.

- Road surface segmentation.

Both of these suffer difficulties. Edge tracking [1] may fail for many reasons, including loss of road edge due to occlusion by other vehicles, its disappearance at junctions or its moving when the road widens or narrows. Road surface segmentation [2] is often unreliable because shadows, other vehicles and changes in surface texture may cause mis-classification.

It is our belief that any effective system will not only need to use several of the above techniques in parallel, but that it will be greatly enhanced by the little-used but very powerful cue of lane markings.

In [3] lane markings are detected in a 'bootstrap' mode using a high level vision approach. Candidate segments are extracted from the image based on their local geometric properties having a high probability of belonging to a lane marking. These segments are then grouped globally to identify possible lane markings.

However approaches such as this are often not robust enough to be relied upon by a road-following vehicle since the local segmentation process is subject to noise and the global grouping has little knowledge of the overall structure of lane markings. Therefore other white objects in the scene i.e. window frames, lettering on signposts or cars may cause mis-tracking to occur.

The approach taken here is to define a model powerful enough to be capable of encapsulating the inherent structure of the lane marking, including its regular mark:space ratio and geometry. The model is then matched to the image and updated in a frame to frame manner using a least squares technique to overcome any problems of noise and to increase robustness.

The theory of the non-linear least squares process is described next, followed by a discussion of the model used and the results obtained using this technique.

2 Theory of Non-Linear Least Squares

A two-dimensional data set $I_{(x,z)}$ may be modelled using the function $S_{(x,z)}(\underline{a})$, where $\underline{a}$ is a set of parameters $a_0 \ldots a_{n-1}$. This n parameter model need not give an exact reconstruction of the original data set but, in general, it should be true that for all points:

$$S_{(x,z)}(\underline{a}) - I_{(x,z)} \approx 0 \tag{1}$$

If we have an approximation $\underline{a}^k$ to the required vector of parameters $\underline{a}$, then we obtain a better estimate $\underline{a}^{k+1}$ using:

$$\underline{a}^{k+1} = \underline{a}^k + \delta\underline{a} \tag{2}$$

To determine the $\delta\underline{a}$ increment in general requires solving a set of non-linear equations. This is avoided by expanding $S_{(x,z)}(\underline{a})$ as a Taylor series about $\underline{a}$ and neglecting all terms which are not linear in δa_i. Hence $\delta\underline{a}$ can be found from:

$$\underline{A}.\delta\underline{a} = \underline{b} \tag{3}$$

where the normal matrix $\underline{A}$ is assumed non-singular and defined by:

$$\underline{A}_{ij} = \sum_{X,Z} \left(\frac{\partial S_{(x,z)}}{\partial a_i} . \frac{\partial S_{(x,z)}}{\partial a_j} \right) \tag{4}$$

and

$$\underline{b}_i = \sum_{X,Z} \left(\frac{\partial S_{(x,z)}}{\partial a_i} [I_{(x,z)} - S_{(x,z)}(\underline{a})] \right) \tag{5}$$

Now that $\underline{a}^{k+1}$ has been calculated, it is refined iteratively using equation (2), the process being repeated until convergence occurs with the desired values $\underline{a}$. This algorithm, known as the Gauss-Newton iterative technique, is a generalised version of the one-dimensional Newton-Raphson. It is suitable for any system in which the model $S_{(x,z)}(\underline{a})$ is smooth and continuously differentiable and for which it is possible to obtain an initial guess 'sufficiently close' to the solution $\underline{a}$ [4].

At convergence, a quantitative measure of the accuracy of the fit over m observations can be obtained from the standard deviations of the n parameters. These are given by the following formula, involving the leading diagonal of the inverse of the normal matrix $\underline{A}$:

$$\sigma^2(a_i) = \underline{A}_{ii}^{-1} \frac{\sum [I_{(x,z)} - S_{(x,z)}(\underline{a})]^2}{(m-n)} \tag{6}$$

3 The Model

Although the number of parameters in our model is large enough to accurately model the marking uniquely, it was necessary to keep the number down to reduce processing time. For this reason the model is of a binary, overhead view of the road as shown in Figure 3, so that perspective and grey level effects can be ignored.

Another factor in choosing our model is that, as mentioned above, the function should be differentiable and this prevents the use of step-edge functions.

The model is split into two parts each having three parameters. The first defines the spatial location and curvature of the marking on the road in the form of a quadratic term similar to that used in [5] to model a road edge. The shape of the lane marking in cross-section, as shown in Figure 1, is approximated

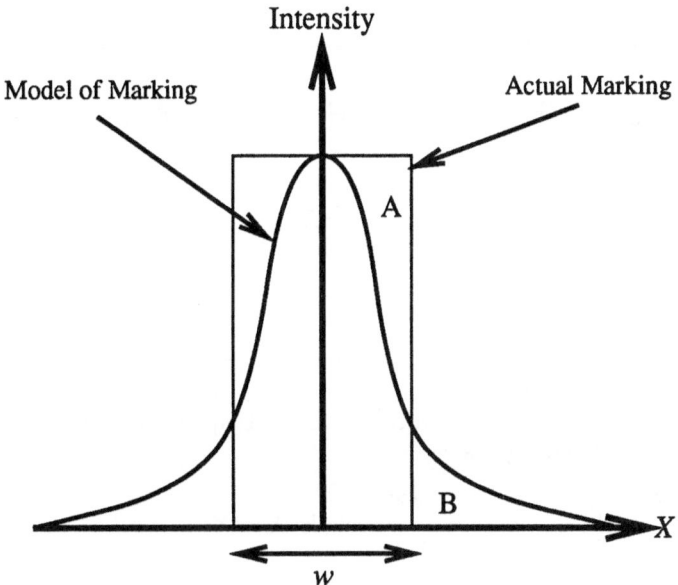

Figure 1: Plot of sech$(K_s X)$, which Models Cross-Section of a Lane Marking

by a sech() term. The long 'tail' on this function will be shown later to be necessary during convergence. The second term models the size and spacing of the lane markings as shown in Figure 2. In plan view (co-ordinates X, Z) a plot of lane marking intensity vs. distance from the camera may be considered as a square-wave having a period $\mathcal{P}$, duty cycle d and phase shift θ.

An example plot of the model is shown in Figure 3 and the model itself is given by:

$$S_{(x,z)}(\underline{a}) = C_{(x,z)}(a_0, a_1, a_2)T_{(z)}(a_3, a_4, a_5) \tag{7}$$

where

$$C_{(x,z)}(a_0, a_1, a_2) = \text{sech}\left(K_s\left[X - a_0 - a_1 Z - a_2 Z^2\right]\right) \tag{8}$$

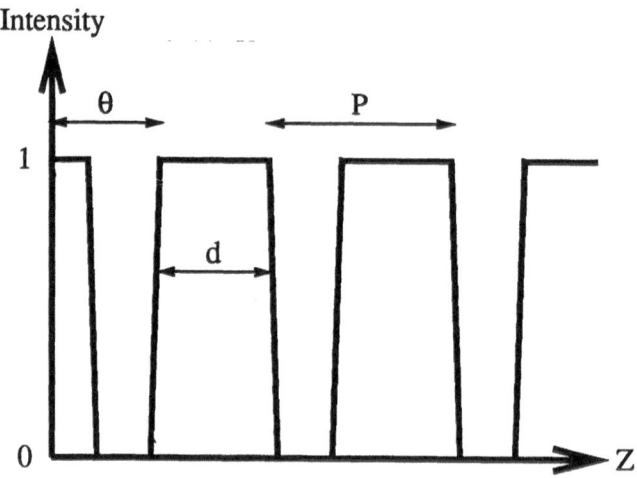

Figure 2: Square Wave Model: Plot of $T_{(z)}(\theta, d, \mathcal{P})$

and

$$T_{(z)}(\theta, d, \mathcal{P}) = \tfrac{1}{2}\tanh\left(K_t\left[\operatorname{asin}(\cos(\frac{2\pi Z}{\mathcal{P}} - \theta - \pi d)) + \pi(d - \tfrac{1}{2})\right]\right) + \tfrac{1}{2} \quad (9)$$

Note that although we use parameters $a_0 \ldots a_5$ it is more meaningful if we use $a_3 \equiv \theta$, $a_4 \equiv d$ and $a_5 \equiv \mathcal{P}$ in equation (9).

The constant K_s in equation (8) relates to the width of a lane marking. We require that the errors incurred by using the $\operatorname{sech}(K_s X)$ term to model the lane marking be the same inside and out, i.e. Area A equals Area B in Figure 1. Equating these two areas gives:

$$\frac{w}{2} - \int_0^{w/2} \operatorname{sech}(K_s X)\, dX = \int_{w/2}^{\infty} \operatorname{sech}(K_s X)\, dX \quad (10)$$

where w is the width of the lane markings in the plan view. This leads to the result: $K_s = \frac{\pi}{w}$. For the images shown here $K_s = 1$ has proven to be satisfactory.

Equation (9) models the square wave shown in Figure 2. The tanh() term models a mark/space boundary in a continuous manner and the asin(cos()) term makes these boundaries periodic. K_t is the constant which changes the slope of the edge of the square-wave so as $K_t \to \infty$ the model more closely resembles a true square-wave. However, if this parameter is too large then the continuous differentiability assumption is violated. A trade-off is required and in current versions of the algorithm $K_t \approx 50$.

4 Processing

The steps required in pre-processing the image into a form suitable for fitting with the model are described here.

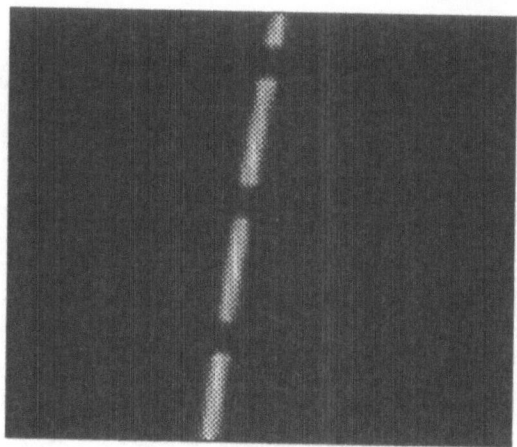

Figure 3: Model of the road scene $S_{(x,z)}(\underline{a})$.

4.1   Plan Transform

Our model of lane markings assumes that the mark:space ratio is constant throughout the image. This is only true when considering the road in plan view since perspective effects are removed [6]. An image (x, y) may be transformed to a 'birds eye' view (X, Z) with knowledge of the camera geometry using:

$$Z = \frac{h(F_y + y\tan\alpha)}{F_y\tan\alpha - y} \qquad X = \frac{(h\sin\alpha + Z\cos\alpha)x}{F_x} \qquad (11)$$

where h is the height of the camera above the road, F_y and F_x are the focal lengths of the camera lens and α is the tilt angle of the camera. This assumes that the tilt angle α remains constant (or is known) for each frame, and that the scene being transformed is locally flat.

4.2   Edge Detection

To fit our model to the image it is necessary that the image be in a similar form i.e. pixel values in $[0, 1]$, where a value of 0 represents road pixels and a value of 1 represents lane markings. Simple thresholding techniques prove too unreliable to achieve this, especially in areas of shadow or where the lane marking becomes obscured by tyre marks or surface dirt. Therefore the image is convolved with an edge detector and the values obtained appropriately scaled so that a typical marking/road boundary has the value 1 and others have smaller values. This method copes well with most shadows since the *magnitude* of edges are similar in shadow or out. The algorithm has worked satisfactorily using many different edge detectors including Canny and Sobel.

A morphological dilate/erode algorithm is now applied so that any gaps between edges, or those caused by noise, are removed. Figure 5 shows the plan-view transformed, pre-processed result of Figure 4.

Figure 4: A Typical Road Scene.

4.3 Solution

A bootstrap algorithm as described in Section 1 obtained initial values for the model at the beginning of every sequence. Then for each frame to be processed the previous frame provided an initial guess of the six parameters in $\underline{a}$. In general, however, this guess was not good enough to give reliable, fast convergence. Any local minima between the initial guess and the global minimum in the 6-dimensional search space lead to an incorrect solution being generated. In practice the approach of updating all six parameters at once failed for this reason. Therefore, as stated in section 2, it is vital that our initial guess be improved before this approach is taken.

In our six parameter model the only parameter almost certain to change from frame to frame is θ, and so it is necessary to estimate this as accurately as possible before any attempt at minimising the other parameters is made. This is achieved by keeping all the parameters of our initial guess, $\underline{a}^k$ say, constant expect for θ and minimising:

$$\sum_{X,Z} \left[I_{(x,z)} - S_{(x,z)}(\underline{a}^k(\theta)) \right]^2 \tag{12}$$

as a function of θ. This new value is used for the parameter a_3 in our refined guess $\underline{a}^{k+1}$.

Now that we have a set of parameters very close to the correct answer, we can update all six parameters using the Gauss-Newton technique to obtain our final answer. By taking advantage of knowledge of the form of the function $S_{(x,z)}(\underline{a})$, convergence was achieved reliably without the need to resort to more

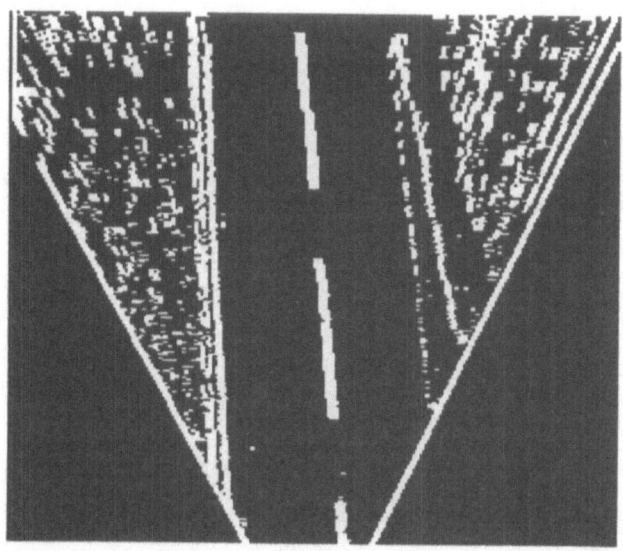

Figure 5: Plan-View Transformed Road Scene after Pre-Processing

computationally expensive methods such as damping [7] or filtering using a spectral analysis of $\underline{A}$.

5 Results

The results of the above techniques are described here.

Figure 6 shows an initial guess made for the road scene and Figure 7 shows the final answer after non-linear least squares fitting. The result obtained is very accurate despite the poor initial guess.

Figure 8 shows the convergence of the model for this image. The shape of this convergence curve and the large number of iterations required is typical for an image with such a poor initial guess. However, it is usual to have a better guess than this, resulting in more rapid convergence, often in ~ 15 iterations. The fact that convergence occurs despite a poor initial guess is due to the tail on the sech() function in equation 8. The value and standard deviation (equation (6)) for each of the six parameters at convergence is shown in Figure 9. The low value obtained for each of the standard deviations shows that the parameters are all well determined, hence justifying the model used.

The algorithm has been run successfully on many different sequences, each of which is hundreds of images long. The only cause of mis-tracking has been found to be if a marking becomes occluded by another vehicle. This situation is readily identifiable from the standard deviations of the parameters, which become very large in relation to the parameters themselves.

One unexpected problem was that any region away from the lane markings that is completely devoid of edges (i.e. black in Figure 5) satisfies the model

Figure 6: Initial Guess

$S_{(x,z)}(\underline{a})$ exactly if $d = 0$. This meant that during convergence after a poor initial guess, $d \to 0$ and the position of the marking was not found correctly. This was solved by making the substitution:

$$d \equiv K_d \sin(d') + K_b + \tfrac{1}{2} \tag{13}$$

in equation (9). For our purposes $K_d = 0.45$ and $K_b = 0.05$ so that d was constrained to lie within the interval $[0.1, 1.0]$.

6 Conclusions

An algorithm has been described which is suitable for use by an autonomous road vehicle to find lane markings in a scene reliably and with high accuracy. It uses a six parameter model to represent the portion of the image around the markings and a non-linear least squares technique is applied to refine these parameters to sub-pixel accuracy.

The algorithm has been tested on many sequences and has coped well with changes in luminance, sharp bends and changes in type of lane marking. It differs from other approaches which are used to guide autonomous vehicles (e.g. [8]) in that the completeness of the model ensures a high degree of confidence in the tracking result. If mis-tracking does occur then it is readily identified using the standard deviations of the parameters.

Further work currently under investigation includes improving the initial guess $\underline{a}^k$ for the current frame by applying a predictive filter to previous values. The method of finding a minimum in a six dimensional search space using the Gauss-Newton method described here is robust but relatively computationally

Figure 7: Final Result

expensive taking around 10 sec. per image on a single transputer. Other, faster techniques to achieve the same result must be investigated if this algorithm is to run in real-time on a small number of processors.

Thanks

We gratefully acknowledge the contribution made by Dr.B.R. Stonebridge to this work.

References

[1] A.D. Morgan, E.L. Dagless, D.J. Milford, and B.T. Thomas. Road edge tracking for robot road following: a real-time implementation. In *Image and Vision Computing*, pages 233–240, 1990.

[2] A. Grunes and J.F. Sherlock. Texture segmentation for defining driveable regions. In *Proceedings of British Machine Vision Conference*, pages 235–239, 1990.

[3] L. Schaaser and B.T. Thomas. Finding road lane boundaries for vision-guided vehicle navigation. In Ichiro Masaki, editor, *Vision Based Vehicle Guidance*, pages 238–254, Springer-Verlag, 1992.

[4] Carl-Erik Fröberg. *Numerical Mathematics : Theory and Computer Applications*, chapter 11. Benjamin/Cummings, 1985.

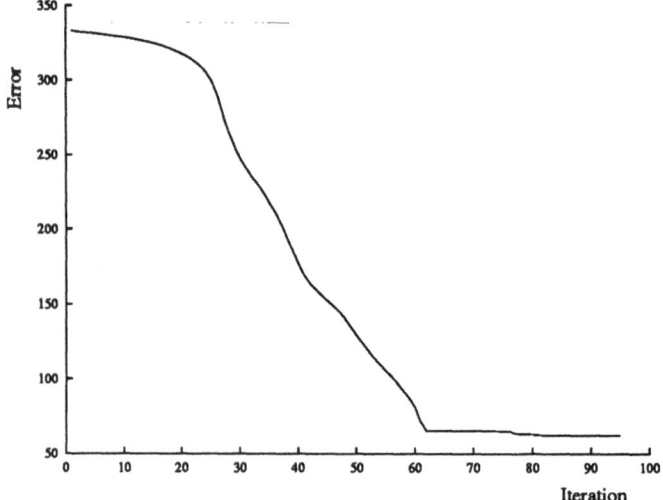

Figure 8: Convergence for One Image

Parameter	Value	$\pm \sigma$
a_0	160.110	$\pm$ 0.1125
a_1	0.0129	$\pm$ 0.0018
a_2	-0.000222	$\pm$ 1e-6
a_3 (i.e. θ)	3.01966	$\pm$ 0.0304
a_4 (i.e. d')	0.139263	$\pm$ 0.0144
a_5 (i.e. $\mathcal{P}$)	90.701	$\pm$ 0.2462

Figure 9: Average Value and Standard Deviation of Parameters

[5] A.D. Morgan. *Application of computer vision techniques to the visual navigation of a road following vehicle.* PhD thesis, University of Bristol, 1991.

[6] R.A. Lotufo, B.T. Thomas, and E.L. Dagless. Road following algorithm using a panned plan-view transformation. In *Proceedings 4th Alvey Vision Conference*, University of Manchester, 1988.

[7] B.R. Stonebridge and R.L. Soulsby. *Damped Nonlinear Least-Squares Computation of a Model for Scouring of the Seabed around a Vertical Cylinder.* Technical Report TR-91-34, University of Bristol, Department of Computer Science, December 1991.

[8] E. D. Dickmanns and B. D. Mysliwetz. Recursive 3-d road and relative ego-state recognition. In *Transactions on Pattern Analysis and Machine Intelligence*, pages 199–213, 1992.

Indexing Two-Dimensional Objects Using Parametrised Geometric Features

C. C. Hand
I.T. Research Institute, Brighton Polytechnic
Brighton, U.K.

Abstract

A new approach to indexing in model-based vision is introduced; using indexing features whose deformed instances are present in a number of different object models. Hypotheses are generated by matching an indexing feature with an image using a viewing transformation incorporating the deformation; the parameter values of the deformational component provide specific indices to those models related to the indexing feature.

1 Introduction

Given a model-based vision (MBV) system which has a number of representations of different objects, how can it access the most appropriate model to match with a scene given that there is no *a priori* knowledge of the objects within it? This is the problem of indexing. One solution is to use indexing features, subsets of model descriptions which are computationally less expensive to match with an image, to generate hypotheses. Two types of indexing feature have been used; 'critical' features - features unique to specific objects [3, 4, 5], and 'similar' features - features common to a number of objects [6, 7]. This paper advocates using deformable indexing features whereby deformed instances are present in a number of quite different object models. As with similar features this has the advantage of representational parsimony. Furthermore, matching a feature with an image using a viewing transformation which has a deformational component can provide different indices for each model related to that feature via the specific parameter values of the deformation. This retains the specificity of the critical features approach.

The structure of this paper is as follows. The relationship between critical and similar features is first clarified and then extended to indexing via the structural deformations of parametrised features using anisotropic scaling. A specific implementation in the domain of 2-D, wire-frame, drawings is then discussed and a representative example is given which demonstrates the system's good performance.

2 Indexing by Critical and Similar Features

Traditionally, object recognition has been regarded as finding the parameters of some viewing transformation between some known object and an image description e.g. [1, 2].

Let I be an image description and M be an object model description, both being comprised of a set of geometrical primitives i.e. $I = \{pi_1, pi_2, ..., pi_r\}$, and $M = \{pm_1, pm_2, ..., pm_s\}$ where pi_j and pm_k are the jth and kth image and model primitives respectively. Object recognition is then characterised by the relation:

$$M \times I \supseteq C = \{(pi_j, pm_k) : T(pi_j) \approx pm_k\} \qquad (1)$$

C is a correspondence between M and I; it is a non-empty subset of their Cartesian Product. T is a geometrical transformation which superimposes a model primitive onto an image primitive. Thus, C is a relation between the two representations such that elements from M and I are related only if they are *geometrically compatible* with each other under some common viewing transformation. The parameters of T must be similar for every element of C so that each pair mutually supports and constrains the solution spaces of every other pair. Though suitable for the 'bin-of-parts' scenario it presents problems when there is a database of models $D = \{M_1, M_2,...\}$. If there is no *a priori* knowledge of what object is present in a scene then one must exhaustively search across D; looking for the model which gives the best measure of fit. This is a major problem for MBV.

Two different types of model-driven indexing schemes have been suggested to reduce this search space. Both use indexing features which are related to particular models but which are simpler structures than the models themselves. They are generally used in hypothesis-verification schemes where a good match with a feature constitutes a good hypothesis which must be verified by a full model match. Such a feature match is less computationally expensive to match with the image than the full model description. The cost of hypothesis verification will also be reduced due to the information provided by the parameters of the viewing transformation of the hypothesis.

The first uses critically distinguishing features for each different object class i.e. there is a subset of each model description which is unique to it e.g. [3, 4, 5]. This gives a set of indexing features $F = \{f_1, f_2,...\}$ where $M_k \supseteq f_k$ and $\forall M_i \in D, i \neq k, M_i \cap T(f_k) = \emptyset$ and T is usually a rigid or a similarity transformation.

The second method uses 'similar' indexing features; features common to a number of object classes e.g. [6, 7]. Each similar feature is a subset of a number of different models; it represents the shared structure between them. Thus a similar feature is a correspondence, as defined in (1) above, but between two or more models i.e.

$$M_i \times M_m \supseteq f_k = \{(pm_{ij}, pm_{mn}) : pm_{ij} \in M_i, pm_{mn} \in M_m, T(pm_{ij}) \approx pm_{mn}\} \qquad (2)$$

where pm_{mn} is the nth primitive of model M_m. Again T acts as a normalisation process. This can be extended to more than two models by recursively applying (2) but with the correspondence being expressed over the indexing feature and some other model. (The only restriction on this process is that the resulting description of the indexing feature should carry enough information such that it can be detected in some reasonably complex image.) A good match with a similar feature indicates a set of possible model hypotheses. Each hypothesis must then be tested by a full model match with each model associated with that feature; though each match will be highly constrained by the parameters of the viewing transformation for the indexing feature.

Both schemes have the disadvantage of exhaustive search across the set of indexing features during hypothesis generation. Though there will be fewer 'similar' indexing features than 'critical' features (due to the sharing of features between different models) this is at the cost of exhaustive search during hypothesis verification. Therefore, one must restrict the number of models that each similar feature indexes; one must find some balance between the generality and uniqueness of a feature [6, 7]. Furthermore, a system must make active use of both the similarities and differences holding between objects. Similar features model similarities for indexing but 'passively' use differences during hypothesis verification and vice-versa for critical features. However, it would be helpful to make use of structural differences and similarities during indexing simultaneously.

3 Indexing via Structural Deformations

In 2D MBV it is generally assumed that the viewing transformation between two representations is either a rigid or a similarity transformation. However, one can use a 'non-similarity' transformation, such as anisotropic scaling or shear, as part of the viewing transformation between two representations. Features can be formed as in (2) above but with a deformational component as part of T. Thus the correspondence between models yields an indexing feature which can map onto different phenomenal forms via the different values of the deformational parameters of T i.e. an indexing feature can be shared across quite different object classes. This 'canonical feature' will be related to each deformed model instance and the index from the feature to each indexed model is given by the parameter values of the deformation. Thus, like 'similar' features, the same indexing feature could index a number of different models. However, the parameter values of the non-similarity part of T can be used to provide critically distinguishing indices. In general there is no need to exhaustively search all of the models associated with an indexing feature. In those cases where the phenomenally same feature exists in more than one model the indexing feature behaves as if it were a similar indexing feature type, but only for those models sharing the same parameter values for the deformation. Thus the similar indexing feature approach is a special case of indexing via structural deformations.

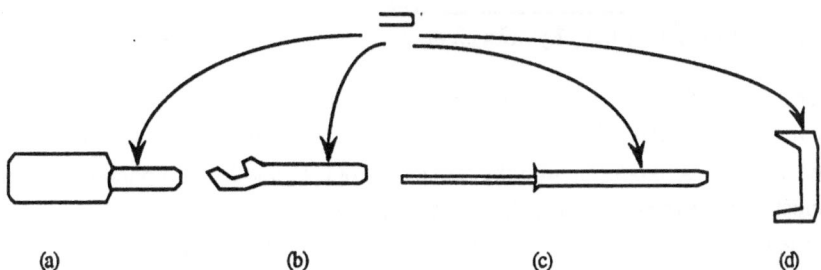

(a)   (b)   (c)   (d)

Figure 1: Deformed instances of the canonical indexing feature 'handle' appear in the object models (a) to (d) which are the 'bottle', 'c-spanner', 'screwdriver', and 'bracket' respectively. The specific parameter values of the four different deformations of the 'handle' provide the indices from that feature to the four models.

Consider the example of the different shapes in Figure 1; there are four different models and a feature. Each model contains an anisotropically scaled instance of the 'handle' - the five line segments furthest to the right in each model. Thus the 'handle' can be thought of as a stretchable feature. Each instance of the 'handle' within the individual models is different, both with respect to the canonical indexing feature and from each other. However, each instance is related to every other as any instance can map onto any other. Thus the 'handle' represents the shared structure between the different object classes whilst the different phenomenal instances resulting from the different parameters used in the deformation represents the differences between the shared structures. This satisfies the condition that an indexing feature simultaneously reflects both the structural similarities and differences existing between forms.

4 Anisotropic Scaling

The 'non-similarity' transformation considered here is that of anisotropic scaling in the context of 2-D wire-frame drawings. The primitives are directed line segments, $\overrightarrow{(x_1,y_1)(x_2,y_2)}$. Let the transformation $T_{\alpha\beta}$ be the mapping:

$T_{\alpha\beta}$: $\Re^2 \to \Re^2$ such that

$$T_{\alpha\beta}(\overrightarrow{(x_1,y_1)(x_2,y_2)}) = (\overrightarrow{(\alpha x_1,\beta y_1)(\alpha x_2,\beta y_2)}) \qquad\qquad \alpha, \beta \in \Re \qquad (3)$$

A transformed feature is simply the set of all directed line segments comprising the structural description of the indexing feature i.e. $f_i = \left\{ (\overrightarrow{(\alpha x_1,\beta y_1)(\alpha x_2,\beta y_2)})_k \right\}$.

Consider the special case $T_{\alpha\beta}$ where $\alpha = \beta$. Here anisotropic scaling mimics the similarity transformation of uniform scaling. To avoid confusion it is best to define this transformation separately:

$T_\lambda = T_{\lambda\lambda}$: $\Re^2 \to \Re^2$ such that

$$T_\lambda(\overrightarrow{(x_1,y_1)(x_2,y_2)}) = (\overrightarrow{(\lambda x_1,\lambda y_1)(\lambda x_2,\lambda y_2)}) \qquad\qquad \lambda \in \Re \qquad (4)$$

Both $T_{\alpha\beta}$ and T_λ can map into the same parameter space P as T_λ is merely a special case of $T_{\alpha\beta}$. Let a feature be deformed according to (3) and then two different uniformly scaled versions be produced with the scaling factors, λ_1 and λ_2. The parameters of the composite transformations give the points $(\lambda_1\alpha,\lambda_1\beta)$ and $(\lambda_2\alpha,\lambda_2\beta)$ in P. There is an equivalence between the two forms as defined by the ratio of the scale factors:

$$\frac{\lambda_1\beta}{\lambda_1\alpha} = \frac{\beta}{\alpha} = \frac{\lambda_2\beta}{\lambda_2\alpha} \qquad (5)$$

Formally, let ~ be an equivalence relation on the set P. We define the equivalence class $\overline{p_1}$ of an element $p_1 \in P$ to be the subset

$$\overline{p_1} = \left\{ p_2 \in P: p_1 \sim p_2 \right\} \qquad (6)$$

where $p_1 = (\alpha_1,\beta_1)$, $p_2 = (\alpha_2,\beta_2)$, α_1 and $\alpha_2 \neq 0$, and ~ is defined as the ratio in (5).

Thus all uniformly scaled instances of some deformed structure form an equivalence class and one can choose a particular instance to act as the class representative. The equivalence class for each deformed instance of a feature forms a line through the origin and the parameter values of the class representative in P. The equivalence classes of the models in Figure 1 are shown in Figure 2.

To index one finds the distance between the point in P defined by the correspondence between an indexing feature and an image, and each point in P defined by the correspondence between the indexing feature and the class representative of each deformed instance of that feature present in some model. This gives a partial ordering of the equivalence classes and, consequently, the models. One can also use a threshold, ε, such that hypothesis verification is conducted only when $d(p_1,p_2) \leq \varepsilon$. This reduces unwarranted verification when the parameters of the transformation between an indexing feature and an image are far from those given by the matches between that indexing feature and the subset of models to which it is related. One metric for P, which uses (6), is the absolute value of the angular difference between two points:

$$d(p_1,p_2) = \left| \cos^{-1}\left(\frac{p_1{}^t p_2}{\|p_1\| \, \|p_2\|} \right) \right| \qquad (7)$$

Where $p_1, p_2 \in P$ and $p_1{}^t p_2$ is the Euclidean inner product. (7) gives the distance around the unit circle and is multi-valued which complicates the metric, but it is well known that:

$$\begin{bmatrix} S_x & 0 \\ 0 & S_y \end{bmatrix}\begin{bmatrix} \cos\pi & -\sin\pi \\ \sin\pi & \cos\pi \end{bmatrix} = \begin{bmatrix} -S_x & 0 \\ 0 & -S_y \end{bmatrix} \text{ and } \begin{bmatrix} -S_x & 0 \\ 0 & S_y \end{bmatrix}\begin{bmatrix} \cos\pi & -\sin\pi \\ \sin\pi & \cos\pi \end{bmatrix} = \begin{bmatrix} S_x & 0 \\ 0 & -S_y \end{bmatrix} \qquad (8)$$

Thus one need only consider the upper half of the plane specified by P to account for all phenomenally different anisotropically scaled instances. This restricts the range of values to $[0, \pi]$ thus avoiding the problem. Also, if an indexing feature is mirror symmetrical, as it is with the 'handle', then only the positive quadrant of P need be considered.

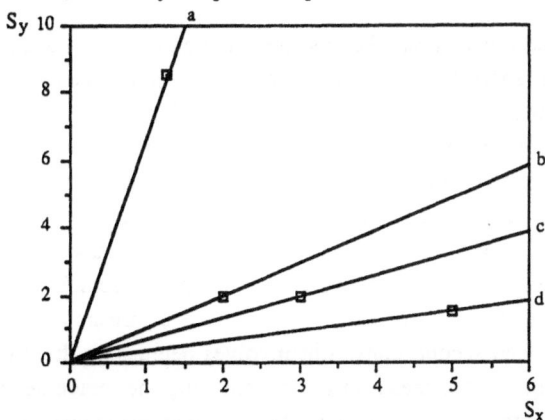

□   Parameter mapping of the class representative for each model equivalence class

Figure 2: The equivalence classes for the four models related to the indexing feature 'handle'. The parameters for the class representatives are (a) 'bracket' (1.25;8.5), (b) 'bottle' (2,2), (c) 'c-spanner' (3,2), and (d) 'screwdriver' (5,1.5).

Anisotropic scaling also deals with reflections. This is a desirable property in certain contexts e.g. the inspection of 2-D machine components which could lie on either side. Separate representations of the indexing feature are not required. Instead the indexing feature will have two indices into the same model, (S_x, S_y) and $(-S_x, S_y)$.

Finally, although the viewing transformation above deals solely with anisotropic scaling, the current work uses a 2-D viewing transformation specified by five parameters; two for scale (S_x, S_y), one for rotation(ψ), and two for translation (T_x, T_y):

$$\begin{bmatrix} x' \\ y' \end{bmatrix} = \begin{bmatrix} \cos\psi & -\sin\psi \\ \sin\psi & \cos\psi \end{bmatrix} \begin{bmatrix} S_x & 0 \\ 0 & S_y \end{bmatrix} \begin{bmatrix} x \\ y \end{bmatrix} + \begin{bmatrix} T_x \\ T_y \end{bmatrix} \tag{9}$$

This presents no problem for the current implementation as the length of a line segment, and hence the scale of an object, is unaffected by rotation and translation.

5 Implementation Details and Performance

The current implementation uses a model-driven hypothesis-verification cycle. It attempts to determine a correspondence between an indexing feature and an image and, if one exists, the scale parameters of the transformation are compared with those of the models associated with the indexing feature as described above. (The current threshold used for (7) is set at 3°.) Each model indexed must then be verified by a full model match. If verified those image primitives forming the correspondence with the *model* primitives are then removed from the image description. This continues recursively until the image is too small to be matched further or until all of the indexing features have been considered.

Because the system is concerned with the parameter values of viewing transformations a natural choice of matching algorithm for both hypothesis generation and verification is the Generalised Hough Transform (GHT) [1, 2]. However, the GHT makes exponential demands upon memory with respect to the number of parameters of the viewing transformation. Two techniques are used to overcome this problem:

1. Parameter space decomposition - Because there is a natural dominance of parameters the 5-D parameter space for the viewing transformation can be decomposed into a linear hierarchy of three sub-spaces; a 2-D scale, 1-D rotation, and 2-D translation parameter space in that order [1, 2]. Possible scale parameters for the viewing transformation are determined and then passed on to determine the orientation parameter and so on. Because the constraining information of the primitive measurements are de-coupled a large number of false local maxima are generated in the scale parameter space. Many of these are eventually disconfirmed by using two simple constraints. The first constraint is to place a threshold on the metric as suggested above. The second constraint involves decreasing the number of false maxima in *subsequent* parameter spaces by reducing the image description. Only those image line segments consistent with at least one indexing feature (or model) line segment under the specific parameter values found in previous parameter spaces are used in determining the parameters in the subsequent parameter spaces.

2. Coarse-fine search within parameter spaces - Hierarchical coarse-fine search is also conducted within each parameter space; this is loosely based upon the Adaptive Hough Transform [8]. It involves using a coarsely quantised accumulator array, incrementing the array according to the normal GHT procedure, and clustering any cells whose count passes a threshold. The parameter values of the *bounds* of each cluster are then dynamically mapped onto the coarse accumulator array and the process continues recursively. The process terminates when the parameter values mapped onto the bounds of the accumulator array reach some pre-defined minimal resolution. The centroids of the clusters of cells found at minimal resolution are taken as possible solutions for the transformation defined by that parameter space.

The current number of intervals used for each parameter is 15. This results in 225 cells for the accumulator array of the scale and translation parameter spaces and only 15 cells for the rotation space, as compared to 15^5 for a single 5-D parameter space.

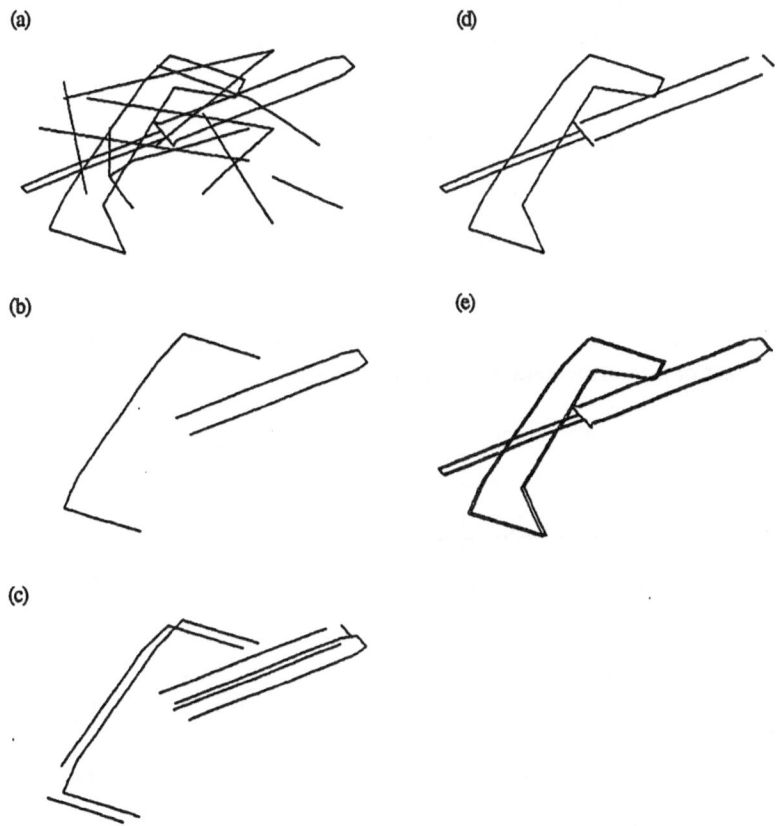

Figure 3: Matching an image containing two different object types.
Hypothesis Generation: (a) The image. (b) The indexing feature hypotheses. (c) The hypothesised instances superimposed upon the reduced image description. Hypothesis Verification: (d) The reduced image description after both hypothesised model matches. (e) The hypothesised models superimposed on the reduced image description of (d).

A typical example of the system at work is given in Figure 3. The 'screwdriver' and 'bracket' were transformed according to the parameters given in Table 1. Noise segments were added giving a total number of 35 line segments, resulting in a signal/noise ratio of 0.1666 for each instance of the indexing feature. (The noise was generated by finding the minimal enclosing rectangle about the objects of interest, and using a uniformly distributed random number generator to set the endpoints of a segment to fall within it).

Table 1: The parameter values of the model transformations used in generating the image.

Model	Scale	Ψ	T_x	T_y
Screwdriver	1.1	30°	20	10
Bracket	1.9	155°	18	9

Table 2 gives the number of hypotheses generated by matching the 'handle' with the image for each parameter space in the hierarchy of sub-spaces. Note the power of the threshold constraint in reducing the number of scale hypotheses and the general reduction in hypotheses as one traverses the hierarchy due to both the constraining nature of the successive transformations and the use of image reduction.

Table 2: The results of matching the indexing feature with the image.

No. Scale Hypotheses	Reduced No. Scale Hypotheses	Rotation Hypotheses	Translation Hypotheses
37	15	3	2

The two translation hypotheses represent the number of correspondences between the image and the 'handle' once scale, orientation, and translation are taken into account. The parameter values of these hypotheses are given in Table 3.

Table 3: The viewing transformation parameter values of the two hypotheses

	S_x	S_y	Ψ	T_x	T_y
First Hypothesis	5.50131	1.63356	30.0267°	20.6667	9.33320
Second Hypothesis	2.38418	16.17490	155.04°	18.6667	9.33332

The first hypothesis indexes the 'screwdriver' whilst the second indexes the 'bracket'. These are the required hypotheses; no others have been made. The distance between the scale parameters and the parameters of the appropriate equivalence classes are 0.16° and 0.13° respectively and fall well within the threshold. Each hypothesis is verified by a match with the appropriate model and the parameters are given in Table 4 (c.f Table 1).

Table 4: The viewing transformation parameter values of the two verified hypotheses

	S_x	S_y	Ψ	T_x	T_y
Screwdriver Hypothesis Verification	1.09927	1.10163	29.9982°	19.9316	10.05130
Bracket Hypothesis Verification	1.90118	1.90018	154.997°	18.1333	9.06666

Several hundred experiments using the database of Figure 1 have been conducted using images containing from one to three objects with varying degrees of noise. The pattern of the results given above is typical of the system's performance. It fails to recognise an object approximately 8% of the time but most of these failures can be avoided by modifications to the algorithms. Furthermore, when the system generates more hypotheses than instances actually present this is often due to slightly different scale parameters. These are still consistent under rotation and translation due to error introduced by quantisation of the accumulator arrays; they effectively represent the same index and this is easily resolved by verification of any one of the similar indexing hypotheses.

6 Conclusions

A new approach to indexing has been presented; indexing by the parameter values of deformable features. This unifies the advantages of both critical and similar indexing features. Like those schemes it is model-driven and thus necessitates exhaustive search at the level of the indexing features. However, unlike those schemes, it does not necessarily entail exhaustive search of those models related to a particular indexing feature. As with similar features a number of different objects can share the same indexing feature; different models can be associated with one another if a deformed instance of an indexing feature is present within them. Furthermore, like critical features, the system gives highly specific indices for each model related to an indexing feature via the parameter values of the deformational component of the viewing transformation.

Finally, further work on the current system includes a larger database, a more detailed analysis of the effect of noise on the system's performance, and an investigation of the effect of error on the image primitives themselves. The latter is of special interest as currently the images are generated by transforming the models and adding noise. Related to this last point is that systems requiring the accurate segmentation of line segments, such as this paper and e.g. [1, 2], may prove to be too 'brittle'. Therefore, some theoretical work has been conducted using collections of vertices for indexing features [9]. The segmentation of vertices may prove to be more reliable than that of line segments. [9] shows that one can recover the ratio of the anisotropic scale factors from the angles of vertices thus satisfying (5) and (6) above. This will be the subject of a future study.

References

[1] D.H. Ballard, and D. Sabbah, On Shapes. In Proc. 7th IJCAI, Morgan Kaufmann, Los Altos, 1981, pp 607-612.

[2] D.H. Ballard, and D. Sabbah, Viewer Independent Shape Recognition. PAMI 1983; 5(6): 653-660.

[3] R.C. Bolles, and R.A. Cain, Recognizing and locating partially visible objects: The local feature focus method. Int. J. of Robotics Research 1982; 1(3): 57-82.

[4] R.C. Bolles, P. Horaud, and M.J. Hannah, 3DPO: A Three-Dimensional Part Orientation System'. In Fischler M.A., and Firschein O, (Eds.) Readings in

Computer Vision: Issues, Problems, Principles, and Paradigms. Morgan Kaufmann, Los Altos , 1987, pp 355-359.

[5] F.P. Sykes, S.B. Pollard, and J.E.W. Mayhew, Hypothesis and Verification in 3-D Model Matching. In Proc. 5th Alvey Vision Conference, University of Sheffield, 1989, pp 7-12.

[6] T.F. Knoll, and R.C. Jain, Recognizing Partially Visible Objects Using Feature Indexed Hypotheses. IEEE J. of Robotics and Automation 1986, RA-2(1): 3-13.

[7] T.F. Knoll, and R.C. Jain, Learning to recognize objects using feature indexed hypotheses. In Proc. First Int. Conf. on Computer Vision, 1987, pp. 552-556.

[8] J. Illingworth, and J. Kittler, The Adaptive Hough Transform. PAMI 1987; 9(5): 690-698.

[9] C.C. Hand, Model Indexing Using Parametrised Features. Phd Thesis (submitted for examination), University of Sussex, 1992.

The Adaptive Bisector Method: Separating Slant and Tilt in Estimating Shape from Texture*

James V. Stone
Department of Computer Science,
University of Wales, Aberystwyth.
JANET: zra@uk.ac.aber

Abstract

Existing techniques for obtaining shape from texture estimate tilt and slant via a single computational mechanism. These techniques do not take advantage of the fact that tilt is relatively easy to estimate, and slant can be estimated more easily once tilt is known. This paper introduces the *adaptive bisector method*, which allows tilt and slant to be calculated separately, and by different computational mechanisms. The method makes minimal assumptions regarding the isotropy of surface textures.

Evidence from psychophysical studies suggests that human observers are able to provide accurate estimates of tilt[8], but are poor at estimating slant[4]. Whilst no psychophysical claims are made regarding the means by which tilt and slant are computed in this paper, it is claimed that a method which depends upon two different mechanisms provides a plausible functional model for how human observers obtain shape from texture.

Results for real and synthetic perspective images of textured planar surfaces are presented.

1 Introduction

Conventional methods for obtaining shape from texture[10, 1, 5, 2] use a single computational mechanism to estimate surface orientation. Such methods ignore constraints implicit in the underlying nature of the problem, namely, that tilt (the direction of surface slope) is relatively easy to compute on its own, and that slant (the amount of surface slope) can be computed more easily once the tilt is known.

The adaptive bisector method (ABM) is unique amongst shape from texture methods in addressing the problems of estimating slant and tilt separately. Evidence from psychophysical studies suggests that human observers are able to provide accurate estimates of tilt[8], but are poor at estimating slant[4]. Reasons for this dichotomy in the perception of slant and tilt are not hard to find. Whereas tilt can be computed without reference to the 3D structure of a surface, slant cannot. Once the tilt is known then slant becomes easier to estimate, as evidenced by the existence of psychophysical models [6, 11] which estimate slant only, and require that the tilt is known before slant can be

---

*This work was undertaken as part of a D.Phil. thesis in the Dept. of Experimental Psychology at the University of Sussex.

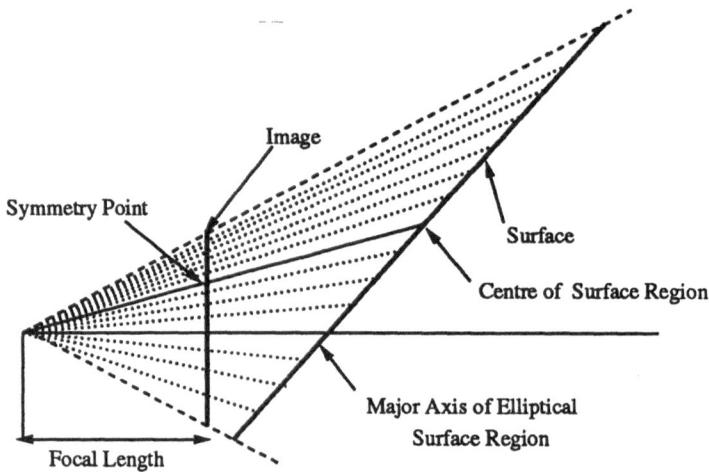

Figure 1: Back-projection of centre of elliptical surface region to the symmetry point in the image.

estimated. Computational results presented here support the implication from psychophysical studies that, whilst it is relatively easy to compute accurate estimates of tilt, estimating slant is both more expensive and less accurate.

As discussed in [7] tilt can be obtained either by finding the image direction in which texture density changes most rapidly, or via the image orientation in which texture density changes least. In practice finding tilt via either of these methods is problematic[10], and neither has been shown to work. Accordingly, we describe a new method for estimating tilt below. The point is that the slant of the surface need not be known in order to estimate tilt. Once the tilt has been established, slant can be estimated by making use of the rate of change of image texture density in the direction of tilt. Thus, the process of estimating surface orientation can be decomposed into two separate processes.

2   The Adaptive Bisector Method

2.1   Preliminary Definitions

The local orientation of a surface is specified by two parameters. Given 'world' coordinates (X, Y, Z) these parameters can be either the partial derivatives $P = -\partial Z/\partial X$, $Q = -\partial Z/\partial Y$, or the more intuitive descriptors, *slant*, $\sigma = tan^{-1}(P^2+Q^2)^{1/2}$, and *tilt*, $\tau = tan^{-1}(Q/P)$. Slant specifies the amount of slope, whereas tilt indicates the direction of slope apparent in an image.

Consider an elliptical region, $\overline{W_0}$, of a planar surface which projects to a circular image, w_0 (see Figures 1 and 2). The major axis of W_0 projects to an image line, the *tilt line*, which passes through the image origin at angle τ. The tilt line bisects w_0, and (by definition) the back-projection of the tilt line bisects the surface region, W_0.

Of the set of image lines orthogonal to the tilt line exactly one back-projects

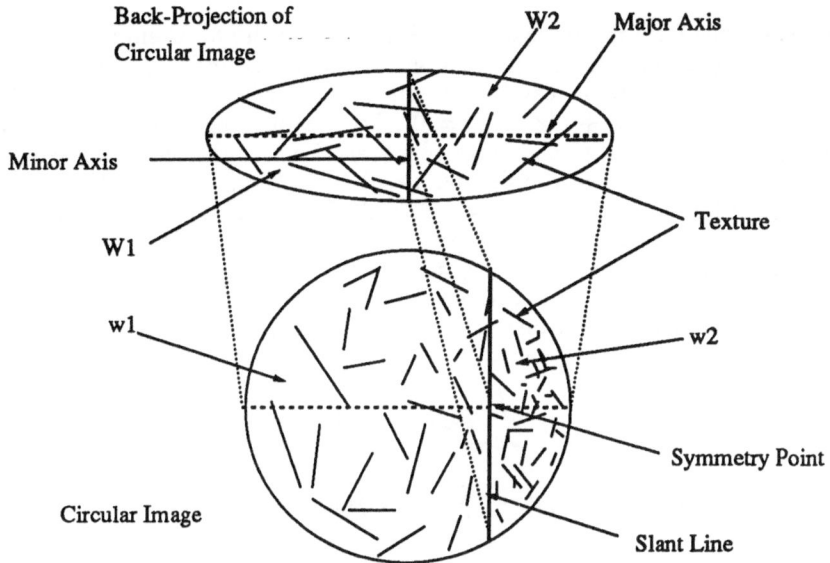

Figure 2: Back-projection of slant line from circular image to elliptical surface region.

to coincide with the minor axis of the surface ellipse. We define this image line as the *slant line*. Like the tilt line, the back-projection of the slant line bisects the surface region, W_0. The intersection of these two lines is the *symmetry point*, $p_C = (x_C, y_C)$. The symmetry point is the projection of the centre of the surface ellipse. Note that the image point p_C and the surface orientation are co-determined. Thus, if the location of p_C is known then it is possible to recover the surface orientation.

The remainder of this section describes how to estimate surface orientation by making use of the definitions given above. A summary of the method is as follows. First, the tilt line is estimated from the direction of the first moment of surface texture around the image origin. This provides an estimate of the tilt of the textured surface. Next, an estimate of the slant line is obtained by finding that image line which is orthogonal to the tilt line and which bisects the total amount of surface texture projecting into the image. The distance of the symmetry point, p_C, from the image origin can then be used to recover surface slant.

The back-projection of the slant line onto the surface divides the surface region, W_0, into two surface regions, W_1 and W_2, with areas, $A_1 = A_2$. These regions correspond to image regions, w_1 and w_2, respectively (see Figure 2). From the definition of the slant line it follows that:

$$A_1 - A_2 = 0 \tag{1}$$

If the amount of texture per unit surface area is constant then the amount of texture, L_1, in W_1 is the same as the amount of texture, L_2, in W_2, so that:

$$L_1 - L_2 = 0 \tag{2}$$

These equations form the basis of the ABM, but in order for them to be useful we need to re-write equations (1,2) in terms of image area and image line length.

The equation of a plane can be written in terms of the 'world' coordinates (X, Y, Z) as $AX + BY + CZ - D = 0$. For an imaging system with focal length, $f = 1$, and with the image plane at $Z = f$, this equation can be written in terms of image coordinates $(x, y) = (x/f, y/f) = (X/Z, Y/Z)$ as $Z = K/(1 + Px + Qy)$, where $P = -\partial Z/\partial X, Q = -\partial Z/\partial Y$, and $K = D/C$ determines the orthogonal distance from the plane to the origin. The term f is omitted hereafter so that $x = x/f, y = y/f$. By using functions, pdf and lcf, that map area and length, respectively, from surface to image we can re-write (1) and (2) in terms of image quantities and the surface orientation parameters, (P, Q):

$$\int_{w_1} pdf(x, y, P, Q, K) \, da - \int_{w_2} pdf(x, y, P, Q, K) \, da = 0 \qquad (3)$$

$$\sum_{l_i \in w_1} \int_{l_i} lcf(x, y, P, Q, \beta) \, ds - \sum_{l_j \in w_2} \int_{l_j} lcf(x, y, P, Q, \beta, K) \, ds = 0 \qquad (4)$$

Where β is the orientation of an image line, l, parameterised by arc-length, s.

The function pdf is a *point density function* that maps the differential of image area, da, to a differential of surface area, dA. The function pdf is defined in [10] as:

$$dA/da = pdf(P, Q, x, y, K) = K^2(P^2 + Q^2)^{1/2}(1 + Px + Qy)^{-3} \qquad (5)$$

The function lcf is a *line compression function* that maps the differential of image length, ds, to a differential of surface length, dS. The function lcf is defined in [10] as:

$$
\begin{aligned}
dS/ds = lcf(P, Q, x, y, K) &= (K/V)(Qx \, sin(\beta) - cos(\beta)(1 + Qy))^2 + \\
&\quad (Py \, cos(\beta) - sin(\beta)(1 + Px))^2 + \\
&\quad (P cos(\beta) + Q sin(\beta))^2)^{1/2}
\end{aligned}
\qquad (6)
$$

Where $V = (1 + Px + Qy)^2$.

For the present let us assume that $Q = 0$ so that the tilt, $\tau = tan^{-1}(Q/P) = 0$, and the slant $\sigma = tan^{-1}(P^2 + Q^2)^{1/2} = tan^{-1}(P)$. Each point, p_m, on the tilt line can be associated with a line which is orthogonal to the tilt line. Each of these lines defines two image regions, w_i and w_j. If a point, p_m, on the tilt line is the symmetry point then its position is associated with a value for the slant $(\sigma = tan^{-1}(P))$. Additionally, if p_m is the symmetry point then w_i and w_j are w_1 and w_2, and their back-projections onto the surface contain the same amount of surface texture. Thus each point, p_m, on the tilt line is associated with a particular value of P, which can be obtained from the inverse of (3). The value of P can be used to compute (from (4)) an estimate of the difference $\delta L = L_1 - L_2$. The value of P (and the corresponding value of p_m with its associated pair of regions) that satisfies (4) is the estimate of the surface slant $(= tan^{-1}(P))$. Thus we are actually seeking a triplet of co-determined entities, σ, p_m, and the image region-pair (w_1, w_2). Before we can set about finding the triplet that satisfies (4) we need to address two remaining problems; how to estimate tilt, and how to obtain P given a value for p_m.

2.2 Estimating Tilt

The back-projection of a circular image, w_0, defines a conic section (an ellipse) on the surface. The image direction, τ, in which the texture density increases most rapidly lies along the tilt line. The value of τ can be obtained from the centre of mass, (x_A, y_A), where:

$$x_A = \frac{\int_{w_0} x\, pdf(x, y, P, Q, K)\, da}{\int_{w_0} pdf(x, y, P, Q, K)\, da}$$

$$y_A = \frac{\int_{w_0} y\, pdf(x, y, P, Q, K)\, da}{\int_{w_0} pdf(x, y, P, Q, K)\, da} \qquad (7)$$

The point (x_A, y_A) is the (normalised) first moment of the surface area around the image origin. For an elliptical surface region, W_0, it can be shown [10] that the direction of tilt, $\tau = tan^{-1}(y_A/x_A)$. If we assume a constant amount of texture (line length) per unit surface area then, for image line segments, $l_i : i = 1..N$, centred at $p_i = (x_i, y_i)$, the point (x_A, y_A) is approximated by:

$$x_L = \frac{\sum_{l_i \in w_0} x_i \int_{l_i} lcf(x, y, P, Q, \beta, K)\, ds}{\sum_{l_i \in w_0} \int_{l_i} lcf(x, y, P, Q, \beta, K)\, ds}$$

$$y_L = \frac{\sum_{l_i \in w_0} y_i \int_{l_i} lcf(x, y, P, Q, \beta, K)\, ds}{\sum_{l_i \in w_0} \int_{l_i} lcf(x, y, P, Q, \beta, K)\, ds} \qquad (8)$$

Therefore τ can be estimated as $\tau \simeq tan^{-1}(y_L/x_L)$.

In attempting to estimate tilt we initially set $(P, Q) = (P_0, Q_0) = (0, 0)$. For isotropic textures $tan^{-1}(y_L/x_L)$ provides an exact estimate of τ if $P_0 = Q_0 = 0$ (or if $Q_0/P_0 = tan(\tau)$). This is because an error generated by the mapping function lcf with respect to an image line at some orientation on one side of the tilt line is guaranteed to be compensated for by an error generated with respect to a corresponding line at a different orientation on the other side of the tilt line.

For anisotropic textures setting $P_0 = Q_0 = 0$ reduces the accuracy of (8) as a positive function of the degree of anisotropy of the surface texture. This is because the error-correcting behaviour described above for isotropic textures diminishes with the degree of anisotropy of the surface texture. The actual method used to compute τ involves iterative improvement such that $P_i \rightarrow P$, $Q_i \rightarrow Q$, and if $P_0 = P$ and $Q_0 = Q$ then (8) provides an exact estimate of τ.

An additional source of error is caused by the differential image resolution associated with surface points at different distances from the image. Image data derived from surface points which are most distant from the focal point have minimal image resolution. Thus, image data becomes increasingly noisy within a given image as distance to the surface increases. As with errors caused by anisotropy of surface texture, errors caused by differential image resolution tend to be symmetric about the tilt line. This is because two image points which are equidistant from the tilt line (and which are on a line orthogonal to the tilt line) back-project to surface points which are the equidistant from the focal point.

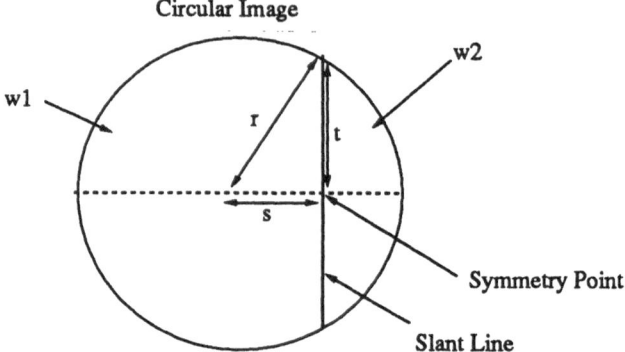

Figure 3: Computing the location of the symmetry point.

From this it follows that, even though lcf makes use of incorrect values for (P, Q), and uses these to back-project noisy image data, the resultant errors tend to be symmetric about the tilt line. Thus, the estimated orientation, $tan^{-1}(y_L/x_L)$, of the tilt line is more robust with respect to these sources of error than is the the length of the vector, (x_L, y_L). Accordingly, errors in the estimated position, (x_L, y_L), of (x_A, y_A) are not necessarily consonant with corresponding errors in the estimated angle, $tan^{-1}(y_L/x_L)$, of τ.

This method for estimating tilt makes use of moments about the image origin, and, in this respect, is similar to the methods described in [5, 10]. However, whereas the orientation of the vector (x_L, y_L) is an accurate estimate of the surface tilt, the length of this vector does not generally provide an accurate estimate of the surface slant (see [10] for a more detailed discussion). By using the quantity (x_L, y_L) to estimate tilt only, we extract information which is most accurately encoded by the measured quantity.

2.3 Estimating Slant

Given an estimate for tilt, the problem of estimating slant consists in finding the image line which passes through w_0, orthogonal to the tilt line, satisfying equation (4). If τ is known then the coordinate system can be rotated around the optic (Z) axis until, in the rotated coordinate system, $(P', Q') = (tan(\sigma), 0)$. Rotating the coordinate system does not affect the value of σ, and is used only as a means of simplifying the equations that follow. Omitting the term Qy from pdf, we now have:

$$\int_{w_0} pdf(x', P', K)\, da' = \int_{w_1} pdf(x', P', K)\, da' + \int_{w_2} pdf(x', P', K)\, da' \quad (9)$$

Combining this with (3):

$$\int_{w_0} pdf(x', P', K)\, da' - 2\int_{w_1} pdf(x', P', K)\, da' = 0 \quad (10)$$

The choice of w_1 in the second term of (10) is arbitrary, and w_2 would do just as well.

 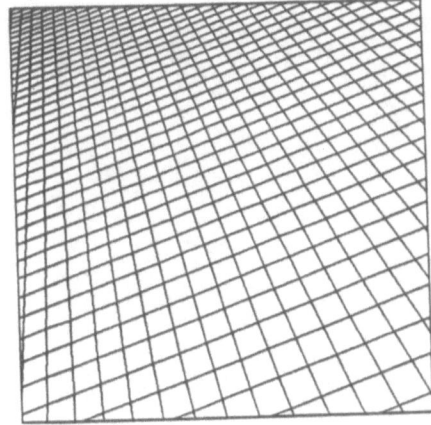

Figure 4: Synthetic image of grid texture.

Figure 5: Line texture derived from
elliptical distribution.

If $(P'x') \ll 1$ then[1] the slant is small relative to the focal length, and
(after [5]) we can approximate (5) by replacing the function *pdf* by the first
two terms of its Taylor expansion:

$$\int_{w_0} 1 + 3P'x' \, da - 2 \int_{w_1} 1 + 3P'x' \, da' = 0 \qquad (11)$$

By evaluating (11) and solving for P' it can be shown [10] that:

$$P' = \frac{st(r^2 cos^{-1}(t/r) - 2)}{2(r^3 - t^3)} \qquad (12)$$

where r is the radius of w_0, s is the distance from p_m to the image origin, and
$t = (r^2 - s^2)^{1/2}$ (see Figure 3). Thus for a putative symmetry point, p_m, we can
compute its distance, s, from the origin and from (12) obtain the corresponding
estimate, P', of $tan(\sigma)$. This estimate is the correct value of $tan(\sigma)$ if p_m is the
symmetry point. Note that, although this derivation was performed assuming
$Q' = 0$, (12) can be used to obtain slant in general if s is known, because (by
definition) $P' = (P^2 + Q^2)^{1/2} = tan^{-1}(\sigma)$.

2.4 Combining Estimates of Slant and Tilt

We are now in a position to define a procedure for estimating (P, Q). We
begin by setting the initial estimate of surface orientation $(P_0, Q_0) = (0, 0)$,
and using the method described earlier to obtain an initial estimate, $\tau_1 = tan^{-1}(y_L/x_L)$, of τ. (Whilst estimating τ the back-projected length of each
image line (evaluated as part of (8)) is recorded, so that it can be used in
estimating slant). Setting $P = P_0, Q = Q_0$ in (4), we then estimate slant, σ_1.
This is achieved by searching along the estimated tilt line for a point, p_m, that

---

[1] Recall that $(x, y) = (x/f, y/f)$.

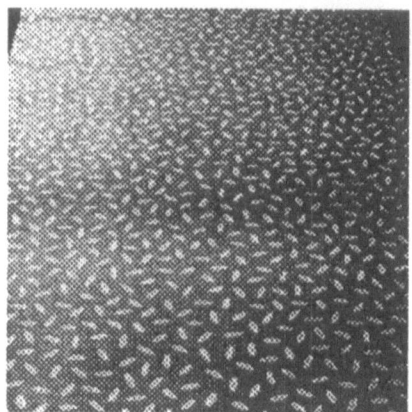

Figure 6: Algorithmically rotated image.

Figure 7: Image of wallpaper.

minimises (4). When this line search is complete the distance, s, of p_m from the image origin can be used to obtain, from (12), an estimate, $\sigma_1 = tan^{-1}(P')$, of the surface slant. We then re-express (σ_1, τ_1) in terms of partial derivatives to obtain (P_1, Q_1). This is used to obtain (from (4)) a new estimate, τ_2, of τ, and then σ_2, as above. This process is repeated until the difference between successive estimates of the surface normal is less than $0.1°$. The procedure for estimating slant involves use of the data recorded whilst computing the tilt, and does not therefore utilise a new value for P (from (12)) for each candidate slant line. In essence, the slant line is found by moving p_m along the tilt line until the pre-computed estimates of surface line length on either side of the candidate slant line (defined by p_m) are the same. Thus, each successive estimate of (P, Q) requires each image edge to be back-projected once only. Re-computing the lengths of back-projected image lines for each candidate slant line would reduce the number of iterations required, but the computational cost of so doing would be prohibitive. Moreover, results indicate that accuracy is not noticeably altered using the form of relaxation described above.

3   RESULTS

Results for the ABM were obtained using synthetic line textures (Figures 4 and 5), an algorithmically rotated grey level image of a photograph (from [3]) of textured surface (Figure 6), and an image of a (real) textured surface. The image area used in each image is defined by a circle with diameter equal to the width of the square images shown in Figures 4-7. The orientation of surfaces in Figures 4-6 is $(\sigma, \tau) = (40°, -20°)$. The orientation of the surface shown in Figure 7 was estimated as $(\sigma, \tau) = (27.23°, 0.050°)$ by manually marking each texel with a dot, and using the method (for dot textures) described in [5]. The focal length of the imaging sytem is equal to half the image height for all images. Each grey-level image consists of a 512x512 array of pixels, with

Figure	Slant	Tilt	Error
4	41.60°	-20.60°	1.66°
5	41.21°	-18.20°	1.70°
6	35.98°	-32.11°	8.46°
7	32.37°	-1.09°	5.41°

Table 1: Results for Figures 4-7.

256 grey levels. The error value given for each image is the angle between the actual and estimated surface normals.

Results for the anisotropic synthetic line textures shown in Figures 4 and 5 are displayed in Table 1. The line segments of the surface texture shown in Figure 5 are derived from an elliptical distribution with ratio of major:minor axes of 5:2; all line segments have the equal length and are randomly distributed on the surface. The initial and final estimates of tilt for these images did not differ by more than 2°, indicating that even when the slant is not known the tilt can be accurately estimated. The number of ABM iterations was 5 for both synthetic images. Results for the grey level images shown in Figures 6 and 7 (see Table 1) were obtained using the ABM with the adaptive multi-scale filtering (AMSF) technique[2][9, 10]. As is usual with AMSF a 'shape from texture' method (e.g. ABM) is used to provide successive estimates of surface orientation, and each of these estimates is used as part of the next filtering operation. This type of nested iteration has been shown [10] to be effective for several different shape from texture methods(e.g. [5, 1])[3]. After each filtering operation of AMSF the estimate of tilt did not vary by more than 2-3°, although (as is to be expected) the estimate of surface orientation varied across successive filtering operations as the AMSF method converged. The number of AMSF iterations required was 15 for Figure 6 and 10 for Figure 7, and the number of ABM iterations (between AMSF filtering operations) was typically around 5.

4 Conclusion

We have demonstrated a method for estimating slant and tilt using separate computational mechanisms. Each of these mechanisms uses quantities ((y_L/x_L) and the origin-symmetry point distance) which robustly index slant and tilt, respectively. Using these two quantities has enable us to utilise two separate

---

[2]This technique iteratively converges to a set of image filters such that all image filters back-project to surface 'filters' with the same centre frequency as each other. This ensures that data from a single band of spatial frequencies on the surface are utilised in estimating surface orientation.

[3]The method used here utilises circular image filters, as described in [10], rather than elliptical image filters as utilised for results presented in [9]; results are not significantly different for these two variants of AMSF.

computational mechanisms which take full advantage of the informational content offered by each of the measured quantities.

Results obtained using real and synthetic images are qualitatively similar to those obtained from psychophysical studies. This indicates that the ABM is at least consistent with the functional characteristics of computational mechanisms responsible for obtaining shape from texture in human observers.

The general principles upon which the ABM is based can be used to estimate surface orientation of curved surfaces. This can be achieved by using multiple circular image regions, and estimating the surface orientation associated with each.

Acknowledgements: Thanks to Raymond Lister and Stephen Isard for comments on drafts of this paper.

References

[1] Aloimonos J, "Shape from Texture", *Biological Cybernetics*, 58, pp 345-360, 1988.

[2] Blake A, Marinos C, "Shape from Texture: Estimation, Isotropy and Moments", *Artificial Intelligence*, 45, pp 323-380, 1990.

[3] Brodatz P, "Textures", Dover Publications Inc., New York, 1986.

[4] Cutting J E, Millard R T, "Three Gradients and the Perception of Flat and Curved Surfaces", *Journal of Experimental Psychology:General*, 113(2), pp 198-216, 1984.

[5] Kanatani K, and Chou T, "Shape From Texture: General Principle", *Artificial Intelligence*, 38, 1, pp 1-49, 1989.

[6] Kube P, "Using Frequency and Orientation Tuned Channels to Determine Surface Slant", *Eigth Annual Conference of the Cognitive Society*, Amhurst, Massechussetts 1986.

[7] Stevens K A, "The Information Content of Texture Gradients", *Biological Cybernetics*, 42, pp 95-105, 1981.

[8] Stevens K A, "Surface Tilt: The Visual Encoding of Surface Orientation", *Biological Cybernetics*, 46, pp 183-195, 1983.

[9] Stone J V, "Shape From Texture: Textural Invariance and the Problem of Scale in Perspective Images of Textured Surfaces", *British Machine Vision Conference*, Oxford, England, pp 181-187, 1990.

[10] Stone J V, "Shape From Texture: A Computational Analysis", *D.Phil. Thesis*, Experimental Psychology, University of Sussex, England, 1991.

[11] Turner MR, Gerstein GL, Bacjsy R, "Underestimation of visual texture slant by human observers: a model", Biological Cybernetics, 65, pp 215-226, 1991.

A Step Towards Efficient Bayesian Signal Reconstruction

J.W. Dickson

IBM UK Scientific Centre,
Winchester, England.

Abstract

This paper presents a theoretical basis for a set of optimal filters for the reconstruction of piecewise-continuous one-dimensional signals, drawing from Bayesian networks and Kalman filters. Results are presented for synthetic and real data, using both the optimal filters and a sub-optimal implementation. The results compare well with linear space invariant filtering or facet fitting approaches, and present a basis for the design of image restoration algorithms.

1 Introduction

This paper addresses the problem of reconstructing a piecewise continuous signal which has been corrupted by noise. This is poorly handled by linear space invariant (LSI) filters as the signal spectrum around a discontinuity is different from the spectrum of a continuous portion of signal. This has led researchers to explore the use of non-linear or adaptive filters.

The approach taken is to derive a particular class of optimal adaptive filters for one-dimensional signals, based on the theory of Bayesian networks [1] and Kalman filters [2, 3]. The signal models on which the optimal filters are based can correspond either to facet-fitting models or deformable membrane models, depending on the choice of parameters. It is recognised that these optimal filters will not be tractable for the ultimate goal of image restoration, so we also investigate the use of a sub-optimal algorithm. We apply optimal and sub-optimal algorithms, using different signal models, to both synthetic and real one-dimensional data; the results in several cases show improvements over traditional facet-fitting or LSI filtering techniques.

This approach brings the power of Bayesian theory to a stage where that theory can be applied to some more concrete problems in an efficient manner. That was the aim in carrying out this work and we see it as one of the important contributions of this paper.

2 Bayes' Rule and Sensors

The purpose of sensing is to combine information from observations with *prior* knowledge of the state of some external object in order to obtain more accurate *posterior* knowledge. We denote the state we wish to determine by a variable h and the observation set by o; the prior knowledge is written probabilistically as a distribution $p(h)$ and the posterior knowledge as a conditional distribution $p(h|o)$. Furthermore we describe the sensor in probabilistic terms by the conditional distribution $p(o|h)$ and the prior knowledge of the observations by $p(o)$; we can then apply Bayes' rule [4]:

$$p(h|o) = \frac{p(h)p(o|h)}{p(o)} \tag{1}$$

In many practical applications we may only wish to find the most likely value of h for a particular observation set o; in these cases we may ignore the (constant) denominator and merely maximise the numerator on the RHS of equation (1). This is the approach taken in this paper.

In order to make the sensor model tractable it is useful to be able to separate the individual members o_i of the observation set o. This can be done if the individual observations can be said to be *conditionally independent*, conditioned upon h [1]. We can then write:

$$p(o_1 \ldots o_n|h) = \prod_{i=1}^{n} p(o_i|h) \tag{2}$$

A particularly simple and common example of this is where the o_i are a sequence of measurements of a constant value h; typically the arithmetic average of these measurements would be taken as the most likely estimate.

3 Recursive Probabilistic Signal Models

In order to make maximum use of the conditional independence assumption, this paper develops recursive probabilistic models for discrete 1D signals. The assumptions that underlie these models are:

1. signal values are corrupted independently by the sensor;
2. the original, uncorrupted values depend only on their direct neighbours' values (Markov assumption).

If we interpret these assumptions in probabilistic terms, then for each discrete value h_i of the uncorrupted signal, we have a single observation o_i and the following distribution functions:

- a prior $p(h_i)$;
- a sensor noise model $p(o_i|h_i)$;
- and a markovian signal model $p(h_{i-1}|h_i)$, $p(h_{i+1}|h_i)$.

The datasets o_i, $o_{j<i}$ and $o_{k>i}$ are assumed to be conditionally independent, conditioned on h_i, so that:

$$p(o|h_i) = p(o_i|h_i)p(o_{j<i}|h_i)p(o_{k>i}|h_i) \tag{3}$$

where each of the last two terms can be expanded recursively as:

$$p(o_{j<i}|h_i) = \oint_{h_{i-1}} p(o_{j<i-1}|h_{i-1})p(o_{i-1}|h_{i-1})p(h_{i-1}|h_i)dh_{i-1} \tag{4}$$

Here the first term in the integral is the same expression as the LHS, evaluated at step $i-1$, the second term is the sensor model at step $i-1$; these two are combined and then the results carried forward to step i via the third term in the integral.

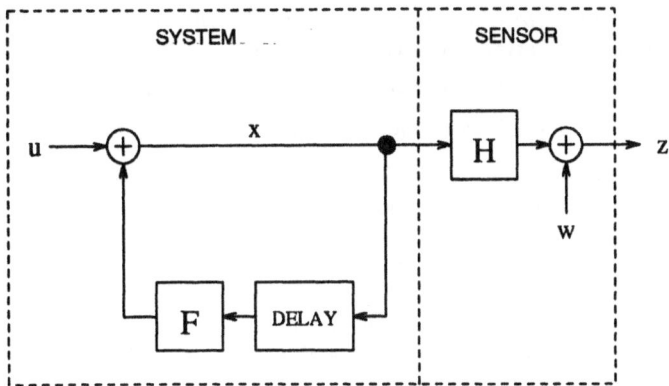

Figure 1: Kalman filter block diagram; **u** is the system noise, **w** is the observation noise.

4   The Kalman Filter

It has been shown ([5], chapter 3) that the Kalman filter [2, 3] is a special case of the class of filter formulated in the previous section, where all the distribution functions are assumed to be Gaussian and the signal can be described by linear state-space equations. Here we will show where the equivalence lies without making a formal proof.

In its usual formulation the discrete Kalman filter step consists of two phases. In the *prediction* phase data up to step $i - 1$, in the form of a state estimate vector $\mathbf{x}(i-1|i-1)$ and a mean squared error (MSE) estimate matrix $\mathbf{P}(i-1|i-1)$, are used to obtain estimates of the state and MSE at the next step i from a state transition matrix $\mathbf{F}(i)$ and a system noise matrix $\mathbf{U}(i)$; the new state estimate is used to obtain an estimate of the corresponding observation $\dot{\mathbf{z}}(i|i-1)$ via the measurement matrix $\mathbf{H}(i)$.

$$\hat{\mathbf{x}}(i|i-1) = \mathbf{F}(i)\mathbf{x}(i-1|i-1) \tag{5}$$
$$\mathbf{P}(i|i-1) = \mathbf{F}(i)\mathbf{P}(i-1|i-1)\mathbf{F}^T(i) + \mathbf{U}(i) \tag{6}$$
$$\mathbf{z}(i|i-1) = \mathbf{H}(i)\mathbf{x}(i|i-1) \tag{7}$$

In the *update* phase, the new measurement vector $\dot{\mathbf{z}}(i)$, along with its expected MSE matrix $\mathbf{R}(i)$, are combined according to filter gain $\mathbf{W}(i)$ with the earlier data to give new estimates of system state and MSE:

$$\mathbf{P}^{-1}(i|i) = \mathbf{P}^{-1}(i|i-1) + \mathbf{H}^T(i)\mathbf{R}^{-1}(i)\mathbf{H}(i) \tag{8}$$
$$\mathbf{W}(i) = \mathbf{P}(i|i)\mathbf{H}^T(i)\mathbf{R}^{-1}(i) \tag{9}$$
$$\dot{\mathbf{x}}(i|i) = \dot{\mathbf{x}}(i|i-1) + \mathbf{W}(i)\left[\mathbf{z}(i) - \hat{\mathbf{z}}(i|i-1)\right] \tag{10}$$

These equations are summarised in the diagram in figure 1, showing the models of the system and sensor.

The equivalence between this formulation and the probabilistic one from equation (4) is as follows:

- observations o_i and $\mathbf{z}(i)$ are equivalent, as are state values h_i and $\mathbf{x}(i)$;

- the expression $p(o_{j<i}|h_i)$ is a Gaussian distribution in h_i with mean $\dot{x}(i|i-1)$ and covariance matrix $\mathbf{P}(i|i-1)$;
- $p(o_i|h_i)$ has mean $\mathbf{H}(i)\mathbf{x}(i)$ and covariance matrix $\mathbf{R}(i)$;
- and finally $p(h_{i-1}|h_i)$ has mean $\mathbf{F}(i)\mathbf{x}(i-1)$ and covariance $\mathbf{U}(i)$.

The prediction phase is the evaluation of $p(o_{j<i}|h_i)$ from $p(o_{j<i}|h_{i-1})$, while the update phase is the combination of $p(o_{j<i}|h_i)$ with $p(o_i|h_i)$ to obtain $p(o_{j<i+1}|h_i)$.

In order to apply this model to the reconstruction of continuous signals, where we can use later observations $o_{k>i}$ as well as past observations $o_{j<i}$, we add a backward "prediction" phase from step $i+1$. This corresponds to $p(o_{k>i}|h_i)$, also evaluated recursively, thus giving all three terms from equation (3).

5   Piecewise Continuous Signals

In order to accommodate piecewise continuous signals we need to extend the model. Preserving the earlier assumptions, we introduce discontinuities by modifying the state transition relationship:

$$p(h_{i-1}|h_i) = \frac{c_i}{c_i+1}p_c(h_{i-1}|h_i) + \frac{1}{c_i+1}p_d(h_{i-1}|h_i) \tag{11}$$

where p_c describes the distribution resulting from a continuous step as before, p_d describes the distribution from a discontinuous step and c_i denotes the odds for the particular step being continuous.

To integrate this into the Kalman filter approach, both p_c and p_d must be Gaussian distributions. In the implementations a discontinuity is simulated by using a system noise matrix $\mathbf{U}_d$ with large covariance values; thus equation (11) corresponds to splitting a single filter into two concurrent filters at each step. As a result we need 2^{n-1} filters for n observation points; the problem is no longer tractable. Two different heuristics to overcome this problem are discussed in a later section.

The other consequence of equation (11) is that each of the filters needs to be ranked by its posterior probability. This too can be done recursively, so that at each step i the distributions $p(o_{j<i}|h_i)$ and $p(o_{k>i}|h_i)$ are represented by a weighted sum of Gaussian distributions. The weights are calculated in the update phase of the Kalman filter, where the Gaussian $p(o_i|h_i)$ is combined with the incoming predictions. This combination corresponds to the multiplication of the distributions, which in turn corresponds to the multiplication of two Gaussian functions (as in equation (4)). The product of two Gaussian functions with means μ_1, μ_2 and covariance matrices $\mathbf{C}_1$ and $\mathbf{C}_2$ respectively is:

$$\mathcal{N}_{\mu_1,\mathbf{C}_1}(\mathbf{x}) \times \mathcal{N}_{\mu_2,\mathbf{C}_2}(\mathbf{x}) = \mathcal{N}_{\mu',\mathbf{C}'}(\mathbf{x}) \times \mathcal{N}_{0,\mathbf{C}''}(\mu_1 - \mu_2) \tag{12}$$

where

$$\begin{aligned}
\mathbf{C}'^{-1} &= \mathbf{C}_1^{-1} + \mathbf{C}_2^{-1} \\
\mu' &= \mathbf{C}'\left(\mathbf{C}_1^{-1}\mu_1 + \mathbf{C}_2^{-1}\mu_2\right) \\
\mathbf{C}'' &= \mathbf{C}_1\left(\mathbf{C}_1^{-1} + \mathbf{C}_2^{-1}\right)\mathbf{C}_2
\end{aligned}$$

In this case,

$$\mathcal{N}_{\boldsymbol{\mu}_1, \mathbf{C}_1}(h_i) = p(o_{j<i}|h_i) \quad \text{and} \quad \mathcal{N}_{\boldsymbol{\mu}_2, \mathbf{C}_2}(h_i) = p(o_i|h_i)$$

As may be seen, the first term on the RHS of equation (12) is of the same form as the terms on the LHS; it is a normal distribution whose parameters are given by the Kalman filter equations ((5) to (10)). The second term on the RHS of equation (12) is a Normal distribution in the difference of the means of the two terms on the LHS; the value of this term is used to rank the individual Gaussian functions that make up $p(o_{j<i}|h_i)$ and $p(o_{k>i}|h_i)$.

Thus the posterior likelihood assigned to each individual Kalman filter is a combination of the prior likelihoods c_i for each step being continuous and the results from equation (12), which is a hypothesis test between $p(o_{j<i}|h_i)$ and $p(o_i|h_i)$.

We can also make use of the hypothesis test to determine the probability of discontinuities. We denote the state of a link from step $i-1$ to step i by the symbol l_i, taking the value 0 for a discontinuity and the value 1 for a continuous step. Then we can evaluate:

$$p(o|l_i) = \oint_{h_i} p(o_{j<i}|h_i, l_i)p(o_i|h_i)p(o_{k>i}|h_i)dh_i \tag{13}$$

which, combined with the prior odds c_i, allows us to calculate the posterior odds for l_i.

6 Algorithms for Reconstruction

The results in this paper were obtained from two different algorithms:

1. a near-optimal algorithm which carries forward all the ranked hypotheses as per equation (11), pruning only those whose likelihood is less than 10^{-3} times the most likely;

2. a sub-optimal algorithm which locally labels each step as continuous or discontinuous and only carries forward the currently-labelled hypothesis.

The second algorithm is a decentralised version of the Highest Confidence First (HCF) algorithm described in [6]. The HCF algorithm makes a single, globally most promising decision at each iteration. The algorithm used here, which could be called Local Highest Confidence First (LHCF) [7], finds the most promising decision over each neighbourhood and takes each of these locally maximal decisions at each step. It might be expected that the performance of the LHCF approach will deteriorate in cases where the decisions made locally have a more global effect.

Four different signal models were used:

- piecewise constant:
  $\mathbf{F}(i) = [1]$, $\mathbf{U}_c(i) = [0]$, $\mathbf{U}_d(i) = \left[\sigma_d^2\right]$

- piecewise constant with system noise (piecewise almost constant):
  $\mathbf{F}(i) = [1]$, $\mathbf{U}_c(i) = \left[\sigma_c^2\right]$, $\mathbf{U}_d(i) = \left[\sigma_d^2\right]$

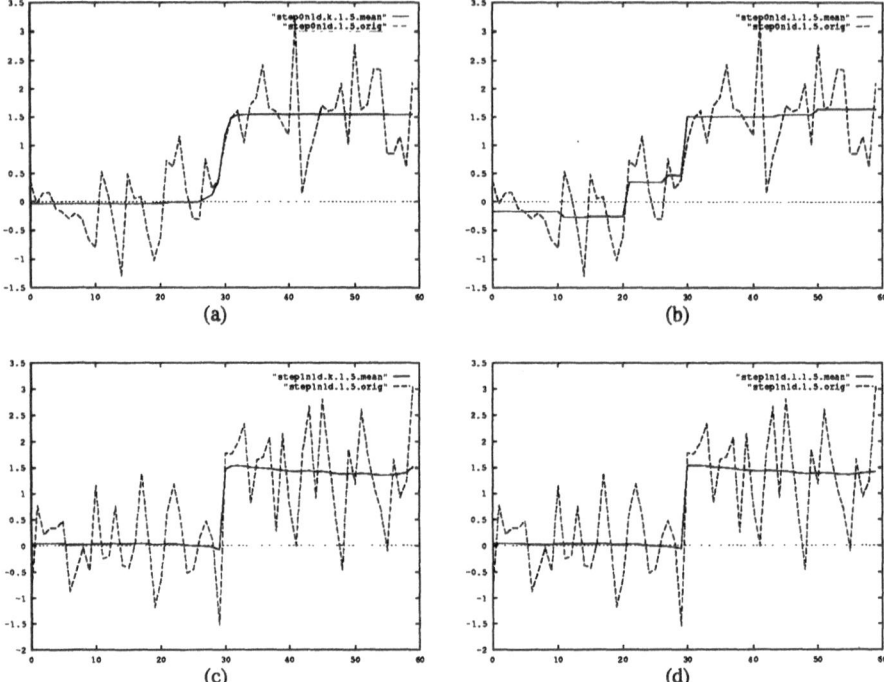

Figure 2: Step size 1.5 in noise $\sigma = 0.707$; (a) piecewise constant, near-optimal algorithm; (b) piecewise constant, LHCF algorithm; (c) piecewise almost constant, near-optimal algorithm, (d) piecewise almost constant, LHCF algorithm

- piecewise linear:
$$\mathbf{F}(i) = \begin{bmatrix} 1 & 1 \\ 0 & 1 \end{bmatrix}, \; \mathbf{U}_c(i) = \begin{bmatrix} 0 & 0 \\ 0 & 0 \end{bmatrix}, \; \mathbf{U}_d(i) = \begin{bmatrix} \sigma_{1d}^2 & 0 \\ 0 & \sigma_{2d}^2 \end{bmatrix}$$

- piecewise almost linear:
$$\mathbf{F}(i) = \begin{bmatrix} 1 & 1 \\ 0 & 1 \end{bmatrix}, \; \mathbf{U}_c(i) = \begin{bmatrix} 0 & 0 \\ 0 & \sigma_{2c}^2 \end{bmatrix}, \; \mathbf{U}_d(i) = \begin{bmatrix} \sigma_{1d}^2 & 0 \\ 0 & \sigma_{2d}^2 \end{bmatrix}$$

7 Experimental Results

We have prepared synthetic data to illustrate the performance of the algorithms and the models with well-controlled inputs. In each case, piecewise-constant or piecewise-linear data was corrupted by additive Gaussian noise and the results processed by the algorithms.

In order to obtain a more realistic assessment of the performance of this approach, we generated 1D datasets from slices through real images. In these cases the actual parameters of the image and noise are not known exactly and have been estimated from plots of the data.

Figure 2 shows the effect of applying the two order zero models to a signal containing a step in noise. The corrupted signal is shown by the dashed line while the reconstructed signal is shown by the solid line. The near-optimal algorithm shows reasonable competence with both models, while the LHCF

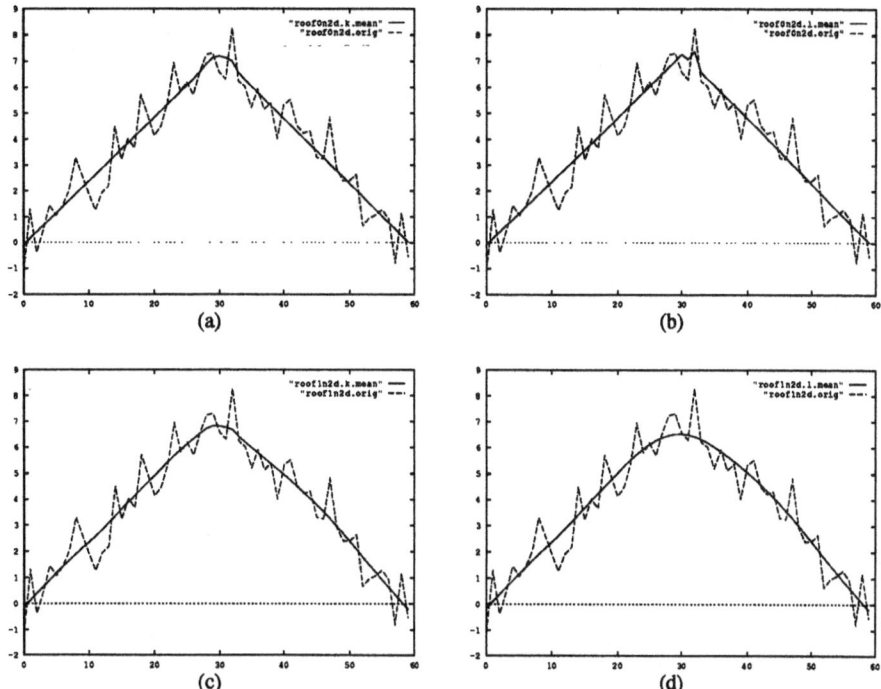

Figure 3: Roof function in noise; (a) piecewise linear, near-optimal algorithm; (b) piecewise linear, LHCF algorithm; (c) piecewise almost linear, near-optimal algorithm, (d) piecewise almost linear, LHCF algorithm

algorithm only performs well with the less rigid signal model; this is as we might expect, since the rigid signal model makes local decisions extremely error-prone. The near-optimal algorithm, using the more rigid model, was able to detect a step size of 1.0 in the same noise.

Figure 3 shows the effect of applying the two first-order models to a roof signal in noise. Once again the near-optimal algorithm shows reasonable competence for both models. The LHCF algorithm appears quite competent with the rigid model, while with the flexible model it fails to detect the discontinuity; however, its competence with the rigid model is likely to derive from the fact that there is only one discontinuity in the dataset.

Finally, figure 4 shows the result of applying the two flexible models to a slice through a real image. As expected the near-optimal algorithm shows better competence than the LHCF one, though the differences are minimal, particularly with the lower pair of plots. The data clearly would not fit either of the rigid models shown in the earlier results. Note also that the discontinuities in the data are not sharp, as per the signal model, but blurred by some form of point spread function.

The results shown here should be compared with either facet-fitting or LSI filtering techniques.

From the results of applying the rigid (facet-like) models to the simulated data, we can see some reasons for the problems encountered with the use of parametric patches to segment images. While the near-optimal algorithm is

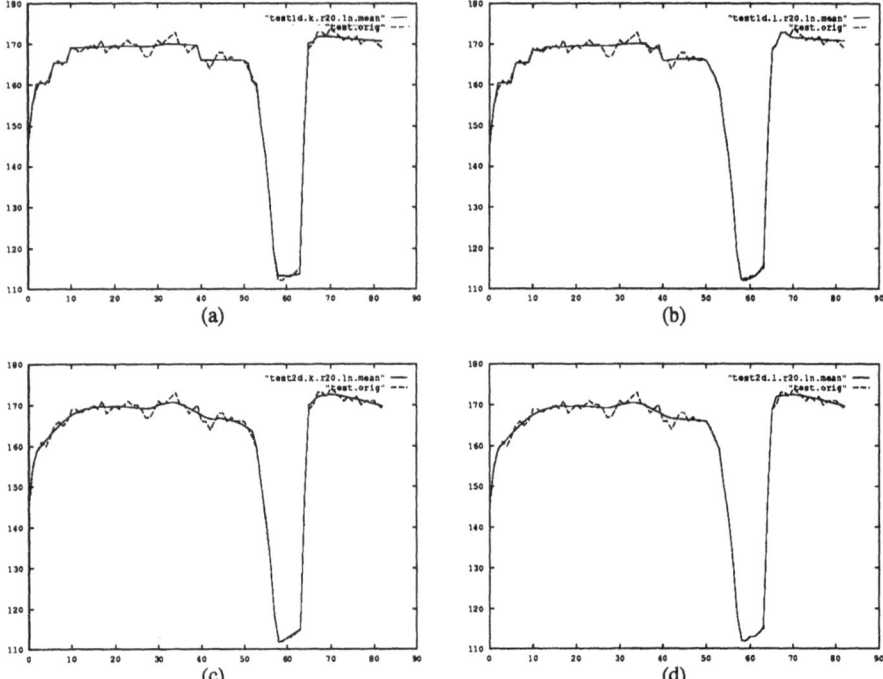

Figure 4: Real data; (a) piecewise almost constant, near-optimal algorithm; (b) piecewise almost constant, LHCF algorithm; (c) piecewise almost linear, near-optimal algorithm, (d) piecewise almost linear, LHCF algorithm

successful in fitting such patches, the sub-optimal algorithm exhibits the fragmentation which is often observed. This is likely to be an unavoidable problem with such models because of the strong dependencies between non-local segmentation decisions. In cases where the optimal approach is not desirable or tractable, sophisticated algorithms will be required to obtain reasonable results.

The flexible models used here provide the closest comparisons with LSI filtering techniques; in fact where there are no edges the sub-optimal filter is equivalent to optimal LSI filtering. Figure 5 shows the application of a standard LSI filter and the corresponding LHCF filter to a step in noise. The use of LSI filters for piecewise continuous data is inevitably a compromise, as outlined in the introduction. Thus while these filters may provide either sharp boundaries or smooth surfaces in between, they cannot provide both simultaneously. The filtering technique developed here is capable of providing them both; where the underlying signal model is sufficiently flexible it can provide these relatively efficiently.

8 Further Work

The task of reconstructing or segmenting images using Bayesian theory has been tackled by several researchers in the past [8, 9, 10]. In many cases the results have proved to be unwieldy.

It is clear that the recursive algorithms and models presented in this paper

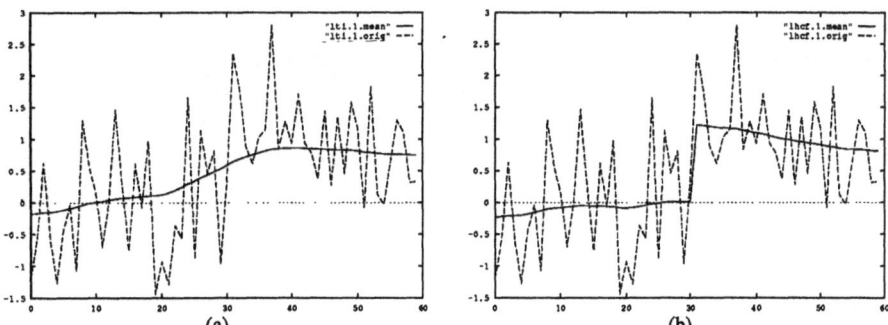

Figure 5: Comparison between LSI and LHCF approaches: Step size 1.0 in noise $\sigma = 0.707$; (a) piecewise almost constant, LSI filter; (b) piecewise almost constant, LHCF algorithm

cannot be directly generalised to the processing of a 2D dataset; the conditional independence assumptions required cannot be justified. However we can show that the recursive models derived here can be expressed equivalently as differential models, which can then be applied to 2D datasets. Though this in itself is not a novel idea, we expect the fusion of the two approaches to provide the inspiration and justification for a more rigorous application of differential models to image restoration and segmentation.

We are currently applying the theory described in this paper to the completion of edge maps, as produced by an edge detector (eg [11]). These edge maps often have gaps where the grey-level contrast is insufficient or where the grey-level surface does not match the edge detector's expectations (for example at junctions). The aim is to use the Kalman filtering technique to estimate the orientation at a terminator so the edges can be extrapolated, and to use the hypothesis testing techniques to determine the best completion for each terminator. This requires the application of some sophisticated search control techniques and decision-theoretical methods. Only preliminary results are available at this time, but these appear encouraging.

9   Conclusions

We have shown how Bayesian theory can be used to design recursive filters for one-dimensional signals. For the special case of Gaussian distribution functions and linear state-space signal models we have shown efficient implementations of this Bayesian theory in terms of Kalman filters, and successfully applied the results to the reconstruction of signals corrupted by additive noise.

A requirement for a more general application of this work is the viability of sub-optimal algorithms; we have compared the performance of one such algorithm with the performance of the optimal filter, showing that in many cases the results are comparable.

The approach described in this paper is currently being applied to a range of problems. Some difficulties still remain:

- the choice of suitable sub-optimal algorithms for both 1D and 2D applications;

- at present the approach does not cater for images which have been blurred as well as corrupted with noise; it is not clear how a blurring model (point spread function) could be integrated with the recursive approach;

- we have not investigated calibration methods; clearly we will need a more rigorous approach to choosing parameters than the current "eyeballing" techniques.

References

[1] Judea Pearl. *Probabilistic Reasoning in Intelligent Systems*. Morgan Kaufmann Publishers, Inc., 1988.

[2] Y. Bar-Shalom and T.E. Fortmann. *Tracking and Data Association*. Academic Press, 1988.

[3] Arthur Gelb. *Applied Optimal Estimation*. MIT Press, 1974.

[4] T. Bayes. An essay towards solving a problem in the doctrine of chances. *Phil. Trans.*, 3:370-418, 1763.

[5] J. W. Dickson. *Image Structure and Model-Based Vision*. PhD thesis, Oxford University, Department of Engineering Science, 1990.

[6] Paul B. Chou and Christopher M. Brown. The Theory and Practice of Bayesian Image Labeling, September 1988.

[7] M.J. Swain and L.E. Wixson. Efficient estimation for Markov random fields. In *Image Understanding and Machine Vision*, 1989.

[8] B.F. Buxton, H. Buxton, and A. Kashko. Optimization, Regularization and Simulated Annealing in Low-Level Computer Vision. In Ian Page, editor, *Parallel Architectures and Computer Vision Workshop*. OUP, 1987.

[9] S. Geeman and D. Geeman. Stochastic relaxation, Gibbs distribution and the Bayesian restoration of images. *IEEE Trans. Pattern Analysis Machine Intell.*, 6:721–741, 1984.

[10] J.F. Silverman and D.B. Cooper. Bayesian Clustering for Unsupervised Estimation of Surface and Texture Models. *IEEE Trans. Pattern Analysis Machine Intell.*, 10(4), July 1988.

[11] J.F. Canny. Finding Edges and Lines. Technical Report 720, Massachusetts Inst. Technol., 1983.

[12] A. Papoulis. *Probability, Random Variables, and Stochastic Processes*. McGraw-Hill, 1984.

Multistage Combined Ellipse and Line Detection

Geoff A.W. West & Paul L. Rosin

Cognitive Systems Group

School of Computing Science Curtin University of Technology

GPO U1987, Perth, 6000, Western Australia

email: geoff@hinault.cs.curtin.edu.au rosin@marsh.cs.curtin.edu.au

Abstract

This paper describes an algorithm for the detection of ellipses and lines in image edge data. Connected edge pixels are transformed into polygonal approximations by a two stage algorithm. Then a second two stage algorithm replaces combinations of lines by ellipses if the ellipse fit is better. For each algorithm a combination of splitting and merging of the data is used to enable global and local constraints to fit different representations to the pixel data. All merging decisions use a significance measure to replace a number of representations by a single representation which removes the need for thresholds in the algorithms. The structure of the algorithm allows any particular representation to describe the data e.g. parabolae, splines, etc. instead of ellipses.

1 Introduction

In computer vision the extraction of meaningful features from images is an important technigue. The most popular approach is based on edge detection. For model based object recognition, edges must be represented in a more manageable form than simply pixels. The type of representation required is highly application dependent but is typically based on a combination of straight line approximations and higher order curves such as arcs, conic sections, spline and curvature primitives.

A number of techniques have been proposed for determining polygonal approximations [5] [10]. However it is only recently that attention has been concentrated on the extraction of higher order representations because of the increased number of parameters or degrees of freedom and the ill-conditioned nature of the problem [8] [4].

This paper describes a technique for generating a higher order description by segmenting the edge data into combinations of ellipses and lines. There are four stages used: (1) lines are fitted to connected lists of edge pixels using Lowe's technique [3], (2) lines are grown by combining adjacent lines, (3) ellipses replace lines and finally (4) ellipses are grown by combining with adjacent lines and/or ellipses. In all stages replacement occurs only if the resultant fit is better. The concept of better fit is that suggested by Lowe which is termed a measure of significance. Significance is the maximum error between the fitted representation and the data (a line fitted to pixels or an ellipse fitted to lines) normalised by the length of the representation. The significance is a scale invariant measure which allows the replacing of (i) pixels by a line, (ii) combinations of lines by a line, (iii) combinations of lines by ellipses and (iv) combinations of lines and ellipses by ellipses. The same measure of significance is used in all four stages removing the requirement for any thresholds.

Previously published results [8] have demonstrated the utility of ellipse and line detection based on stages (1) and (3) of the technique described above. In this paper an improved version of the line and ellipse fitting algorithms is described which overcomes the disadvantages of using a binary search tree for stages (1) and (3) by adding new stages (2) and (4). It is shown that the addition of these stages improves the performance of the algorithm. In addition the improvements have been added to the algorithm for detecting arcs and lines (the LAD algorithm [11]).

2   Finding Representations

The problem of finding the optimum segmentation and hence representation for any curve has resulted in a number of proposed solutions. Simple local techniques such as segmenting at points of high curvature (vertices) and points of change in the rate of change of curvature have been proposed as have more global techniques such as segmenting at the point of maximum distance from the curve to the representation. These usually result in a sub-optimal result. To overcome the problem of using local constraints, context dependent local and global techniques have been proposed. Fischler [2] investigated the performance of people for segmenting curves for a number of objectives and discovered that the points of segmentation varied depending on the objective. From the results he proposed a technique that processed curves using large windows to attempt to capture more global information. In fact it is only possible to correctly segment a curve by taking into account the process that formed the curve e.g. the resulting 2D projections of 3D objects under certain viewing conditions. In addition a large number of curve models need to be available e.g. sine waves, parabolas, etc.

Where the objective is to fit a particular representation, an almost optimal technique can be formulated. All possible combinations of a particular representation such as a line can be fitted to the curve and the combination with the best goodness of fit chosen. Consider the case of dividing a curve up into n segments of equal numbers of pixels. The task is to determine which of the possible segmentation points can be removed, e.g. by combining two adjacent segments to form one line. This is a combinatorially expensive process of order $O(2^{n-1})$. For example, for a curve 40 pixels long where $n=8$, eight 5 pixel long lines is the maximum number of lines that can fit the curve and the minimum is one. To determine the best combination, 128 combinations need to be considered. The combinatorics are compounded if an attempt is made to replace combinations of lines by ellipses or some other representation in a second stage of processing, or if different length segments are considered.

It would appear the best technique would be to restrict the number of combinations that need to be tested. This can be achieved by using a binary search tree as proposed by many researchers [3] [6]. However this has disadvantages as shown in section 3 because some combinations of adjacent representations are never compared as they are in different branches of the tree. To improve

Figure 1. Segmentation of curves

the results a second stage is added that compares these representations.

There are a number of other issues that have to be addressed when considering the segmentation of edge data into higher order representations. The majority of algorithms, with the exception of [1], depend on pre-set parameters to determine such things as the accuracy of fit, the scale at which breakpoints are located, where breakpoints are, and thresholds for selecting the breakpoints. However, the results obtained are dependent on the parameters. For the line and ellipse fitting algorithm [8], breakpoints for the two stages are the well known points of maximum error between the fitted straight line and the data. These have the advantage of not requiring any parameters for their detection. A subset of these become the vertices in the polygonal representation resulting from the first stage. In the second stage groups of lines are replaced by ellipses so the breakpoints between representations are still effectively vertices. However, these may not be optimal breakpoints for higher order representations such as ellipses and it can be argued that other breakpoints such as points of inflection should also be used. Most higher order representations have continuously varying curvature so the detection of breakpoints based on changes in curvature is not valid. Higher order differentials are necessary which are difficult to determine in discrete data.

3 Line Fitting

3.1 Stage (1): binary search tree

The line fitting technique used is the familiar recursive binary tree search. Each curve is hypothesised to be a straight line, figure 1, and segmented at the point of maximum deviation from the curve to the straight line. The process is then repeated for each of the two curves recursively. When the bottom of the tree is reached and the curve cannot be subdivided anymore, tail recursion is used to combine together those lines for which the combined line is a better representation than the other lines. This algorithm has been described in detail elsewhere [11]. The important points to note are (1) the use of points of maximum deviation for breakpoints and (2) the binary search tree preventing all adjacent combinations of lines from being compared. Consider the curve of figure 2a, the result of the algorithm is shown in 2b whereas the intuitively correct result is that in 2c. Figure 3 shows the interpretation tree for this curve. The tree shows that the curve has been segmented into the lines a,b,c,d,e and f, some of which have been combined under tail recursion to give the result of figure 2b.

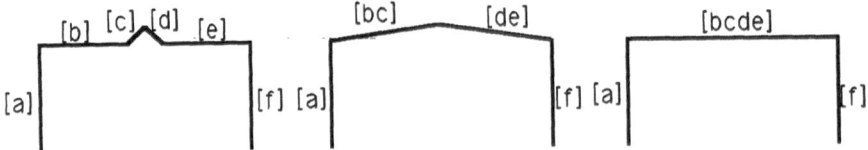

Figure 2. (a) original curve, (b) incorrect interpretation and (c) correct interpretation

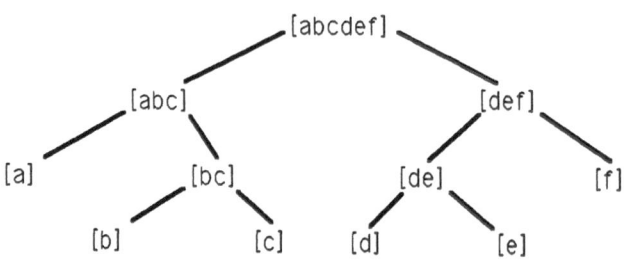

Figure 3. The interpretation tree for figure 2

3.2   Stage (2): combining adjacent lines

To overcome deficiencies such as shown in figure 2 a second stage is used that compares all adjacent lines and combines them if possible. Consider the example of figure 1 again. Figure 3 shows the tree that results from processing this curve. Note the two lines that replace parts of the original curve [bc] and [de] are not compared at all in the tree, hence the incorrect result. The second stage takes the representations [a], [bc], [de] and [f] and computes the goodness of fit for combinations of a selected representation and its neighbours. Possible representations centred on [bc] when compared and possibly combined with its neighbours are: [abcde], [abc]&[de], [a]&[bcde], and [a]&[bc]&[de] (unchanged) where [AB] indicates one representation made up of previous representations [A] and [B]. Note that [abc] has already been tested as [a] and [bc] are adjacent in the tree. The new representation is the one that gives the lowest significance, the same measure used for the first stage. By iterating over the whole curve until no further improvements can be made, all adjacent combinations are tested. Using the second stage on the example of figure 2b should result in representations [bc] and [de] being combined resulting in the result of figure 2c - the intuitively correct result. This is because the combination of [bc] and [de] should result in a lower significance than [bc] and [de] individually.

For part of the curve of figure 2b, segment [bc] and [de] each have a significance of approximately 1/6 and segment [bcde] has a significance of 1/12. Hence [bcde] is a better representation that [bc] or [de]. For a number of real images, table 1 shows results for the total number of lines that result from the original and improved line fitting algorithm. Each of these images contain a number of generalised cylinders. iccv32 is the image for which results have previously been presented for line and ellipse fitting [8].

	Input data	After 1st stage	After 2nd stage
Image name	Number of pixels	Number of lines	Number of lines
mugs6	20870	1195	1109
iccv32	14391	897	728
can3	12887	917	774
cyl21	17659	1257	1166

Table 1. Results for line fitting.

Stage (2) has improved the results in all three images reducing the total number of lines. The use of significance means that the algorithm favours lines that are more significant than others. Note a weighting factor can be used to force the algorithm to favour new combinations i.e. longer lines. Weighting the new combinations by a factor less than 1.0 will give the new line an artificially lower significance forcing the replacement by a longer line.

Figure 4 shows results for the image *cyl21*. For the original image of figure 4a, figure 4b shows the results of the new algorithm. The differences between the results of the old and new algorithms are shown in figures 4c and 4d. Figure 4d shows the lines that replace the lines shown in figure 4c. Note that many old lines have been replaced by fewer longer new lines.

4 Ellipse Fitting

4.1 Stage(3): binary search tree

The basic ellipse fitting algorithm has been presented elsewhere [8]. However a short description is required for completeness. The result of the line fitting algorithm is used as the input of the ellipse fitting algorithm which is, like the line fitting algorithm, based on a binary search. Figure 5a shows the output of line fitting for a curve consisting of two straight lines and one ellipse. In the original algorithm, an iterative Kalman filter was used to fit an ellipse to the endpoints of the curve. A standard least mean square conic fitting algorithm is now used. Although this can generate hyperbolae and parabolae, it usually generates ellipses if the curve is an ellipse. Non-ellipse fits are ignored by giving the result a high significance value so it is always replaced. Replacing the Kalman filter has not altered the results significantly but the algorithm is now faster. Then the list of lines is split into two lists at the vertex of maximum deviation from the ellipse. The algorithm is then repeated for the two lists. Figure 5b shows the resulting interpretation after ellipse fitting which is not correct. The single elliptic arc has been segmented into two elliptic arcs. The intuitively correct result is shown in figure 5c. Figure 6 shows the interpretation tree for figure 5b. Note that the two elliptic arcs [bcd] and [efg] do not get compared in the tree.

When the length of a list of lines reduces to only 5 lines (or 6 vertices) the ellipse fitting is halted and tail recursion used to choose the combination of lines and ellipses that best describe the list of lines in the sense of significance. At each node in the tree the single ellipse describing the list of lines is chosen if it has a better significance than any of the representations below it in the tree.

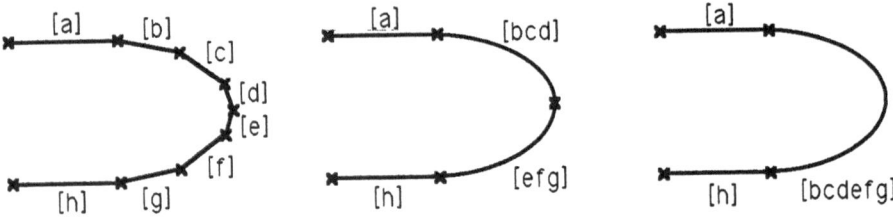

Figure 5. (a) result of line fitting, (b) incorrect oversegmentation, (c) correct segmentation

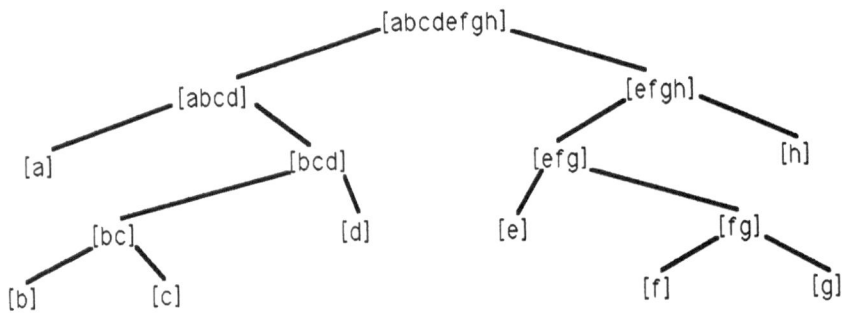

Figure 6. Interpretation tree for figure 5

A number of limitations occur with the algorithm which can be minimised by the following techniques.

4.2 The use of soft breakpoints

Fitting ellipses to line data is more efficient that fitting to pixel data because of the reduction in the number of points for ellipse fitting. However this prevents attempts to fit an ellipse to less that 5 lines (6 vertices) since the fit will be underdetermined. This leads to poor results as ellipses may be missed. It may be that an ellipse would be a better fit to the pixel data than the line. To overcome the problem of not being able to fit an ellipse to less that 5 lines, the concept of soft breakpoints (edge pixel coordinates) is introduced. Consider the hypothesis that an ellipse fits the pixel data better than one line. To confirm the hypothesis 6 data points are required so 4 soft breaks points taken equidistant along the pixel data are used along with the two vertices of the line. If the ellipse is a good fit to the data, a lower significance value should occur. This is the worst case scenario and the algorithm is adaptive such that soft break points are only used when less that 5 lines are used. With three lines only four vertices are available so one additional soft break point is needed for each line. For two lines, two soft breakpoints per line are required. As such it a variable resolution technique that adapts to the amount of data present using the minimum required for the fitting so reducing computation. The alternative of always fitting an ellipse directly to the pixel data is rejected because of the increased computation in the least mean square conic fitting.

4.3   Stage (4): combining ellipses with adjacent ellipses/lines

To overcome the problem of using a binary tree, combinations of adjacent ellipses and lines are tested in a way similar to that described for stage (2) of line fitting. An ellipse will be combined with one or both of its neighbouring lines/ellipses if the significance is better.

Stage (4) is an iterative process that examines each ellipse and determines if it can be extended by combining it with the adjacent representation (line or ellipse). There are four possible outcomes: (i) do not combine, (ii) combine with the left representation, (iii) combine with the right representation and (iv) combine with both representations. For each of these possibilities, the significance measure (as used in the previous three stages) is determined. The resulting representation chosen is the one that results in lower significances (and hence more accurate fits). Iterating over all the ellipses until no further changes can be made results in ellipses growing until further increase in size results in reduced accuracy of fit.

	Input data	After stage 3		After stage 4	
Image name	No. of lines	No. of lines	No. of ellipses	No. of lines	No. of ellipses
mugs6	1108	489	132	447	130
iccv32	728	393	76	381	72
can3	774	451	89	432	88
cyl21	1166	664	142	620	140

Table 2. Results for ellipse fitting.

The results of table 2 show that the additional stage of ellipse growing results in reduced numbers of ellipses and straight lines. Longer ellipses are being generated by combining with lines and with some other ellipses. It is interesting to note that there is an appreciable reduction in the number of lines which implies the use of the point of maximum deviation is not an optimal segmentation method for ellipses. However the shortcomings are reduced by the use of the new stage. The effect of the new algorithm is shown in figure 7 for image *cyl21*. Figure 7a shows the results of the improved algorithm. Figures 7b and 7c show the differences between the two algorithms. Note that in most cases one ellipse has been grown by combining it with lines. However, in some cases, an ellipse has been grown by combining with ellipses and lines.

5   Results

To further demonstrate the performance of the new algorithm, results for a number of images are shown in table 3. The results of the new algorithm are shown along with the differences between the new four stage algorithm and the original two stage algorithm. The differences show the original combinations of features and those replaced by the new stages.

	Two stage algorithm			Four stage algorithm		
	After stage 1	After stage 2	After stage 2	After stages 1 and 2	After stages 3 and 4	After stages 3 and 4
Image name	No. of lines	No. of lines	No. of ellipses	No. of lines	No. of lines	No. of ellipses
mugs6	1195	597	118	1109	447	130
cyl21	1257	924	77	1166	620	140
iccv32	897	550	56	728	381	72
can3	917	626	48	774	432	88
cup3	518	358	33	477	270	53

Table 3. Results for the old two stage algorithm and the new four stage algorithm.

For all the images, the number of lines has been reduced and the number of ellipses has increased. This is expected because more ellipses are being detected at the expense of lines. Figure 8 shows some of a sequence of images of a predominantly circular object. The straight lines and ellipses have been extracted in all the images and can be tracked reasonably well.

6 Conclusion

The new four stage algorithm shows a significant improvement in performance over the original two stage algorithm [8] for a small increase in computation. In the spirit of the original algorithm, the new algorithm avoids the need for any parameters. There are two reasons for the improvement. The first reason is the use of soft breakpoints meaning that ellipses are adaptively fitted to what can be regarded as the minimum number of points. For a curve consisting of a large number of lines, just the vertices are used. For a part of the curve consisting of less than 5 lines, other points are used to allow an ellipse to be fitted in the overdetermined sense. The effect of soft breakpoints is to enable ellipses to be fitted to more parts of the data than before resulting in more ellipses being detected. The second reason is the use of the extra stages to overcome the shortcomings of using a binary search tree. Lines and ellipses already detected can be extended by combining with other ellipses and/or lines. Currently a minor extension to deal with closed boundaries is being carried out. The combining of adjacent lines/ellipses needs to be performed between the first and last elements of a closed list. Although this is a minor extension it is necessary for completeness. A further extension which can be used is to refit the hypothesised ellipses to the original pixel data to get a better estimate of each ellipse. Hence a good estimate of each ellipse can be determined after segmentation.

The algorithm is made up of a number of stages that can be regarded as splits and merges. The sequence of operations is split and merge (stage 1), merge (stage 2), split and merge (stage 3) and merge (stage 4). Each split uses global information to determine breakpoints and each merge uses local information to remove breakpoints.

The improvements increase the usefulness of the algorithm for many applications which require the detection of ellipses in the 2D image (circles in 3D space) such as detection of generalised cylinders [9] and for bottom up perceptual grouping [7] [12].

Finally the algorithm can be easily modified such that any parametric curve can be detected if a fitting algorithm is available e.g. splines and parabolae, and has been extended to 3D curve data [13].

References

[1] C.-H.Teh and R.T. Chin. On the detection of dominant points on digital curves. *IEEE Trans. PAMI*, 11:859–872, 1989.

[2] M.A. Fischler and R.C. Bolles. Perceptual organisation and curve partitioning. *Trans. IEEE PAMI*, 8:100–105, 1986.

[3] D.L. Lowe. Three-dimensional object recognition from single two-dimensional views. *AI*, 31:355–395, 1987.

[4] T. Pavlidis. Curve fitting with conic splines. *ACM Trans. Graphics*, 2:1–32, 1983.

[5] T. Pavlidis and S.L. Horowitz. Segmentation of plane curves. *IEEE Trans. on Computers*, 23:860–870, 1974.

[6] P.L. Rosin and G.A.W. West. Segmentation of edges into lines and arcs. *IVC*, 7:109–114, 1989.

[7] P.L. Rosin and G.A.W. West. Perceptual grouping of circular arcs under projection. *Proc. 1st British Machine Vision Conf.*, pages 379–382, 1990.

[8] P.L. Rosin and G.A.W. West. Segmenting curves into elliptic arcs and straight lines. *Proc. IEEE 3rd Int. Conf. on Comp. Vision*, pages 75–78, 1990.

[9] P.L. Rosin and G.A.W. West. Extracting surfaces of revolution by perceptual grouping of ellipses. *Proc. IEEE Comp. Soc. Conf. on Comp. Vision and Pat. Rec.*, pages 677–678, 1991.

[10] Ramer U. An iterative procedure for the polygonal approximation of plane curves. *CGIP*, 1:244–256, 1972.

[11] G.A.W. West and P.L. Rosin. Techniques for segmenting image curves into meaningful descriptions. *Pattern Recognition*, 24:643–652, 1991.

[12] G.A.W. West and P.L. Rosin. Non-parametric segmentation of 2d and 3d curves into lines and arcs. *Proc. 2nd Int. Conf. of Automation, Robotics and Computer Vision - ICARV92*, 1992.

[13] G.A.W. West and P.L. Rosin. Perceptual grouping of parallel planar ellipses for the recognition of 3d objects. *Proc. 2nd Int. Conf. on Image Processing - ICIP92*, 1992.

Figure 4 (a) Raytraced image of cylinders.

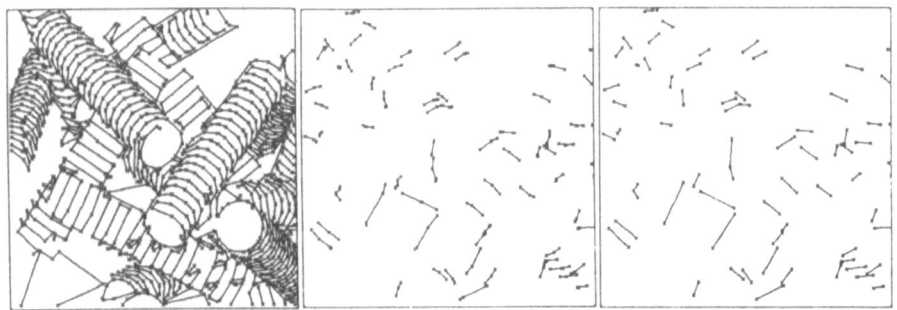

Figure 4 (b) Lines detected by improved algorithm.
Old lines (c) replaced by new algorithm with single lines (d)

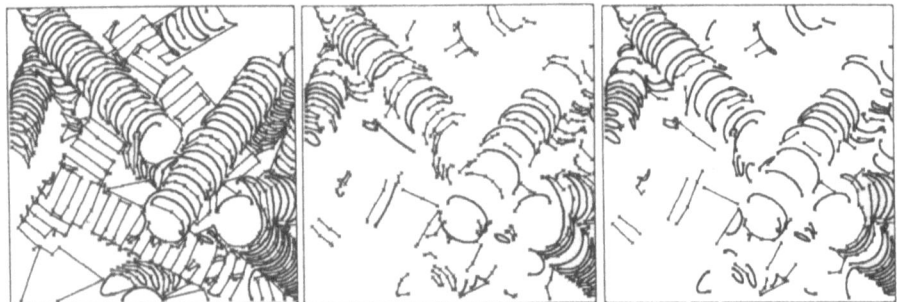

Figure 7 (a) Ellipses and lines detected by improved algorithm
Old lines and ellipses (b) replaced by single ellipses (c).

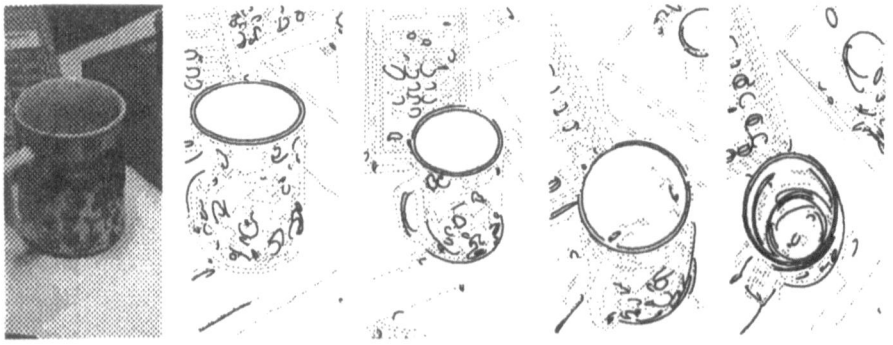

Figure 8. Results for scene from different viewpoints.

Using Colour Templates for Target Identification and Tracking

Simon Brock-Gunn*

Machine Vision Group,
Computer Science Department, City University, London
e-mail: simon@cs.city.ac.uk

Tim Ellis

Machine Vision Group,
Centre for Information Engineering, City University, London
e-mail: t.j.ellis@city.ac.uk

Abstract

A surveillance system is presented which uses colour cues to track people moving in sparse crowd scenes. The erratic motion of the targets, together with their changeable outline, means that they are conventionally difficult to model. However, by examining the colours present in an object in a given frame of a sequence, and looking for these in a later frame, identification and tracking are achieved. The colours are transformed into a template space in which it is easy to match objects to those held in a database, so that even when a person is occluded or disappears totally from view they may be re-located as soon as they can be clearly seen again. A hierarchical approach to template storage and searching reduces the effort required to search through the database and ensures that the system is efficient even with a database containing the templates of hundreds of people. Since work is carried out on a moving sequence, the problems of changing object shape and maintaining colour constancy are minimised by allowing for small changes in these parameters between frames and by continually updating each object's template.

1   Introduction

When examining the problem of following people in sparse crowd scenes, such as in a shopping centre or entering and leaving an aircraft, a number of possible solutions arise. At first, it might seem sensible to model the shape and motion of a person, in order to be able to distinguish them from the rest of the scene [4]. Unfortunately, this task is made difficult by the changing shape of a person, as the camera perceives it, when they move around or change direction and because their movements are influenced by many external and personal factors which are unavailable to a system. Even if we suppose that it is possible to track without modelling the shape [1], any system that follows the target from frame to frame will fail as soon as the targets become occluded when they pass in front of each other or behind stationary objects, such as an advertising

---

*Supported by a SERC award.

hoarding or the aircraft, if it is unable to relocate them when they re-appear. One further consideration which can be overriding, even in current systems, is the storage space and processing power required to maintain large numbers of object representations and to match them between frames in a sequence.

It is clear that if a more general approach is taken to analysing the problem, any practical implementation that arises will be more feasible, but the question remains as to which properties of the application should be examined.

Colour – in machine vision in general and object analysis in particular – is a property which would seem obvious, yet is apparently under-exploited since almost all current machine vision systems involve the use of only monochrome data. When even black and white images take up so much storage space and necessitate so much processing themselves, it cannot seem an attractive option to increase this by at least three-fold in order to utilise the extra information provided by colour data. However, with advances over recent years in computer technology, this is no longer such a concern, and now one of the more salient reasons may be the fact that colour is not an absolute property; rather it depends as much on the frequency content of the illumination as on the surface reflectance. The perceived colours of a target outdoors will change as a cloud passes in front of the sun, or as the target moves into the shadow of a building. The human visual system overcomes some of these problems by using a poorly-understood mechanism referred to as colour constancy [6], whereby the perception of constant colours is maintained under varying illumination levels. However, as yet no widely-accepted algorithms exist in machine vision which will reliably mimic colour constancy and, as a result, colour is seen as an unreliable property.

In this research we are analysing dynamic scenes and so we are able to minimise this problem by allowing the small changes in illumination, and thus the reflected colours, which will occur from frame to frame to fall within a small tolerance level and by continuously updating the object templates to take these changes into account. The way we use colour properties enables us to claim that our method is not sensitive to the small changes in object shape, viewing angle and scale which may also occur between frames.

The use of template definitions naturally avoids the need for large amounts of storage, since a template takes up a fraction of the amount of space that a colour representation of the object requires, and quick object matching within the database is achieved by the use of a hierarchical system of data representation. This ensures that time is not wasted in matching pairs of templates which clearly represent different objects (as will be the case on the majority of occasions) and instead such matches may be dismissed quickly.

Colour templates allow the representation of objects without the need for any type of model. This quality means they can be used in identifying objects of unknown shape and size from a list of possible choices [8], but the principle can be extended so that objects are "learned" by a system, based on the quality of match within frame sequences.

2   Target Identification and Tracking

The underlying technique of the system presented is to look at the objects seen in each frame of a sequence, and learn the combination of colours present

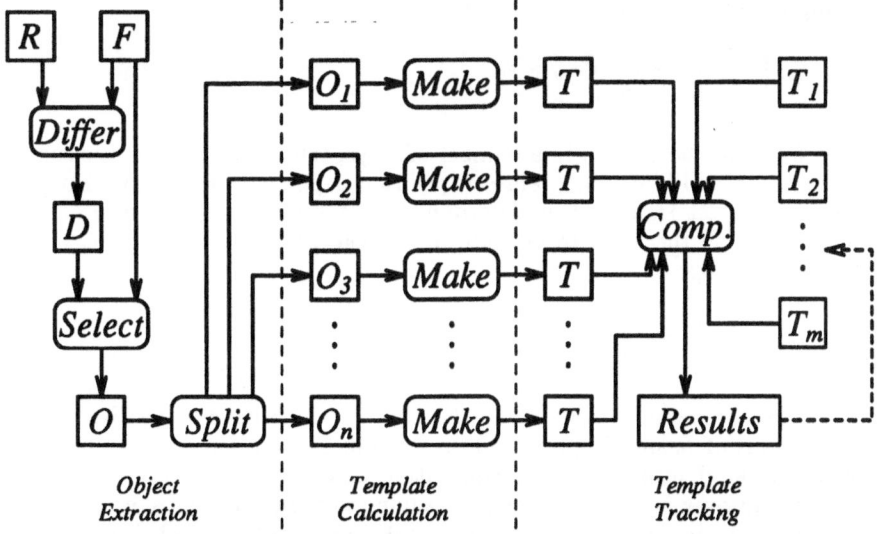

Figure 1: *The Processing Cycle, explained in sections 2.1 – 2.3*

in each object. This combination then becomes associated with the object and is remembered during the rest of the sequence, enabling identification and tracking of targets.

The processing cycle, show in Figure 1, is carried out on each frame and consists of three parts which are detailed below. Firstly, Object Extraction is carried out to locate any moving objects within the field of view and extract them from the image frame. This part of the process is not central to the ideas being presented here and, in our system, a simple and crude technique is used. When the objects have been found, the Template Calculation process examines the colours present within each of them and quantifies these colours by means of a template which can easily be compared to all those previously encountered in the Template Tracking process. The object will be located in a previous frame because of its similar colour "signature", providing a link between the object in two different frames and, therefore, determining its motion.

2.1 Object Extraction

The application uses a static camera, which allows us to assume that there is a one-to-one correspondence between movement in the scene and pixel changes in a frame sequence. This ensures that a simple operation is all that is required in order to extract the objects from a given frame image.

Before the processing begins, a colour reference image, R, is captured and stored. This image represents the background scene, which is composed only of static objects which will remain stationary during processing. Each time around the cycle, this reference image is compared to the incoming colour image frame, F. A resultant binary difference image, D, is then calculated by taking the difference between the intensity images of these two frames and subjecting it to a threshold, t_d, to reduce noise errors:

$$D_{x,y} = \begin{cases} 1 & \text{if } F_{x,y} - R_{x,y} > t_d \\ 0 & \text{otherwise} \end{cases} .$$

$F_{x,y}$ and $R_{x,y}$ may be defined in a number of ways, the simplest being the sum of the red, green and blue point values at position x, y in each respective frame.

Then the object frame, O, is determined by taking an empty image and then copying in all the areas from the frame image where there is an entry in the difference image:

$$O_{x,y} = \begin{cases} F_{x,y} & \text{if } D_{x,y} = 1 \\ 0 & \text{otherwise} \end{cases} .$$

This object frame now contains all the objects which have moved between the acquisitions of R and F, presented on a blank background. These objects are then separated using a simple exploration algorithm which takes all the objects from O over a threshold size, t_o (again, to reduce the effects of noise errors), and relocates them in single object sub-images, $O_1, \ldots, O_n$.

2.2   Template Calculation

There are many different ways of choosing a template, T, such that it represents in some way the colours present in one of these objects, O. Given that the raw data of an object sub-image is generally composed of planes of the three primary colours, red, green and blue, the most obvious and straightforward to calculate is a three-dimensional colour frequency histogram, T_H, which may be created of suitable size $n \times n \times n$ (so n will be the number of bins on each axis), where each $T_{H_{r,g,b}}$ represents the amount of the object which has the colour (r, g, b).

However, we can employ a model that more closely resembles human physiological processes [5] by examining opponent colours. This still gives us three linearly-independent axes, but whereas (r, g, b) represent dimensions of black to red, green and blue respectively, the opponent colour axes of (rg, by, wb) represent dimensions of red to green, blue to yellow and white to black respectively. The transformation [2] from the input data is therefore

$$T_{O_{rg,by,wb}} = \left(r - g, \frac{2b - r - g}{2}, \frac{r + g + b}{3} \right) .$$

This type of template has been used to provide object detection cues [8] for the recognition of static objects, and Figure 2 shows how a template in the rg and by dimensions might look for a person wearing a red shirt and blue trousers. For this template, $n = 16$, so there are sixteen bins on each axis, and regions of intensity (represented by darker areas) are more prevalent in the red and blue parts of the template.

The use of these axes ensures that the template is invariant to rotation, translation and reflection of the object in the image plane since it is only the amount of each colour which is being measured, not its location.

Further analysis of the application shows that it is clear that the objects under study are subject neither to two-dimensional rotation in the image plane

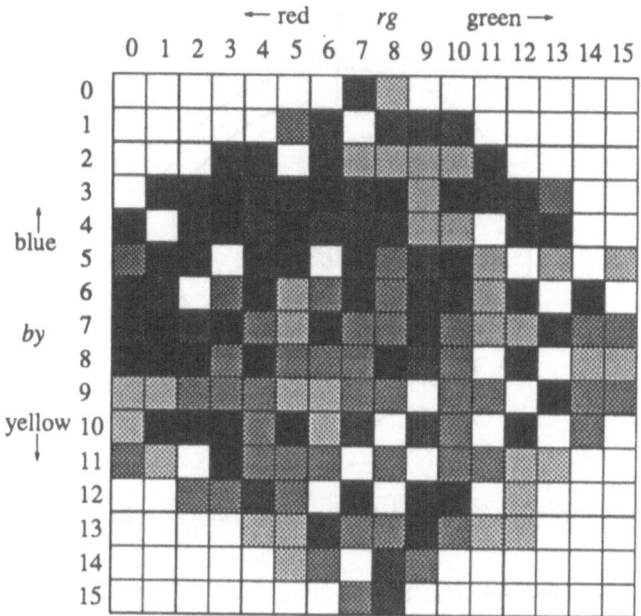

Figure 2: *rg and by plane of a sample template*

nor to reflection. This means that the discriminatory power of the colour may be made greater by weighting the information with respect to its position within the object. When we look at this in the context of our application, it can be seen that the person wearing a red shirt and blue trousers will cause there to be highlights in the appropriate areas of the template. However, someone else wearing a blue shirt and red trousers might be associated with a very similar template, yet it is clear to the observer that these are very different people. By making the spatial distribution of the colours an important factor, it is possible to ensure that such confusion does not arise, without compromising the generality of the system.

To this end, we add two further axes to the template which implement this spatial weighting: r and θ. These are analogous to the polar co-ordinates of a given point, o, in an object, O, such that, as shown in Figure 3, if c is the "centre of gravity" of the object then

$$\theta = \angle vco;$$
$$r = l/r,$$

where v is the point of intersection between the circle, centre c, radius R, which circumscribes the object and a vertical line through c, and l is the distance between o and c.

Finally, it can be recognised that wb is simply a measure of monochrome point intensity, a property which is more susceptible to shadows and changes in lighting than the others being measured. Furthermore, since there are already many methods which make use of monochrome techniques to track objects, and it is our aim to develop the use of a colour technique which may be comple-

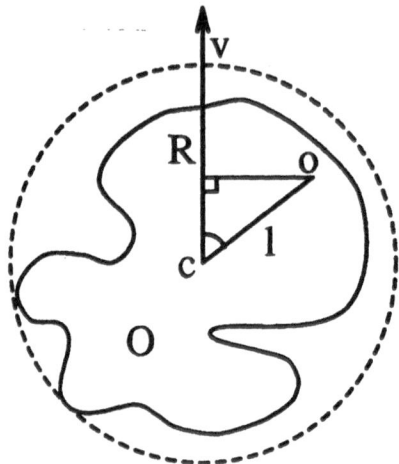

Figure 3: *calculation of the r and θ values*

mentary to these methods, this axis is discarded.

Thus we are left with a template of four dimensions: rg, by, r and θ.

2.3 Template Tracking

The template, T, is subject to a comparison technique together in turn with each of those already stored in the database, $T_1, \ldots, T_m$, where m is the number of templates stored. These T_i are the templates representing all the objects so far encountered. The match coefficient, v, is calculated as

$$v = \sqrt{\sum_{rg=0}^{n-1} \sum_{by=0}^{n-1} \sum_{r=0}^{R} \sum_{\theta=0}^{2\pi} \left(\frac{T_{rg,by,r,\theta}}{S_T} - \frac{Ti_{rg,by,r,\theta}}{S_{Ti}} \right)^2},$$

where

$$S_T = \sum_{rg=0}^{n-1} \sum_{by=0}^{n-1} \sum_{r=0}^{R} \sum_{\theta=0}^{2\pi} T_{rg,by,r,\theta},$$

such that $0 \leq v \leq 1$, where $v = 0$ indicates a perfect match (templates identical) and $v = 1$ indicates a perfect mismatch (templates completely differ). The scaling down by the sum of values in the template, which is the same as the number of image points in the corresponding object, ensures that the coefficient is invariant to scale.

This coefficient is then subject to previously determined minimum and maximum thresholds, t_{min} and t_{max} respectively:

- If $v < t_{min}$ then the two templates are similar enough to say that they represent the same object. In this case, T_i is updated by replacing it with the template definition of T and the object represented by these templates is said to have moved from the position associated with T_i to the position associated with T.

- If $v > t_{max}$ then the two templates are different enought to say that they represent different objects. If $v > t_{max} \forall T_i$ then T is defined as representing an object not previously encountered. A new template, T_{m+1}, is created in the database, consisting of the template definition of T and the object is said to be at the position associated with T.

- If $t_{min} \leq v \leq t_{max}$ the templates are neither sufficiently similar nor sufficiently dissimilar to make a firm decision and the match is said to be inconclusive. This may occur if, for example, T represents a partially-occluded object or two objects occluding each other — situations in which we do not wish to make any judgements about the object associated with T. In this case, T is discarded and there is therefore no information with which to make any decision about the position of any object.

The values of t_{min} and t_{max} are determined by experiment, and will be implementation-independent. Ideally, $t_{min} \to 0$ and $t_{max} \to 1$ and it is only the frame-to-frame change in parameters as well as the usual noise errors which mean that more realistic values need to be chosen. These values will depend on the distribution of unique colours across targets, so that if the objects are all similarly coloured t_{min} and t_{max} will need to be close together in value, whereas when the objects are all distinctly coloured they may be further apart.

3 A Hierarchical Approach

The use of a four-dimensional template means that matching can take a considerable time. A modestly-sized template with sixteen bins on each axis necessitates the comparison of over 65000 pairs of values. In a small application, with 30 different people in the database, millions of calculations would have to be performed in the Template Matching stage for each single frame. Even if each calculation took only one millisecond, Template Matching would take several seconds per frame — clearly unacceptable when the requirement is for a real-time system. The fact that so much time needs to be spent in these comparisons seems particularly absurd since the most cursory inspection by an observer reveals that almost all of the template pairs are clearly different.

The approach taken to alleviate this problem is to manufacture a "pyramid" of templates for each object, instead of just one [3]. This pyramid consists of progressively coarser-resolution templates so, for example, a sixteen-bin template would be accompanied by eight-bin, four-bin and two-bin templates. Although this provides a lot of useful new information, the overheads of this extra work are very low. If using four-dimensional, sixteen-bin templates, the highest-resolution template in a pyramid would hold 65536 values (16^4), whereas the next-highest would only hold 4096 (8^4), the next 256 (4^4) and finally the lowest-resolution which would contain just 16 (2^4) values — a combined total of less than ten percent extra. Furthermore, since each successively lower-resolution template is simply interpolated from the next higher-definition template by averaging the appropriate surrounding values, minimal further processing time is required.

In this hierarchical system, each comparison begins with the lowest definition resolution pair of templates. Of course, in almost all cases, the templates

	Single Template System		Hierarchical System	
Database size	Values (millions)	Calculations (millions)	Values (millions)	Calculations (millions)
30	2.0	2.0	2.1	0.1
100	6.6	6.6	7.0	0.1
1000	65.5	65.5	69.9	0.5

Table 1: *Storage requirements for a template database and processing requirements for comparing a new template with those currently in the database, exemplifying the advantage of a hierarchical system over a single-template system*

will actually be representing different objects, and in the majority of these cases, the difference between the templates is sufficient at this low level to determine that they are not representing the same object. Only if $v \leq T_{max}$, such that the objects are not definitely different, need the process continue with the next higher resolution pair, and so on until the highest resolution pair confirm that the templates are representing the same object (or otherwise). In practice, when trying to match an object's template with 30 others stored in a database, on about 25 occasions the lowest-resolution, two-bin template will provide sufficient information to dismiss the match, and only five will require examination of some of the higher-definition templates, with only one (the correct object match) necessitating a look at the highest-definition, sixteen-bin templates.

As Table 1 shows, the advantages of the pyramid system become abundantly clear as the size of the database increases.

4 Experimental Results

A simple sequence of frames was chosen in order to demonstrate the system in practice. This sequence depicted two people walking towards each other, passing (occluding) and walking away and was augmented with several still frames of other people, in order to provide a database of seven people.

A hierarchical arrangement of four dimensional-templates was used, with sixteen bins on each axis of the highest-resolution template. The system successfully tracked the two moving people and suspended tracking for the period of occlusion, as required. Within the hierarchical implementation, all of the matches into the database were determined to be false at the two-bin or four-bin template level (except, of course, the correct match). This was to be expected since the subjects chosen for the database were all clearly different to the observer, and the choice of more similar-looking subjects would undoubtedly require comparison of the higher-resolution templates.

Figure 4 shows six sample frames from the sequence. Alongside these is shown one of the targets which has been selected to be tracked. During the third frame, this target is lost, as it is occluded. However, the target is matched and found again for the fourth and subsequent frames.

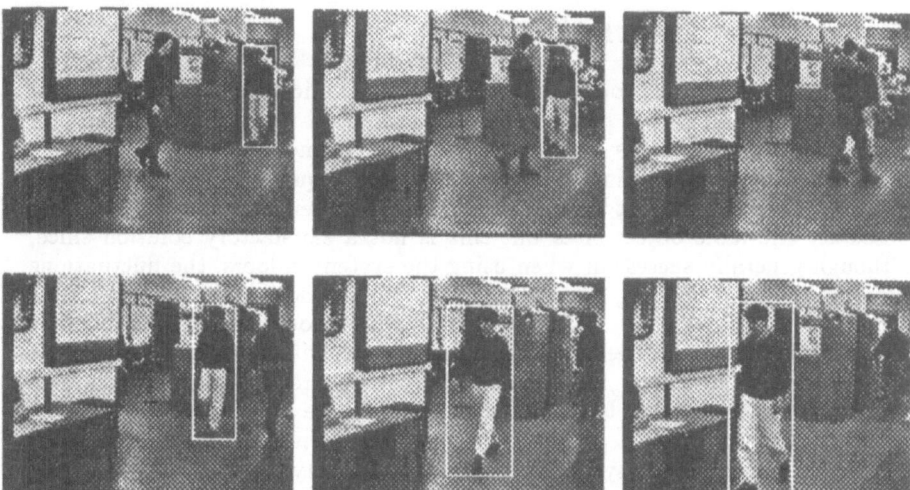

Figure 4: *Six frames from the test sequence and a target which has been tracked*

5 Conclusion

Since the system proposed is very general in nature, it clearly has many potential applications in addition to our own. However, because of its particular advantages, there are three areas of tracking to which it is particularly suited:

- *Where the target objects are of similar shape and/or size.* There is no need to model any object-based parameters such as shape or size so, for example, it would be able to follow people in a street or cars on a road, whereas a system based on distinction by shape and size alone would fail.

- *Where the target objects have irregular motion.* Since there is no predictive element involved, targets may be tracked no matter how they move through the scene since at any given time they may be expected to appear in any part of the image or, indeed, not at all.

- *Where object occlusion occurs.* If an object disappears from view or is occluded for a while it will be re-located when it can be sufficiently seen again as long as the change in colour reflectance over the intervening period falls within the allowable small tolerance level, so there is no need to attempt to track it through every single frame.

- *Where a large database of objects is necessary.* Many database systems suffer because the time taken to search through the database for an object match is proportional to the number of objects held (or even proportional to some higher power of the number of objects held). By using a hierarchical storage and processing system, the time taken to search through the database increases only slowly as the database grows.

It is not suggested that the system presented here is capable of performing in all situations, rather that combination with existing techniques may be the best way of providing a solution to any given application problem.

6 Further Work

The Object Extraction process does not form an explicit part of this work and the simple method chosen has proved to be somewhat unreliable under certain circumstances where areas of the background sometimes show up as "moving" due to a change in lighting conditions between acquisition of the reference image, R, and the current image, F. These are currently removed by a filter based on allowable object sizes but this is not a satisfactory solution since, although generally successful when using the system indoors, the fluctuations sometime cause processing of outside scenes to break down. As one of the aims of this work is to reduce problems associated with colour constancy, a different method of generating R needs to be employed which will ensure the background is constantly updated [7, 1]. Furthermore, there are potential applications where the camera will not be stationary and so a more sophisticated method of Object Extraction must be used which will also take into account ego-motion.

The usefulness of this system depends on the ability to carry out processing in real time and, although the use of hierarchical storage and processing helps enormously in the Template Tracking stage (especially with a large database), the time involved in producing a template from an input frame is still of the order of several seconds (on a Sun SPARCstation). Although the ability to process at frame-rate is not a requirement for real-time processing, some rationalisation of program and storage structures must be undertaken to reduce the program cycle time towards the sub-one second scale. Specialist hardware would probably be required to achieve this but the repetitive nature of the tasks involved make this an ideal candidate for implementation in a parallel environment.

References

[1] A. T. Ali and E. L. Dagless. Computer vision for security surveillance and movement control. In *IEE Colloquium on Electronic Images and Image Processing in Security and Forensic Science*, pages 6/1–6/7. IEE, May 1990.

[2] Dana H. Ballard and Christopher M. Brown. *Computer Vision*. Prentice-Hall, Inc., Eaglewood Hills, NJ, 1982.

[3] Peter J. Burt. Smart sensing within a pyramid vision machine. In *Proceedings of the IEEE*, volume 76, pages 1006–1015. IEEE, August 1988.

[4] David Hogg. Model-based vision: a program to see a walking person. *Image and Vision Computing*, 1(1):5–20, February 1983.

[5] Leo M. Hurvich and Dorothea Jameson. An opponent-process theory of color vision. *Psychological Review*, 64(6):384–404, 1957.

[6] Peter Lennie and Michael D'Zmura. Mechanisms of color vision. *CRC Critical Reviews in Neurobiology*, 3:333–400, 1988.

[7] Simon W. Lu. A multiple target tracking system. In *Proceedings of the SPIE*, volume 1388, pages 299–305, 1991.

[8] Michael J. Swain. Color indexing. Ph.D. thesis, University of Rochester, New York, November 1990.

Machine Vision Inspection of Web Textile Fabric

Dr. L. Norton-Wayne, M. Bradshaw and A.J. Jewell,
Department of Electronic and Electrical
Engineering and CIMTEX Centre,
Leicester Polytechnic,
Leicester, England.

Abstract

The paper describes instrumentation which uses machine vision to inspect rolls of web textile fabric in real time. This involves detection of "message" signals arising from defects buried in noise caused by fabric structure. Analogy with the detection of targets in radar and sonar is exploited to provide effective signal processing. The hardware implementation achieves efficiency with economy by using standard devices wherever possible - such as a CCD linescan camera for sensing and a 486 PC as host processor. A special interface card is provided which compensates for deterministic noise and eliminates more than 99% of the redundant data gathered by the camera.

1. Introduction

Textile fabric is often corrupted by defects introduced during manufacture or subsequent processing. It is customary therefore to inspect the fabric visually, as a moving web; this task is boring and hence inefficient and unreliable. A human inspector notices perhaps 60% of defects present, and copes with 2 metre wide fabric moving at about 30 cm/ second. The performance target for the work reported in this paper is to detect at least 95% of significant defects down to 2 square millimetres area, and to cover fabric 2 metres wide moving at 1 metre per second. It is hoped eventually to identify defects by type so that their causes may be ascertained and corrected. Initial identification work aims merely to specify defects as being from one of four groups based on superficial appearance:- along web, across web, no preferred elongation, slubs.

Though several fabric inspection machines are already being offered commercially, these are much too expensive for application to be widespread. The processing methods used remain undisclosed. Hence the present work.

This paper reports on a research programme aimed at producing an automatic web fabric inspection system whose target cost is perhaps an order of magnitude less than for present systems. Cost effectiveness is achieved by following two strategies:-

1) Using as far as possible hardware components (sensors, processor boards and computers) which are already available commercially and are hence inexpensive.

2) Configuring the processing as a sequence of consecutive stages, in which each stage passes only the small fraction of incoming data likely to include defect information and hence important enough to be delivered to the subsequent more

expensive processing. Processing at each stage can become steadily more elaborate and hence more powerful.

Thus, although the camera must acquire 8 million eight bit pixels each second, only about 100 kilobytes of this (1.25 %) need reach the host computer even in the worst case. Inexpensive, PC-type computers can thus be used as hosts; there is no need for expensive parallel machines.

The research programme aims to produce a configuration (lighting, viewing etc), a processing scheme, and also software algorithms to implement the processing and report the results of the inspection.

2. Previous Work

There have been many attempts to apply machine vision to automate the inspection of moving webs, for a wide range of materials including tinplate [1], cold rolled steel strip [2], sand paper [3], etc in addition to textiles. Detection of defects is generally reasonably easy, provided the defects have sufficient contrast.

Identification of the defects is much more difficult; statistical pattern recognition has often been tried. Promising results have been reported for tinplate; [1] for example quotes 80% correct identification for ten classes using a linear classifier, but tinplate has a smooth and highly reflective surface. Moreover, the results are for simulation only, in which isolated samples of digitised data from real defects were processed off-line. Whether these results would be maintained in on-line operation is hard to judge. Chittineni [3] reports 82% correct identification of defects on sand paper with a linear classifier using only individual scans, which improves to 91% when tentative assignments from successive scans are combined. This is very impressive but the work considered only four classes of defect. Logan and MacLeod [8] reported 80% correct identification for steel strip using linear and quadratic feature space classifiers.

When Hill [3] attempted to apply a linear feature space classifier to identify defects in cold rolled steel strip he was able to achieve only 55% correct in his simulations despite using a least mean square linear classifier which was very carefully designed and thoroughly evaluated. His investigation however considered 37 classes of defect, which were of low contrast compared with the noise arising from surface roughness worsened by laser speckle. Hill claims reasonably that since random classification of defects from 37 equiprobable classes would yield only 2.7% correct, the 55% correctness he obtained is a significant success. However, the end user needed 85%. The inevitable conclusion is that feature space pattern recognition is inappropriate for classifying defects on a moving web unless the number of classes is small, of the order of five. Feature space pattern classifiers are, further, tedious and expensive to design; classification using decision trees seems more promising.

Several publications [9, 10] have appeared which describe signal processing methodologies applicable to fabric inspection, but none provides a comprehensive and detailed technical description of an actual system.

Some [11, 12] have used standard area cameras with associated hardware, but they require many cameras due to their inherent low resolution, making the systems excessively expensive and complicated. They also require either an extremely controlled lighting environment, or a normalisation pre-processing stage to accommodate lighting variations and unevenness.

3. Principle of Operation

The general arrangement is indicated in figure 1. Light returned from the web is sensed using a CCD-linescan electronic camera. The presence of a defect causes the received signal to rise or fall momentarily, and the resultant peaks or troughs are detected by thresholding.

The complete inspection process can be regarded as the sequence of processes shown below:

A) The initial detection stage establishes that a defect of some kind is present.

B) The delineation stage determines the region covered by a defect; it specifies the information which must be used to identify the defect. Delineation also supplies the extent (length, width) and area of a defect.

C) The final stage is identification. This process is the most expensive computationally.

We shall consider detection first. The ideal case is illustrated in figure 2; however, the signals received in practice look like figure 3. Here, two kinds of noise are present which tend to mask the defect indications: low frequency modulations caused mainly by non-uniform illumination, and high frequency modulations. These latter arise partially from variations in responsivity between the photosites in the photodetector array, and partially from the stitch structure of the fabric. Both the illumination variations and the photosite non-uniformity may be compensated following calibration, but the stitch structure noise is effectively a random signal which cannot easily be removed. It is this stitch structure which restricts a pure thresholding operation to detect only defects whose high contrast exceeds the signal excursions due to stitch noise.

To permit low contrast defects to be detected. an analogy is used with the detection of target signals in radar and sonar. Here the problem is to detect a message (of known form) in the presence of noise introduced by clutter, reverberations, and thermal motions. The detection process may be split into three consecutive stages [4], as shown in figure 4. Stage 2, the fundamental decision process implicit in message detection, comprises the thresholding illustrated in figure 5. The incoming signal is analogue but the output is binary; a considerable quantity of data is discarded at this point. Its success can be improved by enhancing the contrast of defects with respect to noise prior to thresholding (stage 1). Stage 1 is most commonly implemented with a matched filter [4], but there are many other possibilities such as the textural filters described in [8]. The form of the message signals generated by defects is very variable, and two dimensional filters may be needed. In certain circumstances, the signal is a vector, and linear transformations may be used to enhance contrast [4].

Stage 3 exploits the property that the false alarm triggers generated by random noise are spaced uniformly and at random over the surface being examined, whereas those due to defects form compact clusters. The map of triggers resulting from stage 2 is therefore scanned and isolated triggers are removed. The trigger clusters which remain are almost invariably due to defects. Many alternative schemes are available for eliminating noise triggers; they differ in efficiency, convenience and in the corruption they cause to genuine defects. Some of these are compared using mathematical analysis in [7], which demonstrates how effective they are quantitatively.

4. Static Experiments

These used a purpose built system, capturing 512*319*8bit images and storing them to disc for subsequent processing. Back lighting was initially used, but due to the open structure of the fabric, an excessive number of noise triggers is generated by the detector. Front lighting was then tried, and although the defect signal amplitude was reduced, its contrast relative to the fabric structure noise was very much improved.

An adaptive threshold (figure 5) is used which compensates for uneven illumination on a pixel by pixel basis, with positive and negative thresholds to detect light and dark defects respectively. Selection of the threshold parameters is automatic, partly due to the wide variations in thresholding requirements of even very similar fabrics, and partly to avoid error prone operator intervention. It involves quantifying the spread of noise in the image, which is Gaussian (Figures 6a and 6b), and calculating the threshold parameters from the equations:-

$$T_{upper} = \mu + k\sigma \qquad \text{and} \qquad T_{lower} = \mu - k\sigma,$$

where k is a constant between 2 and 5. Altering k alters the sensitivity of the detector, as shown in Table 1.

Table 1. Triggers Generated With Varying Detector Sensitivity

k	$P(k)$	Quantity of triggers predicted	Quantity of triggers counted
5	0.0000006	0.1	0-10
4	0.000064	10	30-50
3	0.0027	441	500-1000
2	0.0454	7415	5000-12000

Selection of the optimum value for the threshold in terms of the noise variance (specified by k) is vital to ensure that subsequent stages of noise trigger elimination and defect delineation are effective. It was found that setting the thresholds 4 standard deviations from the mean produced optimal results.

Figure 6a shows the PDF for a region of fabric which is defect free, figure 6b shows the PDF for a region containing a defect. The defect is manifest to the left of the main distribution. This high contrast is typical of fabric defects.

The delineation process then aims to associate triggers arising from the same defect using tests of local adjacency. Initial results using static boundary spatial distance measures are good, correctly clustering all the triggers arising due to a defect with the local triggers arising from fabric structure distortions around the defect. Triggers due to noise are clustered individually in small, localised groups. Needle lines and horizontal defects, which tend to be broken into many smaller clusters by the detector, have their component parts clustered together correctly (Figure 6c). It is envisaged that better results will be obtained by dynamically modifying the adjacency model according to a predetermined plan, biasing it either horizontally or vertically. Research into the use of an optimal estimation filter to implement this is underway.

As the delineation stage works on edge (transition) information only, it is very fast and highly memory efficient, processing many thousands of clusters per second and using around 40 bytes of memory per active cluster.

Placing a size threshold and trigger density threshold on each cluster provides a test for distinguishing a cluster arising from a defect or from noise. The clusters identified as defects are then subjected to a 2 dimensional linear feature space classification system that assigns a cluster to a group based on information obtained from the delineation stage. The catchment rectangle shape factor and cluster size are used to classify a cluster into one of the following classes:-

Horizontal, Vertical, Local, Slubs. (Slubs are essentially horizontal, but have peculiarities that enable a separate classification to horizontal defects.)

A number of images containing various defects from a range of greige fabrics have been subjected to the processing strategy outlined above. In each case all the defects present were correctly detected, delineated and classified, with no false alarm signals being generated. It was found that fabrics with wider stitch spacing resulted in more triggers arising due to noise being present in the signal, but these were very satisfactorily dealt with by the delineation scheme.

5. Experimental Online System

The system used for initial experiments on a moving web (rather than on isolated samples) is based on a commercially available manual tubular fabric inspection machine. The front side only of the web is viewed by a 2048 element linescan camera with the array axis across the fabric perpendicular to the direction of movement. This gives a resolution of 0.5mm across the 1m web. Lighting (figure 1) is provided by a fluorescent tube operating at 35kHz to avoid flicker. The tube is about 25mm away from the fabric; a simple mask with a slit for the camera field of view shields the lens from direct light. Because only the centre portion of the fabric is illuminated only the centre 0.5m width is inspected although the whole width is scanned. The camera is focussed below the tube and perpendicular to the surface; a dark strip is placed under the fabric to maximise the contrast of hole-like defects (since the greige fabric being inspected is light coloured).

The initial system comprised a Fairchild CAM1500R linescan camera, Sentel CCU-M frame store and linescan interface board installed in a 33Mhz 386 AT compatible computer, and a standard Sheltons Tubular Inspection Machine (TIM). A relay driver was incorporated to stop the machine when a defect was found, for evaluation purposes. The set-up was good enough to perform useful inspection, albeit at a slow speed of 5cm/second. The interface board performed no processing, passing the digitised grey scale data to the host computer.

The final system uses a board designed in-house that performs not only the thresholding operation, but binary filtering and data compression as well. The binary filtering examines two adjacent trigger pixels, and outputs a signal based on the rules shown in Table 2.

Table 2. Binary Filter Operating Rules

Input	Output
00	0
01 or 10	No change
11	1

Because of the good noise reduction properties of this filter, the thresholds can be set closer to the signal mean, 3 standard deviations away as opposed to 4 without the filter, thus detecting lower contrast defects. An advantage of using lower thresholds is that more defect information is produced (Table 3). Also, the filter is not destructive to clusters that arise due to a genuine defect (as these are compact), but only to noise triggers which are spread randomly throughout the image. Figure 7a shows the result of using 3 standard deviations for thresholding an image, and figure 7b shows the result of filtering with the binary filter.

Table 3. Comparison of Results With and Without Binary Filter

Binary filtered?	Threshold	Triggers	Individual clusters	Associated clusters
✗	3 SD	633	143	6
✔	3 SD	394	16	1
✗	4 SD	272	16	1
✔	4 SD	NA	Not applicable	Not applicable

Table 3 shows that the defect is correctly delineated using 4 SD's with no binary filtering, and also when using 3 SD's with binary filtering (figure 7c). However, the latter approach retains 45% more information, 394 pixels as opposed to 272, giving a more accurate delineation and classification.

The signal processing methodology described above was successfully implemented into the online system using the following sequence of operations:
do
    grab line of data
    threshold data
    perform binary filtering
    delineate any clusters present
    if cluster complete
      if cluster is due to defect
          classify defect
          output results, or stop machine
while not end of roll
As in the static trials, all defects present were detected. A few false alarm clusters were however generated.

6. Hardware

To achieve the objective of low cost, a commercially available line scan processor board was used originally that was specifically designed for high speed adaptive thresholding. This board could not realise our full operational requirement of 2000 lines/second for a number of reasons, and to achieve the target operating speed a special interface card has been constructed. Its functions are as follows:-
[a] To allow the system to accept and synchronise with exposure and readout pulses generated by the camera, and to receive and condition the video pulses received from the camera.
[b] To correct for deterministic noise corruptions in real time. Individual correction must be provided for each pixel.
[c] To perform the thresholding operation fundamental to defect detection.
[d] To examine the triggers generated by operation [c] and reject as many as possible which have arisen from random noise.
[f] To transition encode the trigger signals selected as being most likely to indicate

defects with their locations over to the PC host, along with certain other information, such as end and edge of web.

Operation [a] is made difficult because there is no standard interface for linescan sensors as there is for two dimensional cameras, and so the interface used in the prototype stage is specific to the IPL linescan camera. Each pixel in the camera signal is digitised to 8 bits and then compared with stored upper and lower threshold values specific to that pixel. This combines thresholding and correction for deterministic noise in one operation. Only those pixels whose values lie outside the thresholding band are retained as potential defects. Some of these are noise pixels arising from the fabric structure, but because they are isolated most can be removed by the simple non-linear binary filter explained previously. It is necessary to choose a form for this filter which is both effective and easy to implement in hardware.

The defect data and end-of-line markers transferred to the computer are further reduced by transition encoding; the four possibilities, transition to light defect, transition to dark defect, end of defect and end of camera scan line are distinguished by two data type bits. These bits together with the pixel address of the transition are passed to the FIFO's for writing by DMA transfer to a circular buffer in the computer.

The development system uses an IPL 5000 series 2048 element line scan camera driven by the camera internal clock; the master clock, exposure strobe and combined video signals are passed to the interface. The exposure pulse is extended to cover the dark reference camera pixels and is used to clamp the video signal and to control the timing. In normal operation the camera scans continuously passing only the essential data to the computer buffer via the FIFO's. The data is processed asynchronously but must be removed from the 64kbyte buffer fast enough to avoid buffer overflow; the processing time increases with the number of the defects present at any given time.

At the start of a fabric roll a number of lines of camera data are read and averaged to provide a relatively noise-free reference line of data. Thresholds for each individual pixel are then computed and written to the threshold RAM's to initiate defect detection. The threshold values reflect changes of pixel sensitivity, uneven light transfer through the lens and non-uniformity in the lighting so that the effects of these anomalies are effectively eliminated. During the inspection process new pixel data is periodically read from the camera and used to update the reference line and the thresholds so that slow changes in lighting and temperature drifts are countered.

All data transfers are 8-bit in the prototype interface which appears as four 8-bit I/O ports and an 8-bit DMA channel to the computer; the 16-bit DMA data needs two 8-bit transfers accessing the FIFO's alternately. A hardware line counter, read as two bytes from separate ports, is provided but its main use has been checking that the number of line markers counted is correct. The prototype is of wirewrapped construction and is based on a standard prototyping card on which buffering and initial decoding are already provided; programmable logic is used for many of the decoding, status and control functions. A 16-bit version is now being developed to improve the data transfer rates. This version will use a pcb and will make extensive use of programmable logic both to reduce the chip count and to simplify modification for different camera parameters.

7. Concluding Remarks

The system has recently been shown in the laboratory to detect all significant defects on greige fabric 1 metre wide moving at 1 metre/second. Doubling of the resolution to

224

enable full width (2 metre) fabric to be inspected to the same specification awaits completion of the updated interface card. Further work is however required, to extend system capability to identify defects by type, to cover patterned and multicoloured material, and to examine "difficult" fabrics such as denim which is dense, dark and has unusual defect types such as loose threads.

Stage (1) of the defect signal enhancement sequence has not as yet been utilised. The analogue processing required is more difficult to implement than the purely digital processing which seems to have been adequate so far. Analogue contrast enhancement may well have to be incorporated to cope with more difficult fabrics.

8. Acknowledgements

This work forms part of the CIMTEX project ongoing at Leicester Polytechnic, whose objective is to apply state-of-the-art technology to apparel manufacture. The consent of the directors of CIMTEX to publication of this paper is gratefully acknowledged, as is the co-operation of Alan Shelton Ltd who are aiming to exploit the results of the work.

9. References

[1] L. Norton-Wayne, W.J.Hill and R.A.Brook, Automated Visual Inspection of Moving Steel Surfaces. Brit. Jnl. of NDT vol.19 no. 5 pp.242-248.
[2] W.J. Hill, L. Norton-Wayne and L. Finkelstein, Signal Processing For Automatic Optical Surface Inspection Of Steel Strip. Trans. Inst. MC vol.5 no.3 1983, pp.137-154.
[3] C.B. Chittineni, Signal Classification for Automatic Industrial Inspection. Proc. IEE. vol.129 pt.E no.3 May 1982 pp.101-106.
[4] L. Norton-Wayne, The Detection of Defects in Automated Visual Inspection. Ph.D. Thesis, The City University, London, 1982.
[5] M.I. Skolnik, Introduction to Radar Systems. 2nd edition, McGraw-Hill, 1980.
[6] F.S. Cohen, Z. Fan and S. Attali, Automated Inspection of Textile Fabrics Using Textural Methods, IEEE Transactions on Pattern Analysis and Machine Intelligence, No. 8, August 1991, p803-808.
[7] L. Norton-Wayne, Non-linear filters for removing noise from binary images. Proc. 3rd IEE Conf. on Image Processing and its Applications, Warwick, July 1989.
[8] I. Logan and J.E.S MacLeod, An Application of Pattern Recognition Algorithms to the Automatic Inspection of Strip Metal Surfaces. Proc. 2nd Intnl. Jnt. Cnf. Pattern Recognition, 1974, p286-290.
[9] I. Tufis, Automated Fabric Inspection Based On A Structural Texture Analysis Method. Pattern Analysis and Recognition, Spring 1989, p377-390.
[10] S. Ribolzi et al, Online Fault Detection on Textile Material by Opto-Electronic Processing. Intelligent Sensor Systems, 1989.
[11] Takatoo, M., Y. Takagi and T. Mori, Automated Fabric Inspection Using Image Processing Techniques. SPIE vol 1004, Automated Inspection and High Speed Vision Architecture II, 1988, p151-158.
[12] Virk, G.S, P.W.Wood, and I.D. Durkacz, Distributed Image Processing for the Quality Control of Industrial Fabrics. Computing and Control Engineering Journal, November 1990, p241-246.

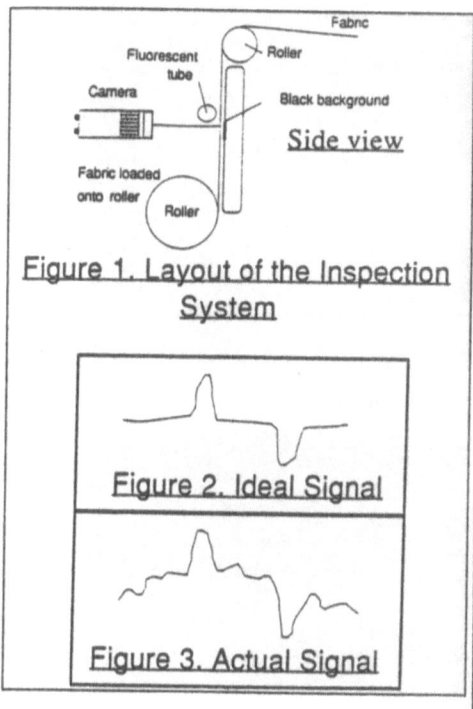

Figure 1. Layout of the Inspection System

Figure 2. Ideal Signal

Figure 3. Actual Signal

Figure 4. Decision Process

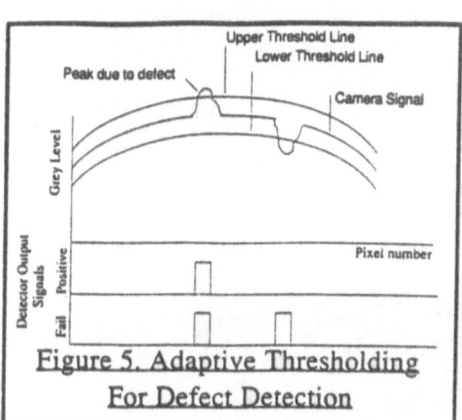

Figure 5. Adaptive Thresholding For Defect Detection

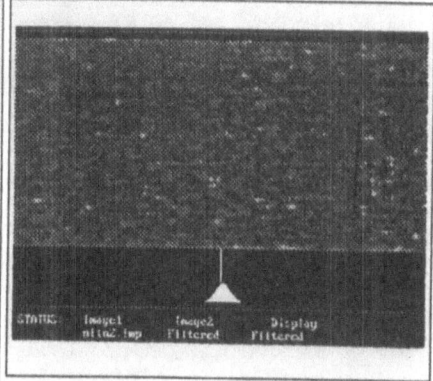

Figure 6a. PDF of whole image.

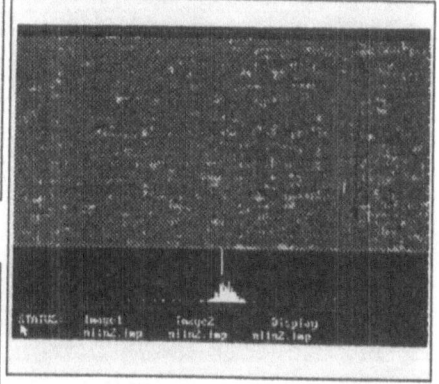

Figure 6b. PDF at the location of a defect.

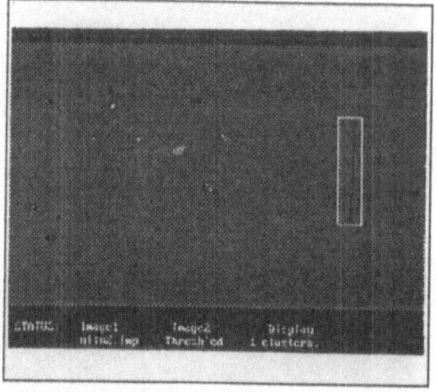

Figure 6c. Noisy binary image successfully delineated.

226

Figure 7(a). Region thresholded at 3SD.

Figure 7(b). 7(a) after binary filtering.

Figure 7(c). Same region thresholded at 4 SD.

Multiresolution Estimation of 2-d Disparity Using a Frequency Domain Approach

A.D. Calway, H. Knutsson* and R. Wilson.

Dept. of Computer Science,
University of Warwick,
Coventry CV4 7AL, England.

* Computer Vision Laboratory,
Linköping University,
S-581 83 Linköping, Sweden.

Abstract

An efficient algorithm for the estimation of the 2-d disparity between a pair of stereo images is presented. Phase based methods are extended to the case of 2-d disparities and shown to correspond to computing local correlation fields. These are derived at multiple scales via the frequency domain and a coarse-to-fine 'focusing' strategy determines the final disparity estimate. Fast implementation is achieved by using a generalised form of wavelet transform, the multiresolution Fourier transform (MFT), which enables efficient calculation of the local correlations. Results from initial experiments on random noise stereo pairs containing both 1-d and 2-d disparities, illustrate the potential of the approach.

1 Introduction

Estimating the disparity between a pair of binocular images in order to determine depth information from a scene has received considerable attention for many years. Essentially a problem of finding corresponding points in the two views of the scene, the complexity of the task is considerable, involving not only the estimation of relative 2-d displacements, but with the added complication of taking into account such effects as geometric transformations and occlusions. This is reflected in the wide range of approaches to solving the problem that have been investigated [1, 2].

Recently, some of the problems have been successfully addressed by the use of frequency domain methods in the form of phase differencing. Using localised frequency representations similar to that proposed by Gabor [3], local phase differences between bandpass filtered versions of the binocular images provide robust estimation of disparity at sub-pixel accuracy [4, 5, 6]. Incorporation of the methods within some form of multiscale framework also allows for efficient matching to be achieved using coarse-to-fine analysis [4, 6]. Nevertheless, these methods are not without their shortcomings. To the authors' knowledge, no straightforward extension to 2-d disparity estimation has been devised and the use of bandpass filters with constant relative bandwidth over scale would appear to be an unwelcome restriction: significant events in a scene are in general broadband and in any case there is no reason why the disparity between the views of a given object should be directly linked to its size or frequency content. In addition, the technique is critically dependent upon the local frequency properties of the images - dominant frequencies significantly different from that of the filter centre frequency can lead to error.

The work reported here is an attempt to address the problems of phase differencing while retaining the advantages of a frequency domain approach. The estimation of 2-d disparity is cast in the form of a least squares minimisation problem over all spatial frequencies and is shown to be equivalent to calculating spatial correlations via the frequency domain. Moreover, by defining the scheme within the framework of a generalised wavelet transform, the multiresolution Fourier transform (MFT) [7, 8], the analysis can be based upon local spatial regions over a range of sizes and so enable a fast coarse-to-fine matching strategy to be adopted. This approach avoids the problem of tying disparity to scale and removes the dependence on a known centre frequency associated with phase differencing. In addition, these advantages are achieved without the high computational cost normally associated with correlation methods. Finally, the unified framework provided by the MFT gives potential for extending the scheme to incorporate feature information to further aid in guiding the disparity estimation and to cope with problems such as local geometric transformations. After outlining the theoretical principles of the algorithm and its implementation using the MFT, results of experiments on random noise stereo pairs with 1-d and 2-d disparities are presented to illustrate the potential of the approach.

2 Multiresolution Disparity Estimation

The purpose of this section is to outline the main features of the algorithm and to indicate its relationship with phase based methods. Towards this end, consider a 2-d image $x(\vec{\xi})$ at a depth Δ from the reference (vergence) plane in a binocular system, where $\vec{\xi} = (\xi_1, \xi_2)$ is the coordinate vector in 2-d space. Ignoring any effects such as scaling, the left and right images in the system are related by

$$x_R(\vec{\xi}) = x_L(\vec{\xi} + \vec{d}) \tag{1}$$

where the 2-d disparity $\vec{d}$ is proportional to the depth Δ. This relationship can also be considered in the Fourier domain as

$$\hat{x}_R(\vec{\omega}) = \hat{x}_L(\vec{\omega}) \exp[\jmath\vec{\omega}.\vec{d}] \tag{2}$$

where '.' denotes scalar product, $\jmath = \sqrt{-1}$ and $\hat{x}(\vec{\omega})$ is the 2-d Fourier transform (FT) of $x(\vec{\xi})$. The significance of (2) is that it suggests a means of estimating $\vec{d}$ using the phase of the inner product between $\hat{x}_L(\vec{\omega})$ and $\hat{x}_R(\vec{\omega})$, ie

$$\arg[\hat{x}_L(\vec{\omega})\hat{x}_R^*(\vec{\omega})] = -(\vec{\omega}.\vec{d}) \tag{3}$$

Thus, providing the direction of $\vec{d}$ is known (eg when considering horizontal disparities) and spectral estimates of the left and right images are obtained at some known frequency $\vec{\omega}$, then estimation of $\vec{d}$ can be made using (3). This is the basis of phase differencing approaches. However, the situation is less clear when the direction of disparity is unknown, as is the case in most natural stereopsis problems [9, 10]. In this instance, single frequency estimates will mean that $\vec{d}$ is indeterminate; to determine $\vec{d}$ requires more than one frequency estimate in different radial directions. This then poses the question as to what is the most appropriate way of estimating $\vec{d}$: to use a subset of frequencies or to devise a method using all frequencies. Given that in most natural scenes it is impossible to predict a priori in which frequency bands significant events will lie, it would seem that the latter should be the preferred option. In fact, such a method of solution can be readily formulated in terms of a least-squares problem and corresponds to selecting $\vec{d}$ to maximise the function

$$\rho(\vec{d}) = \mathcal{F}^{-1}[\hat{x}_L(\vec{\omega})\hat{x}_R^*(\vec{\omega})] = \frac{1}{4\pi^2} \int_{-\infty}^{\infty} \hat{x}_L(\vec{\omega})\hat{x}_R^*(\vec{\omega}) \exp[j\vec{\omega}.\vec{d}] \, d\vec{\omega} \qquad (4)$$

where $\mathcal{F}^{-1}$ denotes the inverse 2-d FT. The maximisation therefore amounts to finding the 'phase correction' term $(\vec{\omega}.\vec{d})$ which maximises the inner product between the spectra of the binocular images, ie $\arg[\hat{x}_R^*(\vec{\omega})]$ is 'rotated' so as to minimise the squared error between the spectra. Moreover, (4) can be written in terms of the spatial domain as [11]

$$\rho(\vec{d}) = \int_{-\infty}^{\infty} x_L(\vec{\xi})x_R(\vec{\xi} + \vec{d}) \, d\vec{\xi} \qquad (5)$$

which is just the cross correlation between $x_L(\vec{\xi})$ and $x_R(\vec{\xi})$. Maximising (4) therefore corresponds to finding the peak in the correlation field and the connection between phased based methods and correlation is made clear: the latter provides a natural extension of the former to deal with 2-d disparities, by making use of the whole frequency domain.

Of course, simply computing the global correlation between the left and right images is inappropriate except in the most trivial of cases. In practice, the interocular disparity will be inherently local: objects in a scene are necessarily confined to some finite spatial region and exist at differing depths, implying that any correspondence measurements must also be based on local properties [6]. This can be achieved in the present case by considering local correlations between neighbourhoods in the binocular images, ie find the $\vec{d}$, denoted by $\vec{d}(\vec{\xi}_1, \vec{\xi}_2)$, which maximises

$$\rho(\vec{\xi}_1, \vec{\xi}_2, \vec{d}) = \frac{1}{4\pi^2} \int_{-\infty}^{\infty} \hat{x}_L(\vec{\xi}_1, \vec{\omega})\hat{x}_R^*(\vec{\xi}_2, \vec{\omega}) \exp[j\vec{\omega}.\vec{d}] \, d\vec{\omega} \qquad (6)$$

where the global spectra in (4) are now replaced by the local spectra $\hat{x}_L(\vec{\xi}_1, \vec{\omega})$ and $\hat{x}_R(\vec{\xi}_2, \vec{\omega})$, centred at $\vec{\xi}_1$ and $\vec{\xi}_2$ respectively, and defined according to

$$\hat{x}(\vec{\xi}, \vec{\omega}) = \int_{-\infty}^{\infty} w(\vec{\chi} - \vec{\xi})x(\vec{\chi}) \exp[-j\vec{\omega}.\vec{\chi}] \, d\vec{\chi} \qquad (7)$$

where $w(\vec{\xi})$ is some appropriate window function, ie $\hat{x}(\vec{\xi}, \vec{\omega})$ is a windowed FT reminiscent of the Gabor representation [3]. It is these equations which underlie the disparity estimation used in the present work: derive local correlation fields $\rho(\vec{\xi}_1, \vec{\xi}_2, \vec{d})$

by computing the inverse FT of $\hat{x}_L(\vec{\xi}_1,\vec{\omega})\hat{x}_R^*(\vec{\xi}_2,\vec{\omega})$ and find the peak to give the disparity $\vec{d}(\vec{\xi}_1,\vec{\xi}_2)$.

The above formulation also suggests a means of overcoming the matching problem, ie selecting $\vec{\xi}_1$ and $\vec{\xi}_2$ in (6). If the neighbourhoods used in the correlations are too small, then finding the best match will involve extensive searching, whereas neighbourhoods which are too large will be susceptible to error due to the presence of more than one disparity. As has been previously noted, eg in [4, 6], the solution is to employ some form of coarse-to-fine analysis so that the disparity estimates can be 'focused' over multiple scales. This approach can be incorporated here by defining the local correlations to be dependent upon a scale parameter σ, ie

$$\rho(\vec{\xi}_1,\vec{\xi}_2,\sigma,\vec{d}) = \frac{1}{4\pi^2} \int_{-\infty}^{\infty} \hat{x}_L(\vec{\xi}_1,\vec{\omega},\sigma)\hat{x}_R^*(\vec{\xi}_2,\vec{\omega},\sigma)\exp[\jmath\vec{\omega}.\vec{d}]\,d\vec{\omega} \qquad (8)$$

where the local spectra are now scale dependent and correspond to multiresolution Fourier transforms (MFT) [8, 7]

$$\hat{x}(\vec{\xi},\vec{\omega},\sigma) = \sigma \int_{-\infty}^{\infty} w(\sigma(\vec{\chi}-\vec{\xi}))x(\vec{\chi})\exp[-\jmath\vec{\omega}.\vec{\chi}]\,d\vec{\chi} \qquad (9)$$

ie a 'stack' of windowed FTs in which the locality of the spectral estimates is varied as a function of σ. Using (8), it is therefore possible to derive local correlation fields, and thus disparity estimates, at multiple spatial resolutions via the Fourier domain. Moreover, there is now no longer a link between those disparity estimates and a specific frequency band as in previous multiscale approaches; in this case, the estimates are based on information from the whole of the frequency domain.

It is now possible to summarise the multiresolution scheme used to derive the required disparity field. Starting at some suitably large scale σ_0, local correlations between neighbourhoods centred at the same spatial positions in the left and right images are formed according to (8) and the disparities $\vec{d}(\vec{\xi},\vec{\xi},\sigma_0)$ found which maximise $\rho(\vec{\xi},\vec{\xi},\sigma_0,\vec{d})$. A disparity field $\vec{D}(\vec{\xi},\sigma_0)$ is then generated such that $\vec{D}(\vec{\xi},\sigma_0) = \vec{d}(\vec{\xi},\vec{\xi},\sigma_0)$, ie it represents the current disparity estimate with respect to the left image at spatial position $\vec{\xi}$ and scale σ_0. The scheme then proceeds through smaller and smaller scales ($\sigma_0 < \sigma_1... < \sigma_{m-1} < \sigma_m$), deriving disparity fields at each scale according to the following update rule

$$\vec{D}(\vec{\xi},\sigma_{k+1}) = \vec{D}(\vec{\xi},\sigma_k) + \vec{d}(\vec{\xi},\vec{\xi}+\vec{D}(\vec{\xi},\sigma_k),\sigma_{k+1}) \qquad 0 \le k < m \qquad (10)$$

where the first term on the rhs serves as both the previous estimate and the 'focusing' term - defining the pair of regions to be correlated - and the second term is the disparity update at scale σ_{k+1} based on the current correlation. The local correlations performed at smaller scales are therefore directed by the disparity estimates obtained at larger scales, producing a more refined estimate at each stage. The final estimate is then given by the disparity field $\vec{D}(\vec{\xi},\sigma_m)$ defined at scale σ_m.

3 Implementation

3.1 The Discrete MFT

The algorithm described above is based upon the MFT as defined by (9). This is a generalised form of wavelet transform designed specifically to enable local Fourier

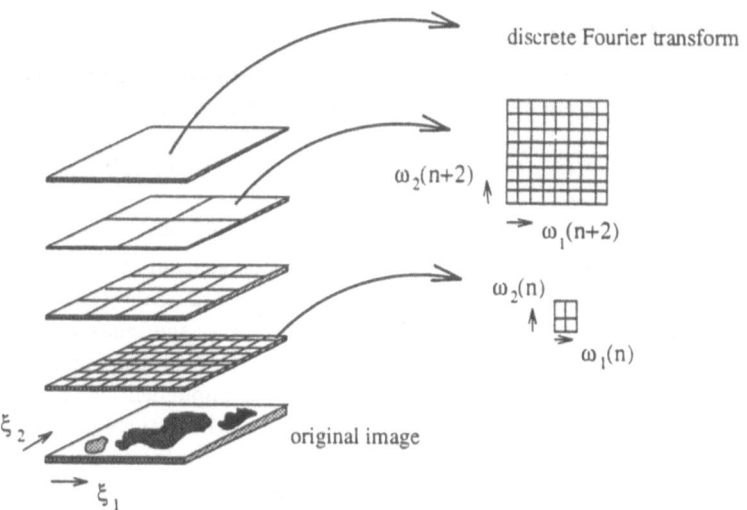

Figure 1: The MFT viewed as a quadtree in which the nodes are assigned the local spectra corresponding to the region "below" the node.

analysis to be performed at multiple scales [8]. A brief summary of the essential properties of the discrete transform is given here. For a discrete 2-d image $x(\vec{\xi}_i)$, its MFT at scale $\sigma(n)$, frequency $\vec{\omega}_j(n)$ and position $\vec{\xi}_i(n)$ is given by

$$\hat{x}(\vec{\xi}_i(n), \vec{\omega}_j(n), \sigma(n)) = \sum_k w_n(\vec{\xi}_k - \vec{\xi}_i(n)) x(\vec{\xi}_k) \exp[-j\vec{\xi}_k.\vec{\omega}_j(n)] \qquad (11)$$

where the discrete window sequence $w_n(\vec{\xi}_i)$ approximates a scaled version of a suitable continuous function $w(\vec{\xi})$, ie $w_n(\vec{\xi}_i) = \sigma(n)w(\sigma(n)\vec{\xi}_i)$. Thus, for some value of $\sigma(n)$, $\hat{x}(\vec{\xi}_i(n), \vec{\omega}_j(n), \sigma(n))$ is a discrete windowed FT of $x(\vec{\xi}_i)$ and corresponds to local frequency estimates centred at spatial positions $\vec{\xi}_i(n)$. As $\sigma(n)$ varies, the spatial and frequency resolution varies, and thus the transform as a whole consists of local estimates over a range of scales.

The two most important factors determining the properties of the MFT are the distribution of the sampling points $\vec{\xi}_i(n)$ and $\vec{\omega}_j(n)$, and the choice of the window sequence. In the present work, the 2-d transform has been formed as the cartesian product of 1-d transforms, and the sampling points in both domains distributed on regularly spaced square lattices of size $N_\xi(n) \times N_\xi(n)$ and $N_\omega(n) \times N_\omega(n)$, where $N_\xi(n)N_\omega(n) = 2N$ for an image of finite size $N \times N$ [8]. The window functions adopted here are bandlimited versions of the prolate spheroidal sequences [11]. These provide maximal spatial localisation and enable efficient computation of the transform using fast Fourier transform techniques [7]. A useful interpretation of the resulting transform is that of a quadtree structure in which the individual nodes are assigned the local spectra referring to the neighbourhood "below" the node and have four associated child nodes whose estimates refer to quadrants of the father's neighbourhood (see Fig. 1). It is this hierarchical framework which forms the basis of the disparity focusing algorithm described below.

3.2 Disparity Focusing

The basic operation employed in the disparity estimation can now be expressed in terms of the discrete MFT coefficients of two binocular images, ie (cf (8))

$$\rho(\vec{\xi}_i(n), \vec{\xi}_k(n), \sigma(n), \vec{d}) = F_{N_\omega(n)}^{-1} \left[\hat{x}_L(\vec{\xi}_i(n), \vec{\omega}_j(n), \sigma(n)) \hat{x}_R^*(\vec{\xi}_k(n), \vec{\omega}_j(n), \sigma(n)) \right] \quad (12)$$

where $F_{N_\omega(n)}^{-1}$ denotes the inverse 2-d discrete FT of size $N_\omega(n) \times N_\omega(n)$ and the correlation is performed between neighbourhoods centred at $\vec{\xi}_i(n)$ and $\vec{\xi}_k(n)$ in the left and right images respectively. A correlation field of size $N_\omega(n) \times N_\omega(n)$ is thereby obtained, in which the position of the peak indicates the relative 2-d displacement between the two neighbourhoods. As a guide to the computational saving obtained by using the MFT, the computational burden of implementing (12) for an $N \times N$ pixel image is in the order of $4N_\omega^2(n) \log_2 2N$ multiplications [7], giving a gain by a factor of around $N_\omega^2(n)/4 \log_2 2N$ over that required for direct calculation of the correlation. For example, for a 256×256 image and a 16×16 neighbourhood, this corresponds to a gain by a factor greater than 5, whilst for neighbourhoods of 32×32 and 64×64, this increases to over 25 and 100 respectively. The saving achieved is therefore considerable, particularly at the larger region sizes.

The disparity focusing algorithm is best described in terms of the quadtree framework discussed above. A level of the MFT, $n = n_0$ say, is chosen as the starting level (typically corresponding to $N_\omega(n) = 64$) and either the left or right channel selected as the reference channel. The algorithm then proceeds as follows (Fig. 2):

1. Cross correlations between corresponding nodes on level n_0 are formed and peak positions in the correlation fields assigned to the relevant nodes in the reference channel.

2. For a father node in the reference channel on level n_0, its child nodes at level n_0+1 are compared with those on the same level in the other channel according to the disparity estimate at the father node. If the estimate is greater than half a block at level $n_0 + 1$ along either or both coordinates, the child nodes are compared with their relevant "neighbours" in the other channel; otherwise they are compared with their corresponding nodes. The peak positions in the resulting correlation fields are used to produce an updated estimate (cf (10)), which is then assigned to the relevant nodes on level $n_0 + 1$ of the reference channel.

3. The process proceeds to level $n_0 + 2$, nodes are compared according to the disparities obtained at the previous level (ie to the nearest block interval) and a new disparity estimate produced. This process then continues through subsequent levels until some final level $n_0 + m$ is reached.

The result of this hierarchical scheme is a set of disparity estimates defined at levels $n_0 \leq n \leq n_0 + m$, with the spatial resolution of each estimate being determined by the corresponding resolution of the MFT level.

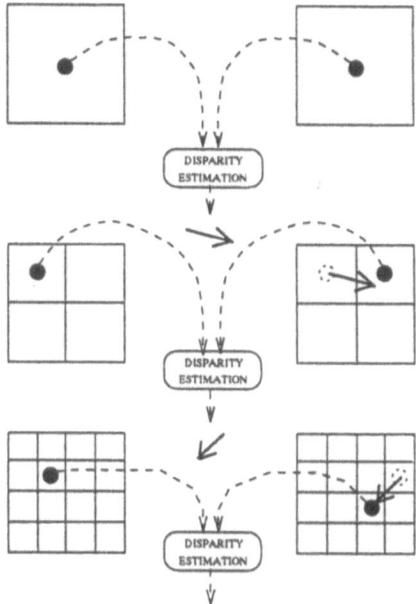

Figure 2: Discrete disparity focusing within the quadtree framework. Estimates obtained at higher nodes dictate which nodes are compared at subsequent levels, yielding an updated and refined disparity field at each scale.

4   Experiments

To test the algorithm, experiments were performed on random noise stereo pairs with horizontal and 2-d disparities. The images were of size 256 × 256 pixels with 8-bit grey level resolution. The MFTs of each image were generated and the levels with $16 \leq N_\omega(n) \leq 64$ used in the focusing algorithm.

The test image pairs are shown in Figs. 4a and 5. The first of these consists of only horizontal disparities, linearly increasing in the positive and negative directions to a peak of ±16 pixels in the centre of the upper and lower halves of the image respectively, ie forming inward and outward projecting peaks when viewed stereoscopically (Fig. 3a). The second pair incorporates 2-d disparities by varying the relative displacement as a function of the radius from the centre of the image. where the disparity at the edges is 16 pixels (Fig. 3b).

Results of the experiments are shown in Figs. 4b and Fig. 6. These show the horizontal and vertical components of the estimates obtained on each of the four levels (only the horizontal component in the case of the first pair, the vertical component being zero). The luminance values (0-255 grey levels) in these images indicates the amount of disparity, where zero disparity corresponds to a grey level value of 128. These results show clearly the focusing steps of the algorithm and the final estimates correspond well to the known disparity variation.

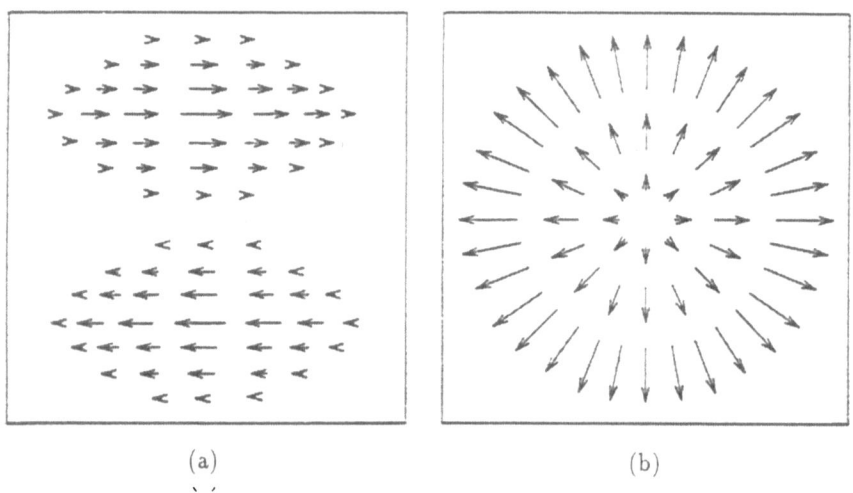

(a)   (b)

Figure 3: Test image disparity variation with (a) horizontal disparities and (b) 2-d disparities.

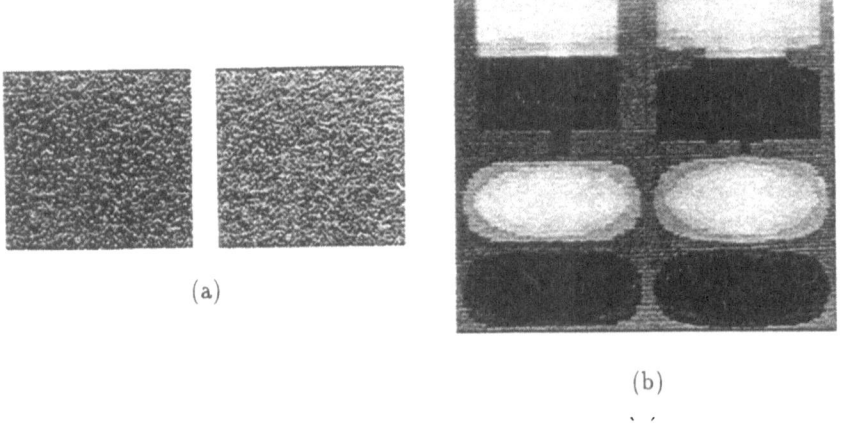

(a)

(b)

Figure 4: (a) Random noise stereo pair containing horizontal disparities. (b) Horizontal component of disparity estimates produced by focusing algorithm from the stereo pair in (a).

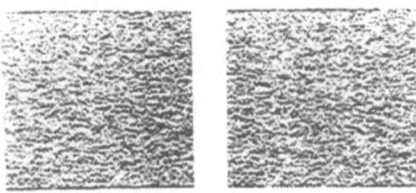

Figure 5: Random noise stereo pair containing 2-d disparities.

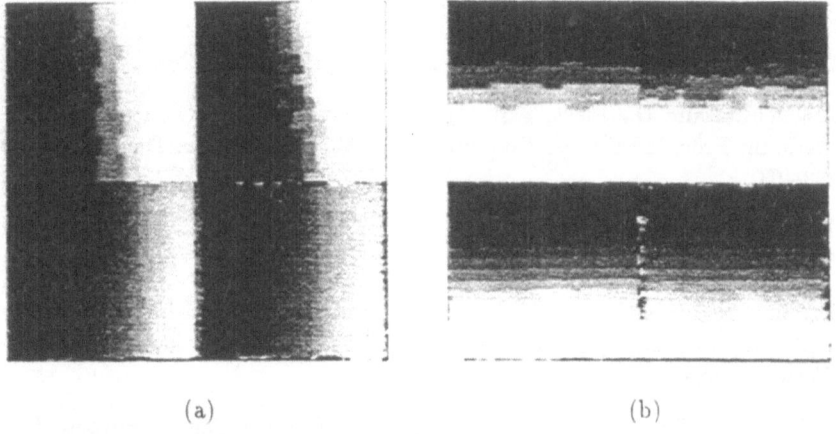

(a)   (b)

Figure 6: (a) Horizontal and (b) vertical component of disparity estimates produced by focusing algorithm from the stereo pair in Fig. 5.

5 Conclusions

An algorithm to compute the 2-d disparity between a pair of binocular images has been presented. The approach is based on the calculation of local correlation fields over multiple scales using a frequency domain method. This has been shown to be a natural extension of phase differencing techniques to deal with 2-d disparities. Efficient implementation of the algorithm is achieved by making use of the MFT. A disparity focusing scheme enables fast matching of corresponding regions in the two images and the results obtained from experiments illustrate the satisfactory performance of the approach.

It should be emphasized, however, that the work presented here is in its preliminary stages. The initial experiments suggest that the approach has considerable potential, although the simplicity of particularly the matching will clearly lead to difficulties when dealing with more complex scenes. Work is under way in extending the approach, most notably on incorporating both local transformation and feature information into the algorithm. Perhaps the most interesting aspect of this work is that due to the flexibility and richness of representation provided by the MFT, the potential exists for incorporating such extensions within the same framework [8].

References

[1] S.T.Barnard and M.A.Fischler, Computational stereo, ACM Computing Surveys 1982; 14, 4: 553-572.

[2] U.R.Dhond and J.K.Aggarwal, Structure from stereo - a review, IEEE Trans. Sys., Man and Cybern. 1989; 19, 6: 1489-1510.

[3] D.Gabor, Theory of communication, Proc. IEE 1946; 93: 429-441.

[4] A.D.Jepson and D.J.Fleet, Fast computation of disparity from phase differences, In: Proc. IEEE Conf. Comput. Vision Patt. Recog., San Diego, 1989, pp 398-403.

[5] T.Sanger, Stereo disparity computation using Gabor filters, Biol. Cybern 1988; 59: 405-418.

[6] R.Wilson and H.Knutsson, A multiresolution stereopsis algorithm based on the Gabor representation, In: Proc. IEE Int. Conf. Image Process. & its Appl., Warwick, 1989, pp 19-22.

[7] A.D. Calway, The Multiresolution Fourier Transform: A General Purpose Tool for Image Analysis, Ph.D. Thesis, Warwick University, 1989.

[8] R.Wilson, A.D.Calway, and E.R.S.Pearson, A generalised wavelet transform for Fourier analysis: the multiresolution Fourier transform and its application to image and audio signal analysis, IEEE Trans. Inform. Th. 1992; 38, 2: 674-690.

[9] H.C.Longuet-Higgins, The role of vertical dimension in stereoscopic vision, Perception 1982; 11: 377-386.

[10] J.Mayhew, The interpretation of stereo-disparity information: the computation of surface orientation and depth, Perception 1982; 11: 387-403.

[11] A. Papoulis, Signal Analysis, McGraw-Hill, New York, 1977.

Estimating mean disparity of stereo images using shift-trials of phase differences

Li-Dong Cai   John Mayhew

Artificial Intelligence Vision Research Unit
University of Sheffield, England

Abstract

Disparity can be estimated using the phase differences between a pair of stereo images after Gabor filtering. In this paper a simple method is proposed for target detection, which uses shift-trials of phase differences to detect the mean disparity over the sampling window up to the precision of one pixel (the sampling interval unit). The process is posed as a problem of minimising the phase differences in terms of their matching residue's norm through direct searching. It involves no computation of the phase derivatives and is therefore robust to noise and phase singularity. Good results are obtained at an acceptable computational cost.

1   Introduction

The estimation of image disparity is a fundamental problem of early biological and computational visual processing. There exists an abundance of literature and methods on this issue. Recently, attention has focused on a new "correspondenceless" technique where disparity is expressed in terms of phase differences in the the left and right views after a local, bandpass filtering. This approach requires neither explicit signal reconstruction or feature detection and localisation. The results may be directly or iteratively used for further depth extraction, texture analysis and motion analysis etc[6, 7].

A kind of widely used methods is to estimate disparity from the local phase differences, as used by Fleet et al. and Sanger[2, 4, 8]. It can be shown that these methods are related to the Newton iterations[1], so the problem of phase singularities remains. To reduce the influence of singular points, edge information can be utilised as proposed in [9], but the problem of how to overcome the phase singularity itself has not been addressed. We propose here a simple method to deal with the singularity problem in purpose of target detection, which uses shift-trials of phase differences over the sampling window to detect the mean disparity up to the precision of one pixel, i.e., the unit of sampling interval. The process is posed as a minimisation of the phase differences in terms of the matching residue's norm through direct searching. It involves no computation of the phase derivatives and is therefore robust to noise and phase singularity. Good results are obtained at an acceptable computational cost.

2   Minimising phase difference by shift-trials

Let $\tilde{L}(x)$ and $\tilde{R}(x)$ be a pair of stereo images of a 1-D signal with a relative shift s:

$$\tilde{L}(x) = \tilde{R}(x + s) \tag{1}$$

This relationship is maintained after a convolution with a function G, such as Gabor filtering [3], over the interval $(-\infty, \infty)$, since

$$L(x) \overset{\text{def}}{=} \tilde{L}(x) * G(x) = \tilde{R}(x + s) * G(x) = \int_{-\infty}^{\infty} \tilde{R}(t + s)G(t - x)dt$$

$$= \int_{-\infty}^{\infty} \tilde{R}(t)G(t - s - x)dt = \tilde{R}(x + s) * G(x + s) \overset{\text{def}}{=} R(x + s) \tag{2}$$

The Fourier components of frequency ω of both images can be presented in the complex forms:

$$L_\omega(x) \overset{\text{def}}{=} \int_{-\infty}^{\infty} \tilde{L}(x)e^{i\omega x}dx = Me^{i\omega x} \overset{\text{def}}{=} Me^{i\Phi_l(x)} \tag{3}$$

$$R_\omega(x) \overset{\text{def}}{=} \int_{-\infty}^{\infty} \tilde{R}(x)e^{i\omega x}dx = Me^{i\omega(x-s)} \overset{\text{def}}{=} Me^{i\Phi_r(x)} \tag{4}$$

where the amplitude M and phase Φ of a complex number $z = a + ib$ is defined as:

$$M \overset{\text{def}}{=} \sqrt{a^2 + b^2} \tag{5}$$

$$\Phi \overset{\text{def}}{=} tan^{-1}\frac{b}{a} \tag{6}$$

In practice, given the (complex) sampling window width $T > 0$, the sampling interval $\Delta = 1$, then the tuning frequency f is limited by the Nyquist frequency $f_c = \frac{1}{2\Delta} = \frac{1}{2}$ to avoid the aliasing effect. That is

$$0 <| f |\le f_c \tag{7}$$

Accordingly, the tuning wavelength λ is limited within the interval:

$$2 \le \lambda \le T \tag{8}$$

Note that any phase $\Phi(x)$ will be modulated into the interval $(-\pi, \pi]$. All detected shifts or disparities s^* will be modulated into the interval $(-\frac{\lambda}{2}, \frac{\lambda}{2}]$ accordingly. That is, any shift s and $\hat{s}$ will be seen as identical under the modulation of λ if they satisfy

$$\lambda - s = \hat{s} \tag{9}$$

$$0 \le s \le \lambda, \quad -\frac{\lambda}{2} < \hat{s} \le \frac{\lambda}{2} \tag{10}$$

This indicates that the range of disparities that can be detected by the phase differencing methods is limited at each tuning wavelength level.

From Eq. (3) and (4), the shift s can be obtained either directly from the phase difference of $\Phi_l(x)$ and $\Phi_r(x)$:

$$s = \frac{\Phi_l(x) - \Phi_r(x)}{\omega} \tag{11}$$

this is the method used in [8], or through Newton iterations of the phase difference converging to zero, such as

$$F(x,s) \stackrel{\text{def}}{=} \Phi_l(x + \frac{s}{2}) - \Phi_r(x - \frac{s}{2}) \rightarrow 0 \tag{12}$$

this is the method used in [2]. Obviously, both methods produce results using local calculations at individual positions and large perturbations may appear at some positions in the presence of singularity of the phase function.

2.1 Problem posed

Disparity estimation is a task dependent problem. For target detection of a moving vehicle, a small sampling window is allowed and the 2-D disparity estimation can be reduced to a 1-D estimation at some scanlines of the image. Meanwhile, disparity estimation can change from local calculations at individual points on $[0, T]$ to global calculations over $[0, T]$ as a whole. These lead to the following shift-trials method, where disparity estimation is posed as a problem of minimising the norm of phase differences over $[0, T]$.

Given a pair of phase functions $\Phi_l(x)$ and $\Phi_r(x)$ over $[0, T]$, let the difference of phases $\Phi_l(x)$ and $\Phi_r(x + s)$ be

$$\Psi(x,s) \stackrel{\text{def}}{=} \Phi_l(x) - \Phi_r(x + s) \tag{13}$$

$\Psi(x,s)$ can be seen as the residue function over $[0, T]$ in the course of $\Phi_l(x)$ being approximated or matched by a family of functions $\Phi_r(x + s)$ which are generated from $\Phi_r(x)$ with (circular) shift trials s over $[0, \lambda]$.

To judge the goodness of approximation, define a norm of the residue $\Psi(x,s)$ as:

$$\nu(s) = \|\Psi(x,s)\| \stackrel{\text{def}}{=} \int_0^T |\Psi(x,s)| dx \tag{14}$$

It is a non-negative functional, whose graph is a continuous curve over $[0, \lambda)$:

$$0 \leq \|\Psi(x,s)\| \qquad s \in [0, \lambda) \tag{15}$$

When the norm $\|\Psi(x,s)\|$ tends to zero, $\Phi_r(x + s)$ converges in *mean* to $\Phi_l(x)$ over $[0, T]$. So the minimum value of $\|\Psi(x,s)\|$ gives the desired shift s^* where the shifted right phase gives the best approximation of the left phase over $[0, T]$, and the resulting shift s^* is thus called the *mean* disparity (to distinguish from the average of disparity values). Hence, by varying s over $[0, \lambda)$ the shift-trials process is virtually a direct search for the minimum phase difference in terms of the norm $\|\Psi(x,s)\|$.

2.2 Norm and singularity

The shift-trials method for target detection has some advantages. First, the disparity detected by the shift-trials method is a global quantity resulting from a function-to-function approximation over the whole sampling window $[0, T]$, rather than a set of local quantities resulting from a pointwise approximation at individual positions $s \in [0, T]$ as in [2, 8].

Second, unlike methods using Newton iterations, the shift-trials method involves no phase derivatives in the process. It is thus free from the pre-conditions of the Newton iteration:

i) A good initial guess $s^{(0)}$ of the real disparity s^*.

ii) An analytic (complex) function $\Phi(x)$.

Notice that it is unlikely for phase differencing methods to satisfy the precondition ii), since the phase function $\Phi(x)$ is not itself analytic everywhere, always leading to large perturbations occurring at some positions due to singularities. In fact, from the ill-defined phase function Φ in Eq. (6), we have[1]:

$$\delta\Phi = \frac{a\delta b - b\delta a}{a^2 + b^2} \tag{16}$$

Hence, when both a and b are small, for example,

$$|a| = |b| < \frac{1}{K} , \qquad K > 10 \tag{17}$$

$$sgn(a\delta b) = sgn(-b\delta a) \tag{18}$$

there must be

$$\frac{|a|}{a^2 + b^2} = \frac{|b|}{a^2 + b^2} > \frac{K}{2} > 5 \tag{19}$$

$$\Phi = tan^{-1}\frac{b}{a} = \frac{\pi}{4} \tag{20}$$

therefore

$$|\delta\Phi| = \frac{|a|}{a^2 + b^2}|\delta b| + \frac{|b|}{a^2 + b^2}|\delta a| > 5(|\delta a| + |\delta b|) \tag{21}$$

$$|\frac{\delta\Phi}{\Phi}| = \frac{4}{\pi}\frac{|ab|}{a^2 + b^2}(|\frac{\delta a}{a}| + |\frac{\delta b}{b}|) > \frac{2}{\pi}(|\frac{\delta a}{a}| + |\frac{\delta b}{b}|) \tag{22}$$

That is, when both a and b are close to zero, any errors δa and δb contained in a and b could be amplified as a fairly large error $\delta\Phi$ contained in Φ, leading to an unstable computation of the phase function. Such severe errors in $\Phi(x)$ result in unreliable disparities.

In contrast, from the definition of the norm of $\Psi(x, s)$ in Eq. (12), it can be seen that singularities will be tolerated in shift-trials as it has no contribution to the integration in the continuous case and only an insignificant contribution in most of the discrete cases.

Hence, it can be expected that the shift-trials method will be more robust to noise and singularities than the Newton iteration methods or other direct searching methods based on local calculations, such as the Golden section method.

2.3   Global minimum and periodic property

As the Fourier components at a specific wavelength λ, both Gabor filtered left and right images are functions with a period λ. So are the left and right phases:

$$\Phi_p(x + \lambda) = \Phi_p(x) \qquad p = l, r \tag{23}$$

Therefore, the norm function of the phase difference $\|\Psi(x, s)\|$ will be a function with the same period λ with respect to shift s, because

$$\|\Psi(x, s + \lambda)\| = \|\Phi_l(x) - \Phi_r(x + s + \lambda)\| = \|\Phi_l(x) - \Phi_r(x + s)\| = \|\Psi(x, s)\| \tag{24}$$

Let the disparity and the shift trial are $s^*, s \in [0, \lambda)$ respectively and suppose the singularities and noise can be ignored. Then from Eq. (3), (4) and (13), we have

$$\Psi(x, s) = \Phi_l(x) - \Phi_r(x + s) = \omega x - \omega(x - s^* + s) = \omega(s^* - s) \qquad (25)$$

So, $\Psi(x, s)$ is a linear, monotonic function over $[0, \lambda)$ in the absence of modulation of phase functions. After being modulated into $(-\pi, \pi]$, $\Psi(x, s)$ becomes discontinuous when it jumps between $-\pi$ and π. However, $|\Psi(x, s)|$ always remains as a piecewise linear, continuous function over $[0, \lambda)$, which has the unique minimum value 0 at the position s^* no matter s^* is zero or not.

Therefore, from the definition in Eq. (14), we have

$$\|\Psi(x, s)\| = \int_0^T |\omega(s^* - s)| dx = |\omega(s^* - s)| T \qquad (26)$$

This shows that with respect to s, the norm $\|\Psi(x, s)\|$ has similar behaviour to $|\Psi(x, s)|$ over $[0, \lambda)$.

Hence, the minimal point s^* is the global minimum point of $\|\Psi(x, s)\|$ over $[0, \lambda)$. Once a minimal point s has been obtained, it must be the desired minimal point s^* of the the norm function.

Note that the singularity of phase functions has insignificant contribution to the norm $\|\Psi(x, s)\|$ and noise has the same effect on each (circular) shift trial $s \in [0, \lambda)$. Therefore it can be expected that the above conclusion can be maintained in practice.

2.4 Computational cost

The computational cost of the shift-trials method relates to the sampling window width T and the disparity s.

Without loss of generality, suppose the shift trials are made over $[0, \lambda)$. When the unknown disparity is zero, there must be $\|\Psi(x, s)\| > \|\Psi(x, 0)\| \ \forall s \in (0, \lambda)$. So, the searching has to run over the whole interval $[0, \lambda)$ as illustrated in Fig. 1.

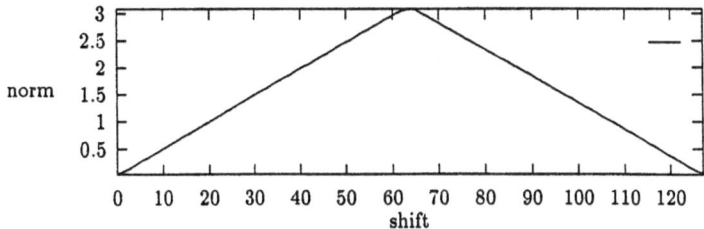

Figure 1: The searching runs over the whole shift range when the minimum point of the norm curve $s^* = 0$ is unknown (wavelength $\lambda = 128$).

On the contrary, when the unknown disparity is non-zero the desired shift s^* must be positive, therefore once a minimal point $s > 0$ has been obtained, further searching becomes redundant. So, when the disparity (*i.e.* the shift s^*) is small, the

242

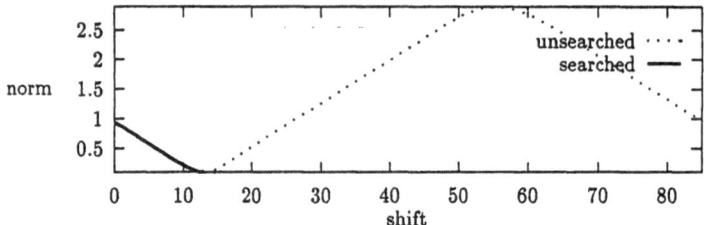

Figure 2: A further searching is unnecessary once a minimal point of the norm curve $s^* = 13$ has been attended.

searching will end at an early stage with a low computational cost as illustrated in Fig. 2.

But worse cases will occur when the disparity $s^* \in [\frac{\lambda}{2}, \lambda)$, as illustrated in Fig 3, where the searching has to finish at a late stage with a higher computational cost.

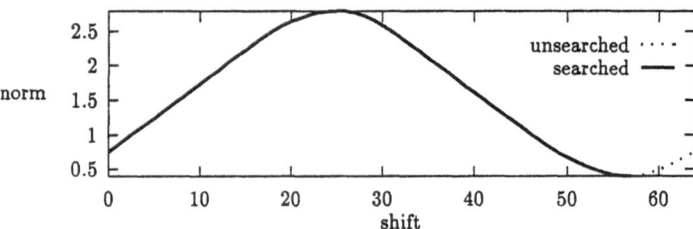

Figure 3: A worse case: more computation is needed in searching for the un-modulated minimum shift point $s^* = 57$ when wavelength $\lambda = 64$.

In this case, according to the periodic property in Eq. (24), $\|\Psi(x, s)\|$ must first increase, then decrease to the minimum point. Hence, a comparison of $\|\Psi(x, s)\|$ at $s = 0$ and $s = 1$ is enough to indicate the tendency of $\|\Psi(x, s)\|$. By letting $\hat{s} = \lambda - s^*$, we get the following results from Eq. (24):

$$\|\Psi(x, s^*)\| = \|\Psi(x, \lambda - \hat{s})\| = \|\Psi(x, -\hat{s})\| \tag{27}$$

where $\hat{s} \in (0, \frac{\lambda}{2}]$ is right the modulated value of s^* (cf. Eq. (10)).

This means that a search along the inverse direction of shifting will lead to the same modulated minimum point $\hat{s} = \lambda - s^*$, but at a reduced computational cost, as illustrated in Fig. 4.

The above treatment also shows that the searching no longer needs to run over the whole range $[0, \lambda)$ as before, a half range run is sufficient. Hence, the actual computational cost in all cases will be proportional to the disparity s^* itself as well as the sampling window width T. The average complexity of the shift trials method is therefore proportional to the wavelength $\frac{\lambda}{4}$ at which it works.

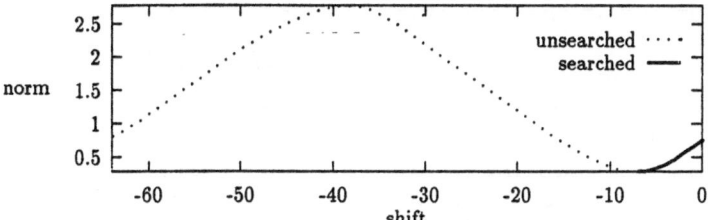

Figure 4: Computational cost is reduced by searching along the inverse direction of shifting for the modulated shift point $\hat{s} = -7$ when $\lambda = 64$.

In this experiment, to reduce the effects of noise, the shift-trial process is applied around the peak frequency of the power spectrum rather than across the whole frequency range $(0, f_c]$. In many cases the frequency range bounded by the left and right image's peak frequencies is a narrow interval, even a single frequency bin. This too reduces significantly the total computational cost at different tuning wavelength levels.

3   Experimental results

The shift-trial method was tested with synthetic and real data. Results of three sets of data are shown in this section. One set of data is a 1-D harmonic signal composed of three frequencies:

$$y = 3\cos\frac{2\pi x}{13} + 4\sin\frac{2\pi x}{13} + 2\cos\frac{2\pi x}{55} + 5\sin\frac{2\pi x}{55} + \cos\frac{2\pi x}{64} + 7\sin\frac{2\pi x}{64} \quad (28)$$

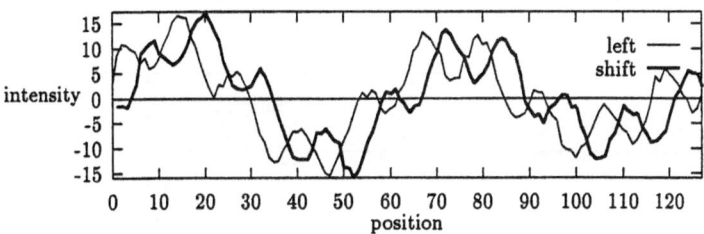

Figure 5: Left and right images of a 1-D harmonic signal.

As shown in Fig. 5 the left image is the signal and the right image its shifted version (circular shift = 5), both images are corrupted by white noise $\sigma(2, 0.7)$.

The shift-trials were made at the tuning wavelength $\lambda = 64$ of the Gabor filtering. The result is illustrated in Fig. 6 along with result yielded by Sanger's method as a comparison.

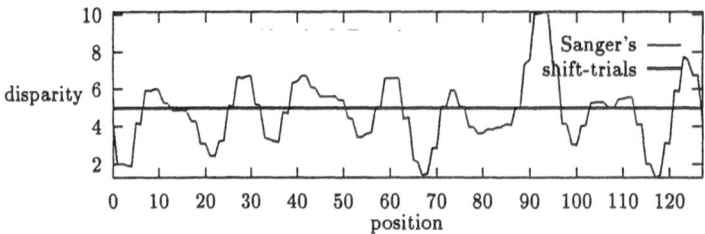

Figure 6: Disparity between the left and right views in Fig. 5 detected by Sanger's and shift-trials' methods respectively.

The second set of data is a 1-D section of a random dot stereogram[5]. The left and right views and the disparities detected by the shift-trial and Sanger's methods (at wavelength = 14) are illustrated in Fig. 7 and Fig. 8 respectively.

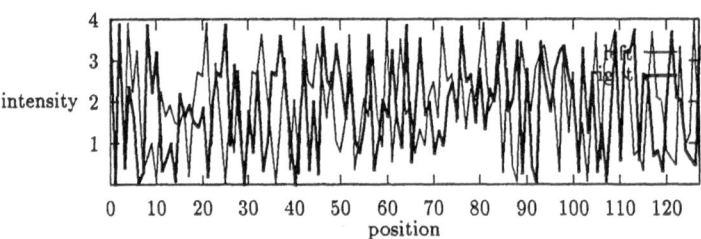

Figure 7: The left and right views of a 1-D random-dot stereogram (target width = 120, image width = 128).

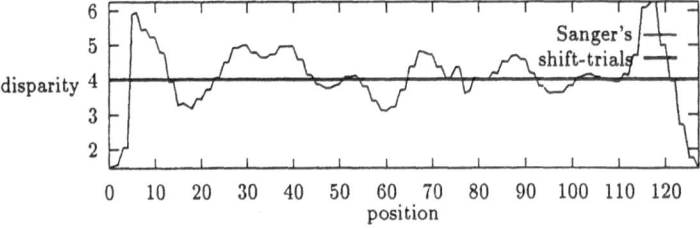

Figure 8: Disparities of the 1-D random-dot stereogram detected by Sanger's and shift-trials' methods respectively.

The third set of data is a pair of stereo images of a shell as shown in Fig. 9. The disparity detection runs for their 1-D section at the scanline = 127 as shown in Fig. 10. The tuning wavelength $\lambda = 85$ is the peak wavelength indicated by a spectral power analysis. Results of both Sanger's and shift-trials' methods are illustrated in Fig. 11.

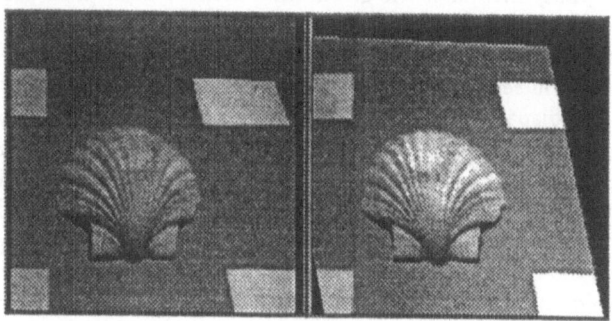

Figure 9: The left and (rectified) right views of a shell.

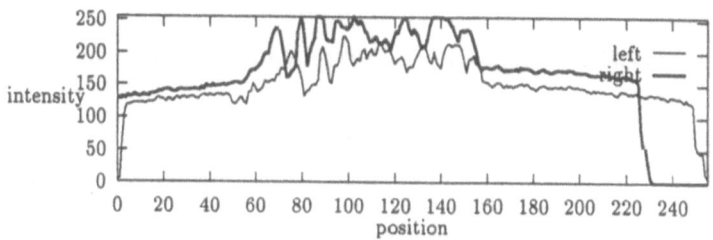

Figure 10: The left and right sections (scan-line=127) in Fig. 9.

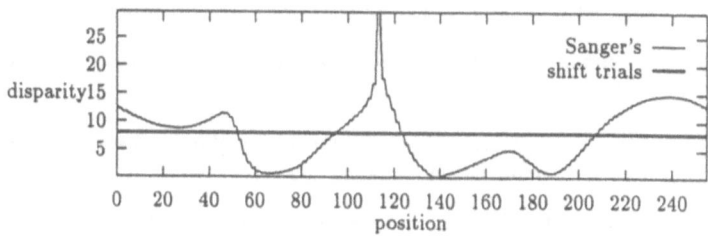

Figure 11: Disparity of the the above sections (scan-line=127) detected by Sanger's and shift-trials' methods respectively.

4 Discussion

So far, the detection of the disparity is up to the precision of one pixel, which is the unit of the sampling interval of the given data. Hence, a sub-pixel precision can be made when a denser data set is produced by using interpolation. However, given that the errors resulting from data noise could be larger than one pixel, a method of one-pixel precision seems enough for target detection in practice.

Note that the disparity detected by the shift-trial method is a global quantity according to the definition of the norm function in Eq. (14). While it is an advantage for shifted harmonic data and real data as shown in Fig. 6 and Fig. 11, it could meet difficulty in some cases of random dot stereograms if the target width is far from the image width, leading to a significant inconsistency between the target's disparity and the background's disparity that is assumed to be zero. As a result the shift-trials method will yield an intermediate value of the above two disparities. However, this problem can be resolved by using a smaller sampling window whose width is close to the size of object width plus margins to account for the expected shift. Thus the effect from the inconsistency can be ignored as shown in Fig. 7.

Acknowledgements

This research was funded by a grant of the ESPRIT VOILA project.

References

[1] Cai, L.D., 1991: A note on Sanger's and Fleet's disparity estimation methods. AIVRU Mem., University of Sheffield, October, 1991.

[2] Fleet, D.J., Jepson, A.D., and Jenkin, M.R.M., 1989: Phase-Based Disparity Measurement. RBCV-TR-89-29, University of Toronto, Nov. 1989.

[3] Gabor, D., 1946: Theory of Communication. *IEE* **93**, 429-459.

[4] Jenkin, M.R.M., Jepson, A.D., and Tsotsos, J.K., 1987: Techniques For Disparity Measurement, RBCV-TR-87-16, University of Toronto, September 1987.

[5] Julsez, B., 1971: *Foundations of cyclopean perception.* University of Chicago Press, Chicago, 1971.

[6] Marr, D., and Poggio, T., 1979: A computational theory of human stereo vision. *Proc. of Royal Society, London* **B204**, 301-328 (1979).

[7] Mayhew, J., and Frisby, J., 1981: Computational studies toward a theory of human stereopsis. *Artificial Intelligence* **17**, 349-385 (1981).

[8] Sanger, T.D., 1988: Stereo Disparity Computation Using Gabor Filters. *Biological Cybernetics* **59**, 405-418 (1988).

[9] Westelius, C-J, Knutsson H. and Wiklund J., 1992: Robust Vergence Control Using Scale-Space Phase Information. LiTH-ISY-I-1363, Linköping University, 1992.

On Transparent Motion Computation

K. Langley

Dept. of Psychology, University College London,
London, U.K. WC1E 6BT

D. J. Fleet

Dept. of Computer Science, Queen's University,
Kingston, Canada K7L 3N6

T. J. Atherton,

Dept. Computer Science, University of Warwick,
Coventry, U.K CV4 7AL.

Abstract

We extend the principle of phase-based techniques for measuring optical
flow and binocular disparity to multiple motion estimation. We analyse
multiple optical flows by estimating phase gradients (instantaneous frequencies) from a set of independent bandpass quadrature filter pairs. Our
approach is similar to that of Shizawa and Mase [22], in which nth-order
differential operators are required to compute n simultaneous velocity estimates. The approach presented here only requires a set of band-pass
filters and their first derivatives.

1 Introduction

Within motion analysis, there are two distinct points of view. There are techniques that detect and track edge or corner features over time (e.g [4]), and those
that compute explicit measurements from the image intensity pattern. Edge-
based approaches must typically deal with the aperture problem, while corner
based approaches generally provide sparse measurements. For intensity-based
approaches this only becomes a problem where the image intensity function has
a one-dimensional structure.

Intensity-based approaches can themselves be subdivided into three different groups. Spatio-temporal energy models[1] compute image velocity from the
relative amplitudes of the outputs of different band-pass filters. This approach,
however, does not perform well when all of the power of the signal lies in the
passband of a single filter. There are differential techniques that measure velocity from spatiotemporal derivatives of intensity of band-pass filter outputs (e.g.
[23]). However, these techniques can be sensitive to noise, geometric deformations between frames, and photometric variations as they assume conservation
of image intensity, or its filtered representation [6].

The approach taken here is phase-based [13, 16]. Measurements are based
on a representation of the signal structure provided by a family of quadrature-
pair bandpass filters. The convolution of an image with a linear bandpass

operator is given by

$$R(\mathbf{x}, t) = I(\mathbf{x}, t) * K(\mathbf{x}, t) = \rho(\mathbf{x}, t) \exp[i\phi(\mathbf{x}, t)] \tag{1}$$

where $K(\mathbf{x}, t)$ is a complex-values bandpass kernel, and $\rho(\mathbf{x}, t)$ and $\phi(\mathbf{x}, t)$ are the amplitude and phase components of the bandpass response. We are assuming that the filters are effectively quadrature pairs and can be expressed as the product of a lowpass envelope and a complex exponential, for example, Gabor filters.

Interestingly, differentiation of the above equation provides two independent equations that can in principle be applied to solve the aperture problem in a local frequency domain model. It should be noted that similar envelope/phase properties are also obtained from differential operators, but the traditional representation of these filter kernels does not pay attention to the bandpass signal. We first consider the motion constraint equation[11] applied to the bandpass signal, assuming $\frac{dR(\mathbf{x},t)}{dt} = 0$:

$$\exp[i\phi(\mathbf{x}, t)]\left[(\rho_x + i\phi_x\rho)v_1 + (\rho_y + i\phi_y\rho)v_2 + (\rho_t + i\phi_t\rho)\right] = 0 \tag{2}$$

where subscripts refer to the direction of partial differentiation with respect to the coordinate frame. This gives two equations:

$$\mathcal{I}m\frac{d\ln R(\mathbf{x}, t)}{dt} = \phi_x v_1 + \phi_y v_2 + \phi_t = 0 \tag{3}$$

$$\mathcal{R}e\frac{d\ln R(\mathbf{x}, t)}{dt} = \frac{\rho_x}{\rho}v_1 + \frac{\rho_y}{\rho}v_2 + \frac{\rho_t}{\rho} = 0 \tag{4}$$

resulting in the following velocity measurement for single motion flow:

$$\begin{bmatrix} v_1 \\ v_2 \end{bmatrix} = \begin{bmatrix} \phi_x & \phi_y \\ \rho_x & \rho_y \end{bmatrix}^{-1} \begin{bmatrix} \phi_t \\ \rho_t \end{bmatrix} \tag{5}$$

providing that the phase and energy derivatives are independent.

The recognition that the human visual system is capable of separately analysing several independent motions at the same point in the image domain, has prompted some authors[20, 3] to investigate the computational rationale behind multiple optical flow analysis. It is hoped that algorithms suitable for measuring multiple image velocities in a single image neighbourhood will be helpful in a wide variety of circumstances, including the superposition of signals, nonlinear transparent phenomena, and several forms of occlusion. Towards this end, several methods have been proposed based on the superposition of two or more translating signals (e.g. [20, 22, 3, 12]). This paper discusses a variation on this theme, using the previous work of Shizawa and Mase[20, 22] as a starting point. It is shown that multiple motions may be computed without the need to estimate second or higher-order derivatives. The problem of computing multiple motion is posed instead in terms of constraints on local measures of instantaneous frequency from different band-pass filter outputs. With a preliminary implementation we find that this method produces reliable estimates of two simultaneous motions of superimposed signals. It also appears to be robust with respect to multiplicative combinations of translating signals, and differences in signal power with respect to transparent surfaces.

The occurrence of more than one legitimate image velocity in a single image neighbourhood may be caused by one of several common phenomena:

- specularities or mirror-like surface reflections like those off a polished floor;

- shadows under diffuse lighting conditions that are seen to move across a stationary surface;

- occlusion such as a single occluding boundary or the fragmented occlusion caused by natural vegetation or certain fences;

- translucency, in which light reflected from one surface is passed through another to the camera, such as stained (or dirty) glass;

- and atmospheric phenomena such as smoke, rain or snow.

In just the past few years several methods that address the problem of multiple image velocities have emerged. For example, some methods compute velocity histograms in relatively local regions of the image [9, 12]. Similarly, Fleet and Jepson[7] showed that local phase information can be used to compute multiple estimates of the normal component of 2-d velocity. However, these techniques do not address the segmentation of the different local measurements to compute separate 2-d velocity estimates. Langley and Fleet[17] have argued that the independence of phase and energy velocity does provide a basis to explain transparent motion to simple signals in human vision. They also noted that the group (energy) velocity of the image signal is not constrained to pass through the origin of the frequency domain, which is one of the properties of multiplicative motion transparency. Bergen et al[3], have derived an iterative method that initially locks onto the one of the motions, allowing it to be cancelled by substrating a deformed version of one frame from another. The same operations can then be applied to the resulting sequence to detect other motions that might exist. However, only Shizawa and Mase[20, 22] have attempted to obtain explicit constraint equations for the analysis of multiple flows from image sequences. Their approach requires that second or higher order derivatives be extracted in space and time, and averaged throughout local apertures in order to estimate the parameters of motion. In order to compute n image velocities simultaneously requires the application of nth-order differential operators.

By contrast, we recast the multiple-flow motion constraint equation in terms of a constraint on instantaneous frequencies of the signals. This as an extension to phase-based methods for measuring image velocity and binocular disparity from the output of band-pass filters [13, 16]. We use the phase gradient, to give a measure of the instantaneous frequency of the filter response as a function of space and time. Instantaneous frequency may be computed from the filter outputs directly, without explicitly representing the phase signal. In addition, we note that the specific form of band-pass filters is not crucial to the approach. Moreover, there exist recent results concerning the general stability of phase information, as well as its potential instabilities, that we may exploit to use instantaneous frequency in a reliable manner [13, 6].

2   Background Theory

With respect to motion transparency, much of the groundwork has already been covered by Shizawa and Mase[20, 22] in terms of understanding the necessary constraints that are required to derive several motions from an image intensity function. Langley and Atherton[14, 15] use a related model to detect corners in images.

The results of Shizawa and Mase[20, 22] are based on the superposition of two translating signals. In this case, not only does the motion constraint equation apply to each of the component signals, but there is a combined constraint that applies to their superposition. For example, let $f(x, t)$ be the sum of two translating signals, $f_1(x, t)$ and $f_2(x, t)$, with velocities $v_1 = (u_1, v_1)$ and $v_2 = (u_2, v_2)$. Individually, the signals satisfy the motion constraint equations:

$$(v_j, 1) \cdot \nabla f_j(x, t) = 0 , \qquad j = 1, 2 , \tag{6}$$

where $\nabla = [\frac{\partial}{\partial x}, \frac{\partial}{\partial y}, \frac{\partial}{\partial t}]$. Their superposition $f(x, t)$ then satisfies:[1]

$$((v_1, 1) \cdot \nabla) ((v_2, 1) \cdot \nabla) f(x, t) = 0 , \tag{7}$$

where $(v, 1) \cdot \nabla = [u\frac{\partial}{\partial x}, v\frac{\partial}{\partial y}, \frac{\partial}{\partial t}]$. When (7) is expanded, the individual terms are found to be :

$$u_1 u_2 f_{xx} + v_1 v_2 f_{yy} + (u_1 v_2 + u_2 v_1) f_{xy} + (u_1 + u_2) f_{xt} +$$
$$(v_1 + v_2) f_{yt} + f_{tt} = 0. \tag{8}$$

Using differential measurements of $f(x, t)$ at five points Shizawa and Mase [21] describe how to compute the individual 2-d velocities. The fitting of three velocity planes through the origin of the frequency domain is a direct extension of this formalism to include higher-order differential operators.

3 Constraints on Instantaneous Frequency

It is well-known that the translation of a 2-d pattern has all its power concentrated on a plane in the frequency domain [5]; that is, the Fourier transform of (6) satisfies:

$$\hat{f}_j(k, \omega) = \hat{h}(k) \delta(v_j \cdot k + \omega) , \tag{9}$$

where k and ω are spatial and temporal frequency variables, $\delta(\cdot)$ is a Dirac delta function, and $\hat{h}(k)$ represents the 2-d Fourier transform of the 2-d pattern that is translating. The velocity constraint in (9) is:

$$v_j \cdot k + \omega = 0 , \tag{10}$$

which also follows from the Fourier transform of (6).

In these terms, finding a solution to (8) for the two velocities amounts to simultaneously fitting two planes to the power of $f(x, t)$ in the frequency domain. Towards this end, note that the Fourier transform of (8) is given by:

$$iu_1 u_2 k_1^2 \hat{f}(k, \omega) + iv_1 v_2 k_2^2 \hat{f}(k, \omega) + i(u_1 v_2 + u_2 v_1) k_1 k_2 \hat{f}(k, \omega) + \tag{11}$$
$$i(u_1 + u_2) k_1 \omega \hat{f}(k, \omega) + i(v_1 + v_2) k_2 \omega \hat{f}(k, \omega) + i\omega^2 \hat{f}(k, \omega) = 0 .$$

---

[1] This derivation assumes more than the conservation of f_1 and f_2, as would be required by (6) alone. In (7), because of the cascaded differentiation, it is important that the two velocities v_1 and v_2 be constant as a functions of space and time.

If we factor out $i\hat{f}(\mathbf{k}, \omega)$ from (12), we are left with the constraint:

$$u_1 u_2, k_1^2 + v_1 v_2 k_2^2 + (u_1 v_2 + u_2 v_1)k_1 k_2 + (u_1 + u_2)k_1\omega +$$
$$(v_1 + v_2)k_2\omega + \omega^2 = 0 . \tag{12}$$

In effect, (12) constrains the locations of nonzero power in the frequency domain $(\mathbf{k}, \omega)$ to lie on one of two planes.

Following from equation (3) in the signal domain and noting that the expectation of phase derivatives (instantaneous frequency) under certain circumstances relate to Fourier frequencies[18], we are assuming that a number of independent measurements can be obtained from bandpass filters:

$$\mathcal{I}m[\frac{d\ln R}{dt}] = (\phi_x u_i + \phi_y v_i + \phi_t) = 0 \text{ for } i = 1, 2. \tag{13}$$

when combined for two flows gives:

$$(\phi_x u_1 + \phi_y v_1 + \phi_t)(\phi_x u_2 + \phi_y v_2 + \phi_t) = 0 \tag{14}$$

such that each bandpass filter selectively responds to an individual component of the multiple flow field. Independent measurements are possible by using filters tuned to different scales and orientation. The assumption that individual filters are individually selective to components of the image velocity is not without problems. For example, when transparent image sequences are defined within the passband of an individual filter kernel, reliable discrimination cannot be expected. Further, marked differences in the signal power of transparent sequences may also restrict velocity estimates to the dominant signal using the approach presented here.

Our approach to solving for the two velocities involves finding a solution to the coefficients in (12) that contain the components of $\mathbf{v}_1$ and $\mathbf{v}_2$. We then compute the individual velocities from these terms. For convenience, we rewrite (12) directly in vector form as:

$$\mathbf{a}^T\mathbf{m} = 0 , \tag{15}$$

where $\mathbf{m} = (\phi_x^2, \phi_y^2, \phi_x\phi_y, \phi_x\phi_t, \phi_y\phi_t, \phi_t^2)^T$, and $\mathbf{a} = (u_1 u_2, v_1 v_2, u_1 v_2 + u_2 v_1, u_1 \cdot u_2, v_1 + v_2, 1)^T$. At least 5 independent measurements of instantaneous frequency (at which there is significant power) are required to solve for the five unknown elements of $\mathbf{a}$. Given six or more measurements of instantaneous frequency we have an overconstrained system, and can solve for the elements of $\mathbf{a}$ more robustly. We do this by minimizing the squared error between the model and the instantaneous frequencies; that is, we minimize

$$\sum_j (\mathbf{a}^T\mathbf{m}_j)^2 \tag{16}$$

with respect to $\mathbf{a}$. Differentiating (16) with respect to $\mathbf{a}$, and setting the result to zero produces the linear normal equations:

$$M\mathbf{a} = 0 , \quad \text{where } M \equiv \sum_j \mathbf{m}_j\mathbf{m}_j^T . \tag{17}$$

Equation (17) constrains a to lie in the null space of M, the matrix of outer products. Ideally, in the case of two motions, M has a rank of 5, with a 5-dimensional column space and a 1-dimensional null space. Following Barman et al[2] the goodness of fit of the model may be determined by ordering the eigenvalues of M ($\lambda_1 > \lambda_2... > \lambda_6$) and comparing the ratio of $\frac{\lambda_5 - \lambda_6}{\lambda_1}$ to unity. Note if M has a rank of 3 or 4, the flow fields are ambiguous and cannot be individually determined. The null space is spanned by the eigenvector corresponding to the zero eigenvalue. Therefore, to solve for the elements of a, we compute the smallest eigenvalue of M (which should be zero for two motions), and its corresponding eigenvector. We then scale the eigenvector so that its last element is unity, which produces our least squares estimate of a.

From the elements of a, as described in [21], the individual velocities are found as follows: For convenience, let the computed elements of a be denoted a_j, $j = 1, ..., 5$. Then, the two components of velocity are given by solutions to:

$$u_j = \frac{1}{2}a_4 \pm \sqrt{\frac{1}{4}a_4^2 - a_1}, \quad v_j = \frac{1}{2}a_5 \pm \sqrt{\frac{1}{4}a_5^2 - a_2}. \tag{18}$$

The correct combinations of these roots to one another to obtain estimates of v_1 and v_2 are then determined by a_3. In particular, note that their are only two ways to combine the different estimates of velocity in the x and y directions. Only one of these two will equal the third component of a.

4   Computing Instantaneous Frequency

In order to compute various measurements of instantaneous frequency, we assume that a family of band-pass filters, such as those used by Heeger, or Fleet and Jepson [10, 5], are applied to the image sequence. If the tuning of the filters is sufficiently different, we can assume that the frequency measurements represent independent degrees of freedom of instantaneous frequency measurements.

Instantaneous frequency is defined as the spatiotemporal phase gradient [8, 19] It gives a local approximation to the structure of the filter response in terms of an amplitude-modulated, sinusoidal signal. We measure the instantaneous frequency of the filter output $R(x, t)$ using the identity:

$$\phi_x(x, t) = \frac{Im[R^*(x, t)\, R_x(x, t)]}{|R(x, t)|^2}, \tag{19}$$

where $R^*(x, t)$ is the complex conjugate of $R(x, t)$.

But not all phase gradients are useful in constraining the multiple motions that may exist in the image. First, it is important that more weight be given to those frequencies that correspond to greater amounts of local energy, given by the amplitude of the filter output $|R|$. Second, it is important that the measurements of instantaneous frequency be ignored in regions where the phase of the filter output is overly sensitive to small variations in spatial position of the scale of input. These are detected using the theory of phase singularities described by Jepson and Fleet [13, 5]. They occur because of interference between energy maxima in the power spectrum. Finally, we only expect to be able to isolate transparent motion when there is some parameter (scale, orientation or speed) that can be used to distinguish the different motions. We require at least 5 independent measurements of instantaneous frequencies from filters that respond primarily to only one of the two motions.

5   Implementation and Results

We have completed an implementation of this approach that works for 1-d and 2-d signals. One-dimensional signals allow phase gradients to be displayed as images, an aid to an intuitive grasp of the approach. That is we are using the properties of the bandpass filter to discriminate the phase velocity of transparent motion fields. The details are as follows: Figure 1 shows a 1-d space-time image of superimposed random dots moving at $\pm 18.26°$; velocity is conveniently viewed here as orientation. The energy and phase responses of the bandpass filters are then shown. In the first case (top-right and bottom-left) the filter responds mainly to the rightward moving stimulus. Then we show the phase response of a filter whose tuning bisects the two stimuli. This shows some of the distortion that occurs when applying filters that respond in part to both patterns. For the signal in Figure 1, the mean velocity (orientation) error was $-0.327°$ for leftward motion, and $-0.046°$ for rightward motion, with standard deviations $1.63°$ and $1.58°$.

The experiment was repeated (see Figure 2) with the same stimuli multiplicatively combined rather than superimposed. In this case, we find mean errors of $0.398°$ for left and $0.053°$ for rightward motion with standard deviations of $3.98°$ and $5.1°$. At present, we lack an explanation for the stability of phase information in cases of multiplicative transparency, except that the local support of the filters appears to play a major role in separating the two signals, whether combined linearly or not. We also find that this method is stable with respect to 66% differences in contrast between the two signals. The final results to 1-d sequences (Figure 2) are presented to a translating and dilating noise pattern. In particular, the phase contours from this sequence provide some indication of the problems that arise when two independent flow fields are defined within the neighborhood of support of individual bandpass filter kernels.

The final example in figure 3 shows the velocity and error field in the case of 2-d motion, with two an added random noise pattern superimposed upon the translating tree sequence. Mean errors to both flow fields were found to be $1.2°$ with a standard deviation of $7.0°$. A total of 16 independent Gabor filter kernels and their derivatives were used, defined over a 20x20x20 neighborhood of support and tuned to a frequency magnitude of 0.2 cycles per pixel.

6   Conclusion

This paper outlines a new method for computing multiple optical flows using quadrature-pair filters and their first-order derivatives. The approach is extendable to several independent velocities by increasing the number of filters and modifying our constraint equation. The basic approach is a variation on the theme discussed in detail by Shizawa and Mase. But it offers a substantially different perspective, since it requires only first-order filters, and a mechanism

Figure 1: *Phase and energy contours for 1-d motion sequences (a)* **Left** *Additive transparency (b)* **Right** *Multiplicative transparency. For each image;* **Top left** *Image intensity sequence* **Top right** *phase and* **Bottom left** *amplitude response for a filter tuned to the motion sequence,* **Bottom right** *Phase contours for a filter equally sensitive to both components of motion sequence.*

to estimate instantaneous frequency.

There are a number of advantages in the approach that we have chosen. In particular, our processing paradigm allows higher order (deformation, dilation, rotation) properties of the optic flow field to be derived from further differentiation of the bandpass signal representation ($\nabla \frac{dR(\mathbf{x}, t)}{dt} = 0$).

However, the approach presented here still retains a number of difficulties. The foremost problem is the ability to determine precisely when reliable measurements can be obtained from bandpass filters under transparent motion. Our preliminary results suggest that the energy derivative may play an important role in supporting phase measurements from similar bandpass filters when both provide similar velocity measurements.

References

[1] Adelson.E.H and Bergen.J.R. Spatiotemporal energy models for the perception of motion. *J.Opt. Soc. Am. A*, 2,2:284–299, 1985.

[2] Barman.H, Haglund.L, Knutsson.H, and Granlund.G.H. Estimation of Velocity, Acceleration and Disparity in Time sequences. *IEEE motion workshop*, 44–51, 1991.

[3] Bergen.J.R, Burt.P, Hingorani.R, and Peleg.S. Computing two motions from three frames. *Proc. 3rd ICCV, Osaka,Japan*, 27–32, 1990.

[4] Deriche.R and Faugeras.O. Tracking line elements. In Faugeras.O, editor, *Computer Vision-ECCV 90*, pages 341–345, Antibes,France, 1989. Springer-Verlag.

[5] Fleet.D.J. PhD thesis, University of Toronto, 1990.

[6] Fleet.D.J and Jepson.A. Stability of Phase information. *IEEE workshop on Visual Motion, Princeton.*, 1991.

[7] Fleet.D.J and Jepson.A.D. The computation of normal velocity from local phase information. *IJCV*, 5:77–104, 1990.

[8] Franks.L. *Signal Analysis.* Prentice-Hall,Inc,N.J., 1969.

[9] Girod.B and Kuo.D. Direct Estimation of Displacement Histograms. *Proc. Image Understanding and Machine Vision.*, pages 73–76, 1989.

[10] Heeger.D.J. A model for the extraction of image flow. *J.Opt.Soc.Am*, 4:1455–1471, 1987.

[11] Horn.B.K.P and Schunk.B.G. Determining optic flow. *Artificial Intelligence*, 17:185–204, 1981.

[12] Jasinschi.R and Rosenfield.A. Sumi.K. The perception of visual motion coherence and transparency: a statististical model. Technical report, Univ. Maryland, 1990.

[13] Jepson.A.D. and Fleet.D.J. *Scale-Space Singularities.* (Ed.) Faugeras.O, Proc. ECCV,Antibes , Springer-Verlag, 1990.

[14] Langley.K. Ph.D thesis. *University of Warwick*, 1990.

[15] Langley.K and Atherton.T.J. Inferring the structure of images using multi-local filters. *BMVC91 University of Glasgow*, pages 111–118, 1991.

[16] Langley.K., Atherton.T.J., Wilson.R.G., and M.H.E.Larcombe. *Vertical and Horizontal Disparities from Phase.* (Ed.) Faugeras.O, Proc. ECCV,Antibes , Springer-Verlag, 1990.

[17] Langley.K. and D.J. Fleet. Using group and phase velocity to explain coherent and transparent motion. *AVA conference, manchester*, 1–2, 1992.

[18] Mandel.L. Interpretation of Instantaneous Frequency. *AJP*, 42:840–845, 1974.

[19] Papoulis.A. *Systems and Transforms with Applications in Optics.* McGraw-Hill, New York, 1968.

[20] Shizawa.M and Mase.K. Determining multiple optic flow using spatio-temporal filters. *ICPR90, atlantic City*, pages 274–278, 1990.

[21] Shizawa.M and Mase.K. Principle of Superposition: A common Computational Framework for Analysis of Multiple Motion. *IEEE workshop on visual motion,Princeton,N.J.*, 1991.

[22] Shizawa.M and Mase.K. A unified computational theory for motion transparency and motion boundaries based on eigenenergy analysis. *CVPR91, Hawaii*, pages 289–294, 1991.

[23] Uras.S, Girosi.F, Verri.A, and Torre.V. A Computational Approach to Motion Perception. *Biological Cybernetics*, 60:79–87, 1988.

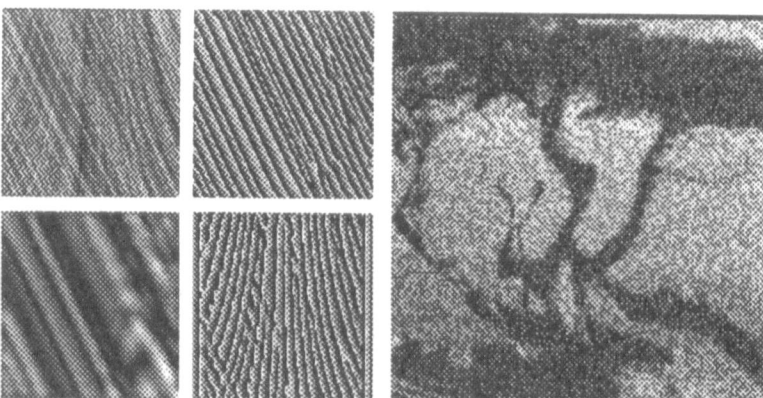

Figure 2: **Left** *A 1-d motion sequence consisting of a dilating and translating noise pattern additively combined. Phase contours are shown for a filter tuned to the translating component and a filter tuned to zero velocity.* **Right** *Frame 15 from a noise pattern moving vertically at one pixel per frame added to a translating tree sequence moving with a mean velocity of two pixels per frame to the right.*

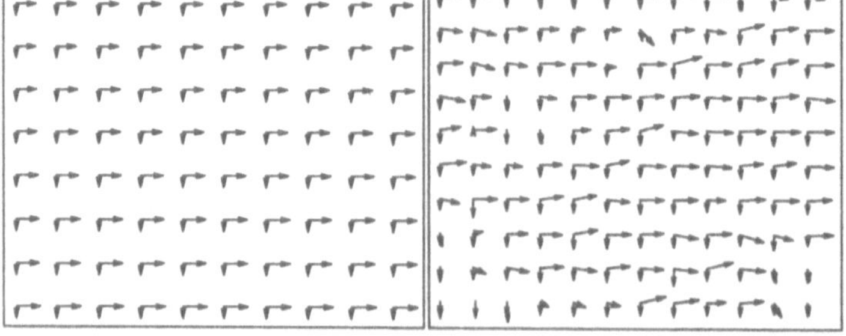

Figure 3: **Left** *True flow field from figure 2.* **Right** *Measured flow field.*

A Neural Network Approach to Recognition of Structural Aberrations in Chromosomes[1]

M. Turner[a], J. Austin[a], N. Allinson[b], P. Thompson[c].
Departments of Computer Science[a], Electronics[b] and Psychology[c],
York University,
York, UK.

Abstract

Neural networks are applied to the problem of detecting structural aberrations
in chromosomes from shape. We present a simple technique for initial location
of scattered chromosomal objects within multi-resolution images of human
blood cells. A system for classifying located objects is also described. It is
proposed that the system be applied to multi-resolution images. Application to
low resolution images is illustrated.

1. Introduction

The rapid development of human cytogenetics in recent years has produced an increased
awareness of the relationships between chromosomal abnormalities and medical disor-
ders, which in turn has lead to a large growth in the volume of clinical microscope work.
There is now an great demand for rapid, accurate and cheap methods for analysing chromosome spreads. Currently, semi-automated systems are available for clinical karyotyping
(1), prevalent in pre-natal screening, where the task is to determine the characteristic chromosome constitution of an individual. However, no systems are available for chromosome
aberration scoring - the search for specific structural abnormalities within a cell.

A number of physical and chemical agents can induce chromosome structural aberrations - breakage and rearrangement of chromosomal material. Affected cells may contain
unusually shaped objects, or have abnormal chromosome constitutions, and this may be
spotted by a skilled cytotechnician using a microscope at high magnification. There is,
however, a shortage of skilled technicians and their training - which for the most part consists of viewing many examples of damaged cells - is time consuming. Attempts at automation have focused on image segmentation, measurement of predefined features, and
classification based upon these measurements (2-4). However, this approach has failed to
cope with the general variability in chromosome appearance. Chromosomes are often
bent, or they may touch or overlap. Diffraction effects make objects appear fuzzy under a
light microscope. In addition, the existence of staining gradients and artefacts is troublesome.

We use neural network systems acting on multi-resolution images in an attempt tackle
the problem of variability. Neural networks are adaptive systems of interconnected units
which acquire knowledge through experience rather than preprogramming. Knowledge is

1. Supported by SERC grant 06R00174

stored implicitly in the connection strengths, and recalled in response to the presentation of cues. We hope that by presenting many examples of image portions to neural networks that they can learn through experience to locate chromosomal objects, and to extract image features which permit centring and classification of located objects. Simple, supervised, single-layer networks are employed in the initial location of scattered chromosomal objects within a cell. Kohonen self-organising maps (5) are used to extract salient image features within windows placed in the neighbourhood of located objects. Feature extraction forms the first stage in a centring and classification system.

We have been fortunate to have had assistance from a cytogenetics company in our work. As a result we have been able to learn much about how cytotechnicians recognise structural aberrations, and compare their classification of microscope cell images and digitised (computer) images. In addition, access to a large set of mutagen-treated blood cells has been provided.

2. Initial Location of Chromosomal Material

2.1 Data Set.

Photographs of mutagen-treated human blood cells at metaphase were taken using a microscope-camera system, with enlargements produced and scanned. The resolution of digitised images obtained ranged from 750x750 to 1200x1200 pixels. 256 grey levels were available. Figure 1 shows a portion from a cell image.

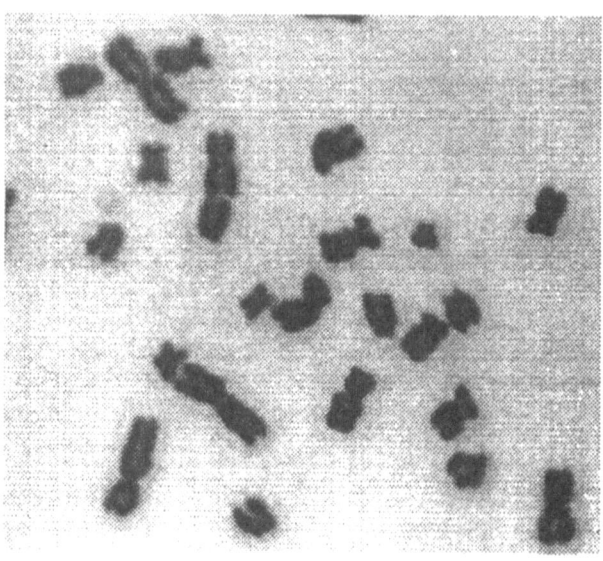

Figure 1. Portion from a cell image. Cells usually contain 46 chromosomes of varying size and orientation.

A number of images were chosen from the collected set and rescaled by linear interpolation to be 512x512 (the originals were approximately square), and a lowpass circular filter of diameter 128 pixels applied (in the frequency domain). Each resultant image formed the base for a 3-layer Gaussian pyramid. We convolved 4x4 neighbourhoods of base level images with Gaussians of standard deviation 1.0, to produce pixel values for the pyramid layer above. By only choosing neighbourhoods centred on every other pixel in every other row a reduction of 2 in each dimension for the next layer was achieved. By repeating the process 512-256-128 Gaussian pyramids were created.

2.2 Network Training.

One simple, supervised, single layer neural network was associated with each of the three levels in our pyramids. Units in a network were arranged as a 2-D sheet. All units had an associated adjustable reference vector $m(t)$, where t is a time step, and all received identical inputs $x(t)$, namely the set of normalised image portions grabbed by a moveable window on an image . Each unit in a network was designated responsibility for signalling the presence of chromosomal material within a particular subregion of the window; if a network consisted of n x m units then the window would be divided into n x m regions, and unit (i,j) in the network would be responsible for signalling the presence of material centred within window subregion (i,j). For each pyramid level many image portions containing chromosomal material were presented to a network. Each time, the reference vector of the responsible, or required winning unit, w, and its neighbours, (following Kohonen, see section 3), were adjusted to be closer to the image input $x(t)$,

$$m_i(t+1) = \frac{m_i(t) + \alpha(t) x(t)}{\| m_i(t) + \alpha(t) x(t) \|} \quad (i \in N_w(t)) \dots\dots\dots\dots\dots\dots (1)$$

$$m_i(t+1) = m_i(t) \quad (i \notin N_w(t))$$

Each of the three networks contained 5x5 units and input images were 64x64 pixels. During training the neighbourhood, $N_w(t)$, a square centred on the required winning unit, decreased from 5x5 units to 1x1 units (i.e., just the required winner) over the first 500 steps. Thereafter it remained at 1x1 units. Since neighbourhoods encompassed different subsets of units the net effect of adjustments for each unit in the network tended to be smoothed out over time. (So, for example, if a unit was seldom chosen as the required winner early in training, its reference vector would tend towards an averaged version of the vectors of units in its neighbourhood).

We took $\alpha(t) = 100/t$, so that as the neighbourhood decreased so did the magnitude of reference vector adjustments. Over 1000 steps were carried out for each network. Units in each network developed into something like blob detectors whose receptive fields overlap. Network reference vectors for each pyramid level are shown in figure 2.

2.3 Recall.

Once trained the three networks could be used in a recall phase to locate chromosomal material. Gaussian pyramids were constructed from cells not in the training set.Units were assigned activations according to the inner product between their reference vectors and the input (both normalised), and the unit with the greatest activation was deemed to best indicate the position of a chromosomal object. The best guess location of an object was taken as the centre of the window region which had the same row-column coordinates relative to the window as the winning unit had relative to the network. Once found, the best guess location at one level determined the position at which to centre the window at the level below. As we moved down from the top pyramid level it was possible to locate chromosomal material with increased precision. Furthermore, by gradually reducing the window's size at the pyramid base, and only matching with reference vectors over this region, we could land on chromosomal material with great reliability even in the presence of clutter. A test on 8 pyramids constructed from 8 cell images not in the training set produced one "missed" chromosome at the base level from a total of 374 start positions at the top pyramid level. Here the size of base level window was gradually shrunk from 64x64 to 9x9 (about a chromosome's width) in steps of 5 pixels.

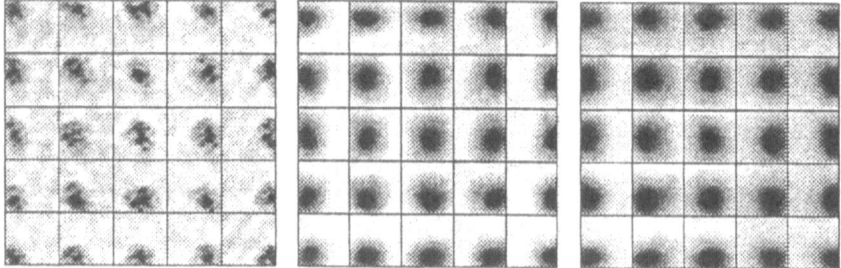

Figure 2. Reference vectors learnt by the networks for the top, middle and bottom pyramid levels.

3. Classification of Low Resolution Images

Given that we can land on chromosomal objects we would like to centre on them (the pyramid system lands on chromosomal objects but does not necessarily find their centres), and perform some classification. Due to the varibility in chromosome appearance we are initially only interested in estimating the rough position, size and orientation of objects within low resolution cell images. This information may then be used to constrain chromosome appearance at higher resolutions. The centring and classification system employed on low resolution images (128x128 pixels) is shown schematically in figure 3. The system is centred around the use of a Kohonen network for feature extraction.

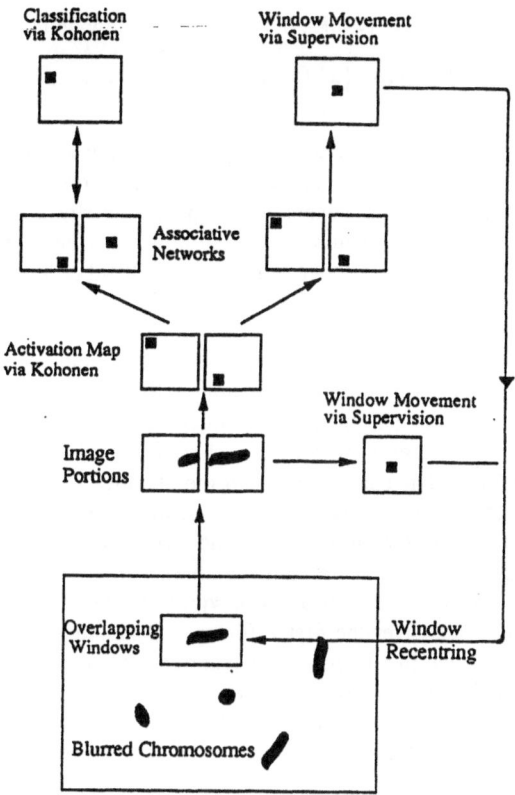

Figure 3. Schematic diagram of centring and classification system.

3.1 Feature Extraction.

The Kohonen Network or Self-Organising Map is a self-organising neural network whose units become specifically tuned to various features present in a set of input signals through a period of unsupervised training. Moreover, the locations of responses on the network tend to become ordered such that nearby units respond to similar input features.

Our supervised networks in section 2 were based upon the Kohonen network. During training we specified a required winning unit for each input image, and adjusted the reference vector of the required winner and its neighbours accordingly. As stated, the Kohonen network itself is unsupervised. Each time an input vector is presented to the network the winning unit is not specified by the user, rather it is taken as that unit whose reference vector is closest by some distance measure to the input vector. When input and reference vectors are normalised a suitable updating algorithm for reference vectors is as given in

262

equation (1) for our supervised networks, but with the winning unit, w, now such that

$$x^T(t)\,m_w(t) \;=\; max\{x^T(t)\,m_i(t)\} \quad\ldots\ldots\ldots\ldots\ldots\ldots\ldots\ldots\ldots (2)$$

The effect of equation (2) is that, over many time steps, units tend to become specifically tuned to particular domains of the input space. Furthermore, because reference vectors in the neighbourhood of the winner are adjusted, nearby units on the Kohonen network respond to similar input features. At small time steps the neighbourhood is chosen to be relatively large in size, encouraging a rough global ordering of reference vectors. Gradually the neighbourhood is decreased in size, usually until just the best-matching vector is updated.

Our Kohonen network is arranged as a 2-D sheet, 8x8 units in size. Inputs are normalised image portions grabbed from each window within a set of 5x5 overlapping windows placed on low resolution cell images. The window set is centred on positions located by the pyramid system described in section 2. Each window is 8 x 8 pixels. Neighbouring windows overlap by 60%. (The characteristics of the window set are chosen after consideration of: the need to encode similarity between features without losing the ability to express differences; the requirement that the window set covers chromosomes; the need for computations to be performed in a reasonable time).

During training we took $\alpha(t)=100/t$. The square neighbourhood, $N_w(t)$ decreased from 7x7 units to 1x1 units over the first 1000 steps, remaining at 1x1 thereafter. Over 10,000 steps were performed in all, where a step corresponds to the presentation of a single image portion. The feature map for the trained network is shown in figure 4. It consists of blob and bar-like features. (The map provides for good reconstruction of images from their extracted features, indicating that the window set chosen is reasonable).

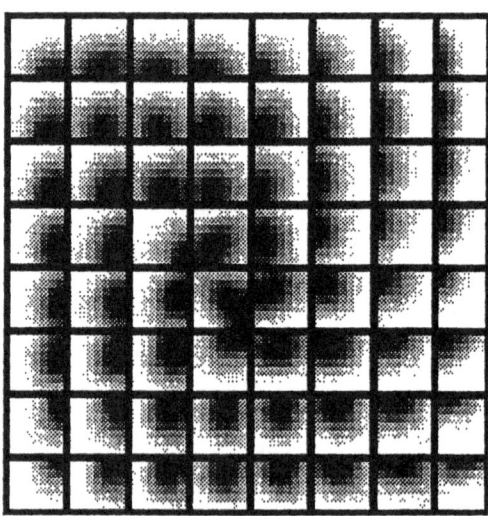

Figure 4. Feature map for low resolution images.

It is the responses of the trained Kohonen network to each image portion that determine the values on our activation map. The activation map is a 2-D sheet of binary units. The sheet is divided into 5x5 regions to mirror the organisation of the window set. We transform the Kohonen network responses produced for each image portion, setting the most responsive Kohonen unit to 1, all others to 0, and copy these responses onto the appropriate region of the map. Thus, the activation map identifies the best-matching feature for each of the image portions grabbed by the window set.

3.2 Auto-associative Networks.

The activation map is the input layer of two auto-associative neural networks - one involved in centring, the other in classification of centred objects (see figure 3). Associative networks learn to store input-output pairs of patterns during a period of training. When a noisy version of an input pattern is then presented to a network, the associated (noise-free) output pattern is retrieved (under certain conditions relating to storage capacity of the network). For auto-associative networks the input and output patterns are the same - networks learn to associate patterns with themselves. When a noisy pattern is presented to such a network the noise-free pattern is retrieved.

As stated, our auto-associative networks (which are similar to the Willshaw network (6)) have the activation map as their input layer. Output layers are identical in structure to the activation map (i.e., 2-D sheets of 40x40 binary units, divided into 5x5 regions). We taught networks to store activation map states produced from images of "good" chromosomes (i.e., straight, uncluttered and non-overlapping) . During training we simply clamped a network's output layer into the state of its input layer (i.e., the activation map), and created connections between an input unit-output unit pair if the units were on together.

On recall, we wish a network to output a stored activation map state given a noisy version as input. We set the output state of unit i which is in region S, to 1 if

$$\sum_j T_{ij} I_j > \sum_j T_{kj} I_j \, (k \in S) \dots\dots\dots\dots\dots\dots\dots\dots\dots\dots\dots\dots \, (3)$$

where I_j is the state of input unit j and $T_{ij}=1$ if input unit i and output unit j are connected, $T_{ij}=0$ otherwise. If the inequality is not satisfied the output of unit i is set to 0.

In effect, each of the best-matching features encoded in the input state votes for compatible features to be present in the output (or retrieved) state. For each region of the output layer the feature with the most votes is chosen. While our auto-associative memories can retrieve states which permit centring on, and classification of, relatively straight chromsomes in clutter, we are interested in their use as part of a technique for bent and overlapping chromosomes (see section 4).

3.3 Window Centring.

Initially, the activation map sets the input state of the associative network employed in centring. The output state of the network passes directly as input to a single-layer, supervised network of the type described in section 2. In this case each unit in the supervised network is responsible for signalling the presence of a chromosome's central region within a particular window. The network works in tandem with a second supervised network taking input directly from the image portions. This second network tries to ensure that the new centre for the window set suggested by the first is actually on chromosomal material rather than just to one side. Attempts are made to locate chromosome centres on new images by repeatedly cycling through the Kohonen, associative and supervised networks until a stable position is found (see figure 3). Each time the window set is moved a new activation map is produced, and fed on to the associative network.

3.4 Classification.

When centring has been achieved the current activation map statte forms the input to the associative network associated with classification. The network output state is passed on as input to a "semi-supervised" Kohonen network for classification. By semi-supervised we mean that for each presented input we specify which row we would like the most responsive unit in the network to belong to. The specification is based upon the length of the chromosome within the window set. The network is 6x6 units. (We require classification into one of six lengths at this resolution). During training the network self-orgainsed such that the locations of the responses within a row became ordered with respect to chromosome orientation.

4. Discussion.

Our system is capable of making good estimates of chromosome position, length and orientation from low resolution images of chromosomes provided chromosomes are reasonably straight. This applies in the presence of clutter. However, cells examined for structural aberrations usually contain at least some bent chromosomes, and possibly overlaps. We are currently addressing the problems of overlapping and bent chromosomes. We are investigating a technique which prevents features from appearing in the output of an associative network unless they are "similar enough" to the features extracted directly from the image. The similarity measure is a distance on the Kohonen map, and increases over time. Also, we are looking at ways of suppressing appropriate image features once one classification of an overlap has been performed in order that a second classification may be made.

We intend to apply the centring and classification system described here to higher resolution images. The appearance of chromosomal objects at such resolutions would be constrained by the estimates of position, size and orientation obtained at the low resolution. We would expect a Kohonen network to extract features associated with structurally aberrant chromosomes, such as gaps and breaks.

References

[1] M.G. Daker. Automation in cytogenetics. Medical Laboratory Sciences, V45, 324-332, 1988.

[2] M.Turner, J.Austin, N.Allinson, P.Thompson. Automated Chromosome Analysis. ACAG Memo 36, Computer Science Department, York University, 1991.

[3] J.Bille, A.Erhardt, G.Johannsen and B.Ueberreite. Biological dosimetry by aberration scoring using POLYP. Proceedings of the 6th International Conference on Pattern Recognition, 1200, IEEE Computer Society, Munich, 1982.

[4] R.K.Aggarwal and K.S.Fu. Automated Recognition of Irradiated Chromosomes. Journal of Histochemistry and Cytochemistry, V22, N7, 561-568, 1974.

[5] T.Kohonen. The Self-Organising Map. Proceedings of the IEEE, V78, N9, 1464-1480, 1990.

[6] D.J.Willshaw, O.P.Buneman, H.C.Longuet-Higgins. Non-Holographic Associative Memory. Nature V222, 960-962, 1969.

Active Shape Models - 'Smart Snakes'

T.F.Cootes and C.J.Taylor

Department of Medical Biophysics
University of Manchester
Oxford Road
Manchester M13 9PT
email: bim@wiau.mb.man.ac.uk

Abstract

We describe 'Active Shape Models' which iteratively adapt to refine estimates of the pose, scale and shape of models of image objects. The
method uses flexible models derived from sets of training examples.
These models, known as Point Distribution Models, represent objects as
sets of labelled points. An initial estimate of the location of the model
points in an image is improved by attempting to move each point to a
better position nearby. Adjustments to the pose variables and shape parameters are calculated. Limits are placed on the shape parameters ensuring that the example can only deform into shapes conforming to global
constraints imposed by the training set. An iterative procedure deforms
the model example to find the best fit to the image object. Results of applying the method are described. The technique is shown to be a powerful
method for refining estimates of object shape and location.

1 Introduction

Flexible models can represent classes of objects whose shape can vary, and can be
used to recognise examples of the class in an image. Various authors have described
iterative techniques for fitting flexible models to image objects. Kass, Witkin and
Terzopoulos [1] described 'Active Contour Models', flexible snakes which can stretch
and deform to fit image features to which they are attracted. The iterative energy
minimisation technique used is a powerful one, but only simple, local shape constraints are applied. Yuille *et al* [2] describe hand built models consisting of various
geometric parts designed to represent image features; they also describe methods
for adjusting their models to best fit an image. Unfortunately both the models and
the optimisation techniques have to be individually tailored for each application.
Staib and Duncan [3] use a Fourier shape model, representing a closed boundary as
a sum of trigonometric functions of various frequencies. They too use a form of iterative energy minimisation technique to fit a model to an image. However, using trigonometric functions does not always provide an appropriate basis for capturing
shape variability, and is limited to closed boundaries. Lowe [4] describes a technique
for fitting projections of three–dimensional parameterised models to two dimensional images by iteratively minimising the distance between lines in the projected
model and those in the image.

We have developed a method of building flexible models by representing the
objects as sets of labelled points and examining the statistics of their co–ordinates
over a number of training shapes – Point Distribution Models (PDMs) [5]. In this

paper we describe an iterative optimisation scheme for PDMs allowing initial estimates of the pose, scale and shape of an object in an image to be refined. The linear
nature of the model leads to simple mathematics allowing rapid execution. Because
the models can accurately represent the modes of shape variation of a class of objects
they are compact and prevent 'implausible' shapes from occurring. Since PDMs can
represent a wide variety of objects the same modelling and refinement framework
can be applied in many different applications.

Given an estimate of the position, orientation, scale and shape parameters of
an example in an image, adjustments to the parameters can be calculated which give
a better fit to the image. Suggested movements are calculated at each model point,
giving the displacement required to get to a better location. These movements are
transformed to suggested adjustments of the parameters, giving a better overall fit
of the model instance to the data. By applying limits to the ranges of the parameters
it can be ensured that the shape of the instance remains similar to the original training examples. Enforcing these limits applies global shape constraints, allowing only
certain deformations to occur. Because the models attempt to deform to better fit
the data, but only in ways which are consistent with the shapes found in the training
set we call them 'Active Shape Models' or 'Smart Snakes'.

2 The Point Distribution Model

The Point Distribution Model (PDM) is a way of representing a class of shapes using
a flexible model of the position of labelled points placed on examples of the class
[5]. The points can represent the boundary or significant internal locations of an object (Figure 1).

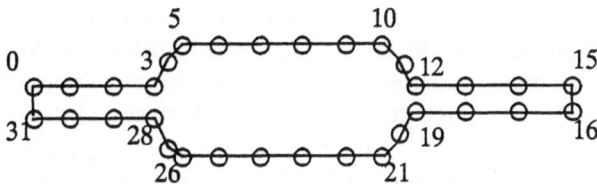

Figure 1 : 32 point model of the boundary of a
resistor.

The model consists of the mean positions of these points and the main modes
of variation describing how the points tend to move from the mean;

$$\mathbf{x} = \bar{\mathbf{x}} + \mathbf{Pb} \tag{1}$$

where x represents the n points of the shape,

$$\mathbf{x} = (x_0,\ y_0,\ x_1,\ y_1,\ \ldots,\ x_k,\ y_k,\ \ldots,\ x_{n-1},\ y_{n-1})^T$$

(x_k, y_k) is the position of point k

$\bar{\mathbf{x}}$ is the mean position of the points

$\mathbf{P} = (\mathbf{p}_1\ \mathbf{p}_2\ \ldots\ \mathbf{p}_t)$ is the matrix of the first t modes of variation, $\mathbf{p}_i$,
corresponding to the most significant eigenvec-
tors in a Principal Component Decomposition
of the position variables.

$\mathbf{b} = (b_1 \; b_2 \; ... \; b_t)^T$ is a vector of weights for each mode.

The columns of $\mathbf{P}$ are orthogonal so $\mathbf{P}^T\mathbf{P} = \mathbf{I}$ and

$$\mathbf{b} = \mathbf{P}^T(\mathbf{x} - \bar{\mathbf{x}}) \qquad (2)$$

The mean and linearly independent modes of variation are estimated from a set of training examples. The above equations allow us to generate new examples from the class of shapes by varying the parameters (b_i) within suitable limits. The limits are derived by examining the distributions of the parameter values required to generate the training set (typically three standard deviations from the mean). Each parameter varies the global properties of the shape reconstructed.

We can define the shape of a model object, in an object centred co-ordinate frame, by choosing values for $\mathbf{b}$. We can then create an instance, $\mathbf{X}$, of the model in the image frame by defining the position, orientation and scale;

$$\mathbf{X} = M(s, \theta)[\mathbf{x}] + \mathbf{X}_c \qquad (3)$$

where $\quad \mathbf{X}_c = (X_c, \; Y_c, \; X_c, \; Y_c, \; ..., \; X_c, \; Y_c)^T$

$M(s, \theta)[\;]$ is a rotation by θ and a scaling by s.

(X_c, Y_c) is the position of the centre of the model in the image frame.

3 Using the PDM as a Local Optimiser – Active Shape Models

Suppose we have a PDM of an object, and we have an estimate of the position, orientation, scale and shape parameters of an example of the object in an image. We would like to improve our estimate, updating the pose and shape parameters to make the model instance fit more accurately to the image evidence. The approach we use is as follows: at each point in the model we calculate a suggested movement required to displace the point to a better position; we calculate the changes to the overall position, orientation and scale of the model which best satisfy the displacements; any residual differences are used to deform the shape of the model object by calculating the required adjustments to the shape parameters. The global shape constraints are enforced by ensuring that the shape parameters remain within appropriate limits.

3.1 Calculating The Suggested Movement of Each Model Point

Given an initial estimate of the positions of a set of model boundary points which we are attempting to fit to the outline of an image object (Figure 2) we need to estimate an adjustment to apply to move each boundary point toward the edge of the image object. There are various approaches that could be taken. In the examples we describe later we use an adjustment along a normal to the model boundary towards the strongest image edge, with a magnitude proportional to the strength of the edge (Figure 3).

A set of adjustments can be calculated, one for each point of the shape (Figure 4). We denote such a set as a vector $d\mathbf{X}$, where

$$d\mathbf{X} = (dX_0, dY_0, \; ..., \; dX_{n-1}, \; dY_{n-1})^T$$

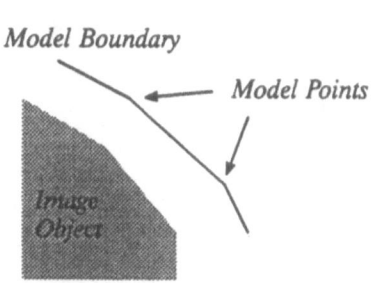

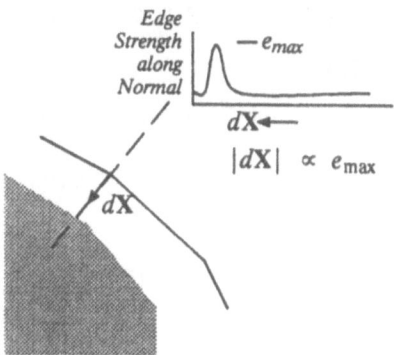

Figure 2 : Part of a model boundary approximating to the edge of an image object.

Figure 3 : Suggested movement of point is along normal to boundary, proportional to maximum edge strength on normal.

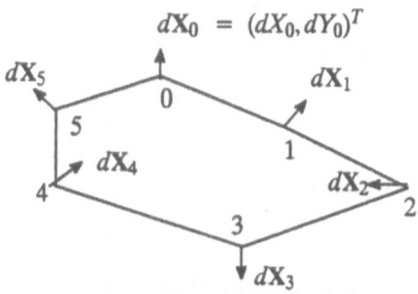

Figure 4 : Adjustments to a set of points

We aim to adjust the pose and shape parameters to move the points from their current locations in the image frame, X, to be as close to the suggested new locations $(X + dX)$ as can be arranged and whilst still satisfying the shape constraints of the model. If the current estimate of the model is centred at (X_c, Y_c) with orientation θ and scale s we would like first to calculate how to update these parameters to better fit the image. One way is to find the translation (dX_c, dY_c), rotation $d\theta$ and scaling factor $(1 + ds)$ which best map the current set of points, X, onto the set of points given by $(X + dX)$. Although exact solutions for dX_c, dY_c, $d\theta$ and ds are possible [5], we have used the approximation method given in Appendix A, which is quick to calculate and adequate given the iterative nature of the overall scheme.

Having adjusted the pose variables there remain residual adjustments which can only be satisfied by deforming the shape of the model. We wish to calculate the adjustments to the original model points in the local co-ordinate frame x required to cause the scaled, rotated and translated points X to move by dX when combined with the new scale, rotation and translation variables.

The initial position of the points in the image frame is given by

$$X = M(s, \theta)[x] + X_c \qquad (3)$$

We wish to calculate a set of residual adjustments dx in the local model co-ordinate frame such that

$$M(s(1 + ds), \theta + d\theta)[x + dx] + (X_c + dX_c) = (X + dX) \tag{4}$$

Thus

$$M(s(1 + ds), \theta + d\theta)[x + dx] = (M(s, \theta)[x] + dX) - (X_c + dX_c)$$

and since

$$M^{-1}(s, \theta)[\ ] = M(s^{-1}, - \theta)[\ ]$$

we obtain

$$dx = M((s(1 + ds))^{-1}, - (\theta + d\theta))[M(s, \theta)[x] + dX - dX_c] - x \tag{5}$$

Equation 5 gives a way of calculating the suggested movements to the points x in the local model co-ordinate frame. These movements are not in general consistent with our shape model. In order to apply the shape constraints we transform dx into model parameter space giving db, the changes in model parameters required to adjust the model points as closely to dx as is allowed by the model. Equation 1 gives

$$x = \bar{x} + Pb \tag{1}$$

We wish to find db such that

$$x + dx \approx \bar{x} + P(b + db) \tag{6}$$

Since there are only t ($< 2n$) modes of variation available and dx can move the points in $2n$ different degrees of freedom, in general we can only achieve an approximation to the deformation required, since we only allow deformation in the most significant modes observed in the training set. This truncating of the modes of variation is equivalent to setting limits of zero on the parameters controlling other modes of variation. Applying such limits and truncation enforces the global shape constraints.

Subtracting (1) from (6) gives

$$dx \approx P(db)$$

$$db = P^T dx \tag{7}$$

It can be shown that Equation 7 is equivalent to using a least squares approximation to calculate the shape parameter adjustments, db.

3.2 Updating the Pose and Shape Parameters

The equations above allow us to calculate changes to the pose variables, dX_c, dY_c, $d\theta$ and ds, and adjustments to the shape parameters db required to improve the match between an object model and image evidence. We have applied these to update the parameters in an iterative scheme as follows;

$$X_c \rightarrow X_c + w_t\, dX_c \tag{8}$$

$$Y_c \rightarrow Y_c + w_t\, dY_c \tag{9}$$

$$\theta \rightarrow \theta + w_\theta\, d\theta \tag{10}$$

$$s \rightarrow s(1 + w_s\, ds) \tag{11}$$

$$\mathbf{b} \rightarrow \mathbf{b} + \mathbf{W}_b \, d\mathbf{b} \tag{12}$$

Where w_t, w_s and w_θ are scalar weights, and $\mathbf{W}_b$ is a diagonal matrix of weights for each mode. This can either be the identity, or each weight can be proportional to the standard deviation of the corresponding shape parameter over the training set. The latter allows more rapid movement in modes in which there tends to be larger shape variation.

In order to ensure that the new shape is plausible it is necessary to apply limits to the b–parameters. If the variance about the origin of the ith parameter over the training set is λ_i then a shape can be considered acceptable if the Mahalanobis distance D_m is less than a suitable constant, D_{max} (for instance 3.0) ;

$$D_m = \sum_{i=1}^{t} \left(\frac{b_i^2}{\lambda_i} \right) \leq D_{\max} \tag{13}$$

(The vector $\mathbf{b}$ lies within a hyper–ellipsoid about the origin.) If updating $\mathbf{b}$ using (12) leads to an implausible shape, ie (13) is violated, it can be re–scaled to lie on the closest point of the hyper–ellipsoid using

$$b_i \rightarrow b_i \cdot \frac{D_{\max}}{D_m} \quad (i = 1..t) \tag{14}$$

Once the parameters have been updated, and limits applied where necessary, a new example can be calculated, and new suggested movements derived for each point. The procedure is repeated until no significant change results.

4 Examples Using Active Shape Models

The techniques described above have been used successfully in a number of applications, both industrial and medical [8]. Here we show results obtained using the resistor and hand models described in the companion paper [5].

In both cases initial estimates of the position, orientation and scale are made, and the shape parameters are all initialised at zero ($b_i = 0$ $(i = 1..t)$). Suggested movements for each model point are calculated by finding the strongest edge (in the correct direction) along the normal to the boundary at the point (See 3.1 and Figure 3). Adjustments to the parameters are calculated and applied, and the process repeated.

4.1 Finding Resistors with an ASM

We have constructed a Point Distribution Model of a resistor representing its boundary using 32 points (Figure 1). Figure 5 shows an image of part of a printed circuit board with the resistor boundary model superimposed as it iterates towards the boundary of a component in the image. We interpolate an additional 32 points, one between each pair of model points around the boundary, and calculate adjustments to each point by finding the strongest edge along profiles 20 pixels long centred at each point. We use a shape model with 5 degrees of freedom. Each iteration takes about 0.025 seconds on a Sun Sparc Workstation.

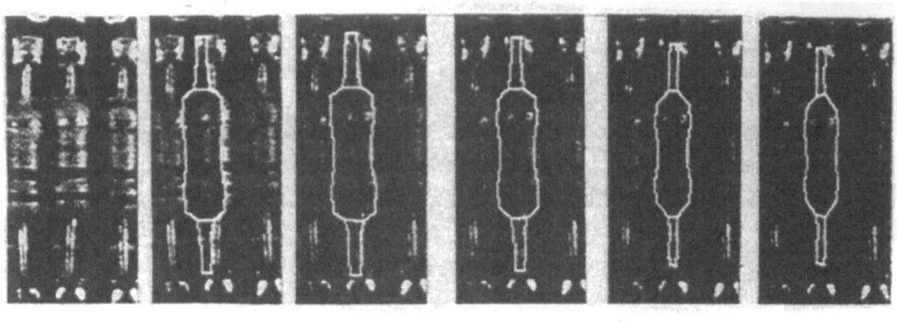

| (a) Original Image | (b) Initial Position | (c) After 30 iterations | (d) After 60 iterations | (e) After 90 iterations | (f) After 120 iterations |

Figure 5 : Section of Printed Circuit Board with resistor model superimposed, showing its initial position and its location after 30, 60, 90 and 120 iterations.

The ends of the wires are not found correctly since they are not well defined – there is little edge evidence to latch on to. We intend to produce a better model by including the square solder pads. The method is effective in maintaining the global shape constraints of the model and works well given a sufficiently good starting approximation; we discuss methods of obtaining such initial hypotheses elsewhere [6,7,8].

The relatively simple method of calculating the movement of each point, looking for a strong nearby edge, can cause problems when a model is initialised some distance from a component. Highlights and the banding patterns on the resistors can attract the boundary of the model, pulling it away from the true edge. A more sophisticated technique which modelled the banding and possible highlights would be required to overcome this.

4.2 Finding Hands with an ASM

We have constructed a Point Distribution Model of a hand representing the boundary using 72 points. Figure 6 shows an image of the author's hand and an example of the model iterating towards it. We calculate adjustments to each point by finding the strongest edge on a profile 35 pixels long centred on the point. The shape model has 8 degrees of freedom, and each iteration takes about 0.03 seconds on a Sun Sparc Workstation. The result demonstrates that the method can deal will limited occlusion.

As in the previous example the method works reliably, given a reasonable starting approximation. The example shows that the method is tolerant to quite serious errors in the starting approximation, though this depends on the amount of clutter in the image.

5 Discussion and Conclusions

The iterative approach described above, using image evidence to deform a Point Distribution Model, is effective at locating objects, given an initial estimate of their position, scale and orientation. How good an estimate is required will depend on how cluttered the image is and how well the model describes the object in the image.

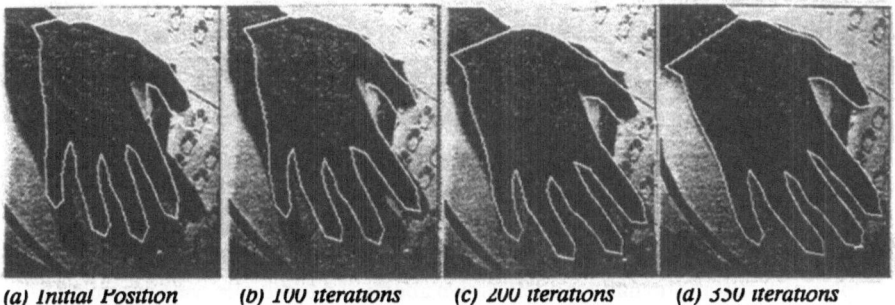

(a) Initial Position   (b) 100 iterations   (c) 200 iterations   (d) 350 iterations

Figure 6 : Image of authors hand with hand model superimposed, showing its initial position and its location after 100,200 and 350 iterations.

How suggested adjustments are found for each point is important. Calculating the suggested movement by looking for strong nearby edges is simple and has proved effective in many cases. However, when searching for more complex objects, where the model points do not necessarily lie on strong edges, more sophisticated algorithms are required. Potential maps can be derived, describing how likely each point in an image is to be a particular model point. During a search each model point attempts to move to more likely locations, climbing hills in the potential map. Alternatively a model of the expected grey levels around each model point can be generated from the training examples, and each point moved toward areas which best match its local grey level model. Preliminary experiments using both these techniques have proved promising [9]. The model points do not have to lie only on the boundary of objects, they can represent internal features, and even sub–components of a complex assembly. In the latter case the model describes both the variations in the shapes of the sub–components and the geometric relationships between components. The refinement technique can be applied as easily in this situation as to a model of a single boundary.

By allowing the model to deform, but only in ways seen in the class of examples used as a training set, we have a powerful technique for refinement. The constraints on the shape of the model are applied by the limits on the shape parameters. The $2n{-}t$ unrepresented modes of variation effectively have limits of zero on their parameters. Rather than fixed limits being used to enforce shape constraints, restoring forces in the parameter space could be applied, pulling the parameters back towards zero against the external 'forces' from the image;

$$\mathbf{b} \rightarrow \mathbf{b} + \mathbf{W}_b d\mathbf{b} - k_b \mathbf{W}_b \mathbf{b} \qquad (0 < k_b < 1) \qquad (15)$$

This would give more weight to solutions closer to the mean shape, and require strong evidence for shapes which are considerably deformed. However, this would be likely to lead to compromise solutions between image data and model.

The work we present here can be thought of as a two dimensional application of Lowe's refinement technique [4]. Because of the linear nature of the Point Distribution Model, the mathematics is considerably simpler and can lead to rapid execution.

We have conducted experiments which suggest that the local optimisation method described can be fruitfully used in conjunction with a Genetic Algorithm (GA) search [8]. The GA can be run as a cue generator to produce a number of object hypotheses, which can be refined using the Active Shape Model. Alternatively the ASM can be combined with the GA search, applying one iteration at each generation of the Genetic Algorithm. Both techniques appear very promising.

The method of calculating the parameter changes is straightforward, and new examples of a model can be generated rapidly using linear algebra. As well as the examples of resistors and hand models shown above, the technique has been successfully used in a variety of applications and has great potential for image search in many image analysis domains.

Acknowledgements

This work is funded by SERC under the IEATP Initiative (Project Number 3/2114). The authors would like to thank the other members of the Wolfson Image Analysis Unit for their help and advice, particularly D.H.Cooper, J.Graham, D.Bailes and A.Hill.

Appendix A : Estimating the Pose Parameter Adjustments

Suppose we have a shape defined by the n points in the vector $\mathbf{x}$ relative to the centre of the model, and we wish to find the translation (dX_c, dY_c), rotation about (X_c, Y_c), $d\theta$ and scaling factor $(1+ds)$ which best maps the current set of points, $\mathbf{X}$, onto the set of points given by $(\mathbf{X} + d\mathbf{X})$.

The translation is given by

$$dX_c = \frac{1}{n}\sum_{i=0}^{n-1} dX_i \qquad dY_c = \frac{1}{n}\sum_{i=0}^{n-1} dY_i \tag{16}$$

If we now remove the effects of the translation, letting

$$dX'_i = dX_i - dX_c \qquad\qquad dY'_i = dY_i - dY_c \tag{17}$$

$$dX'_i = (dX'_i, \ dY'_i)^T$$
$$\mathbf{X'} = \mathbf{X} - \mathbf{X}_c \tag{18}$$

then the problem becomes one of finding the rotation $d\theta$ and scaling factor $(1+ds)$ which best maps $\mathbf{X'}$ onto the set of points given by $(\mathbf{X'} + d\mathbf{X'})$.

Consider point i. We wish to move it to point i' (Figure 7)

It is (relatively) easy to show that

$$dX_{ir} = \frac{X'_i dX'_i + Y'_i dY'_i}{(X'^2_i + Y'^2_i)}\begin{pmatrix} X'_i \\ Y'_i \end{pmatrix} \tag{19}$$

$$dX_{ia} = dX'_i - dX_{ir} \tag{20}$$

$$ds_i = \frac{|dX_{ir}|}{\sqrt{X'^2_i + Y'^2_i}} = \frac{X'_i dX'_i + Y'_i dY'_i}{(X'^2_i + Y'^2_i)} \tag{21}$$

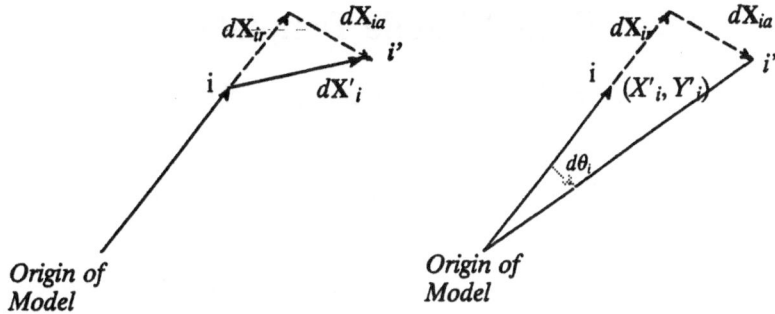

Figure 7 : Estimating the angle and scale changes require to map one point to a new position.

$$d\theta_i = \frac{|dX_{ia}|}{\sqrt{X'^2_i + Y'^2_i}} \tag{22}$$

Then

$$d\theta \approx \frac{1}{n} \sum_{i=0}^{n-1} d\theta_i \tag{23}$$

$$ds \approx \frac{1}{n} \sum_{i=0}^{n-1} ds_i \tag{24}$$

6 References

[1] M. Kass, A. Witkin and D. Terzopoulos, Snakes: Active Contour Models. First International Conference on Computer Vision, pub. IEEE Computer Society Press, 1987; pp 259–268.

[2] A.L. Yuille, D.S. Cohen and P. Hallinan, Feature extraction from faces using deformable templates. Proc. Comp. Vision Patt. Rec. 1989; pp104–109.

[3] L.H. Staib and J.S Duncan, Parametrically Deformable Contour Models. IEEE Computer Society conference on Computer Vision and Pattern Recognition, San Diego, 1989.

[4] D.G. Lowe, Fitting Parameterized Three Dimensional Models to Images. IEEE PAMI 1991; 5, pp441–450.

[5] T.F. Cootes, C.J. Taylor, D.H. Cooper and J. Graham, Training Models of Shape from Sets of Examples. This Volume.

[6] A. Hill and C.J. Taylor, Model–Based Interpretation using Genetic Algorithms. Proc. British Machine Vision Conference, Glasgow, 1991, pub. Springer Verlag, pp266–274.

[7] A. Hill, C.J. Taylor and T.F. Cootes, Object Recognition by Flexible Template Matching using Genetic Algorithms. Proc. European Conference on Computer Vision, Genoa, Italy, 1992.

[8] A. Hill, T.F. Cootes and C.J. Taylor, A Generic System for Image Interpretation Using Flexible Templates. This Volume.

[9] A. Lanitis, Modelling Faces. Internal Progress Report, Wolfson Image Analysis Unit, Manchenster University, 1992.

A Generic System for Image Interpretation Using Flexible Templates

A. Hill, T. F. Cootes and C. J. Taylor

Department of Medical Biophysics, University of Manchester,
Oxford Road, Manchester M13 9PT, England

Abstract

We describe a generic approach to image interpretation, based on combining a general method of building flexible template models with Genetic Algorithm (GA) search. The method can be applied to a given image interpretation problem simply by training a Point Distribution Model (PDM), using a set of examples of the image structure to be located. A local optimisation technique, developed for use with PDMs, has been incorporated into the GA search with the aim of improving the speed of convergence and optimality of solution. We present results, from three practical applications, demonstrating that the new method offers significant improvements when compared to previously reported approaches to flexible template matching. The benefits include the ability to deal with different domains of application using a standard method, the ability to deal with complex multi-part models and improved search performance.

1 Introduction

Flexible templates have been employed widely as a means for model–based image interpretation. In most cases, however, the flexible template employed is *hand–crafted* and the search strategy used to locate instances of the template within a given image is problem specific [11,12,13,14]. In a previous publication [8] we described an approach to flexible template matching using a generic search strategy – Genetic Algorithms (GAs) [4,6,9]. The method generates good interpretation hypotheses robustly and at moderate computational cost. We presented results obtained using a system for automatically delineating the left ventricle of the heart in echocardiograms, employing a flexible template model of the left ventricle which was hand–crafted. We present here a development of that work which employs a generic technique for flexible template construction based on Point Distribution Models (PDMs) [2]. Several improvements to the original GA method result from the use of the PDM approach :

- The method is now entirely generic; it can be applied to new problems by training a PDM using examples of the image structures to be located.
- The ability of a PDM to capture both the variability in shape of an object and the spatial relationships between a number of different objects enables complicated biological structures to be modelled and subsequently located using a GA search.
- By incorporating into the GA a local optimisation technique developed for use with PDMs [3], significant improvements in the performance of the GA

search can be achieved both in terms of speed of convergence and optimality of solution; the ability of a GA search to extract multiple candidate interpretations from an image is also enhanced by the use of the local optimiser.

Our results demonstrate how complex biological structures can be automatically located using the combination of PDMs and a GA search. The delineation of the left ventricle in echocardiograms previously reported [8] has been enhanced by employing a more complex, PDM model of the heart. We also present results of locating the first and second ventricles of the brain in Magnetic Resonance images. The improved performance of the GA search when incorporating the PDM local optimiser is also demonstrated. The effect of the local optimiser when using a GA to extract multiple interpretation hypotheses is discussed and results are presented for locating simultaneously many resistors on a printed circuit board.

2 Genetic Algorithms

GAs employ mechanisms analogous to those involved in natural selection to conduct a search through a given parameter space for the global optimum of some objective function. The main features of the approach are as follows :

- A point in the search space is encoded as a *chromosome*.
- A *population* of N chromosomes/search points is maintained.
- New points are generated by probabilistically combining existing solutions.
- Optimal solutions are *evolved* by iteratively producing new *generations* of chromosomes using a *selective breeding* strategy based on the relative values of the objective function for the different members of the population.

A solution, $z = (z_1, z_2, .., z_n)$, is encoded as a string of *genes* to form a *chromosome* representing an *individual*. In many applications the gene values are [0,1] and the chromosomes are simply bit strings. An objective function, f, is supplied which can decode the chromosome and assign a *fitness value* to the individual a chromosome represents. In our case the z_i are model parameters which define the shape and pose of possible image objects. The objective function measures the extent to which the potential interpretation represented by a particular chromosome is supported by image evidence.

Given a population of chromosomes the genetic operators *crossover* and *mutation* can be applied in order to propagate *variation* within the population. Crossover takes two *parent* chromosomes, cuts them at some random gene/bit position and recombines the opposing sections to create two *children* e.g. crossing the chromosomes 010–11010 and 100–00101 at position 3–4 gives 010–00101 and 100–11010. Mutation is a background operator which selects a gene at random on a given individual and mutates the value for that gene (for bit strings the bit is complemented).

The search for an optimal solution starts with a randomly generated population of chromosomes; an iterative procedure is used to conduct the search. For each iteration a process of *selection* from the current *generation* of chromosomes is followed by application of the genetic operators and re–evaluation of the resulting chromosomes. *Selection* allocates a number of trials to each individual according to its *relative fitness value* $f_i/\bar{f}$, $\bar{f} = 1/N\ \{f_1 + f_2 + .. + f_N\}$. The *fitter* an individual the more trials it will be allocated and vice versa. Average individuals are allocated a single trial.

Trials are conducted by applying the genetic operators (in particular crossover) to selected individuals, thus producing a new generation of chromosomes. The algorithm progresses by allocating, at each iteration, ever more trials to the high performance areas of the search space under the assumption that these areas are associated with short sub–sections of chromosomes which can be recombined using the random cut–and–mix of crossover to generate even better solutions.

The major feature of GAs which makes them attractive for object location is the use of a population of solutions, allowing competition between alternative interpretations. If there are several possible candidates, within an image; for the object we wish to locate, either because there are several instances of the object itself or instances of similar objects, the manner in which a GA search is conducted allows the various plausible interpretations to compete with one another, the strongest solution having the greatest probability of success. We have shown previously [8] how this facet of a GA search can be exploited to extract multiple plausible interpretations from an image by allowing separate *species* to adapt to various *niches* within the search space.

3 Point Distribution Models

We have employed the method described by Cootes et al [2] for constructing flexible templates. The technique captures the statistical variation in the distribution of sets of points to produce Point Distribution Models (PDMs). We describe the construction of PDMs for a chamber of the heart in echocardiograms and structures in the brain in Magnetic Resonance images. We also describe briefly *Active Shape Models* : instances of PDMs which deform to improve their fit to image data.

3.1 Constructing Point Distribution Models

The key steps in constructing a PDM from a set of examples of an object to be modelled are as follows :

- Generate a set of object descriptions upon which the model is to be trained. An object description is simply a labelled set of points $(x_0, y_0;\ x_1, y_1; ...)$.

  Each labelled point represents a particular position on the object (for example, the corner of a boundary). Corresponding points on different objects represent equivalent locations on each object.

- Align the sets of points and perform a Principle Components Analysis to the locations of the points. This involves finding the mean position of each point and the co–variance matrix of the position variables $(x_0, y_0;\ x_1, y_1; ...)$. The principle eigenvectors of the matrix give the main modes of variation of the training set.

This procedure results in a model with a small set of parameters $\mathbf{b} = (b_1, b_2, .., b_m)$ which act as weights for the major m eigenvectors of the co–variance matrix. These weights can be manipulated to create new instances from the class of objects modelled.

Because the technique is concerned only with the statistical variations of sets of points it can be employed not only to model objects which are, for example, closed boundaries but can also be used to capture the spatial relationships between objects. In this paper we have constructed PDMs of two complex biological structures : the left ventricle, septum and mitral valve of the heart (as imaged in echocardiography – see figure 3) and the first and second ventricles of the brain (as imaged in Magnetic Resonance Imaging – see figure 4). Both of these models exhibit complex structure (several parts) as well as variability in shape (see figure 1).

Heart   *Brain*

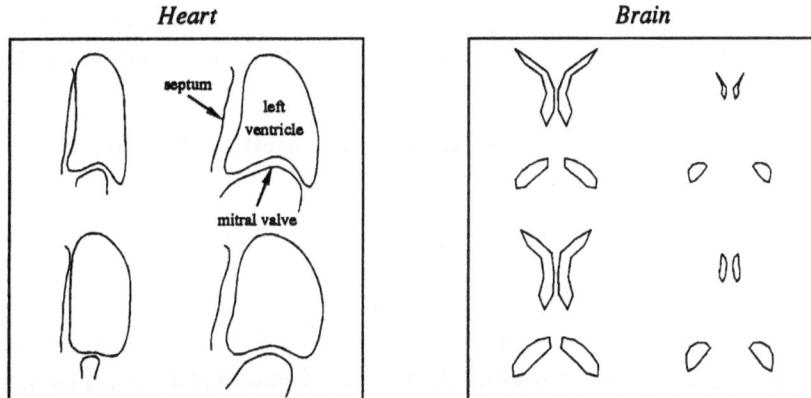

Figure 1 : Examples from Heart and Brain Ventricle Models.

For the PDM of the heart we employed echocardiogram time sequences from 33 individuals. From each of these sequences two images were selected which showed the left ventricle in its most contracted and extended states, giving a training set of 66 images. These images were labelled by an expert. For the ventricles of the brain we employed 3D Magnetic Resonance data sets from ten individuals. Because the ventricles of the brain are 3D structures we were able to employ several "slices" from each data set to give variability in shape due to slice position as well as variability in shape between individuals. On average, nine images were used from each sequence giving a total training set of 88 images.

3.2 Active Shape Models

Cootes and Taylor [3] have developed a local optimisation technique which can deform a PDM to fit image data. The technique is similar to that employed in the so called "Snake" approach presented by Kass et al [10] in that image evidence is used to suggest deformations of the model in order to improve the correspondence between model and data. An important property of the method presented by Cootes and Taylor, however, is that the shape constraints learned during the training of the PDM are never violated during model deformation i.e. the shape of the model is always legal during the iterative deformation process.

Each iteration of the procedure is as follows :

- Place the current instance of the model onto the image and interrogate the image data locally to deduce the deformation of each point required to improve the correspondence between model and data.
- Compute changes in the translation (t_x, t_y) scale (s) and orientation (θ) of the model from these local deformations. Apply these changes to $(t_x, t_y), s, \theta$ and evaluate any residual deformations which remain.
- The residual deformations are now employed to suggest changes to the model parameters which control shape, **b**. The current model parameters **b** are updated to reflect these changes in shape.

By applying this procedure iteratively the position, scale, orientation and shape of the model which best fit the data in the locality of the original estimate can be determined.

4 Combining Active Shape Models and Genetic Algorithms

4.1 Possible Frameworks

There are two possible ways we might combine Active Shape Models (ASMs) and GAs:

- Consider the techniques as separate but complementary. Here the GA search would be conducted as normal and the ASM applied to the solutions suggested by the GA. This uses the ASM as a *refinement* procedure; if the GA suggests a solution in a non–optimal area of the search space, the ASM can do no better than locate the local optimum for that area of the search space.
- Incorporate the ASM directly into the GA search; in the GA literature it has been suggested that incorporating heuristic information and local optimisation techniques within a GA search can improve performance significantly (see the discussion of the Travelling Salesman Problem in [4]). The basis of this approach is that the GA can locate the *hills* in the search space while the local optimiser embedded within the GA can climb to the top of these hills.

We are particularly interested in the second of these approaches which allows a more thorough investigation of the search space, especially when a *speciated* version of the GA is employed.

4.2 Genetic Algorithms and Active Shape Mutation

In a single iteration of the ASM procedure, a set of local deformations are computed from the image data and applied to the model to generate suggested changes to $(t_x, t_y), s, \theta$, and **b** in order to improve the correspondence between model and data. When using a chromosomal representation of the parameters these changes can be realised by first decoding the chromosome to generate the current values of $(t_x, t_y), s, \theta, \mathbf{b}$, then updating the parameter values and re–encoding the parameters to produce an updated chromosome. In effect, the genes on the chromosome are mutated to reflect the suggested changes in the parameters. This mutation will, in general, be beneficial to the individual concerned because the template which the chromosome encodes will represent a better fit to the image data; consequently the objective function value associated with the template will be improved. There are two parameters which control the incorporation of the ASM mutation into a GA :

- The rate at which the mutation is applied (M_{asm}); each individual in the current population is mutated with probability M_{asm}.
- The number of iterations of the ASM technique, M_i, to be applied for each mutation.

It will often be the case that the local deformations applied to the model can be generated as a by–product of evaluating the objective function which we are attempting to minimise/maximise i.e. when estimating how well the model fits the data, estimates of improved positions of model points suggest themselves naturally. This means that we can apply a single iteration of the ASM for "free" every time the objective function is evaluated – the only additional computing requirements being those to calculate the model transformation from the given local deformations. Taking this into consideration, together with the fact that the ASM mutation is generally beneficial, we suggest the values $M_{asm} = 1$ and $M_i = 1$ i.e. a single iteration of the local optimiser is applied every time the objective function is evaluated.

5 Results

The results we present below were obtained using a GA with "standard" parameter values [7] (unless otherwise stated): rate of crossover = 0.6, population size = 50, rate of random mutation = 0.005. The crossover operator employed was the constrained, two–point version suggested by Booker in [4]. The Remainder Stochastic Independent Sampling (RSIS) algorithm suggested by Baker [1] was used for selection. All model parameters were encoded as unsigned gray–code binary integers as suggested by Fitzpatrick et al [5]; 8 bits were used to represent each parameter. The same objective function was employed in all cases. The function was constructed in such a manner that it was minimised when strong edges of similar magnitude were located within the image close to the boundary of a given instance of a PDM (see [8] for more detail).

5.1 Improved Performance of GAs using ASM Mutations

In [8] we showed how GAs could be employed together with a flexible template model of the left ventricle of the heart to locate left ventricular boundaries in echocardiograms (see figure 3). We have compared the performance of GAs with and without ASM mutations using this exemplar. We applied a GA search to 5 echocardiograms 10 times, starting from different (random) initial populations. Two measures were recorded at each iteration of the GA procedure and average values of these measures computed for all 50 applications. The measures used were :

- the value of the objective function averaged over the current population.
- the best objective function value in the current population.

On average the performance of the GA incorporating the ASM was significantly better than that of the GA without the ASM. Convergence was more rapid and the best solution found was also improved. The results are shown in figure 2.

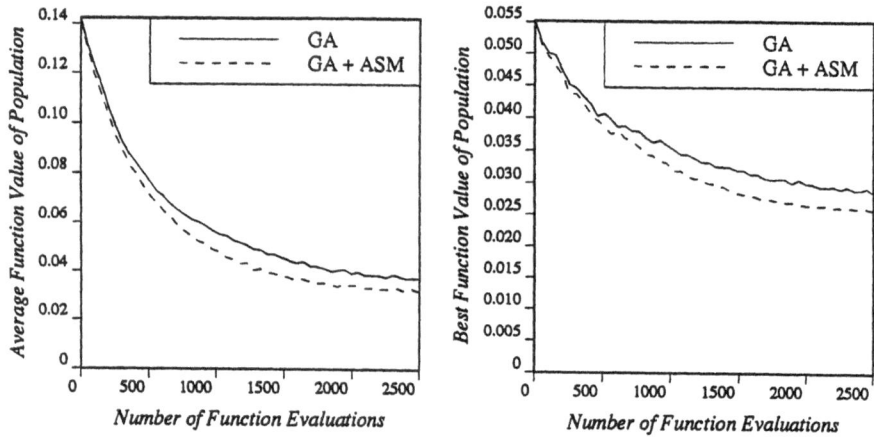

Figure 2 : Comparison of GA and GA+ASM Searches.

5.2 Objects with Multiple Parts

As we have already stated, employing the PDM method as a generic flexible template construction technique extends greatly the applicability of image interpretation using GA search. Objects with many variable parts can be modelled using a single PDM and the search technique applied as before. We have found that the increased specificity of such models can help resolve potentially ambiguous interpretations of image data. For example, in the model of the left ventricle of the heart we have included both sides of the septum and mitral valve in the model (see figure 1) placing stricter constraints on possible interpretations of the image data than if only the boundary of the left ventricle itself had been modelled. Figure 3 shows the automatic location of the septum, mitral valve and left ventricle in an apical 4–chamber echocardiogram using the GA+ASM technique. Figure 4 shows the automatic delineation of the first and second ventricles in a Magnetic Resonance image of the brain. The PDM of the ventricles of the brain was trained on images from different individuals and also from images at various positions in a 3D data set; the 2D shapes and locations of the ventricles vary considerably from slice to slice. The image shown was selected at random from a 3D data set and no indication of the "slice index" was employed in the interpretation process.

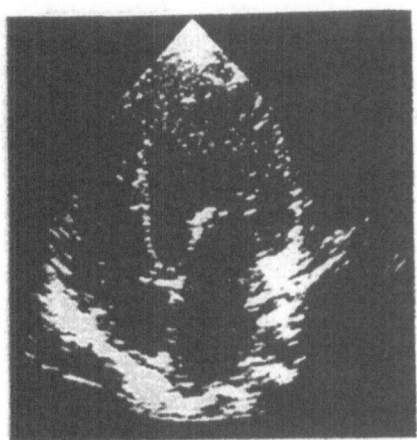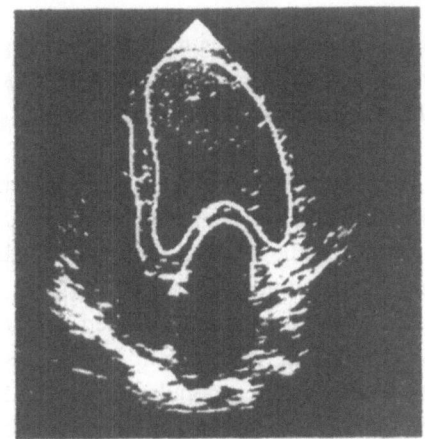

*Figure 3 : Automatic Delineation of the Left Ventricle,
Septum and Mitral Valve in the Heart.*

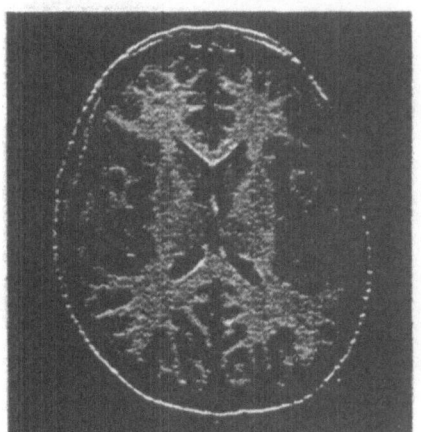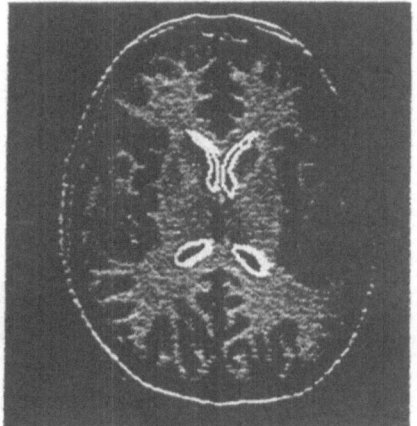

*Figure 4 : Automatic Delineation of the First and
Second Ventricles in the Brain.*

5.3 Extracting Multiple Plausible Interpretations using Speciation

In [8] we showed how a *speciated* version of the GA could be employed to extract multiple candidates for a given object within an image. In this case the GA is forced to spread its effort over different areas of the search space, rather than converging upon one particular area. This is accomplished by penalising individuals that reside in over-crowded areas of the search space in order to force migration to less crowded, yet still promising, areas. Speciation is also promoted by encouraging individuals to mate (crossover) with nearby rather than distant individuals. One problem with this approach is that adaptation of any particular species to the particular niche it occupies in the search space can be quite poor. This is because only a small gene pool is available for any given species due to the low numbers of individuals belonging to each species (given a population of 100 individuals and just 3 or 4 possible interpretations, species might be represented by groups as small as 10 individuals). The danger here is that a species will become extinct in a promising area of the search space.

What is required is the ability to optimise locally with only a small number of individuals. This is exactly what the ASM mutation incorporated within the GA search can achieve. An example of the ability of the GA+ASM to maintain stable, small sub-populations is shown in figure 5 where a number of resistors have been located automatically on a printed circuit board using a PDM resistor model and a speciated GA+ASM search. The 7 resistors identified (5 "real" and 2 spurious) were located using a population size of 100 individuals and the 7 species shown were stable i.e. the GA maintained the sub-populations indefinitely without species becoming extinct.

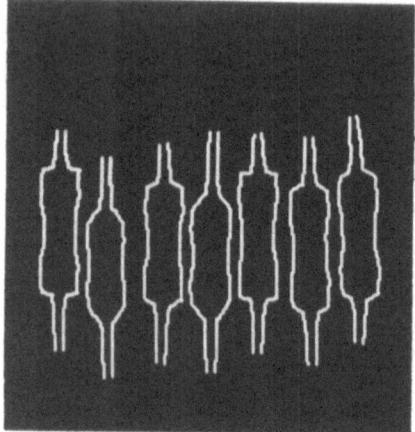

Figure 5 : Multiple Plausible Interpretations Employing a Speciated GA+ASM (the component on the left was excluded from the search area).

6 Conclusions

Combining a generic method of flexible template construction with the GA search technique has been shown to produce a very powerful and generally applicable method of model-based image interpretation. The ability of the PDM to capture both the variation in shape of single objects and the spatial relationships between different objects enables models of complex structures to be derived. We have shown that the GA search technique can be successfully employed to locate instances of these complex models within image data. Furthermore, the incorporation of a local optimisation technique for PDM models within the GA framework as an *Active Shape Mutation* improves considerably the performance of the GA search both in terms of speed of convergence and optimality of solution. The active shape mutation also improves the *speciated* version of the GA, in which many solutions are extracted simultaneously, by enabling small sub-populations to survive and improving the adaptation of each species to its particular environmental niche.

7 Acknowledgements

This research was funded by the UK Science & Engineering Research Council and Department of Trade & Industry. Dr Hill is seconded from C. N. Software Ltd.

8 References

[1]   J. E. Baker, Reducing Bias and Inefficiency in the Selection Algorithm, Proc. of the 2nd Int. Conf. on Genetic Algorithms, Hillsdale, NJ, USA, 1987; 14–21.

[2]   T. F. Cootes, D. H. Cooper, C. J. Taylor, and J. Graham, Training Models of Shape from Sets of Examples, Proc. British Machine Vision Conference, Leeds, 1992.

[3]   Cootes T. F., and Taylor C. J., Active Shape Models – Smart Snakes, Proc. British Machine Vision Conference, Leeds, 1992.

[4]   L. Davis, Genetic Algorithms and Simulated Annealing, Pitman, London, 1987.

[5]   J. M. Fitzpatrick, J. J. Grefenstette, and D. Van Gucht, Image Registration by Genetic Search, Proc. IEEE Southeastcon, Louisville 1984; 460–464.

[6]   D. E. Goldberg, Genetic Algorithms in Search, Optimisation and Machine Learning, Addison–Wesley, 1989.

[7]   J. J. Grefenstette, Optimisation of Control Parameters for Genetic Algorithms, IEEE Trans. on Systems, Man and Cybernetics 1986; 16(1):122–128.

[8]   A. Hill, and C. J. Taylor, Model–Based Interpretation using Genetic Algorithms, Proc. British Machine Vision Conference, Glasgow, 1991, Springer Verlag; 266–274.

[9]   J. H. Holland, Adaptation in Natural and Artificial Systems, University of Michegan Press, Ann Arbor, 1975.

[10] M. Kass, A. Witkin, and D. Terzopoulos, Snakes: Active Contour Models, 1st International Conference on Computer Vision, pub. IEEE Computer Society Press (1987) pp 259–268.

[11] P. Lilly, J. Jenkins, and P. Bourdillon, Automatic Contour Definition on Left Ventriculograms by Image Evidence and a Multiple Template–Based Model, IEEE Trans. on Medical Imaging (1989); 8(2):173–185.

[12] P. Lipson, A. L. Yuille, D. O'Keeffe, J. Cavanaugh, J. Taaffe, and D. Rosenthal, Deformable Templates for Feature Extraction from Medical Images, Proc. 1st European Conference on Computer Vision, Lecture Notes in Computer Science, Springer–Verlag, 1990; 413–417.

[13] L. H. Staib, and J. S. Duncan, Parametrically Deformable Contour Models, IEEE Computer Society Conference on Computer Vision and Pattern Recognition, San Diego, 1989.

[14] A. L. Yuille, D. S. Cohen, and P. Hallinan, Feature Extraction from Faces using Deformable Templates, Proc. Computer Vision, San Diego, 1989; 104–109.

Recognition of Volcanoes on Venus using Correlation Methods

Charles R. Wiles[1,2]

M.R.B. Forshaw[1]

[1]Image Processing Group, [2]Planetary Science Group
Department of Physics and Astronomy, University College London
London WC1E 6BT, UK

Abstract

Radar images of 95% of the surface of Venus have been obtained by the *Magellan* spacecraft at resolutions of 100-300 m. The surface area covered is 3 times the total land-mass area of the Earth; this corresponds to a data volume of about 10^{11} bytes.

A large population of volcanoes has been observed in this data set. Measurements of these features are essential for a full understanding of Venusian geology. The scale of the task, however, precludes the use of manual methods to make these measurements. An algorithm for the automated location and counting of these volcanoes is therefore being developed.

The noisy nature of the data makes it appropriate to use correlation-based techniques to recognise the features. A least-squares-error template matching algorithm has been implemented, which includes local DC removal and contrast normalisation.

Preliminary experimental results from running the algorithm on Magellan data are presented, along with the corresponding measurements of expert human observers. Because there is no ground truth information for Venus, it has also been necessary to undertake a control experiment, using simulated radar images of artificial terrain. The results of this experiment are also included and compared with theoretical predictions: their implications for the calibration of both human and automated measurements are discussed.

1   Introduction

The main objective of the Magellan mission is to increase knowledge about the geological history and geophysics of the planet Venus, by mapping its surface [1]. To accomplish this, a spacecraft containing a synthetic aperture radar (SAR) instrument was inserted into orbit about Venus in August 1990. Since then, the spacecraft has returned SAR images of 95% of the planet's surface at resolutions of between 100 and 300 m. Because Venus does not possess any oceans, the surface area mapped by Magellan is over three times the total area of all the land masses on Earth; this represents an image data volume of almost 100 Gb [2].

The fact that its bulk properties are so similar to Earth's makes Venus a unique and important planetary comparison. By studying its geology and surface processes it is hoped that our understanding of the formation and evolution of the Earth will be improved. Unfortunately, it is impossible to observe Venus' surface using conventional remote sensing techniques because it is continuously obscured by optically opaque clouds. However, Magellan's radar, operating at a wavelength of 12.6 cm, is able to penetrate the cloud cover.

Both the Magellan and earlier (low-resolution) Venera 15/16 SAR data have revealed that small volcanoes are a prevalent feature on the Venusian surface. These edifices, which are predominantly low shield volcanoes <15 km in diameter, have been extensively described by several earlier workers (eg. [3,4]). Their abundance indicates that volcanism has been of fundamental importance to Venus' evolution. The fact that they are also extremely widespread makes the volcanoes an important statistical tool for studying Venus' geological history. Measuring their global population, spatial distribution and size-frequency distribution would provide important information about both the geological and geophysical processes on Venus.

However, from previous work it is estimated that the global population of small volcanic features on Venus is of the order of 5×10^6 [3]. This, together with the vast size of the Magellan data set (described above), means that even a preliminary visual survey would take a trained observer at least 5 years. To make detailed, global measurements would therefore be extremely time-consuming and prone to error. It is thus desirable that an automated method be developed for the consistent identification and measurement of these features.

This paper discusses one such method which is based on correlation techniques, specifically a template matching algorithm. There is an additional problem associated with this task, however: the total lack of ground truth information against which to calibrate the algorithm. Consequently, the paper also describes a control experiment which involved generating artificial SAR images of synthetic terrain, designed to emulate closely real Magellan imagery of the small volcanoes.

2 Magellan data characteristics

Magellan's SAR transmits pulses of coherent, microwave radiation perpendicular to its flight-path and records the back-scattered echoes. These echoes are then processed to form images by analysing their intensity, time-delay and frequency-shift. Multiple sampling allows the spatial resolution of a larger antenna to be synthesised.

This image formation mechanism is fundamentally different to the way in which images at visible wavelengths are produced. Consequently, SAR images possess certain properties which set them apart from conventional images. The slant-range/doppler SAR coordinate system is very different from the coordinate system of a traditional image and leads to different types of geometric distortions. The much longer wavelengths involved mean that radar reflectance characteristics are also different: radar back-scatter strength is governed by the topography, roughness and electrical properties of a surface in a complex and incompletely understood manner. Finally, since they are produced from coherent illumination, SAR data inherently suffer from the effects of speckle noise, giving them a grainy texture.

These characteristics can place additional demands on pattern recognition algorithms used to analyse SAR data. For example, both the degree of geometric distortion and the dominant back-scatter mechanism observed, depend strongly on the radar incidence angle, which may not remain constant (as is the case with Magellan).

Due to technical constraints, Magellan's polar orbit is highly elliptical, which causes significant variation in the SAR imaging parameters. The range resolution varies from 110 to 280 m, the nominal incidence angle changes from 17° to 47° and the number of looks (which affects the speckle noise) goes from 5 to 17. Thus, in the Magellan data, similar features can appear quite different, depending on which combination of parameters were in force at that stage of the mapping sequence.

Every orbit, the spacecraft maps a swath that is 16000 km long and 25 km wide. The SAR data undergo several stages of processing to convert them into grey-level images. This processing includes both geometric and radiometric corrections, followed by resampling to give a uniform pixel spacing of 75 m. Finally, the image swaths are mosaicked together into data products known as *MIDRs*. Each MIDR contains 56 Mb of data, in the form of an 8192x7168 array of 8-bit pixels. To cover the entire planet, 1650 such MIDRs are required. (A full description of the Magellan mission is given in [5]).

3 Description of algorithm

Because the SAR images are inherently noisy, it is not possible to employ feature-detection methods which are based on edges or contours. Instead area-based methods must be used, foremost amongst which is the classical correlation of a template with the scene. In practice, classical correlation is unsatisfactory because it breaks down when faced with variations in background brightness or structures whose contrast differs from that of the template.

The correlation process must be carried out in spatial coordinates to allow for the locally-varying DC removal and contrast normalisation. These corrections are essential if low-contrast features are to be detected. Although correlation could be carried out using global FFT techniques it would be extremely inefficient to implement such local corrections using these methods.

The algorithm which is used for this work is known to be statistically robust and relatively insensitive to noise [6]. For a given position within an image (p,q), using a template of size $N \times N$, the correlation function $C(p,q)$ is defined as follows

$$C(p,q) = \sum_{i,j=0}^{N-1} \left[\left[\frac{(t(i,j) - \bar{t})}{\left[\sum_{k,l=0}^{N-1} (t(k,l) - \bar{t})^2 \right]^{0.5}} \right] - \left[\frac{(I(i+p,j+q) - \bar{I})}{\left[\sum_{m,n=0}^{N-1} (I(m+p,n+q) - \bar{I})^2 \right]^{0.5}} \right] \right]^2 \quad (1)$$

where: t and I represent the pixel values of the template and the image respectively; and $\bar{t}$ and $\bar{I}$ are the mean template value and the local mean image value. The t and $\bar{t}$ values are independent of p and q, and are calculated prior to correlation. Here, (p,q) defines the top-left corner of the portion of the image covered by the template. This would produce an offset in the final detected position, so, in practice, a correction is applied to centre the coordinates recorded by the algorithm.

The features being searched for, small shield volcanoes, exhibit a wide range of morphological types (see fig. 1). For each volcanic type, a distinct set of templates would have to be employed, to ensure successful matching. This is in addition to the usual requirement of using a range of template sizes to permit recognition of features at differing scales. Together with the variable imaging geometry already described, these considerations make the problem of automatically recognising all of the volcanoes in the Magellan data set very demanding.

Template-matching is a numerically intensive process. For correlations undertaken in the spatial domain, the number of calculations required for a template of size N, in a scene of size M, is proportional to $N^2 M^2$. The small volcanoes are typically 3-4 km in diameter, or 40-50 pixels in the Magellan images. However, it is estimated

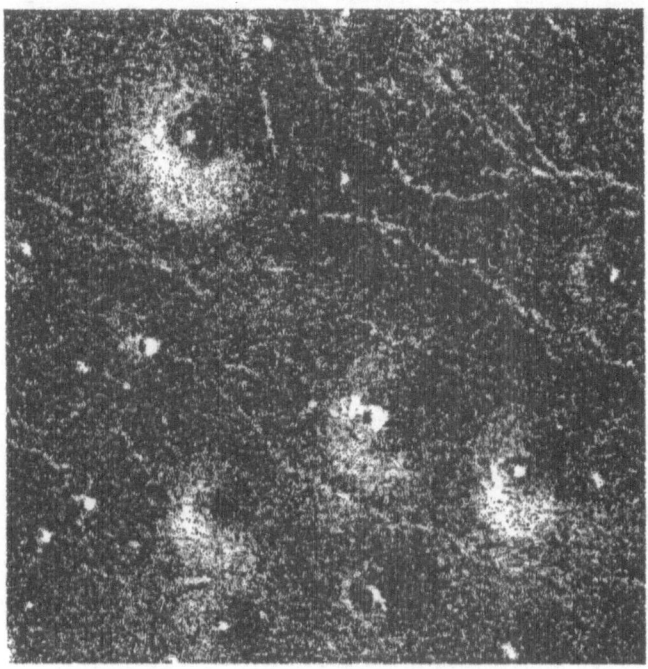

Fig. 1 Example of a Magellan SAR image showing various
types of small volcanoes: many possess summit pits.

Fig. 2 Radar image simulation of artificial pits of
several sizes, prior to addition of speckle noise.
Note, however, inclusion of random surface texture.

that 85% of the volcanoes possess central features known as summit pits [4]. Because these pits have high visual contrast, they are discernible at sizes of about 5 pixels, though they are typically about 10 pixels across, with maximum diameters of 20 pixels or more. This means that they can be matched much more rapidly than can the (considerably larger) complete edifices (since $N_{pit}^2 \ll N_{volcano}^2$). The pits are also more regular in appearance than the volcanic edifices. For these reasons it was decided to concentrate initially on recognising the summit pits. To further reduce processing times, it was also decided that preliminary experiments should be performed using sub-scenes from MIDRs of just 512x512 pixels (ie. to restrict M^2).

4   Generation of synthetic data

There are two types of error to which all forms of feature identification are subject: failures, or 'false negatives' (ie. the number of real features missed); and false alarms, or 'false positives' (ie. the number of non-existent features identified). When dealing with real, noisy data it is unlikely that pattern recognition algorithms can ever achieve 100% success rates (ie. zero false negatives) whilst making no spurious identifications (ie. zero false positives). It is therefore highly desirable to be able to assess the absolute accuracy of any chosen algorithm.

With Magellan data, the lack of ground truth forces a dependence upon uncalibrated human observations against which to compare recognition algorithms. For this work, it was therefore decided to undertake a control experiment, in order to calibrate the ability both of humans and the machine to identify small 'pit-like' features, in the presence of speckle noise. To achieve this, it has been necessary to produce simulated radar images of synthetic terrain, designed to resemble Magellan imagery as closely as possible. This procedure involved several stages –

1. Production of artificial terrain:
The requirement was for a digital elevation model (DEM) to be produced which would closely resemble the morphology of volcanic summit pits on Venus. The characteristics of the shape required were estimated by expert interpretation of Magellan images of typical pits and by using knowledge of analogous features on Earth. A 3-D DEM was then generated using B-splines with ten free parameters, which were varied until the correct shape was obtained. The parameters controlled morphometric quantities such as pit depth and diameter. Once a satisfactory DEM was obtained, simulated surface texture was included by adding small random height perturbations. To reduce large discontinuities, the roughened surface was then smoothed using 3x3 height-averaging.

2. Radar image simulation:
There are several means by which simulated radar images of a DEM can be obtained. For this project it was decided to employ a radar image simulation (see [7]). This technique involved several phases. First, an illumination vector was defined according to the nominal radar incidence angle that was being used for the simulation. By considering the DEM to consist of surface facets, a radar reflectance map was then obtained using simple vector geometry and a suitable, standard back-scatter model. This map contained the radiometric components of the simulated image. However, it was also necessary to model the geometric properties of radar imagery. This was achieved by simulating the effects of slant-range binning. The bin sizes were made to be significantly larger than the reflectance map pixels, since in a real radar image, the back-scattered signal is considered to be produced by returns from many individual scatterers within a

resolution element.

Binning occurred in the axis of the reflectance map that was parallel to the illumination vector (the *range* direction), utilising the DEM once again to ensure correct simulation of the geometric distortions. Since the range resolution was now considerably coarser than that in the other axis, cross-range averaging was performed to make the resolutions in both axes comparable; a similar procedure is used in real SAR processing to reduce the effects of speckle. Finally, the pixel brightness values were adjusted to emulate the specific effects of the Magellan SAR processing sequence (fig. 2 shows examples of the simulated radar images of artificial pits of various sizes).

The radar image simulation was now complete, except for the effects of speckle. These were modelled separately and added to the final image (see below).

3. Scene generation:

Simulated radar images were generated for a whole range of pit diameters, from 2-16 pixels. Pre-determined numbers of these different-sized images were then added to a uniform background (possessing the mean Magellan pixel brightness) to yield scenes with a variety of pit size distributions. The distributions used included exponential, 1/radius and uniform.

The algorithm placed each pit at a random location within the 512x512 scene and stored in a table both the coordinates and the size of the pit. A check was first made, however, to ensure that the randomly generated location for the pit, was entirely within the scene and also that another pit had not been previously placed there. The second restriction was enforced to avoid small pits being covered by larger ones (fig. 3 shows one such artificial scene).

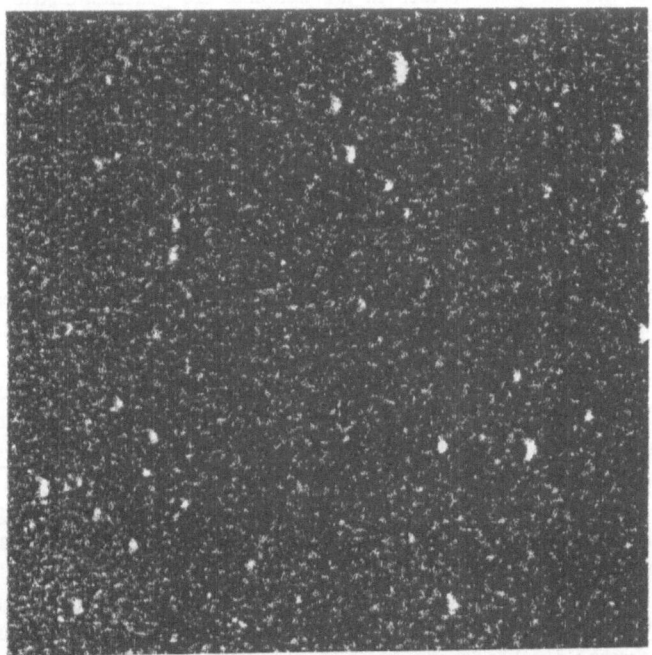

Fig. 3 Simulated radar scene containing pits of various sizes: includes speckle noise.

4. Addition of speckle:

Speckle noise is observed in any images produced by coherent radiation. With SAR images, it results from the interference of returns from many radar scatterers within a resolution element. Speckle can be reduced by making several statistically independent observations, or *looks*, of a target and averaging them. Magellan always obtains images comprising at least five looks.

SAR speckle can be modelled as a random multiplicative noise source with a Rayleigh distribution. The effects of speckle were therefore simulated by using a Rayleigh random number generator to produce a 512x512 'speckle' image, with a mean value of 1.0. To obtain 'multiple looks', several such images were generated using different random number seeds. For this experiment, five of the speckle images were averaged together. Finally, since speckle noise is multiplicative, the artificial scene was multiplied by the five-average speckle image, pixel by pixel.

5. Resolution degradation:

The last stage of the data simulation, was to emulate the resampling of Magellan images, described in section 2. To achieve this it was decided to employ 3x3 local-neighbourhood blurring. The blurring kernel was chosen so as to mimic the resampling of a typical Magellan resolution cell (150x110 m) into 75x75 m pixels.

Fig. 3 is the final product of one such scene simulation, which was used in the control experiments described in section 6.1.

5 Theoretical correlation performance

This section briefly describes a theory which has been developed to predict the performance of the correlation algorithm with the test data generated using the procedure outlined above. Consider a template of area A pixels, containing the noise-free image of a pit: assume that the pixel values in the template are gaussian-distributed with zero mean and some known standard deviation. The template is shifted across the scene and is correlated with the scene data according to expression (1). Note that both the scene data lying within the current window and the template data are first normalised to unit variance. At points in the scene where there is only a uniform background which is corrupted by speckle, then the correlation function may be written in the simplified form:

$$C = \sum_A (t_{norm} - B_{norm})^2 \qquad (2)$$

where t_{norm} represents the normalised template data values and B_{norm} represents the independently normalised, noisy background values. C is a sum of A independent samples, each of variance 2, since both t_{norm} and B_{norm} have unit variance. The expectation value for C , <C>, is therefore 2A and its standard deviation, σ, is approximately $\sqrt{2A}$. By making similar approximations it can also be shown that when the template overlays a portion of the scene which contains a noisy version of the template data, the value for <C> is ≈ A, with a standard deviation of $\sqrt{2A}$.

A further assumption is made, that the background values are also gaussian-distributed (in fact they follow an approximate Rayleigh distribution). There are therefore two sets of gaussian-distributed correlation values: one with a mean value of two and a standard deviation of $\sqrt{2A}$ (the "background" correlation values); the other with a mean of unity and the same standard deviation (the "signal" correlation values).

A suitable threshold value is now chosen, such that only rarely is a background correlation value obtained which falls below this threshold, giving rise to a false match. Setting the threshold at 2.5σ below the background value has been found to be convenient in practice. Once the threshold has been fixed in this way, one may readily deduce how often a "true" match will be detected (ie. a value from the signal distribution which falls below the threshold).

The main error is that the distribution of intensity values in the template is not gaussian. However, a correction factor may be applied which *does* produce an approximately gaussian distribution in the template. Inspection of the distribution of values shows that a majority (70%) of the pixel values in the template lie very close to the mean, giving rise to a spike in the distribution. Removing these pixels from the correlation, results in a probability of detecting a pit which varies with the pit diameter (as shown in fig. 4).

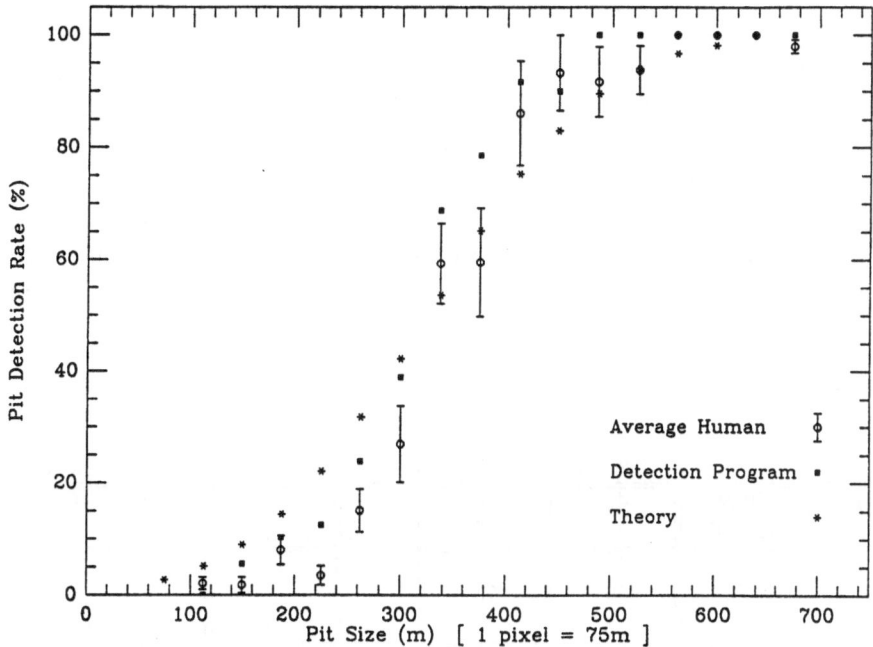

Fig. 4 Comparison of detection rates for simulated pits: humans, machine and theory.

6 Analysis of results

In this section the results of two distinct sets of experiments are presented. First, the outcome of the control experiment involving both human and machine measurements of artificial images is analysed. Second, the results of running the correlation algorithm on Magellan data are assessed, by comparing them with human measurements of the same data.

6.1 Synthetic data

A variety of tests were carried out with different distributions of simulated pit sizes and speckle noise levels. Software routines were developed which logged the (x,y) coordinates where human observers believed that they had detected a pit. Since the coordinates of the pits were already known, it was possible to calculate the detection rate (and false alarm rate) as a function of pit size. Modified versions of these routines were used to calculate the same quantities for the pit-detection algorithm.

Fig. 4 summarises the results of some of these tests. It shows how the probability of detecting a pit image varies with the pit diameter. Three sets of points are shown: one set is the average for all observers and all pit size distributions; one set is the equivalent for the detection algorithm; and the third set is derived from the theoretical prediction outlined in section 5.

It can be seen that the three sets of data are very similar to each other (the differences between the human results and the machine results are almost within the error bounds for each data point). The departure of the theoretical curve from the machine results is readily explained, as being due to the approximations used in deriving the theory. What is perhaps of more interest, is the similarity between the machine results and the human results, which suggests that some underlying similarity may exist between two apparently disparate detection mechanisms. In addition, results such as those shown in fig. 4 can be used to provide corrections to pit size-distributions derived previously by human observers (eg. [3]).

6.2 Magellan data

To date, the algorithm has only been run with a very limited number of Magellan images for which human observations are available (eg. fig. 1). The results of this comparison between the pit detection rates of humans and the machine, using real Magellan images, are shown in table 1.

Table 1	Pit Detection Results for Real Data					
MGN Image	Observer	No. seen by observer	Seen by observer only	Seen by machine & observer	Seen by machine only	No. seen by machine
1	CRW	42	4	38	3	41
	MRBF	48	10	38	3	41
2	CRW	19	8	11	4	15
	MRBF	37	26	11	4	15
	JEG	18	9	9	6	15
3	CRW	26	7	19	6	25
	MRBF	36	19	17	8	25
	JEG	41	22	19	6	25

From this table it is apparent that there are significant variations in the observations made by humans. It can also be seen that, under certain circumstances, the machine results are very promising: the algorithm performs well on images where the background is relatively uniform (although it has yet to be modified to take account of variations in the background texture which require a variable threshold). Finally, it was also noted that, whilst some of the pits seen by the machine alone were false positives, many represented real features that the humans had missed.

7   Discussion

In this paper, we have described the preliminary results of an investigation into the automatic detection of volcanic features in noisy, radar images. Whilst some progress has been made towards accomplishing this task, we have had to limit ourselves to detecting the small pits which many, but not all of the volcanoes possess. We have also encountered difficulties when using our algorithm in regions where the background texture is non-uniform.

In the immediate future, we aim to increase the reliability of the algorithm at detecting the pits. It is then intended that as much Magellan data as possible is processed using the improved technique. At the same time, modifications will be made to enable detection of the volcanoes themselves, probably using reduced-resolution Magellan data.

This leads to the question of processing times. At present, running on a Sun SPARCstation 1+, the algorithm is able to scan one 512x512 pixel image, making several passes with different-sized templates, in approximately 30 minutes. For regions with high volcano-densities, this time is comparable to that required by a trained geologist to undertake the same task of recognising and recording the coordinates of all the volcanic edifices. The machine, however, has certain advantages over the human observer. In particular, it could, in principle, process data 24 hours per day with little human intervention. Furthermore, from the experiments conducted with artificial data, it has been discovered that humans occasionally miss large features for which the machine has a consistent, 100% success rate.

In conclusion, therefore, whilst still at a relatively early stage of development, the methods described here appear to have great potential in the fields of planetary science and remote sensing. Data sets in these areas are already reaching sizes that make comprehensive analysis by trained observers impossible. As the processing speeds of relatively inexpensive computers continue to increase, however, automated image-data analysis and reduction will become an increasingly important and effective scientific tool.

Acknowledgements

C.R. Wiles has been supported by a NERC Ph.D. studentship during the course of this research.

We acknowledge the Magellan project at the NASA Jet Propulsion Laboratory, for the provision of radar image data of Venus.

References

[1] JPL D-6724, Magellan science requirements document, 630-6 (Rev. D). California Institute of Technology, Pasadena, 1991, pg 2-1.

[2] G.H. Pettengill, P.G. Ford, W.T.K. Johnson, R.K. Raney and L.A. Soderblom, Magellan: radar performance and data products. *Science* 1991; 252: 260-265.

[3] J.C. Aubele and E.N. Slyuta, Small domes on Venus: characteristics and origin. *Earth, Moon and Planets* 1990; 50/51: 493-532.

[4] J.E. Guest, M.H. Bulmer, J.C. Aubele, et al., Small volcanic edifices and volcanism in the plains of Venus. *J. Geophys. Res.* 1992; (in press).

[5] R.S. Saunders, G.H. Pettengill, R.E. Arvidson, et al., The Magellan Venus radar mapping mission. *J. Geophys. Res.* 1990; 95: 8339-8355.

[6] J.P. Secilla, N. Garcia and J.L. Carrascosa, Template location in noisy pictures. *Signal Processing* 1988; 14: 347-361.

[7] J. Thomas, Radar image simulation. In: Leberl F.W. Radargrammetric image processing. Artech House, Boston London, 1990, pp 165-187.

The Use of Symmetry Chords for Expressing Grey Level Constraints

D. R. Bailes and C. J. Taylor

Dept. of Medical Biophysics, Manchester University,
Stopford Building, Oxford Road,
Manchester M13 9PT.

Abstract

In many object recognition systems only geometric constraints on the boundaries of the objects are used. Many authors have shown the merit in determining the distinctiveness of boundary constraints, and using these to speed up and improve the robustness of object recognition algorithms. However, if the objects have variable shapes or the image is cluttered, then the use of boundary constraints alone can cause object recognition to be be both slow and lacking in robustness. We present a method for expressing grey level constraints using the grey levels along symmetry chords. We describe a method for forming groups of symmetry chords, and calculating their distinctiveness on the basis of the grey levels along the chords in training images. We show how distinct groups of symmetry chords can be used in object recognition by creating evidence images for specific groups of boundary points. Our initial results are promising.

1 Introduction

Most object recognition methods for 2D objects fall into the following framework. The object is modelled by its boundary, which is divided into segments, and the object is found by finding part or whole boundary segments. Geometric constraints are used for the matching of image segments to model segments, and to ensure consistency between pairs of matched image and boundary segments. Many authors have shown that using the distinctiveness of a segment or a group of segments can speed up and improve the robustness of an object recognition algorithm [1, 2, 3].

However, if the objects have variable shapes or the image is cluttered, then the use of boundary constraints alone can cause object recognition to be both slow, and lacking in robustness. The following example illustrates the problem. Two image boundary segments may match two model boundary segments which belong to the same model, and the pair of matches may be geometrically consistent, but the image boundary segments may in fact belong to different objects. These bad matches can significantly reduce the speed and robustness of object recognition. The additional constraint of the expected grey levels between these segments can often exclude such bad matches.

We are interested in finding ways of expressing grey level constraints which satisfy the following criteria:

1. They must model variability, which should be learnt from training data.
2. Be able to cope to some extent with partial occlusion.
3. Be unaffected by scaling the grey levels in the image.
4. Be able to cope with variations in spatial size.
5. Allow the calculation of distinctiveness from training data.
6. Be reasonably general.

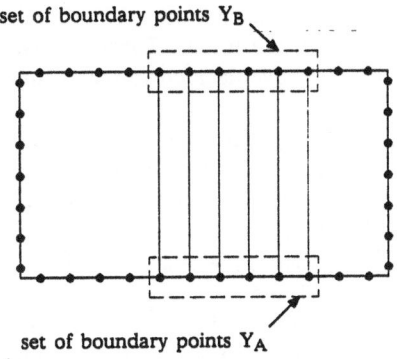

set of boundary points Y_B

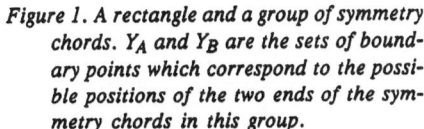

set of boundary points Y_A

Figure 1. A rectangle and a group of symmetry chords. Y_A and Y_B are the sets of boundary points which correspond to the possible positions of the two ends of the symmetry chords in this group.

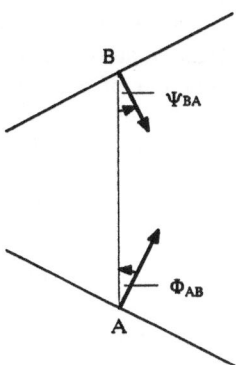

Figure 2. Definition of a symmetry chord.

We have explored the possibility of using the grey level profiles along symmetry chords as grey level constraints, and will give some justification for this in section 2. In many cases, symmetry chords belonging to the same symmetry region have similar grey level profiles, and so these are grouped together to form groups of symmetry chords on the basis of the grey levels along the chords in a set of training images. We then calculate the distinctiveness of each group of chords with respect to all the other groups of chords.

A distinct group of symmetry chords G can be used for object recognition by creating evidence images for the two sets of boundary points Y_A and Y_B which correspond to the possible positions of the two ends of the symmetry chords in G (Fig 1). The edge points in the image are found using the method of Canny[4]. All the possible symmetry chords between pairs of edge points which have a length which lies within the range of lengths of the chords in G are found. For each chord, the probability that it belongs to G is calculated using the grey levels along the chord. Using these probabilities, we calculate for each edge point, the probability that it belongs to the set of boundary points Y_A, and the probability that it belongs to the set of boundary points Y_B, thus creating two probability images. These probability images contain far fewer unwanted points than the original edge image, and can easily be incorporated into any of the standard object recognition methods.

2 Grey levels along symmetry chords

In this section we will explain why we have chosen the grey level profiles along symmetry chords as a way of expressing grey level constraints. We will also describe how the grey levels are represented.

We are trying to find distinctive constraints which apply to parts of the grey level landscape. There are a number of options as to what kind of parts of the landscape to use: 2d patches of any shape; curves of any shape; or chords. The advantage of using 2d patches rather than curves or chords is that there is a better chance of finding patches which are very distinctive, since they contain more information. However, the extremely large number of possible patches makes it difficult to construct an automatic method for finding distinctive constraints. There are also a large number of possible curves, and therefore we have decided to try to find distinctive chords.

We will consider only the chords which begin and end on object boundaries. This has the advantage that it greatly reduces the number of possible chords. One option would be to use all such chords, and parameterise each chord by the following parameters: length; the angles between the boundary normals and the chord; and the grey levels along the chord. However, the number of possible chords is still very large. If we model the boundaries in an image by N discrete points, then there are $O(N^2)$ possible chords. For each of these chords we would have to collect the grey level profiles from several training images. N does not have to be too large before this results in an unmanageable amount of data which has to be stored and processed.

The number of of chords can be reduced by putting a constraint on the angles between the chord and the boundary normals. We have investigated the constraint that a chord must be a symmetry chord [5, 6]. In Fig 2 Φ_{AB} is the angle of the symmetry chord AB with respect to the boundary normal at A, and Ψ_{BA} is the angle of the boundary normal at point B with respect to the chord BA. The chord AB is a symmetry chord if $\Phi_{AB} = \Psi_{BA}$.

The locus of the midpoint of the symmetry chords of a boundary takes the form of a number of disconnected symmetry axes. The set of chords associated with a symmetry axis can be considered to 'cover' a region, and we refer to such a region as a symmetry region. For many objects, symmetry chords in the same symmetry region have similar grey level profiles. Thus we are unlikely to be able to find a symmetry chord which is distinct with respect to all other chords. Rather, we hope to find groups of chords which are distinct with respect to all other groups. By moving from trying to find a distinct chord to finding a distinct group we are reducing the specificity of the method, but we are increasing its robustness to occlusion. A symmetry region can be partially occluded, and still have some of its symmetry chords unoccluded.

We have investigated representing the grey levels along a symmetry chord in four different ways: either the grey levels or the values of the component of the gradient in the direction of the chord are used; in each case either the unnormalised values are used, or the values are normalised by dividing the values by the sum of the absolute values along the chord. The values of the component of the gradient in the direction of a chord are obtained from the x and y components of the gradient of the Gaussian smoothed images which are produced by the first stage of the Canny edge finding scheme. To ensure a fair comparison, the values of the grey values are taken from images which have been smoothed by a Gaussian filter with the same standard deviation as was used for the edge detection. If it is required that the constraints be unaffected by scaling the grey levels in the image, then the values must be normalised. The advantage of using the component of the gradient in the direction of the chord is that the constraints are also unaffected by a uniform constant being added to the grey levels of the image.

For convenience we will often refer to the values of any of these representations of the grey levels along a chord as 'the grey levels along the chord', and denote their values by the vector **g**. The grey levels along a chord are not independent, and so we

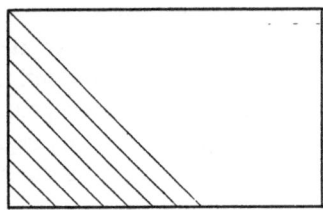

Figure 3. An example of an unwanted symmetry region.

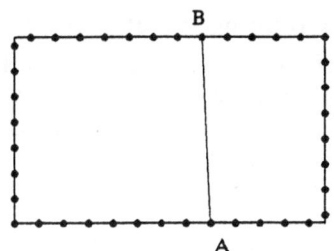

Figure 4. A symmetry chord for a discrete boundary.

describe the distribution of the grey levels by their mean $\bar{g}$, and their covariance matrix S.

3 Finding distinct groups of chords from training data

We now describe an automatic method for finding distinct groups of chords. For each symmetry region we need to end up with a discrete set of chords, so that the grey levels along each chord can be sampled from the training images. We have therefore decided to model the object boundary as a discrete set of points, and to calculate the symmetry chords only between these points. Initially, we will only consider boundaries which are slightly variable.

3.1 Calculating the symmetry chords

We do not want to calculate all the symmetry chords for two reasons. Firstly, the set of chords in the symmetry region of the rectangle shown in Fig 3 have widely varying lengths and are unlikely to have similar grey level profiles. We are therefore not interested in such symmetry regions. Secondly, there is the well known problem that all the chords of a circle are symmetry chords. We have decided to find only the chords which are diameters.

In the continuous case, the condition for a symmetry chord is that $\Phi_{AB} = \Psi_{BA}$ exactly. However, in the discrete case, this has to be relaxed. Consider point A in Fig 4 which lies on the boundary of a rectangle. In general, there will not be a point on the opposite side of the rectangle which lies directly opposite point A. The condition for a symmetry chord has to become $|\Phi_{AB} - \Psi_{BA}| < t_1$, where t_1 depends on the spacing of the points along the boundary. There may be more than one point opposite point A which satisfies our relaxed condition, and therefore we choose the chord for which $|\Phi_{AB} - \Psi_{BA}|$ is a minimum with respect to B.

By introducing a second condition that $|\Phi_{AB}| + |\Psi_{BA}|$ is a local minimum with respect to B, we find only the symmetry chords in circles which are diameters. A third condition that $|\Phi_{AB}| < t_2$ rules out symmetry regions like the example in Fig 3. Thus our set of conditions for a symmetry chord are:

$$|\Phi_{AB} - \Psi_{BA}| < t_1 \ ,$$

$$|\Phi_{AB}| < t_2 \ ,$$

$$|\Phi_{AB} - \Psi_{BA}| + |\Phi_{AB}| + |\Psi_{BA}| \text{ is a local minimum with respect to B.}$$

The third condition is valid for both axial and circular regions.

3.2 Grouping the chords into symmetry region chord sets

We refer to the set of chords which belong to a symmetry region as a symmetry region chord set. For the symmetry chords of a discrete boundary, we define a symmetry region chord set as follows. Each chord is specified by its start point a, and its end point b, where $a = 1 .. N_{bound}$, and $b = 1 ..N_{bound}$. A symmetry region chord set is an ordered set of symmetry chords (a_j, b_j), $j = 1 .. N_{region}$, such that:

1. a_j is monotonically increasing.
   b_j is monotonically increasing or decreasing.
2. $a_j - a_{j-1} <= 1$
   $| b_j - b_{j-1} | <= 1$

3.3 Getting grey level data from the training images

For each image in the training set, the model boundary is matched to the image manually, and for each of the symmetry chords the grey levels are sampled along the chord. For a given symmetry chord we use the same number of equally spaced sampling points for each image. This ensures that the method can cope with variations in spatial size. The number of sampling points is determined by the maximum length of the chord in the training images, and the standard deviation of the Gaussian filter used to smooth the images. We currently sub-sample by a factor of 2 times the standard deviation. We compute the mean $\bar{g}$, and the covariance matrix S for each chord.

3.4 Forming groups of similar chords

We currently choose groups of similar chords manually. However we envisage an automatic method of grouping which consists of two stages: for each symmetry region we form groups of chords, such that within groups there is little variation in $\bar{g}$ and S; then, within an object, similar groups are merged. For each group, the overall values of $\bar{g}$ and S are calculated, and these values are used to calculate how distinct each group is with respect to all the other groups. Thus each group has the following parameters:

1. A measure of distinctiveness.
2. Overall values of $\bar{g}$ and S.
3. The range of angles of the symmetry chords with respect to the boundary normal at A, Φ_{AB_1} to Φ_{AB_2}.
4. The range of lengths of the chords, l_{min} to l_{max}.
5. The number of sampling points for the chord n_s.
6. The set of boundary points Y_A which corresponds to the possible positions of point A.
7. The set of boundary points Y_B which corresponds to the possible positions of point B.

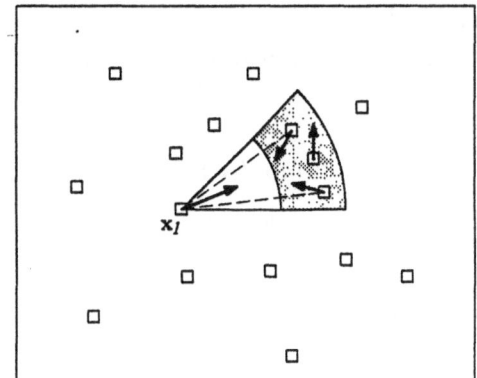

Figure 5. Finding the edge points which form symmetry chords with the point x_1.

4 An application of grey level constraints

If an object contains one or more distinct groups of chords, then we can usefully create evidence images for specific groups of boundary points. Given an image in which we want to find the object, the edge points in the image are found using Canny's method, except that we do not apply any thresholding. For a given group of symmetry chords G_i, we find all the possible symmetry chords between pairs of edge points which have a length in the range l_{min} to l_{max}, and an angle with respect to the boundary normal at A in the range Φ_{AB_1} to Φ_{AB_2}. For each symmetry chord, the probability that it belongs to G_i is calculated using the grey levels along the chord, which are sampled at the same number of equally spaced points for each chord. Using these probabilities, we calculate for each edge point, the probability that it belongs to the set of boundary points Y_{A_i}, and the probability that it belongs to Y_{B_i}, thus creating two probability images.

The probabilities that an edge point belongs to Y_{A_i}, or Y_{B_i} are calculated as follows. If a point x_1 is hypothesised as a member of Y_{A_i}, then the other end of the symmetry chord must lie within the sector of an annulus which has radii l_{min} and l_{max}, and range of angles $\Theta_A + \Phi_{AB_1}$ to $\Theta_A + \Phi_{AB_2}$, where Θ_A is the angle of the gradient at x_1. In general, there will be a number of edge points within this sector, and of these, a number will form symmetry chords with point x_1, as shown in Fig 5. Let set X_B be the set of edge points within the sector which form a symmetry chord with point x_1. The point x_1 is a member of Y_{A_i} and the point x_2 is a member of Y_{B_i} if, and only if the chord (x_1, x_2) belongs to the chosen group G_i on the basis of its grey level profile. Therefore the point x_2 in set X_B which is most likely to belong to Y_{B_i} is the point for which the probability that the chord (x_1, x_2) belongs to the group of chords is a maximum. Hence the probability that point x_1 is a member of Y_{A_i} is given by:

$$P(\ x_1 \ is \ a \ member \ of \ Y_{A_i} \mid x_1, \ X_B\ ) \ = \ \max_{x_2 \in X_B} P(\ G_i \mid g(\ x_1, \ x_2\ )\ )$$

And similarly:

$$P(\ x_2 \ is \ a \ member \ of \ Y_{B_i} \mid x_2, \ X_A\ ) \ = \ \max_{x_1 \in X_A} P(\ G_i \mid g(\ x_1, \ x_2\ )\ )$$

4.1 Probabilities

It can be seen that we need to calculate the probability that a particular chord comes from the chosen group of chords. Using Bayes theorem:

$$P(\ G_i\ |\ \mathbf{g}\ ) = \frac{P(\ G_i\ )P(\ \mathbf{g}\ |\ G_i\ )}{P(\ G_i\ )P(\ \mathbf{g}\ |\ G_i\ ) + P(\ G_i^C\ )P(\ \mathbf{g}\ |\ G_i^C\ )}$$

$$= \frac{p_i\ f_i(\ \mathbf{g}\ )}{p_i\ f_i(\ \mathbf{g}\ ) + p_C\ f_C(\ \mathbf{g}\ )}$$

where G_i^C is the complement of G_i, p_i and p_C are the prior probabilities that a chord belongs to G_i and G_i^C respectively, and $f_i(\ \mathbf{g}\ )$ and $f_C(\ \mathbf{g}\ )$ are the probability densities of G_i and G_i^C respectively. The main decision is what form the distribution for $f_C(\ \mathbf{g}\ )$ should take. We have made the assumption that this is a normal distribution, even though we know that there will be spikes in the distribution due to other distinct groups of chords.

It is computationally efficient to compute the probability via the odds ratio:

$$P(\ G_i\ |\ \mathbf{g}\ ) = \frac{odds}{1 + odds}$$

$$odds = \frac{P(\ G_i\ |\ \mathbf{g}\ )}{P(\ G_i^C\ |\ \mathbf{g}\ )} = \frac{p_i\ f_i(\ \mathbf{g}\ )}{p_C\ f_C(\ \mathbf{g}\ )}$$

$$= \frac{\dfrac{p_i}{(2\pi)^{\frac{n_g}{2}}\ |S_i|^{\frac{1}{2}}}\exp(\ -(\ \mathbf{g} - \bar{\mathbf{g}}_i\ )'S_i^{-1}(\ \mathbf{g} - \bar{\mathbf{g}}_i\ )/2\ )}{\dfrac{p_C}{(2\pi)^{\frac{n_g}{2}}\ |S_C|^{\frac{1}{2}}}\exp(\ -(\ \mathbf{g} - \bar{\mathbf{g}}_C\ )'S_C^{-1}(\ \mathbf{g} - \bar{\mathbf{g}}_C\ )/2\ )}$$

$$= \left(\frac{p_i}{p_C}\right)\left(\frac{|S_C|}{|S_i|}\right)^{\frac{1}{2}}\exp(\ -\mathbf{g}'(\ S_i^{-1} - S_C^{-1}\ )\mathbf{g}/2 + (\ \bar{\mathbf{g}}_i'S_i^{-1} - \bar{\mathbf{g}}_C'S_C^{-1}\ )\mathbf{g} - (\ \bar{\mathbf{g}}_i'S_i^{-1}\bar{\mathbf{g}}_i - \bar{\mathbf{g}}_C'S_C^{-1}\bar{\mathbf{g}}_C\ )/2\ )$$

5 Results

We have applied the method described above to both printed circuit board (PCB) and brake assembly images, and have obtained promising results with both sets of images. The method for automatically finding the distinct groups of chords is not complete, and therefore we have chosen groups manually which look as if they are distinct.

We present the results of one of our experiments on PCB images. Both the training set and the validation set contain 8 images, and an example of one of the training images is shown in Fig 6. We have chosen the group of symmetry chords which lie across the bodies of the resistors (there are 41 and 38 resistors in the training and validation sets respectively). The edge points which belong to the sets Y_A and Y_B were manually labelled, and these are shown in Fig 6c. The symmetry chords which were used for collecting training data from this image are shown in Fig 6d. An example of a validation image, its edge magnitude image, and the set of points which were manually labelled as belonging to Y_A or Y_B are shown in Fig 7 a,

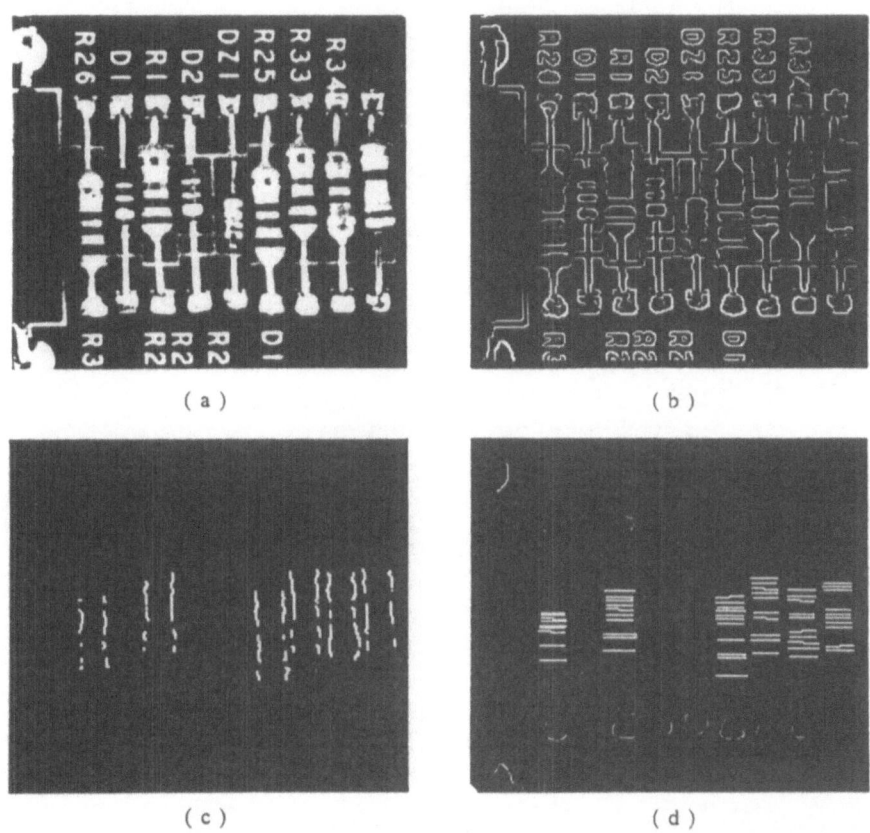

Figure 6. An example of a training image. a) the original image, b) edge magnitude image, c) edge points which were manually labelled as belonging to sets Y_A or Y_B, d) symmetry chords which were used for collecting the training data.

b and c respectively. Fig 7d shows the edge points for which the computed probability that they belong to the set Y_A is greater than 0.5. Because the grey level profiles along each of the symmetry chords in the chosen group is roughly symmetric about the midpoint of the chord, the probability that a point belongs to the set Y_A is roughly the same as the probability that it belongs to the set Y_B. Hence this image corresponds to the membership of the union of the sets Y_A and Y_B.

The method was quantitatively evaluated by computing how well it classified the edge points in the validation set. If the value of a pixel in a probability map is greater than a threshold, then we consider it to be classified as a member of Y_A or Y_B of the chosen group of chords. ROC curves for the four methods of representing the grey levels were computed by varying this threshold, and are shown in Fig 8. The curves are averages over the validation set, and only edge points which satisfied the geometric constraint of being an end of a symmetry chord with length in the range l_{min} to l_{max}, and angle with respect to the boundary normal at A in the range Φ_{AB_1} to Φ_{AB_2} were classified. The average number of boundary points per image which belonged to the sides of the resistors was 449, and of these 412 satisfied the geometric constraint; the average number of other boundary points was 12623, and

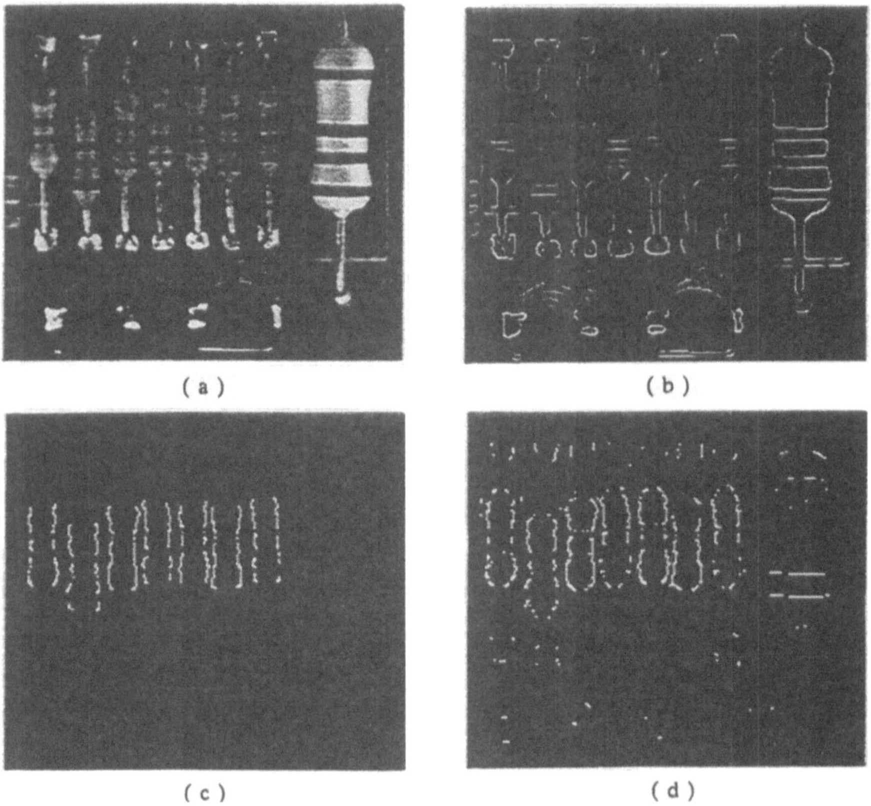

304

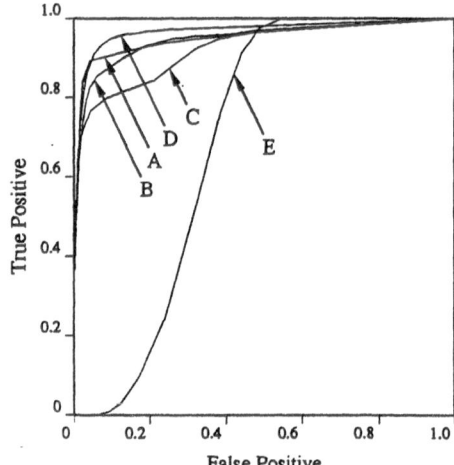

Figure 7. An example of a validation image. a) original image, b) edge magnitude image, c) edge points which were manually labelled as belonging to Y_A or Y_B, d) edge points for which the computed probability that they belong to Y_A is greater then 0.5.

Figure 8. Classification results: ROC curves for the 4 ways of representing the grey level profiles, and for the geometric constraints alone:

A: grey levels
B: normalised grey levels
C: gradient
D: normalised gradient
E: geometric constraint alone

of these 6967 satisfied the geometric constraint. It can be seen that the normalised gradient representation has the best ROC curve, and is the preferred representation. We have also given the classification results for the use of the geometric constraint alone: the ROC curve was obtained by thresholding the values of edge magnitude of the edge points which satisfied the geometric constraint. It can be seen that using the grey level constraint as well as the geometric constraint greatly decreases the number of misclassifications.

For each 256 x 256 PCB image the algorithm takes about 17 seconds on a Sun Sparc 2 workstation. For the chosen group of chords: l_{min} = 14, l_{max} = 20, Φ_{AB_1} = $-18°$, and Φ_{AB_2} = $18°$. No prior knowledge of the orientation of the symmetry chords is used.

6 Conclusions

We have developed a method of expressing grey level constraints which uses the grey level profiles along symmetry chords. We represent a grey level profile by the normalised values of the component of the gradient in the direction of the chord. These grey level constraints satisfy the criteria given in the introduction: they model the variability in the training data; are able to cope to some extent with partial occlusion; they are unaffected by scaling the grey levels in an image; they are able to cope with variations in spatial size; they allow the calculation of distinctiveness from training data; and are reasonably general. We have outlined an automatic method for grouping the chords, but the implementation of this needs completing.

We have shown how distinct groups of symmetry chords can be used in object recognition by creating evidence images for specific groups of boundary points. These evidence images contain far fewer unwanted points than the original edge image. The results are encouraging.

There are two straightforward generalisations of the method. Firstly, rather than the grey levels along symmetry chords, the values of any intrinsic image, eg texture, could be used. Secondly, instead of the symmetry chords between edge points, the symmetry chords between ridge points or trough points could be used.

7 Acknowledgements

Various members of the Wolfson Unit, and especially Dave Cooper, Tim Cootes, and Jim Graham for their helpful comments. This work was funded by SERC project grant no. GR/F63633.

8 References

[1] R.C. Bolles, and R.A. Cain, Recognising and locating partially visible objects: the local-feature-focus method. Int. J. Rob. Res. 1982; 1: 57–82.
[2] J.L. Turney, T.N. Mudge, and R.A. Voltz, Recognizing partially occluded parts. IEEE Trans. PAMI 1985; 7: 410–421.
[3] T.N. Mudge, J.L. Turney, and R.A. Voltz, Automatic generation of salient features for the recognition of partially occluded parts. Robotica 1987; 5: 117–127.
[4] J. Canny, A computational approach to edge detection. IEEE Trans PAMI 1986; 8: 679–698.
[5] H. Blum, Biological shape and visual science (part 1) . J. Theor. Biol. 1973; 38: 205–287.
[6] J. M. Brady, and H. Asada, Smoothed local symmetries and their implementations. Intern. J. Robot. Res. 1984; 3: 36–61.

A Matching and Tracking Strategy for Independently Moving Objects

Larry S. Shapiro, Han Wang and J. Michael Brady

Robotics Research Group,
Department of Engineering Science,
19 Parks Road, Oxford University, OX1 3PJ, U.K.

Abstract

We present a robust and inherently parallel strategy for tracking "corner" features on independently moving (and possibly non–rigid) objects. The system operates over long, monocular image sequences and comprises two main parts. A *matcher* performs two–frame correspondence based on spatial proximity and similarity in local image structure, while a *tracker* maintains an image trajectory (and predictor) for every feature. The use of low–level features ensures an opportunistic and widely applicable algorithm. Moreover, the system copes with noisy data, predictor failure, and occlusion and disocclusion of scene structure. Motion and scene analysis modules can then be built onto this framework. The algorithm is aimed at applications with small inter–frame motion, such as videoconferencing.

1 Introduction

This paper addresses the problem of robustly extracting the image motion of independently moving (and possibly non–rigid) objects over long periods of time. The application motivating this research is model–based coding of facial image sequences [1], which requires fast, reliable motion estimation on lengthy monocular sequences. Here, we focus on the low–level "front–end" of such a system, and show that the use of general–purpose "corner" features enables us to cope with a wide range of facial variations and accessories (e.g., beards and glasses). The image trajectories that emerge are intended to drive "higher–level" modules, which will group features into coherently moving objects and estimate their 3D motion. Further details of the research described here can be found in [6].

We build on recent work at Oxford by Wang and Brady [11], who developed a "corner" finder that runs at 14Hz on T800 transputers. These "corners" are curvature extrema in the image intensity surface, and have already served successfully as features in a stereo–matching algorithm [10]. We extend this work to the general motion case, which introduces both benefits and complications: the former because temporal integration facilitates noise resistance and allows ambiguity to be resolved in time, and the latter because objects can change over time in ways they can't over space alone. As in the DROID system [3], we employ a single tracker and predictor per feature.

Our framework has two parts: a *matcher*, to perform two–frame correspondence, and a *tracker*, to maintain the trajectories and perform prediction. The utility of corners as correspondence tokens is demonstrated in Section 2, and their extraction mechanism described. The matcher and tracker subsystems are then discussed separately in Sections 3 and 4, and we conclude with directions for future research. Results on real imagery are given throughout.

2   Corner detection

In order to match different views of an object, one must first obtain a set of reliable features from each view. (The explicit extraction of "correspondence tokens" was supported by Ullman [8] and further justified by Verri and Poggio [9].) We employ the term "corners" to refer to distinctive feature points such as discontinuities, points of occlusion, and various intensity curvature maxima (e.g., surface markings). These appear in the image as loci of two–dimensional intensity change (i.e., *second–order features*), and impose more constraint on visual processes than edges (which encode only one–dimensional change) [2]. This is particularly true of visual motion; various authors (e.g., [4]) have shown that the *full* optic flow field μ is recoverable at corner points, whilst only the normal component $\mu^\perp$ can be recovered locally along an edge (owing to the aperture problem). Furthermore, corner tokens are discrete and distinguishable, so can be explicitly tracked over time; arbitrarily curving edges are difficult to describe and hence to track [3].

To date, corners have mainly been used where there is an abundance of physical corners arising from man–made objects (e.g., factory environments). We contend, however, that corner features are equally useful in many natural scenes. In a human face, for instance, there are no right–angles or sharp discontinuities; facial features are rounded and the skin surface is smooth. Nonetheless, our experiments on many images [6] have shown that there *are* numerous "corner points" in the image of a human face (Figure 1). Often they correspond to salient anatomical features (e.g., nostrils or corners of the eyes), but more importantly, they are *stable, robust* beacons which can be tracked as the head moves.

Furthermore, by using corner features we gain the important advantages of *generality* and *opportunism*. Videoconferencing researchers in particular have largely overlooked this low–level approach, aiming directly for high–level facial features. (An exception was So et al. [7] who used centres–of–gravity of iso–density contours.) Our tracking system can

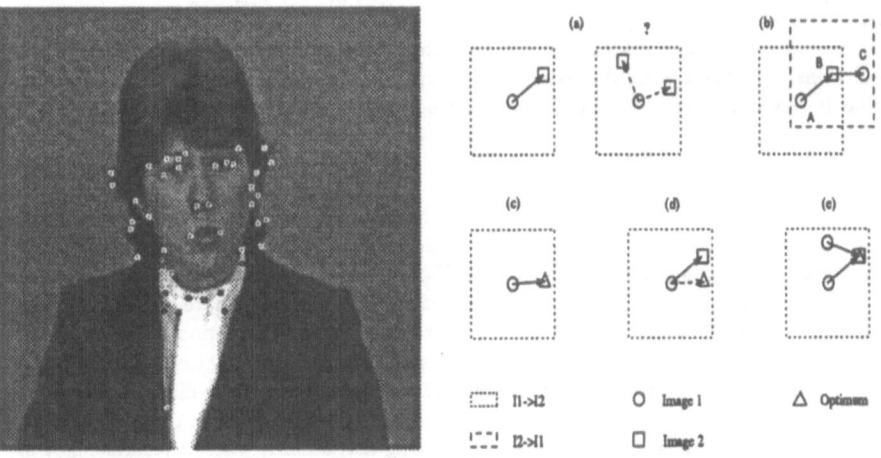

Figure 1: [Left] *CCITT sequence "Clair" with corners superimposed;* [a] I_2 *features (squares) falling in a search window centred on an I_1 feature (circle) are match candidates;* [b] *A "love triangle": feature A in I_1 chooses B in I_2, but B prefers C to A;* [c] *An unmatched I_1 feature finds a ghost match (triangle) by correlation search;* [d] *A nearby unclaimed feature in I_2 is accepted in a compromise match;* [e] *A ghost match coincides with a claimed corner, so the paths merge.*

cope with male and female subjects, glasses and eye–patches, plasters and nose–casts, and even facial hair (including beards, mustaches and sideburns). Moreover, it *uses* to its advantage any distinctive features that deviate from the norm, such as earrings, acne and facial scars. This opportunism applies both statically *and* dynamically; if the speaker dimples when smiling, or wrinkles his forehead when frowning, the system will utilise the corners while they are available.

Corners are detected using a *corner operator* [11], which involves only local operations and generates a response Γ which attains a local maximum at the corner location:

$$\Gamma = \left(\frac{\partial^2 I}{\partial t^2}\right)^2 - S\,|\nabla I|^2 \longrightarrow \max, \quad \Gamma > R, \quad |\nabla I|^2 > E. \tag{1}$$

There are four parameters to specify: scale S, edge strength E, corner response R and mask size M. Details appear in [11, 6], along with discussion on automatic parameter setting, the effects of Gaussian smoothing, and the problem of false corners (arising from profile edges, conjunctions of edges lying at different depths, shadow lines and specularities).

3 The matcher

The two–frame correspondence problem is familiar in computer vision. Seminal work in this area was done by Ullman [8], and numerous algorithms have since been proposed in an extensive body of literature. The demand for eventual frame–rate processing places constraints on our matcher: the algorithm must be parallel in nature and computationally cheap. The fact that inter–frame motion is small, however, ensures that the intensity pattern around a corner (its local *shape*) doesn't change much between frames.

The matcher receives as input two grey–scale images (I_1 and I_2) along with their respective sets of corners (having image coordinates $p_{i,1}$ and $p_{j,2}$ respectively). These corners are generated automatically as images enter sequentially in time. This section describes the two–frame matcher in the absence of predictions; the modifications when predictions are available are discussed in Section 4. An important feature of our matcher is that it leaves no corner in I_1 without a pairing, and uses only local operations.

3.1 Strong matches

Consider two sets of corners superimposed on the same system of image axes. For every corner in I_1, we construct a search window centred on it, and all corner points from I_2 lying in this window are candidates for the match (Figure 1(a)). We then perform a *local patch correlation* between the corner in I_1 (the "template") and each candidate corner in I_2 (the "patch"). The winning candidate is the one with the highest correlation value, provided it surpasses a certain minimum threshold; this is necessary since the "best" corner in the window need not be the correct one (e.g., the actual feature may have disappeared).

We then repeat the procedure, working back from the second image to the first. This widely–used technique (e.g., [3]) resolves conflicting attractions, where the preference of one feature for another is not reciprocated (Figure 1(b)). We only accept matches which concur in both directions, and discard the rest. A match that survives this pruning is a *strong* (or *natural*) match, with a confidence value c. The correlation metric we use is the standard *product moment coefficient*,

$$c = \frac{\sum_{i=1}^{n}(t_i - \bar{t})(p_i - \bar{p})}{\sqrt{\sum_{i=1}^{n}(t_i - \bar{t})^2}\sqrt{\sum_{i=1}^{n}(p_i - \bar{p})^2}}, \quad -1 \le c \le 1,$$

where $\{t_i\}$ and $\{p_i\}$ are the intensity values of the template and patch, $\bar{t}$ and $\bar{p}$ are their means, and we raster–scan pixels in the blocks of interest to give the two n–point data-sets. This metric is invariant to a linear change between the data–sets, i.e., $c = 1$ when $p_i = at_i + b$, with constants $a, b \geq 0$. Hence, c compares the *structure* of the patches, rather than their absolute intensities[1]. Only positive correlation values are considered; negative c indicates inversion of the intensity values. Since perfect correlation obtains when $c = 1$, c serves as a measure of confidence in the match.

3.2 Ghost and compromise matches

There are several reasons why there may still be unmatched corners:

1. A feature may permanently disappear from sight, due to occlusion. Hence, it will appear in I_1 but not in I_2 (a *ghost*).

2. A previously obscured feature may become visible as new structure sweeps into view (or previously–seen structure becomes visible once more). Hence, the feature will appear in I_2 but not in I_1 (an *intruder*).

3. A feature may appear intermittently ("flash" on and off) due to instability in the corner detector (e.g., the response Γ oscillates about the cutoff threshold). This can result in both ghosts and intruders.

4. A feature may appear once and then disappear, due to noise in the signal. This leads first to an intruder and then to a ghost.

Although it is impossible to distinguish these scenarios on the basis of only two frames (and it is precisely the last two problems which have made corner detection unattractive in the past), sustained observation of the features (coupled with prediction) makes this task simple.

It is, however, important that the matcher doesn't deprive the tracker of potentially useful information (which could always be overridden or discarded later). We therefore require the matcher to generate a best position estimate for *every remaining corner* in I_1. The assumption here is that the third scenario has happened, i.e., the corner has flashed off and will soon flash on again. If indeed this *has* happened, the "bridging" estimate will be good, since the feature is still visible and does have some second–order structure. If the assumption is incorrect and the feature has disappeared, the tracker will soon realise this (since the corner won't reappear and the ghost matches will be poor).

The best position estimate is computed via a correlation test over the whole search window, using every location as a candidate. The location with highest correlation (c_{max}) is accepted (Figure 1(c)). If there is a real corner close to the estimated position (with correlation c_{ok}), we accept it if it is sufficiently similar to c_{max} (Figure 1(d)). This is a *compromise* match, and its objective is to reduce the number of intruders and ghosts. If the ghost point coincides with an I_2 corner already claimed by a different point, we merge their paths, destroying the ghost and its trajectory (Figure 1(e)).

If there is no corner nearby the optimum correlation position, we settle for the *ghost* match. The new position assigned to the unmatched corner is thus decided purely by correlation, *with no actual corner being there*. On the next cycle ($I_2 \rightarrow I_3$), this ghost corner will be treated as a real corner in I_2, in the hope that it will find a strong match

---

[1] Assuming the albedo of the patch doesn't change over time, a accounts for automatic gain control of the camera, uniform changes in scene lighting and changes of object pose relative to a constant light source, while b accounts for a uniform intensity offset [6].

Figure 2: *Two-frame matches (markers indicate corners in the first frame, and vectors are drawn double their true length for clarity):* [top to bottom, left to right] *"Curl"; "Div"; "Salesman" (CCITT); "Car".*

in I_3 (signalling reappearance). We refer collectively to ghost, compromise and merge matches as *forced* matches.

This search procedure is expensive, since an $n \times n$ window yields n^2 possible positions. However, the operation can be done in parallel, and is performed only for the unmatched points. Moreover, n^2 is the worst–case scenario; when predictions are available, the search space is substantially reduced (see Section 4.2).

3.3   Results

The algorithm was implemented in sequential form in C and run on SUN SPARC–1 work-stations. An 11×11 search window was used for initial candidates, and a 3×3 window for compromise matches(see [6]). We have tested this algorithm on a wide range of sequences, under different lighting conditions and facial poses.

Figure 2 shows several two–frame matches, combining the natural and forced pairings (camera is stationary). The "Curl" sequence shows a head (LSS) rotating about an axis parallel to the optic axis, while the "Div" sequence shows a subject looming towards the camera amidst a cluttered background (a diverging flow pattern). The CCITT "Salesman" sequence is a challenging one, for the speaker moves his arms, flexes his fingers, turns the object he is demonstrating and ripples his shirt. The corners accurately reflect this movement; his head moves left while his right arm moves right and upwards. Finally, we show a car accelerating forwards, indicating that the algorithm is transportable to other application domains. Similar results have been obtained when the camera moves as well as the scene (paper in preparation).

These results illustrate that the motion vectors give a clear indication of where in the image (and in what direction) movement occurs. This testifies both to the temporal consistency of the corners (strong and compromise matches), and to the suitability of corner locations for computing flow (ghost matches). The accuracy of the motion vectors despite the small motion indicates how well the corners are localised; this will prove a solid foundation for trajectories spanning multiple frames.

4 The tracker

The tracker has two responsibilities. Firstly, it maintains an image trajectory for each feature, charting its motion through successive frames. Secondly, it oversees the matcher, feeding it predictions and obtaining a set of matches in return. This also involves supervising the initial startup (*boot mode*) and securing the transition to normal operation (*run mode*).

4.1 Trajectory maintenance

Every feature has a record in the "world" database, describing its general details (e.g., when it first appeared) and its frame–specific information (e.g., position and velocity). Maintaining these spatio–temporal trajectories comprises various subtasks, e.g., *instantiate* new features, *retire* features which have disappeared, and *update* the records of tracked features. The number of tracked corners grows for the first few frames and then reaches an approximate equilibrium, once the instantiation of new points is offset by the retirement program.

4.2 Correspondence control strategy

When there are n frames in the sequence rather than just two, there are $n-1$ pairs of images to process in temporal order: I_1 and I_2, I_2 and I_3, etc. The algorithm in Section 3 operates until the predictor kicks in, whereafter the algorithm presented below is used. Hereafter, I_1 will refer to the "previous" image, and I_2 to the "current" image.

4.2.1 The prediction philosophy

Finding a suitable role for prediction in long–term tracking is a tricky problem. On the one hand, past behaviour can be a valuable indicator of future behaviour, since physical objects moving in the world build up inertia; to ignore these "motion trends" is therefore to discard useful information. On the other hand, predictors require a model of object motion: when the model is valid, the predictor works well, but when the model fails, do does the predictor (often badly). Typically, tracking systems match directly from predicted

312

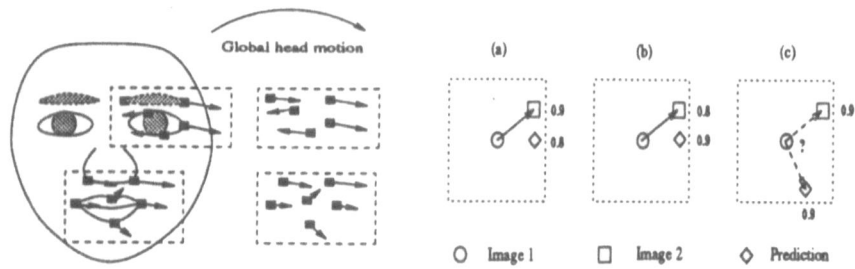

Figure 3: [left] *Valid facial motion patterns, which one might reject under the rigidity assumption (eyes move and mouth opens);* [right] *Strong matches versus predictions:* [a] *Match wins;* [b] *Prediction wins and is nearby, so we get a* compromise *match;* [c] *Prediction wins and is far away, so the corner is discarded (a* wimp*).*

positions to the data in the new frame. This approach only has merit when the prediction takes you *closer* to the true position; when it moves you further away, you are worse off than if you had used the *raw* data. Often, the problem of what to do when prediction fails is bypassed by using other cues [3], solved via adaptive filters [5], or even ignored.

However, the question is not *whether* to use the predictor, but *when* to use it; we need to distinguish between a predictor which is working well (and can be trusted) and a predictor which is failing (and should be restarted). Occasional predictor failure is inevitable; in videophony, for instance, heads can change direction rapidly (e.g., a nodding action). A mechanism for *graceful degradation* is thus of fundamental importance.

Our solution is to maintain a predictor (in image coordinates) for every corner, and use local patch correlation as an indicator of predictor success. When the predictor fails, we simply revert to the original image data. We use a simple model of feature motion, eschewing the use of (2D or 3D) global (or semi-global) motion models at this early stage of processing; such motion models can slot in at a higher stage. Thus, we avoid having to first segment the scene into differently moving objects; indeed, this is one purpose for which the trajectories are intended. Furthermore, we are able to track independent and non-rigid motions. Faces, for instance, are not only non-rigid themselves, but also *contain* non-rigid objects (e.g., the mouth). Consequently, the corners simply *don't* move in a locally consistent way (Figure 3).

Although our system has much in common with DROID [3] (which broke new ground in tracking corner features), we differ from it in several important ways. For example, DROID catered only for an observer moving through a static (hence rigid) world. A detailed comparison between the systems is provided in [6].

4.2.2 Strong matches

Predictions are initially held in abeyance while we obtain a set of mutually consenting matches from the raw data (Section 3.1). For every corner in I_1, we then compare its prediction against its strong match. If the strong match wins (Figure 3(a)), it is accepted as the correct solution. If the prediction is better than the match, then the course of action depends on how far apart these two locations are. If they are close together, we accept the corner but *downgrade* the match to a *compromise* (Figure 3(b)). If, they are far apart, we have an I_1 corner being strongly pulled by two very different image locations (Figure 3(c)). Such *wimp* corners arise in areas of uniform texture, caused either by a poor corner in a region without much structure (e.g., a cheek), or a good corner in a highly (but similarly) textured area (e.g., chin stubble or hair). Either way, the corner is an unsuitable feature and is destroyed.

4.2.3 Compromise and ghost matches

Now the unmatched corners remain. If a prediction is available, we search a small region around the prediction and see whether the best correlation value there is good enough to be accepted. If not, we do a full correlation search starting from the original corner. Thus, when the predictor works, we save greatly on time, but when it fails, we revert to the full search method. The reason for not searching in a gradually expanding region around the prediction (e.g., radially) is that once the predictor has failed, a more accurate result obtains by ignoring it completely. Note that if the prediction correlates well but is incorrect, this becomes apparent in several frames' time when no match reappears, and the corner is retired.

A comparison (over a long sequence) between the cases with and without a predictor show interesting results. Firstly, the total number of corners being tracked is very similar in the two cases, and we also get almost identical matches; prediction simply *speeds up the process*. This differs from many prediction schemes where, in the absence of prediction, the number of unmatched points grows due to the large uncertainty. Because we force unmatched corners in I_1 to find "virtual" partners in I_2, we contain the uncertainty in the system. Secondly, the number of strong matches is fairly constant over time, suggesting that a fixed percentage of the corners are very robust [6].

4.2.4 Prediction strategies

We examined two fixed–coefficient predictors (constant velocity and constant acceleration). At best these model could only be approximate since they operate on *image* data, without modelling 3D motion (or camera projection). However, our goal here is *not* to deduce the world motion parameters, but rather to utilise trends in image motion to improve efficiency and reject unsuitable matches. For the constant velocity case (in finite difference notation), $\dot{x}(k+1) = \dot{x}(k)$, so $x(k+1) = 2x(k) - x(k-1)$. Two frames are needed for a prediction, and distance changes linearly in time. (We treat x and y coordinates independently.)

For the constant acceleration case, $\ddot{x}(k+1) = \ddot{x}(k)$, so $x(k+1) = 3x(k) - 3x(k-1) + x(k-2)$. Here, three frames are needed for prediction, and distance changes quadratically in time. Li et al. [5] also used these filters but favoured the adaptive coefficient forms. However, accurate tracking was crucial to their system since they matched from predictions to data; it is far less critical in our approach.

Experiments have shown that the "linear" predictor often outperforms the "quadratic" one, because the smallness of the inter–frame motion often leads to locally linear trajectories. Also, since the motion is small relative to the quantisation errors, the noise introduced by "second temporal derivatives" has a detrimental effect.

4.3 Results

Figure 4 shows image trajectories obtained over several frames. "Richard" (bespectacled) nods his head downwards, "Clair" moves her head round in a circular motion, and "Dave" (bearded) performs a "curl" motion while his body sways slightly in the opposite direction.

When the camera is still, it is simple to distinguish stationary from moving points on the basis of their velocity history. For each feature, we compute the mean ($\bar{s}$) and standard deviation (σ_s) of its speed over several frames. Classification as a stationary point requires small $\bar{s}$ and small σ_s (with a minimum number of sightings). Figure 4 shows the segmentation for the "Dave" sequence, with the stationary points removed.

5 Conclusions

We have presented an algorithm to track moving objects in the image, using local operations. A key strength of this algorithm is the use of low-level "corner" features. These corners are stable and well-localised, making them suitable for tracking – even in applications where there aren't "physical" corners, such as human faces. Furthermore, being entirely image–driven, the corners differ from scene to scene, giving the powerful advantages of generality and opportunism. We further ensure robustness by means of temporal

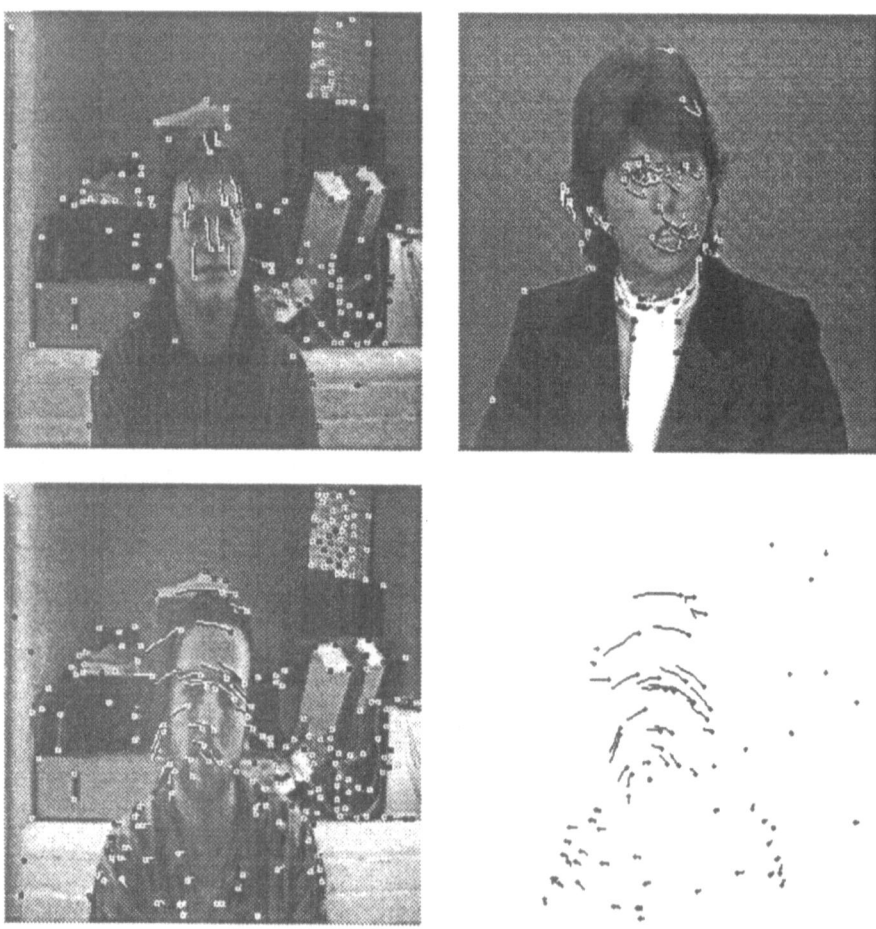

Figure 4: *Image trajectories, each spanning an equal number of frames (true–length vectors, markers show corner positions in final frame):* [top to bottom, left to right] *"Richard" (8 frames); "Clair" (6 frames); "Dave" (6 frames); moving points for "Dave".*

integration, which overcomes the problems of "flashing" corners, noise, and occlusion and disocclusion of 3D structure.

Our matcher–tracker verifies matches by correlating local image structure, and limits uncertainty by assigning *ghost* matches to unmatched points. Use of a simple predictor (per point) speeds up the matching process significantly, though we take care to degrade gracefully when predictors fail (by reverting to the raw data). The image trajectories that finally emerge give a strong impression of "what motion occurs where".

The matcher–tracker forms only the first level of our motion analysis system, and there are several directions of research to pursue. Firstly, by forming clusters of corners having similar motion, we aim to segment the scene into coherently–moving objects and then compute 3D motion parameters. Secondly, as the "Salesman" sequence illustrates, edge information will be very useful for eliciting motion boundaries. We therefore plan to combine edge motion with corner motion. Thirdly, the usefulness of point features other than the ones we have described here (e.g., distinguished points from invariant theory) will be explored. Finally, we plan to implement the tracker in parallel.

Acknowledgments

LSS thanks Rami Guissin and Andrew Zisserman for useful discussions. Ian Reid, Dave Dyer and Richard Lewis willingly sat before our camera, and Bill Welsh (of British Telecom Research Labs) provided the CCITT sequences. LSS is supported by an ORS award (UK) and by the Foundation for Research Development (RSA).

References

[1] K. Aizawa, H. Harashima and T. Saito, "Model–based analysis synthesis image coding (MBA-SIC) system for a person's face", *Signal Processing: Image Communication*, Vol. 1, No. 2, Oct. 1989, pp. 139–152.

[2] J.M. Brady, "Seeds of perception", *Proc. 3rd Alvey Vision Conference*, Cambridge University, Sept 1987, pp. 259–265.

[3] D. Charnley, C. Harris, M. Pike, E. Sparks and M. Stephens, "The DROID 3D vision system: algorithms for geometric integration", Plessey Research, Roke Manor, Technical Note 72/88/N488U, Dec. 1988.

[4] L. Dreschler and H. Nagel, "Volumetric model and 3D trajectory of a moving car derived from monocular TV–frame sequence of a street scene", *Computer Vision, Graphics and Image Processing*, Vol. 20, No. 3, Nov. 1982, pp. 199–228.

[5] H. Li, P. Roivainen and R. Forchheimer, "3D motion estimation in model–based facial image coding", Report LITH–ISY–I–1278, Dept. Electrical Engineering, Linköping University, 1991.

[6] L.S. Shapiro, H. Wang and J.M. Brady, "A matching and tracking strategy applied to videophony", Report OUEL 1933/92, Dept. Engineering Science, Oxford, May 1992.

[7] I. So, O. Nakamura and T. Minami, "A study on a model–based coding system based on isodensity maps of facial images", *Picture Coding Symposium (PCS–91)*, 1991, pp. 299–302.

[8] S. Ullman, *The Interpretation of Visual Motion*, MIT Press, USA, 1979.

[9] A. Verri and T. Poggio, "Against quantitative optic flow", *Proceedings of the International Conference on Computer Vision (ICCV–1)*, London, UK, May 1987, pp. 171–180.

[10] H. Wang and J.M. Brady, "A structure–from–motion vision algorithm for robot guidance" in I. Masaki (ed.), *Proc. IEEE Symposium on Intelligent Vehicles*, Detroit, June 1992.

[11] H. Wang and J.M. Brady, "Corner detection with subpixel accuracy", Report OUEL 1925/92, Dept. Engineering Science, Oxford, 1992.

Statistical Analysis of a Stereo Matching Algorithm

N. A. Thacker* and P. Courtney.
AI Vision Research Unit, University of Sheffield
Sheffield, UK
(* now at the Department of Electrical and Electronic Engineering,
University of Sheffield, Sheffield, UK)

Abstract

This paper discusses the problems of image processing algorithm design
and comparison and suggests that a suitable approach may be to model
algorithms. We introduce the corner matching algorithm which we have
used to provide reliable data for 3D computation modules [5][6]. The
development of a simple model of the matching process permits the understanding of the influence of various parameters in the matching algorithm. This model also allows optimisation of the algorithm using data
distributions obtained from representative scenes.

1 Introduction

The vision literature contains many algorithms for image processing involving feature extraction and matching. Often these algorithms take the form of
heuristic solutions which attempt to exploit natural properties of generic images. Such algorithms are rarely perfect and peformance on real images is often
degraded compared to simulated images due to the presence of noise. Clearly
some algorithms must be better than others at doing a particular task, but
how can we determine which? The conventional method for algorithm evaluation, demonstration and comparison seems to be to show the results on a set
of "standard" images. This is useful to show that the algorithm has been successfully implemented and will work on real data [10]. However, the validity of
the assumptions and heuristics underlying the algorithm can rarely be seriously
tested in this way. There are several reasons for this: algorithm performance
is often determined by the specific images for which it has been developed.
Secondly, it is nearly always impossible to obtain an absolute measure of performance (often one needs to be defined and this can be a matter of subjective
choice). Finally, correct implementation and use of other people's algorithms
is often difficult [1], due to a lack of information about control parameters.

It has been suggested [8] that many implementations of vision algorithms
lack a stability analysis (see for example [9]). Fundamentally, the only way to
evaluate the quality of output data from an algorithm is on the basis of how
well the data is suited to a particular application [3]. However, we need to
get away from the dependence of the evaluation of the algorithm on a specific
source of input data. One alternative involves developing a model of the effects
of an algorithm on input data distributions. This achieves two things, firstly

the statistical properties of the input data are specified and secondly the validity of any heuristics are made explicit so that any data independent behaviour can be identified. Also, by developing such a model the effects of algorithm modification may be directly assessed. Different algorithms can then be compared, either directly on the basis of their models, or on results predicted by their models for specific data distributions. Thus algorithm modeling can be fundamentally useful in understanding, optimising and comparing image processing algorithms. This paper has thus adopted this approach to describe the performance of a corner matching algorithm. Although this algorithm is relatively simple, we believe that the same basic analysis strategy could be applied to any other algorithm.

2 Feature Matching

The robust matching of any image features obtained from a pair of grey level images involves the use of a limited set of heuristics.
(a) local image similarity (eg image correlation).
(b) restricted search strategies (eg epi-polars in the case of stereo).
(c) disparity gradient (or smoothness) constraints.
(d) one to one matching.
(e) reliability.

The relative merits of any matching algorithm will be determined by the extent to which these heuristics are utilised.

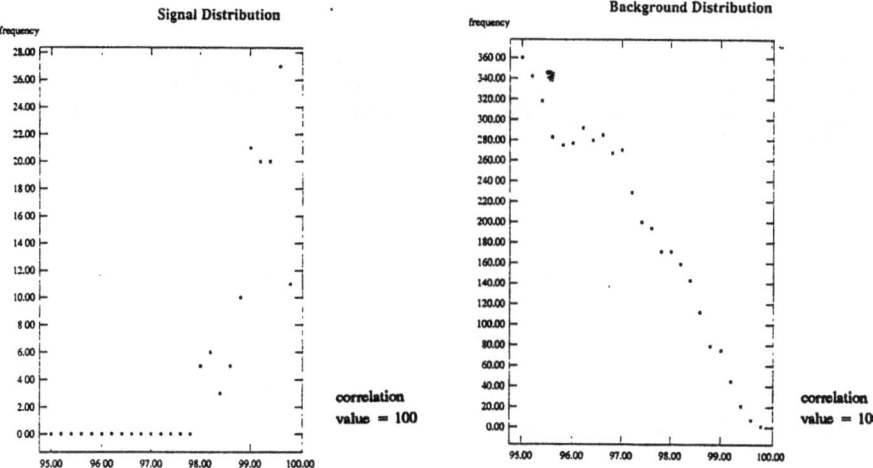

Fig 1(a): Cross correlation distribution for known correct matches.

Fig 1(b): Cross correlation distribution for incorrect matches.

We use the corner detection algorithm of Harris and Stephens [2]. Our corner matching algorithm makes use of all but (c) as corners are generally too sparse to formulate a sensible constraint (but see [7]). Such information may be available from edge based matching algorithms but is not considered further here.

The basic matching algorithm we use for corners is as follows: (a) Construct a list of possible matches on the basis of limited search. This involves choosing a valid search area A. (b) Order the list according to a cross correlation measure c. (c) Select good matches on the basis of: (i) a threshold ρ on the minimum acceptable cross correlation $c < \rho$. (ii) a threshold ω on the ratio of absolute corner strengths $\frac{(c_1-c_2)^2}{(c_1+c_2)^2} < \omega$. (iii) reliability of the best candidate match c_m compared to the next best match c_n on the basis of a uniqueness parameter δ : eg: $c_m - c_n < \delta$. (iv) the same best candidate match must be obtained when matching from image 1 to image 2 and image 2 to image 1 (this enforces one to one matching). We have previously given reports on the performance of this algorithm for stereo/temporal matching for use in ego-motion determination and camera calibration [5],[6], but what we are aiming for here is a more systematic model of the effects of the parameters which control the matching process. We can consider these rules and control parameters as a prune to select a valid set of candidate matches followed by selection on the basis of image cross correlation. If the prune results in completely unambiguous assignment then the result of the matching process is already determined. If however, there are still several candidates for matching then the success or failure of the algorithm is determined by the extent to which the image correlation measure separates correct from incorrect matches. We start by justifying our image correlation measure as the best measure of its sort for choosing appropriate matches. By modeling the distribution of this measure for correct and incorrect matches we are able to assess the effects of the algorithms matching parameters.

2.1   What should we use as our match strength measure?

Given that corners are defined as the peak of an auto-correlation function it makes sense to use cross-correlation. There are many ways to construct the correlation function but we will assume that the function should be radially symmetric so we choose a function of a similar form to the corner detection definition.

$$c = \int_{-\infty}^{\infty} A^{-2} w_{uv} I_{uv} I'_{uv} \partial u \partial v$$

where I_{uv} is the image, $w_u v$ is a gaussian weighting function and with

$$A = \int_{-\infty}^{\infty} w_{uv} I_{uv}^2 \partial u \partial v \int_{-\infty}^{\infty} w_{uv} I'^2_{uv} \partial u \partial v$$

We have hand selected a set of correct matches from several images of different objects including simple widget like objects, complicated machine castings, plastic childrens toys and cassette and lightbulb boxes. For this the cameras were configured at a typical verge angle of 0.1 radians and data at a distance of 5-10 inter-occular separations from the camera. The distribution of the correlation measure for these correct matches is shown in figure 1 (a). This distribution is generated by the differences in local image formation between the two images due to lighting, sensor differences and surface orientation. By computing the cross-correlation for incorrect matches we can get an idea of the shape of the background underneath the correlation signal when using this measure for matching (figure 1 (b)).

We have control over two aspects of the nature of this correlation measure the first is the range of the correlation. This clearly should be large compared to the localisation accuracy (0.3 pixels) but not so large that the correlation computation is costly or that we demand image similarity on a scale which is unrealistic. We use a range of 3 pixels.

Secondly we are free to choose the form for our cross-correlation measure by any non-linear rescaling of the image values:

$$I_u v := I_u v^n, I'_u v := I'_u v^n$$

We find that the best signal to noise ratio is given when $n = 1$ (figure 2). This is presumably because the corner detection auto correlation is also defined on the original image ($n = 1$) and this is therefore the correlation measure that we us in the matching algorithm.

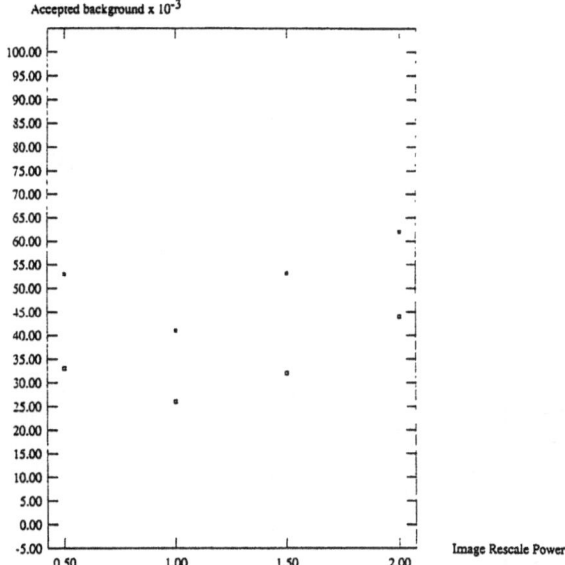

Figure 2: Incorrect cross correlations above 90% and 95% signal cut.

3   Getting an Incorrect Match

The possible forms of mismatch and signal rejection are determined by the reliability of corner detection process. An inefficient corner detector and occlusion will generate cases where some corners do not have a detected partner and can only be matched incorrectly. We can analyse the conditions under which we will get a mismatch and reject a correct match by considering each corner feature and its available match candidates in turn. We will thus show how the probabilities of accepting noise and rejecting signal can be controlled by the parameters ρ, ω, A and δ in the matching algorithm. In the following analysis we assume that the cross correlation distributions for correct and incorrect matches are independant of the detection process.

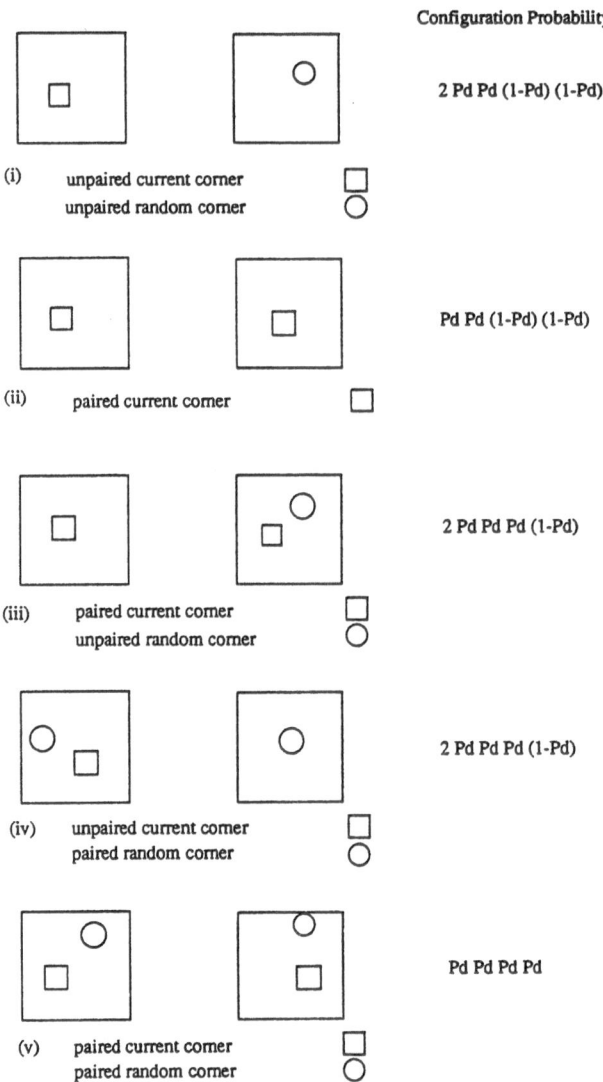

Figure 3: Basic Detection Configurations.

Case (a): The first matching case we consider is that of finding that the cross-correlation for the best candidate match x_m to the corner under consideration ("current") is incorrect in the case where neither candidate has their pair detected in the other image (Figure 3 (i)). The probability P_m^a for the best candidate match to the current corner being one of n_u unpaired random corners can be computed as a product of several factors. These are: the probability of getting the configuration shown in figure 3 (i) and the probability $P_{I(\rho)}$ that a random match drawn from the cross correlation for incorrect matches P_N will have a value greater than ρ:

$$P_m^a = 2n_u P_d^2 (1 - P_d)^2 P_I(\rho)$$

where P_d is the probability of finding a corner again given that it has already been detected in one image and

$$P_I(x) = \int_x^1 P_N(a) da$$

Notice that this source of matching error is very sensitive to the reliability of corner detection P_d, and the only way to remove such matches is to increase the minimum accepted cross-correlation ρ.

Case (b): The next case we consider is that of obtaining an incorrect match with one of n_u unpaired random corners when the correct match to one of the corners was also present (Figure 3 (iii) & (iv)). This is slightly more complicated than the previous case because the existence of the correct match in the matching list may still prevent this getting accepted as a match due to the uniqueness parameter δ. The probability of this happening P_m^b is given by:

$$P_m^b = 4n_u P_d^3 (1 - P_d) P_n(\delta, \rho)$$

where $P_n(\delta, \rho)$ is the probability that an incorrect match can be chosen even when the correct match is present in the match list above a value of ρ. Given that P_N and P_S (the cross correlation distribution for correct matches) are uncorrelated this is given by:

$$P_n(\delta, \rho) = \int_\rho^{1-\delta} P_S(x) \int_x^{1-\delta} P_N(a - \delta) da dx$$

We can see from this that as the uniqueness factor increases the probability of keeping a noisy match of this type is reduced.

Case (c): Finally we consider the case where the current corner is paired but has been matched incorrectly with one of $2n_p$ paired random corners (figure 3 iii). We may wish to write the probability for the acceptance rate for mismatches P_m^c as:

$$P_m^c = 2n_p P_d^4 P_n(\delta, \rho)^2$$

This equation assumes that the two probabilities for mismatch $P_n(\delta, \rho)$ are uncorrelated which is unrealistic, as the two corners must be figurally similar if they are to have mismatched in one matching direction. Thus it is better to write this as:

$$P_m^c = 2n_p P_d^4 P_n(\delta, \rho) P_k(\delta, \rho)$$

where $P_k(\delta, \rho)$ is the probability that the cross correlation value for the complementary pair of the original random match will also be bigger than the correlation value for the correct match.

4 Rejected Signal

We now consider ways in which corner pairs are rejected by the matching process.

Case (a): The first case we consider is when the current corner has been detected in both images (ie paired) and a random matching feature has not been detected in either image (Figure 3 (ii). The probability of rejecting this match is given by:

$$P_r^a = P_d^2(1 - P_d)^2 P_J(\rho)$$

where

$$P_J(\rho) = \int_0^\rho P_S(x)dx$$

Case (b): When the current match is paired and there is a random unpaired match present in either image (Figure 3 (iii)). The probability of rejecting a correct match due to the presence of n_u unpaired random corners P_r is given by:

$$P_r^b = 2P_d^2(1 - P_d)^2(P_J(\rho) + n_u P_l(\delta, \rho))$$

where $P_l(\delta, \rho)$ is the probability of rejecting a correct match due to the proximity of a random corner.

$$P_l(\delta, \rho) = \int_\rho^1 P_S(x) \int_x^{1+\delta} P_N(a + \delta)da dx$$

Case (c): The final case for consideration is when the current match is paired and there is a random paired match (Figure 3 (v)). The rejection rate for good corner matches P_r^c is given by

$$P_r^c = P_d^4(P_J(\rho) + 2n_p P_l(\delta, \rho) - n_p^2 P_l(\delta, \rho)^2)$$

5 Algorithm Analysis and Conclusions

Although the above probabilities can be computed directly from examples of the cross correlation histograms for signal and background there are advantages to modelling the data as a functional form. With an analytic model of the matching process we can compute directly the effects of varying parameters in the matching algorithm and can thus minimise the number of background matches obtained, for various amounts of signal, using standard numerical minimisation methods.

The cross-correlation distributions $P_N(x)$, $P_S(x)$ and $P_k(\delta, \rho)$ can be approximated by triangular distributions. The remaining unknown parameters are the detection efficiency P_d and the numbers of paired and unpaired random corners n_p and n_u. In some ways these values are closely related as the ratio $n_p : n_u$ has a maximum value of $P_d : 1 - P_d$. In an application where the full image contains several hundred corners and the search regions are of the order of a few percent of the image we estimate these values as;

$$P_d = 0.85, n_u = 0.75, n_p = 4.25$$

With these values we can now compute typical signal rejection and noise acceptance curves for the matching algorithm as a function of the matching parameters ρ and δ (Figures 4 and 5).

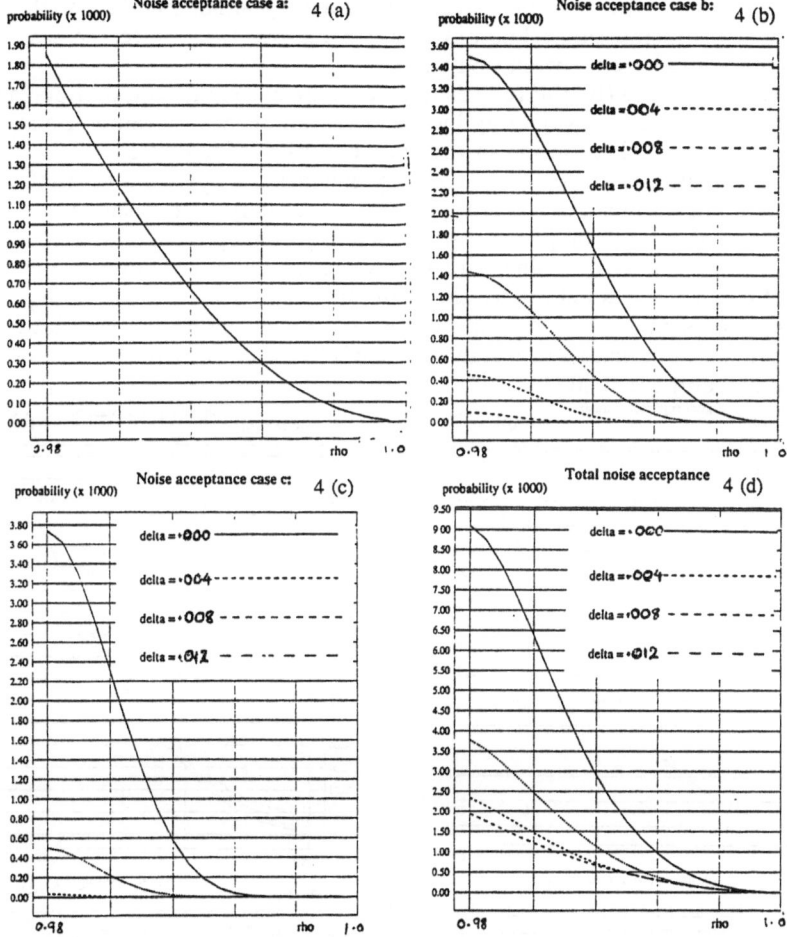

We can thus approximate the total number of incorrect matches (Figure 4 (d)) by

$$P_m^T = P_m^a + P_m^b + P_m^c$$

and the total rejection rate for paired corners (Figure 5 (d)) as

$$P_r^T = P_r^a + P_r^b + P_r^c$$

In specific applications where the detected corners have correlated properties the probability distributions for cross correlations of signal and background may be significantly different. In these cases the probabilities for mismatch and signal rejection would also be different. However, we can still draw some qualitative conclusions about the generic case of corner matching which must be true regardless of the signal and background distributions. These are:

1) All terms in P_m^T are proportional to the mean number of candidate matches, thus we would expect the total number of mismatches to vary proportionately with the search area A.

2) We expect type (a) mismatches to be a very small fraction of the total number of mismatches. The only way to remove these is to increase the minimum required cross correlation value ρ.

3) We expect type (b) and (c) mismatches to be of roughly equal importance and both are reduced considerably by use of the uniqueness parameter δ at the cost of only marginal reduction in the overal number of matches.

4) There is no improvement obtained by increasing δ beyond a value of $1-\rho$ as at this point all mismatches of type (b) have already been rejected.

5) There is no set of parameters which give an optimal signal to noise ratio, this value keeps on rising with increasing ρ. There are however optimal values of ρ and δ corresponding to the minimum noise obtainable for a required proportion of signal. For example using the above model for the data the minimum noise obtainable at a signal level of 60% is 0.2% at parameter values of $\rho = 0.985$ and $\delta = 0.0032$.

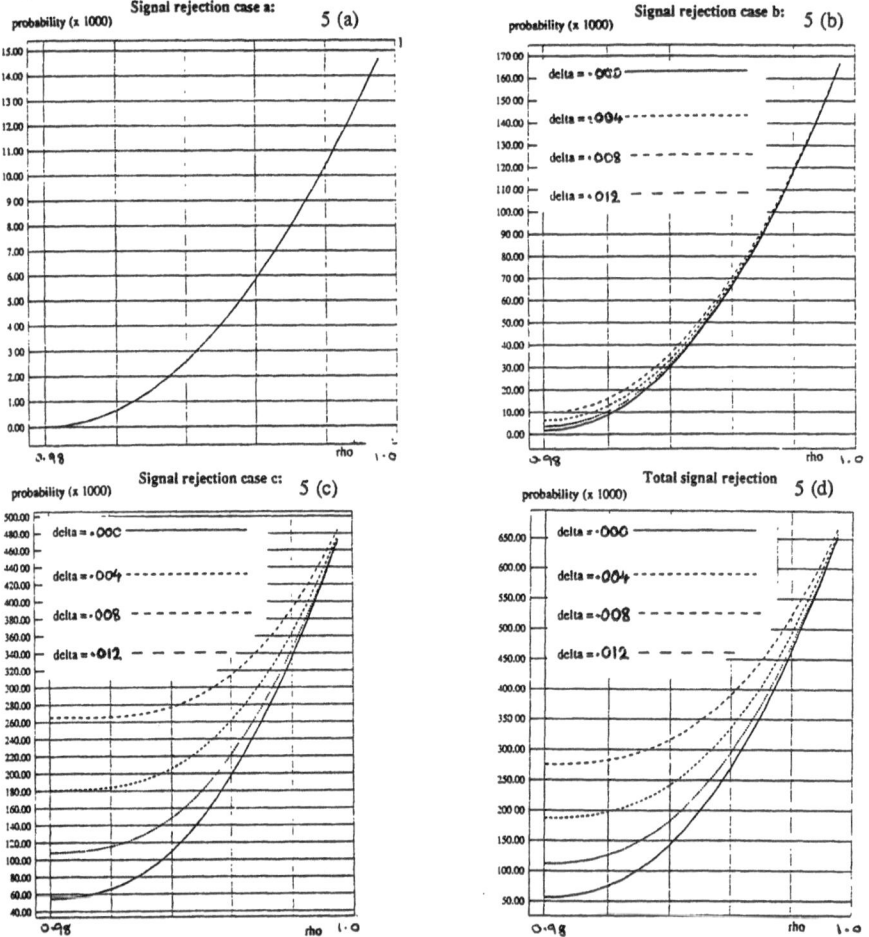

6) Even in very severe cases we expect this matching algorithm to have a

signal to noise ratio in excess of 100:1.

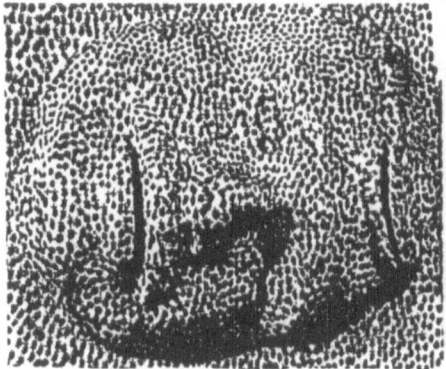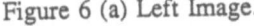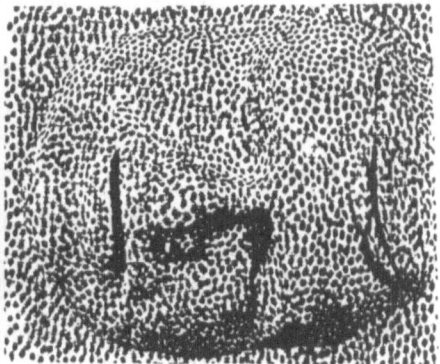

Figure 6 (a) Left Image.   Figure 6 (b) Right Image.

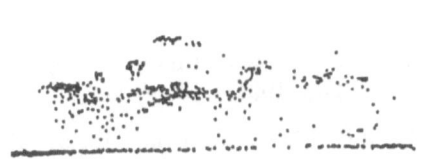

Figure 6 (c) Reprojected 3D data from stereo.

The algorithm is demonstrated here with stereo images of a highly textured head (figure 6(a) and (b)). In this case there were in excess of 1000 corners detected with an average of 36 candidate matches for each corner. For parameter values of $\rho = 0.99$ and $\delta = 0.002$ the predicted number of correct and incorrect matches of 600 and 7. As can be seem from the reprojection of the reconstructed 3D data (figure 6(c)), this is close to what is observed.

The discussion of the matching algorithm has centred on selecting a set of candidate matches and then choosing between them. Given that the cross correlation distributions can be replaced with any relevant similarity measure this work can be considered as a general model of constrained feature matching. This model puts us on a sound footing for considering potential modifications to the corner detection and matching processes.

326

Acknowledgements

We gratefully acknowledge the grant holders Prof. John E.W. Mayhew Dr. Paul Dean and Prof. John Frisby, the support of ESRC/MRC/SERC and the EEC, under the ESPRIT VOILA project, for the funding of this work. We would also like to thank Prof. David Hogg for letting us have this extra page for our paper.

References

[1] Day, T. and J.P.Muller, Digital Evaluation Model Production by Stereo-Matching SPOT Image Pairs: A comparison of Algorithms, Proc. 4th Alvey Vision Conference, pp.117-122, 1988.

[2] Harris, C. and M.Stephens, A Combined Corner and Edge Detector, Proc. 4th Alvey Vision Conference, pp.147-151, 1988.

[3] McLauchlan, P.F., J.E.W.Mayhew and J.P.Frisby, Stereoscopic Recovery and Description of Smooth Textured Surfaces, Proc. BMVC, pp.199-204, 1990.

[4] Moravec, H.P., Obstacle avoidance and navigation in the real world by a seeing robot rover, Ph.D Thesis, Stanford Univ., Sept. 1980.

[5] Thacker, N.A., Y.Zheng and R. Blackbourn, Using a Combined Stereo/Temporal Matcher to Determine Ego-motion, Proc. BMVC, pp.121-126, 1990.

[6] Thacker, N.A. and J.E.W.Mayhew, Optimal Combination of Stereo Camera Calibration from Arbitrary Stereo Images, Image and Vision Computing, pp.27-32, vol 9 no 1, Feb. 1991.

[7] Wang, H. and J.M.Brady, A Structure-from-Motion Algorithm for Robot Vehicle Guidance, Proc. IEE Symp. Intel. Vehic., Detroit, June 1992.

[8] Aloimonos, J., Integration of Visual Modules, Academic Press, 1989.

[9] Grimson, W.E.L., D.P.Huttenlocher and D.W.Jacobs, A Study of Affine Matching with Bounded Sensor Error, pp.291-306, ECCV, Genoa, Italy, May 1992.

[10] Ferrari, D., G.Garibotto and S.Masciangelo, Towards Experimental Computer Vision, Performance Assessment of a Trinocular Stereo System, EEC ESPRIT Day, Santa Margherita Ligure, Genoa, Italy, 18 May, 1992.

Line Based Trinocular Stereo

D. Yang and J. Illingworth

Dept. of Electronic and Electrical Engineering,

University of Surrey,

Guildford, Surrey GU2 5XH, United Kingdom

Abstract

An approach to solving the stereo correspondence problem in trinocular stereo vision is described. It is based on geometric matching constraints relating the orientation of lines extracted in three images taken from different viewpoints. These novel constraints are termed *unary orientation* and *binary orientation* constraints. Matching is achieved within an optimisation framework in which the constraints are encoded into a cost function that is optimised using the *simulated annealing* method. Results are demonstrated and the characteristics of the approach are explored on both synthetic and real[1] trinocular images.

1 Introduction

A fundamental problem in computer vision is the inference of 3D structure from 2D images. An important machine vision technique for this is stereo vision. In stereo vision several images of a scene are taken from different viewpoints and an attempt is made to find the correct correspondence of image features that are projections of the same physical entity. If the geometric relationship between viewpoints is well known then accurate 3D structure can be recovered using the measured image disparity (or shift) between matched features. The problem addressed in this paper is that of finding the correct feature correspondences in a set of three stereo images.

Previous work in stereo vision has involved matching of point or line features. Line matching schemes in binocular and trinocular vision [1, 2, 3, 6, 7, 8] have been primarily based both on similarity of feature attributes and on using point-based representations of lines (midpoints or endpoints) that allow adoption of epipolar constraints developed for point based stereo matching. Unfortunately, matching based on similarity of features is limited to cases where the angle between views is small, while point-based matching is not robust as occlusion effects or line fragmentation artifacts mean that midpoints or endpoints may not correspond to the same 3D point in a scene.

In this paper we attempt to exploit fully the geometric constraint information inherent in lines among three views. We suggest more direct geometric constraints based on the relationship between the orientation of lines in three distinct images. Two main results are exploited: the first uses the orientation

---

[1]The real trinocular data was kindly supplied by Dr N. Ayache of INRIA, Rocquenfort, France and is the data used in his recent book "Artificial Vision for Mobile Robots: stereo vision and multisensory perception"

328

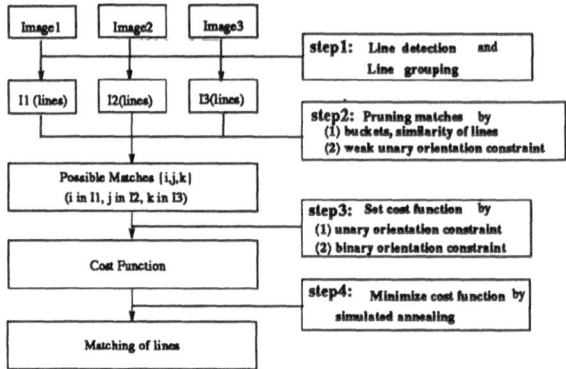

Figure 1: Overview of the Algorithm

of lines in three images and is called the *unary orientation* constraint while the second constraint relates pairs of lines matched in three images and is called the *binary orientation* constraint. These constraints are viewpoint independent and it is the distinctive feature of our method. These constraints are encoded into an optimisation matching algorithm.

An overview of the proposed algorithm is shown in Figure 1. Lines segments are extracted in three images taken from different viewpoints. Segments which are collinear are grouped into a single line as our algorithm primarily uses only orientation information. Each line segment is represented by an N-vector [4], which is the unit normal to the plane that contains the image line and the optical centre of the viewing camera. The second stage of the algorithm is a simple pruning step that discards inadmissible matches using heuristics such bucketing of lines and line similarity information which are commonly used by other authors [1, 2, 3, 6, 7, 8]. In addition, the proposed weak unary orientation constraint is also used in this step. Note that they are applied with loose thresholds so that they throw out wrong matches but retain the subset that includes the correct matches. The objective of the second stage is to quickly reduce the size of the matching problem. In the third step of our method the proposed unary and binary orientation constraints are encoded into a cost matching function. The configuration of matches represented by the minimum of the cost function correspond to the best correspondence of the image lines. Any suitable optimisation algorithm could be used for minimisation but in this work we have used simulated annealing in order to find a global minimum. The computational cost of the process kept manageably low as the number of matches has been reduced by the preceding grouping and pruning stages.

The rest of the paper is organised as follows: section 2 discusses in detail the geometric constraints for the orientation of lines viewed from three different views and derives the unary and binary constraints. Section 3 shows how the constraints may be encoded into a matching cost function and briefly discusses the simulated annealing method that is used to find the minimum cost solution. Section 4 shows experimental results of tests of the algorithm on both real and synthetic data while Section 5 summarises the contribution of the work and offers suggestions for its future development.

2 Line Orientation Constraints

In this section, the representation of 2D and 3D lines is considered and the relation between 2D lines and 3D lines is established. Geometric constraints for orientation of lines among three different views are explored. Two kinds of orientation constraints for matching lines in three images are introduced: *unary* and *binary* orientation constraints.

The camera model used is the pinhole model and perspective projection is adopted as the model of the image formation process. The relation between a 3D point and its 2D projection is expressed by the following:

$$\begin{pmatrix} \omega u \\ \omega v \\ \omega \end{pmatrix} = T \begin{pmatrix} x \\ y \\ z \\ 1 \end{pmatrix}$$

where (u, v) are image coordinates, ω is a scalar, (x, y, z) are 3D point coordinates, T is a 3×4 matrix called the perspective transformation matrix and is defined up to a scale factor.

Points and lines in the image plane are represented (uniquely up to a sign) by unit vectors of homogeneous coordinate called *N-vectors* [4]. The unit vector starting from the optical center and pointing toward a point P in the image plane is called the *N-vector of point P*. The unit vector normal to the plane passing through the optical center C and intersecting an image plane along a line l is called the *N-vector of line l*. Figure 2(a) shows an example of representation of a point and a line by N-vectors.

For 3D vectors a, b and c the following notations are used throughout the paper: $a.b$ denotes the inner or dot product, $a \times b$ denotes the cross product, $< a, b, c >= (a \times b).c = (b \times c).a = (c \times a).b$ denotes the scalar triple product, $[a] = a/\|a\|$ denotes the normalization of vector a and $\|a\|$ denotes the norm of vector a.

The N-vector, m, of a point $P(u, v)$ can be obtained from the perspective projection matrix T as $m = [(t_1 - ut_3) \times (t_2 - vt_3)]$ where the vector $t_i = (t_{i1}, t_{i2}, t_{i3})^t$ and t_{ij} is an element of T. The N-vector, n, of a line l is then simply given by the cross product of the N-vectors of any two points which lie on the image line i.e. $n = m_1 \times m_2$.

A 2D line and the optical centre define a plane in 3D space and the N-vector describes the normal to that plane. If a line is taken from each of two distinct views then the two planes generated may intersect in 3D to define a 3D line. The equation of the 3D line is easily calculated from the N-vectors, n_1, n_2 of the two lines. The 3D scene line can be parameterised by a unit direction vector v and the foot of the normal point, Q, on the 3D line. v and Q can be determined using the relations (see Figure 2(b)):

$$v = n_1 \times n_2, \qquad \vec{OQ}.v = 0, \qquad \vec{QC_1}.n_1 = 0, \qquad \vec{QC_2}.n_2 = 0$$

However, taking a line from each of two views is not a sufficient condition to guarantee the existence of a physical line in 3D. Any matched pair of lines will generate a hypothesis for a 3D scene line but it can only be verified as a

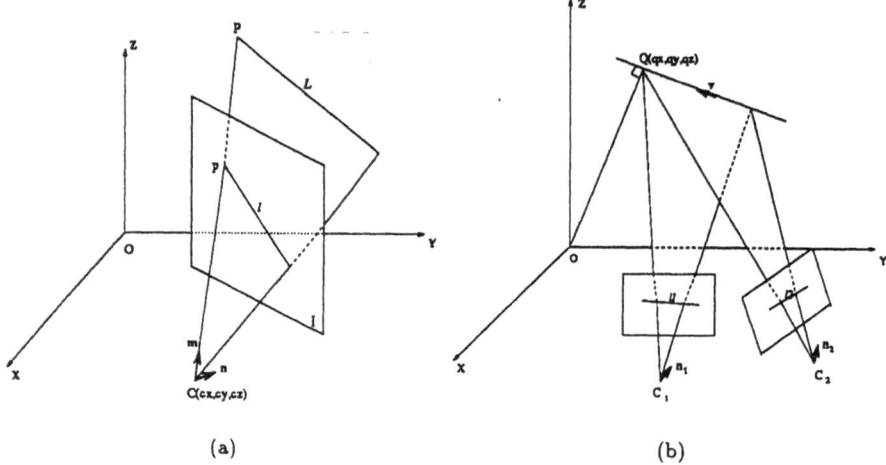

(a)   (b)

Figure 2: (a) N-vector **m** representing a point p and **n** representing line l on the image plane I, where C is the optical center, P and L are the corresponding 3D point and line, respectively (b) Reconstruct 3D line by two views

physically realisable line by finding a match with a geometrically consistent line in a third view. The geometric constraints on lines in three views will now be explored and used to define criteria for identifying correct line correspondences.

Consider images taken from three different viewpoints in which lines can be represented by N-vectors as shown in Fig. 3. If the N-vectors of the projection of a given 3D line in each of the 3 views are denoted as n_1, n_2 and n_3 then two unary orientation constraints (so called as they relate to the match of a single 3D line) can be defined as follows:

Weak Unary Orientation Constraint:

$$f_{wu}\left(n_1, n_2, n_3\right) = <n_1, n_2, n_3> = 0 \tag{1}$$

This constraint is called a weak constraint as it is a necessary but not sufficient condition for a correct match. It is used to discard inadmissible matches in the pruning step in our scheme. A single match $m = \{n_1, n_2, n_3\}$ is correct iff a pair of 2D lines $\{n_1, n_2\}$ reconstruct the same 3D line as a pair of 2D lines $\{n_2, n_3\}$. Thus a stronger unary constraint is:

Unary Orientation Constraint:

$$f_u\left(n_1, n_2, n_3\right) = (Q_{12}x - C_3x, Q_{12}y - C_3y, Q_{12}z - C_3z).n_3 = 0 \tag{2}$$

where Q_{12} denote the foot of the normal on the 3D line reconstructed by $\{n_1, n_2\}$ and C_3 is the viewpoint of the third camera (see Figure 3).

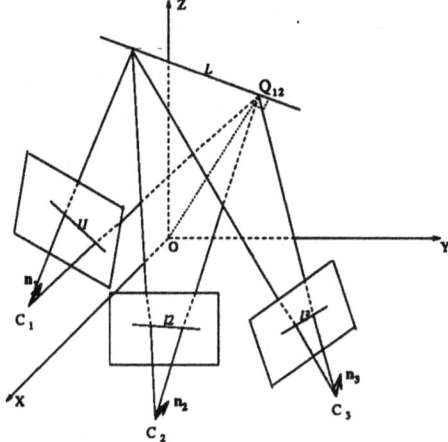

Figure 3: N-vectors representing lines in trinocular vision

Further constraints can be found by considering the geometric relation for a pair of matches $m_1 = \{n_{i1}, n_{j1}, n_{k1}\}$ and $m_2 = \{n_{i2}, n_{j2}, n_{k2}\}$. A 3D line can be constructed by two 2D lines in different viewpoint images. For three views, there are two independent combinational two 2D lines, say, (n_{i1}, n_{j1}) and (n_{j1}, n_{k1}). Since m_1 is supposed to be a correct match the constructed 3D lines $v_1 = [n_{i1} \times n_{j1}]$ and $v_2 = [n_{j1} \times n_{k1}]$ are supposed to be the same line. Similarly, the 3D line $u_1 = [n_{i2} \times n_{j2}]$ and $u_2 = [n_{j2} \times n_{k2}]$ are supposed to be the same line. Thus, the angle between line v_1 and line u_1 should be equal to the angle between line v_2 and line u_2. Thus we have the following binary constraint (so called as it involves pairs of matches):

Binary Orientation Constraint:

$$
\begin{aligned}
f_b &\left((n_{i1}, n_{j1}, n_{k1}), (n_{i2}, n_{j2}, n_{k2})\right) \\
&= ([n_{i1} \times n_{j1}].[n_{i2} \times n_{j2}]) - ([n_{j1} \times n_{k1}].[n_{j2} \times n_{k2}]) \\
&= 0
\end{aligned}
\tag{3}
$$

This constraint can be used to enforce mutually support between correct matches

It is worth noting that the proposed orientation constraints are viewpoint independent. This is in contrast to other methods that use orientation information primarily as a similarity measure for matching. Similarity approaches only work where the angle between views are small and are therefore viewpoint dependent. The proposed orientation constraints make extensive use of the third view and are valid for all viewing positions.

It is also worth noting that the proposed constraints are local constraints. in the next section we discuss how they can be incorporated into a matching framework where the correct matches are determined by the global minimum of a cost function. The minimum is found by making local changes to search through the solution space.

3 Matching as Optimization

In this section, we concentrate on the representation of the solution space for the three view matching problem and then formulate a cost function which aggregates local costs to quantify the global goodness of the solution. The three view stereo matching problem then corresponds to locating the minimum cost solution and this is achieved via local changes using simulated annealing.

A $N_1 \times N_2 \times N_3$ binary matrix $\boldsymbol{P}$, where N_1, N_2 and N_3 are respectively the numbers of lines in each of the three images, is used to represent a solution of the stereo matching problem. An element of $\boldsymbol{P}$ is defined by:

$$p(i, j, k) = \begin{cases} 1 & \text{if } (i, j, k) \text{ is a match} \\ 0 & \text{otherwise} \end{cases} \tag{4}$$

Since a line in an image can match at most one line in the other images or no lines due to occlusion, $\boldsymbol{P}$ is constrained to have the following property:

$$\sum_{i=1}^{N_1} p(i, j, k) = 1 \text{ or } 0 \text{ , } \sum_{j=1}^{N_2} p(i, j, k) = 1 \text{ or } 0 \text{ and } \sum_{k=1}^{N_3} p(i, j, k) = 1 \text{ or } 0 \tag{5}$$

In principle, the solution space of the problem, $\boldsymbol{S}$, is the set of all possible values of the matrix $\boldsymbol{P}$ subject to the constraint defined by the above equation. However this is extremely large and therefore it is necessary to reduce the size of $\boldsymbol{S}$ by pruning the set of lines and matches using simple heuristics based on factors such as grouping of collinear segments in a single image and applying loose similarity constraint and weak orientation constraint to exclude infeasible matches between images.

The cost of a particular feasible solution can be expressed a linear sum over all admissble matches in solution space of two terms:

$$E(\boldsymbol{P}) = \lambda_1 \sum_{p \in P} p(i, j, k) E_1(i, j, k) + \lambda_2 \sum_{p_1 \in P} \sum_{p_2 \in P(p_1 \neq p_2)}$$
$$p_1(i_1, j_1, k_1) p_2(i_2, j_2, k_2) E_2((i_1, j_1, k_1), (i_2, j_2, k_2)) \tag{6}$$

The first term relates to the degree of satisfaction of the unary orientation condition while the second term relates to the binary orientation constraint. λ_1 and λ_2 are weight factors controlling the relative influence of the two constraints. In our work so far we have used $\lambda_1 = \lambda_2 = 0.5$ throughout although this choice warrants further detailed investigation. The particular form adopted for energy terms, E_1 and E_2, is

$$E_1 = \exp\left(\frac{|f_u|}{\sigma_1}\right), \qquad E_2 = \exp\left(\frac{|f_b|}{\sigma_2}\right) \tag{7}$$

where σ_1 and σ_2 is the control parameters set to values related to the expected variance of the values of the orientation constraints. In all our work the values of σ_1 and σ_2 were fixed.

In order to find the minimum cost solution of the matching function we have applied the method of simulated annealing [5]. This is a stochastic search method in which state changes that lead to cost decreases are always accepted but local changes that result in a cost increase may be accepted with a small random probability. This strategy allows the search to jump out of local minimum and asymptotically guarantees convergence to the global minimum. The probability with which cost increases are accepted is varied as the process continues, being initially high but then decreasing later. The parameter that controls this probability is known as the "temperature" following the analogy of annealing in physical systems.

4   Experimental Results and Discussion

In this section results are presented of the algorithms performance on both synthetic images and real trinocular stereo data. All algorithms were implemented in C and run on a Sun Sparcstation.

Synthetic images were generated using the Rayshade software package. This permits scenes composed of simple geometric primitives to be constructed and allows explicit control and knowledge of factors such as viewing parameters and noise. Parts (a), (b) and (c) of Figure 4 shows a typical triplet of images of the corner of a laboratory. These images were processed by an implementation of the Canny edge detector and lines were extracted using a Hough Transform. It should be noted that matching methods based on line attribute similarity would fail to match several of the lines in these images as the viewpoint changes are relatively large. The results of the algorithm for this test image are shown in parts (d) through (f) of Figure 4, where correct matches are shown via line numbers placed adjacent to the lines i.e. line number 1 in (d) is matched to line number 1 in (e) and line number 1 in (f). In this example all 15 lines are successfully matched. On a larger selection of synthetic images the average correct matching percentage was about 90%. Incorrect matches can generally be attributed to not having obtained a global minimum in optimization. The computation time for these images was less than 1.5 minute of CPU.

To test the robustness of our method synthetic image data was corrupted by adding uniform random noise to each of the three components of the original line N-vectors. The amount of noise added to each component was a constant fraction of the size of the component. The method was found to work reliably up to values of noise where the maximum deviation in orientation was 2.5 degrees. The average correct matching percentage was about 85%. We found that presence of noise in these cases results in more local minimums in cost function, but global minimum of cost function still corresponds to the best matching. It means that the proposed unary and binary orientation constraints have a fairly stringent tolerance on noise.

The algorithm was also tested on real data which was supplied to us by Dr Ayache of INRIA. This data has been extensively used by him and his colleagues for experiments in trinocular stereo matching. Figure 5 shows one of the four triplets which were analysed. Processed line data was also supplied by Dr Ayache based on line detection using a recursive line splitting routine. The characteristics of this line detector are slightly different from that of the Hough based detector. The three images contained 282, 300 and 307 respectively and our algorithm found 156 correct line correspondences (lines with length less than 10 pixels were excluded). This is better than the 129 line matches found by Ayaches's method [2].

Simulated annealing has been adopted in the initial proving stages of our algorithm development as it provides a relatively simple, general purpose and reliable method of finding a global minimum. However it is a computationally intensive method and therefore it is entended to explore more specialised optimisation methods which will better exploit the structure of the problem and lead to faster solutions.

A consequence of the slowness of the optimisation method used was the adoption of heuristic methods to reduce the search space via grouping of all line segments in the same straight line as one feature at an early stage of the process. This potentially introduces a weakness into the method as two lines which physically unrelated but accidentally collinear in one but not all views will be represented by different structures in different images and as the matching scheme enforces unique matches between lines in all frames this may cause some lines to remain unmatched. However, to date we have not found these accidental alignments a common situation.

5 Conclusions

In this paper a novel method of finding correspondences between line segments in trinocular stereo images has been developed. The method exploits constraints based on the orientation of lines and encodes them into a matching function. The method has advantages over previous methods: (1) it works for arbitrary viewing positions whereas previous methods often make the assumption of similarity of viewing positions; (2) it can deal with occlusion and line fragmentation whereas many line based methods fail because calculated endpoints or midpoints of image line segments don't correspond to the corresponding points on the 3D line.

Experimental results show the method is effective for both synthetic and real trinocular image data. Future work will include more detailed studies of the method on a greater variety of images and the development of better computational methods to solve the global optimisation problem.

References

[1] N. Ayache and B. Faverjon. "Efficient registration of stereo images by matching graph descriptions of edge segments". *International Journal of Computer Vision*, 1(2):107–131, 1987.

[2] N. Ayache and F. Lustman. "Trinocular stereo vision for robotics". *IEEE Transactions on Pattern Analysis and Machine Intelligence*, PAMI-13(1):73–85, January 1991.

[3] R. Horaud and T. Skordas. "Stereo correspondence through feature grouping and maximal cliques". *IEEE Transactions on Pattern Analysis and Machine Intelligence*, PAMI-11(11):1168–1180, November 1989.

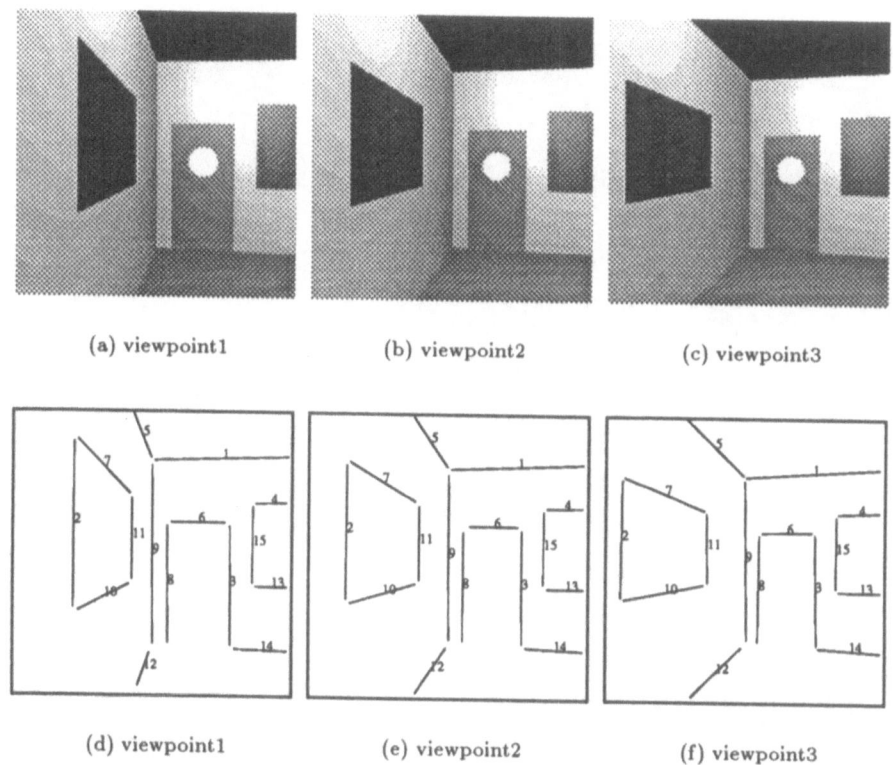

(a) viewpoint1   (b) viewpoint2   (c) viewpoint3

(d) viewpoint1   (e) viewpoint2   (f) viewpoint3

Figure 4: Syntectic Trinocular Stereo Data and Results

[4] K. Kanatani. "Computational projective geometry". *CVGIP: Image Understanding*, IU-54(3):333–348, 1991.

[5] S. Kirkpatrick, C. D. Gellatt, and M. P. Vecchi. "Optimization by simulated annealing". *Science*, 220:671–680, 1983.

[6] J. H. McIntosh and K. M. Mutch. "Matching straight lines". *Computer Vision, Graphics and Image Processing*, 43:386–408, 1988.

[7] G. Medioni and R. Nevatia. "Segment-based stereo matching". *Computer Vision, Graphics and Image Processing*, 31:2–18, 1985.

[8] R. L. Vergnet, S. B. Pollard, and J. E. W. Mayhew. "Stereo-matching of line-segments based on a 3-dimensional heuristic with potential for parallel implementation". In *Proceedings of Alvey Vision Conference*, pages 181–186, 1989.

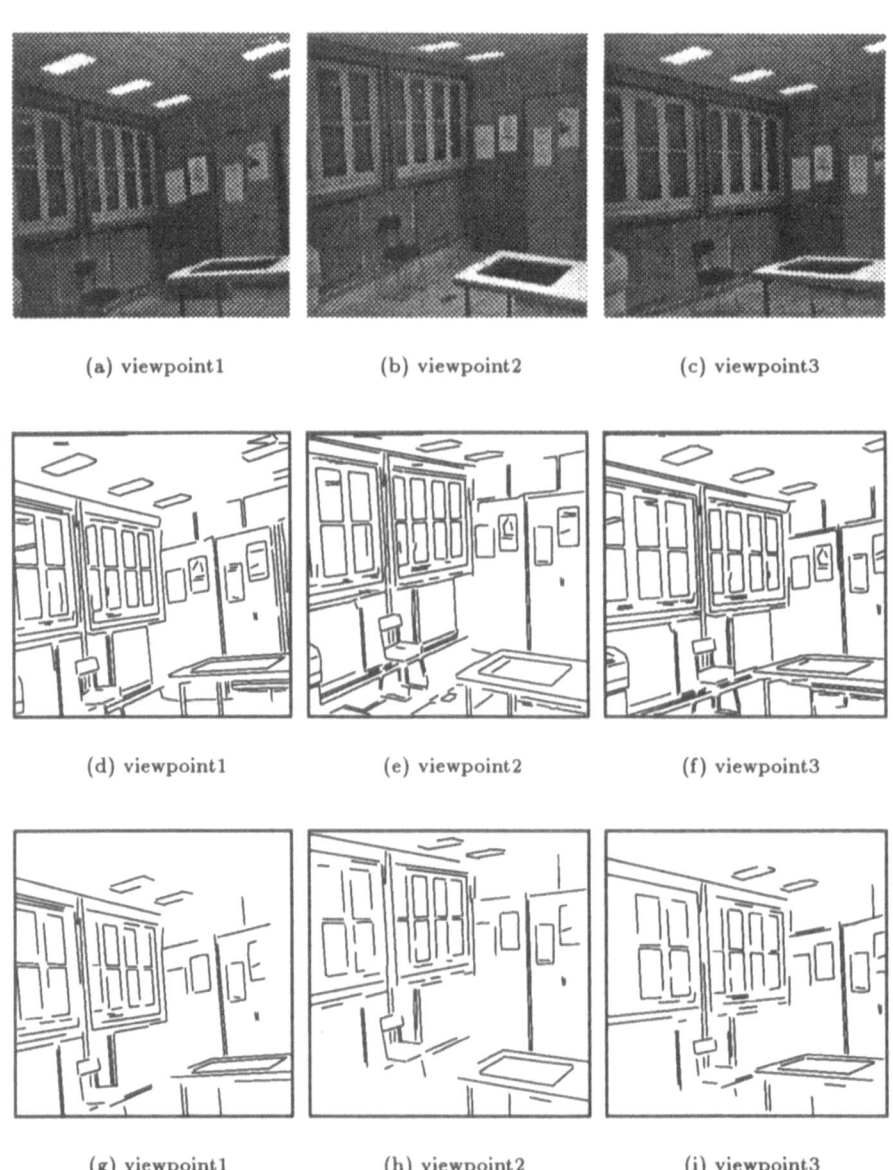

(a) viewpoint 1   (b) viewpoint 2   (c) viewpoint 3

(d) viewpoint 1   (e) viewpoint 2   (f) viewpoint 3

(g) viewpoint 1   (h) viewpoint 2   (i) viewpoint 3

Figure 5: Real Trinocular Stereo Data and Results

Stereo Without Disparity Gradient Smoothing: a Bayesian Sensor Fusion Solution

Ingemar J. Cox, Sunita Hingorani,
Bruce M. Maggs and Satish B. Rao

NEC Research Institute, 4 Independence Way, Princeton, NJ 08540, U.S.A.

Abstract

A maximum likelihood stereo algorithm is presented that avoids the need for smoothing based on disparity gradients, provided that the common uniqueness and monotonic ordering constraints are applied. A dynamic programming algorithm allows matching of the two epipolar lines of length N and M respectively in $O(NM)$ time and in $O(N)$ time if a disparity limit is set. The stereo algorithm is independent of the matching primitives. A high percentage of correct matches and little smearing of depth discontinuities is obtained based on matching individual pixel intensities. Because feature extraction and windowing are unnecessary, a very fast implementation is possible.

Experiments reveal that multiple global minima can exist. The dynamic programming algorithm is guaranteed to find one, but not necessarily the same one for each epipolar scanline. Consequently, there may be small local differences between neighboring scanlines.

1 Introduction

Stereo algorithms seek to find corresponding features between a *pair* of images. Stereo algorithms can be characterized by (1) the primitive features that are matched, (2) the *local* cost of matching two features and (3) the *global* cost function and associated constraints. The stereo framework presented here is, at the algorithmic level, independent of the feature primitives. However, for the experimental results of Section (3), matching was performed directly on the scalar intensity values of the individual pixels. Matching occurs along epipolar lines which are assumed, for convenience, to be coplanar with the image scanlines. The epipolar constraint reduces the stereo correspondence problem from two to one dimension. Most, if not all, previous stereo algorithms include a cost based on the disparity gradient [1, 2, 4, 5, 11], i.e., the difference in depth between two pixels divided by their distance apart. This cost can be thought of as a regularization factor [10] which serves to constrain surfaces to be smooth. However, surfaces are not smooth at depth discontinuities which are the most important features of depth maps. One contribution of this paper is to show that penalizing disparity gradients is unnecessary, provided that the common assumptions of uniqueness and monotonic ordering are made. This is detailed in Section (2), in which stereo is formulated as a Bayesian sensor fusion problem. A local cost function is derived that does not penalize disparity gradients. Section (2.1) then describes how a global minima can be found using a dynamic programming algorithm that enforces the uniqueness and monotonicity constraints. The experiments described in Section (3) reveal that multiple *global* minimum may exist. This can give rise to (minor) artifacts in the disparity map. Similar multiple global minima may exist for other stereo algorithms. Results for several natural scenes are included. Finally, Section (4) concludes with a discussion of the advantages and disadvantages of this algorithm and possible future work.

2   Deriving Cost Functions

In this section, the cost of matching two features, or declaring a feature occluded is first derived, then a global cost function that must be minimized is derived. To begin, we introduce some terminology as developed by Pattipati *et al* [9]. Let the two cameras be denoted by $s = \{1, 2\}$ and let $\mathbf{Z}_s$ represent the set of measurements obtained by each camera along corresponding epipolar lines: $\mathbf{Z}_s = \{z_{s,i_s}\}_{i_s=0}^{m_s}$ where m_s is the number of measurements from camera s and $z_{s,0}$ is a dummy measurement, the matching to which indicates no corresponding point. For epipolar alignment of the scanlines, $\mathbf{Z}_s$ is the set of measurements along a scanline of camera s. The measurements z_{s,i_s} might be simple scalar intensity values or higher level features. Each measurement z_{s,i_s} is assumed to be corrupted by additive, white noise.

The condition that measurement z_{1,i_1} from camera 1, and measurement z_{2,i_2} from camera 2 originate from the same location, x_k, in space, i.e. that z_{1,i_1} and z_{2,i_2} correspond to each other is denoted by Z_{i_1,i_2}. The condition in which measurement z_{1,i_1} from camera 1 has no corresponding measurement in camera 2 is denoted by $Z_{i_1,0}$ and similarly for measurements in camera 2. Thus, $Z_{i_1,0}$ denotes occlusion of feature z_{1,i_1} in camera 2.

Next, we need to calculate the *local* cost of matching two points z_{1,i_1} and z_{2,i_2}. The likelihood that the measurement pair Z_{i_1,i_2} originated from the same point x_k is denoted by $\Lambda(Z_{i_1,i_2} \mid x_k)$ and is given by

$$\Lambda(Z_{i_1,i_2} \mid x_k) = \prod_{s=1}^{2} [P_{D_s} p(z_{s,i_s} \mid x_k)]^{1-\delta_{i_s}} [1 - P_{D_s}]^{\delta_{i_s}} \qquad (1)$$

where δ_{i_s} is an indicator variable that is unity if a measurement is not assigned a corresponding point, i.e. is occluded, and zero otherwise. The term $p(z \mid x)$ is a probability density distribution that represents the likelihood of measurement z assuming it originated from a point x in the scene. The parameter P_{D_s} represents the probability of detecting a measurement originating from x_k at sensor s. This parameter is a function of the number of occlusions, noise etc. Conversely, $(1 - P_D)$ may be viewed as the probability of occlusion. If it is assumed that the measurements vectors z_{s,i_s} are normally distributed about their ideal value z, then

$$p(z_{s,i_s} \mid x_k) = \mid (2\pi)^d S_s \mid^{-\frac{1}{2}} exp\left\{ -\frac{1}{2}(z - z_{s,i_s})' S_s^{-1}(z - z_{s,i_s}) \right\} \qquad (2)$$

where d is the dimension of the measurement vectors z_{s,i_s} and S_s is the co-variance martix associated with the error $(z - z_{s,i_s})$. Since the true value, z, is unknown we approximate it by maximum likelihood estimate $\hat{z}$ obtained from the measurement pair Z_{i_1,i_2} and given by

$$z \approx \hat{z} = S_{2,i_2}(S_{1,i_1} + S_2)^{-1} z_{1,i_1} + S_{1,i_1}(S_{1,i_1} + S_{2,i_2})^{-1} z_{2,i_2} \qquad (3)$$

where S_{s,i_s} is the covariance associated with measurement z_{s,i_s}.

Now that we have established the cost of the individual pairings Z_{i_1,i_2}, it is necessary to determine the total cost of all pairs. Denote by γ a feasible pairing of all measurements and let Γ be the set of all feasible partitions, i.e. $\Gamma = \{\gamma\}$. If γ_0 denotes the case where all measurements are unmatched, i.e., the case in which there are no corresponding points in the left and right images, then we wish to find the pairings or partition γ that maximizes $L(\gamma)/L(\gamma_0)$ where the likelihood $L(\gamma)$ of a partition is defined as

$$L(\gamma) = p(Z_1, Z_2 \mid \gamma) = \prod_{Z_{i_1,i_2} \in \gamma} \Lambda(Z_{i_1,i_2} \mid x) \left(\frac{1}{\phi_1}\right)^{n_1} \left(\frac{1}{\phi_2}\right)^{n_2} \qquad (4)$$

where ϕ_s is the field of view of camera s and n_s is the number of unmatched measurements from camera s in partition γ. The likelihood of no matches, $L(\gamma_0)$ is therefore given by $L(\gamma_0) = 1/(\phi_1^{n_1}\phi_2^{n_2})$

The maximization of $L(\gamma)/L(\gamma_0)$ is eqivalent to

$$\min_{\gamma \epsilon \Gamma} J(\gamma) = \min_{\gamma \epsilon \Gamma} \left[\ln(L(\gamma_0)) - \ln(L(\gamma))\right] \tag{5}$$

which leads to

$$\min_{\gamma \epsilon \Gamma} J(\gamma) = \min_{\gamma \epsilon \Gamma} \sum_{Z_{i_1, i_2} \epsilon \gamma} \left\{ \sum_{s=1}^{2} \left\{ (1 - \delta_{i_s}) \left[\frac{1}{2}(\hat{z} - z_s)'S_s^{-1}(\hat{z} - z_s)\right] + \delta_{i_s} \left[\ln\left(\frac{P_{D_s}}{1 - P_{D_s}} \frac{1}{|(2\pi)^d S_s^{-1}|^{\frac{1}{2}}}\right)\right] \right\} \right\} \tag{6}$$

The first term in the inner summation of Equation (6) is the cost of matching two features while the second term is the cost of an occlusion/disparity discontinuity. Clearly, as the probability of occlusion $(1 - P_{D_s})$ becomes small the cost of not matching a feature increases, as expected.

2.1 Dynamic Programming Solution

The minimization of Equation (6) is a classical weighted matching or assignment problem [8]. There exist well known algorithms for solving this with polynomial complexity $O(N^3)$ [7]. If the assignment problem is applied to the stereo matching problem directly, non-physical solutions are obtained. This is because Equation (6) does not constrain a match at z_{i_s} to be close to the match for $z_{(i-1)_s}$, yet surfaces are usually smooth, except at depth discontinuities. In order to impose this smoothness condition, previous researchers have included a disparity gradient term to their cost function [1, 4, 5, 11, 12]. The problem with this approach is that it tends to blur the depth discontinuities as well as introduce additional free parameters that must be adjusted.

Instead, we assume as in [6] (1) *uniqueness*, i.e. a feature in the left image can match to no more than one feature in the right image and vice versa and (2) monotonic ordering, i.e. if z_{i_1} is matched to z_{i_2} then the subsequent measurement z_{i_1+1} may only match measurements z_{i_2+j} for which $j > 0$. The minimization of Equation (6) subject to these constraints can be solved by dynamic programming. If there are N and M measurements in each of the two epipolar scanlines, respectively, then Ohta and Kanade [6] presented a solution with complexity $O(N^2M^2)$. We have improved this minimization procedure to $O(NM)$:

```
Occlusion = [ln ( PDₛ/(1-PDₛ)  1/|(2π)ᵈSᵣ⁻¹|½ )]
for (i=1;i≤ N;i++){ C(i,0) = i*Occlusion }
for (i=1;i≤ M;i++) { C(0,i) = i*Occlusion}
for(i=1;i≤ N;i++){
    for(j=1;j≤ M;j++){
        C(i,j) = min (C(i-1,j-1)+c(z1,i,z2,j), C(i,j-1)+Occlusion,
                      C(i-1,j)+Occlusion) } }
```

where $C(i,j)$ represents the cost of matching the first i features in the left image with the first j features in the right image and $c(z_{1.i}, z_{2,j})$ is the cost of matching the two features $z_{1.i}, z_{2,j}$ as shown in Equation (6).

Of course, this general solution can be further improved by realizing that there is a practical limit to the disparity between two measurements. This is also true for human stereo, the region of allowable disparity being referred to as Panum's fusional area [3]. If a measurement z_{i_1} is constrained to match only measurements z_{i_2} for which $i_1 - \Delta x \le i_2 \le i_1 + \Delta x$ then the time required by dynamic programming algorithm can be reduced to linear complexity $O(N)$.

3 Experimental Results

Unless otherwise stated, all experiments described here were performed with scalar measurement vectors representing the intensity values of the individual pixels, i.e. $z_{i_s} = I_{i_s}$. The field of view of each camera, ϕ_s, is assumed to be π and the measurements are assumed to be corrupted with white noise of variance $\sigma^2 = 16$. Finally, the probability of detection P_{D_s} is assumed to be 0.9 so that the cost of an occlusion is 3.8.

3.1 Random Dot Stereograms

Figure (1) shows the depth map obtained from the left image of a "wedding cake" random dot stereogram - three rectangular regions one above the other. Note that black pixel values indicate no match with pixels in the right image. While the number of correct matches is 95.4%, it is interesting to examine why the correct depth estimates have not been found at every point on every line. In particular, since the RDS pair is noise free, a perfect match is expected, so the right side of each rectangle should exhibit a depth discontinuity that is aligned with neighboring scanlines. This is not the case in practice. Close examination of this phenomenon revealed there are multiple *global* minima! Dynamic programming is guaranteed to find a global minima but not necessarily the same one for each scanline. Hence, the misalignment of the vertical depth discontinuities. This is a problem. Note however, that the jagged vertical discontinuity caused by the multiple global minima is a characteristic of other stereo algorithms [2, 6] and may be indicative of the presence of multiple global minima in other stereo algorithms.

Rather than choose an arbitrary solution from amongst the set of global minimum, a second optimization can be performed that selects from the set of solutions, that solution which contains the least number of discontinuities. Performing this minimization *after* first finding all maximum likelihood solutions is very different from incorporating the discontinuity penalty into the original cost. The second level of minimization can be easily accomplished as part of the dynamic programming algorithm without having to enumerate all maximum likelihood solutions.. The result of applying the maximum likelihood minimum discontinuity algorithm to the random dot stereogram is shown in Figure (2). A significant improvement is evident, with the percentage of correct matches increasing to 98.7%. Once again, multiple global minima are evident but their number is far fewer.

Note that using the dynamic programming algorithm with a disparity limit of 25 pixels a 256x256 pixel image pixel scanline takes approximately 11 seconds on a SGI Personal Iris. Each scanline therefore takes 0.04 seconds which is very close to video rates of 0.033 seconds per frame, if all scanlines are processed in parallel.

3.2 Natural Scenes

Figure (3) is the left image of the "Pentagon" stereogram. Figures (4) and (5) shows the resulting disparity maps for the maximum likelihood (ML) and ML with minimum discontinuities (MLMD) algorithms. The MLMD provides a qualitative improvement. Note that for display purposes, those pixels that were not matched are assigned the disparity value of whichever of the left or right neigboring pixel is furthest away. Once again, vertical depth discontinuities exhibit some misalignment between scanlines. Nevertheless, significant detail is obtained, as is evident from the overpasses and freeways in the upper right corner of the image. Figure (6) shows the result of applying the MLMD algorithm for $P_D = 0.99$ and supports our observation that the algorithm is stable for reasonable variations in the free parameter value.

Figures (8) and (9) show the results of applying the ML and MLMD algorithm to a stereo pair, the left image of which is shown in Figure (7). Especially noteworthy is the narrow sign pole in the middle right of the image which illustrates the sharp depth detail that is extracted.

The algorithm was tested on other stereograms and similar performance was obtained. However page restrictions, prevent further examples.

4   Conclusion

Determining the correspondence between two stereo images was formulated as a Bayesian sensor fusion problem. A local cost function was derived that consists of (1) a normalized squared error term that represents the cost of matching two features and (2) a fixed penalty for an unmatched measurement that is a function of the probability of occlusion. These two terms are common to other stereo algorithms, but the additional smoothing term based on disparity gradients is avoided. Instead, uniqueness and monotonicity constraints, imposed via a dynamic programming algorithm constrain the solution to be physically sensible.

The dynamic programming algorithm has complexity $O(NM)$ which reduces to $O(N)$ if a disparity limit is set. The algorithm is potentially very fast. especially since a high percentage of correct matches were obtained on intensity based matching primitives that require no feature extraction.

Experimental results were presented for RDS and natural images with good results. The random dot stereograms revealed that multiple *global* minima may exist. Consequently, there may be small local differences between neighboring scanlines. Similar differences are visible for other stereo algorithms which may indicate that multiple global minima are a problem for these algorithms as well. A more detailed study of this phenomenon is needed. In particular, does a sensible cost function with only a single global minima exist?

The experimental results described here do *not* use any information between scanlines. This is somewhat surprising, but was a concious decision to avoid blurring horizontal depth discontinuities. The maximum likelihood minimum horizontal discontinuities (MLMD) also suffers from multple global minima, though far fewer than the maximum likelihood algorithm alone. A third level of optimization should be investigated that maximizes the continuity between scanlines. This is being examined.

Acknowledgements

Thanks to Yaakov Bar-Shalom and Davi Geiger for valuable discussion on issues related to this paper. Also thanks to Takeo Kanade and Tomoharu Nakahara for supplying several stereo images. Special thanks to K.G. Lim of Cambridge University for the interest shown in this algorithm. Thanks to George V. Paul for implementation assistance.

References

[1] A. Blake and A. Zisserman. *Visual Reconstruction.* MIT Press, 1987.

[2] D. Geiger, B. Ladendorf, and A. L. Yuille. Binocular stereo with occlusions. In *Second European Conference on Computer Vision*, 1992.

[3] D. Marr. *Vision.* W. H. Freeman & Co., 1982.

[4] D. Marr and T. Poggio. A cooperative stereo algorithm. *Science*, 194, 1976.

[5] J. E. W. Mayhew and J. P. Frisby. Psychophysical and computational studies towards a theory of human stereopsis. *Artificial Intelligence*, 17, 1981.

[6] Y. Ohta and T. Kanade. Stereo by intra- and inter- scanline search using dynamic programming. *IEEE Trans. Pattern Analysis and Machine Intelligence*, PAMI-7(2):139–154, 1985.

[7] C. H. Papadimitriou and K. Steiglitz. *Combinatorial Optimization.* Prentice Hall, 1982.

[8] C. H. Papadimitriou and K. Steiglitz. *Combinatorial Optimization: Algorithms and Complexity.* Prentice Hall, 1982.

[9] K. R. Pattipati, S. Deb, and Y. Bar-Shalom. Passive multisensor data association using a new relaxation algorithm. In *Multitarget-Multisensor Tracking: Advanced Applications*, pages 219–246. Artech House, 1990.

[10] T. Poggio, V. Torre, and C. Koch. Computational vision and regularization theory. *Nature*, 317:638–643, 1985.

[11] K. Prazdny. Detection of binocular disparities. *Biological Cybernetics*, 52, 1985.

[12] A. L. Yuille, D. Geiger, and H. Bulthoff. Stereo integration, mean field theory and psychophysics. In *First European Conference on Computer Vision*, pages 73–82, 1990.

Fig 1: Maximum likelihood disparity map for random dot stereogram with $P_D = 0.9$.

Fig 2: Maximum likelihood minimum discontinuity disparity map for rds.

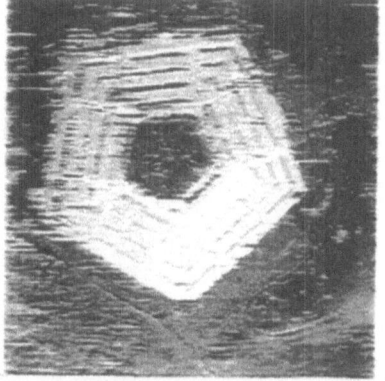

Fig 4: Maximum likelihood disparity map for the Pentagon for $P_D = 0.9$.

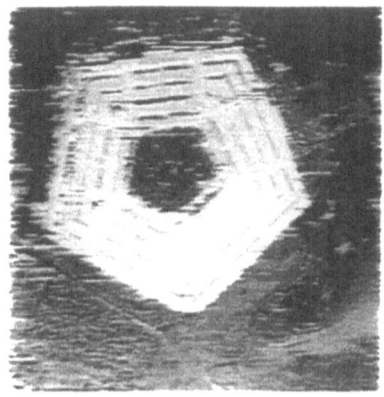

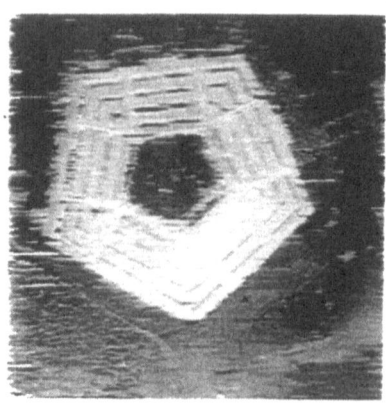

Fig 5: Maximum likelihood minimum discontinuity disparity map for the Pentagon for $P_D = 0.9$.

Fig 6: Maximum likelihood minimum discontinuity disparity map for the Pentagon for $P_D = 0.99$.

Fig 7: Left image of the "parked car" stereo pair.

Fig 8: Maximum likelihood disparity map for the "Parked car".

Fig 9: Maximum likelihood minimum discontinuity disparity map for the "parked Car".

On Local Matching of Free-Form Curves

Zhengyou Zhang

INRIA Sophia-Antipolis, 2004 route des Lucioles
BP 93, F-06902 Sophia-Antipolis Cedex (France)
E-Mail: zzhang@sophia.inria.fr

Abstract

Geometric matching in general is a difficult unsolved problem in computer vision. Fortunately, in many practical applications, some a priori knowledge exists which considerably simplifies the problem. In visual navigation, for example, the motion between successive positions is usually either small or approximately known, but a more precise registration is required for environment modeling. The algorithm described in this paper meets this need. Objects are represented by free-form curves, i.e., arbitrary space curves of the type found in practice. A curve is available in the form of a set of chained points. The proposed algorithm is based on iteratively matching points on one curve to the closest points on the other. A least-squares technique is used to estimate 3-D motion from the point correspondences, which reduces the average distance between curves in the two sets. Both synthetic and real data have been used to test the algorithm, and the results show that it is efficient and robust, and yields an accurate motion estimate.

Keywords: Free-Form Curve Matching, 3-D registration, Motion Estimation, 3-D Vision

1 Introduction

Most of the previous work on geometric matching focused on polyhedral objects; geometric primitives such as points, lines and planar patches were usually used. This is of course very limited compared with the real world we live in. Recently, curved objects have attracted the attention of many researchers in computer vision. This paper deals with objects represented by curves, particularly free-form curves, i.e., arbitrary space curves of the type found in practice.

A free-form curve is represented by a set of chained points. Several matching techniques for free-form curves have been proposed in the literature. In the first category of techniques, curvature extrema are detected and then used in matching [1]. However, it is difficult to localize precisely curvature extrema [2, 3], especially when the curves are smooth. Very small variations in the curves can change the number of curvature extrema and their positions on the curves. Thus, matching based on curvature extrema is highly sensitive to noise. In the second category, a curve is transformed into a sequence of local, rotationally and translationally invariant features (e.g., curvature and torsion). The curve matching problem is then reduced to a 1-D string matching problem [4, 5, 6]. As more information is used, the methods in this category tend to be more robust than those in the first category. However, these methods are still subject to noise disturbance because they use arclength sampling of the curves to obtain point sets. The arclength itself is sensitive to noise.

The methods cited above exploit global matching criteria in the sense that they can deal with two sets of free-form curves which differ by a large motion/transformation. This ability to deal with large motions is usually essential for applications to object recognition. In many other applications, for example, visual navigation, the motion between curves in successive frames is in general either small (because the maximum velocity of an object is limited and the sample frequency is high) or known within a reasonable precision (because a mobile vehicle is usually equipped with several instruments such as odometric and inertial systems which can provide such information). In the latter case, we can first apply the given estimate of the motion to the first frame to produce an intermediate frame; then the motion between the intermediate frame and the second frame can be considered to be small. In this paper we propose a new method for the registration of curves undergoing small motion.

The key idea underlying our approach is the following. Given that the motion between two successive frames is small, a curve in the first frame is close to the corresponding curve in the second frame. By matching points on the curves in the first frame to their closest points on the curves in the second, we can find a motion that brings the curves in the two frames closer (i.e., the distance between the two curves becomes smaller). Iteratively applying this procedure, the algorithm yields a better and better motion estimate. Interestingly enough, during the preparation of this paper Besl and McKay published a paper in PAMI (issue February 1992) which exploited the same idea [7]. Our work is an independent and much improved treatment. See [8] for a more detailed comparison.

2 Problem Statement

A 3-D (space) curve segment $\mathcal{C}$ is a vector function $\mathbf{x} : [a, b] \to \mathbb{R}^3$, where a and b are scalar. In computer vision applications, the data of a space curve are available in the form of a set of chained 3-D points from either a stereo algorithm [9] or a range imaging sensor [10]. If we know the type of the curve, we can obtain its description $\mathbf{x}$ by fitting, say, conics to the point data [11, 12]. In this work, we shall use directly the chained points, i.e., we are interested in free-form space curves without regard to particular curve primitives.

The use of chained points is equivalent to a piecewise linear approximation to a curve. Let $\mathbf{x}_{i,j}$ $(j = 1, \ldots, N_i)$ be the N_i chained points on the curve $\mathcal{C}_i$. The approximation error can be made arbitrarily small by increasing N_i and decreasing the distances $\|\mathbf{x}_{i,j} - \mathbf{x}_{i,j+1}\|$. At every point $\mathbf{x}_{i,j}$, we compute the tangent direction $\mathbf{u}_{i,j}$ which will be used in the matching procedure. It is not necessary in our algorithm to know precisely the tangent directions. We use the simple estimate

$$\mathbf{u}_{i,j} = (\mathbf{x}_{i,j+1} - \mathbf{x}_{i,j-1})/\|\mathbf{x}_{i,j+1} - \mathbf{x}_{i,j-1}\|,$$

except at the beginning and end points where

$$\mathbf{u}_{i,1} = (\mathbf{x}_{i,2} - \mathbf{x}_{i,1})/\|\mathbf{x}_{i,2} - \mathbf{x}_{i,1}\|, \mathbf{u}_{i,N_i} = (\mathbf{x}_{i,N_i} - \mathbf{x}_{i,N_i-1})/\|\mathbf{x}_{i,N_i} - \mathbf{x}_{i,N_i-1}\|.$$

Given two 3-D frames of a scene observed at two different positions, each containing a set of curves. Let $\mathcal{C}_i$ $(i = 1, \ldots, m)$ and $\mathcal{C}'_k$ $(k = 1, \ldots, n)$ be the curves observed in the first and second frames, respectively. Let $\mathbf{x}_{i,j}$ $(j = 1, \ldots, N_i)$ and $\mathbf{x}'_{k,l}$ $(l = 1, \ldots, N_k)$ be the points on the curves $\mathcal{C}_i$ and $\mathcal{C}'_j$, respectively. The objective is to find the motion between the two frames, i.e., **R** for rotation and **t** for translation, such that the following criterion

$$\mathcal{F}(\mathbf{R}, \mathbf{t}) = \frac{1}{\sum_{i=1}^{m} \sum_{j=1}^{N_i} p_{i,j}} \sum_{i=1}^{m} \sum_{j=1}^{N_i} p_{i,j} \, d^2(\mathbf{R}\mathbf{x}_{i,j} + \mathbf{t}, C_k') . \tag{1}$$

is minimized, where $d(\mathbf{x}, C)$ denotes the distance of the point $\mathbf{x}$ to the curve C (to be defined below), $p_{i,j}$ takes value 1 if the point $\mathbf{x}_{i,j}$ can be matched to a point on the curve C_k' in the second frame and takes value 0 otherwise.

Furthermore, we assume the motion between the two frames is small or approximately known. In the latter case, we can first apply the approximate estimate of the motion between the two frames to the first one to produce an intermediate frame; then the motion between the intermediate frame and the second frame can be considered to be small.

3   Iterative Pseudo Point Matching Algorithm

We describe in this section an iterative algorithm for curve registration by matching points in the first frame, after applying the previously recovered motion estimate $(\mathbf{R}, \mathbf{t})$, with their closest points in the second. A least-squares estimation reduces the average distance between curves in the two frames. As a point in one frame and its closest point in the other do not necessarily correspond to a single point in space, several iterations are indispensable. Hence the name of the algorithm.

3.1   Finding Closest Points

Let us first define the distance $d(\mathbf{x}, C_k')$ between point $\mathbf{x}$ and curve C_k', which is used in Eq. (1), the criterion defined in the last section. If C_k' is a parametric curve $(\mathbf{x}_k' : [a, b] \to \mathbb{R}^3)$, then $d(\mathbf{x}, C_k') = \min_{u \in [a,b]} d(\mathbf{x}, \mathbf{x}_k'(u))$, where $d(\mathbf{x}_1, \mathbf{x}_2)$ is the Euclidean distance between the two points $\mathbf{x}_1$ and $\mathbf{x}_2$, i.e., $d(\mathbf{x}_1, \mathbf{x}_2) = \|\mathbf{x}_1 - \mathbf{x}_2\|$. In our case, C_k' is given as a set of chained points $\mathbf{x}_{k,l}'$ $(l = 1, \ldots, N_k)$. We simply define $d(\mathbf{x}, C_k') = \min_{l \in \{1, \ldots, N_k\}} d(\mathbf{x}, \mathbf{x}_{k,l}')$. See [8] for more discussions on the distance.

The closest point $\mathbf{y}$ in the second frame to a given point $\mathbf{x}$ is the one satisfying

$$d(\mathbf{x}, \mathbf{y}) = \min_{k \in \{1, \ldots, n\}} d(\mathbf{x}, C_k') = \min_{k \in \{1, \ldots, n\}} \min_{l \in \{1, \ldots, N_k\}} d(\mathbf{x}, \mathbf{x}_{k,l}') .$$

The worst case cost of finding the closest point is $O(N_k^n)$, where N_k^n is the total number of points in the second frame. The total cost while performing the above computation for each point in the first frame is $O(N_i^m N_k^n)$, where N_i^m is the total number of points in the first frame. The use of k-D trees can considerably speed up this process, see the next section.

3.2   Pseudo Point Matching

For each point $\mathbf{x}$ we can always find a closest point $\mathbf{y}$. However, because there are some spurious points in both frames due to sensor capability, or because some points visible in one frame are not in the other due to sensor/object

350

motion, it probably does not make any sense to pair **x** with **y**. Many constraints can be imposed to remove such spurious pairings. For example, distance continuity along a curve, which is similar to the figural continuity in stereo matching [13, 14], should be very useful to discard the false matches. These constraints are not incorporated in our algorithm in order to maintain the algorithm in its simplest form. Instead, we impose the following two simple constraints, which are all unary.

The first is the maximum tolerance for distance. If the distance between a point $x_{i,j}$ and its closest one $y_{i,j}$, denoted by $d(x_{i,j}, y_{i,j})$, is bigger than the maximum tolerable distance $D_{\max}$, then we set $p_{i,j} = 0$ in Eq. (1), i.e., we cannot pair a reasonable point in the second frame with the point $x_{i,j}$. This constraint is easily justified for we know that the motion between the two frames is small and hence the distance between two points reasonably paired cannot be very big. In our algorithm, $D_{\max}$ is set adaptively and in a robust manner during each iteration by analyzing distances statistics, as described below.

The second is the orientation consistency. It can be easily shown that the angle between the tangent of point **x** and that of its closest point **y** can not go beyond the rotation angle between the two frames [15]. Therefore, we can impose that the angle between the tangents of two paired points should not be bigger than a prefixed value Θ, which is the maximum of the rotation angle expected between the two frames. In our implementation, we set $\Theta = 60°$ to take into account noise effect in the tangent computation. If the tangents can be precisely computed, Θ can be set to a smaller value. This constraint is especially useful when the motion is relatively big.

3.3 Updating the Matching

Instead of using all matches recovered so far, we exploit a robust technique to discard several of them by analyzing the statistics of the distances. To this end, one parameter, denoted by $\mathfrak{D}$, needs to be set by the user, which indicates when he considers the registration between two frames is good. See the next section for the choice of the value $\mathfrak{D}$.

Let $D_{\max}^I$ denote the maximum tolerable distance in iteration I. At this point, each point in the first frame (after applying the previously recovered motion) whose distance to its closest point is less than $D_{\max}^{I-1}$ is retained, together with its closest point and their distance. Let $\{x_i\}$, $\{y_i\}$, and $\{d_i\}$ be the resulting sets of original points, closest points, and their distances after the pseudo point matching, and let N be the cardinal of the sets. Now compute the mean μ and the sample deviation σ of the distances, which are given by

$$\mu = \frac{1}{N}\sum_{i=1}^{N} d_i , \quad \text{and} \quad \sigma = \sqrt{\frac{1}{N}\sum_{i=1}^{N}(d_i - \mu)^2} .$$

Depending on the value of μ, we adaptively set the maximum tolerable distance $D_{\max}^I$ as shown below*:

if $\mu < \mathfrak{D}$ then $D_{\max}^I = \mu + 3\sigma$ /* the regist. is quite good */

*Here we assume the distribution of distances is approximately Gaussian when the registration is good. This has been confirmed by experiments. A typical histogram is shown in Fig. 1.

```
else if μ < 3𝔇  then D^I_max = μ + 2σ  /* the regist. is still good */
else if μ < 6𝔇  then D^I_max = μ + σ   /* the regist. is not too bad */
else            D^I_max = ξ             /* the regist. is really bad */
endif
```

Here, ξ is the median of all the distances. That is, the number of d_i's less than ξ is approximately equal to the number of d_i's larger than ξ.

At this point, we use the newly set D^I_{max} to update the matching previously recovered: a paring between x_i and y_i is removed if their distance d_i is bigger than D^I_{max}. The remaining pairings are used to compute the motion between the two frames, as to be described below.

Because D_{max} is adaptively set based on the statistics of the distances, our algorithm is rather robust to relatively big motion and to gross outliers (as to be shown in the experiment section). For example, when the registration is really bad, only half of the originally recovered matches are retained. Even if there remain several false matches in the retained set, the use of least-squares technique yields still a reasonable motion estimate, which is sufficient for the algorithm to converge to the correct solution.

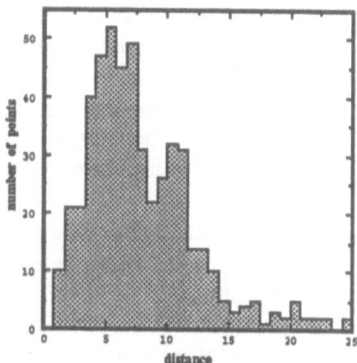

Fig. 1. A histogram of distances

3.4 Computing Motion

At this point, we have a set of 3-D points which have been reasonably paired with a set of closest points, denoted respectively by $\{x_i\}$ and $\{y_i\}$. Let N be the number of pairs. Because N is usually much greater than 3 (three points are the minimum for the computed rigid motion to be unique), it is necessary to devise a procedure for computing the motion by minimizing the following mean-squares objective function

$$\mathcal{F}(\mathbf{R}, \mathbf{t}) = \frac{1}{N} \sum_{i=1}^{N} \|\mathbf{R}x_i + \mathbf{t} - y_i\|^2, \tag{2}$$

which is the direct result of Eq. (1) with the definition of distance given above. Any optimization method, such as steepest descent, conjugate gradient, or complex, can be used to find the least-squares rotation and translation. Fortunately, several much more efficient algorithms exist for solving this particular problem. They include quaternion method [16], singular value decomposition [17] and dual number quaternion method [18]. We have implemented both the quaternion method and the dual number quaternion one. They yield exactly the same motion estimate. See [8] for more details.

3.5 Summary

We can now summarize the iterative pseudo point matching algorithm as follows:

- **input:** Two 3D frames containing m curves C_i and n curves C'_k, respectively. Each curve C is a set of chained 3D points x_j.
- **output:** The optimal motion between the two frames.
- **procedure:**
  a) **initialization**
     $D^0_{\max}$ is set to $20\mathfrak{D}$, which implies that every point in the first frame whose distance to its closest point in the second frame is bigger than $D^0_{\max}$ is discarded from consideration during the first iteration. The number 20 is not crucial in the algorithm, and can be replaced by a larger one.
  b) **preprocessing**
     (i) Compute the tangent at each point of the first frame.
     (ii) Compute the tangent at each point of the second frame.
     (iii) Build the k-D tree representation of the second frame (see the next section).
  c) **iteration** until convergence of the computed motion
     (i) Finding the closest points satisfying the distance and orientation constraints.
     (ii) Update the matching through statistical analysis of distances.
     (iii) Compute the motion between the two frames from the updated matches
     (iv) Apply the motion to all points and their tangents in the first frame.

Several remarks should be made here. The construction and the use of k-D trees for finding closest points will be described in the next section. The motion is computed between the original points in the first frame and the points in the second frame. Therefore, the final motion given by the algorithm represents the transformation between the original first frame and the second frame. The iteration-termination condition is defined as the change in the motion estimate between two successive iterations. The change in translation at iteration I is defined as $\delta t = \|t_I - t_{I-1}\|/\|t_I\|$. To measure the change in rotation, we use the rotation axis representation, which is a 3-D vector, denoted by $\mathbf{r}$. Let $\theta = \|\mathbf{r}\|$ and $\mathbf{n} = \mathbf{r}/\|\mathbf{r}\|$, the relation between $\mathbf{r}$ and the quaternion $\mathbf{q}$ is $\mathbf{q} = \begin{bmatrix} \sin(\theta/2)\mathbf{n} \\ \cos(\theta/2) \end{bmatrix}$. We do not use the quaternions because their difference does not make much sense. We then define the change in rotation at iteration I as $\delta\mathbf{r} = \|\mathbf{r}_I - \mathbf{r}_{I-1}\|/\|\mathbf{r}_I\|$. We terminate the iteration when both $\delta\mathbf{r}$ and δt are less than 1%.

4 Practical Considerations

In this section, we consider several important aspects in practice, including search for closest points, choice of the parameter $\mathfrak{D}$, and coarse-to-fine search strategy.

4.1 Search for Closest Points

As can be observed in the last section, the search for the closest point to a given point is $O(N)$ in time, where $N = N^n_k$ is the total number of points in the second frame. Several methods [19] exist to speed up the search process, including bucketing techniques and k-D trees (abbreviation for *k-dimensional binary search tree*). We have chosen k-D trees, because the curves we have form chained points which are sparse in space. It is not efficient enough to use bucketing techniques because only a few buckets would contain many points, and many others nothing. The reader is referred to [8] for more details of the implementation.

4.2 Choice of the Parameter $\mathfrak{D}$

The only parameter needed to be supplied by the user is $\mathfrak{D}$, which indicates when the registration between two frames can be considered to be good. In other words, the value of $\mathfrak{D}$ should correspond to the expected average distance when the registration is good. When the motion is big, $\mathfrak{D}$ should not be very small. Because we set $D^0_{max} = 20\mathfrak{D}$, if $\mathfrak{D}$ is very small we cannot find any matches in the first iteration and of course we cannot improve the motion estimate. (A solution to this is to set D^0_{max} bigger, say $30\mathfrak{D}$).

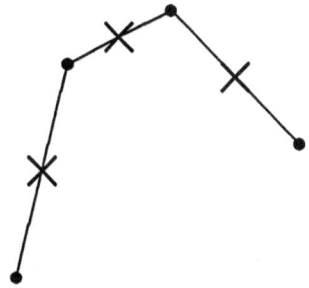

Fig. 2. Illustration of a perfect registration to show how to choose $\mathfrak{D}$

The value of $\mathfrak{D}$ has an impact on the convergence of the algorithm. If $\mathfrak{D}$ is smaller than necessary, then more iterations are required for the algorithm to converge because many good matches will be discarded at the step of matching update. On the other hand, if $\mathfrak{D}$ is much bigger than necessary, it is possible for the algorithm not to converge to the correct solution because possibly many false matches will not be discarded. Thus, to be prudent, it is better to choose a small value for $\mathfrak{D}$.

We have worked out a better solution to $\mathfrak{D}$ instead of an ad hoc choice. Let $\bar{D}$ be the average distance between successive points in the second frame, that is

$$\bar{D} = \frac{\sum_{k=1}^{n} \sum_{l=1}^{N_k-1} \|x_{k,l} - x_{k,l+1}\|}{\sum_{k=1}^{n}(N_k - 1)} .$$

Consider a perfect registration shown in Fig. 2. Points from the first frame are marked by a cross and those from the second, by a dot. Assume that a cross is located in the middle of two dots. Then in this case, the mean μ of the distances between two sets of points is equal to $\bar{D}/2$. Therefore, we can expect $\mu > \bar{D}/2$ when the registration is not perfect. In our implementation, we set $\mathfrak{D} = \bar{D}$ which gives us satisfactory results.

4.3 Coarse-to-Fine Strategy

As to be shown in the next section, we find fast convergence of the algorithm during the first few iterations that slows down as it approaches the local minimum. We find also that more search time is required during the first few iterations because the search space is larger at the beginning, as described above. Since the total search time is linear in the number of points in the first frame, it is natural to exploit a coarse-to-fine strategy. During the first few iterations, we can use coarser samples (e.g., every five) instead of all sample points on the curve. When the algorithm almost converges, we use all available points in order to obtain a precise estimation.

5   Experimental Results

The proposed algorithm has been implemented in C. In order to maintain the modularity, the code is not optimized. In all the experiments described below, the same parameters are used. The program is run on a SUN 4/60 workstation, and any quoted times are given for execution on that machine.

We have tested our algorithm using computer-generated data with different levels of noise. The results show that it is quite robust to noise. Due to space limitation, we only provide in this section the experimental results with real data. The reader is referred to [8] for more examples.

A trinocular stereo system mounted on our mobile vehicle is used to take images of a chair scene (the scene is static but the robot moves). We show in Fig. 3 two images taken by the first camera from two different positions. The displacement between the two positions is about 4 degrees in rotation and 100 millimeters in translation. The chair is about 3 meters from the mobile vehicle.

Fig. 3. Images of a chair scene taken by the first camera from two different positions

The curve-based trinocular stereo algorithm developed in our laboratory [9] is used to reconstruct the 3-D frames corresponding to the two positions. There are 36 curves and 588 points in the first frame, and 48 curves and 763 points in the second frame. We show in the two left-hand pictures of Fig. 4 the front view and the top view of the superposition of the two 3-D frames. The curves in the first frame is displayed in solid lines while those in the second frames, in dashed lines. We apply the algorithm to the two frames. The algorithm converges after 12 iterations. It takes in total 32.5 seconds on a SUN 4/60 workstation and half of the time is spent in the first iteration (so we could speed up the process by setting D_{max}^0 to a smaller value). The final motion estimate is

$$\hat{\mathbf{r}} = [-1.527 \times 10^{-3}, 6.639 \times 10^{-2}, 2.894 \times 10^{-3}]^T ,$$
$$\hat{\mathbf{t}} = [-4.266 \times 10^0, -1.586 \times 10^0, -1.009 \times 10^2]^T .$$

The motion change is: $\delta r = 0.78\%$ and $t = 0.53\%$. The result is shown in the two right-hand pictures of Fig. 4 where we have applied the estimated motion to the first frame. Excellent registration is observed for the chair. The registration of the border of the wall is a little bit worse because more error is introduced during the 3-D reconstruction, for it is far away from the cameras.

 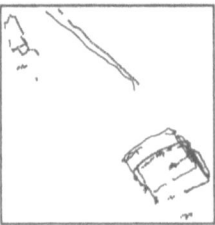  

(front view)before registration(top view)   (front view) after registration (top view)

Fig. 4. Superposition of two 3-D frames before and after registration: front and top views

Now we exploit the coarse-to-fine strategy. We do coarse matching in the first five iterations by sampling evenly one out of every five points on the curves in the first frame, followed by fine matching using all points. The algorithm converges after 12 iterations and yields exactly the same motion estimation as when only doing fine matching. The execution time, however, decreases from 32.5 seconds to 10.5 seconds, about three times faster. If now we sample evenly one out of every *ten* points on the curves in the first frame, and do coarse matching in the first five iterations and fine matching in the subsequent ones, the algorithm converges after 13 iterations (one iteration more), and the final motion estimate is

$$\hat{\mathbf{r}} = [-1.438 \times 10^{-3}, 6.653 \times 10^{-2}, 2.995 \times 10^{-3}]^T,$$

$$\hat{\mathbf{t}} = [-4.282 \times 10^{0}, -1.637 \times 10^{0}, -1.007 \times 10^{2}]^T,$$

which is almost the same as the one estimated using directly all points. The motion change is: $\delta\mathbf{r} = 0.71\%$ and $\mathbf{t} = 0.50\%$. The execution time is now 8.8 seconds.

6   Conclusions

We have described an algorithm for the registration of free-form curves, i.e., arbitrary space curves of the type found in practice. We have used the assumption that the motion between two frames is small or approximately known, a realistic assumption in many practical applications including visual navigation. A number of experiments have been carried out and good results have been obtained.

Our algorithm has the following features:

- It is simple. The reader can easily reproduce the algorithm.
- It is extensible. More sophisticated strategies such as figural continuity can be easily integrated in the algorithm.
- It is general. First, the representation used is general for representing arbitrary space curves of the type found in practice. Second, the ideas behind the algorithm are applicable to (many) other matching problems. The algorithm can easily be adapted to solve for example 2-D curve matching and 3-D surface matching.
- It is efficient. The most expensive computation is the process of finding closest points, which has a complexity $O(N \log N)$. Exploiting the coarse-to-fine strategy considerably speeds up the algorithm with only a small change in the precision of the final estimation.
- It is robust to gross errors and can deal with appearance, disappearance and occlusion of objects. This is achieved by analyzing dynamically the statistics of the distances.
- It yields an accurate estimation because all available information is used in the algorithm.
- It does not require any preprocessing of 3-D point data such as for example smoothing. The data are used as is. That is, there is no approximation error.
- It does not require any derivative estimation (which is sensitive to noise), in contrast with many other feature-based or string-based matching methods.

Our algorithm converges to the closest local minimum, and thus is not appropriate for solving large motion problems. Two possible extensions of the algorithm to deal with large motions have been described in [8]: coupling with a global method or sampling the motion space.

In our algorithm, one parameter, the parameter $\mathfrak{D}$, needs to be set by the user. It indicates when the registration can be considered to be good. It has an impact on the convergence rate. In our implementation, $\mathfrak{D}$ is automatically computed using the intervals of chained points. This method works well for all experiments we have carried out. However, a better method probably exists. Intuitively, the parameter $\mathfrak{D}$ is related not only to the intervals of chained points but also to the shape of the curves. $\mathfrak{D}$ should be smaller for rough curves than for smooth ones. We are currently investigating this issue.

We are currently extending the algorithm to solve surface matching problems arising in navigation. When a mobile vehicle navigates in a natural environment, a correlation-based stereo algorithm or a range finder provides a sequence of dense 3-D maps. Only minor modifications are needed in order to produce an algorithm for registering successive 3-D maps.

Acknowledgment: The author would like to thank Olivier Faugeras for stimulating discussions during the work, and Steve Maybank for carefully reading the draft version.

References

[1] R. Bolles and R. Cain, "Recognizing and locating partially visible objects, the local-feature-focus method," *Int'l J. Robotics Res.*, vol. 1, no. 3, pp. 57–82, 1982.

[2] D. Walters, "Selection of image primitives for general-purpose visual processing," *Comput. Vision, Graphics Image Process.*, vol. 37, no. 3, pp. 261–298, 1987.

[3] E. E. Milios, "Shape matching using curvature processes," *Comput. Vision, Graphics Image Process.*, vol. 47, pp. 203–226, 1989.

[4] T. Pavlidis, "Algorithms for shape analysis of contours and waveforms," *IEEE Trans. PAMI*, vol. 2, no. 4, pp. 301–312, 1980.

[5] J. T. Schwartz and M. Sharir, "Identification of partially obscured objects in two and three dimensions by matching noisy characteristic curves," *Int'l J. Robotics Res.*, vol. 6, no. 2, pp. 29–44, 1987.

[6] H. Wolfson, "On curve matching," *IEEE Trans. PAMI*, vol. 12, no. 5, pp. 483–489, 1990.

[7] P. J. Besl and N. D. McKay, "A method for registration of 3-D shapes," *IEEE Trans. PAMI*, vol. 14, pp. 239–256, February 1992.

[8] Z. Zhang, "Iterative point matching for registration of free-form curves," Research Report 1658, INRIA Sophia-Antipolis, March 1992.

[9] L. Robert and O. Faugeras, "Curve-based stereo: Figural continuity and curvature," in *Proc. IEEE Conf. Comput. Vision Pattern Recog.*, (Maui, Hawaii), pp. 57–62, June 1991.

[10] R. E. Sampson, "3D range sensor-phase shift detection," *Computer*, no. 20, pp. 23–24, 1987.

[11] R. Safaee-Rad, I. Tchoukanov, B. Benhabib, and K. C. Smith, "Accurate parameter estimation of quadratic curves from grey-level images," *CVGIP: Image Understanding*, vol. 54, pp. 259–274, September 1991.

[12] G. Taubin, "Estimation of planar curves, surfaces, and nonplanar space curves defined by implicit equations with applications to edge and range image segmentation," *IEEE Trans. PAMI*, vol. 13, pp. 1115–1138, November 1991.

[13] J. E. W. Mayhew and J. P. Frisby, "Psychophysical and computational studies towards a theory of human stereopsis," *Artif. Intell.*, vol. 17, pp. 349–385, 1981.

[14] S. Pollard, J. Mayhew, and J. Frisby, "PMF: A stereo correspondence algorithm using a disparity gradient limit," *Perception*, vol. 14, pp. 449–470, 1985.

[15] Z. Zhang, O. Faugeras, and N. Ayache, "Analysis of a sequence of stereo scenes containing multiple moving objects using rigidity constraints," in *Proc. Second Int'l Conf. Comput. Vision*, (Tampa, FL), pp. 177–186, December 1988.

[16] O. Faugeras and M. Hebert, "The representation, recognition, and locating of 3D shapes from range data," *Int'l J. Robotics Res.*, vol. 5, no. 3, pp. 27–52, 1986.

[17] K. Arun, T. Huang, and S. Blostein, "Least-squares fitting of two 3-D point sets," *IEEE Trans. PAMI*, vol. 9, pp. 698–700, September 1987.

[18] M. W. Walker, L. Shao, and R. A. Volz, "Estimating 3-D location parameters using dual number quaternions," *CVGIP: Image Understanding*, vol. 54, pp. 358–367, November 1991.

[19] F. Preparata and M. Shamos, *Computational Geometry, An Introduction.* New-York: Springer, Berlin, Heidelberg, 1986.

Coarse Image Motion for Saccade Control

Philip F. McLauchlan, Ian Reid and David W. Murray
Robotics Research Group,
Department of Engineering Science, University of Oxford,
Oxford, U.K.

Abstract

We describe a 2D vision module that estimates the motion of moving objects for the purpose of driving saccadic head/eye motions to fixate them. Robustness and a fast reaction time are the main requirements of the module. The apparently moving background is segmented from the moving objects in the scene using a prediction of the background flow obtained from head odometry. Subsequently the velocities of the detected objects are determined using a least-square method to solve the aperture problem. The algorithm has been implemented at frame rate (25Hz) on a network of five transputers, and has a latency of approximately 0.06s.

1 Introduction

We are developing a reactive vision system to investigate real-time gaze control. The mount consists of a 4 degree-of-freedom stereo head (pan, elevation, independent vergence), of which elevation and vergence for one camera are currently implemented. Given the fast reaction times required of a useful gaze control system, we have required high quality engineering of the mount to supply the speed and precision required, with real time control and vision systems. The latter is made up of a combination of DataCube pipelined image processing boards and Transputers, with high speed links for communication with the mount. Details of the head design can be found in [1], the overall structure and aims of the project are described in [2], and details of the information processing system are given in [3].

This paper describes the implementation of a 2D vision module that will supply signals to drive saccadic camera motions for the purpose of obtaining fixation on a moving object. The image motion will be used to compensate for the inevitable processing and head motion delay by predicting the new position of the object. We wish to estimate large motions (up to 15 pixels per frame, corresponding to 30° s^{-1} for the cameras we currently use) for this purpose. Given the inertia of the head we require motion to be computed with a latency of about 0.1s. Other modules will maintain fixation on (track) the object of interest, and our initial experiments with the head will investigate the interaction of these modules in conjunction with position and velocity head control.

Nelson [4] has described an algorithm designed to solve the background/moving object segmentation problem (but not the object velocity estimation problem) for the translating camera case as well as the case of rotation. Burt and his co-workers [5] have developed a multi-scale pyramidal motion segmentation algorithm designed for use in conjunction with control of the sensor and the parameters of the algorithm. François and Bouthemy [6] have designed an algorithm that uses qualitative information about the camera motion to aid motion segmentation, using a Markov Random Field (MRF) approach to segment the scene into regions with common affine flow field.

Our approach differs from the previous work in two main ways: firstly we must have an algorithm that can recover moving objects at the frame rate of 25Hz and, just as importantly, a latency below 0.1s. This eliminates iterative minimisation approaches like the MRF and the pyramidal multi-scale method of Burt. Secondly, we can take advantage of the precisely known head motion and use quantitative estimates of head rotation to predict the background motion and hence aid segmentation. Areas in the image not in agreement with the background motion can be flagged as being due to independent motion. This is *only* possible with a real-time gaze control system that allows effectively instantaneous determination of head odometry for each image, which explains why this method has not been tried before. On the assumption of zero camera translation and a static background, the image motion is constrained to be independent of the scene geometry. Because the rotation axes of the camera do not coincide with the optical centre, there is a small error which varies inversely with the distance of the object from the camera. The worst case error for an object at 2 metres distance is negligible for the vergence axis, 0.5 pixels for the elevation axis and 1.5 pixels per frame for the pan axis, undetectable by the coarse motion algorithm. Indeed it appears not to be necessary to align elevation and vergences axes precisely with the camera optical centres for the visual tasks we are concerned with.

We use normal image flow estimates directly for the first stage of the algorithm, segmenting the background from the moving objects. The normal flow is calculated from spatial and temporal image derivatives using the motion constraint equation [7]. The flow due to the known camera angular velocity is then "subtracted" from the normal flow vectors in the sense described in section 3. This yields image regions whose non-zero residual motion is incompatible with the background. These are analysed individually on the assumption that their motion can be approximated as a constant flow. A least-squares method is used to find the best-fitting full flow vector to the set of normal flow vectors, and also provides a measure of the error via the RMS residual. Thus the aperture problem is solved within each segmented region. The algorithm is not designed to segment two moving objects that happen to appear at adjacent positions in the image. In that case the motion will be flagged with a large error in the velocity estimate.

We have implemented the algorithm on our head/eye platform at frame rate (25 Hz) with a latency of 0.06s using a network of five transputers plus a transputer frame-grabber. Details of the real-time implementation can be found in [3].

The following sections 2 to 5 describe the various stages of the algorithm in more detail. There follow some results on moving imagery and a discussion of the future development of the project.

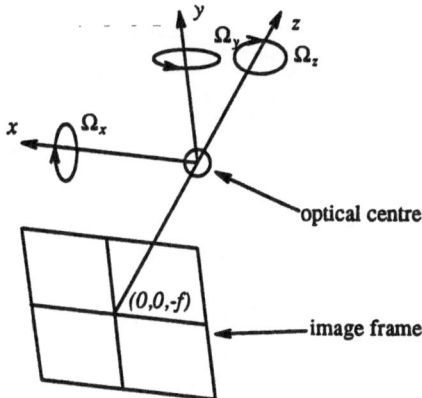

Figure 1: The camera coordinate frame.

2 Calculation of Normal Image Flow

Let the image be $E(x, y)$, x and y being the pixel coordinates. The motion constraint equation [7] is

$$\frac{dE}{dt} = \frac{\partial E}{\partial t} + u\frac{\partial E}{\partial x} + v\frac{\partial E}{\partial y} = 0, \quad \text{where } u = \dot{x}, \; v = \dot{y} \tag{1}$$

Calculation of the spatial gradients $\partial E/\partial x$, $\partial E/\partial y$ and temporal gradient $\partial E/\partial t$ employs gaussian convolution and is described in section 5. The image flow $\mathbf{v} = (u \; v)^T$ is approximately the projected motion of the moving object, although Verri and Poggio have shown [8] that equation 1 does not hold exactly, but increases in accuracy as the spatial image gradients increase. We place a threshold on $|\nabla E| = \sqrt{(\partial E/\partial x)^2 + (\partial E/\partial y)^2}$, and only use points where the threshold is exceeded.

Equation 1 is the equation of a line in (u, v) space, corresponding to the set of image flow vectors consistent with equation 1. This *constraint line* is perpendicular to ∇E. The *normal flow* is defined as the vector $\mathbf{v}_\perp = (u_\perp \; v_\perp)^T$ that satisfies equation 1 and is parallel to ∇E:

$$\mathbf{v}_\perp = -\frac{\partial E}{\partial t}\frac{\nabla E}{|\nabla E|^2} \tag{2}$$

$\mathbf{v}_\perp$ is related to $\mathbf{v}$ by the equation $\mathbf{v} \cdot \mathbf{v}_\perp = |\mathbf{v}_\perp|^2$. That we can calculate only the component of the image flow parallel to the gradient is of course a result of the *aperture problem* [9].

3 Segmenting the Background

Let us place a coordinate frame at the camera's optical centre, and assume an ideal pinhole camera model with x, y, z axes as shown in figure 1, the z axis aligned with the principal axis of the lens, and the image origin at the point $(0, 0, -f)^T$. This frame is rotating with angular velocity $\Omega = (\Omega_x, \Omega_y, \Omega_z)^T$, related to the pan, elevation and vergence angles. This gives rise to motion in the image:

$$\mathbf{u} = \begin{pmatrix} \dot{x} \\ \dot{y} \end{pmatrix} = \frac{1}{f}\begin{pmatrix} xy\Omega_x - (f^2 + x^2)\Omega_y - fy\Omega_z \\ (f^2 + y^2)\Omega_x - xy\Omega_y + fx\Omega_z \end{pmatrix}. \tag{3}$$

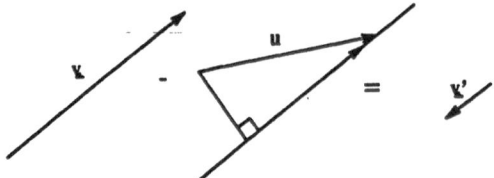

Figure 2: The method of nulling the background motion. Its projection onto the normal flow is subtracted from the normal flow.

We wish to "subtract" the effects of this from the calculated normal flow $\mathbf{v}_\perp$, forming a new normal flow estimate $\mathbf{v}'_\perp$, which will represent the motion of the independent objects. This is done by subtracting from $\mathbf{v}_\perp$ the projection of $\mathbf{u}$ onto $\mathbf{v}_\perp$, as illustrated in figure 2:

$$\mathbf{v}'_\perp = \mathbf{v}_\perp \left(1 - \frac{\mathbf{u} \cdot \mathbf{v}_\perp}{|\mathbf{v}_\perp|^2}\right). \tag{4}$$

In this way we reconstruct the normal flow field that would have arisen if the camera had not been rotating. Combining equations 2 and 4 allows us to simplify the process to the following, computing image flow and background compensation in a single stage:

$$\mathbf{v}'_\perp = -\frac{\nabla E}{|\nabla E|^2}\left(\frac{\partial E}{\partial t} + \mathbf{u} \cdot \nabla E\right). \tag{5}$$

Note that the method can be applied whatever the form of the predicted flow $\mathbf{u}$.

The next step is to decide, on the basis of the new normal flow $\mathbf{v}'_\perp$, which image regions are background and which are not. The simplest method, and the one we use at present, is to threshold $|\mathbf{v}'_\perp|$, i.e. label residual velocities greater than a certain value as due to independently moving objects. In future versions of the algorithm we hope to integrate the results over time to obtain a more robust segmentation.

Lastly, we enforce spatial coherence on the moving objects found. This is done by searching for small square patches in the image dominated by non-background flow vectors. The patches are arranged in two sets each of which covers the image as shown in figure 3. Thus diagonally adjacent patches overlap over a quarter of their area, while laterally adjacent patches do not overlap. Then, if the total number of vectors in a patch is n_{total}, and the number of non-background vectors is $n_{non-back}$, we label a patch as non-background if:

1. $n_{total} > T1$, and

2. $n_{non-back}/n_{total} > T2$

where $T1$ and $T2$ are constant thresholds.

Adjacent non-background patches are then connected on the assumption that they are part of the same moving object. A region-growing procedure then finds all distinct sets of mutually connected patches. These are the image regions corresponding to the moving objects.

4 Calculating the Independent Motion

We assume that a moving object will give rise to a spatially constant image flow. The algorithm will fail to deliver accurate velocity information if:

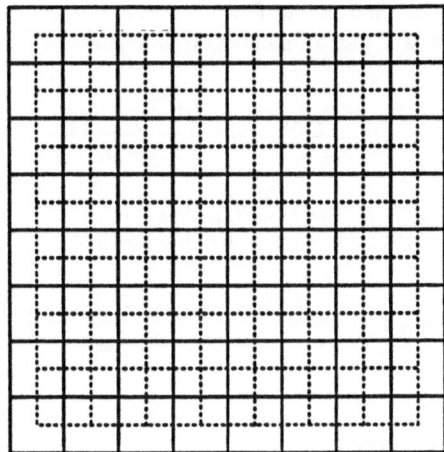

Figure 3: The arrangement of the square image patches. One set is drawn in solid lines, the other dotted.

1. The visible part of the object has a large extent in depth (z).

2. The component of motion perpendicular to the image plane is large.

3. The object is rotating significantly.

4. The velocity of the object is too great (see below).

In these cases the algorithm will only be able to label the regions as moving, without any specific velocity, essentially flagging a region of non-zero image difference [10]. We do not think that this is a drawback of the method, which is intended as a crude but robust way of labelling moving regions in an image; complex motions can be recovered after the moving region is stabilised in the centre of the image. We feel that robust, real-time algorithms should have simple, specific tasks to perform, and be able to report when they need help. We have developed a separate module that detects strong flow divergence [11] which is an "alarm" cue that an object may be on collision course with the cameras and hence deals with case 2 in the above list. We use a generalisation of the current method using a linear flow field approximation, as suggested by Campani and Verri [12] as a good approximation for the flow field generated by a moving planar surface.

Finding the Full Motion

Given regions of non-background normal flow vectors we wish to find the image velocities of moving objects. This involves solving the aperture problem, which until this stage we have been able to avoid. For each labelled moving object we find a single full flow vector $\mathbf{v}$ which is closest, in a least-squares sense, to the constraint lines generated by the background-compensated normal flow vectors (see equation 5) making up the object. The distance between the constraint line represented by $\mathbf{v}'_\perp$ and $\mathbf{v}$ is $(1 - \mathbf{v} \cdot \mathbf{v}'_\perp / |\mathbf{v}'_\perp|^2)|\mathbf{v}'_\perp|$. Thus we minimise the expression

$$\sum_{i=1}^{N} \left(1 - \frac{\mathbf{v} \cdot \mathbf{v}'_{\perp i}}{|\mathbf{v}'_{\perp i}|^2}\right)^2 |\mathbf{v}'_{\perp i}|^2 \tag{6}$$

in $\mathbf{v} = (u, v)^T$ where N is the number of flow vectors and $\mathbf{v}'_{\perp i}$ is the value of $\mathbf{v}'_\perp$ for the ith point. The result is a pair of simultaneous equations from which u and v can be obtained as:

$$u = \frac{S_y S_{xy} - S_x S_{yy}}{S_{xy}^2 - S_{xx} S_{yy}}, \quad v = \frac{S_x S_{xy} - S_y S_{xx}}{S_{xy}^2 - S_{xx} S_{yy}}$$

where $S_x = \sum v'_{\perp ix}$, $S_y = \sum v'_{\perp iy}$, $S_{xx} = \sum v'^2_{\perp ix}/|\mathbf{v}'_{\perp i}|^2$, $S_{xy} = \sum v'_{\perp ix} v'_{\perp iy}/|\mathbf{v}'_{\perp i}|^2$ and $S_{yy} = \sum v'^2_{\perp iy}/|\mathbf{v}'_{\perp i}|^2$. The root-mean-square residual is then

$$\text{RMS residual} = \frac{1}{\sqrt{N}} \left[S_{xy2} + \frac{S_x^2 S_{yy} - 2 S_x S_y S_{xy} + S_y^2 S_{xx}}{S_{xy}^2 - S_{xx} S_{yy}} \right]^{1/2} \quad \text{pixels}$$

where $S_{xy2} = \sum |\mathbf{v}'_{\perp i}|^2$.

5 Implementation

We use a 512×256 pixel field of an interlaced frame subsampled to 64×32. All subsequent processing takes place at the coarse resolution. The subsampled images are smoothed using a gaussian convolution. The temporal image gradient $\partial E/\partial t$ is then simply the difference between the current and previous images, and the spatial gradients $\partial E/\partial x$ and $\partial E/\partial y$ are calculated by applying the gradient masks $[-1\ 0\ 1]$ and $[-1\ 0\ 1]^T$ to the average of the previous and current smoothed images, as in [4]. The parameter and threshold values used were:

1. Standard deviation of gaussian convolution: 1.5 pixels. The larger this value the larger the velocities that can be measured. This current value gives us the ability to measure velocities up to about 12 pixels per frame (1.5×8). Larger masks would allow us to measure greater velocities, but at greater computational cost.

2. Threshold on $|\nabla E|$: 4 grey levels per pixel.

3. Threshold on residual velocity $|\mathbf{v}'_\perp|$: 0.3 pixels per frame.

4. Segmentation patch size: 8×8 pixels. This specifies the smallest size of moving object that can be located by the method.

5. Threshold on total patch vectors $T1$: 20.

6. Proportion threshold on non-background to total vectors ratio $T2$: 0.7. Thus 70% of a patch must be non-background for itself to be labelled as non-background.

The transputer implementation runs at frame-rate (25Hz) on the even field of each interlaced image, and the latency (pipeline delay) is 0.06 sec.

6 Results

Initial tests were made using a camera carried on a robot arm. The camera was rotated about its x-axis by known amounts between frames, while an object, a white head, was moved to the right. The problem is to recover the motion of the object in the apparently moving background. The first two frames in the sequence at 512×512 pixel resolution are shown in figure 4a. They may be fused stereoscopically.

Figure 4: a) The first two frames of an image sequence. b) The same frames after subsampling.

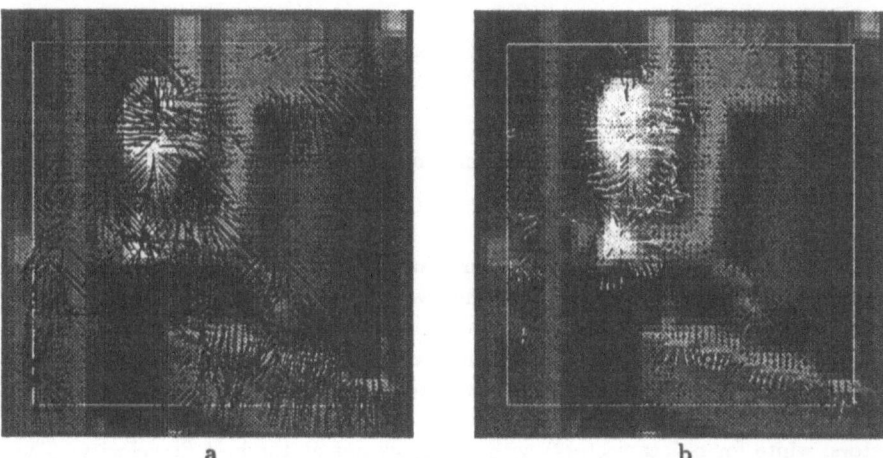

Figure 5: a) Normal image flow vectors. b) The result of subtracting the background flow.

When block-averaged and subsampled down to 64×64 pixels the result is as shown in figure 4b. From these the normal flow vectors are calculated. The vectors in figure 5a found from the images in figure 4b are displayed six times their actual length to aid visibility.

To predict the background flow in this simple case we use a small field of view approximation to equation 3, justified since the field of view of the camera is only 20°. The quadratic terms in x and y disappear. Ω_y and Ω_z are both zero, so we have

$$\mathbf{u} = \left(\begin{array}{c} 0 \\ f\Omega_x \end{array} \right)$$

Ω_x is the known x-axis rotation, and f is known to be approximately $25mm$. Thus the predicted flow $\mathbf{u}$ is a constant vector (w.r.t. x and y) in the y-direction.

The result of subtracting $\mathbf{u}$ from the normal flow vectors is shown in figure 5b. The background vectors are clearly smaller than those in figure 5a, while those due to the moving object have remained about the same size. The vectors shown as black are those that have been labelled as background according to the $|\mathbf{v}'_\perp|$ threshold criterion (page 4). The vectors labelled as part of a moving object are shown white.

364

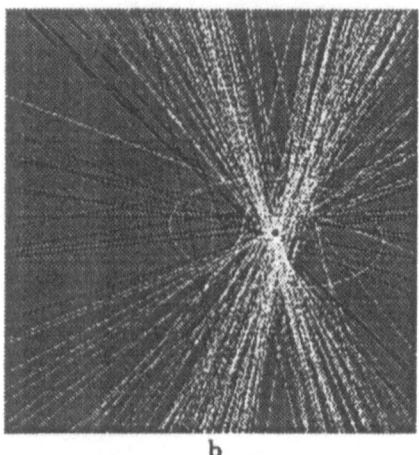

a   b

Figure 6: a) The result of segmenting the motion field. b) The least-squares fit to the constraint lines.

Figure 6a illustrates the flow segmentation and the results of the least-square velocity fit. Four adjacent square patches were found to be non-background and these were connected and the least-squares fit method described in section 4 applied to the non-background (white) normal flow vectors. The long black vector, displayed at ten times its actual length, is the estimated object velocity, correctly found to be a sideways motion. In figure 6b are the constraint lines (black for background vectors, white for object vectors) with the black dot at the best fit velocity. The ellipse gives an indication of the fit error and the amount of anisotropy in the fit, given by the ellipse eccentricity. If r is the RMS residual of the least-squares fit then the major and minor axes of the ellipse are set to ϵr and $\frac{r}{\epsilon}$, where ϵ is such that the ellipse takes on the shape of the cross-section through the quadratic sum-of-squares function in equation 6. Thus the ellipse shows the goodness of fit and how it changes with orientation in (u, v) space, allowing the common case of the motion being constrained mainly in one direction (for instance due to a moving straight edge) to be detected as a large major axis, indicating the lack of constraint in that direction. The ellipse is magnified 8 times to aid visibility.

Results for the Head/Eye Platform

The second sequence was produced by the real-time transputer implementation. Shown in figure 7 are 16 frames from a sequence showing two people walking past each other. There was no camera motion in this case. We used only the even field of each frame to avoid motion effects of interlacing. The outline of each detected moving region is shown along with the velocity vector and the error ellipse, both magnified six times. The sequence shows that the people are initially detected as one moving region, the nearer person moving to the right dominating the result (frames 1-4), while as they separate the estimated error increases greatly since there is an equal amount of image data travelling in opposite directions (frames 5-8). At that point the algorithm separates the two people and the velocities are estimated with greater subjective precision, i.e. the algorithm "knows" when it has good velocity data. The real-time implementation allows us to obtain such sequences routinely.

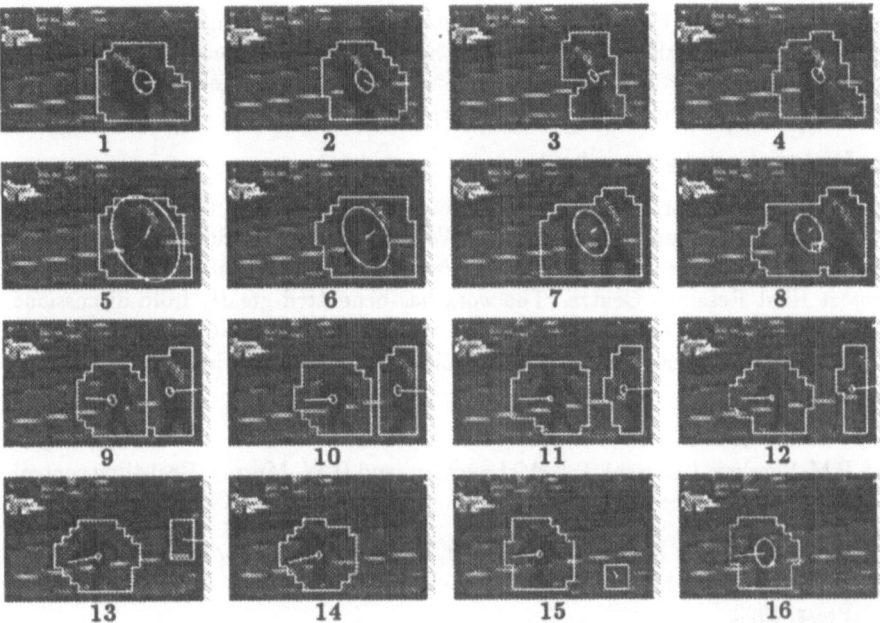

Figure 7: Results for real-time sequence of two people passing on a pavement next to the road outside out laboratory.

7   Conclusions

The algorithm we have developed can quite robustly calculate the position, extent and velocity of isolated moving objects against a stationary background in the presence of known head rotation, in real time (25Hz) and with short latency (0.06s). As it stands this will allow us to perform experiments with the head. We need to analyse the accuracy of the algorithm for different velocities and types of object. The basic test that the algorithm must pass is to be accurate enough that using the position and velocity data it provides the head will be able to saccade, fixate on the object, and start tracking it in a small foveal window. We are currently working on a Kalman filter-based tracking module that will allow the coherence of the temporal image to be made explicit, and to determine the trajectory of moving objects. With this module we wish to maintain an updateable memory of the moving objects in the scene, so that while tracking one object knowledge is retained of others, enabling change of attention if an another object becomes more "interesting" according to the size, velocity, persistence, etc. of the object. We have also extended the method to deal with more general types of object motion. For instance, an object moving towards the cameras gives rise to a strongly divergent flow field. This divergence can be used as an "alarm" signal that the object is about to hit the camera. The time-to-contact can be determined from the divergence (see [13]). Using a linear flow approximation [12] enables us to calculate the flow divergence. This work will be reported in future pulications.

This algorithm should be seen in context as a single module of a complex vision system, much of which will be bootstrapped from the results of this module. As such, robustness has been of greatest importance, and given higher priority than the ability to deal with complex scenes. The important outcome in the context of a real-time gaze control system is not the "perfection" of the flow vectors, but the number of correct *actions* that the module gives rise to.

Acknowledgements

This work was supported by Esprit (Project 5390: Real Time Gaze Control) and the SERC (grant number GR/G30003). We have had fruitful discussions with our RTGC partners, especially Paul Sharkey here at Oxford, and Donald Weir and Nigel Gent at Hirst Research Centre. The work has benefitted greatly from discussions within the Oxford Robotics group, and we would especially like to thank Phil Torr, Paul Beardsley, Rupert Curwen, Andrew Zisserman and Mike Brady.

References

[1] P.M. Sharkey, I.D. Reid, P.F. McLauchlan, and D.W. Murray. Real-time control of an active stereo head/eye platform. In *Proc. 2nd ICARCV, Singapore*, 1992.

[2] D.W. Murray, F. Du, P.F. Mclauchlan, I.D. Reid, P.M. Sharkey, and M. Brady. Design of stereo heads. In A. Blake and A. Yuille, editors, *Active vision*. MIT Press, 1992.

[3] I.R. Reid, P.M. Sharkey, P.F. McLauchlan, and D.W. Murray. A modular head-eye platform for real-time reactive vision. Technical Report 1941/92, Dept. of Engineering Science, Oxford University, 1992.

[4] R.C. Nelson. Qualitative detection of motion by a moving observer. *IJCV*, 7(1):33–46, 1991.

[5] P.J. Burt. Image motion analysis made simple and fast, one component at a time. In *Proc. 2nd BMVC*, pages 1–8, 1991.

[6] E. François and P. Bouthemy. Multiframe-based identification of mobile components of a scene with a moving camera. In *Proc. CVPR*, 1991.

[7] B.K.P. Horn. *Robot vision*. MIT Press, 1986.

[8] A. Verri and T. Poggio. Against quantitative optical flow. In *Proc. 1st ICCV*, pages 171–180, 1987.

[9] E. Hildreth. *The Measurement of Visual Motion*. MIT Press, 1984.

[10] Y.Z. Hsu, H. Nagel, and G. Rekers. New likelihood test methods for change detection in image sequences. *CVGIP*, 26:73–106, 1982.

[11] J.J. Koenderinck. Optic flow. *Vision Res.*, 26(1):161–180, 1986.

[12] M. Campani and A. Verri. Computing optical flow from an overconstrained system of linear algebraic equations. In *Proc. 3rd ICCV*, pages 22–26, 1990.

[13] D.N. Lee. The optic flow field: the foundation of vision. *Phil. Trans. R. Soc. Lond.*, 290, 1980.

Vergence Micromovements and Depth Perception

Antônio Francisco *

CVAP, Royal Institute of Technology (KTH)

S-100 44 Stockholm, Sweden

Abstract

A new approach in stereo vision is proposed in which 3D depth information is recovered using *continuous vergence angle control* with simultaneous local correspondence response. This technique relates elements with the *same* relative position in the left and right images for a continuous sequence of vergence angles. The approach considers the extremely fine vergence movements (micromovements) about a given fixation point within the depth of field boundaries. It allows the recovery of 3D depth information given the knowledge of the geometry of the system and a sequence of pairs $[\alpha_i, \mathbf{C}_i]$, where α_i is the ith vergence angle and $\mathbf{C}_i$ is the ith matrix of correspondence responses. Due to its local operation characteristics, the resulting algorithms are implemented in a modular hardware scheme using transputers. Unlike currently used algorithms, there is no need to compute depth from disparity values; at the cost of the acquisition of a sequence of images during the micromovements. Experimental results from physiology and psychophysics suggest that the approach is biologically plausible. Therefore, the approach proposes a functional correlation between the vergence micromovements, depth perception, stereo acuity and stereo fusion.

The perception of the 3D-distance, depth, of objects using stereo images have been studied by many researchers for a long time. Some of these studies use vergence camera systems [1] integrating position control, image acquisition and depth processing on the modality of vision system named "active vision" [2]. Following this line of research, the present work analyses the correlation between the real time depth acquisition and the extremely fine vergence movements (micromovements) of the cameras about the fixation point. We assume that these movements are synchronized between the two cameras.

The *continuous vergence micromovements* differ from the vergence, translation and rotation movements used on the other methods to fixate the cameras on a new fixation point. The previous methods (using particular techniques as multi-resolution) compute some depth-map or depth directly from the acquisition of the left and right images at this fixation point, i.e., using two images and some stored information (estimation) about the depth-map they are able to infer the current depth at the correctly matched image points. Generally,

---

*Researcher at the National Institute of Space Research (INPE), São José dos Campos, São Paulo, Brazil. The support from the Swedish National Board for Industrial and Technical Development, NUTEK, is gratefully acknowledged. I would like to thank Prof. Ruzena Bajcsy and Prof. Jan-Olof Eklundh for the support to the development of this work as well as Kourosh Pahlavan, Akihiro Horii and Thomas Uhlin for valuable help when using the KTH head-eye system.

the strategies used is such methods have the search space for correspondence matches along epipolar lines. Therefore the depth is calculated using the disparity information between the left-right matched points and the geometry of the camera system.

The current approach uses neither epipolar lines nor disparities to calculate the depth of any 3D point. The depth is determined by the geometry of the camera system (mainly, the vergence angle) and by the relative position of a pixel with respect to the image plane. The procedure can be simply described as following: micromovements of two cameras occur about the fixation point. For each left-image point, on the left image plane, the vergence angle and the "correspondence response" of this point and the right-image point at the same relative position on the right image plane are stored. For each left-point, using these "correspondence response" signals and the camera geometry, the depth of the 3D points where the correspondence response reach the highest level are calculated.

Therefore, the approach is functionally different from the previous ones in the sense that the depth is calculated locally for each point (without searching epipolar lines) with the necessity of acquiring a sequence of images during the micromovements. The objective here is to clarify how to calculate depth using micromovements.

The paper covers the theoretical background, the experimental results and the biological support for depth acquisition from the vergence micromovents approach. The theoretical and simulations parts of this work were developed [3] during my stay at the General Robotics and Active Sensory Perception (GRASP) laboratory, University of Pennsylvania, USA. The experiments to validate the approach have been done using the KTH head-eye system [4].

1 The stereo vision system and the horopter

Each lens of the right and left camera is considered to be thin and *ideal*, in the sense that an object at a distance d_{out} (the *object distance*) from the principal plane has its image (with inverted direction) at distance d_{in} (the *image distance*) from this plane. The relationship of these two distances and the focal length of the lens is given by the *Gaussian Lens Equation:*

$$\frac{1}{f} = \frac{1}{d_{in}} + \frac{1}{d_{out}} \tag{1}$$

With respect to the camera platform, a symmetric fixation in the visual plane is assumed. Therefore the vergence angles of the two cameras have the same value α and the point being fixated is in the visual (horizontal) plane of the cameras. With this assumption, any camera torsion (about the axis connecting the lens center and the image plane center) is considered to be zero. According to the symmetric fixation model (figure 1) associated with each image plane there is a coordinate system with its origin at the image plane center. The lens centers are separated by a baseline b. These coordinate systems define the left projection (x_{pl}, y_{pl}) and the right projection (x_{pr}, y_{pr}) of a point in space. To identify the 3D position $(X_o, Y_o, Z_o)^T$ of a point $\vec{P_o}$ in space a global coordinate system xyz is used (figure 1).

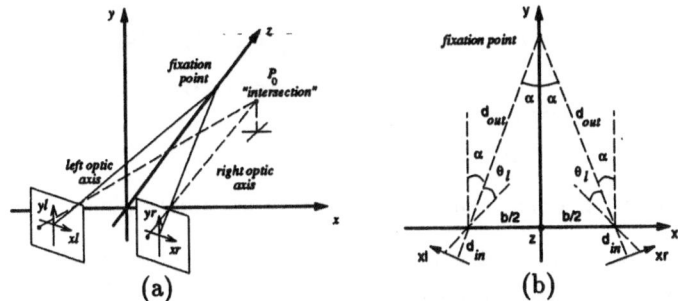

Figure 1: Stereo camera geometry: (a) Perspective view (b) Top view

In order to adopt the same terminology used in the human vision field we will review some concepts from the physiology and psychophysics sciences. The **horopter** defines the set of points in space for which the binocular disparity is zero [5]. The **point horopter** is the locus of zero disparities for the point stimulus where *both* horizontal and vertical disparities are zero. (There has been a considerable amount of confusion in the literature caused by the laxity in defining the horopter [5].)

We consider the ideal point horopter composed of points with zero horizontal and vertical disparities for the symmetric fixation in the visual plane, with any position, torsion and optical aberrations assumed to be absent. As described in [5], any point off the *point horopter* in space (off-axis points) projects to the two image planes with horizontal and vertical disparities. With vergence, the points at the distance corresponding to the point horopter would nullify the horizontal disparity. Note that in the ideal case nothing can be done to nullify the vertical disparity produced by off-axis points being necessarily closer to one eye than the other, with a resulting difference in the projection angle in the two eyes. The present analysis concerning the horopter is based on *zero horizontal disparity*.

For the *zero horizontal disparity* case we can define a 3D "intersection" point of the left and right optic axes (same x and z and different ys) passing through the same corresponding element $(x_p, y_p)^T$ on the image planes coordinate systems. The coordinate $(x_i, y_i, z_i)^T$ of this "intersection" point [3] is:

$$\left(\frac{b}{2}\frac{\tan(\alpha+\theta_l)-\tan(\alpha-\theta_l)}{\tan(\alpha+\theta_l)+\tan(\alpha-\theta_l)}, \quad -\frac{y_p}{d_{in}}\frac{b}{2}\frac{\cos^2(\theta_l)}{\sin(\alpha)}, \quad \frac{b}{\tan(\alpha+\theta_l)+\tan(\alpha-\theta_l)}\right)^T \tag{2}$$

where,

$$\alpha = \arctan(\frac{b}{2d_{out}}), \quad \theta_l = -\arctan(\frac{x_p}{d_{in}}), \quad \text{and (eq. 1)} \quad d_{in} = \frac{d_{out}f}{d_{out}-f} \tag{3}$$

The above equation for y_i was deduced considering the average of the left and right y coordinates of the left and right optic axes at the "intersection" point. This assumption includes an error in the present analysis. In order to evaluate the dimension of this error with respect to the length of the photo-receptor element, the difference between the projections on both image planes of a point $\vec{P}_o$ in $(x_i, y_i, z_i)^T$ (eq. 2) is analyzed. Note that the projections are determined by the intersection of the right and left optic axes, passing through the left and right optic lens center respectively, with the image planes. Let us denote the intersection of the left optic axis with the left image plane as $(x_{pl}, y_{pl})^T$ and the intersection of the right optic axis with the respective

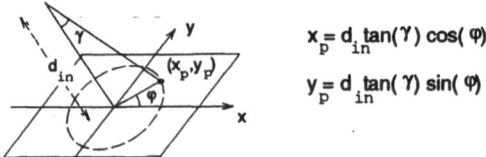

$$x_p = d_{in} \tan(\gamma) \cos(\varphi)$$

$$y_p = d_{in} \tan(\gamma) \sin(\varphi)$$

Figure 2: Polar coordinate system of the image plane

image plane as $(x_{pr}, y_{pr})^T$. It is possible [3] to define the following euclidian *projections deviation (dev)* for a given point $\vec{P}_o$ as:

$$dev = \sqrt{(x_{pl} - x_{pr})^2 + (y_{pl} - y_{pr})^2} \qquad (4)$$

where,

$$x_{pl} = -d_{in} \tan(+\arctan(\tfrac{\frac{b}{2}+X_o}{Z_o}) - \alpha), \quad y_{pl} = \frac{-Y_o d_{in} \sin(\alpha+\theta_l)}{\cos(\theta_l)(X_o+\frac{b}{2})} \qquad (5)$$

$$x_{pr} = -d_{in} \tan(-\arctan(\tfrac{\frac{b}{2}-X_o}{Z_o}) + \alpha), \quad y_{pr} = \frac{Y_o d_{in} \sin(\alpha+\theta_r)}{\cos(\theta_r)(X_o-\frac{b}{2})} \qquad (6)$$

$$\theta_l = -\arctan(\tfrac{x_{pl}}{d_{in}}), \qquad \theta_r = \arctan(\tfrac{x_{pr}}{d_{in}}) \qquad (7)$$

The *dev* analysis is done considering the object distance as a multiple of the baseline $(d_{obj} = k_b b)$ and the image planes mapped by a polar coordinate system (figure 2). Having deduced all needed equations for the *dev* analysis, it is time to show some simulated results about the human visual system and the GRASP platform system. The procedure to accomplish the *dev* analysis can be synthesized in the following simulation steps:

- for a given: k_b, γ, φ, f , b ;   compute: d_{obj} , α , d_{in} (eq. 3), x_p, y_p (figure 2), $(x_\iota, y_\iota, z_\iota)^T$ (eq. 2),
- using $\vec{P}_o = (x_\iota, y_\iota, z_\iota)^T$, compute: $x_{pl}, y_{pl}, x_{pr}, y_{pr}$ (éq.s 5 to 7) and then *dev* (eq. 4).
- plot: the results of the *dev* normalized with respect to the distance between centers of adjacent photo-receptor elements *dce*. Note that *dce* is a constant for most machine vision systems $(dce(.))$ and is a function of γ for the human visual system $(dce(\gamma))$ [3].

The results of the simulation shown in Figure 3.a imply that the highest *dev* occurs when $\varphi = k$ 90 degree + 45 degree (k = 0,1,..). Therefore all other simulations are done with the value of φ equal 45 degree. A conclusion from Figure 3.a, is that the normalized *dev* is smaller for the human visual system since $dce(\gamma)$ increases on the periphery. This characteristic implicit in the human visual system tends to diminish *dev*. Another feature of the human visual system that tends to diminish *dev* is the known difference between nasal and temporal retina eccentricity (if nasal is larger than temporal) for every pair of corresponding points. This difference in eccentricity could explain the deviation of the empirical horopter [5] from the V-M circle as well as the necessity to diminish *dev*.

For the GRASP system we want to know how *dev* varies with object distance. From the analysis of Figure 3.b we see that *dev* decreases with object

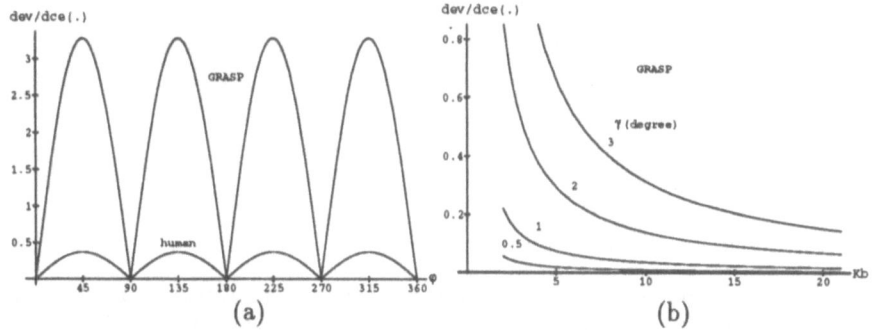

Figure 3: Normalized deviation as a function of: (a) φ ($\gamma = 2°$, b = 65 mm, f = 17 mm, $d_{obj} = 2b$), (b) k_b ($d_{obj} = k_b b$, $\varphi = 45°$, b = 128 mm, f = 65 mm)

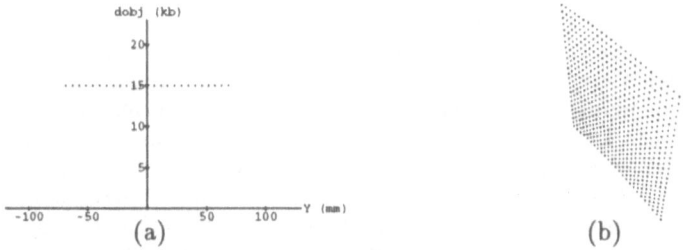

Figure 4: Point horopter for GRASP platform system (b = 128 mm, f = 65 mm, $d_{obj} = 15$ b, dce(.) = 30.0 10^{-3} mm): (a) Top view, (b) Perspective view

distance, therefore the next set of simulations are done with $d_{obj} = 15b$ which ensures a small dev for $\varphi = 45$ degree. The dev has been investigated for its maximum value, in spite of the zero value of this deviation on the image planes coordinate axes ($\varphi = k$ 90 degree, k = 0,1,..) for any value of γ, d_{obj}, b and f.

Having shown that the normalized dev is very small when d_{obj} is greater than fifteen times the length of the baseline, the point horopter is plotted using equation 2 for this value of d_{obj}. Almost all the parameters of the above equations can be computed directly (like d_{in} and α) from the defined values of d_{obj}, b and f. The only two parameters that do not have a defined range are x_p and y_p. In the present paper, 80 photo-receptor elements are used as the distance from the image plane center to the periphery of the workspace being analyzed. Therefore, a *square workspace* of side size equal to 160 pixels centered in the image plane is assumed. This range of x_p and y_p gives the point horopter plotted in Figure 4 for the GRASP platform system. It can be seen that the point horopter is a surface in space.

2 Micromovements

The shape of the point horopter has been analyzed for a given vergence angle calculated from the object distance under fixation. The main analysis now is conducted for a number of vergence angles α_i about the *fixation point in the visual plane*, described by the following equation:

$$\alpha_i = \arctan(\frac{b}{2d_{obj}}) + \varepsilon_i \tag{8}$$

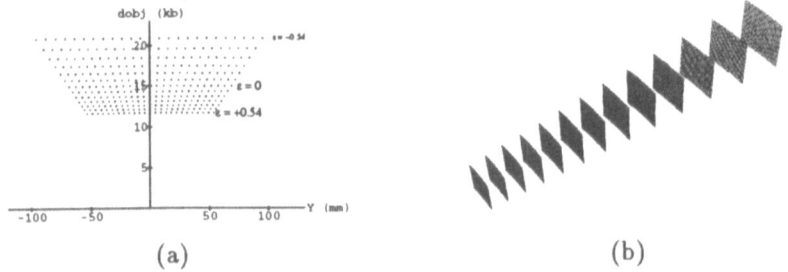

Figure 5: Set of point horopter surfaces for GRASP platform system ($\varepsilon_i \in$ [$-0.54°$, $0.54°$]): (a) Top view, (b) Expanded perspective view

where ε_i is a small angle increment (positive or negative) that describes the micromovement about the *fixation point* (first part of equation 8). The set of α_i about a given fixation point determines a complete *micromovement cycle*.

Figure 5 shows the locus of "intersection" points in 3D space of the GRASP platform system for a given micromovement cycle. The surfaces shown correspond to the set of point horopter surface generated for each vergence angle α_i. As can be seen in Figure 5, the locus of all the "intersection" points form a volume in the 3D space. Therefore, any object inside this volume can have its depth measurements determined by the response of a *local correspondence operator* to the *continuous vergence angle control*. Remember that this operator relates elements (image plane points) with the *same* horizontal and vertical distance from the center of the left and right image planes. It is possible to use a local correspondence operator since we assume that d_{obj} is greater than fifteen baselines, implying a small *dev* (see previous section).

The errors in stereo (along z axis) with the present approach are due the vergence angle quantization (angle steps fixated by ε_i). These errors differ from the quantization errors due to discrete photo-elements in cameras, that are a common characteristic of other stereoscopic methods. As described in [6], the errors due the photo-receptor quantization are significant and increase with the distance from the object to the cameras system. The present approach allows us to overcome the photo-receptor quantization limitation by using a sequence of pairs $[\alpha_i, C_i]$, where α_i is the ith vergence angle and C_i is the ith matrix of correspondence responses. Although the present analysis considers only the micromovements in the visual plane (horizontal micromovements), the human eye system performs micromovements in a vertical plane including the visual axis as well as rotations about the visual axis itself [7].

3 Biological support of the micromovements

The following discussion of eye-movements according to physiological and psychophysical experiments is offered as a working hypothesis, useful for the understanding the role of the micromovements on depth perception. Physiological results [8, 7] show that the human eye performs fine movements during the process of fixation on a single point, which are collectively called *physiological nystagmus*. Physiological nystagmus is composed of three different kinds of movements: (1) high-frequency tremor, (2) slow drifts, and (3) rapid *binocular flicks*. The drift and flick movements occur in opposing directions and produce convergence/divergence waves of the eyes on a similar way as the micromove-

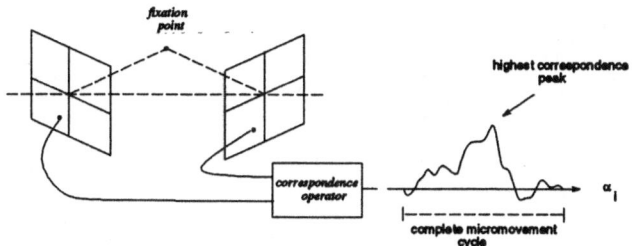

Figure 6: Correspondence operator response

ments studied in previous sections.

Assuming the vergence micromovements mechanism as the basis of the depth perception, it is easy to understand the phenomena of *stereoacuity* (depth or stereoscopic acuity, stereopsis). As well described in [8, 5], it is almost incredible that most observers under normal conditions can discriminate a difference in depth corresponding to an *angular disparity* (interocular disparity) of about 10 arc sec. The best values reported in the literature have been obtained by the apparatus called the *Howard-Dolman apparatus*, devised by Howard in 1919. The best observers achieve a 75% discrimination level close to 2 arc-seconds in that experiment. The most incredible fact is that this disparity value is much smaller than the distance between the cones' centers at the *central part* of the *fovea* ($\approx$ 22 arc sec).

We suggest that the high sensitivity to slight disparity can be explained by the correlation between depth perception and the vergence eyes micromovements and not by the capacity of the human visual system to spatially detect disparity on the retinas. Therefore the idea of an *angular disparity* that can be detected *spatially* by the visual system is substituted by a *local approach* where the human visual system determines the depth values by the highest peak of correspondence response (figure 6) during a complete *micromovement cycle* (section 2). The highest peak of correspondence occurs when there is no spatial disparity between the left and right stimulus of elements with the same relative position on both retinas, i.e., when the spatial disparity is cancelled for a given vergence angle.

Another phenomenon that can be explained by the present approach is known in the literature [8] as *Panum's fusional area*: the range of interocular disparities within which objects viewed with both eyes on corresponding retinal regions appear single. This area is such that *fusion* occurs, only one dot is seen, when two points that are perceived in different eyes fall closer together in the combined view. Note that these two points can be seen through an uncrossed (left and right optic axis do not cross) or crossed disparity. The classical static limits for Panum's area, the mean crossed to uncrossed range of horizontal disparities, is reported as being 14 arc min. The experiments described in [9] support the existence of *binocular fusion* as a *unique* category of sensory performance, disconfirming several non fusional explanations of single vision. While the range of binocular disparities allowing fusion (Panum's fusional horizontal diameter) is typically in the region of 14 arc min, stereoscopic depth can be perceived from a disparity 500 times smaller.

In the present approach, the phenomena of binocular fusion and stereoscopic depth are assumed to be supported by the mechanism of vergence eye micromovements about a fixation point. In this way, the fusion area dimension

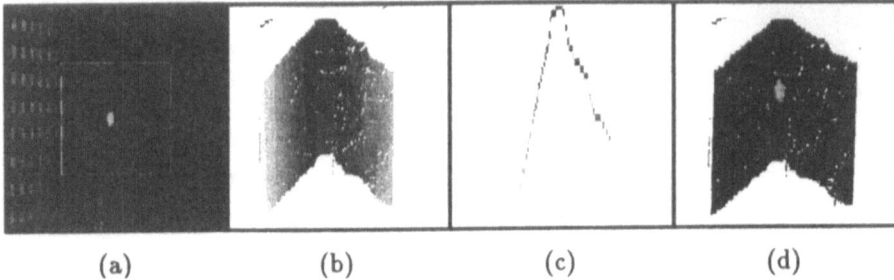

(a)	(b)	(c)	(d)

Figure 7: Two planes experiment (33 vergence steps): (a) original image, (b) perspective view of the acquired object, (c) horizontal cut of the acquired object, (d) perspective view pasted with real grey value from original image

is determined by the range of a complete *micromovement cycle* (section 2). It is important to point out that the *classical* value of the Panum's fusional *horizontal radius* (average of the crossed and uncrossed disparities), 7.0 arc min, coincides with the micromovement range value described in [7]. Note that the Panum's fusional horizontal radius must be compared to the total range of a *monocular* micromovement reported in [7] to be coherent to both definitions. In the present analysis the *vertical* fusion radius is not considered since this radius follows the monocular spatial resolution limit of the retina [10]. As a conclusion, the "real" binocular fusion is assumed to occur only between cells adjacent on the horizontal axis of the retinas, and that binocular vertical fusion is a result of the monocular fusion mechanism.

4   Experimental results

In order to validate experimentally the micromovements approach, a practical implementation was done using the KTH head-eye system. This active vision system is composed of several motors, two cameras and two camera lens controllers. The system is connected on the VME-bus of a SUN-SPARC station via a transputer board. Our main control was over the vergence motors, image acquisition, zoom and focus. At the beginning, we do not use the transputer board to execute the algorithms. Instead of that we did the experiments acquiring and storaging a sequence of images pairs in the SPARC station for further processing. The two experiments that will be described below consist of an object located in front of the head-eye system around 3000 mm from the baseline. The set up is done by choosing a value for the focal length (zoom) and adjusting manually the focus and the initial fixation point over the central part of the object surface. These values of zoom and focus are kept constant during the experiment. A program makes the sweeping of the object by changing the vergence angle between the initial vergence value and a final value determined by the number of steps specified. For each vergence angle, two images (from the left and right cameras) are stored on the SPARC station. After the acquisition of the desired number of images pairs, we compute the correspondence response for every pixel (x,y) on the acquired left and right images, using the

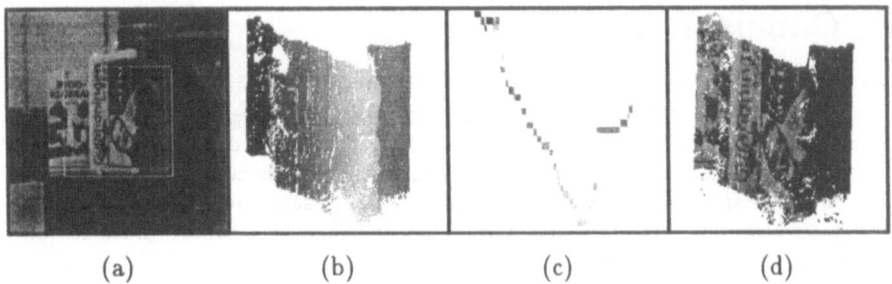

| (a) | (b) | (c) | (d) |

Figure 8: The box-plane-cylinder experiment (41 vergence steps): (a) original image, (b) perspective view of the acquired object, (c) horizontal cut of the object, (d) perspective view pasted with real grey value from original image

following correlation operator:

$$Corr(x, y) = \frac{E[Left(x, y)Right(x, y)] - E[Left(x, y)]E[Right(x, y)]}{\sigma[Left(x, y)]\sigma[Right(x, y)]} \quad (9)$$

where,

$$E[X(x, y)] = \frac{\sum_{i=-w/2}^{w/2}\sum_{j=-w/2}^{w/2} X(x+i, y+j)}{2(w+1)}, \quad (\sigma[X])^2 = E[X^2] - (E[X])^2$$

For every pixel, using the vergence angle α_i that gives the highest correspondence response for that pixel and using (8) we compute the object depth d_{obj} for that pixel.

The following parameters are common in both experiments: b = 200 mm, f = 30 mm, d_{obj} = 15 b, the work window size of 200x200 pixels, the operator size of 21x21 pixels and the vergence step resolution of 26 arc sec (imposed by the head-eye system). Our first experiment was done using two vertical planes as the object being viewed. In Figure 7.a the left image of the object before the vergence sweeping is shown. Figure 7.b shows the perspective view of the acquired object after the vergence sweeping and the use of equation 8. The darker grey patches of the object are farther from the baseline than the lighter grey patches. On Figure 7.c, a horizontal cut of the acquired object is shown permitting us to have the correct idea of the object being viewed. The last picture is the perspective view pasted with the real grey values instead of the depth represented grey values. The second experiment is shown in Figure 8. The object is composed of a box (left side of the object) a cylinder (right side) and an inclined newspaper. Figure 8.c gives the notion of the object used.

The step-shape wave seen on Figures 7.c and 8.c is a consequence of the vergence step resolution. The processing time using the previous scheme was about one hour for the processing of the entire 200x200 pixel being viewed. Actually the correspondence operation is being executed on the transputer board. The entire image was split in four transputers before the correspondence operation. Using this new scheme the experiment took around 20 seconds. We are not using the entire power of our transputer board since we did not have time enough to implement it. In spite of the great improvement using the transputer board our goal was not reach yet, since we want to process depth at the frame rate.

5   Conclusion

The *continuous vergence micromovements* approach permits to overcome the physical limitation of the photo-receptor dimension (CCD element or cone) on the depth perception. Moreover, there is no need to compute depth from disparity values since the disparity is cancelled by the vergence micromovements. Note that the *stereoscopic matching problem* still exists, since there is the possibility to have two or more correspondence peaks with similar values for an element of the correspondence matrix.

The highlight of this new approach is the vergence micromovements as a mechanism to nullify the disparity between the left and right visual stimulus at the same retina locus. Therefore, the concept of a "neural structure spread spatially" in the visual system to perceive depth via measurement of disparity is substituted by a "neural structure connected locally with the neighborhood" of each retina locus.

References

[1] E. P. Krotkov. *Exploratory visual sensing for determining spatial layout with an agile stereo camera system.* PhD thesis, School of Engineering and Applied Science, University of Pennsylvania, Philadelphia, PA, USA, 1987.

[2] R. Bajcsy. Active perception vs. passive perception. In *Proc. Workshop on Computer Vision*, pages 55–59, Bellaire,MI, October 1985.

[3] A. Francisco. The role of vergence micromovements on depth perception. Technical Report MS-CIS-91-37, GRASP LAB, CIS, University of Pennsylvania, Philadelphia, PA, USA, 1991.

[4] K. Pahlavan and J.O. Eklundh. A head-eye system - analysis and design. In *Computer Vision, Graphics, and Image Processing: Image Understanding*, page (To appear.), July 1992.

[5] C. M. Schor and K. J. Giuffreda. *Vergence eye movements: basic and clinical aspects.* Butterworth, 1983.

[6] F. Solina. Errors in stereo due to quantization. Technical Report MS-CIS-85-34, GRASP LAB, CIS, University of Pennsylvania, Philadelphia, PA, USA, 1985.

[7] R. W. Ditchburn. Eye-movements in relation to retinal action. *Optica Acta*, 1(4):171–176, 1955.

[8] J. W. Kling and L. A. Riggs. *Experimental psychology.* Holt, Rinehart and Winston, Inc., 1971.

[9] T. Heckmann and C. M. Schor. Panum's fusional area estimated with a criterion-free technique. *Perception & Psychophysics*, 45(4):297–306, 1989.

[10] C. Schor, I. Wood, and J. Ogawa. Binocular sensory fusion is limited by spatial resolution. *Vision Res.*, 24(7):661–665, 1984.

Layered Architecture for the Control of Micro Saccadic Tracking of a Stereo Camera Head

J.E.W.Mayhew

Artificial Intelligence Vision Research Unit,
University of Sheffield, Sheffield, S10 2TN,England

Y.Zheng, S.A.Billings

Department of Automatic Control and Systems Engineering,
University of Sheffield, Sheffield, S1 4DU,England

Abstract

The paper describes a 3-layered architecture for the control of the stereo-scopic eye-saccade system of a stereo-camera head [1] mounted on an autonomous vehicle.

The 0-level is a proportional feedback controller providing a micro-saccadic [2] control for eye movements enabling the head to foveate and track targets but requiring iteration through the vision system with the attendant computational overhead.

The 1-level provides the feedforward inverse kinematics for saccadic eye movements allowing a ballistic movement to replace the 0-level control loop. The training data is provided by the feedback error signal from the 0-level controller.

The 2-level is an adaptive lattice filter which is used to track moving targets. The filter is 'trained' using vision error-feedback from previous saccades. The filter learns to predict the future target position in the next image. This is used by the inverse kinematics module to generate the eye movement commands for the appropriate predictive saccade.

---

[1] The stereo camera rig used for this work comprises a 3-link kinematic chain, whose degrees of freedom are rotations around the following axes: i) Pan: a vertical axis corresponding to the 'neck'; ii) Tilt: an axis at right angles to the neck; and iii) Verge: each camera ('eye') can rotate independently around an axis at right angles to the tilt axis. The rig has been constructed so that the centres of rotation of the tilt and pan links coincide, and the centres of rotation of left and right verge and the tilt links coincide. The length of the tilt link is approximately 12.5 cm for each eye (i.e. the head is about 25 cm wide); the length of the verge link (i.e. approximately how far the centre of rotation is from the focal centre of the camera) is 5cm so that tilting the eye also produces a small translation. It is also of note that the right camera has been mounted with a 5 degree heterophoria and about 2.5 degrees of cyclotorsion. Stepper motors control the head and give a maximum saccade velocity of 50 degrees/second.

[2] Microsaccades are generally used to refer to the very small saccades which, if they have any function at all, may be used to correct errors arising from drift during fixation of a stationary target (Carpenter, 1988). We use the term microsaccadic tracking to describe a form of tracking which uses small vergence saccades (ranging in size from a few minutes of arc to two degrees), characterised by a fast movement stage, followed by a 80ms image capture stage during which the eyes remain stationary. In humans, this form of tracking may not normally occur in isolation but seems to be an important component of pursuit movements (Carpenter, 1988, page 55).

1 Introduction

We describe the implementation of a 3-layered architecture for the control of the stereoscopic eye-saccade system of a stereo-camera head mounted on an autonomous vehicle. This system is shown in figure 1 as a functional block diagram and has been implemented on a 4x4 transputer network (see legend for details).

0-level: This layer is a proportional feedback controller providing a microsaccadic control for eye movements enabling the head to foveate and track targets but requiring iteration through the vision system with the attendant heavy image processing overhead. The latter processing in the current implementation is a simple centre-of-gravity blob tracker. This rather crude level of image processing is driven by the real-time demands of the task and current equipment constraints. The image capture has three modes:

1. Tracking: a 3.75 degree square 'foveal' region of interest (64 x 64 pixels) is used when a target has been located and is being tracked. It may be of interest to note that the location of a small target in this fovea takes at least 20 ms (and more depending on the size of the blob).

2. Recovery: a 7.5 degree square region of interest (ROI) is used to recover when the target is temporally lost when tracking.

3. Initialisation: the full 30 degree square image for initialisation of target tracking.

In the tracking mode, image processing is done concurrently and independently in the two images; in the recovery and initialisation modes a subsampling strategy is used to locate a target in one image and then focus the search around the corresponding point in the other image.

The details of the implementation are unimportant but a principle may be worth elaborating. A tracking competence working on primitive, fast and even crude vision processing can provide the 'temporal glue' by which "the thing you saw then is the object you recognise now". Thus during tracking the foveal ROI is distributed as a continuous stream to another image processing system, completely independent of the tracking system, which samples the image stream at a very different and much slower rate. Currently this system is used only to display the images, but the direction of future system evolution is obvious.

1-level: This layer provides the feedforward inverse kinematics for saccadic eye movements allowing a ballistic movement to replace the 0-level control loop. Only a brief description of this level is given because the work has been described elsewhere (Dean et al 1991; Mayhew et al, 1992). They used adaptive PILUTs (Parameterised Interpolating Look-Up Tables) as the architecture to learn the state dependent correction to the 0-level controller. Following Kawato et al (1989) the feedback error signal from the 0-level simple proportional controller was used to provide the training data.

2-level: This is an adaptive lattice filter which is used to track moving targets. The filter is trained using error feedback from previous saccades within the current tracking sequence, so that the filter learns to predict the future target position in the next image. This is used by the inverse kinematics module to generate the eye movement commands for the appropriate predictive saccade.

For the 2-level layer we wished to develop a tracking prediction module with the following properties: i) it should be as general as possible, making minimal assumptions about the complexity and stationarity of the target trajectory; ii) it should adapt in very few time steps or samples, both to the onset of motion and to any discontinuities in the trajectory, yet at the same time it should be robust over sequences of missing data such as frequently result from occlusions and low level image processing infelicities; and iii) the implementation of the predictor should be computationally inexpensive. We describe below the details of the experimental evaluation of this module, using both simulated data, and to control the real stereo while tracking a moving light source.

2   Lattice Predictor

The use of multi-stage lattice filters (Goodwin and Sin, 1984; Alexander, 1986) for prediction is commonplace especially in the speech processing domain (Makoul,1975). The general principle underlying their design is that the successive stages of the filter compute the partial correlations (or regressions) at different delays. We have explored several different adaptive algorithms for doing this. As expected, we found gradient methods of training inefficient compared to recursive data projection algorithms. However, an alternative method has been implemented that calculates the reflection coefficients of the lattice filter directly using a decaying running average of the smoothed partial correlations. By controlling the time constant of the estimator of the partial correlations, both the requirements of fast adaptive response and relative robustness to missing data can be satisfied. Because successive stages of the filter are orthogonal and independent it is easily adapted on-line to the complexity of the signal by the simple expedient of adding or deleting stages of the lattice in response to variations in the partial correlations of the last stage. (See figure 2. For further experimental details see Zheng, et al 1991).

In implementing the adaptive lattice predictor in a simulation environment, we have found that the choice of initial conditions can significantly influence the rate of convergence of the reflection coefficients. If the reflection coefficients are initialised to zero this has the effect of introducing an artificial discontinuity in the input data which would propagate through the stages of the lattice predictor influencing the calculations of all reflection coefficients, resulting in delayed convergence and poor predictive performance.

We noticed that the first stage of a lattice predictor is very similar to a differentiator. Thus an appropriate initial condition for the reflection coefficient of the first stage should be -1. When so initialised a very significant increase in the rate of convergence of all the reflection coefficients is obtained with a much improved predictive performance.

This is of particular importance when the trajectory to be modeled contains discontinuities such as a sudden step change in velocity and/or change in direction. These occurrences can be readily recognised by monitoring the prediction error. Unless dealt with appropriately these discontinuities corrupt the future tracking behaviour. The strategy we have adopted is to maintain a running estimate of the standard deviation of prediction errors assuming they were normally distributed. If the current prediction error exceeds the 95% confidence limit, the memory is immediately flushed and all the stages of the lattice are reinitialised. The strategy is effective only because of the rapid convergence obtainable when correctly initialised.

We have compared the performance of a two-stage lattice filter, a Kalman filter of the same order, and a non-predictive tracker. The performance of the predictor is significantly better than the Kalman filter. This is because the lattice filter is optimal in the least squares sense and, unlike the Kalman filter, incorporates no assumptions about the structure of the trajectory. Another attractive feature of the lattice filter is that, because successive stages are orthogonal, it is very simple to adapt its length on-line as the complexity of the target trajectory increases or decreases. This can be done by monitoring the residuals. The Kalman filter does not have this degree of flexibility.

3   Online Saccadic Tracking

We have evaluated the lattice predictor in several modes:

1. Relative mode: visual target prediction using fixation error feedback. The filter was used to generate predictions of the future retinal coordinates of the target with respect to the fovea. The prediction was then used (via the kinematics) to move the head to a position which nulled off the predicted retinal error. The predictor has no access to the actual target trajectory but must estimate it in the context of its own saccades and the measured retinal errors. Four independent filters are used, one to track each of the retinal coordinates of the target in the left and right images.

2. Absolute mode: motor state prediction using fixation error feedback. The filter was used to generate predictions of the future motor states which would foveate the target. The predictor has access to the absolute motor states at which the image was taken, and the error measured in retinal coordinates is converted via the inverse kinematics to motor commands. Three filters are required: one for each of the verge motors and the other to control the tilt.

3. Image capture modes: serial and pipeline. The above tracking task can be broken into the following four stages: a) image capture; b) image processing; c) prediction and inverse kinematics; and d) head motion.

Pipelining is a form of parallelism which is appropriate when a repetitive activity consists of a sequence of stages. The strategy is to overlap the processing of the stages so that while stage n is being processed, stages n-1 and n-2 etc of successive instances of the action are processed concurrently. Figure 3 shows how it is possible to pipeline the components of the microsaccadic tracking task.

The advantage of pipelining is clear: it increases through-put of a processing stream. Here, the important difference from serial processing is that in pipeline mode the target-locked image sampling frequency is maximised. Furthermore, while maintaining the same sampling frequency or image capture rate, it is possible to treble the amount of time available for the image processing and inverse kinematic stages. Also, because the number of head motion stages has doubled, the maximum target velocity can be increased proportionately. From this it follows that a pipeline tracker is much less vulnerable to temporal noise than a tracker operating in serial mode. There is some potential for oscillation because the sequence involves a two-step lag but this danger is reduced by using the lattice filter to generate 2-step ahead predictions. This stabilises the system and reduces the tracking errors. Figure 4 shows the effect of using the filter to model the trajectory and the advantages over a simple non-predictive pipelined tracker in terms of the off-fovea retinal error.

4 Conclusions

This study has shown several attractive features of the lattice predictor as a component of an architecture for microsaccadic tracking. i) The order of the lattice predictor can be changed by simply adding on or taking off stages, making it easy to adapt to changes in the complexity of the input signal process. ii) The lattice predictor is capable of providing robust several-step-ahead predictions. These may be used to bridge sequences of missing data and the gap produced by sensor action delays. iii) It is robust to discontinuities in the target trajectory. iv) The lattice filter implementation is extremely economical and computationally efficient. v) It plays an important stabilising role in pipelining.

Pipelining and the resulting maintenance of the maximum image sampling frequency is potentially important, perhaps less for its effects on the tracking performance per se, but because a pipelined tracker can support other concurrently operating vision processes which themselves require high sampling frequencies with limited temporal noise (eg modelling object deformations as a time series, or building a model of the target trajectory in world geometry coordinates in order to evaluate the risk of collision). That the tracking competence can be subsumed by other vision processes is an important consideration for the long term development and evolution of the system.

5 Acknowledgements

Dr. Y. Zheng is supported by a grant from RSRE awarded to Prof. Frisby, Prof. Billings and Prof. Mayhew. The authors wish to acknowledge their debt to all the members of AIVRU.

References

[1] Alexander, S. T., (1986) **Adaptive Signal Processing**, Springer-Varlag, New York.

[2] Carpenter, R. H. S., (1988) **Movements of the Eyes**. Pion, London.

[3] Dean, P., Mayhew, J. E. W., Thacker, N., and Langdon, P. M. (1991), Saccade control in a simulated robot camera-head system: neural net architectures for efficient learning of inverse kinematics. Biological Cybernetics, 66, 27-36.

[4] Goodwin, G. C., Sin, K. S., (1984) **Adaptive Filtering, Prediction and Control**, Prentice-Hall, New Jersey.

[5] Kawato, M., (1989) Neural network models for formation and control of multijoint arm trajectory. In: Ito M. (ed) Neural programming. Taniguchi Symposia on Brain Sciences No 12. Japan Scientific Society Press/Karger, Basel, pp 189-201.

[6] Makhoul, J., (April, 1975) Linear Prediction: A Tutorial Review, Proc. IEEE, Vol. 63, No.4.

[7] Mayhew, J. E. W., (1992) ANIT: Architecture for navigation and intelligent tracking (in preparation).

[8] Mayhew, J. E. W., Dean, P., Langdon, P., (1992) Artificial neural networks for the kinematic control of a stereo camera head (in preparation).

[9] Widrow, B. and Stearns, S. D., (1985) **Adaptive Signal Processing**, Prentice-Hall, New Jersey.

[10] Zheng, Y., Mayhew, J. E. W., Billings, S. A., and Frisby, J. P., (1991) Lattice predictor for 3D vision and intelligent tracking. AIVRU Memo No 67.

LAYERED ARCHITECTURE FOR SACCADE CONTROL

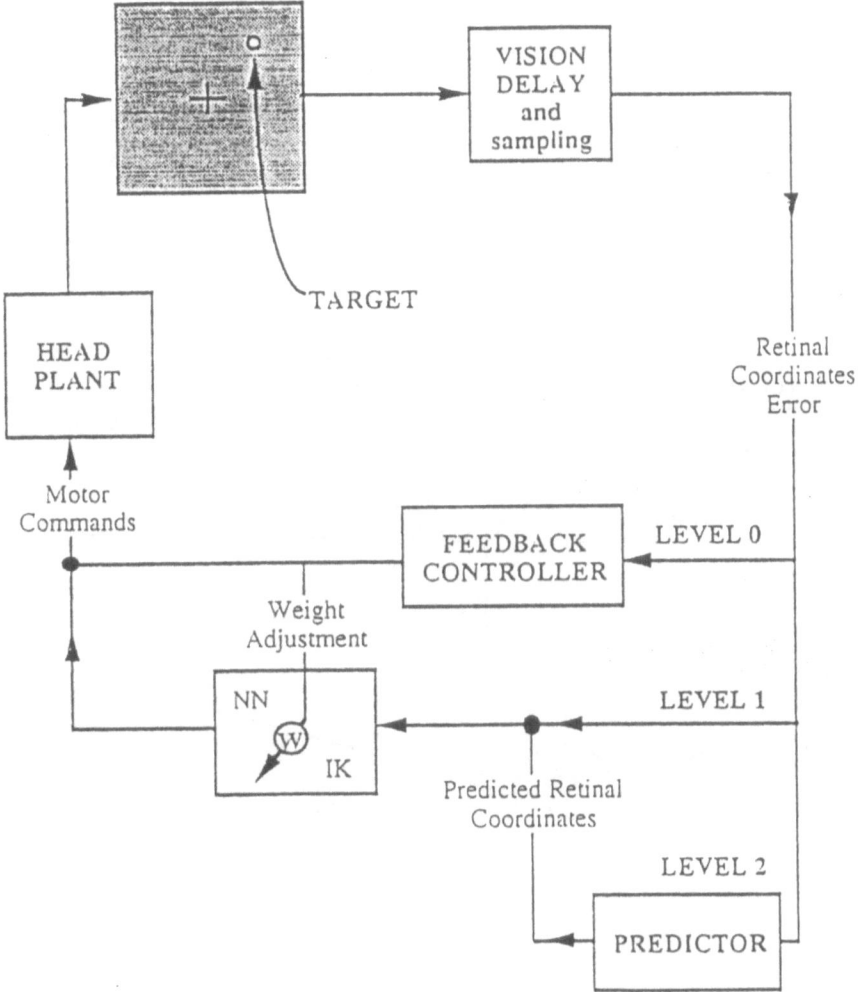

Figure 1. The philosophy of subsumption applied to a perceptual-motor task: three layers of competences for a microsaccadic tracking system able to maintain zero fixation error. Level-0 is the basic competence, a proportional feedback controller providing a stable starting point that is enhanced by the addition of two further layers of visuo-motor competence. Level-1 subsumes the Level-0 competence, and improves it to provide a single saccade to achieve fixation of a stationary target. Level-2 subsumes both the lower level competences and augments them by providing zero fixation under conditions when the target is moving. See text for details.

LATTICE FILTER

$$y_{t+1} = a_0 y_t + a_1 y_{t-1} + a_2 y_{t-2} \ldots \text{etc.}$$

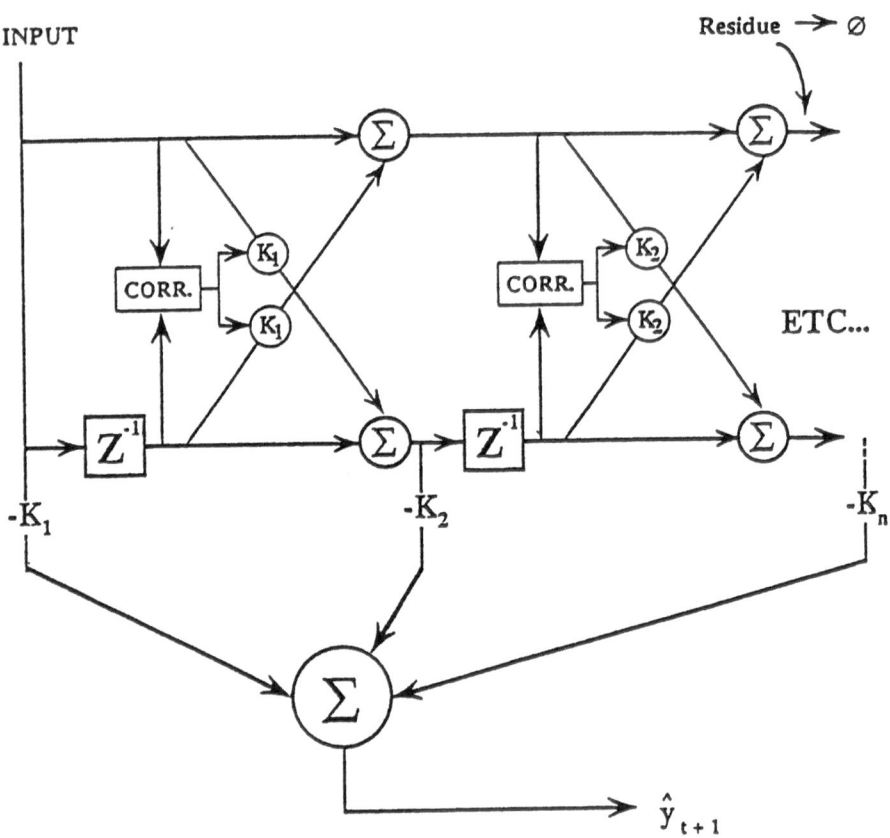

Figure 2. The lattice filter implementation of a transversal filter. The filter uses the current and past observations to form a prediction of the future output. It makes no assumption about the signal being linear and finite dimensional, it simply uses an auto-regressive model for the structure of the filter (which may not be optimal) and chooses the coefficients to minimise the mean square prediction error. K_1, K_2 etc are the reflection or partial correlation coefficients computed between the top and bottom 'delay' lines. (K_1 is generally negative, and initialised to -1 to give rapid convergence). Z^{-1} is a delay operator. Successive stages are delayed by increments of the sampling interval. The one-step-ahead prediction is given by summing the negated reflection coefficients K_1, $K_2 \ldots K_n$.

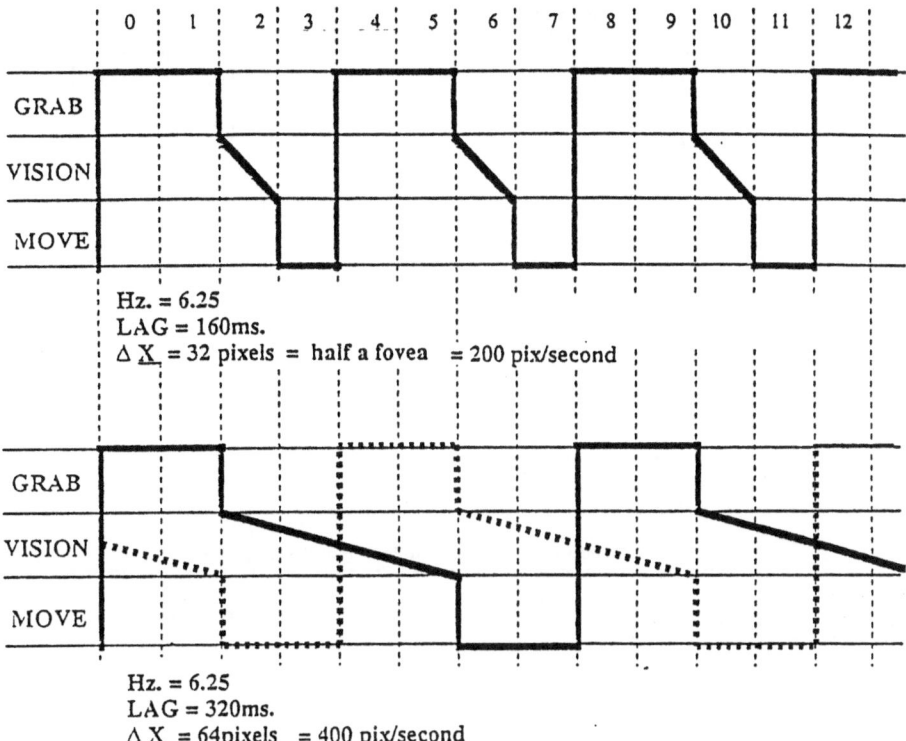

Figure 3. Serial and pipeline modes for micro saccadic tracking. Time is shown left-to-right quantised into 12 chunks, each of 40ms (set by the image frame grabbing rate). The vertical axis depicts the image capture (GRAB), image processing (VISION) and head movement (MOVE) stages. For clarity of exposition, it is assumed that the target velocity is constant and, at the maximum consonant with the time allotted to the head movement stage, is sufficient to maintain tracking without 'slipping' a frame. To minimise blur, the image capture stage is triggered by completion of the head movement stage. In the upper half of the figure (serial mode), the thick line shows which stage is in operation at any given time. In the lower half (pipeline mode), the thick line and the dotted line show the simultaneous operation of different stages.

a) Serial mode: The maximum sampling rate is 6.25 Hz, the samples have a 160 ms lag, and at a maximum head velocity of 50 degrees a second the retinal target velocity is 200 pixels a second. This is equivalent to a displacement across the image of half the 'foveal ROI' per sample. The critical feature is that to maintain this rate the demand on visual processing stage is maintained at 40 ms. This provides serious constraints on both the size of the region that can be processed and the complexity of the algorithms that can be used to support the tracking.

b) Pipeline mode: The sampling rate is maintained at 6.25 Hz. The samples lag by 320 ms, and the maximum allowed average velocity is 400 pixels a second. The feature is that if the lag does not cause instability (a function of the complexity of the target trajectory), the pipelining mode increases the time allowed for image processing by a factor of four. This provides an important buffer for maintaining the temporal stability of the tracking sample frequency.

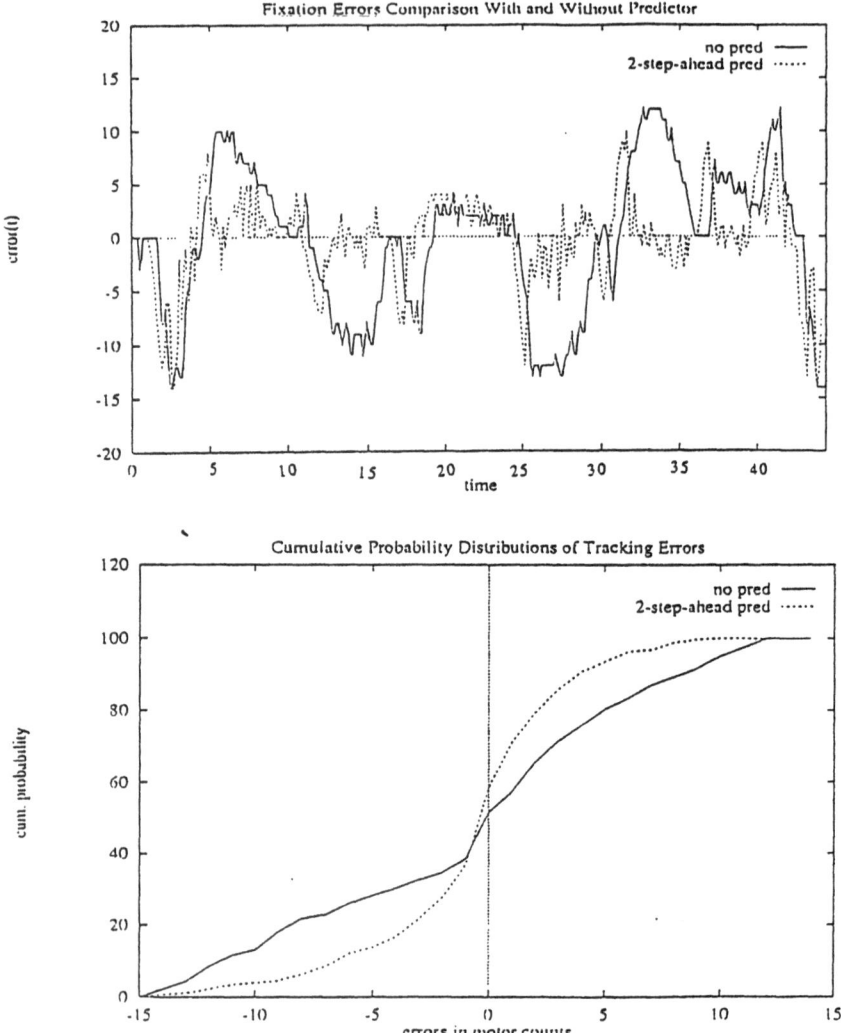

Figure 4. The improvement in microsaccadic tracking performance provided by a 4-stage, 2-step-ahead lattice predictor (absolute mode using vision error feedback) compared with a non-predictive tracker, both running in pipelining mode. The target was a small light source moved by a robot arm over the ground plane in front of the vehicle. (a) Fixation errors (represented as left verge motor counts). (b) Normalised cumulative frequency distribution of the fixation errors. The important point to be noted is the rapidity with which the errors of the predictive tracker quickly return to near-zero after major error excursions caused by a sudden change in the direction of target trajectory. The predictor rapidly learns the trajectory, the non-predictive tracker is one step behind. The difference between the two modes is statistically very marked as can be seen from (b). The rms errors are: predictor 4.16, non-predictive tracker 6.84. Each motor count corresponds to about 7.7 min visual arc so the residual tracking errors of the predictor are very small and roughly of the same order as human vision (Carpenter, 1988, p.125).

Image Tracking in Real-Time: a Transputer Emulation of some Early Mammalian Vision Processes

P.H.Welch and D.C.Wood

Computing Laboratory, University of Kent at Canterbury,
Canterbury, Kent, England

Abstract

By emulating some of the 'early' vision processes believed to occur in the visual cortex of mammals, dramatic savings can be made in the computational efforts usually associated with image processing. This paper describes the design and implementation of a low-cost visual tracking system that uses no special signal processing, vector processing nor floating-point hardware. Its effectiveness relies only on the parallel replication of fast scaler integer processors with low latency communications. The current implementation uses 10 T425 transputers, processes 544 × 544 byte-pixel images at camera frame rates and provides 100 Hz. tracking feedback signals to the camera pan-and-tilt motors with a latency of no more than one-fifth of a second. The system will track any object that occupies the (majority of the) centre of the field of view moving against any background — no special lighting conditions or artificial scenery are required. The design and implementation follow naturally parallel structures and are expressed at all levels in *occam*.

1   Introduction

Biological systems have evolved highly efficient mechanisms for the accurate tracking of moving objects. Rather than wielding a computational sledge-hammer, it is a good idea for computer systems designers — charged with solving the same problem — to look at nature and see what can be 'borrowed'.

The algorithms and data-structures employed within KITTEN (the Kent Intelligent Target Tracker) exploit the following features of mammalian vision:-

- *the fovea:* light receptors are concentrated at the centre of the retina;

- *adaptation:* an individual receptor quickly stops firing if the light level falling upon it remains constant;

- *rapid eye movement:* the eye never remains still — there are always 'tremors', even when apparently at rest;

- *foveal tracking:* only objects maintained in the centre of the field-of-view are tracked — we cannot track objects using peripheral vision;

- *velocity modelling:* there is some physiological evidence [1] that suggests that eye movement during object tracking is controlled by modelling eye velocities rather than absolute eye position.

The differing densities of retinal light receptors from the fovea (the highest) to the periphery (the lowest) allow the same scene to be processed with a range of views and resolutions. Each view — from the narrow angled, but highly resolved, foveal patch to the wide angled, but low resolution, full scene — demands a similar quantity of memory and processing power. Each view offers differing, but complementary, information to higher level tasks (such as tracking). All these views can be processed in parallel.

The adaptation of retinal light receptors to constant stimuli mean that they are naturally sensitive to moving objects — this allows moving objects to be detected and 'acquired' (i.e. the eye is moved to locate the detected object on its fovea). During tracking, this same effect enables objects that change their (angular) velocities, with respect to the eye, to be detected and for correcting signals to be sent to the eye muscles.

The rapid movement (or 'dithering') of an apparently resting eye continuously shakes the image falling on its retina. If this did not happen, the adaptation effects. would cause a stationary eye to become blind. The effects of adaptation and dithering need to be well-balanced!

The eye can only track objects if they can be kept 'locked' in the fovea. During tracking, the tracked objects will appear to be stationary, but the background will be moving. If the tracking response mechanisms for the eye operated across the whole scene, correction signals generated by the moving background (usually the majority of the scene) would quickly bring the eye to a halt. Tracking would only be possible on objects that occupied the majority of the whole scene or on smaller objects moving across a uniform background. Clearly, the eye does much better than this!

Finally, there is some debate as to whether a tracking eye is controlled by sending it change of velocity information (i.e. object accelerations) or change of position information (i.e. object velocities). The latter would imply that the eye needs to be repeatedly brought to rest momentarily, whilst the continuing velocity of the tracked object was observed. The former implies that observations can continue 'on-the-fly'. Because the motors in the pan-and-tilt unit on which our camera is mounted have considerably greater inertia than the muscles that control eyeball (or head) movement, we choose the former model. It also seems more simple and elegant.

2　Top-Level Design

Figure 1 shows the basic feedback control loop for the tracker. Images acquired from the camera are processed into control signals for operating the pan-and-tilt motors on which the camera is mounted. Objects moving in the centre of the field of view cause the camera to move to try and follow it — this, of course, impacts back upon the image acquired. Clearly, the lower the time around this feedback loop, the better will be the quality of the tracking response.

During tracking, the movements of the target and the angular velocities of the camera pan-and-tilt unit need to be reasonably matched. Any discrepancies — due to errors in processing or real motion changes by the target — cause deviation of the target from the centre of the acquired image. These deviations must be measured in real-time and correction signals returned to the pan-and-tilt motors to bring the target image back on track.

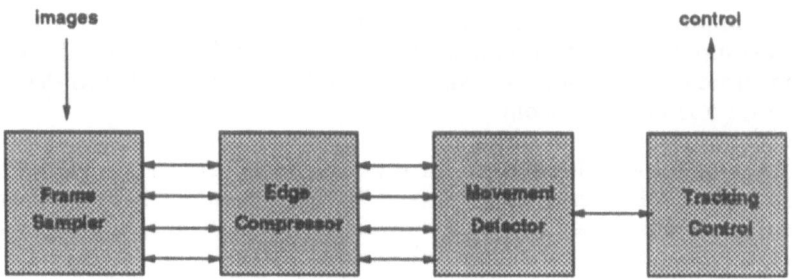

images control

Frame Sampler | Edge Compressor | Movement Detector | Tracking Control

Figure 1: Image Processing Pipeline

Figure 1 also shows the top-level processing structure — a four-stage pipeline. Currently, the `Frame Sampler` and `Tracking Control` use one transputer each, whilst the `Edge Compressor` and `Movement Detector` both use four. The `images` input represents frames generated by the camera at 25 cycles per second. The `control` outputs are digital/analogue signals for operating the pan-and-tilt motors — these are generated at 100 cycles per second. The internal channels represent *transputer* links.

Information flows continuously from `images` through to `control` feedback. Occasional control message flow in the reverse direction — these are to change the mode of operation of certain processes and for the fine-tuning of various parameters.

3   Frame Sampling and Dithering

Our *transputer*-controlled frame-grabber delivers one 544 × 544 8-bit greyscale image into video-RAM every 40 milli-seconds (i.e. 25 frames/second). This represents a data bandwidth of about 7 Mbytes/second, whereas the capacity of the four links leaving this transputer (and routed, in our prototype system, through electronic switches) is only about 5 Mbytes/second.

Each frame-grabbed 544 × 544 image, therefore, is reduced to four 68 × 68 images called `layer[0]` through to `layer[3]`:-

- `layer[0]` is just the middle 68 × 68 square of the original image. It corresponds to the 'foveal patch' on the retina, where image sampling density is greatest;

- `layer[1]` is the middle 136 × 136 square of the original image sampled every other pixel in each dimension — i.e. from each 2 × 2 square in this patch, one pixel value is chosen;

- `layer[2]` is the middle 272 × 272 square sampled every fourth pixel — i.e. from each 4 × 4 square, one pixel is selected;

- `layer[3]` is the original 544 × 544 image sampled every eighth pixel. This corresponds to the 'widest angle' view of the scene, but has the lowest resolution.

This sampling pattern was previously described in [2], where it was used to provide input to the 'Self-Similar Stack' — see also [3, 4, 5, 6]. Using the language from this stack model of vision, we shall refer to layer[0] as the 'top' and to layer[3] as the 'bottom'.

Rapid eye movement (or dithering of the sampled image) is a by-product of the pixel sampling algorithm used. On layer[1], layer[2] and layer[3], when pixels are selected from each (respectively 2×2, 4×4 and 8×8) sub-square, a different location is chosen in successive frames. The dithering sequence is not a straight row-by-row scan of each sub-square. Instead, the sequence jumps around the square, choosing for each successive location a pixel from an area not recently visited. On layer[1], the sampling sequence for each 2×2 sub-square (of the 136×136 inner square) is given by Figure 2(a):-

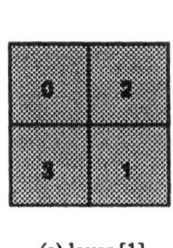

(a) layer [1]

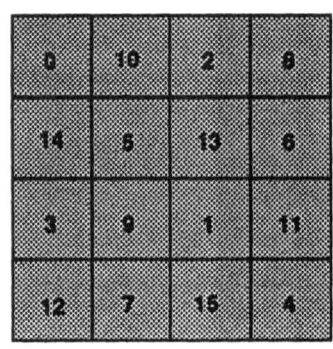

(b) layer [2]

Figure 2: Dithered Sampling Sequences

The layer[1] sequence is used recursively to generate the rather longer sequence for layer[2] (Figure 2(b)) and layer[3] (not shown). The aim of this dithering is to ensure that a 'good' representation of each sub-square is chosen over the last n samplings — for any value of n.

Viewing the result of this dithered sampling in real-time (i.e. 25 frames/sec.), the images from the lower layers (i.e. layer[1], layer[2] and layer[3]) show gradually increasing amounts of 'tremor'. These tremoring effects are very useful for the next stage in processing — edge extraction.

The reduced images from each layer are output down the corresponding link. At 25 images/second, each link carries approximately 100 Kbytes/second — well within its capacity.

4 Edge Compression

Before movement detection is attempted, the images are reduced still further by the removal of texture information. This edge-filtering operation needs to be computationally light in order to cycle at camera frame rates.

In [5, 6], the implementation of algorithms based on the Self-Similar Stack, Hierarchical Discrete Correlation and Difference of Gaussians [2, 3, 4] are described. Using a pipe-line of 3 T800 transputers per level (i.e. floating-point arithmetic is necessary), the necessary computational bandwidth can be achieved — but at some cost in latency (about 200 milli-seconds for just this stage).

In KITTEN, a much lighter approach is used. Instead of trying to extract edges from each image separately, we make use of the fact that they form part of a moving sequence. Following (loosely) the adaptation characteristics of retinal light receptors, successive images are simply differenced (absolutely) and then thresholded. Parts of the image that have not changed simply disappear. Objects that have *moved* (slightly) are reduced to just their edges with varying levels of thickness (from greatest along the axis perpendicular to the direction of movement, down to zero along the axis parallel to the direction of movement). These filtering effects work well for natural objects regardless of their orientation or direction of movement — i.e. we are not restricted to just the major points of the compass.

Because of the dithered image sampling, all *stationary* images in the bottom three layers have an apparent tremor. This is sufficient for the differencing computations to yield their edges. If the system were perfectly locked on to a (constant velocity) *moving* object, its image would be stationary — this way, we can still see it!

Images from each layer are processed in parallel by separate *transputers*. The thresholding level that determines whether the difference between sampled pixel values (in consecutive frames) are sufficiently different may be adjusted dynamically by (backwards flowing) control signals — these are 'higher level' decisions.

After thresholding, the 68×68 byte-pixel images have only binary values — 'edge' or 'not-an-edge'. These are now compressed into 68×68 bit-pixel images. Each row is squashed on to three (32-bit) integers with a generous overlap.

The output image from each layer has now been reduced to just 816 bytes, lowering the data-flow on each link to only 20 Kbytes/second.

5 Movement Detection

The method for extracting movement information from the sequence of images is the same as that reported in [2, 6]. However, with the bit-compressed representation of edge-only images now available, the required computational effort is more than one order of magnitude less.

For tracking purposes, movement detection should only be sought within the 'foveal patch' of each image. Four streams of images, layer[0] through layer[3], are output from the Edge Compressor. The whole of each (68×68) layer[0] image represents the foveal patch from the original (544×544) image captured by the camera. On layer[1], this foveal patch is only the inner 34×34 square. On layer[2] and layer[3], respectively, the foveal patches reduce to the inner 17×17 and 9×9 squares.

Because of concern that the amount of information available in these fovea might be too low to yield accurate tracking information, the KITTEN prototype arbitrarily defines an extended fovea to include the whole (68×68) layer[1] image. Thus, on layer[2] and layer[3], the actual foveal patches searched are the inner 34×34 and 17×17 squares.

Movement detection operates independently (and in parallel) on each of the four layers. For each frame at time t, two copies shifted respectively left and right by one pixel are made. These three images (the original plus the two copies) are compared against three other images: the frame at time $(t - 1)$ plus two copies shifted (respectively) up and down by one pixel.

In the *occam* KITTEN implementation, the horizontal shifts are achieved by logical one-bit shifts. The duplicated boundary regions in the compressed bit-mapping for each line mean that no bits need to be shifted between words. Vertical shifts are implemented by *occam* abbreviations. The comparisons are (one-cycle) *exclusive-or* operations followed by full-word BITCOUNT instructions. The extra-foveal regions are excluded by logical *and*-masks (horizontally) and by suitable index boundaries (vertically). The *and*-masks also exclude unwanted (horizontal) duplicates. All this is easy and efficient to express in *occam*, with full anti-alias checking and no assembler inserts.

A 'significant' minimum from the 3×3 comparisons between successive frames indicates the movement of a 'significant' object. A *one*-pixel movement (up, down, left or right) detected on layer i corresponds to a 2^i-pixel movement in the original camera image. Thus, the different layers of image processing respond to different scales of movement.

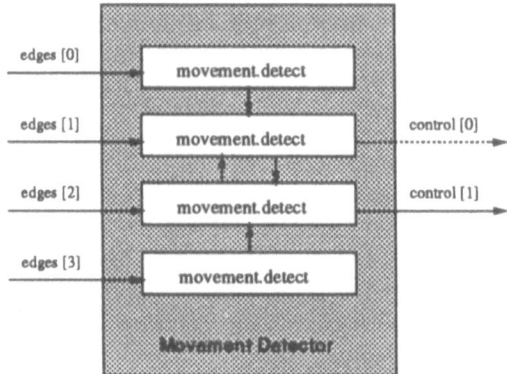

Figure 3: Movement Detection Architecture

The movement detectors for each layer are handled by separate transputers — see Figure 3. Because we only restrict our attention to the foveal region in each image, layer[0] and layer[1] have the same work-load, but the duties become increasingly lighter lower down the stack. To make up for this, layer[2] and layer[3] are programmed to conduct wider searches for movement. In fact, layer[3] has time to shift each foveal image by (plus and minus) three pixels in each axis and compute 7×7 comparisons.

It is important not to accept just any minimum from the computed comparison matrix at each layer. A minimum sufficiently close to the average may just be the result of 'noise' in the images and not indicative of any real object movement. The test for ' significance' adopted is only to accept a minimum that is more than k standard deviations below the average for the matrix, where k is a 'tuning' parameter set by experiment. During tracking, this threshold needs to be highest for layer[3] (which has the least amount of information from which to compute movements). Again, Figure 3 has been simplified by omitting the reverse-flowing control channels that are used to adjust k (or pass back other control signals destined for earlier stages in the pipeline).

The movements detected by layer[0] (± 1 pixel) are passed down to layer[1]. The movements from layer[3] (± 8, ± 16 and ± 24 pixels) are passed up to layer[2]. Layer[1] (detecting ± 2 pixels) and layer[2] (detecting ± 4 pixels) then exchange all their information, select the largest absolute movements horizontally and vertically and output these. These outputs are, of course, identical. One of these is connected to the next stage in the pipeline (Tracking Control — see Figure 1). The other (dashed in Figure 3 and not shown in Figure 1) is connected in our prototype system to a display manager process (so we can see what the system is doing!).

6 Tracking Control

Tracking is implemented using just one *transputer* that manages the hardware interface to the camera control system. DACs and ADCs (for feedback information) are memory mapped on to *occam* ports that enable absolute positioning of pan and tilt settings. Camera zoom, focus and gain may also be adjusted.

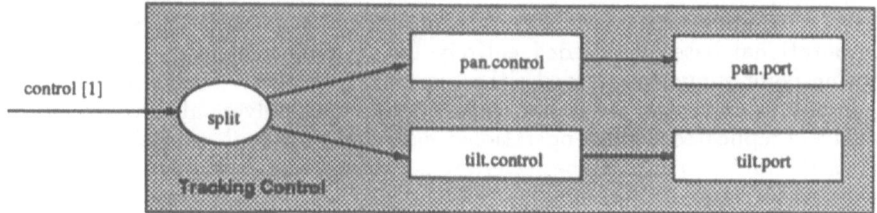

Figure 4: Tracking Architecture

Figure 4 shows the software architecture that controls the pan-and-tilt motors. The split process de-multiplexes the incoming movement (or initial position) information on to separate pan-and-tilt channels. The pan.control and tilt.control processes are identical (modulo some 'gearing' tables), as are the pan.port and tilt.port processes (modulo the actual DAC/ADC ports).

The gearing ratio depends on the level of zoom applied to the camera lens — the greater the zoom, the smaller must be the camera movement that corresponds to a detected object movement (measured in pixel displacements). The movement spans of our pan and tilt motors are not identical — i.e. the pan and tilt gearings need to be different.

An estimate is maintained of the current (angular) pan-and-tilt velocities of the tracked object. Change reports, arriving 25 times per second from the Movement Detector, are treated as accelerations and integrated to yield current velocities. These are, in turn, integrated once more to produce absolute pan-and-tilt coordinates that are output to the DACs. This final integration cycles at 100 outputs per second because this gives smoother pan-and-tilt motion, especially for slower velocities.

An object whose velocity has been correctly estimated and which does not accelerate or brake will be held in the centre of the field of view (the fovea). So long as this object occupies the majority of the fovea, the movement detectors will report zero accelerations (i.e. the moving background will be ignored) and tracking will be maintained. If the estimate were too high or the object brakes, negative accelerations will arrive to correct the tracking. If the estimate were too low or the object accelerates, the tracking will again be suitably modified. Pan and tilt movements are tracked independently and in parallel.

The accelerations reported by the movement detectors are 'noisy' approximations to the true changes in velocity of the object. The control processes only receive discrete values from the set $\{0, \pm1, \pm2, \pm4, \pm8, \pm16, \pm24\}$ and a percentage of these will be completely wrong! When an incorrect value is processed, the camera starts to move away from the tracked object. The image of this object starts to move in the fovea and this generates correcting signals that are fed back to Tracking Control. Depending on the latency within the loop and the percentage of erroneous movement detections, this feed-back control can maintain tracking despite the apparent paucity of information. In KITTEN, the latency and error rate are sufficiently low for the tracking to work well — see [7].

7 Discussion — a Lesson From Biology

KITTEN demonstrates a practical real-time image processing application whose computational base is provided entirely by a small *transputer* system. No floating-point support is needed either from DSP or vector-processing engines — or even from T800s! The bulk of the computation involves only low-level, but massively replicated, logical operations: *addition, greater-than, and, exclusive-or, shift* and *bit-count*. Higher-level integer arithmetic, like *multiplication*, is relatively infrequent.

The design follows physiological models evident within the mammalian retina and visual cortex. The data from each captured image is (intelligently) reduced to about 1% of its original volume, before being analysed for movement. The reduction method exploits the fact that the images form a moving sequence and does not attempt to operate on individual frames. The detected movements are fairly 'noisy', but the overall feedback-loop ensures that errors are self-correcting and allows (reasonable) tracking to be maintained.

It is interesting to note how certain characteristics (such as rapid eye *dithering*), that appear at first glance to be physiological 'errors' that we should not bother to emulate, turn out to play a useful role. *Dithering* allows a cheap way for extracting edge information from stationary objects (e.g. it enables us to 'see' the object being tracked). It also provides a more complete response from naturally-shaped moving objects (by showing up edges even if they are parallel to the direction of movement). When tracking a target that has stopped moving, *dithering* helps keep the camera 'locked on' (because having the edge-information on all objects continuously available means that the correction response to dampen out erroneously reported movements — or deliberate 'feints' — can be much faster).

This paper describes only the *tracking* algorithm used within KITTEN. All the control processes actually operate in two modes — *passive* and *tracking*. In *passive* mode, the camera is held still and movement is sought from all parts of the images — including peripheral fields of view. Once a sufficiently large movement is located, the camera is moved to 'acquire' the target on its fovea and the system switches to *tracking* mode. Further details on these mode switches may be found in [8].

The 'intelligence' within KITTEN is at a fairly low level. The system is, after all, only emulating 'early' vision processing within the visual cortex. Tracking may well benefit from further correction signals fed back from 'higher-level thoughts' (e.g. for locating the tracked object more centrally in the fovea, choosing which of several moving objects to track, deciding when to lose interest in a tracked object, ...).

Clearly, the edge-filtered images could be delivered to other processing streams (e.g. for object recognition), that run on other *transputers* in parallel with this tracker pipline. In fact, the KITTEN prototype already delivers these images to a graphics processor with no degradation to its tracking capability. The message here is that extra functionality can be grafted on to the system by applying extra processors, without weakening the performance of its original task. Indeed, a response such as tracking may well benefit from multiple sources of feedback control — even if these signals are sometimes contradictory. Of course, the highest-level image processing system available to us is human. KITTEN could be made to interact well with such an operator.

At present, the implementation of the KITTEN prototype is largely unoptimised. We are using several — possibly 4 — more *transputers* than will finally be needed just to manage the tracking response! The prototype is heavily instrumented with information gathering (and filtering) processes that allow us to observe in real-time what is really happening inside the system. We also have on-line manual control of various system parameters (e.g. edge-detection thresholds, minimum-significance thresholds, gearing ratios, ...) to allow us to determine their best settings. Because we did not want such instrumentation to affect unduly the real-time performance of the system, some spare processing capacity was reserved in each processor. This instrumentation will not be needed in production versions of KITTEN.

The availability of T9000 *transputers* will have a considerable impact on the performance and economics of KITTEN. Its current capabilities could be performed by just two processors — one of which would require no external memory! Because of the physically shortened pipeline, latency would automatically be (at least) halved. Alternatively, we could revert to six processors, possibly five of which would need no external memory, and go for higher quality tracking (by processing more information to achieve lower error rates and detect higher accelerations of the target).

The aim is to produce a KITTEN system with tracking abilities that match those of a real kitten! At present, KITTEN will follow a person walking at a distance of about ten feet from the camera — or a coffee cup at a distance of three feet (moving with similar angular velocities/accelerations to the walker) [7]. In both cases, the camera zoom level is set to render the size of the tracked object to be about one fifth of the camera image.

Applications for such tracking systems are widespread. Potential large markets include (semi-)intelligent TV cameras, automobile guidance systems, surveillance cameras, docking control for in-flight refuelling, general target interception, video telephones and novel man-machine interfaces.

An important feature of all these image-tracking applications is that, given an external light source, they are all *passive* systems. Neither the tracker nor the object being tracked needs to be equipped with active radiation emitters (e.g. light, heat, radio, radar, sound, ...). This considerably enhances their practicality and security and lowers their costs.

8   Acknowledgements

The work reported in this paper complements research into Active Vision Systems by the University of Kent at Canterbury (Computing Laboratory) and the Defence Research Agency (RARDE, Fort Halstead, Kent). The basic notions of layered image sampling and motion detection may be found in [2, 3, 4]. with direct implementation details in [5, 6]. The 'short-cutting' ideas of dithered sampling, difference edge-filtering, bit-compression and velocity (rather than position) tracking are outside the standard stack model of vision. Further details may be found in [8].

We are especially grateful to Nigel Haig, Ian Moorhead and Richard Clement (from RARDE) and to Andrew Smith, Vedat Demiralp, Colin Willcock, Gordon Makinson and John Southall (from UKC) for countless lengthy debates about all these matters. A recent devious idea to use the *transputers*'s '2-D block move' instruction to speed up dithered image sampling (and, thereby, reduce feedback latency) is due to Tony Debling of INMOS — to whom thanks!

The prototype KITTEN system has been developed and implemented on the MEiKO Computing Surface at the University of Kent.

References

[1] R.H.S.Carpenter, **Movements of the Eyes**; published by Pion Ltd., London; ISBN 0-85086-109-8; 1988.

[2] P.J.Burt, **The Laplacian Pyramid as a Compact Image Code**; IEEE Transactions on Communications, pp. 532-540; 1983.

[3] G.J.Burton, N.D.Haig and I.R.Moorhead, **A Self-Similar Stack Model for Human and Machine Vision**; in Biological Cybernetics (53), pp. 397-403; 1986.

[4] I.R.Moorhead and N.D.Haig, **A Stack Model of Vision with some Pre-attentive Properties**; in 'Conference Proceedings of MARI 1987'; AGPB, Paris; 1987.

[5] A.B.Smith and P.H.Welch, **Real-Time Transputer Models of Low-Level Primate Vision**; in Proc. of the 11th OUG Tech. Conf., Univ. of Edinburgh, Scotland; edited by J.Wexler; pp.171-181; IOS Press, the Netherlands; ISBN 90 5199 11 1; Sept., 1989.

[6] A.B.Smith and P.H.Welch, **A Transputer Based Active Vision System**; in Proc. of the 15th WoTUG Technical Conference, University of Aberdeen, Scotland; edited by A.Allen; pp.112-121; IOS Press; ISBN 90 5199 085 5; April, 1992.

[7] P.H.Welch, **Silicon Retina II: a Foveal Tracking System (KITTEN)**; video (running time 25 minutes); the University of Kent; October, 1991.

[8] P.H.Welch and D.C.Wood, **KITTEN — A Foveal Image Tracker**; in 'Image Processing and Transputers'; edited by H.Webber, IOS Press; July, 1992.

A CURVATURE SENSITIVE FILTER AND ITS APPLICATION IN MICROFOSSIL IMAGE·CHARACTERISATION

J. P. Oakley and R. T. Shann

Department of Electrical Engineering
University of Manchester
Manchester M13 9PL U.K.

Abstract

A new class of oriented, curvature sensitive filters are introduced. These filters provide a low–level detection facility for noisy curves without a prior edge extraction stage. The application of these filters to the detection of Carboniferous Foraminifers (a type of microfossil found in plane rock sections) is described. A symbolic representation of the detected curves is stored in a database which is then queried to recover the required structures. We show that the curves identified by the filter correspond to salient features of the microfossil evidence in the image.

1 Introduction

We report here on a research project on techniques for content addressing of image databases. The aim is to provide the equivalent of "free–text" document retrieval for real image data. This is achieved by extracting some representation of image content using image analysis and then storing this information in a database together with the raw image data. The user interacts with the system to define a query set of interesting images which can be inspected on–screen.

One important application which is under investigation for this system is the dating of rock samples using microfossil evidence (a procedure known as biostratigraphy). Work has been reported on automatic or semi–automatic biostratigraphy of 3–D microfossil samples [1] but the systems described rely on graphical input from the user rather than image analysis. Here we are looking at a different type of microfossil (the Carboniferous Foraminifera), which because their hardness is similar to that of the surrounding rock can only be viewed as 2–D sections from thin sections. A reasonably clear example of this type of microfossil is shown in figure 2 and a less clear example in figure 3. It can be seen that the actual microfossils can be distinguished by eye from the surrounding rock because of a pronounced geometrical structure. The level of visual noise is quite high, and this factor is inherent in the nature of the material, not an artifact of the imaging technique.

We have looked at the Canny–type edge maps for these images but the amount of data is huge and very few of the edge segments relate to significant object features. This has lead us to look towards robust methods for detecting noisy curves, which are the predominant visual features in this application.

Low level schemes for curve detection have been reported [2] [3] but the work seems to have been motivated by physiological considerations rather than by a need for robust curve detection. We feel that any scheme which relies on bottom–up grouping of edge information is doomed to failure on images containing this level of visual noise. Instead we take a top–down approach and fit a local model of a curved boundary directly to the grey–level image itself. This is achieved by a process based on linear filtering of the image data with a specialised filter.

This filter is a variation of that reported in [4] as a method for precise localisation of curved boundaries, although the actual problem addressed here is quite different.

In an effort to quantify the performance of these filters we will show, using actual microfossil images:

(a) In section 3.1, how the detection performance of the operator varies with σ_β (the filter arc length);

(b) In section 3.2, that the curvature correction (i.e. the Newton step in κ) is self consistent;

(c) In section 3.3, that the curvature of the filter has a strong influence on which curves are detected and that the curvature selectivity can be increased by placing a restriction on the size of the Newton step in κ; and

(d) In section 3.4, that a simple Hough–like centre binning algorithm provides a reliable method for detecting microfossil structures.

2 Technical Approach

2.1 The Filter

The basic approach is to firstly filter the image with a specialised linear filter and then to detect local intensity extrema (minima and maxima) in the filtered image. There are typically about 200 of these local extrema (which we call "hits") in each image; this is a much more manageable amount of data than an edge map. The hits are then subject to further selection based on:
(a) parametric information and
(b) intensity ranking.

The parametric information quantifies the behaviour of the hit as the filter parameters are varied (this information is obtained from partial derivatives).

The filters are highly direction sensitive and so the whole process is repeated for several different orientations.

The filter used for detecting curved edges is illustrated in fig (1). The filter kernel is generated starting from a simple Gaussian shape which is elongated and then curved and finally differentiated in the direction perpendicular to the elongation so as to give an curved edge sensitive detector. The elongated curved Gaussian $G(x,y)$ is the product of two terms:

$$e^{-\frac{1}{2}\left[\frac{((\kappa x-\cos\theta)^2 + (\kappa y-\sin\theta)^2)^{\frac{1}{2}} -1}{\sigma_\alpha \kappa}\right]^2}$$

and

$$e^{-\left[\frac{x^2+y^2}{2\sigma_\beta^2}\right]}$$

Here θ is the angle of the filter and κ is the curvature. The first of these terms decays (via σ_α) with distance from the circle centred on $\cos(\theta)/\kappa$, $\sin(\theta)/\kappa$. The second term decays more gradually (via σ_β) with distance from the filter centre. The directional derivative $K(x,y)$ is given in terms of the derivatives with respect to the spatial directions x and y by

$$K(x,y) = -\{ G_x \cos\theta + G_y \sin\theta \}$$

where the minus signs cause the sense of differentiation to be radially outwards at the filter origin. This is our filter kernel (see fig 1). Derivatives of this expression with respect to x,y and κ up to second order were calculated using the symbolic algebra package Macsyma. By the usual argument [5] the derivatives of the filtered image with respect to these parameters can be obtained by filtering the image with the differentiated kernels. The filtered images are computed at every point on the pixel grid using an FFT convolution routine. At this stage we could detect local maxima by examining adjacent pixels. However, this is not a wholly reliable method, marking too few or two many points at places where the image is changing rapidly. Instead by computing the Newton step (a term from optimisation theory –see for example [6]) we are able to locate extrema of the filtered image and get a sub–pixel estimate for the position. The Newton step in x,y is given by

$$(\delta x, \ \delta y) = -H^{-1} \quad (G_x, \ G_y)$$

where (G_x, G_y) is the vector of first derivatives (the gradient) and H is the matrix of second derivatives (the Hessian).

In the output of this filter, local maxima correspond to the edges of locally convex bright objects, while local minima correspond to the edges of locally convex dark objects. The Newton step in θ and κ can also be calculated and this gives an estimate for the optimum values for θ and κ.

For θ we used the step in (x,y,θ) space

$$(\delta x, \ \delta y, \ \delta \theta) = - \begin{bmatrix} G_{xx} & G_{xy} & G_{x\theta} \\ G_{xy} & G_{yy} & G_{y\theta} \\ G_{x\theta} & G_{y\theta} & G_{\theta\theta} \end{bmatrix}^{-1} (G_x, \ G_y, \ G_\theta)$$

and similarly for κ. It would be possible to compute a four dimensional Newton step in (x,y, θ, κ) space, but previous experience with this sort of filter has suggested there may be numerical problems doing that.

Obviously this estimate is short–range in that if the filter parameters are such that the filter is poorly matched to the visual object sought then the Newton estimate is unreliable.

Tests on phantom images of curved objects have been carried out and these experiments show the expected results and are not reported here. The key issue is the performance of the operators described when applied to noisy image data. This issue has been addressed by the experiments reported in section 3.

2.2 The Database

The database infrastructure for this project is a simple prototype triple store, along the lines of the Essex Sierra IFS simulator [7], with each triple representing a relationship between two entities. Lexical items are either encoded directly into the identifiers (eg integers, floats) or stored separately (strings). Unlike a relational database where the access programs need to know the names of tables and fields in the database in order to access the data, in the triple store the lowest level of storage is more uniform (a triple of indentifiers) and doesn't change when new relationships are introduced. The relational model prejudges (to some extent) the queries to be made (and in compensation offers efficient access), whereas the triple store allows the database schema to be represented in the database itself, so that it can easily evolve.

For the fossil database data items were computed and stored as follows: for a given orientation and curvature the filter output was computed, along with first and second derivatives with respect to position, angle and filter curvature. From these extrema of the filter output were computed together with estimations for the true values of the curvature and orientation of the curved edge.

Experiments can now be expressed as computations on the result of queries to the database: eg "what is the mean position of hits which have their centres within a given distance of a given point?". At present these queries are embedded in programs, but the plan is to implement them in a special-purpose language.

3 Experimental Results

An example of the 5 strongest hits from a single oriented filter is shown in figure 4. The curves are marked over a range $\pm 2\sigma_\beta$ and follow the centre line of the filter. In figure 5 we show the two strongest hits from each of 20 directions for the image in figure 2.

3.1 Choice of parameters

The best choice of σ_α and σ_β for a given domain depends on the noise in the image [8] which in this application precludes the use of very small values. The upper limit on σ_α is dictated by the proximity of clutter. The filter arc length parameter σ_β is more model–dependent: for example in most cases we are interested in longer curves of low curvature and shorter curves of high curvature, so σ_β should be chosen longer or shorter to match. Apart from this the expected amount of deviation from circularity will also influence σ_β. Table (1) shows the variation in the number of hits on the spiral fossil of image (fig 1) as σ_β is varied – in this case the best value is about 30 pixels. The next two tables (2,3) show the results for two rather more circular spiral fossil images extracted from published photographs. σ_α was kept constant at 4.4 pixels, which is about half the thickness of the fossil walls, and so avoids interference between hits arising from the inside and from the outside walls of the fossil.

Table 1 –   Fossil 1   Table 2 – Fossil 2

σ_β	hits on fossil	% total hits
11	16	8
22	20	10
29	22	11
45	20	10

σ_β	hits on fossil	% total hits
11	13	11
22	18	13
29	17	13.5
45	21	15

Table 3   –   Fossil 3

σ_β	hits on fossil	% total hits
11	10	5
22	14	7
29	15	7.5
45	15	7.5

3.2 Curvature correction

To get a feel for the extent to which we can predict the curvature of an edge from the Newton step in curvature we give three examples of points where filters of different radius have hits. (These points were chosen at random from the spiral image fig 1). The best matching filters (the consensus of the estimated radii) are marked * in tables 4 to 6 below.

Table 4 – point 1

filter radius	estimated radius of curve
40	55.23
50	47.62*
60	47.09*
70	44.28
80	3.39
90	89.35

Table 5 – point 2

filter radius	estimated radius of curve
40	30.71
50	-131.12
60	1050.50
70	109.33
80	85.19*
90	84.97*

Table 6 – point 3

filter radius	estimated radius of curve
40	24.42
50	71.1
60	60.3
70	58.7*
80	51.2*
90	26.8

3.3 Curvature Selectivity

To show curvature selectivity, six filters with a radius ranging from 40 to 90 pixels were applied to the image in fig 2. The twenty strongest hits from each of twenty directions were taken. If the spiral were perfectly smooth then as we travel out from the centre along the curve of the spiral we would expect to encounter hits from filters of a given radius only in one section of the spiral.

In practice with the fossil shown (fig 1) the curvature is quite irregular, but a relatively uncluttered part of the spiral occurs at a distance of 50 to 70 pixels from the fossil centre and plotting the density of hits in this area against filter radius reveals a peak at a filter radius of 60 to 70 pixels, shown as the upper line in figure 6.

This shows that the geometrical form of the filter introduces a selectivity or bias towards specific curvatures in the image.

The curvature selectivity of the filter may be increased by insisting that the detected curves should "match" the filter in the sense that the Newton step should be less than some given threshold. The lower trace of figure 6 shows the result of discarding hits whose Newton step in curvature is greater than 0.01 pixel^{-1}. In principle the lower the threshold the greater the selectivity although in practice the effect of very small thresholds is to eliminate almost all the hits.

3.4 Microfossil detection

To show the usefulness of the hit evidence from these microfossil images we used a simple binning method for the centres of curvature of the hits on spiral fossil images. In a perfect spiral such centres would lie on a tight curve near the centre of the fossil. Thus dividing the image up into square bins and counting the number of centres in each bin should yield peaks centred on spiral fossils. To test this a series of 14 images containing spiral fossils and 4 images with no fossils were taken and the strongest ten hits from each of twenty angles was computed. For various possible bin sizes the highest occupancy bin was found, and the density of hits (per square pixel) recorded. For very large bin sizes this will tend to a constant value, the average density, while for very small sizes it will be inversely proportional to the bin area (with the bin holding just one hit). However, in between we expect images with spiral fossils to have higher densities, and this effect can be seen if fig 7 where the average density is plotted against the bin width for the fossil bearing and non-fossil bearing images. We can use this effect to detect microfossil evidence simply on the basis of a fixed bin count threshold.

5 Conclusions

The methods employed yield reliable cues for locating spiral microfossil structures in plane rock samples. The curve finding algorithm described here provides a rich source of visual evidence from quite noisy images. The ability of the algorithm to output points rather than curves is a marked advantage in our database application since this simplifies the construction of visual models/queries. We hope that the set of points resulting from a well directed query will be useful for characterisation of microfossils as well as detection. The operator has several novel aspects, in particular the exclusive processing of local extrema and the use of partial derivatives for parametric estimation. Much work remains to be done on how this new type of feature evidence can be used effectively.

We have found that a symbolic representation for the curve evidence gives us the ability to experiment quickly by simply changing a database query without re-coding a program. Using this tool we have been able to explore the area of visual querying in one application.

6 Acknowledgements

This work is supported by the SERC. The fossil samples were provided by Fiona White of the Department of Geology, and we would like to thank her for sharing her expertise in foraminifera. The Sierra database simulator software was kindly made available by Prof. Lavington and his group at the University of Essex.

7 References

[1] Swaby P. A. "A Graphical Expert system for Microfossil Identification", BCS Expert Systems Conference, London, 20-22 September 1989.

[2] Parent P. and Zucker S.W. "Trace Inference, curvature consistency and curve detection", IEEE PAMI, Vol II, No 8 1989 pp 823-839.

[3] Zucker S.W. David C. Dobbins A. and Iverson L. "The organisation of Curve Detection: Coarse tangent fields and fine spline coverings", Proceedings of 2nd ICCV, pp 568-577, 1988.

[4] Oakley, Robinson & Shann, "An Efficient and Robust Local Boundary Operator", BMVC 91.

[5] Shann & Oakley "Novel Approach to Boundary Finding" IVC 8 p32 (1990).

[6] Murray Gill and Wright "Practical Optimisation".

[7] Lavington & Wang "The External Procedural Interface for the IFS/2" Dept Computer Science, University of Essex report CSM 164, June 1991.

[8] J.P. Oakley and R. T. Shann "An efficient method for finding the position of object boundaries to sub-pixel precision", Image and Vision Computing, Volume 9, no 4, August 1991.

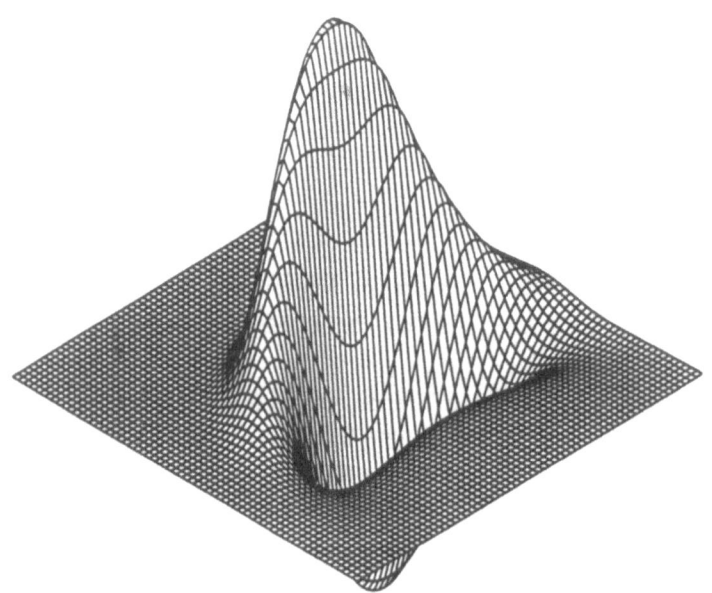

Figure 1 FILTER KERNEL

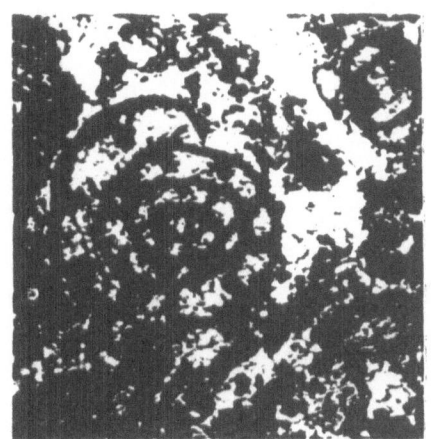

Figure 2 SPIRAL FOSSIL

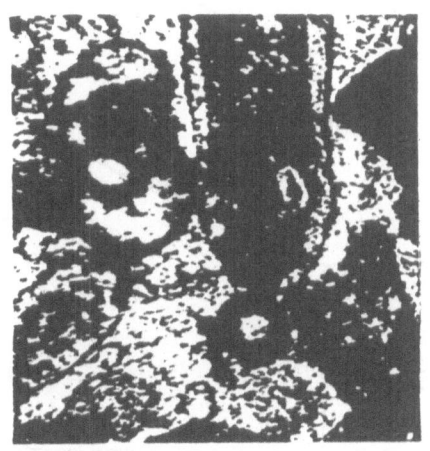

Figure 3 NOISIER FOSSIL

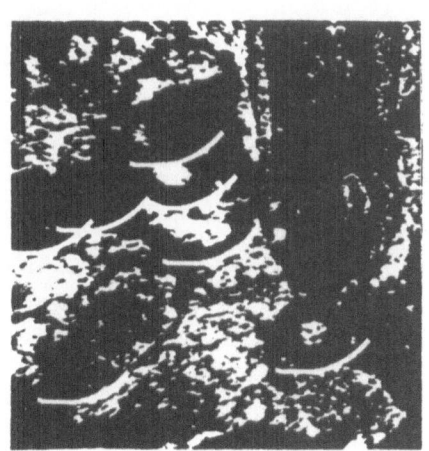

Figure 4 HITS FROM ONE DIRECTION

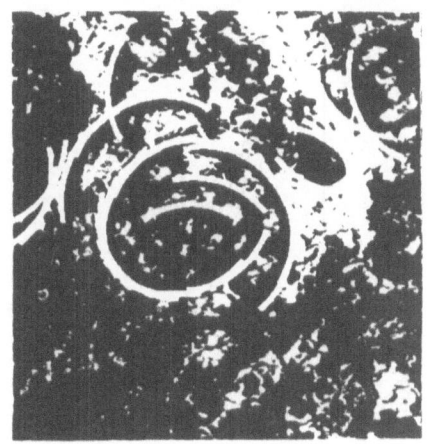

Figure 5 HITS FROM SEVERAL DIRECTIONS

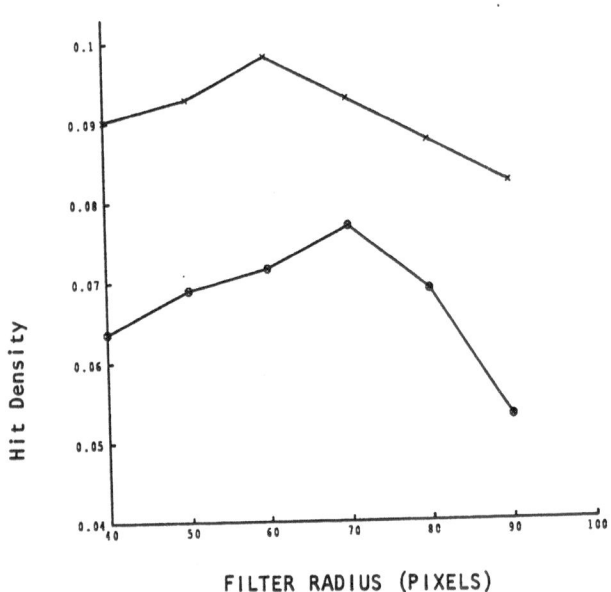

Figure 6 EFFECT OF FILTER CURVATURE

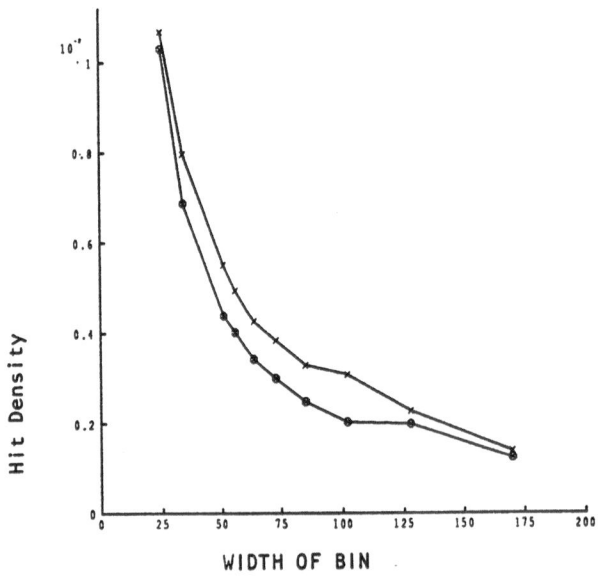

Figure 7 MICROFOSSIL DETECTION PERFORMANCE

Measuring Geometrical Parameters of Involute Spur Gears to Sub-pixel Resolution.

Mark J Robinson *   John P Oakley
Dept. of Electrical Engineering
University of Manchester
Manchester M13 9PL
email mark-r@spec0.ee.man.ac.uk

Abstract

The problem of accurate determination of the position, orientation and width of involute spur gear teeth using a single front-lit image is addressed. The information is used to compute global measures of tooth concentricity and pitch variations across a gear wheel. The approach is based on the use of specialised filters which make maximum use of the grey level information available in the image. Tooth location consistent to ±0.04 pixels has been achieved in realistic images.

1   Introduction

Gears are manufactured to very close tolerance, better than 10μm for high quality components ([1], [2]). The current methods of testing these tolerances is to use co-ordinate measuring machines, or a shadowgraph (an optical projection system which casts a magnified shadow of the tooth onto a calibrated screen). Both these techniques have inherent disadvantages; the former is expensive while the latter is time consuming. Contact with industry has revealed that there would be demand for a reliable and low-cost automatic inspection system based around image analysis. Measurement using image analysis is difficult, however. because the high precision required suggests that a "close-up" image of each tooth should be used, but the need to measure in a global co-ordinate system would then require accurate registration of a number of these sub-images. This implies that an array of cameras, or a mechanical system to move the part under a single camera, would be required. An image analysis system which uses one image would be simpler. cheaper and more reliable.

The aim of this study is to determine the measurement accuracy which is possible in this type of application from a single image.

---

*Funded by the SERC

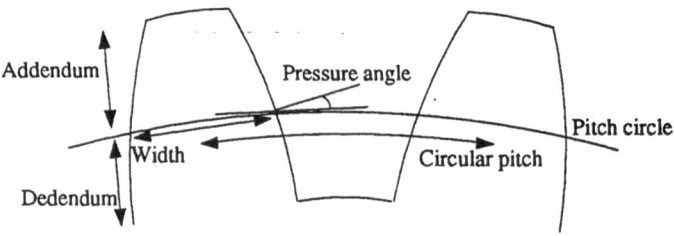

Figure 1: A gear profile

The definitions of dimensions associated with a gear tooth are shown in figure 1. The sides of the teeth must accurately match a well defined curve called an involute to a circle. A complete inspection system would need to measure

- Concentricity of the teeth.

- Error in the pitch of the teeth.

- Error in the involute profile and pressure angle.

The system discussed here takes an initial estimate of the position of a tooth from a cue generator, and iteratively refines it to return the position of the centre of the tooth and its orientation, in a global co-ordinate system, and the width of the tooth across the pitch circle.

In [3] we show that a specialised filter, the Curved Iterative Boundary Locator (CIBL), makes more effective use of image information than model fitting procedures based on edge data. When applied to the gear images, however, the small length of available boundary and the inaccurate match between the filter shape and the boundary shape led to two difficulties. The first was that the final position of the operator was affected by changes in grey level along the tooth boundary. Also, the operator was not completely reliable in that highlights on the tooth could cause the operator to align itself with the top of the tooth rather than the side, even when the initial position was very close to the required boundary.

This disappointing result led us to define more specialised operators for this particular visual task. These operators, CIBL-2, CIBL-3 and INVOLUTE-2 are described in the following sections.

2   CIBL-2 and CIBL-3

A new operator, CIBL-2, was developed to make use of prior knowledge of the structure of the gear tooth. This was based on the sum of two CIBL filters, one for each side of the tooth. The filters were free to move and rotate as a unit, but their orientation was fixed relative to each other. The position, orientation and separation of these filters were varied by an optimisation routine to maximise the filter output.

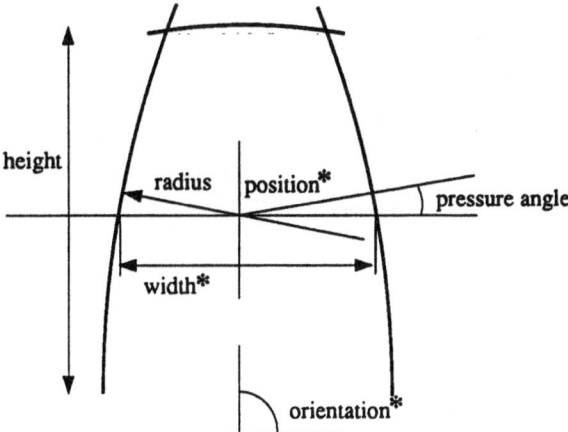

Figure 2: The composite operator. Parameters marked * are under the control of the optimisation routine.

This led to a significant improvement over the single filter, but it was still easily confused by highlights on the edges of teeth, where one of the filters would centre itself on the highlight. In order to counteract this problem a third filter was introduced to the operator, aligned with the outside edge of the tooth. This filter's output was weighted so as to keep the operator locked onto a tooth, while not dominating the operator output. The geometry and parameters of this operator, CIBL-3, are shown in figure 2.

Although this operator worked well, its performance is compromised by a number of factors, the choice of filter radius to give the best approximation to the required involute is not obvious, and the need for the operator to use the outer (non-contacting) surface of the tooth is undesirable because this face is not subject to any critical tolerance specification.

3   INVOLUTE-2

To circumvent these problems, a new filter which modelled the involute profile of the tooth more closely was developed.

The parametric equations for the involute to a circle of radius a, starting from the origin, are

$$x_i = a(\cos t + t \sin t - 1)$$

$$y_i = a(\sin t - t \cos t)$$

(1)

This curve is shown in figure 3.

The distance, $r(t)$, of a general point (x_p, y_p) from this involute is given by

$$r(t)^2 = (x_i - x_p)^2 + (y_i - y_p)^2$$

(2)

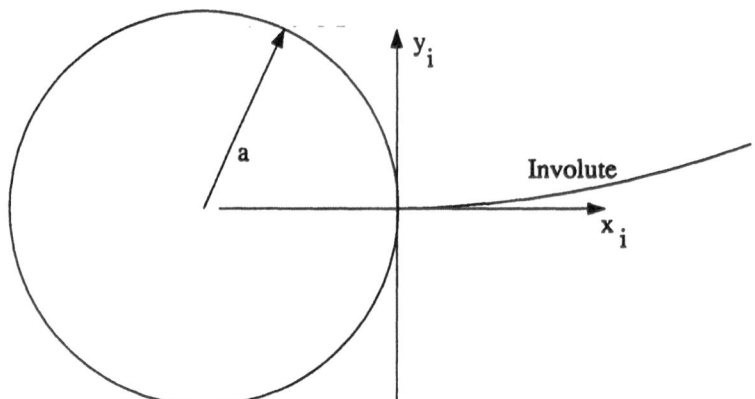

Figure 3: The involute to a circle of radius a.

This distance is minimised when

$$\frac{\partial r(t)^2}{\partial t} = 0 \tag{3}$$

which implies that

$$y_p \sin t + (x_p + a) \cos t - a = 0 \tag{4}$$

Solving 4 gives t_{min}, the value of t at which the involute passes closest to (x_p, y_p) and by substituting into 1 we can find $r(t_{min})$, the minimum distance. The basic gaussian filter is then defined by

$$f = \exp(\frac{r(t_{min})^2}{\sigma_d^2} + \frac{p^2}{\sigma_p^2}) \tag{5}$$

where σ_d and σ_p are the filter extent in the detector and projector directions respectively, and the projector function p is defined as

$$p = (x_p - x_{cen})^2 \tag{6}$$

which assumes that the filter is approximately straight. x_{cen} is then the distance of the peak in the projector Gaussian function from the origin, $3\sigma_p$ is a reasonable value. A typical filter defined in this way is shown in figure 4.

The derivative of this filter normal to the involute is used in the operator, and it is moved to its required position by applying an affine transform to it.

The INVOLUTE-2 operator was built using two of these filters, one to model each side of the tooth. Again, the filters are free to move and rotate as a unit, and to move apart to fit the width of the tooth.

The partial derivatives of the operator required by the optimisation routine were estimated using finite differences because of the complexity of these functions.

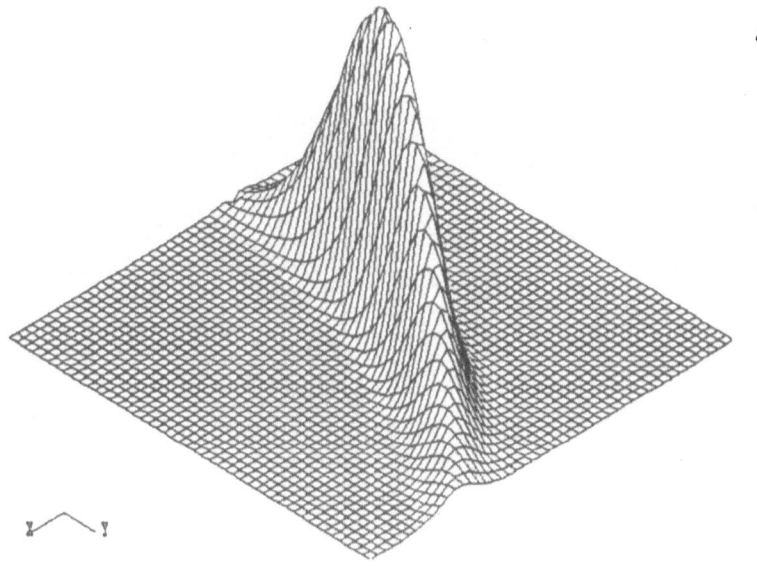

Figure 4: A typical involute filter shape.

4 Testing

Figure 5 shows part of an image of a gear wheel, magnified to show a single tooth, with the final position of the operator overlayed, for both the CIBL-3 and INVOLUTE-2 operators.

The gear locators were tested by measuring the distance between the centres of three adjacent teeth in six images of the same gear wheel. A tooth occupied an area of around 40 by 40 pixels, where one pixel corresponded to about 0.1mm. In each successive image the gear wheel was moved very slightly, and the direction of the lighting changed, in order to make the images as different as possible without affecting the geometry. The values of distance measured should be invariant between the images, but changes in noise and shading result in discrepancies in the position which maximises the operator output.

The RMS variation between the results was used as an indication of the random error in the operator. although systematic errors cannot be quantified without a calibrated reference.

The results for both operators are summarised in table 1.

It can be seen that both operators produced similar results. of around 0.04 pixels, which corresponds to 4μm. The INVOLUTE-2 operator is still affected by highlights and noise to some extent. although it is considerably better than CIBL-2. To improve the performance of INVOLUTE-2, a cost function based on the expected position of the base circle or outer surface of the tooth could be included. Requesting an accuracy better than around 0.005 pixels caused the optimisation routine to fail. or return a very inconsistent result, implying

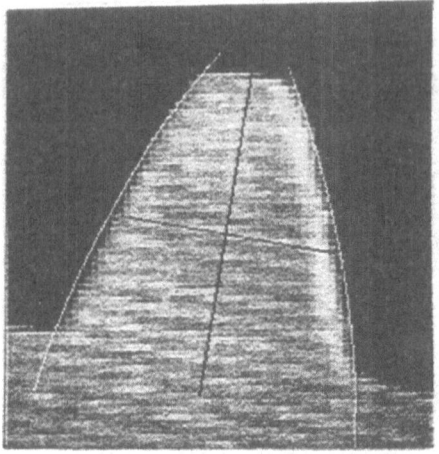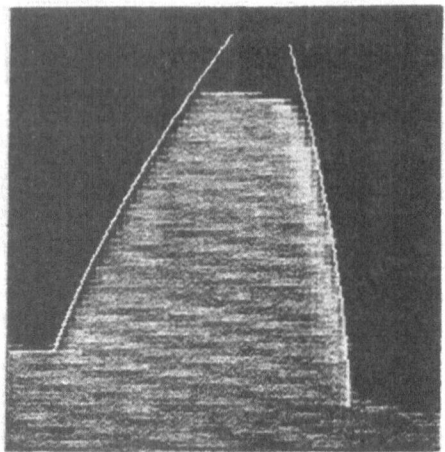

Figure 5: Magnified views of the final tooth position measured by (a) the CIBL-3 operator, and (b) the INVOLUTE-2 operator.

that this is the limit of accuracy available with these test images.

5 Conclusion

Results from the initial trials of the tooth locator are very encouraging, achieving the accuracy required for even the closest tolerance gears from a single image. Obviously the final accuracy achievable by the system will depend on the camera calibration, but the requirements of the calibration procedure and image acquisition system are greatly simplified by the ability to work from a single image.

There are other limitations on the accuracy available from the system apart from the image analysis algorithm, particularly from the camera optics, and this is an area currently under investigation.

6 Acknowledgements

The authors would like to acknowledge the assistance of the following companies, who provided useful information about the gear manufacturing process and samples of their products.

- Newmont Engineering Co. Ltd.

- Portescap UK Ltd.

- S.H. Muffett Ltd.

- Spline Gauges Ltd.

- University of Newcastle.

The project is funded by the SERC.

CIBL-3			INVOLUTE-2		
distance (pixels)		error	distance (pixels)		error
image 1	image 2	(pixels)	image 1	image 2	(pixels)
60.4003	60.3819	0.0184	62.7561	62.6994	0.0567
60.3819	60.3943	0.0124	62.6994	62.6988	0.0006
60.3843	60.4022	0.0179	62.6988	62.6564	0.0423
60.3922	60.4003	0.0081	62.6664	62.7344	0.0670
60.4003	60.4334	0.0330	62.7344	62.7629	0.0285
60.7703	60.7923	0.0220	60.8294	60.8709	0.0415
60.7923	60.8059	0.0135	60.8709	60.8548	0.0161
60.8059	60.9102	0.1043	60.9548	60.9072	0.0476
60.9102	60.8829	0.0273	60.9072	60.8539	0.0532
60.8829	60.8983	0.0155	60.8539	60.8981	0.0442
RMS error = 0.035			RMS error = 0.040		

Table 1: Summary of test results.

References

[1] British Standard 436 part 2 *The British Standards Institute 1970*

[2] British Standard 4582 part 1 *The British Standards Institute 1970*

[3] Robinson M J, Oakley J P and Shann R T. *An efficient and Robust Local Boundary Operator*. Proceedings of the BMVC 1991.

Camera Calibration using Vanishing Points

Paul Beardsley and David Murray *

Department of Engineering Science, University of Oxford, Oxford OX1 3PJ, UK

Abstract

This paper describes a method for measuring the intrinsic parameters of a camera. If aspect ratio is already known, the method requires two images of a plane; if aspect ratio is not known, the method requires three images of the plane. (Equivalently, one image of n planes could be used instead of n images of one plane). The plane must be marked with four points in known configuration.

1 Introduction

This paper addresses the measurement of a camera's intrinsic parameters, Figure 1. Main references for camera calibration include [7], [4]. The underlying theme in the calibration processing described here is the use of vanishing point and vanishing line information. Related methods are [3], [8], [2].

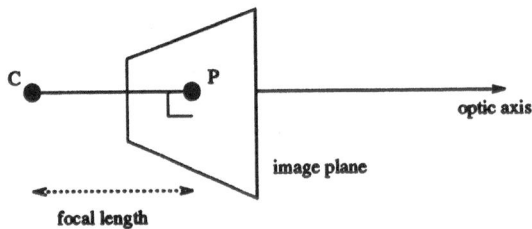

Figure 1: *Intrinsic parameters. The camera's focal point is C. The perpendicular projection of C onto the image plane is the principal point P. The distance CP is the focal length. On the image plane, the ratio of unit lengths along the y- and x-axes is the aspect ratio.*

Section 2 describes some key concepts used in the calibration, sections 3 to 6 cover the calibration itself, and section 7 contains the results.

2 Key Concepts

Some important ideas from projective geometry are reviewed here. See [6] for fuller discussion.

Consider a set of parallel lines on a projective plane, Figure 2 - the lines intersect at infinity, at an *ideal point*. There is a unique association between parallel lines of a given direction and the corresponding ideal point (in this sense, an ideal point can be regarded as specifying a direction). The set of all the ideal points on a projective plane constitutes an ideal line.

*This work was supported by SERC Grant No GR/G30003. PAB is in receipt of a SERC studentship.

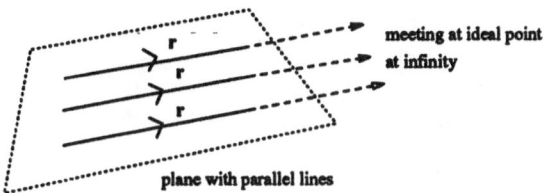

Figure 2: *Parallel lines on a projective plane have a well-defined intersection point at infinity, at an* **ideal point**.

The image of an ideal point is a vanishing point, Figure 3. This figure also shows a standard result from projective geometry - the line CV through the camera focal point and the vanishing point is parallel to the lines in the scene which have given rise to the vanishing point.

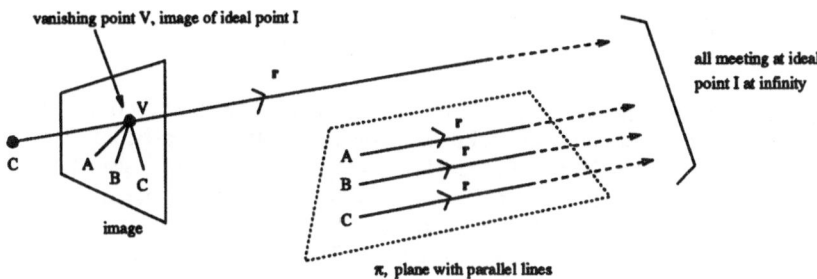

Figure 3: *A camera with focal point* C *views a set of parallel lines on the plane* π. *The parallel lines intersect at the ideal point* I. *The corresponding image lines converge to the vanishing point* V, *which is the image of* I. *Note that the ray* CV *is parallel to the lines in the scene.*

The other main idea in the work concerns the choice of coordinate frame for the processing. Figure 4 shows a set of points q_i on the image plane and the corresponding points Q_i on a plane π in the scene. The projectivity from the image plane to π is denoted α and the inverse projectivity from π to the image plane is α^{-1}. The coordinates of an image point q_n and the coordinates of the corresponding scene point Q_n are related by $Q_n = \alpha q_n$ and $q_n = \alpha^{-1} Q_n$. This can be treated conceptually as if there is a single point which has coordinates q_n in the image coordinate frame, and Q_n in a coordinate frame based on π. Thus α and α^{-1} can be regarded as operators which describe a *change of coordinate frame*. This conceptual treatment is the one adopted for the rest of the paper.

3   Finding the Principal Point - Overview

The intrinsic parameters of the camera are found sequentially, starting with the principal point. This section is an overview of the method for finding the principal point, while a step-by-step description is given in Section 4.

Consider a camera with focal point C and principal point P viewing a plane π.

418

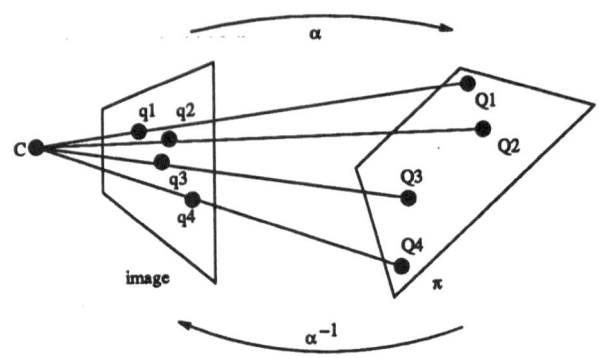

Figure 4: *A camera views a plane π in the scene. The projectivities between the image plane and π are α and α⁻¹, in the directions shown. As described in the text, α and α⁻¹ can be regarded as operators which transform the coordinates of a point between the image coordinate frame and a coordinate frame based on π.*

The vanishing line of π on the image plane is L, Figure 5a. The question is whether it is possible to use L to determine the unknown principal point P. Clearly, line L on its own is insufficient for this. However, there is a distinguished point on L also shown in Figure 5a - this is the vanishing point V_A on L at infinity on the image plane. It will be shown that V_A can be used to compute a second vanishing point V_B on L, Figure 5b, and the perpendicular to L at V_B passes through P. Thus, given one image of a plane, the position of P is constrained to lie upon a line M; given two or more images of planes in different orientations, the position of P is given by the point of intersection of the set of lines M_i. (Equivalently, one image of n planes can be used instead of n images of one plane).

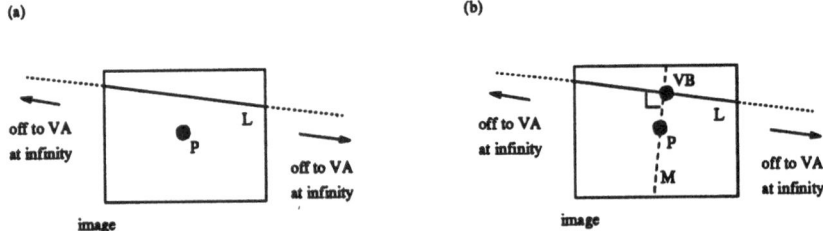

Figure 5: *(a) The image contains the principal point* P, *and the vanishing line* L *measured for a plane in the scene. Any point on* L *is a vanishing point - the vanishing point which is on* L *at infinity on the image plane is* V_A. *(As shown, one can travel to* V_A *at infinity by going in either direction along* L.)
(b) An important vanishing point on L *is* V_B *which has the property that the perpendicular to* L *at* V_B *passes through the principal point* P.

4   Finding the Principal Point - Step-By-Step

(a) Take an image of a plane as shown in Figure 4. Compute the projectivity α from the image plane to π, and the inverse projectivity α^{-1} from π to the image plane. A method for computing a projectivity is given in appendix A. The computation requires four image points, together with the coordinates of the corresponding scene points in a coordinate frame based on π (the *scale* of the configuration of the scene points is immaterial), no three of the four points to be collinear. Apply α^T to the coordinates of the ideal line of π in order to obtain the coordinates of the vanishing line L of π on the image plane.[1] See Figure 6.

(b) Find the intersection point V_A of L with the ideal line of the image plane.

(c) Apply α to the coordinates of V_A to obtain the coordinates of the corresponding ideal point I_A on the plane π.

(d) As pointed out in Section 2, an ideal point such as I_A can be associated with a *direction* on the projective plane. Call the direction d_A. Use I_A to compute I_B with associated direction d_B, such that d_A and d_B are orthogonal. Specifically, if the homogeneous coordinates of I_A are $(a, b, 0)$, then a convenient choice for the homogeneous coordinates of I_B is $(-b, a, 0)$.

(e) Apply α^{-1} to the coordinates of I_B to obtain the coordinates of the corresponding vanishing point V_B on the image plane.

To see that V_B has the property shown in Figure 5, that the perpendicular to L at V_B passes through the principal point P, consider the following argument using Figure 6.

(i) Line L is perpendicular to CP, which follows from the definition of P (Section 1) and the fact that L lies on the image plane.

(ii) Line L passes through V_A and therefore through I_A (because V_A is the image of I_A). Thus L has direction d_A, the direction associated with I_A. Line CV_B passes through V_B and therefore through I_B (because V_B is the image of I_B). Thus CV_B has direction d_B, the direction associated with I_B. It follows that L is perpendicular to CV_B.

(iii) The plane CPV_B is perpendicular to L because both CP and CV_B are perpendicular to L. Therefore, *any* perpendicular to L at V_B is in the plane CPV_B. Therefore, the perpendicular to L at V_B which is on the image plane passes through P.

In summary, note that a key step in the processing is the generation of the point V_B such that the line CV_B is perpendicular to L. The route to finding V_B might seem tortuous. It is necessary because, when working projectively on the image plane, it is not possible to compute a perpendicular such as L and CV_B by making use of a point C which is *off* the image plane. By shifting the coordinate frame to π, it is possible to bring in the perpendicularity condition using only ideal points which are on the plane π, and then to transform back to the image plane where the required condition that L is perpendicular to CV_B will hold.

5   Finding the Focal Length

Figure 7a shows the information available when viewing a plane, following the processing in Section 4 - the vanishing line L of the plane, the special vanishing point

---

[1] α^T is the transpose of α. Note that the projectivity α^{-1} between π and the image plane is a *point* projectivity; when operating on a line, as here, the corresponding line projectivity must be used and this is α^T. See [6].

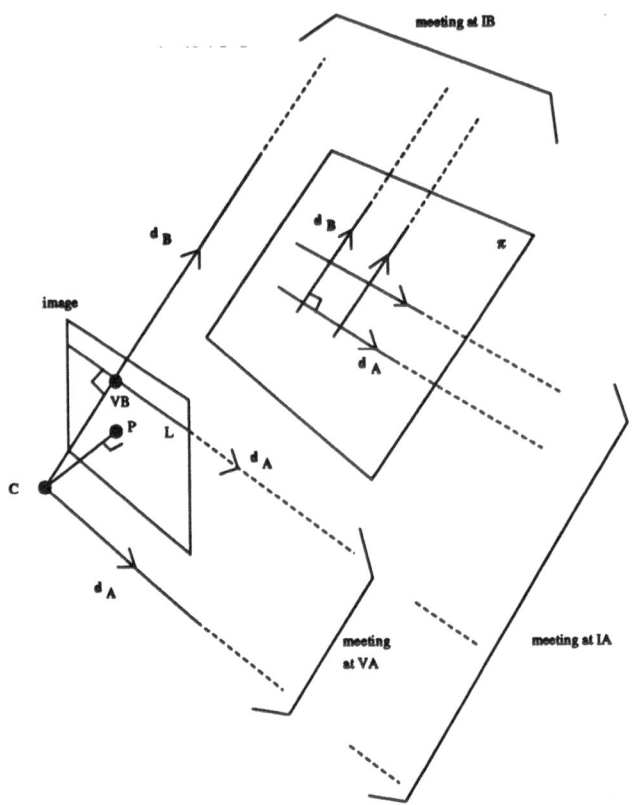

Figure 6: *This figure shows all the information involved in computing the vanishing point* V_B. *First vanishing point* V_A *is found. This is the image of the ideal point* I_A, *so* V_A *can be used to compute the coordinates of* I_A. *Then* I_B *is found using the constraint that the directions associated with* I_A *and* I_B *are perpendicular. The image of* I_B *is* V_B, *so* I_B *can be used to compute the coordinates of* V_B.

V_B described in Sections 3 and 4, and the principal point P. Figure 7b includes one further piece of information, a vanishing point V_C lying on L such that angle $V_B C V_C$ is 45°, to be used in the computation of the focal length.

V_C is found by the same method for computing V_B from V_A in Section 4 - recall the three stages (a) V_A was used to compute the coordinates of the corresponding ideal point I_A, (b) I_A was used to compute I_B such that the directions associated with I_A and I_B were perpendicular, and (c) I_B was used to compute the coordinates of the corresponding vanishing point V_B. We apply the same method here to compute V_C from V_B, the only difference lying in the second stage - the ideal point I_B is used to compute I_C such that the directions associated with I_B and I_C are at 45° (this is just as straightforward as computing perpendicular directions).

Once V_C is known, the focal length can be found from simple trigonometry. The distance $V_B V_C$ is known, and this is equal to the distance $C V_B$. Thus, the lengths of sides $C V_B$ and $V_B P$ of the right-angle triangle $C P V_B$ are known, and the length $C P$

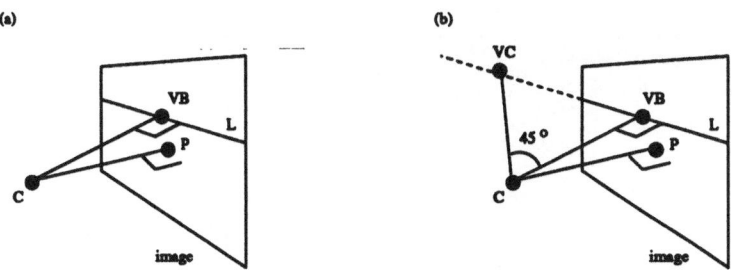

Figure 7: *(a) This figure shows just that part of the information in Figure 6 which is to be used in the computation of focal length. (b) One extra piece of information is also used when computing focal length - the point V_C which lies on L such that angle $V_B C V_C$ is 45°.*

(which is the focal length) can be computed.

6 Finding the Aspect Ratio

It has been assumed so far that the aspect ratio of the image has been corrected. If aspect ratio is unknown, it can be computed in the following way.

Consider the method for computing the principal point in Sections 3 and 4 - a set of "constraint lines" is generated, one from each image, and these lines are concurrent at the principal point. If the aspect ratio has not been corrected, however, the constraint lines will not be concurrent (the constraint lines are generated by finding a perpendicular to the vanishing line L, but perpendicularity is not preserved under changes of aspect ratio). Starting from an initial aspect ratio which is near the true value, aspect ratio is iteratively adjusted until the constraint lines are as close as possible to concurrency - at this stage, the value of aspect ratio is recorded as its true value. The concurrency test can be carried out given three or more constraint lines.

7 Results

Results are given for two cameras which are part of a stereo system. The cameras are both of the same type, so principal point and focal length measurements are expected to be similar, and the aspect ratio is identical. A typical stereo pair of a calibration plane (dimensions about 4ft-by-4ft) is shown in Figure 8.

The images are 256-by-256 pixels. The first stage of the processing is to remove radial distortion. The radial distortion correction factor is calibrated separately, by a method [1] which does not require full knowledge of the camera parameters because it assumes approximate values for the radial distortion centre and the aspect ratio[2]. Canny edge detection to sub-pixel accuracy is run on the images, and straight lines are fitted by orthogonal regression along the eight horizontal and eight vertical lines available on the calibration plane. These lines are used to generate 64 vertex positions,

---

[2]It is assumed that the centre of the image array is a reasonable approximation to the radial distortion centre, while a reasonable approximation to the aspect ratio is available from the specification of the digitisation equipment.

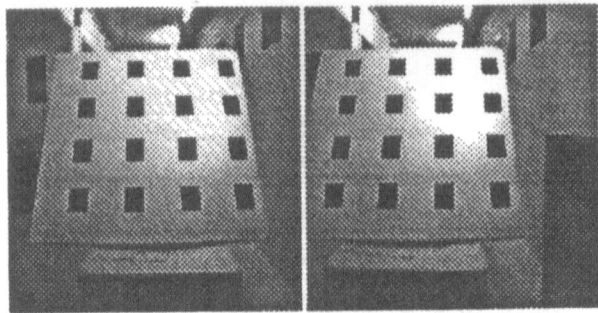

Figure 8: *A stereo pair of the calibration plane. (Bowing of straight lines is just discernible - this is due to radial distortion).*

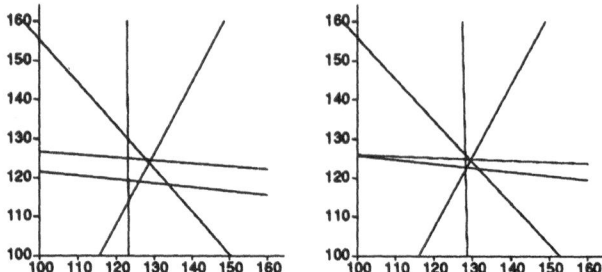

Figure 9: *Left and right cameras - the constraint lines meeting at the principal point. The axes are labelled in pixels.*

and the vertices are used to determine the projectivity between the image and the calibration plane by the method in appendix A.

Results are shown for five images for the left and right cameras - the calibration plane is repositioned for each image so that the constraint lines used to compute the principal point vary significantly in gradient. Figure 9 shows the five constraint lines meeting at the principal point, when the aspect ratio has been adjusted to its correct value. Table 1 shows the focal lengths computed from the five images. Figure 10 shows the way in which the concurrency of the five constraint lines varies with aspect ratio - the concurrent point of the constraint lines is measured using orthogonal regression, and the residual obtained from this process is used as the measure of concurrency.

Table 2 shows a summary of the results. Obtaining a ground-truth to check camera calibration is difficult - we have so far carried out only one main test. A tape measure is placed parallel to the image plane at a measured distance in front of the camera (the parallelism and the measurement can only be approximate). By comparing the ratio of the tape length to the tape-camera distance, and the ratio of the number of image pixels spanned by the tape to the focal length, it is possible to compute a value for the focal length. The result was within 1% of the calibrated focal length. This indicates that the calibrated value is reasonable, although it can only be treated as a precursor to more rigorous cross-checking of the calibration.

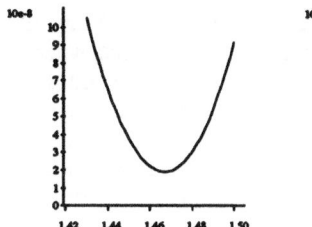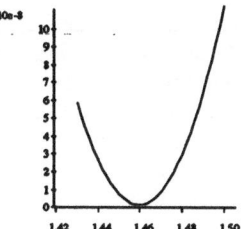

Figure 10: *Left and right cameras - a plot of the residual from the computation of the concurrency of lines at the principal point, against aspect ratio. The minimum of the graph provides the estimate of the actual aspect ratio.*

Image number	Left camera focal length	Right camera focal length
1	267.7	264.3
2	262.8	266.6
3	261.5	266.6
4	262.6	268.0
5	270.0	266.2
Mean	265	265
Standard deviation	4	1

Table 1: *Measurement of focal length in pixels. A focal length of 265 pixels is about 6mm (using the pixel size from the camera specification).*

8 Evaluation and Conclusion

As shown in Section 5, the constraint lines which are used to compute the principal point are reasonably close to concurrency, with an error area of about ±2.5 pixels in the worst case (the left camera). The values of focal length, one from each image, show only small variation.

The results are for two short focal length wide-angle cameras. We have found that the method degrades in longer focal length cameras. This is probably because it relies on perspective effects, and these decrease as the focal length increases.

Finally, regarding the application of the method, note that the processing has no particular requirement for a symmetrical calibration pattern like the one used - the minimum requirement on the data is in fact four points in a known configuration on a

Parameter	Left camera	Right camera
Principal point	(127,122)	(129,124)
Focal length	265	265
Aspect ratio	1.466	1.460

Table 2: *Summary of results.*

plane (the actual scale of the configuration is immaterial). This raises the possibility of extending the method to work in natural scenes - in indoor scenes, for instance, where the four corners of doorways or windows of known configuration would provide one way of obtaining the required input data. Calibrating from a natural scene is increasingly desirable with the advent of active cameras which can continually change focus and zoom - clearly though, there is a huge gulf between obtaining results using a calibration plane as presented here and the ability to calibrate from a natural scene.

Acknowledgements.

Thanks to Bill Triggs for useful insights, to Andrew Zisserman and Charlie Rothwell for their help with projective geometry, and to Han Wang for his tape calibration results.

References

[1] P.A. Beardsley. The correction of radial distortion in images. Technical report 1896/91, Department of Engineering Science, University of Oxford, 1991.

[2] P.A. Beardsley, D.W. Murray, and A.P. Zisserman. Camera calibration using multiple images. In *Proc. 2nd European Conf. on Computer Vision*, pages 312–320. Springer-Verlag, 1992.

[3] B. Caprile and V. Torre. Using vanishing points for camera calibration. *Int. Journal of Computer Vision*, pages 127–140, 1990.

[4] O.D. Faugeras and G. Toscani. The calibration problem for stereo. *Proc. Conf. Computer Vision and Pattern Recognition*, pages 15–20, 1986.

[5] C.A. Rothwell, A. Zisserman, C.I. Marinos, D.A. Forsyth, and J.L. Mundy. Relative motion and pose from arbitrary plane curves. *Image and Vision Computing*, 10(4):250–262, 1992.

[6] J.G. Semple and G.T. Kneebone. *Algebraic projective geometry*. Oxford University Press, 1952.

[7] R.Y. Tsai. An efficient and accurate camera calibration technique for 3d machine vision. In *Proc. IEEE CVPR*, 1986.

[8] L. Wang and W. Tsai. Computing camera parameters using vanishing line information from a rectangular parallelepiped. *Machine Vision and Applications*, 3:129–141, 1990.

A. Determining a Projectivity

Figure 11(a) shows the layout of the calibration plane used in the experiments. Figure 11(b) shows the typical appearance of this calibration plane in an image, with distortion due to perspective effects. This appendix describes how to determine a linear transformation, a *projectivity*, which maps corresponding features between Figures 11(a) and (b).

When working with projectivities, points and lines are represented in homogeneous coordinates. Let the vertices of the calibration squares in the image be represented by $(x_1, y_1, 1)...(x_n, y_n, 1)$ where (x, y) is the image position in pixels and the third component is set to 1. Let the corresponding vertices in a coordinate frame based on the calibration plane (hereafter, just the "calibration frame") be $(X_1, Y_1, 1)...(X_n, Y_n, 1)$ where (X, Y) are Euclidian coordinates and the third component is again set to 1.

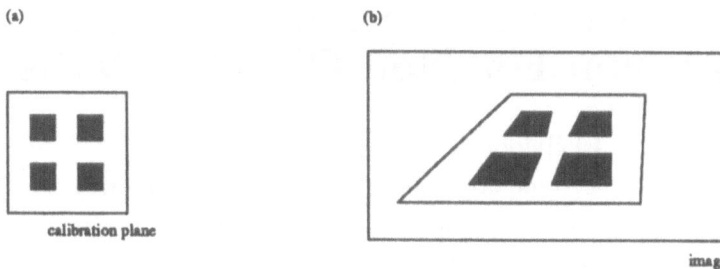

(a)   (b)

calibration plane

image

Figure 11: *(a) A fronto-parallel view of the calibration plane (the plane used in the actual experiments had 4-by-4 squares). (b) The calibration plane as it appears in a typical image, with distortion due to perspective.*

The choice of origin and axes for the calibration frame is arbitrary, but it is convenient to align the $x-$ and $y-$axes with the horizontal and vertical directions of the calibration squares. Note that there is no requirement here that the calibration pattern consists of squares.

The projectivity is determined by a method in [5]. Each vertex in the image is related to its corresponding point in the calibration frame by

$$
\begin{bmatrix} X_i \\ Y_i \\ 1 \end{bmatrix} = k_i \begin{bmatrix} \alpha_{11} & \alpha_{12} & \alpha_{13} \\ \alpha_{21} & \alpha_{22} & \alpha_{23} \\ \alpha_{31} & \alpha_{32} & 1 \end{bmatrix} \begin{bmatrix} x_i \\ y_i \\ 1 \end{bmatrix} \tag{1}
$$

where the matrix α is the required projectivity, $(x_i, y_i, 1)$ and $(X_i, Y_i, 1)$ are corresponding points in the image and calibration frame respectively, and k_i is a scale factor which varies with each particular pair of points. Eliminating k_i from the three equations present in matrix equation (1) leaves two linear equations. Thus, given four pairs of points, there are eight linear equations in the eight unknowns $\alpha_{11}...\alpha_{32}$, and α can be found[3]. Given n points, there are $(n * 2)$ equations in the eight unknowns, and a method such as the pseudo-inverse can be used to find a solution.

It is shown in [6] that if α maps points from the image to the calibration frame, then α^{-1} is guaranteed to exist and maps points from the calibration frame to the image, $\alpha^{-\mathsf{T}}$ (the inverse transpose of α) maps *lines* from the image to the calibration frame, and α^{T} maps lines from the calibration frame to the image.

---

[3]The condition for linear independence of the equations is that no three of the four points are collinear.

Ground Plane Motion Parameter Estimation For Non Circular Paths

G.J.Ellwood Y.Zheng S.A.Billings

Department of Automatic Control and Systems Engineering
University of Sheffield, Sheffield, UK

J.E.W.Mayhew J.P.Frisby

Artificial Intelligence Vision Research Unit,
University of Sheffield, Sheffield, UK

Abstract

A method of motion parameter estimation for AGV's is developed based
on a new trajectory constraint algorithm. It is assumed that the vehicle will follow a circular path over the vision sampling interval. The
algorithm has been found to be more effective and consistent than least
squares estimators when the motion obeys the trajectory constraint. Reliable estimates of the motion parameters can even be made from individual data pairs. In this way, a static world can be segmented from
moving objects, and so the motion parameters can be obtained using the
stationary points alone.

In practice the vehicle will not necessarily follow a circular path, and
hence there may be a bias in the parameter estimates. Experiments
were carried out using simulated data, where the true trajectory was a
clothoid, to investigate the robustness of the algorithm when the trajectory constraint is violated.

1   Motion Parameter Recovery

The prime objective of the work is to estimate the rotation and translation
parameters of a vehicle using a pair of stereo images. Corners are found in the
images using the Plessey group algorithm [1]. These are matched to form a 3D
point based map of the robot's environment [6]. The vehicle then moves in the
ground plane. The procedure is repeated for the new coordinate frame of the
AGV.

A linked list of the corners q_i observed at time t and their corresponding
coordinates p_i observed at $t + 1$ is formed.

The motion parameters R and T are found by solution of the equation

$$q_i = Rp_i + T \tag{1}$$

Most algorithms achieve this using the principle of least squares to estimate the rotation matrix R. The translation vector T is then obtained from
equation(1).

It has been observed by many authors [3, 5] that although the vehicle moves
smoothly, the estimated paths are very erratic and noisy. Such trajectories are
impossible. This is due to the sensitivity of least squares methods in general to

noise, causing the estimate of R to be significantly degraded. This is magnified when solving for T using equation (1). The points $\mathbf{q_i}$ and $\mathbf{p_i}$ are usually distant from the vehicle, so a small error in R will be amplified through to yield a large error in T.

Our proposal was to constrain the parameters by some assumptions [5] regarding the dynamics of the vehicle. A clear constraint on the motion would be that, if it does not skid, the vehicle will always move in the direction it is heading. It seems reasonable to assume also that the rate of change of steering is small. So, provided the vision sampling rate is sufficiently fast, one can assume that the rate of change of steering is approximately zero over the sampling interval. This means the vehicle is travelling along a path of constant curvature, or a circle, between samples.

A geometrical analysis [5] shows that, when moving along a circular arc, the three motion parameters can be related by

$$t_x = t_z tan \frac{\theta}{2} \tag{2}$$

Expanding equation (1), there are three equations and three unknowns. The parameters can be found analytically.

This method has been found to be a considerable improvement on least squares in the estimation of lateral motion and rotation for motion obeying the circular constraint. There is little change in the accuracy of the depth motion parameter estimate.

The motion parameters can be estimated from each individual data pair. This property can be used to segment [4] moving objects from a static background. The estimates made using stationary points will form a cluster in the motion parameter space. The corners on an object moving arbitrarily in the scene will not in general satisfy the trajectory constraint. The motion parameters obtained using these points will be scattered widely in the motion parameter space.

Each of the motion parameters are arranged in ascending order of the heading parameter θ. A threshold based on the variance of θ is set up such that if two adjacent motion parameters in the ordered list represent the vehicle's motion, their difference in heading is almost certainly below the threshold. Another threshold L_{th} is set up on the minimum length of a group. This is to avoid groupings between consecutive points which coincidentally have similar headings rather than because they are coherent. If both θ_{th} and L_{th} are set up appropriately, only one cluster should be obtained with members representing the motion parameters calculated from the static world.

For a vehicle to follow a planned trajectory, obeying the trajectory constraint, would require instananeous steering at the moment of sampling. This is, of course, impossible. There will, therefore, always be a bias in the estimates of the motion parameters. It is necessary therefore to investigate the effect that violation of this assumption has upon parameter estimation and scene segmentation.

2 Simulated Experiments

2.1 Trajectory Generation

A series of experiments were carried out in which the "true" motion of the vehicle was a clothoid. This a function [2] whose curvature c varies linearly with respect to the distance s moved along it.

$$c(s) = ks + c_0 \tag{3}$$

where k is known as the sharpness coefficient. The direction of heading is therefore

$$\theta(s) \quad = \quad \int_0^s (ks + c_0)ds = \frac{ks^2}{2} + c_0 s + \theta_0 \tag{4}$$

A set of clothoids is illustrated in figure (1). The larger the coefficient k, the tighter will be the curve.

The parameter k can be thought of as representing, physically, the rate of steering. It is however quite hard to grasp what any particular value of k will mean.

A commonly desired trajectory would be to join two straight lines at right angles to each other. A solution is a clothoid pair. This is a trajectory that starts with a coefficient k until a turn of $45°$ is achieved, it then continues with sharpness $-k$ until the total change in heading is $90°$. The distance t_z moved to join P_0 to P_1 for a certain sharpness k gives a better intuitive grasp of the sharpness of the curve than the value of k itself.

It is noted that a clothoid will not in reality be the true trajectory. It is used here, simply because it represents a trajectory that does not obey the constraint of a constant curvature.

2.2 Data Generation

At time t, a set of 3D points were generated in front of the vehicle. All the coordinates were generated using independent Gaussian distributions.

Assume that at time $t+1$ the vehicle has moved to a new location. The motion parameters were calculated. Using these parameters, the points generated at time t were mapped onto the vehicles new local coordinate frame at time $t + 1$. Each point at $t + 1$ had a known correspondence with a point at time t. Points which could not be seen by the vehicle at both times were rejected. Data were generated until a predetermined number, in this case 20, were visible to the vehicle before and after motion.

The two sets of points were then transformed into left and right camera images for times t and $t + 1$, using the exact calibration file. Each image was then blurred to produce two sets of 3D data points, with known correspondences. The motion parameters were then estimated using both the trajectory constraint and the least squares algorithms.

2.3   Experimental Results

2.3.1   The Effect of Violating the Trajectory Constraint on Estimating Motion Parameters

Experiments were carried out to investigate the effect of violating the trajectory constraint on parameter estimation. In each of the experiments the vehicle moved a distance of $250mm$ along the curve. This is equivalent to moving at $1.25m/s$ with a vision sampling rate of $5Hz$

The vehicle turned through $5.73°$ over the sampling interval, following several trajectories. These had sharpness coefficients (see equation (3)) of 0 (ie a circle), $1 \times 10^{-6}, 2 \times 10^{-6}, 3 \times 10^{-6}$ and 4×10^{-6}. Clothoid pairs of such sharpness would join two perpendicular straight lines in $\infty m, 1.5m, 1.2m, 1.0m$ and $0.8m$ respectively. Clothoids with corresponding positive sharpness coefficients are illustrated in figure 1. Estimates of the bias and standard error were made using 50 identical samples of each movement.

The bias and standard error for the parameter estimates for these experiments can be seen in Tables 1 to 3. The results of an individual experiment where $k = 4 \times 10^{-6}$ can be seen in figures 2, 3 and 4.

The estimates obtained using least squares appear to be unbiased, but consistently noisier than those obtained using the trajectory constraint method.

It is noted that there is a bias in the estimates of θ (see table 1 and figure 2) for the trajectory constraint method. This increases with the degree to which the circular assumption is violated and arises because equation (2) is no longer valid. For motion along a clothoid trajectory the direction of heading is given by

$$\theta(s) = \frac{ks^2}{2} + c_0 s \tag{5}$$

The Trajectory Constraint Method assumes that θ is given by

$$\theta(s)_{est} = c's \tag{6}$$

where c' is some constant value. Therefore, the bias is given by

$$(\theta - \theta_{est}) = \frac{ks^2}{2} + (c_0 - c')s \tag{7}$$

One would therefore expect the bias to be linear with respect to k This is shown in figure 5.

Results for estimating the depth parameter can be seen in table 2 and figure 3. There was little difference in the estimates of t_z for either technique. For small θ, the error in t_z is given by

$$\tilde{t}_z \approx (c_{px} + c_{pz}\theta)\tilde{\theta} + \tilde{c}_{qz} + \tilde{c}_{px}\theta - \tilde{c}_{pz} \tag{8}$$

The bias is caused by the terms in $\tilde{\theta}$, but these are small, of the order 10^{-1}. $\tilde{c}_{qz}$ and $\tilde{c}_{pz}$ will be of the order 10^1 to 10^2 and since the error in each estimate will be dominated by these terms. t_z is effectively unbiased.

Any improvement in the estimates of t_x and θ will have little effect on the t_z estimate. The accuracy of t_z will only will be significantly improved by better error modelling of the vision data.

Sharpness	θ^o true	LS2D bias	sd	traj bias	sd
0	5.73	-0.01	0.08	0.0005	0.008
1×10^{-6}	5.73	-0.005	0.039	-0.043	0.013
2×10^{-6}	5.73	0.0037	0.062	-0.077	0.016
3×10^{-6}	5.73	0.0039	0.065	-0.122	0.023
4×10^{-6}	5.73	0.014	0.063	-0.159	0.029

Table 1: Bias and Standard Deviation for Estimates of θ for Increasing Violation of Trajectory Constraint, $\theta_{true} \approx 0.1 rad$

Sharpness	$t_z(mm)$ true	LS2D bias	sd	traj bias	sd
0	249.6	0.78	2.27	0.74	2.17
1×10^{-6}	249.7	0.85	2.46	0.90	2.48
2×10^{-6}	249.8	-0.33	2.57	-0.36	2.69
3×10^{-6}	249.8	-0.59	2.37	-0.60	2.46
4×10^{-6}	249.9	0.03	2.73	-0.17	2.85

Table 2: Bias and Standard Deviation for Estimates of t_z for Increasing Violation of Trajectory Constraint, $\theta_{true} \approx 0.1 rad$

Sharpness	$t_x(mm)$ true	LS2D bias	sd	traj bias	sd
0	12.50	0.29	2.97	0.005	0.11
1×10^{-6}	11.20	0.20	1.33	1.21	0.098
2×10^{-6}	9.90	-0.097	2.58	2.43	0.11
3×10^{-6}	8.60	-0.15	2.55	3.64	0.13
4×10^{-6}	7.30	-0.58	2.47	4.87	0.14

Table 3: Bias and Standard Deviation for Estimates of t_x for Increasing Violation of Trajectory Constraint, $\theta_{true} \approx 0.1 rad$

Sharpness	θ^o true	LS2D bias	sd	traj bias	sd
0	16.12	-0.0003	0.05	0.0008	0.0043
1×10^{-6}	16.12	0.004	0.057	-0.030	0.0059
2×10^{-6}	16.12	-0.0019	0.068	-0.062	0.0072
3×10^{-6}	16.12	0.0039	0.046	-0.095	0.0078
4×10^{-6}	16.12	-0.009	0.044	-0.13	0.010

Table 4: Bias and Standard Deviation for Estimates of θ for Increasing Violation of Trajectory Constraint, $\theta_{true} \approx 0.281 rad$

Sharpness	$t_x(mm)$ true	LS2D bias	sd	traj bias	sd
0	34.96	-0.12	2.17	-0.13	0.37
1×10^{-6}	33.67	-0.18	2.45	1.23	0.47
2×10^{-6}	32.39	0.20	3.07	2.51	0.47
3×10^{-6}	31.10	-0.07	2.06	3.73	0.38
4×10^{-6}	29.80	0.37	2.05	4.83	0.34

Table 5: Bias and Standard Deviation for Estimates of t_x for Increasing Violation of Trajectory Constraint, $\theta_{true} \approx 0.281 rad$

$\theta°$		LS2D		traj	
Distance	true	bias	sd	bias	sd
100	16.11	0.004	0.03	-0.005	0.004
200	16.11	0.0062	0.032	-0.033	0.004
300	16.10	0.0008	0.06	-0.11	0.01
400	16.10	0.001	0.037	-0.26	0.021

Table 6: Bias and Standard Deviation for Estimates of θ for Increasing Distance Moved over Sampling Interval , $\theta_{true} \approx 0.281 rad$

$t_x mm$		LS2D		traj	
Distance	true	bias	sd	bias	sd
100	13.82	-0.20	1.43	0.13	0.40
200	26.62	-0.21	1.41	1.36	0.42
300	37.45	-0.21	2.56	4.02	0.34
400	45.30	-0.016	1.56	9.68	0.39

Table 7: Bias and Standard Deviation for Estimates of t_x for Increasing Distance Moved over Sampling Interval , $\theta_{true} \approx 0.281 rad$

The estimation of the lateral motion parameter was then examined. The results can be seen in table 3 and figure 4. The estimates made using the trajectory constraint method were biased. Again though, they were smoother than those made using least squares. For small θ, the error in t_x is given by

$$\tilde{t}_x = \tilde{c}_{qx} - \tilde{c}_{px} - \tilde{c}_{pz}\theta - c_{qz}\tilde{\theta} \tag{9}$$

The only term with bias is $\tilde{\theta}$, therefore the bias in t_x is given by $c_{pz}\tilde{\theta}$. The bias in the estimates of lateral motion is therefore linear with respect to the bias in θ. This is illustrated in figure 6. The bias appears large with respect to t_x. It is however small with respect to the total distance travelled over the sampling interval. When viewed in this context the bias is very small.

A similar set of experiments was carried out for a rotation of approximately $16°$ over the same sampling interval as before. The bias and standard deviation for both methods of estimation can be seen in tables 4 and 5. The Least Squares estimates are more noisy but unbiased. There seems to be little change in the accuracy in the estimates obtained for the trajectory constraint method for the larger rotation, although the absolute value of the parameters has changed. The bias in terms of the total distance moved is the same as for the smaller turn. This suggests that the bias in θ and t_x is a function of the degree of violation and sampling rate alone.

2.3.2 The Effect of Sampling Rate on Parameter Estimation for Motion Violating the Trajectory Constraint

In these experiments a clothoid of sharpness 2×10^{-6} was used. The vehicle turned through approximately $16°$ over the sampling interval. In each experiment the total distance s moved over a sampling interval was different. As $s = Vt$, this investigates the effect of the sampling rate and vehicle speed on parameter estimation.

The results for θ and lateral motion can be seen in tables 6 and 7. The least squares estimates seem to be unaffected by the sampling interval. However, the bias in the estimates obtained using the trajectory constraint method grows sharply with the distance moved between samples. This is illustrated in figure 7. The bias in t_x is small though, compared to the distance moved over the sampling interval.

2.3.3 The Effect of Violation of the Trajectory Constraint on Scene Segmentation

The trajectory constrained method yields very accurate estimates of the motion even when applied to single static data pairs. This property can be exploited to segment moving objects from a static background. It is necessary to investigate the effect a non circular trajectory has upon scene segmentation.

The background data, consisting of 20 points, were generated as before. The moving object was generated using a 3D Gaussian generator, typically centred between $2000\text{-}2500mm$ in front of the vehicle. The object was represented by 10 points. Given the movement of the object, a corresponding set was obtained with respect to the vehicle's new position. A linked list was then produced including both the background and the object points. To illustrate how the segmentation was achieved, only one vehicle movement was recorded.

In the first experiment the vehicle moved along a clothoid of sharpness $k = 2 \times 10^{-6}$. The motion parameters were $\theta = 5.73^o, t_x = 9.90mm$ and $t_z = 249.8mm$. The object moved $100mm$ horizontally with respect to the position of the vehicle at time t. The estimates of of t_x and t_z obtained by applying the trajectory constrained method to each individual data pair can be seen in figure 8. It is clear that a group of the points lie in a straight line.

The clustering algorithm described in [4] was applied to these estimates with $\theta_{th} = 0.03^o$ and $L_{th} = 5$. All 20 of the stationary points in the scene formed a cluster, and the trajectory constraint method was used to estimate the motion parameters using these alone.

The data is clustered on the basis that the variance of the parameter θ is small. An experiment was carried out to investigate how the variance of the estimates of θ from individual data points was affected by increased violation of the trajectory consraint. The standard deviation plotted against sharpness coefficient can be seen in figure 9. The standard deviation does not seem to be increased greatly by violation of the trajectory constraint. So, for reasonable trajectories successful segmentation can be achieved.

3 Conclusions

A slight bias is inherent in the trajectory constraint method when the vehicle is not following a circular path as equation (2) is no longer strictly valid. It is however, extremely small for reasonable sampling rates.

Segmentation was only very mildly affected by violation of the trajectory constraint. For clothoids of reasonable sharpness, segmentation is successful, and so very accurate estimates of the motion parameters can still be obtained in the presence of moving objects.

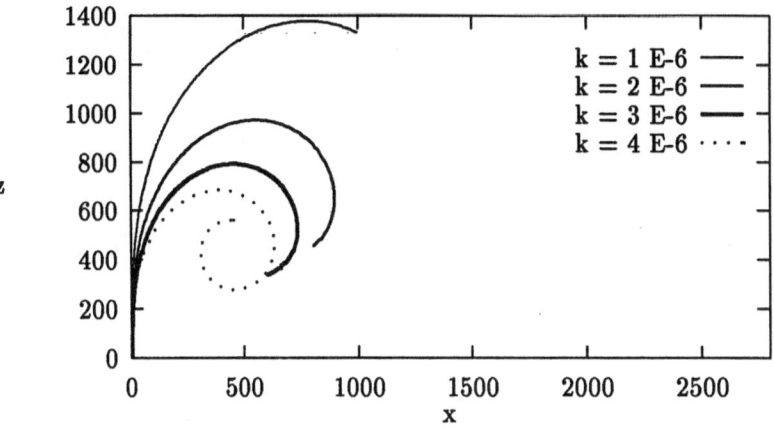

Figure 1: A Set of Clothoids

4 Acknowledgements

Dr Y. Zheng is supported by a grant from RSRE awarded to Prof. J. Frisby, Prof. S Billings and Prof J. Mayhew. The authors wish to acknowledge their debt to all the members of AIVRU.

References

[1] C. Harris and M. Stephens. A combined corner and edge detector. In *Proceedings of the Fourth Alvey Vision Conference*, pages 147–151, August 1988.

[2] Y. Kanayama and N. Miyake. Trajectory generation for mobile robots. In *Robotics Research*, volume 3, pages 333–340, 1986.

[3] L. Matthies and T. Kanade. The cycle of uncertainty and constraint in robot perception. In *Robotics Research, The Fourth International Symposium*, pages 327–335, 1988.

[4] Y. Zheng N.A. Thacker S.A. Billings J.E.W. Mayhew and J.P. Frisby. 3d scene segmentation using a constrained motion parameter recovery algorithm. Technical report, 1991.

[5] Y. Zheng N.A. Thacker S.A. Billings J.E.W. Mayhew and J.P. Frisby. Ground plane motion parameter recovery and 3d segmentation using vehicle trajectory constraints. Technical report, 1991.

[6] N.A. Thacker. Corner detection and matching. Memo 58, AIVRU, 1990.

434

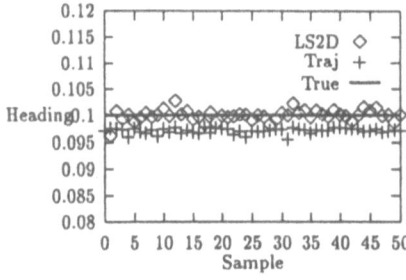

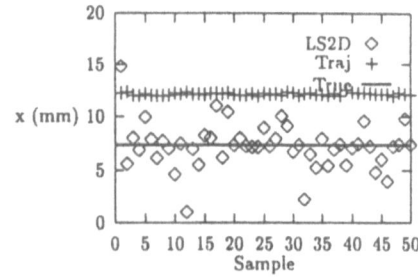

Figure 2: Estimates of θ in radians for Least Squares and Trajectory Methods for Motion Along a Clothoid

Figure 4: Estimates of $t_x(mm)$ for Least Squares and Trajectory Methods for Motion Along a Clothoid

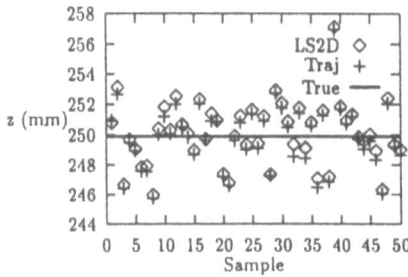

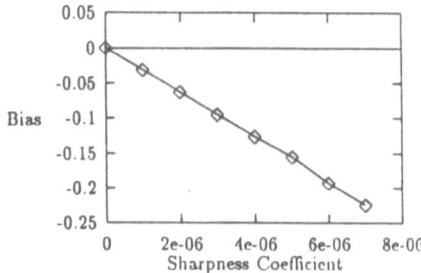

Figure 3: Estimates of $t_z(mm)$ for Least Squares and Trajectory Methods for Motion Along a Clothoid

Figure 5: Bias in $\theta(deg)$ for the Trajectory Method for Motion Along Clothoids of Increasing Sharpness

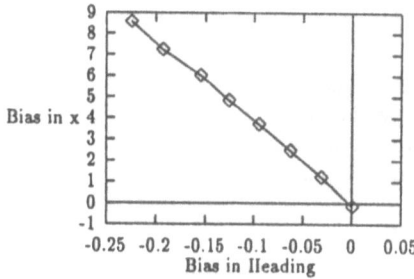

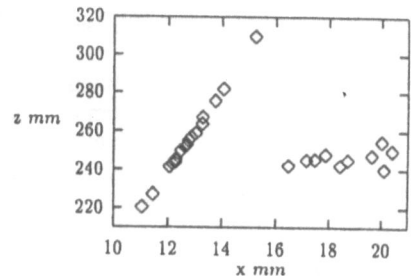

Figure 6: Bias in $t_z(mm)$ against Bias in $\theta(deg)$ for the Trajectory Method

Figure 8: Estimates of t_x and t_z from Individual Data Points with Object Moving Horizontally

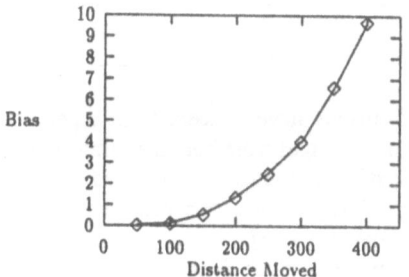

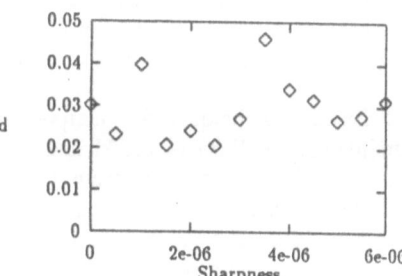

Figure 7: Bias in $t_z(mm)$ for the Trajectory Method for Different Sampling Rates

Figure 9: Standard Deviation of θ for Motion along Clothoids of Increasing Sharpness

Estimation of Cloud Cover using Colour and Texture

K. Richards and G.D. Sullivan

Intelligent Systems Group,
Department of Computer Science
University of Reading, RG6 2AY.
K.Richards@reading.ac.uk.

Abstract

We describe methods for using colour and texture to discriminate cloud and sky in images captured using a ground based colour camera. Neither method alone has proved sufficient to distinguish between different types of cloud, and between cloud and sky in general. Classification can be improved by combining the features using a Bayesian scheme.

1   Introduction

In order to provide accurate local forecasts, meteorologists require good quality observations, including those of cloud. The increased use of automatic stations means that valuable visual assessment of weather conditions is lost. This paper considers colour and texture as possible features to provide automatic ground-based estimation of cloud cover.

Existing techniques for interpreting cloud images have concentrated largely on satellite images. Work by Parikh and Rosenfeld used images from both the infra-red and visible regions of the spectrum [1]. Their system extracted spectral and textural information, and reached a classification using a complex selection procedure. Other papers in this field have used these features independently [2,3].

Ground-based assessment of cloud cover provides a much finer resolution in both space and time than is possible by satellite, and can provide valuable information concerning cloud at different heights. Coomes and Harrison developed a cloud estimation system using ground-based radiometric measurements [4], but although this method provides good estimates of total cover, it was unable to distinguish between different classes of cloud.

In this paper we report experiments with colour and texture analysis on cloud images obtained from a ground-based colour camera, and demonstrate that although each provides good discrimination neither is sufficient for complete analysis. A combination of these features using a simple Bayesian scheme provides improved classification.

2   Colour processing using normalised colour space

Colour video cameras provide an inexpensive and easy means for capturing colour images, although the stability of the colour signal is uncertain. We captured a sample set of images under varying conditions of brightness, magnification and weather in order to

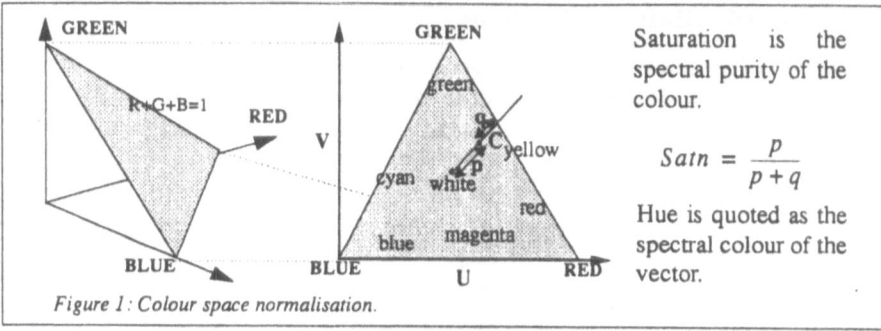

Saturation is the spectral purity of the colour.

$$Satn = \frac{p}{p+q}$$

Hue is quoted as the spectral colour of the vector.

Figure 1: Colour space normalisation.

find (i) what consistency can be expected from such cameras under different viewing conditions, and (ii) the separation of cloud and sky in colour space.

Early work compared a number of colour metrics and found little difference in their discriminatory abilities; consequently we have adopted a simple colour normalisation based on the CIE colour transform [5]. Each channel is normalised with respect to the total signal from all three channels, effectively projecting RGB colour space from the origin onto the plane $R+G+B=1$ (see Figure 1).

2.1 Training set distributions

A total of thirteen images were used for the colour training set and covered a range of weather conditions, from overcast to very sunny. For each image, masks were manually created to identify regions as cloud or sky. This was critical to our analysis since mislabelling would introduce errors into the final colour distributions, consequently classification was conservative and ambiguous regions were excluded from further analysis.

The pixels within each classified region were used to generate colour space distributions for cloud and sky. On an image-by-image basis it was evident that cloud and sky were distinct. To examine the *class* distributions, the results for single images were combined. Three dimensional plots of the class distributions are shown in Figure 2.

Clearly there is some overlap between sky and cloud, and to estimate the extent of overlap a confusion matrix was generated (Table in Figure 2). From these distributions the confusion between cloud and sky appears to be less than 3%, therefore the CCD colour camera is sufficiently stable to offer good distinction between cloud and sky over the range of conditions in this sample set.

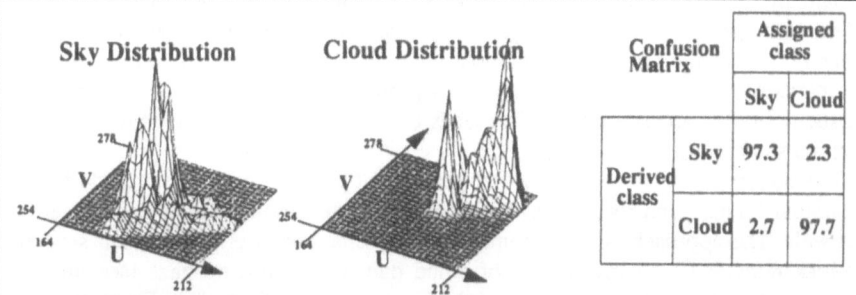

Confusion Matrix		Assigned class	
		Sky	Cloud
Derived class	Sky	97.3	2.3
	Cloud	2.7	97.7

Figure 2: A section of colour space showing the sample sky and cloud distributions. The table shows the resulting confusion matrix, with manually assigned class compared against the class derived automatically.

2.2 Classification using colour

A further set of images was captured under a wider range of conditions than with the original training set. These were classified pixel-by-pixel by looking up the most probable classification based upon the sample distributions. Typical results are shown in Figure 3, where the relative probability of cloud is shown as intensity.

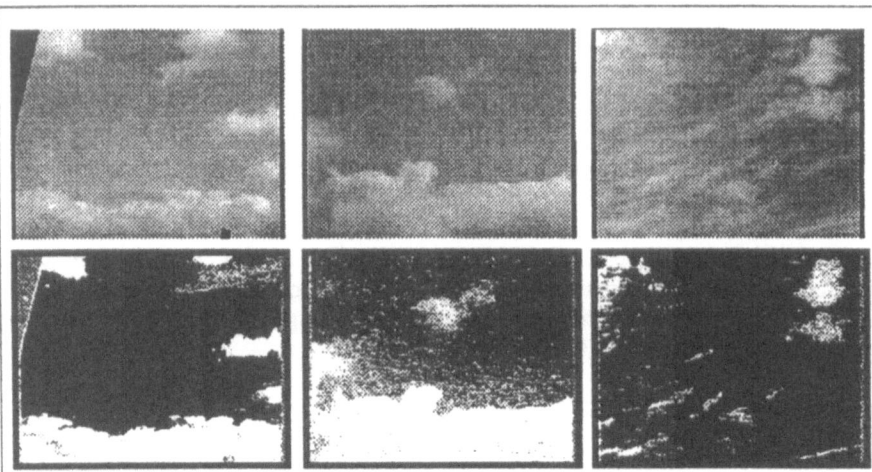

Figure 3: Colour Segmentation of cloud images. In the lower images the intensity indicates the relative probability of cloud against sky (white = cloud, black = sky)

The left image shows a clear sky bordered by several cumulus clouds of varying size. The top right corner of this image contains a shadowed base of some cumulus surrounded by a small amount of haze. For the purposes of this report, haze is defined as whitish sky arising from illumination of pollution or other particles in the atmosphere. The results of segmentation are clearly good; all the cumulus clouds are located including the shadowed section, although there is some confusion shown by the grey areas (top right of left image in Figure 3).

The image in the middle column was taken with a telephoto lens and shows a border of cumulus at the bottom, with a fragment of cumulus slightly above, and a small degree of haze near the borders. There also appears to be some surrounding fibrous elements of cloud. This classification highlights the problem of *haze* in colour analysis. Chromatically it is very similar to cloud and results in the mis-classification of sky.

The right image contains a large amount of cirrus, which appears wispy and transparent. The image also contains obvious cumulus in the foreground, to the right and left corners. Again this image demonstrates a problem with colour segmentation; thin cloud such as cirrus appears bluish. This leads to an obvious colour ambiguity between such cloud and clear sky.

Nevertheless we have found that inexpensive CCD technology can provide fairly stable colour images, and for the purpose of cloud segmentation a simple colour metric works well. The approach is largely unaffected by changes in brightness, but several difficulties remain: (i) overestimation of cloud due to surrounding haze (see middle Figure 3), and (ii) highly textured cloud such as cirrus is missed (see right Figure 3).

3   Texture analysis of cloud images

Texture has been successfully employed to classify satellite images of cloud, and we have seen that on the basis of intensity and colour there is little difference between transparent cloud (such as cirrus) and sky. Texture may provide a means to distinguish between them, and may also be useful for recognising different classes of cloud such as cumulus, stratus and cirrus.

3.1   Laws' texture description

We have considered a number of approaches including second order statistical moments and fractal texture models [8,9]. A texture model which has received considerable interest over recent years is based on *local linear transforms*. We have found this approach provides much better classification than with either of the other two approaches, although it does result in an n-fold increase in computation (where n is the number of local linear transforms).

The image is first transformed using 9 local linear operators which were defined by Laws [6]. These make use of a number of empirical linear transformations which are intended to extract elementary underlying structures such flats, spots and edges. For any given point an energy measure, such as the variance or some higher order statistical measure, is calculated over a local neighbourhood in each of the transformed images. The local texture properties of that point are described by the vector of feature energy measures. Figure 4 shows Laws' spatial-statistical texture model diagrammatically.

Unser extends this work, attempting to provide a more formal basis for such texture analysis [7]. He demonstrates how optimal transformation filters may be calculated for two textural classes and also proposes sub-optimal transforms, such as the Discrete Sine Transform (DST) and the Discrete Cosine Transform (DCT), which provide good discrimination over a broad range of textures.

For the purposes of comparison we applied both Laws' 3x3 linear transforms and Unser's sub-optimal DST to the data. The DST performed very well on a number of images, but Laws' transforms consistently produced more accurate classification.

3.2   Bayesian classification using feature vectors

Assuming the class conditional probability distributions to be multivariate gaussian distributions, Unser uses a Bayesian classifier which minimises the possibility of mis-classification. Given a set of class-mean vectors v, and their covariance matrices C, a

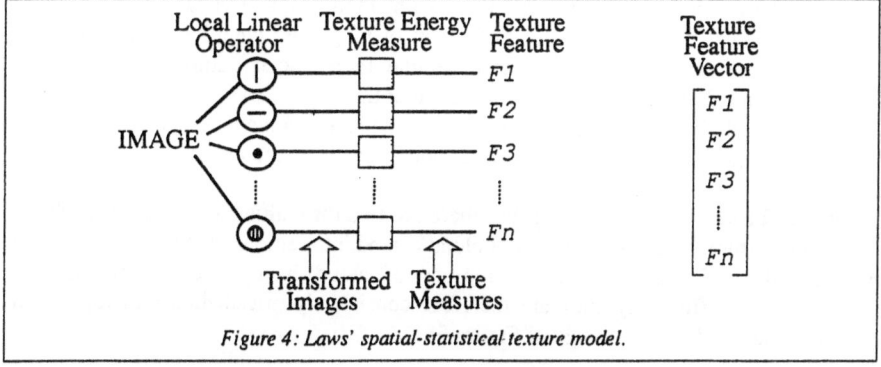

Figure 4: Laws' spatial-statistical texture model.

feature vector f can be classified by calculating the distance to each of the class-mean vectors and selecting the closest. Using a Bayesian classifier, the distance measure reduces to a simple matrix multiplication [7].

$$d_i = ((f - v_i)^T \bullet C_i^{-1} \bullet (f - v_i)) + \log \{ |C_{ii}| \}$$

[i=1.......n] $\qquad\qquad\qquad\qquad\qquad\qquad\qquad\qquad\qquad\qquad (1)$

The original colour training set was expanded to include a much wider range of cloud types, providing a number of texture samples for cirrus, haze, and cumulus. A total of twenty-eight images were used for this second training set. The class-mean vectors and covariance matrices were calculated using Laws' 3x3 linear transforms followed by a moving 8x8 window to calculate the energy measure. Classification used the same process, with the Bayesian classifier labelling each point with the nearest class (either sky, haze, cirrus or cumulus).

3.3   Classification using Laws' texture measures

The images classified earlier using colour were reclassified using Laws' texture measure (see Figure 5). The black regions indicate the class sky, the dark grey regions represent haze, light grey regions represent cirrus, and the white regions are cumulus.

Figure 5: Images labelled using nine of laws 3x3 transform functions, and energy measured over an 8x8 moving window. Black is sky, dark grey is haze, mid-grey is cirrus and white is cumulus.

The left image seems to be quite accurately classified: the central region of sky has now been divided into two regions of sky and haze, with evidence of some small regions of cirrus. The major bodies of cumulus seem to be labelled accurately, although small parts of cumulus are mis-classified (see bottom and top right section of the image in Figure 5). These are very smooth areas of cloud and are consequently very difficult to distinguish from sky using texture.

The middle image is encouraging since it shows correctly labelled magnified regions of cloud. The majority of cloud is cumulus (with small wisps of cirrus and haze in the surrounding sky). The successful classification of magnified cumulus using texture is somewhat surprising, but may possibly be due to a fractal nature of such cloud. However, such a discussion is beyond the scope of this paper.

Finally, with the right image we can see that the image containing a large quantity of high cirrus is now classified fairly well, although some of the cirrus is mislabelled as cumulus.

In conclusion, texture is useful, but there are two difficulties with the approach: (i) sky has no texture, so that the borders of cloud extend into regions of sky causing a small overestimation of cloud; (ii) some regions of cloud have little texture and are indistinguishable from sky, they are therefore consistently mislabelled (see regions of cumulus in left and middle image of Figure 5).

4 Using colour and texture

Both colour and texture have been shown to be important indicators to the presence of cloud, but neither is sufficient for complete segmentation. Chromatically, haze is indistinguishable from cloud, and cirrus is indistinguishable from sky. Texturally, smooth regions of cloud are frequently mis-labelled as sky. There is a clear attraction in using a combination of both methods.

4.1 Combining colour and texture

Normalised colour space produces two chromaticity measures. As we have already discussed texture can be described by a vector of spatio-statistical measures. By creating a combined feature vector using colour metrics and texture features, a description can be made in terms of both colour and texture. As with the texture classification a Bayesian classifier can be employed to minimise the probability of mis-classification.

The second training set of images was reprocessed to calculate the new class-mean feature vectors consisting of texture and colour measures. The classification procedure was based on that used for texture analysis. The results of classification using the combined colour and texture approach can be seen in Figure 5.

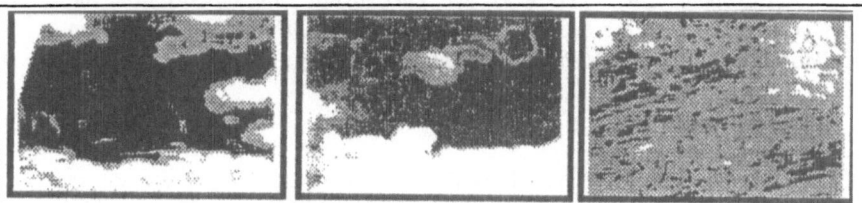

Figure 6: Images labelled on the basis of chromaticity and texture features.

4.2 Classification using colour and texture

The segmentation of the left image again seems good; the regions of haze located using texture are now correctly labelled as sky. The kind of improvement possible from using both texture and colour is demonstrated with the right image. Even using texture classification a number of cumulus patches were mistakenly found within cirrus. Using colour and texture these have been reduced considerably, the few remaining small patches most probably do belong to the lower cumulus clouds.

The centre image in Figure 5 is still troubled by noise, which is surprising since texture processing produced very clear regions of sky. It is difficult to determine whether this noise is an image artifact from magnification or an accurate interpretation of the data. At this degree of magnification small elements of cloud are likely to exist near cloud borders. This difficulty highlights one of the problems with interpreting cloud images, both by Man and machine.

This combination of colour and texture can greatly improve classification but is still imperfect. Obvious errors which persist from texture classification are the mis-classification of smooth regions of cloud (see the top right of first image in Figure 5), and the extended classification of cloud at borders with sky.

5   Conclusion and future work

It has been demonstrated that both colour and texture are useful features in the differentiation of cloud from sky, although neither approach has proved sufficient for complete analysis. A simple method to combine the features using a Bayesian classifier has shown an improvement in classification over the two independent approaches, but there are still problems to be resolved concerning: (i) cloud border accuracy, and (ii) the weighting of the classifier towards texture.

The approach presented here is suitable for simple models, but part of the success relies on the feature distributions being normal. This cannot be guaranteed in a problem domain such as this, since the radiation passing through the atmosphere is known to change throughout the day. In order to compensate for this, and deal with extreme situations such as sun-rise and sun-set, a more elaborate technique for combining colour and texture is required.

Despite these shortcomings, the automatic analysis of cloud cover does seem feasible. A full field trial using a portable system is planned for later this year which will provide an opportunity to collect a large body of classified data for future study and to compare the present system against the judgement of a trained meteorological observer.

Acknowledgements: This work was carried out with support from the Science and Engineering Research Council, under CASE award 90593345, in collaboration with the British Meteorological Office.

6   References

[1]   J.A. Parikh and A. Rosenfeld, "Automatic Segmentation and Classification of Infrared Meteorological Satellite Data", IEEE Transactions on Systems, Man, and Cybernetics, SMC-8, No.10, 1978., pp 736-743.

[2]   K.S. Lau and G. Wade, "Spatial-spectral clustering using recursive spanning trees", IEE Proceedings-I, Vol. 138, No. 4, August 1991, pp

[3]   Z.Q. Gu et al, "Comparison of techniques for measuring cloud texture in remotely sensed satellite meteorological image data", IEE Proceedings, Vol. 136, Pt F, No. 5, October 1989, pp 232-238.

[4]   C.A. Coomes and A.W. Harrison, "Radiometric Estimation of Cloud Cover", Journal of Atmospheric and Oceanic Technology, Vol. 2, 1984, pp 482-490.

[5]   R.J. Schalkoff, "Digital Image Processing and Computer Vision", John Wiley & sons INC. 1989.

[6]   J. Foley, A. van Dam, S. Feiner and J. Hughes, "Computer Graphics: principles and practice", 2nd Edition, Addison-Wesley 1990, 579-585.

[7]   A.P. Pentland, "Fractal based description of natural scenes", IEEE Transactions, 1984, PAMI-6, pp 661-674.

[8]   T.J. Dennis and N.G. Dessipris, "Fractal modelling in image texture analysis", IEE Proceedings, 1989, Vol. 136, No.5, pp 227-235.

[9]   K.I. Laws, "Textured Images Segmentation", PhD dissertation, University of Southern California, 1980.

[10]   M. Unser, "Local linear transforms for texture measurements", Signal Processing 11, 1986, pp. 61-79.

Building a Model of a Road Junction Using Moving Vehicle Information

Xu Li-Qun
David Young

School of Cognitive and Computing Sciences, University of Sussex
Brighton BN1 9QH, UK

David Hogg

School of Computer Studies, University of Leeds
Leeds LS2 9JT, UK

Abstract

We describe a program to construct a model of a road junction using data from a single camera. The model specifies the ground plane orientation in camera coordinates and the positions of traffic lanes, and is obtained entirely from observations of vehicle movements, with no static image analysis. At present, the model is restricted to representing straight lane segments. We describe our methods for segmentation, object tracking, ground plane estimation and lane identification. Throughout, we emphasise techniques which are computationally cheap and can be used with fairly low resolution data and a low frame rate. We nevertheless obtain a reliable model by using the statistics of large numbers of vehicle movements.

1 Introduction

In the course of a project aimed at producing reliable information about vehicle movements through a road junction, using computer vision, we have tackled the problem of building a model of the junction. The model, which specifies the position of the ground plane in camera coordinates, and the positions of vehicle lanes in the ground plane, is produced using measurements of moving vehicles in a single camera's view. In this paper, we report our overall strategy and some of the technical problems and our choice of solutions to them, and we propose how the work could usefully be developed. There has been much other work on the analysis of road traffic scenes, e.g. [1-7]; our aim here is to demonstrate how some computationally straightforward techniques, applied to visual data with coarse spatial and temporal sampling, can give us a reliable site model on which to base more detailed analysis of individual vehicle movements.

Specifically, we examine: segmentation of moving vehicles from the static background; tracking of multiple vehicles from frame to frame; estimation of ground plane parameters; and estimation of the positions of lane boundaries in the ground plane.

The approach taken was motivated by two complementary factors. First, automatic model generation means that the need for a site survey, or any manual intervention, is avoided, allowing a complete system to be set up quickly, and recalibrated quickly following any movement of the cameras. Second, producing a good model does not necessarily require tracking every vehicle accurately; the model can be based on the statistics of many vehicle movements. Once a model has been generated, it will clearly assist in making more reliable measurements of individual vehicles' movements. For present purposes, then, we require reasonably cheap computational methods, but we can afford to use methods that make occasional mistakes.

The traffic scene domain, and the practicalities of data collection, introduced some specific challenges. These included: a large inter-frame interval (determined by the low digitisation rate of our equipment, but not unrealistic for a practical application); variations in illumination (real, and also due to camera exposure compensation); static or slowly moving vehicles, producing difficulties with segmentation based on dynamic information; overlapping and very close images of vehicles, in queues and when travelling close together in parallel lanes; unconventional and unexpected behaviour by drivers; occlusion of vehicles by street furniture. Despite these problems, we chose to base our approach on dynamic information, since the difficulties and ambiguities of static scene analysis are likely to be harder to overcome. Since the purpose of the model is to provide a predictive framework within which individual vehicles can be tracked accurately, it is sensible to base the model on exactly this kind of information.

To test our ideas, we collected data from an intersection on a four-lane road, with two lanes in each direction, the two directions partially separated by islands. This was crossed by an ordinary two-lane road; the junction was controlled by traffic lights. We made a video recording using a conventional camcorder with automatic electronic exposure control from a foot bridge about 25 m from the junction. Traffic was moderately heavy, and contained a wide spread of vehicle sizes from motorcycles to large articulated trucks, travelling at speeds up to about 80 km/h. When the lights were against the traffic on the main road, queues formed, and there were frequently vehicles waiting to make right turns. Fig. 1 shows the view obtained by the camera.

We digitised about 30 minutes of the recording at a rate of about 3.3 frames/s, at a spatial resolution of 180 × 143 pixels, so the data rate was only about 85 kilobytes/s. Efficient algorithms which can produce useful results from these data therefore stand a good change of being applicable in practice. The algorithms described below were implemented in POP-11 and C using the POPLOG system.

2 Segmentation

The segmentation scheme described here is based on the combination of motion and contrast cues. First, difference information from successive frames of an image sequence is the key to detecting regions in which motion is taking place [9, 10]. Second, edges derived from individual frames indicate the surface boundaries of possible objects. Combining the filtering effect of the former and the localisation of the latter, the individual moving objects (vehicles or pedestrians), irrespective of their sizes, are isolated. The inspiration for this scheme can be traced to [11], [12] and [13], though the exact operation adopted differs from all these.

Let f_c and f_p denote the current and previous images respectively. The operation s

Figure 1. Typical views of the road junction.

corresponds to taking the sum of the absolute values of the convolutions with the two 3 ×
3 Sobel edge templates; b_T is a threshold operation with threshold T; and ∧ means the
operation of coincidence or logical AND. The operation we used is then described as

$$B = b_{T1}(s(f_c)) \wedge b_{T2}(s(|f_p - f_c|))$$ (1)

The thresholds $T1$ and $T2$ are in fact identical, and are chosen to be the mean of the local
minima in the smoothed histogram of the values in $s(|f_p - f_c|)$.

Fig. 2b shows a typical binary image of the traffic scene after this processing. Moving objects are clearly picked out, although some failures still occur. In general, the most
significant problems are the fragmentation of large slow-moving vehicles and the grouping together of close vehicles.

As a kind of domain-independent low-level processing, this approach can consistently remove all the static parts of the background scene, including of course temporarily
stopped vehicles. This makes the method inadequate for atomic event recognition, but
entirely suitable for the present purpose of collecting sufficient statistical information
about vehicle movements to generate a site model, without the cost of maintaining a reference image.

We adopt a simple approach to segmenting the moving region into individual vehicles. Each vehicle's image is represented as a rectangle with its sides parallel to the
image axes. These are extracted by recursively grouping set pixels in the binary image,
using 8-connectivity, and recording the bounding rectangle for each connected blob.
Rectangles which are close together relative to their sizes are merged. More sophisticated techniques such as chain codes [14] or medial axis transforms [15] would allow
more exact representation of blob shape, at greater computational expense.

Fig. 2c shows the results of this processing. Generally speaking, the fit between the
vehicle images and the rectangles is adequate, despite the coarse representation.

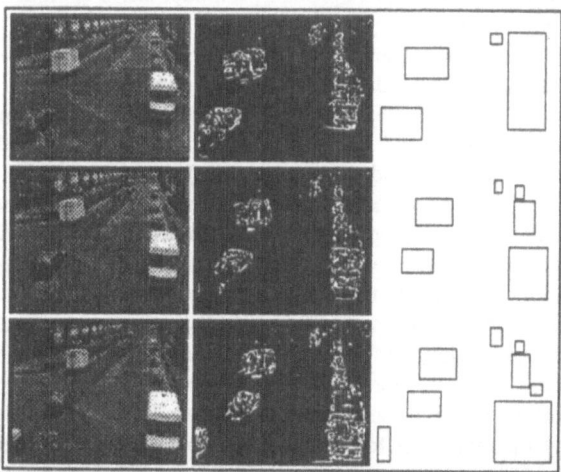

Figure 2. (a) Three images from the traffic sequence. (b) Results of applying the combined
motion and texture detection operator. (c) Segmentation into rectangles.

3 Object tracking

The establishment of feature correspondence between two related images has been a challenging area in computer vision research. Examples can be found in stereopsis algorithms [16, 17]and in motion understanding systems [7, 18-22]. Furthermore, Jenkin & Tsotos [23], performing stereo matching simultaneously with temporal feature tracking, presented a successful method of attacking the dynamic stereo problem. Important theoretical and practical developments included Ullman's minimal mapping approach [24], based on the 2-D distance between tokens, relaxation-based algorithms [25, 26], and Scott and Longuet-Higgins' eigendecomposition method [27-29]. An important constraint in tracking moving objects, when more than two frames are available, is that of motion smoothness, and this was notably exploited in [23] and [30]. The relationship to human visual perception has been elucidated by many experiments [31-36].

Early work in the road traffic domain was well summarised by Dreschler & Nagel [37]. These authors tried firstly to detect and track the *prominent points* of a single moving car, then to recover 3-D positions of these points in object coordinates, and finally to approximate a convex hull description of the car by using a minimisation approach. The assumptions introduced include a known camera tilt angle and planar road surface. Recent work can be found in [38] and [39], where a model-based paradigm is adopted. Here, tracking is carried out in 3-D world coordinates rather than 2-D image coordinates; this has clear advantages for any predictive scheme, but requires prior knowledge of object and world models.

Since our aim is to construct a site description, we carry out tracking in 2-D, and retaining our criteria of simplicity and efficiency we adopt a method in which path coherence and motion smoothness are embedded in the formulation of a similarity matrix for inter-frame matches. The most important constraint on our choice is our large inter-frame interval of about 0.3 s, which results in large image motions between frames. In addition, many vehicles (up to about 15) are often visible and moving in assorted directions in the same frame, and frequently occlude or pass close to one another. These factors mean that strong assumptions, and probably a model-based approach, are necessary to get highly reliable results.

For each image frame, the processing in section 2 results in a list of *rectangles*. Each of these is supposed to represent a different vehicle or pedestrian, and we are required to track these to produce a list of *trajectories*, each consisting of a list of linked rectangles extending over a number of frames. Each trajectory is taken to correspond to one *object*. We make the assumption that it is adequate to take each frame sequentially and to match the rectangles in it to the current set of active trajectories; this avoids the need for any backtracking through previous frames. The first requirement is then a *similarity measure*: a function from trajectory-rectangle pairs to positive real numbers; given this, we also need an algorithm for deciding unique matches given a matrix of the similarity measures for the rectangles in the current frame and the currently active trajectories.

Our similarity measure involves the sizes, aspect ratios and image velocities of the rectangles. For each active trajectory, a predicted rectangle, or *template*, is created, whose height, width and position are calculated using polynomial extrapolation of the same parameters of the rectangles forming the trajectory [40]. (The "position" of a rectangle here is that of the centre of the base of the rectangle.) Taking a rectangle in the current frame, we then calculate the relative difference in width, height and position between this rectangle and the template (e.g. in the case of width we find $(2|w_t - w_r|) / (w_t + w_r)$, where w_t is the width of the template and w_r the width of the current rectangle). If all three of these measures are less than some thresholds, a potential match is recorded in the similarity matrix, with the distance between the positions of the template and rectangle stored as a measure of dissimilarity. Otherwise, we rule out a match between this template and this rectangle. We thus end up with a sparse {no of active trajectories} × {no of rectangles} matrix for the current frame.

The important problem of discontinuous trajectories is tackled in building the matrix. Our low-level processing often causes objects to fail to be represented in a given frame; for example, two close vehicles may be merged into a single object, or one vehicle crossing the path of another may occlude it. In principle, an arbitrarily long gap can be tackled by keeping a trajectory active even when no rectangle has been appended to it in the current frame, and extrapolating over more than one frame interval in generating subsequent templates. In practice, we have found it sufficient to deal with gaps of one frame and no more. Thus a trajectory becomes inactive (i.e. is simply stored) when no rectangle has been appended to it for two frames. In matching a rectangle to a trajectory which has been extrapolated for two frame intervals, the thresholds mentioned in the last paragraph are increased.

Ideally, the sparse (dis)similarity matrix would have no more than one entry in each row or column, indicating unique matches. In practice, this will not be the case. Further processing is needed to establish the final matches. Our algorithm is straightforward: for each row of the matrix (i.e. for each trajectory) we find the minimum value, which corresponds to the closest rectangle, and remove all the other entries. Then for each column (i.e. for each rectangle), we find the minimum value amongst the remaining entries, and we note the corresponding match. This is a conservative approach, in the sense that matches made are likely to be good ones, but some matches can be missed.

Finally, if a frame cannot be matched to a trajectory, it is regarded as the start of a new trajectory.

This approach has proved quite robust to noise in our data, producing trajectories reliable enough for subsequent work. Fig. 3 shows some of the trajectories obtained. Errors arise mainly from occlusion between vehicles, resulting in merging or splitting mistakes in the segmentation. We have applied the method of Scott & Longuet-Higgins [28] to our similarity matrix, but the results, assessed in term s of mean trajectory smoothness, were not significantly better. The high noise levels and large inter-frame interval for our rectangle data mean that improved computational methods have strongly diminishing returns. Our simple method captures the essentials for adequate, if not optimal, tracking: trajectory prediction and a robust similarity measure.

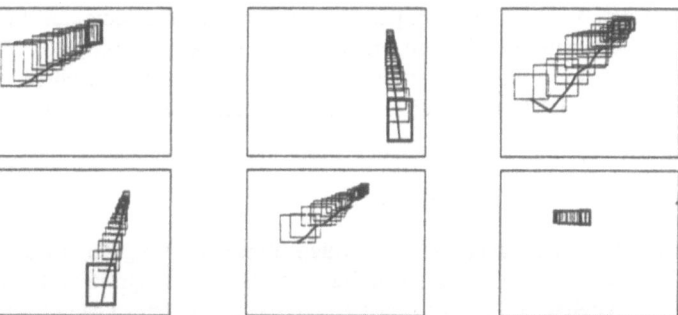

Figure 3. Some trajectories found by the method described. The top right shows a trajectory affected by the proximity of other vehicles. The bottom right shows a pedestrian crossing the road.

448

4 Ground plane estimation

For simplicity, we treat the ground surface as a plane. This is reasonable in many urban environments, but we expect to need to use a more accurate model in the future. The problem is to find the orientation of this plane in camera coordinates. Although several methods are possible, for example using the image size of vehicles, one which we have found successful and which we describe here is based on finding vanishing points for parallel vehicle trajectories. Although this requires the assumption that some parts of the vehicle trajectories lie in families of parallel straight lines, this will be true for many sites. The technique turns out to be robust against noise, largely because it is based on non-local measurements: the parameters of each trajectory are derived from a significant number of images and from a large region of the image. Furthermore, a large number of trajectories can be analysed in order to derive the ground plane parameters, using a form of the Hough Transform. Other advantages are the possibility of ready integration with information from the analysis of static features such as the road boundaries and lane markings, and the fact that the segmentation into groups of parallel trajectories can be used immediately in finding traffic lane positions.

The basis of the method is the fact that under perspective projection, parallel lines in a 3-D scene form a fan of lines in a 2-D image, all intersecting at a common vanishing point. If the 3-D lines lie in a plane, two vanishing points are sufficient to determine the rotational part of the transformation from the camera coordinate system to a coordinate system with two axes lying in this plane. In our case, given straight segments extracted from traffic trajectories, the problem of finding two vanishing points, and hence the road surface parameters, reduces to a problem already well studied in, e.g., [41-44]. We extend this work by applying it to trajectories rather than to static features, although we note also the effective application of vanishing points in autonomous vehicle road following by Liou & Jain [45].

The first stage of the process is to segment the trajectories into straight line segments, using recursive subdivision with a test for goodness of fit based on maximal normal distance of the trajectory segment from the straight line joining its ends [46]. Segments shorter than a threshold are discarded.

Each straight segment is then represented by the unit normal n of an *interpretation plane* which contains the segment and the focal point of the camera, O. If this normal is mapped onto the Gaussian sphere centred on O, the great circle normal to n contains all the directions which could be parallel to 3-D trajectory segment, given the evidence of the line segment in the 2-D image. The intersection of more than one such great circle gives the vanishing point for a group of parallel segments, and their common direction in 3-space. For further details, see the references cited above, [47] and [48].

In our implementation, we parameterise the forward Gaussian hemisphere using azimuth α and elevation β. These are related to image coordinates (x,y) and camera focal length f by

$$x = f\tan\alpha \qquad y = f(\tan\beta/\cos\alpha) \qquad (2)$$

A hierarchical Hough transform [43], which utilises an irregular division of the Gaussian sphere, was used to find the vanishing points. In this algorithm, the sphere is divided into four patches in terms of α and β, and the line segments vote for the patch through which their great circles pass. Those patches with votes exceeding a user-defined threshold are further subdivided and the process proceeds recursively, using a breadth-first strategy on the quadtree, until the patch size is as small as the expected scatter in the data. (We used a 1.5 degree × 1.5 degree limiting region.) In order to reduce the effects of noisy short segments, we weighted each line segment by a linear combination of its duration in time and its length in the image.

The procedure outputs a small patch of the Gaussian sphere and a list of the segments voting for it. After the dominant vanishing point has been found, the contributing line segments are removed from the whole data set, and the process repeated to find subsequent vanishing points.

Results obtained from the traffic sequence are shown in Fig. 4. The first vanishing point along the main road is shown at the top right of Fig. 4a by a small blob, whilst the second vanishing point along the crossing road is outside the frame to the left, at the intersection of the roughly horizontal lines in Fig. 4b. The line joining the two vanishing points (the estimated horizon) is shown along the top of Fig. 4a. Although our data are dominated by vehicles moving along the dual carriageway, and trajectories in the direction across the line of sight are under-represented, the method nonetheless finds two groups of parallel trajectories. It can be seen that both the vanishing points found lie close to the apparent position of the horizon (which is not actually visible in the image), indicating qualitative success for the method.

Given two vanishing points in the image, it is straightforward to obtain the ground plane parameters, up to a scale factor. The two lines in 3-space through O and each of the vanishing points are both parallel to the ground plane. The normal to the ground plane in camera coordinates is therefore given by the cross product of these two vectors, and the rotation matrix follows easily.

5 Identification of traffic lanes

Since our goal is to provide a basis for event recognition in traffic scenes, it is valuable to have a description in terms of traffic lanes, which provide primary expectations for traffic behaviour, and summarise large numbers of vehicle movements. We begin with the assumption that the lanes are straight, which gives strong computational advantages, though clearly will not be adequate for all sites. We choose to carry out the lane analysis in image coordinates, since the vanishing point method for the ground plane leads naturally to an approach to lane segmentation, the direction histogram method.

Given a vanishing point, the line of each associated trajectory segment can be described in terms of a single parameter - its orientation in the image - since it must

Figure 4. (a) Trajectory line segments and estimated horizon position.. (b) The lane structure superimposed on an image from the original sequence.

450

(approximately) pass through the vanishing point. We measure this orientation as the angle between the horizontal and the mid-point of the trajectory segment. The set of trajectories can then be described by the histogram of their orientations. In practice, we weight each contribution to the histogram by the duration in time of the trajectory segment.

The histogram clearly contains information about the lane structure for the vehicles whose paths contributed to the particular vanishing point, for instance those moving along the dual carriageway road. In order to find the lane boundaries, we smoothed the histogram with a 3-point average, and found local maxima. These were taken to correspond to lane centres; we simply took bisectors of these directions as the lane boundaries. This is repeated for each vanishing point. Once found, the lane boundaries can be projected onto the image plane, or onto the ground plane.

Results are shown in Fig. 4b, by projection of the lane boundaries onto an image. It can be seen that the actual lanes are fitted quite well in this case, despite the fact that the main road description uses over 400 segments and the cross road only about 30.

6 Discussion

We have demonstrated that by making suitable assumptions and adopting appropriate algorithms, a site model can be generated from coarsely sampled image data, without great computational cost. For a computer vision system designed to interpret a traffic scene, this is a valuable ability. Our method relies on utilising the dynamic information in the scene, and on measuring reasonably large numbers of vehicle trajectories so that statistical techniques can overcome high noise levels. We have obtained promising results from challenging data.

Several improvements are possible to this initial system. The representation of vehicle images using an "upright" rectangle is probably over-simple, and some more degrees of freedom would be useful. Certainly one extra parameter specifying the orientation of the major axis of the blob would be worthwhile. The tracking algorithm could possibly benefit from more sophisticated prediction: Kalman filter techniques rather than the polynomial extrapolation would undoubtedly give improvements at some computational cost. The vanishing point method is sensitive to the thresholds used, and needs to be set automatically. Finally, we need to be able to deal properly with vehicles turning at the junction; one way to do this is to regard their movements as transitions from one lane to another; alternatively we could introduce curved lane segments into our model, extending the Hough Transform method to deal with circular trajectories.

The main aim, however, is to provide a relatively simple model to bootstrap more sophisticated processing. Once an approximate model has been obtained, image positions can be projected into 3-D, and various possibilities immediately arise, for example Kalman filtering on 3-D trajectories, which will allow much better tracking. The model can then be refined in parallel with the acquisition of information about individual traffic events.

Acknowledgments

We are grateful to NTT-Data of Japan for supporting this work.

References

[1]   D.C. Hogg, G.D. Sullivan, K.D. Baker & D.H. Mott, Recognition of vehicles in traffic scenes using geometric models. In: IEE First Conf. Road Traffic Data Collection. London, 1984, pp 115-119.

[2]   A. Houghton, G.S. Hobson, L. Seed & R.C. Tozer, Automatic monitoring of vehicles at road junctions. Traffic Engineering and Control 1987; 28: 541-543.

[3]   K.W. Dickinson, Traffic data capture and analysis using video image processing. PhD Thesis, Sheffield University, 1986.

[4]   K.W. Dickinson & C.L. Wan, Road traffic monitoring using the TRIP II system. In: IEE Second Conf. Road Traffic Monitoring. London, 1989.

[5]   A.T. Ali & E.L. Dagless, Computer vision for automatic road traffic analysis. In: Proc. Int. Conf. 'Automation Robotics and Computer Vision'. Singapore, 1990.

[6]   L. Du, G.D. Sullivan & K.D. Baker, 3-D grouping by viewpoint consistency ascent. In: Proc. British Machine Vision Conf. Glasgow, 1991, pp 45-53.

[7]   T.N. Tan, G.D. Sullivan & K.D. Baker, Structure from constrained motion using point correspondences. In: Proc. British Machine Vision Conf. Glasgow, 1991, pp 301-309.

[8]   R. Marslin, G.D. Sullivan & K.D. Baker, Kalman filters in constrained model based tracking. In: Proc. British Machine Vision Conf. Glasgow, 1991, pp 371-374.

[9]   D.C. Hogg, Model-based vision: a program to see a walking person. Image and Vision Computing 1983; 1: 5-20.

[10]   R.C. Gonzalez & P.A. Wintz, Digital Image Processing. Addison-Wesley, Reading Mass., 1987.

[11]   R. Jain, W.N. Martin & J.K. Aggarwal, Segmentation through the detection of changes due to motion. Computer Graphics and Image Processing 1979; 11: 13-34.

[12]   W.B. Thompson, Combining motion and contrast for segmentation. IEEE Trans. PAMI 1980; 2: 543-549.

[13]   S.M. Haynes & R. Jain, Detection of moving edges. Computer Vision, Graphics and Image Processing 1983; 21: 345-367.

[14]   H. Freeman, Computer processing of line-drawing images. Computing Surveys 1974; 6: 57-97.

[15]   D.H. Ballard & C.M. Brown, Computer Vision. Prentice Hall, Englewood Cliffs, 1982.

[16]   W.E.L. Grimson, Computational experiments with a feature based stereo algorithm. IEEE Trans. PAMI 1985; 7: 17-34.

[17]   N. Ayache & F. Lustman, Fast and reliable trinocular stereo vision. In: Proc. First Int. Conf. Computer Vision. London, 1987, pp 422-427.

[18]   J.K. Aggarwal, L.S. Davis & W.N. Martin, Correspondence processes in dynamic scene analysis. Proc. IEEE 1981; 69: 562-572.

[19]   O.D. Faugeras, F. Lustman & G. Toscani, Motion and structure from motion from point and line matches. In: Proc. First Int. Conf. Computer Vision. London, 1987, pp 25-34.

[20]   J.K. Aggarwal & N. Nandhakumar, On the computation of motion from sequences of images - a review. Proc. IEEE 1988; 76: 917-935.

[21]   H. Shariat & K.E. Price, Motion estimation with more than two frames. IEEE Trans. PAMI 1990; 12: 417-434.

[22]   M.E. Spetsakis & J. Aloimonos, A multi-frame approach to visual motion perception. Int. J. Comp. Vision 1991; 6: 245-255.

[23]   M. Jenkin & J.K. Tsotsos, Applying temporal constraints to the dynamic stereo problem. Computer Vision, Graphics and Image Processing 1986; 33: 16-32.

[24]   S. Ullman, The Interpretation of Visual Motion. MIT Press, Cambridge MA, 1979.

[25]   S.T. Barnard & W.B. Thompson, Disparity analysis of images. IEEE Trans. PAMI 1980; 2:

333-340.

[26] K.E. Price, Relaxation matching techniques - a comparison. IEEE Trans. PAMI 1985; 7: 617-623.

[27] G.L. Scott & H.C. Longuet-Higgins, Feature grouping by "relocalisation" of eigenvectors of the proximity matrix. In: Proc. British Machine Vision Conf. Oxford, 1990, pp 103-108.

[28] G.L. Scott & H.C. Longuet-Higgins, An algorithm for associating the features of two images. Proc. Royal Soc. London B 1991; 244: 21-26.

[29] L.S. Shapiro & J.M. Brady, A modal approach to feature-based correspondence. Proc. British Machine Vision Conf. Glasgow, 1991, pp 78-85.

[30] I.K. Sethi & R. Jain, Finding trajectories of feature points in a monocular image sequence. IEEE Trans. PAMI 1987; 9: 56-73.

[31] J.T. Todd, Visual information about rigid and non-rigid motion: a geometric analysis. J. Experimental Psychology: Human Perception & Performance 1982; 8: 238-252.

[32] V.S. Ramachandran & S.M. Anstis, Extrapolation of motion path in human visual perception. Vision Research 1984; 23: 83-85.

[33] M.R.W. Dawson, The cooperative application of multiple natural constraints to the motion correspondence problem. In: Proc. Seventh Biennial Conf. Canadian Soc. for Comp. Studies of Intelligence. Edmonton, 1988, pp 140-147.

[34] M.R.W. Dawson, Apparent motion and element connectedness. Spatial Vision 1989; 4: 241-251.

[35] M.R.W. Dawson, The how and why of what went where in apparent motion: modeling solution to motion correspondence problem. Psychological Review 1991; 98: 569-603.

[36] M.R.W. Dawson & R.D. Wright, The consistency of element transformations affects the visibility but not the direction of illusory motion. Spatial Vision 1989; 4: 17-29.

[37] L. Dreschler & H.-H. Nagel, Volumetric model and 3D-trajectory of a moving car derived from monocular TV-frame sequences of a street scene. Computer Graphics and Image Processing 1982: 20: 199-228.

[38] A.D. Worrall, R.F. Marslin, G.D. Sullivan & K.D. Baker, Model-based tracking. In: Proc. British Machine Vision Conf. Glasgow, 1991, pp 310-318.

[39] D. Koller, K. Daniilidis, T. Thórhallson & H.-H. Nagel, Model-based object tracking in traffic scenes. In: Proc. Second European Conf. on Computer Vision. Santa Margherita Ligure, 1992, pp 437-452.

[40] J. Stoer & R. Bulirsch, Introduction to Numerical Analysis. Springer-Verlag, New York, 1980.

[41] M.J. Magee & J.K. Aggarwal, Determining vanishing points from perspective images. Computer Vision, Graphics and Image Processing 1984; 26: 256-267.

[42] H. Li M.A. Lavin & R.J.L. Master, Fast Hough transform: a hierarchical approach. Computer Vision, Graphics and Image Processing 1986; 36: 139-161.

[43] L. Quan & R. Mohr, Determining perspective structures using hierarchical Hough transform. Pattern Recognition Letters 1989; 9: 279-286.

[44] B. Brillault, Parallel and perpendicular line grouping in a 3-D scene from a single view. In: Proc. British Machine Vision Conf. Oxford, 1990, pp 257-292.

[45] S.P. Liou & R.C. Jain, Road following using vanishing points. Computer Vision, Graphics and Image Processing 1987: 39: 116-130.

[46] R.O. Duda & P.E. Hart, Pattern Recognition and Scene Analysis. Wiley, New York, 1973.

[47] S.T. Barnard, Interpreting perspective images. Artificial Intelligence 1983; 21: 435-462.

[48] T. Kanade, Geometrical aspects of interpreting images as a three-dimensional scene. Proc. IEEE 1983; 71: 409-460.

Off-line Handwriting Recognition by Recurrent Error Propagation Networks

A.W.Senior*   F.Fallside

Cambridge University Engineering Department
Trumpington Street,
Cambridge,
CB2 1PZ.

Abstract

Recent years have seen an upsurge of interest in computer handwriting recognition as a means of making computers accessible to a wider range of people. A complete system for off-line, automatic recognition of handwriting is described, which takes word images scanned from a handwritten page and produces word-level output. Normalisation and preprocessing methods are described and details of the recurrent error propagation network and Viterbi decoder used for recognition are given. Results are reported and compared with those presented by researchers using other methods.

1   Introduction

As computers become more powerful, and take a greater rôle in everyday life, the search continues for ways of integrating them into workplaces by making them conform to existing conventions of human communication. Though automatic speech recognition holds out some hope as a natural method of human to computer communication, for many applications it is unsuited and some of its promises are slow to be fulfilled. To fill these gaps, much interest is being shown in handwriting recognition, which offers applications as a direct input medium and in automatic document processing. Accordingly it is convenient to divide handwriting recognition systems into *on-line*, in which the writing is input directly via an electronic pen, and *off-line* in which words are written normally on paper and subsequently input using an optical scanner. This paper investigates the latter, more general, problem in which the data is in the form of a noisy image written with wide, overlapping pen strokes, rather than a time-ordered sequence of precisely known pen co-ordinates.

So far, most researchers have concentrated on on-line recognition (now seeing commercial application in pen computers) or the recognition of isolated characters, written separately and usually in capitals (as used in automatic postcode readers). Božinović and Srihari [1,2] have described a system for off-line cursive script word recognition which uses a feature-based approach, segmenting the word at potential character boundaries and decoding words by

---

*E-mail: aws@eng.cam.ac.uk

a probabilistic method. Edelman, Ullman and Flash [3] describe a more recent system, based on 'alignment of letter prototypes', which seeks to parametrise the strokes as splines and compare these with letter prototypes, allowing affine transformations of the models.

Here we have used a recurrent error propagation network to tackle word recognition, a technique which avoids the difficult problem of finding a valid segmentation of the word into letters, and naturally fits a dynamic programming procedure which constrains the search to a lexicon of permitted words. Real, scanned handwritten data is used in a whole word, single writer system. Earlier authors who have studied the off-line problem have in fact used data collected from a digitising tablet.

2 Database Used

The images used were written by a single writer, scanned on a 300dpi, 8 bit flatbed scanner and transferred to a workstation which was used for all of the subsequent processing. Words are written surrounded by white space so that a simple program can search for them before extraction into separate files. Figure 2 shows one such word.

For this study, the corpus chosen was the set of numbers 'one' to 'twenty', tens from 'thirty' to 'hundred' plus 'thousand', 'million' and 'zero' all written in words. These thirty-one words were chosen because, though constituting a limited vocabulary, they might be considered to form a useful corpus for an application such as cheque amount verification. Seven exemplars of each of these words were taken: three to serve as a training set and four as test data. Subsequently to achieve consistent training of the network, a validation set was collected which consisted of three exemplars of each word.

3 Preprocessing

To prepare the data for recognition by a network, extensive preprocessing is required to normalise the data and present it in an appropriate form. A number of operations are involved in this sifting process. Figure 1 illustrates the whole process for which each operation is described below.

3.1 Histograms

An important tool used throughout the preprocessing is that of the density histogram. These are formed by counting the number of black pixels in each vertical or horizontal line in the image, and plotting these counts as histograms which are used to identify the baselines of the word. (See figure 2.)

3.2 Baseline Detection

Among the most fundamental features extracted from the raw image data are the base lines (the lines which run along the top and bottom of a lower case 'o', above an 'l' or below a 'y'). It is with these lines that one identifies the ascenders and descenders so important in determining word shape. We shall

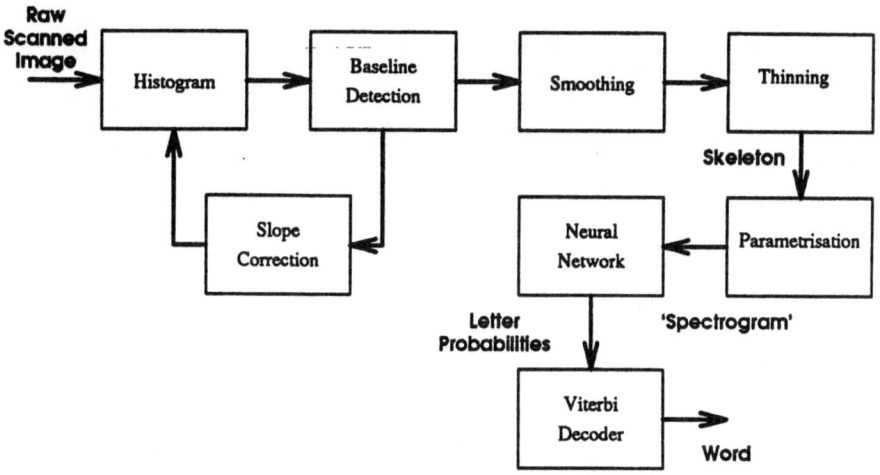

Figure 1: Schematic of the Recognition System

refer to these as the half, base, upper and lower lines, with a centre line midway between the half and base lines, as shown in figure 2.

The heuristic used for base line detection, is :

1. Calculate the vertical density histogram.

2. Reject that part of an image likely to be a descender.

3. Find the lowest pixel in each vertical scan line.

4. Carry out moment analysis to find the line of best fit.

5. Reject all outliers, and calculate the new line of best fit, which is considered to be the base line of the character.

3.3   Slope Correction

Given the best estimate of the base line, we can attempt to straighten the writing to give it a horizontal baseline. The image is straightened by application of a shear transform parallel to the y axis. (See figure 3b.) Now, we can re-calculate the base line (using a more accurate histogram) and the half line. These two lines are fundamental to the normalisation used in subsequent processes since the distance between the two is taken to be the character height, and for a given writer assumed to be proportional to the character width.

Figure 2: Histograms and Half, Centre and Base Lines.

3.4 Thinning

Having de-skewed and smoothed the image (by convolution with a 2-dimensional filter), it is thresholded to leave every pixel in one of two states. Then, an iterative thinning algorithm is applied to reduce the lines in the writing to one-pixel width so that the strokes can be followed later. Algorithms due to Gonzalez and Wintz [4] and Davies [5] were used to generate lines one pixel wide. (See figure 3c.)

4 Recognition

Having reduced the image to a standard form, which highlights invariants of the words and suppresses irrelevant variations, we come to the main task — that of actually extracting the word information from the image. As described earlier, a number of approaches have been taken to this problem, but the one used here is that of error propagation networks, using variations of the basic backpropagation algorithm [6] to learn the shapes of letters and to give probabilistic estimates of letter and word identities. We have investigated two architectures, but here we shall only describe the more successful recurrent error propagation network whose design as a time-invariant pattern recogniser makes it suitable for this application, with the the x axis substituted for the time axis.

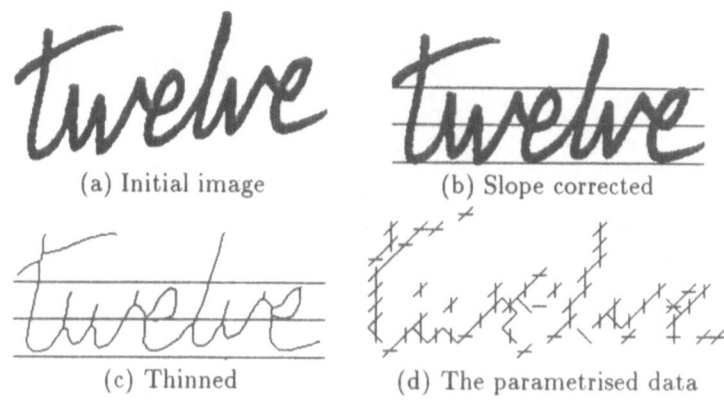

(a) Initial image   (b) Slope corrected

(c) Thinned   (d) The parametrised data

Figure 3: Successive stages in the preprocessing.

4.1 Parametrisation

Having chosen this recognition method, the preprocessed image needs to be parametrised in an appropriate manner for input to the network. The parametrisation scheme chosen is as follows:

We first divide the area covered by the word into rectangles. There are sixteen horizontal rows and a variable number of vertical frames of a width proportional to the estimated character width. Thus long words have more frames than short words, but, a given character will always occupy approximately the same number. For each of these rectangles, we allocate four bins representing line angles (vertical, horizontal, and the lines 45 degrees from these.) Given this framework, we coarse code the lines of the skeleton image:

Working from left to right in the thinned image, we take each stroke (a stroke being taken to mean a one-pixel wide line between junctions or end points) and divide it into equal length segments (the length being determined by the character size) for which the centroid position and angle are calculated. We now 'fill' the box associated with this segment's (x, y, θ) values. Segments which are not perfectly aligned with the angles of the bins, are coarse coded — that is they contribute to the two bins representing the closest orientations.

Figure 3d shows the input pattern schematically. In each square, a line indicates the presence of a line segment at approximately that angle in that bin. Because of the coarse coding, some line segments contribute to two bins and this is seen on the 'l' stroke which is between the vertical and 45 degrees so both these lines are shown in the corresponding boxes in Figure 3d.

4.2 Recurrent Networks

Recurrent error propagation networks have been successfully applied to speech recognition and other dynamic problems which treat a time-varying signal [7, 8]. A recurrent network is well suited to time-invariant pattern recognition because the same processing is performed on each section of the input

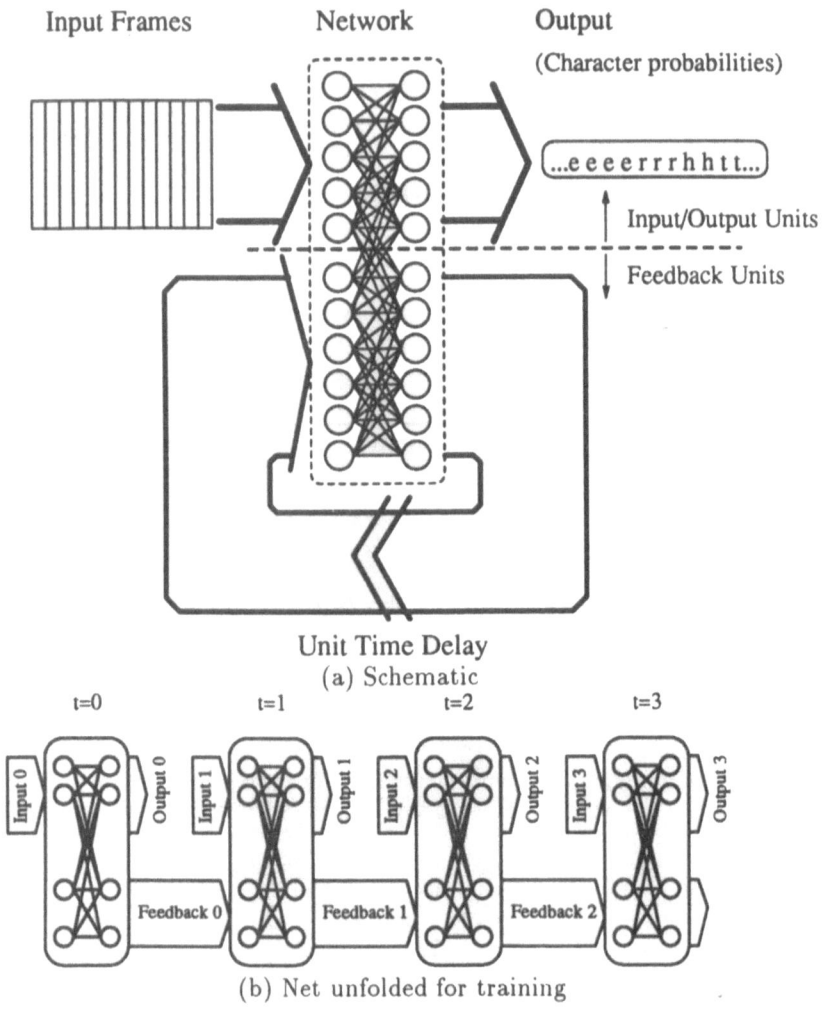

Input Frames   Network   Output

(Character probabilities)

...e e e e r r r h h t t...

Input/Output Units

Feedback Units

Unit Time Delay
(a) Schematic

t=0   t=1   t=2   t=3

Input 0 Output 0 Input 1 Output 1 Input 2 Output 2 Input 3 Output 3

Feedback 0 Feedback 1 Feedback 2

(b) Net unfolded for training

Figure 4: Recurrent Error Propagation Network

stream, but hidden units are available to encode context information about the preceding signal.

The network is in the form of a single layer of standard perceptrons with sigmoid activation functions (as described by Rumelhart, Hinton and Williams [6]). However some of the output units are connected one-to-one to some of the input units with a unit time-delay. (See figure 4a.) The remaining input units accept the parametrised input described above, one frame at a time, and the 27 output units give letter probabilities for the 26 lower case letters plus a space character. During the forward pass, successive frames of input data are presented to the network and results are fed back through the time delay, while letter probabilities are read off from the other outputs. To allow the network to assimilate context information, a delay of several frames is permitted before reading off the probabilities.

To be able to obtain the correct outputs from the network, we have to determine the correct values of the internal parameters ('weights') by a training procedure. Training the network requires 'unfolding' it in time. During training on a word, the inputs, outputs and feedback activations are stored for each frame. Then the errors are propagated back using the generalised delta rule [6], treating the network at different times as different layers of a multi-layer network (Figure 4b). To give target values for the training, we require a segmentation of the word into characters. For simplicity, an 'equal length' segmentation is used initially, where letters are assumed to be of the same length, though this is not true in practice. When the network performs reasonably well with this segmentation, we can use its own estimation of the optimal segmentation, which will be more accurate and enhances subsequent performance.

Since the network is being used to estimate probabilities, Bridle's [9] Softmax formulation is used on the output units. To improve performance and speed of convergence of the network we have made use of Jacobs' delta-bar-delta update rule [10], which gives each weight a variable training rate. Further improvements were obtained when Robinson and Fallside's [11] modifications to this rule were used. In order to achieve the optimum training time for the network, a validation procedure is used, whereby the network is tested on the validation set between periods of training. When recognition rates start to tail off, the network is considered trained and is tested on the test set.

5 Post-processing

The output of the network is in the form of probability estimates for each of the 27 categories in each of the frames. From this data we need to decide which word was written. This is achieved using dynamic programming [12] which relies on the principal of optimality, in the Viterbi algorithm to find a best match. We create a Markov model for each word in our vocabulary, with one state per letter, each state containing the probability that the data so far was generated by the preceding letters in that model. For each new frame, we take the network's probability estimates and multiply each state's probability by the probability of the corresponding letter. Transitions are allowed only from one state to itself or to the next state. After the last frame of data has been processed, the final state of each word model contains the probability that the observed data was produced by that model. By choosing the maximum of these likelihood estimators, if our model is good, we have a good estimate of the identity of the original word.

The Viterbi decoder can be improved by allocating transition probabilities which affect the average length of each letter. A simple durational model is included in the Viterbi decoder used, by giving each letter two successive identical states and thus forcing the minimum duration to be two frames.

6 Results

The system performance is judged by the number of words in the test set which are correctly identified, and these are quoted as a percentage of the 124 test words used. Using a network with 40 feedback units and 64 inputs/frame, a 73.4% recognition was obtained. Though the corpus used so far has only a small (31 word) vocabulary, we can investigate the system's scaling performance by testing the same data with a larger lexicon. This extended lexicon is created using the most common words in the LOB (Lancaster–Oslo/Bergen) corpus [13] to supplement the numbers lexicon. Using this lexicon the recognition rate was 60.4%.

7 Conclusions

The above results show that the method of recurrent error propagation networks can be applied successfully to the task of off-line cursive script recognition. Though comparison of results with other researchers is difficult (because of the difference in experimental details, the actual handwriting used and the method of data collection) we might note here the results which have been published for this problem . Božinović and Srihari achieved a 78% recognition rate with a 780 word lexicon [1] and Edelman, Ullman and Flash [3] quote a 50% recognition rate for whole words on a 30,000 word vocabulary.

Further improvements can be expected by adjustments in the preprocessing and parametrisation and also by the introduction of a language model, such as those already used successfully in the domain of speech recognition.

References

[1] R.M. Božinović and S.N. Srihari. Off-line cursive word recognition. *IEEE PAMI*, 11(1):68–83, January 1989.

[2] Sargur N. Srihari and Radmilo M. Božinović. A multi-level perception approach to reading cursive script. *Artificial Intelligence*, 33:217–255, 1987.

[3] S. Edelman, T. Flash, and S. Ullman. Reading cursive script by alignment of letter prototypes. *International Journal of Computer Vision*, 5(3):303–331, 1990.

[4] Rafael C. Gonzalez and Paul A. Wintz. *Digital Image Processing*. Addison Wesley, Reading, Mass., 1977.

[5] E.R. Davies. *Machine Vision: theory algorithms, practicalities*. Microelectrics and signal processing Number 9. London Academic, 1990.

[6] D.E. Rumelhart, G.E. Hinton, and R.J. Williams. Learning internal representations by error propagation. In D.E. Rumelhart and J.L. McClelland, editors, *Parallel Distributed Processing: Explorations in the Microstructure of Cognition*, volume 1, chapter 8, pages 318–362. Bradford Books, 1986.

[7] Barak A. Pearlmutter. Dynamic recurrent neural networks. Technical Report CMU-CS-88-191, CMU, School of Computer Science, Pittsburgh, PA15213, December 1990.

[8] Tony Robinson and Frank Fallside. A recurrent error propagation network speech recognition system. *Computer Speech and Language*, 5:259-274, 1991.

[9] John S. Bridle. Probabilistic interpretation of feedforward classification network outputs, with relationships to statistical pattern recognition. *Neurocomputing*, F 68:227-236, 1990.

[10] Robert A. Jacobs. Increased rates of convergence through learning rate adaptation. *Neural Networks*, 1:295-307, 1988.

[11] A.J. Robinson and F. Fallside. Phoneme recognition from the TIMIT database using recurrent error propagation networks. Technical Report TR 42, Cambridge University Engineering Department, Cambridge, UK., March 1990.

[12] H.F. Silverman and D.P. Morgan. The application of dynamic programming to connected speech recognition. *IEEE ASSP Magazine*, pages 6-25, July 1990.

[13] Stig Johansson, Roger Garside, Knut Hofland, and Geoffrey Leech. The tagged LOB corpus vertical/horizontal version. Technical report, Norwegian Computing Centre for The Humanities, 1986.

Evaluating a Hidden Markov Model Of Syntax In A Text Recognition System

Stephen Hanlon and Roger Boyle,

Division of Artificial Intelligence, School of Computer Studies,

The University of Leeds.

Abstract

Recognition of text by whole word shapes generates a set of candidate words for each printed word. A Hidden Markov Model (HMM) of syntax may be used to find the most probable sequence of syntactic tags for a sentence given the sequence of candidate sets. Candidate sets are then reduced by removing all words which are not associated with the chosen tag. We show that the tagging performance of the HMM does not deteriorate despite an increasing proportion of mis-classified words. We also show that using the model significantly reduces the number of candidates.

1 Introduction

Whole word recognition has been proposed as one solution to the problem of off-line interpretation of handwritten script [1, 6]. An unknown word has a set of features extracted describing the whole word shape and a set of words with similar features are generated as possible candidates. Such a set is called the candidate set for the printed word.

Post-processing of candidate sets aims to reduce the membership to one correct word. One method is to use a word's surrounding context in order to remove unlikely choices; previous methods have used both syntactic and semantic constraints [11, 8, 5, 4]. Syntactic constraints can be determined by the frequency of adjacent pairs or triplets of syntactic tags in an annotated corpus of text; tuples of tags with relatively high frequency imply a stronger contextual relationship compared to tuples with low frequency. If the tuple frequencies are represented as probabilities the language syntax can be approximated as a Markov process.

Hanlon and Boyle [4] and Hull [5] independently demonstrated that a Hidden Markov Model (HMM) of syntax can be used in a system where visual features of the words are synthesised. In both cases, a sentence was represented as a sequence of candidate sets and the most probable sequence of tags was calculated using the Viterbi algorithm [3]. Although these experiments did not use images of words, the results were promising with Hull showing that the candidate sets could be reduced by up to about 84%. A possible (and detrimental) consequence of removing words from the candidate sets is that when a word is mis-tagged the correct word may be removed from the candidate set.

This paper describes a set of experiments evaluating the use of a HMM of syntax in a text recognition system using off-line images of text. These experiments differ from those using synthesised features since noise effects in scanning and classification can generate erroneous candidate sets which do not contain the correct word.

Data used in the experiments was extracted from the tagged LOB (Lancaster Oslo Bergen) corpus [7]. Recognising words by their shape and not internal letters places a restriction on the number of words that can be recognised; for these experiments we chose to start with a small lexicon of 100 words and increase this to 1000 words in steps on 100 words to evaluate the effect of an increasing lexicon. We note that a lexicon of 1000 words is sufficient for many practical domains. The words chosen for the lexicon were the most frequent words in the LOB corpus removing digits and separately tagged suffixes such as 's. Test sentences extracted from the corpus were constrained to be single clause sentences containing only those words in the lexicon. Consequently as the lexicon grows, the length and complexity of the sentences increases.

2 Word Recognition

Words were recognised using vector quantisation; features were extracted from a training set of words and were clustered in feature space. Unknown words were classified by extracting features and finding the closest cluster centre in feature space. The members of the closest cluster were used to generate the candidate set.

Two different sets of features were used, Fourier based descriptors and a set based on ascenders, descenders and moment based features. Further, the training set of words used to generate the clusters was printed twice to determine the influence of different noise effects. One set contained each word printed four times in a roman font, the other set contained each word printed in four different fonts.

2.1 Features used

2.1.1 Word shape features

A nine dimensional vector of word shape features consisted of three moment based features, four counts of ascender and descender and the aspect ratio of the word image. Since words were not segmented into individual characters, a letter count was impossible to determine. Instead a relative measure of word length was given by the aspect-ratio, i.e. *length/height*.

Ascender and descender counts were taken by first splitting the word image into top, middle and bottom sections, where the top section contained any ascenders of the word and the bottom section contained any descenders. The word was then split down the vertical centre of gravity and ascenders and descenders were counted in the left and right regions.

A measure of the internal structure of the word image was calculated using moment based features. These were taken from [2] and have been used in character recognition and other applications [9]. These features are invariant over translation, rotation and scaling. The measures used were M_2, M_3 and

M_4:

$$\mu_{pq} = \frac{1}{N} \sum_{i=1}^{N} (u_i - \bar{u})^p (v_i - \bar{v})^q$$

$$r = (\mu_{20} + \mu_{02})^{1/2}$$

$$M_2 = [(\mu_{20} - \mu_{02})^2 + 4\mu_{11}^2]/r^4$$

$$M_3 = [(\mu_{30} - 3\mu_{12})^2 + (3\mu_{21} - \mu_{03})^2]/r^6$$

$$M_4 = [(\mu_{30} + \mu_{12})^2 + (\mu_{21} + \mu_{03})^2]/r^6$$

2.1.2 Fourier features

Fourier descriptors have been used to classify whole words by taking a two dimensional Fourier transform of the word image [10], and individual characters by taking a one dimensional Fourier transform of the outline of the letter [12]. One-dimensional Fourier features were chosen to describe the shape of the word envelope.

In order to extract one dimensional Fourier features of a word image, a chain code of the word envelope must first be taken. Letters in the word were then joined to create a connected region from which a chain code could be extracted.

Figure 1 : Word image and word envelope

The chain code was then represented as a set of points (x_i, y_i). The sequences $x_1, x_2 \ldots x_n$ and $y_1, y_2 \ldots y_n$ are interpreted as two one-dimensional signals, $x(m)$ and $y(m)$ where $x(L) = x(0)$ and $y(L) = y(0)$. A one-dimensional Fourier transform of each is taken and then normalised, i.e. from [12].

$$a(n) = \frac{1}{L-1} \sum_{m=1}^{L-1} x(m)e^{inw_0 m}$$

$$b(n) = \frac{1}{L-1} \sum_{m=1}^{L-1} y(m)e^{inw_0 m}$$

These descriptors are not invariant to rotation, shift or size, so the following normalisations are done.

$$r(n) = [|a(n)|^2 + |b(n)|^2]^{1/2}$$

$$s(n) = r(n)/r(1)$$

The power spectrum showed that the most information was represented by approximately fifteen low frequency descriptors and a small number of high frequency descriptors. The shape of the word was then stored as a 15 dimensional feature vector containing the low frequency descriptors $s(2) \ldots s(16)$.

2.2 Clustering

The lexicon is generated with each word printed four times, which is then scanned and a set of features extracted from each word image. The resulting feature vectors are then clustered using an adapted K-means clustering algorithm based on [13]. The number of clusters for these experiments was defined to be $n - 30$ where n is the number of words in the lexicon.

The result of the clustering algorithm is a number of clusters each containing a set of words as members. If the image acquisition and feature extraction were perfect, clustering would result with all four instances of a word in the same cluster. However, in practice, we find that noise effects cause some words to appear in more than one cluster. This has the effect of distributing the probabilities of words across different clusters and hence affects the tagging.

3 A Hidden Markov Model For Tagging Candidate Sets

A Hidden Markov Model is a probabilistic model consisting of a state transition matrix, a confusion matrix and a set of initial probabilities. The hidden states are the features of the model which are generated by a Markov process, that is, the probability of the system being in a particular state at time t depends only on the immediately preceding states. The observations are events probabilistically related to the hidden states and are those events of the system which can be measured. Usual HMM problems include finding the probability a model generated a sequence of observations, finding the most probable sequence of hidden states given a sequence of observations and generating a HMM given a large sequence of observations.

When tagging candidate sets, the hidden states are the syntactic tags and the observations are the candidate sets. The probabilities for the tag transition matrix and initial tag probabilities are calculated from the whole LOB corpus. First order probabilities are calculated from the frequency of tag bigrams and second order probabilities from the frequency of tag trigrams. The confusion matrix probabilities, $Pr(observation|tag)$ are calculated by summing $\frac{m}{4}Pr(word|tag)$ for all tags in the cluster where $word$ appears in the cluster m times (up to a maximum of four printed words).

A set of 22 tags were used, which were logical groupings of the 134 tags used in the LOB corpus.

Thus a sentence, $S = w_1, w_2 \ldots w_n$ is represented as a list of candidate sets, $\hat{S} = c_1, c_2 \ldots c_n$, for which we wish to find the sequence of tags, $T = t_1, t_2 \ldots t_n$, which maximises :

$$\max_T [Pr(t_1)Pr(c_1|t_1) \prod_{i=2}^{n} Pr(t_i|t_{i-1})Pr(c_i|t_i)]$$

This is calculated using the Viterbi algorithm using the standard HMM approach [3].

4 Results

Results were measured in three ways. The tagging performance was measured by comparing the Viterbi output with the tags in the LOB corpus. Two measures proposed and used by Hull [5] to measure the candidate set reduction and the error rate in the tagging were also used.

Measuring the tagging performance indicates how well the Hidden Markov Model tags the candidate sets in contrast to candidate set reduction and error rate which indicates the consequence of using syntactic tags as information to constrain possible candidates.

4.1 Tagging performance

The simplest measure of results is to calculate the percentage of tags generated by the Viterbi algorithm which correspond to the sentences in the LOB corpus. The graph in figure 2 shows the tagging performance for the first and second order experiments.

These results show that the choice of fonts in the training set can dramatically improve performance. The system tagged significantly better when trained with one font rather than four different fonts.

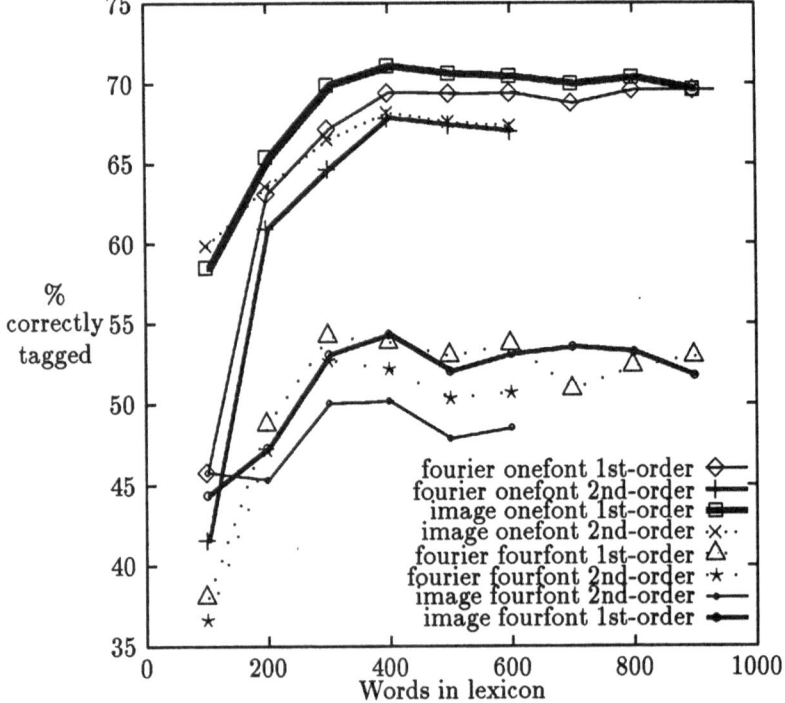

Figure 2 : Percentage of words correctly tagged

Perhaps the most interesting result is that the second order results were marginally worse than the first order results. In order to find why this was so, the probability distribution of tag bigrams in the test sentences was compared to the distribution in the tag transition matrix of the HMMs. This gave a measure of how well the tag transition matrix modelled the structure of the test sentences. It was found that the structure of the test sentences were better represented by the first order transition matrix than the second order matrix. This was due mainly to the choice of test sentences; by constraining the words in the test sentences, the structure of the test sentences were found to be much different to the structure of sentences in the overall LOB corpus. Hence we would expect sentences with richer syntactic structure to be better modelled by a second-order model.

It was also found by comparing transition probability distributions that the tag transition matrix better represents sentences with a larger lexicon than those with a small lexicon. Figure 3 shows a χ^2 measure of the test sentences and the first order transition matrix and it shows the two distributions converging in an inverse curve to the results. So, the results appear to be better as the transition matrix better approximates the test sentences.

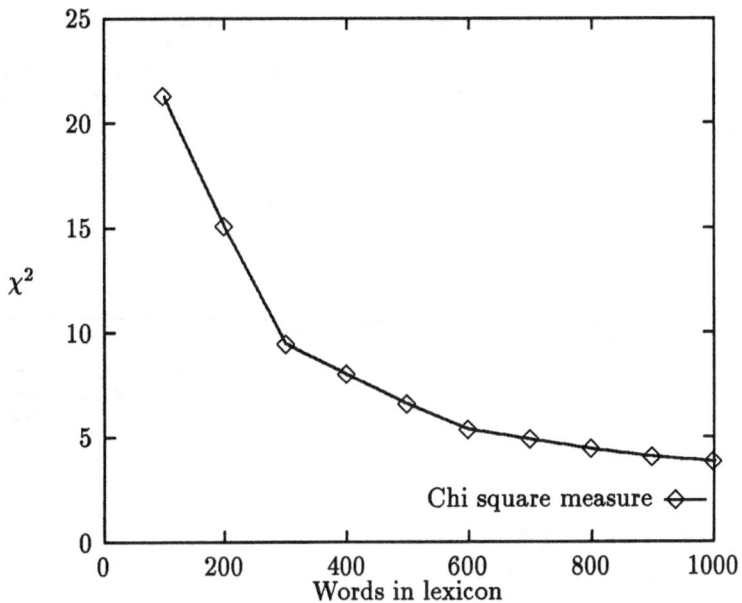

Figure 3 : χ^2 comparison of transition matrix and test sentences

4.2   Error rate

The error rate is defined as the percentage of candidate sets which do not contain the correct word. When tagging word images, this error occurs in classification where a word may be mis-classified, and after mis-tagging where the correct word may be removed from the candidate set. Consequently we use two error rate measures with word images, *classification error* is the percentage of candidate sets in error before tagging and *reduction error* is the additional percentage of candidate sets in error after tagging.

Figure 4 shows the error rate measures for the first order experiments, classification errors for second order experiments were identical and reduction errors slightly higher than those shown for first order.

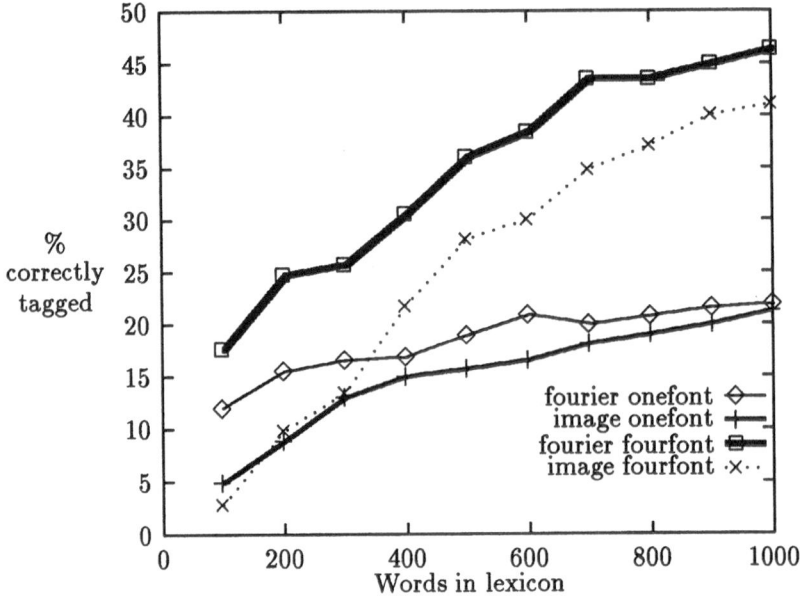

Figure 4a : First order classification results

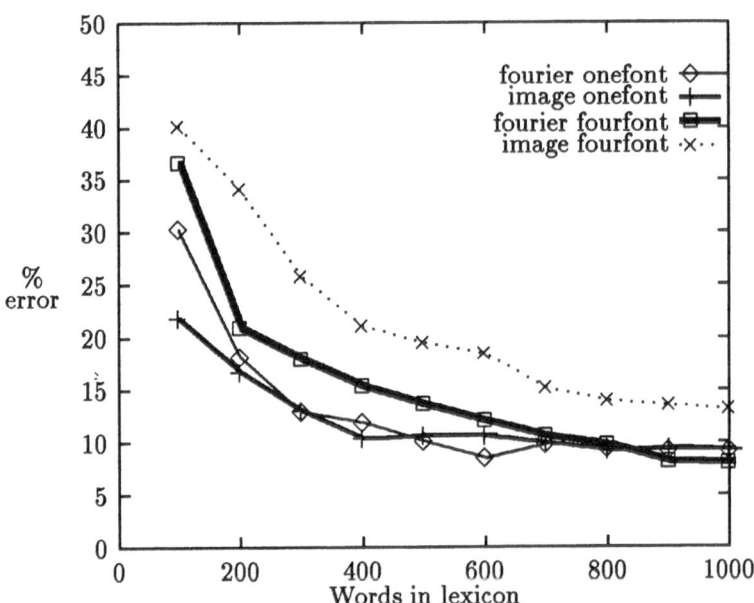

Figure 4b : First order reduction errors

The classification error shows that classification routines are introducing more errors as the number of words in the lexicon increases, reflecting the intuitive result that whole word features become less effective as the lexicon increases. If a large lexicon is to be used, a more robust set of features would have to be chosen. However, reduction errors are decreasing as the lexicon increases showing that when a word image is correctly classified, syntactic reduction of candidates introduces only a small amount of error.

Considering the increasing error in classifying words, the tagging performance is encouraging and shows that the HMM approach to tagging candidate sets is robust when coping with high levels of noise in the observation sequence.

4.3 Reduction of candidates

The percentage reduction of candidate sets shows how useful the tagging is when used in a text recognition system. This reduction is defined as the percentage reduction in the average candidate set size before and after tagging. Figure 5 shows the candidate set reduction for first order experiments. Second order results showed a slightly larger reduction.

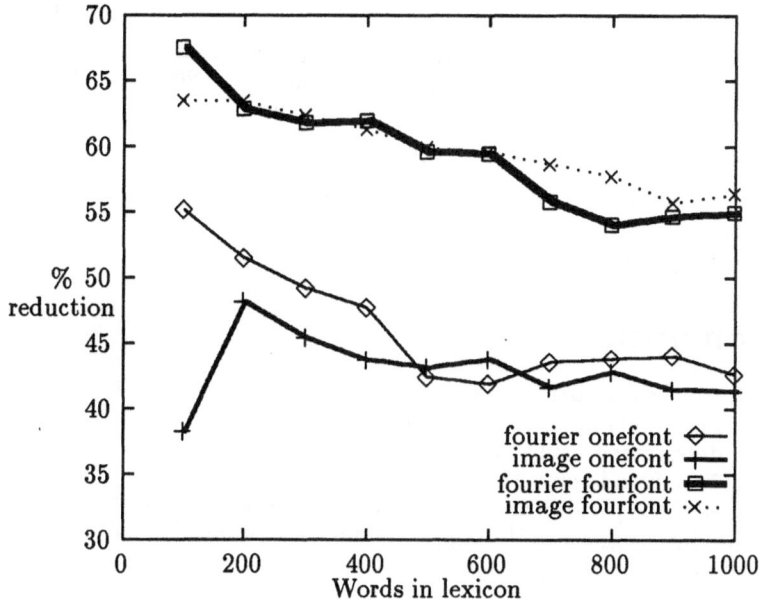

Figure 5 : Candidate set reduction

These results show that the set reduction tends to become worse as the lexicon increases. That is, the model is choosing tags which represent more candidate words and hence less words are removed. This suggests that perhaps for larger lexicons, the model chooses tags which are more probable given the observation, rather than tags which are less probable and promoted because of surrounding context.

We measure the number of times the Viterbi algorithm chooses the most probable tag given the observation, this is a crude measure of how the Viterbi algorithm is choosing the tags. The graph in figure 6 shows the percentage of most probable tags chosen.

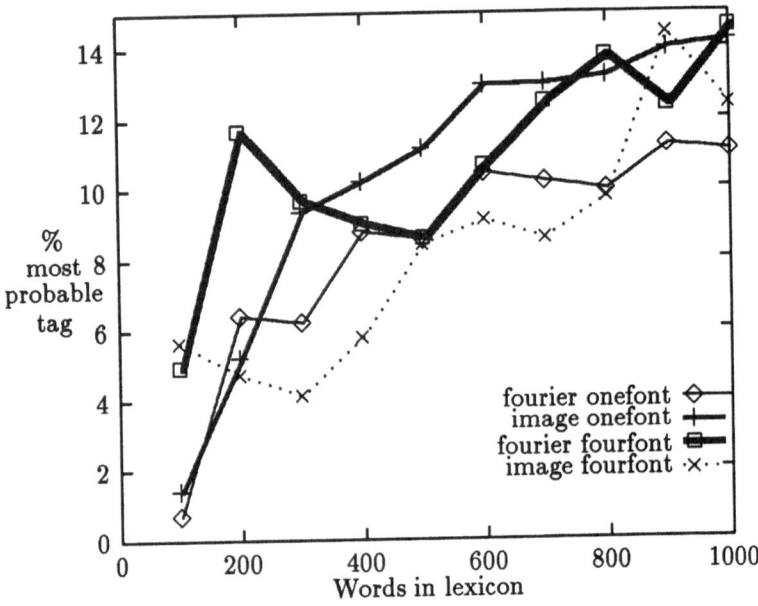

Figure 6 : Percentage of most probable tags chosen by Viterbi algorithm

The trend shown in the graph is that the Viterbi algorithm increasingly chooses the most probable tag for an observation as the lexicon grows. This is then reflected partly in the falling candidate set reduction.

5 Conclusion

A Hidden Markov Model was added to a text recognition system to provide syntactic constraints on word candidates. A syntactic tag for each set of candidates was determined using the Viterbi algorithm and the candidate sets were constrained to include only those words that could exist with the chosen tag.

Analysis of the tagging and reduction results were promising. The percentage of candidate tags correctly tagged did not deteriorate despite an increasing amount of errors in classification. We conclude that a Hidden Markov Model of syntax can introduce a large amount of contextual information when tagging scanned images of printed sentences.

References

[1] R D Boyle and R C Thomas. Interpretation of cursive script at the word level. Technical report, School of Computer Studies, University of Leeds, June 1990.

[2] S A Dudani, K J Breeding, and R B Mcghee. Aircraft identification by moment invariants. *IEEE Transactions on Computers*, 26:39–45, 1977.

[3] G D Forney. The Viterbi algorithm. *Proceedings of the IEEE*, 61:268–278, March 1973.

[4] S J Hanlon and R D Boyle. Syntactic knowledge in word level text recognition. In R Beale and J Finlay, editors, *Neural Networks and Pattern Recognition in Human–Computer Interaction*. Ellis Horwood, 1992.

[5] J Hull. Incorporation of a Markov Model of language syntax in a text recognition algorithm. In *Symposium on Document Analysis and Information Retrieval*, University of Nevada, Las Vegas. 16th – 18th March, 1992.

[6] J Hull. *A computational theory of visual word recognition*. Report number 88-07, University of NY at Buffalo, February 1988.

[7] S Johansson, E Atwell, R Garside, and G Leech. *The tagged LOB corpus.* Norwegian Computing Centre for the Humanities, Bergen, 1986.

[8] F G Keenan, L J Evett, and R J Whitrow. A large vocabulary stochastic syntax analyser for handwriting recognition. In *First International Conference on Document Analysis and Recognition*, Saint-Malo, France. September 30 – October 2, 1991.

[9] A Kundu, Y He, and P Bahl. Recognition of handwritten word : First and second order Hidden Markov Model based approach. *Pattern Recognition*, 22(3):283–297, 1989.

[10] M A O'Hair and M Kabrinsky. Beyond the OCR: reading whole words as single symbols based on the two dimensional, low frequency Fourier transform. In *First International Conference on Document Analysis and Recognition*, Saint-Malo, France. September 30 – October 2, 1991.

[11] T G Rose, L J Evett, and R J Whitrow. The use of semantic information as an aid to handwriting recognition. In *First International Conference on Document Analysis and Recognition*, Saint-Malo, France. September 30 – October 2, 1991.

[12] M Shridhar and A Badreldin. High accuracy character recognition algorithm using Fourier and topological descriptors. *Pattern Recognition*, 17:515–524, 1984.

[13] Q Zhang and R Boyle. A new clustering algorithm with multiple runs of iterative procedures. *Pattern Recognition*, 24(9):835–848, 1991.

Segmentation of Music Primitives

K.C.Ng and R.D.Boyle

Division of Artificial Intelligence, School of Computer Studies,
The University of Leeds,
Leeds LS2 9JT, United Kingdom

Abstract

In this paper, low-level knowledge directed pre-processing and segmentation of music scores are presented. We discuss some of the problems that have been overlooked by existing research but have proved to be major obstacles for robust optical music recognisers [1] to help entering music into a computer, including sub-segmentation of interconnected primitives and identification of nonstraight stave lines, and present solutions to these problems. We conclude that, with knowledge, a significant improvement in low-level segmentations can be achieved.

1 Introduction

Computers are being increasingly used for musical applications, and numerous available musical software package require a machine representation of music to perform their task. Currently, input methods are very time consuming and require some musical knowledge. Optical musical score recognition, especially if able to analyze handwritten scores, would be an interesting and time saving input technique. This is similar to the job of an 'Engraver' who reads a handwritten music score and with specific knowledge transforms it into an engraved music score for printing [5].

This visual problem might seem simple, since writing is black on white paper. Unfortunately, many of the symbols are highly interconnected (Figure 1). Only with musical knowledge can the meaningful figures be discerned.

Figure 1: A number of features which are interconnected.

Interpreting the handwritten notation of a composer is even more difficult because sloppy handwriting results in unclosed and ambiguous note heads, stems not attached to note heads, beams looking similar to slurs, with some phrase marks tying a number of stems and joining up all these features.

The overall target process is thus as follow :

- A score is scanned optically, and the digitised image fed into the computer as a raster image.

- The computer performs low-level processing; thresholding and deskewing.

- Segmentation; locate and erase the staves and decompose any composite features until they are recognisable as a primitive.

- Recognition; classified a primitive feature, and

- output it in some appropriate representation, of which standard MIDI file [6] is perhaps the most popular, being understood by most currently available software, although it is not ideal as an internal representation during processing.

In this paper, we concentrate on the issue of low level primitive segmentation, especially confusion introduced by slurs, ties, phrase marks and beams. This problem has received attention before - in particular, [4] notes that most musical score recognition is amenable to attack by examining vertical and horizontal projection histograms of suitably chosen windows of an image. Our approach is based on this observation, and our results tend to confirm the view that such simply derived observations carry a wealth information (often sufficient) in this domain.

2 Pre-processing

2.1 Thresholding

The continuous-tone image from the scanner is converted into a binary (black and white) image. For this purpose the Iterative Threshold Selection Method of Ridler and Calvard [10] with Lloyd's modification [8] is used.

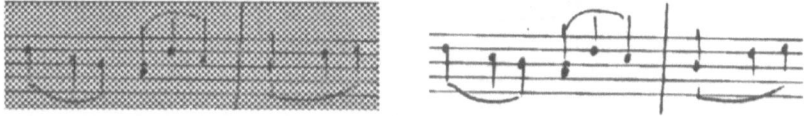

Figure 2: An example input grey tone digital image and the thresholded output.

The threshold method works well and the threshold value converges, usually in less than eight iterations. It produces clean output with little or no noise (Figure 2). Usually, such noise is just some isolated points and can be identified easily.

2.2 Skew correction

At this stage, musical information is contained in the black pixels. In music scores, horizontal alignment is an important property and an indispensable clue during recognition, permitting projection techniques to be used to detect feature position. Symbols are aligned on a five lines (stave lines) staff, and results will only be correct if the staves are horizontal at acquisition time. In

474

practice there is always a slight skew (characteristically less than two degrees) when the image is captured. We adopt a modification of the approach used by Martin and Bellissant [9] to find the skew angle.

First, we compute a measure of horizontality at some range of possible skew angles with a fixed step, for example, −5° to +5° at 0.1° intervals. The middle column of the image, which is the most likely to cut through any horizontal line, is scanned from the top to the bottom row. When we find a foreground (black) pixel, a line template, computed using Bresenham's discrete line algorithm [2], at each possible skew is offered to it, and the number of foreground pixels which fall on the template line counted. The count for each angle is then accumulated for each row, after which the angle with the highest count provides the skew angle. Frequently, there is no clear single peak to determine the skew

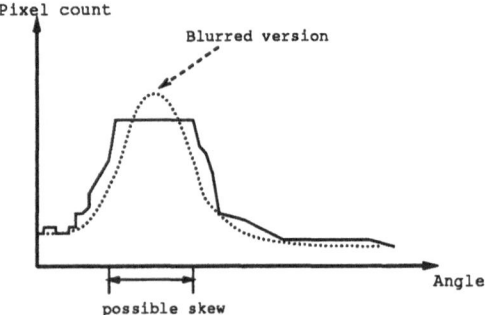

Figure 3: Skew measurement.

(Figure 3). In this eventuality, a Gaussian blur is applied to the responses with sufficiently high σ to make it unimodal - the resulting mode is then accepted as the skew.

Figure 4: An input image with skew, and its deskewed version.

The original 256 grey-level image is then rotated , after which the thresholding algorithm is reapplied. This deskewing process is certainly worthwhile (Figure 4). Deskewed images have more even and smooth stave thickness and lighten the effort of locating and removing them at a later stage. Often, the results are not pixel-perfect, but are at least good enough.

3 Locating and erasing the staves

The stave is the fundamental element of a musical score. A note head by itself may represent the duration of a note, but it has to associate with the stave line to gain its pitch. A staff is a group of five stave lines which are equally spaced, the stave line thickness and the distant between two stave lines being important parameters at all later stages of recognition. Once we know the position of these lines, they become distractions when we try to recognise the features which have been engraved on or around the staff. Hence, the stave must first be located and measured before erasing it to isolate the musical features. Unfortunately, the stave lines often pass through musical symbols, and so they must be erased selectively in order not to disconnect these symbols.

A histogram of the horizontal projection is generated in which graphs of five equally spaced peaks are usually clear. If they are not (due, for example, to inter-staff text) this pattern can be observed by a suitable blurring of this histogram. From this information, the stave line position, average line width and the space between lines are extracted and recorded.

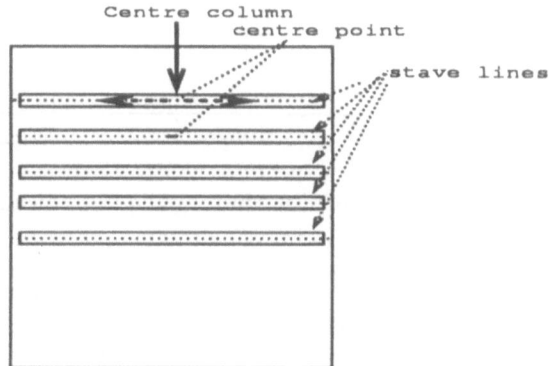

Figure 5: Tracing the stave line.

In practice, the stave lines are not found to be completely straight or of even thickness, causing the parameters for each stave line to differ considerably. Thus we have to repeat the process for each stave line.

For each stave line, we start from its centre column at its central (vertical) position and trace it right and left (Figure 5). Assuming that the line is straight and horizontal, for each column the pixel in the indicated vertical position is taken, and the height of the foreground feature in that column of which it is part is recorded. The distribution of these heights provides a clear mode which provides the 'usual' stave line thickness. Since in practice the line thickness is seen to fluctuate, we choose as the highest acceptable width (W_L), the point after this mode with the distribution gradient almost equal to zero (Figure 6).

Each stave line is then scanned again; commencing from the centre column, the predicted centre pixel is inspected. If it is background, the closest foreground pixel (within the range of $2 * W_L$) to the predicted position in that column is located. If there is no foreground pixel in that range, we assume that the line may be disconnected and go on to scan the next column. If the height of the connected foreground strip so defined does not exceed W_L, its vertical position centre is recorded as the correct best estimate of the line centre position and the strip is deleted. If the strip exceeds W_L in height, it is allowed to remain and the centre estimate not amended. This procedure is then iteratively repeated in neighbouring columns.

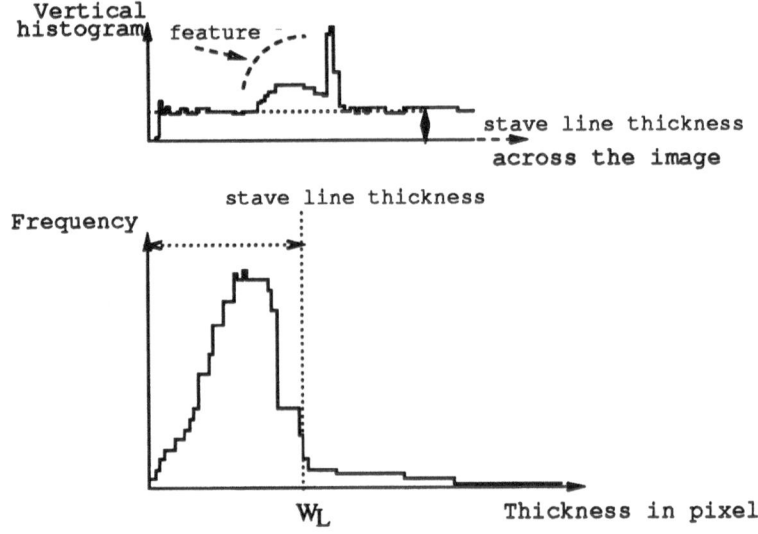

Figure 6: Determining the line thickness.

The output is good but not perfect (Figure 7). Features that were engraved on the staff inherit some noise from the stave lines when they are removed, while thin and long feature such as slurs, ties and phrase marks tend to be disconnected. At this stage, perfect erasing is very unlikely, but these imperfections are not important as they can be overcome during recognition.

Figure 7: An example of stave line removal.

4 Segmentation

The primitives with which the composer deals (crochets, rests, slurs etc.) are often not as simple for an automatic system to identify as their component parts. For this reason, henceforward we refer to 'primitives' as graphical primitives on the page which sometimes (but not always) do not correspond to the musical primitives which the destination representation will expect. In particular, stems will be regarded as features to be recognised independently of the note head to which they are connected, and beams connecting quavers (for example) will likewise be regarded as primitives in their own right. Given accurate recognition of these low level primitives, a reliable reconstruction of the derived musical symbols should be straightforward.

When the stave lines are erased, the image will be left with blocks of connected foreground pixels which may be recognisable primitives, such as note heads or stems, or composite objects, such as a group of four semi-quavers, or noise or part of a stave. These are inspected sequentially to determine whether they are primitive features or need further segmentation.

4.1 Primitive sub-segmentation

From the object segmentation, if the object is too 'large' as a primitive feature relative to the staff, it must be a composite made up of a number of connected primitive features (Figure 8).

Figure 8: Examples composite objects.

In practice, the connections are frequently straight lines (beams) or curves (phrase marks, slurs, ties) which cut through the other note stems or connected note heads. Within such composites, a sudden change in vertical projection histogram usually suggests a possible junction point of two separate features.

In a similar application (separation of merged characters during OCR), Kahan et al. [7] observe that maxima in the absolute value of the second difference of the projection is a good indicator of these positions, and that, since 'break points' may be expected to be thin, the ratio of this difference to the projection height is a better measure still. Consequently, we evaluate the measure $(V(x-1)-2V(x)+V(x+1))/V(x)$ across the vertical projection $V(x)$, (Figure 9).

When the horizontal position (X) with the maximum value of this function is found, we assume that this is a junction point at which the object may be separated into two or more smaller features which are connected by a possibly long and relatively thin feature. Instead of just separating the object into two, we attempt to trace and extract the connecting feature. Starting from X, we trace to its left and right until the image boundary is met, or there is no foreground connected pixel ahead. First, find the centre position (interpreted as above) of the connector at X and get its thickness. For the next column to trace, the first guess of the centre will be that of the preceding one; if this prediction is background, 8-neighbour connectivity is used to find any possible immediately connected pixel - this may occur during a sharp turning point on a curve. If the thickness of the column is less than half of the space between two stave lines, the foreground pixels which connect with the centre pixel are marked as an unambiguous part of a connecting feature; otherwise the column must be shared by two closely neighbouring features and is preserved.

478

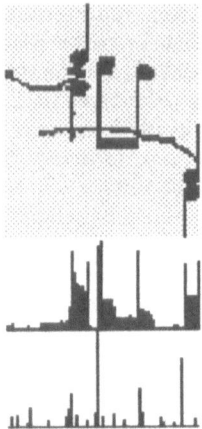

Figure 9: A composite feature, its vertical projection and the ratio of second difference to the projection.

The features we are tracing are either linear (in the case of beams) or approximately quadratic (in the case of slurs, phrase marks and ties). After a suitable number of columns (characteristically 10) have been processed, we can, via a least square estimate, fit a polynomial $y = a + bx + cx^2$ to the observed centre points which is used to predict the likely feature position in the next column. This permits a sufficiently accurate prediction of feature position 'through' objects such as stems within which accurate measurements cannot be made. Figure 10 shows that the connector, in this case a long slur, was identified and when we separate the slur, other primitives are not disturbed. Figure 11 shows the separation of a phrase mark joined with a note head and an accent sign. Notice that some of the estimated centre points of the segmented phrase

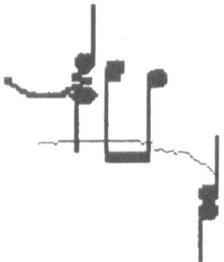

Figure 10: Connector was identified (thin line).

mark are not continuous with their neighbours; this is due to the least squares estimate being insufficiently accurate and falling into background. When this happens, we try to reuse the previous centre position. It is possible that this problem would be solved by a higher order curve approximation.

This process is repeated until the output sub-segment is a possible primitive feature. The termination criteria is the feature having density within its

Figure 11: The sub-segmentation routine, segment out the connector.

bounding box higher than 75%, or being recognisable as a basic primitive such as a note head.

This works very well for phrase marks, slurs, ties and beams if a good break point can be identified. In practice, noise by the side of a stem or bar line may be indicated as the break point; this happens rarely, and is identifiable from its size (very small relative to stave line thickness) and we may simply continue the process.

For vertically connected features, such as a chord, we try to apply the same technique to the horizontal histogram, but the response is not so clear. By making use of the knowledge of the inter-stave line distance and the estimated break points we can deduce good estimates of the location of note heads; possible break points which fall on a stave line or halfway between two stave lines are likely candidates (Figure 12). A complete and robust segmentation of such connected features is likely to require fuller interaction with the recognition phase, and suitable feedback, and this represents work in hand.

Figure 12: Example input with some estimated break points and its original staff position. The three strong peaks are in 'good' positions.

5 Conclusion

In this paper, we have discussed some potential problems encountered in the early processing of musical scores, and proposed some solutions to them. We have chosen to interpret symbols at the most primitive of levels, and have attempted to segment out such primitives from features which are often highly interconnected, and have demonstrated success at extracting long connecting features such as phrase marks, ties, slurs and beams, leaving the primitives isolated for a subsequent recogniser to work on. Features closely connected in a vertical direction respond to a similar approach, exploiting knowledge of the score geometry. When this approach is combined with a higher-level process providing recognition, we expect to be able to take advantage of the musical syntax [3] to verify the identity of features, or to provide feedback to questionable segmentation, thereby making the whole system very robust and reliable.

References

[1] M Brown A Clarke and M Thome. Problems to be faced by developers of computer based automatic music recognisers. In *International Computer Music Conference*, pages 345–347, 1990.

[2] J E Bresenham. Algorithm for computer control of a digital plotter. *IBM Systems Journal*, 4(1):25–30, 1965.

[3] H Fahmy and D Blostein. A graph grammar for high-level recognition of music notation. In *First International Conference Document Analysis and Recognition*, pages 70–78, September 1991. Sep 30 - Oct 2.

[4] I Fujinaga. Optical music recognition using projections. Master's thesis, Ma McGill University, 1988.

[5] W Gamble. *Music Engraving and Printing*. Da Capo Press Music Reprint Series. Da Capo Press, 1923.

[6] International MIDI Association. *Standard Musical Instrument Digital Interface Files 1.0*, July 1988.

[7] S Kahan, T Pavlidis, and H S Baird. On the recognition of printed characters of any font and size. *IEEE Transactions on Pattern Analysis and Machine Intelligence*, PAMI-9(2):274–288, March 1987.

[8] D E Lloyd. Automatic target classification using moment invariants of image shapes. Technical Report RAE IDN AW126, Farnborough, U.K., December 1985.

[9] P Martin and C Bellissant. Low-level analysis of music drawing images. In *First International Conference Document Analysis and Recognition*, pages 417–425, September 1991. Sep 30 - Oct 2.

[10] T W Ridler and S Calvard. Picture thresholding using an iterative selection method. *IEEE Transactions SMC*, 8(8):630–632, August 1978.

Active Contours using Finite Elements to Control Local Scale

Persephoni Karaolani,
G D Sullivan and K D Baker
Intelligent Systems Group,
Department of Computer Science
University of Reading, RG6 2AY, UK.
p.karaolani@reading.ac.uk.

Abstract

Finite elements allow smoothness to be enforced on the measurement of the image-dependent term in active contours. This improves the stability of the solution, with less computational cost than is incurred by increasing the number of elements. Performance is best when the size of the element matches the scale of the image detail sought.

This property of Finite elements can be used deliberately to select the scale of the image detail recovered by the active contour. The method offers a way to reduce the sensitivity of active contours to localised noise.

1   Introduction.

Active contours ("snakes" and "bubbles") provide global solutions to the problem of finding continuous edges in images [1, 2, 3, 4, 5, 10]. Instead of building connected edges from point evidence, obtained by purely local operators, the active contour has a pre-established structure (stiff, elastic strings in the case of "snakes"; elastic loops, under internal pressure forces, in the case of "bubbles"). Starting from an initial state, the contour moves under the influence of local forces derived from the image, until these are in equilibrium with the contour's internal structural forces.

Equilibria occur at minima of the energy potential of the forces acting on the contour. The energy comprises internal and external factors. In typical applications of active contours in vision, the internal forces represent the contour's elasticity and stiffness; the external force is provided by the negative of the magnitude of the gradient image, acting along the contour. The former forces tend to make the contour short and smooth; the latter deflects it towards regions of the image with high gradient (i.e. edges). The energy term to be minimised can be expressed as:

$$E(v(s)) = \int_0^1 \left(\alpha \left| \frac{dv}{ds} \right|^2 + \beta \left| \frac{d^2v}{ds^2} \right|^2 + \gamma P_{im}(v) \right) ds \tag{1}$$

where: v(s) = (x(s),y(s)) defines a parametric curve in the image plane.

The three terms represent the elastic energy, the stiffness energy, and the potential due to external forces, which in our models is represented by the image intensity function. The parameters α, β, and γ are weights provided to balance the relative importance of the terms in (1).

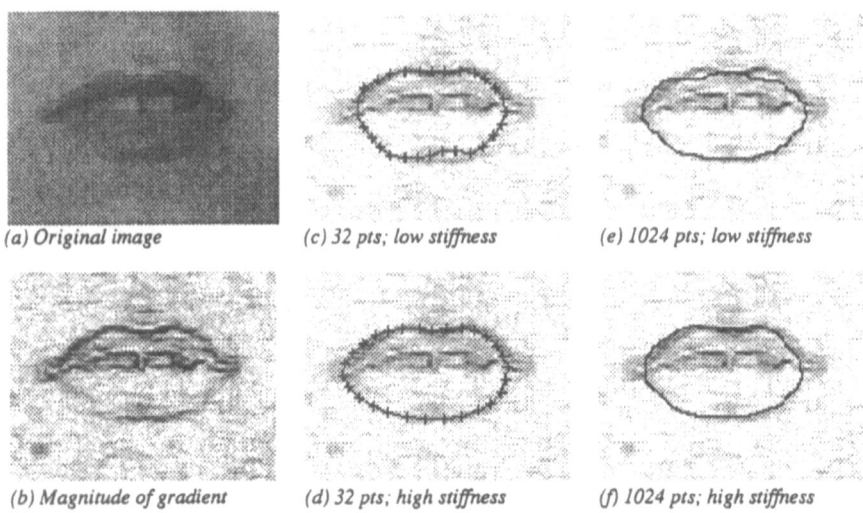

(a) Original image   (c) 32 pts; low stiffness   (e) 1024 pts; low stiffness

(b) Magnitude of gradient   (d) 32 pts; high stiffness   (f) 1024 pts; high stiffness

Figure 1: "Bubbles" solved by Finite Differences

1.1 Finite differences solution

The original method to minimise the energy in Equation (1) was developed using the Euler method with Finite Differences (FD) [1, 2]. This represents the contour as a string of control points, connected to each other by an elastic, stiff link. The image forces act at the control points, where the local derivative of the potential (itself usually a derivative - to emphasise edges) provides a force, deflecting the contour from its natural shape. Each control point is therefore subject to a force dependent on the image at the control point, together with forces due to its 2 immediate neighbours (representing elasticity) and its nearest 4 neighbours (representing stiffness). The global solution to the set of local problems is given by solving two sets of n_d equations in n_d unknowns (the x & y coordinates of the n_d control points). Each equation involves 5 unknowns, which leads to a penta-diagonal system of equations. The problem is profoundly non-linear, in that the image forces are strictly local, and change in an uncontrolled way as the contour moves. The system of equations is therefore repeated iteratively, each iteration moving a little way towards the preferred solution, until stability results.

There are many difficulties with this approach which have limited its application in practice. Firstly, the image forces are only sampled at the control points; therefore in order to make the contour sensitive to fine detail in the image, n_d must be large. Secondly, with large n_d, the gap between control points is small, so that the effect of stiffness (spreading only over 5 points) is only felt on a local scale; this means that a contour is easily caught by local details in an image, which may not be part of the desired global structure - e.g. random noise.

These defects of the Finite Differences method are illustrated in Figure 1, using the implementation reported in [10]. Figure 1(a) shows an original image of a mouth; our purpose is to find and outline the outer edge of the lips. Figure 1(b) shows the negative of the magnitude of the gradient image: we seek to identify the irregular ridge of dark points around the lips. An elliptical "bubble" was initially located just inside the line of the lips, and the balance of elastic and pressure forces caused it to expand, until the control points were caught on strong local minima (Figures 1(c-f)).

In Figure 1(c), there are 32 control points (shown as +) which have found good individual minima, but the bubble ignores the image lying between them. In Figure 1(e) there are 1024 control points (not shown) and the solution follows the local edges more accurately, but the contour easily becomes snagged on local detail. The problem of snagging can be overcome in part by increasing the weight of the stiffness term (Figure 1(d & f)), but then fine detail (e.g. at the corner of the mouth) becomes even poorer.

The problem is the familiar one of spatial scale. The image force is proportional to a second order derivative of the image (change in the gradient image), which is inherently noisy. Noise can only be combated successfully by integrating information using a scale of smoothness appropriate to the structure being sought; this is highly context-dependent, e.g. the corners of the mouth require finer scale than the lips. As normally implemented, the stiffness and elasticity coefficients in Equation (1) apply globally (though in our implementation a simple method of biasing the equilibrium shape away from symmetry is available [10]). The finite differences method gives no natural way to impose a sense of scale locally.

1.2 Finite element method

An alternative approach to solving the non-linear energy minimisation problem relies on the use of the Finite Element Method (FEM) [9]. Here, the domain s of the contour $v(s)$ is divided into a number of elements, in which the elastic, stiffness and image forces seek their own local minima. Within each element the position of the contour $(x(s),y(s))$ is described by a low order polynomial in s; successive elements are connected together by constraints which seek to enforce low-order continuity in x & y at their links. A popular choice of function is given by Hermite Cubics, third order curves having fixed end-points and fixed tangents at their ends, which can therefore be connected to each other with first order continuity, while retaining second order continuity within each element [6]. This approach leads to the problem of solving a system of $2n_e$ simultaneous equations in $2n_e$ unknowns, where n_e is the number of elements. An important difference between the FEM and FD approaches is that the image forces act along the entire element, not merely at discrete control points.

The global minimum is sought (as with FD) by moving in the direction of the solution of a system of linear equations determined by the current position of the contour, and iterating. At each iteration, the force acting on each element is recomputed, according to the new position in the image of the element. As before, the process is iterated until the contour is stable.

A key issue in the FEM is the method used to integrate the image forces along an element. A simple scheme ([5]) merely approximated the integral by sampling the image along the element at a fixed number of points. However, this method runs into the same problems encountered by FD: to be sensitive to fine scale in the image, we need many small elements, and in coarser scale parts of the image these easily become tangled by irrelevant local image detail.

More accurate ways to integrate the image forces numerically are available. A widely used scheme is the Gauss-Legendre n-point rule for choosing the sample points for numerical integration, and their relative weights [7]. The order of the rule refers to the number of sample points. If this is high then the estimate of the integral takes account of more image detail, so that noise has a smaller effect on the overall shape of the element (which is still constrained to the cubic form).

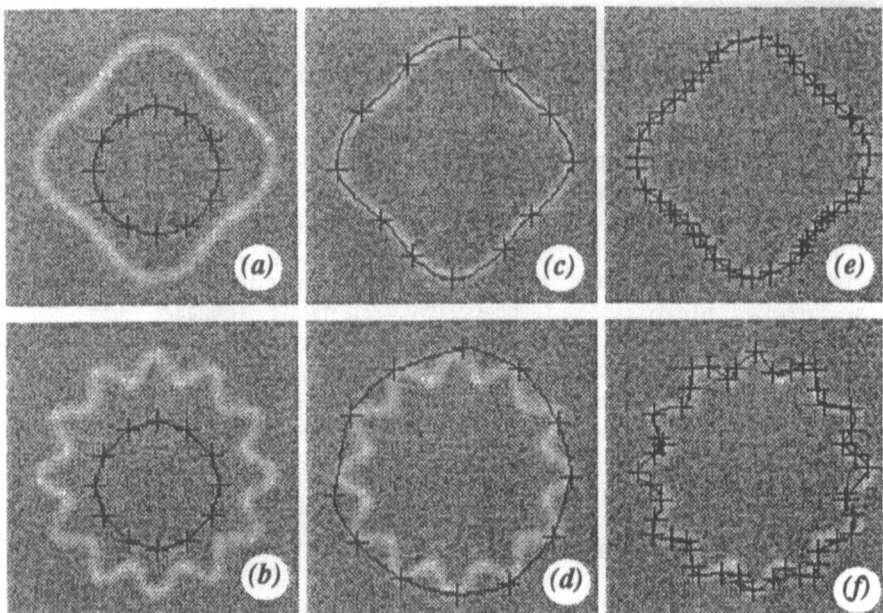

Figure 2: Examples of FEM bubbles applied to images having coarse (top) and fine (bottom) scales. Left: Initial position; Middle: 12 elements; Right: 48 elements.

The order can be varied according to the length of the elements: in a simple scheme described here, we make it proportional to the length of the element computed in pixel units, thus (approximately) ensuring regular steps along the contour of approximately 1 pixel distance. This provides a natural way to guarantee that the image forces pay good attention to all the detail along the locally-cubic description of the contour. We also weight the contribution of each element in proportion to its length; this helps to ensure that elements do not form clusters. By deliberately varying the length of the elements it therefore becomes possible to control the scale of smoothness adopted by the contour - and this can be done on a local basis.

2   FEM bubble with adaptable integration.

2.1   Outline of method

The function $v(s)$ is represented as a Hermite Cubic polynomial over each element. The nodes between elements are shown as + in the relevant Figures. The image energy term in Equation (1) is derived by numerical integration along the Hermite Cubic, using the Gauss-Legendre rule [7]. In each element an approximation to the length of the element (in the $(x.y)$ space) is computed, and the number of sample points is made approximately proportional to the length in pixels. The gaussian rule determines the sample points (in the s space) for optimal sampling. The derivative of the image is then computed at these points, and integrated over s. It is known that the Gauss-Legendre rule is optimal for approximating integrals since among all integration rules requiring n_g integrand evaluations, it has the highest possible order of accuracy $(2n_g-1)$ [7].

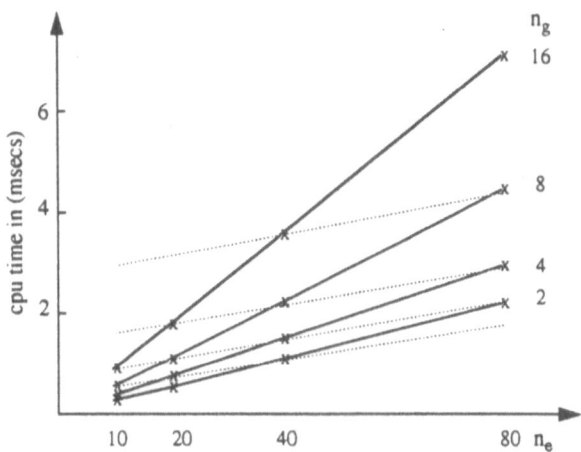

Figure 3: Cpu times [SPARC2] for one iteration of the energy minimisation calculation, using different number of elements (n_e) and gauss-points (n_g). The dashed lines link cases with equal numbers of integration points (n_g*n_e)

At each iteration, the method leads to a hepta-diagonal system of equations in $2n_e$ unknowns, which is currently solved by a conjugate gradient algorithm, though linear methods would be faster (e.g. Cholesky [8]). A scaling factor is introduced to weight elements in proportion to their lengths in the iterated problem, which prevents elements which collapse from having excessive influence.

2.2 Imposing local smoothness

Figure 2 illustrates bubbles consisting of 12 and 48 elements, applied to synthetic images of a blurred modulated circle with added gaussian noise (peak signal is 72, and the noise sd was 20 intensity levels). In each case, the bubble was initially positioned as shown in Figure 2(a & b) for the 12 element version. The images illustrate two different scales of spatial detail. Best performance is obtained (Figure 2(c & f)) when the scale of the element matches the scale of the image. If elements are too coarse (Figure 2(d)), then the bubble cannot follow the image; if elements are too fine, (Figure 2(e)) then the system of equations is unnecessarily large, and the elements become over-sensitive to local noise (note the irregular spacing, as elements have collapsed onto noise points).

2.3 Computational costs

Integration along the elements is less costly than increasing the number of elements. Figure 3 shows the results of a simulated experiment in which the computational cost per iteration (averaged over 1000 iterations) was measured as a function of the number of elements and the number of gauss points used to compute the integral. The algorithm is close to linear on the number of elements. However, the number of image points used in the computation (= n_g*n_e) - and hence the opportunity to integrate out image noise - can be doubled with significantly less cpu time cost, by increasing the number of gauss points (moving vertically to the next dashed line in Figure 3).

486

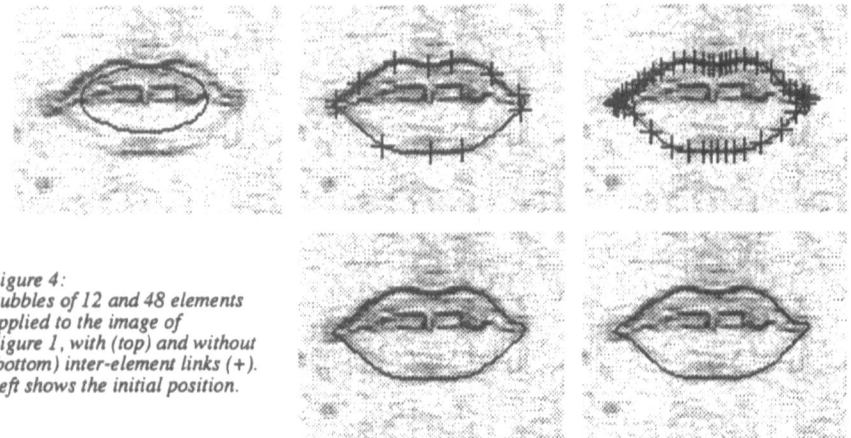

Figure 4:
Bubbles of 12 and 48 elements
applied to the image of
Figure 1, with (top) and without
(bottom) inter-element links (+).
Left shows the initial position.

2.4 Demonstration on real images

Figure 4 shows the 12 and 48 element bubbles applied to the magnitude of the gradient image of lips in Figure 1(b), from the initial starting position shown. In comparison with the FD method of Figure 1, the FEM seems more accurate; it is also far less sensitive to initial conditions. The 48 element bubble has more accurate performance at high curvature points (e.g. the corners of the mouth), but the 12 element bubble recovers a smoother description of the slowly varying regions (edges of lips).

3 Conclusions.

We have demonstrated a new approach to active contours, using FEM with an adaptable order of integration. This provides a mechanism to integrate the local forces accurately. A major benefit is that fuller use is made of the local information, and smoothness can be enforced, without becoming excessively sensitive to noise; a secondary benefit is that the scale of the detail sought can be imposed on a local basis.

The scheme is also efficient. The well-known FD approach [1, 2] has a computational cost that increases linearly with the number of control points (those points where the image is sampled). If the size of an element is well-matched to the local spatial scale, then many fewer elements are needed than FD control points. The information loss is avoided by interpolating the element, within the constraint of the cubic form; this represents a smaller computational burden, since finer sampling does not increase the number of unknowns in the linearised problem. In iterative schemes, computing speed is critical. We have found that the FEM requires far fewer iterations than the FD method for convergence, and in our implementations (the FD on an 1024 processor SIMD multi-processor, the FEM on a solitary SUN), the FEM is also far faster, and produces better results from simple initial conditions, though direct comparisons on more similar implementations have not been made.

The effect of the scaling of the forces by the length of the element is to prevent very small elements (perhaps contracting onto a purely local feature) from contributing as much to the global solution as longer elements. This encourages elements to preserve their lengths, and to avoid elements collapsing onto isolated noise points. In

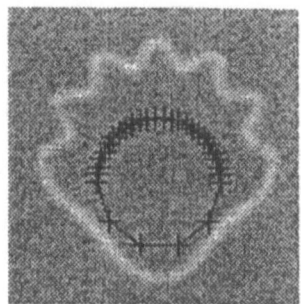 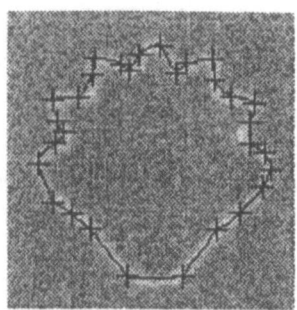

Figure 5: Multi-scale FEM with 30 elements applied to a synthetic image:
Left: starting position; Right: final position.

consequence, it is possible to prescribe the initial scale of the contour locally. Figure 5 illustrates performance on a synthetic multi-scale image.

In practical uses of bubbles for image analysis, it will often be the case that some knowledge is available of the expected structure of the objects of interest. The two greatest determinants of the form adopted by an active contour are its preferred (blank image) state, and its initial conditions. The adaptable integration scheme provides a convenient way to create active models which mimic the expected smoothness of images of objects as well as their initial expected positions and shapes.

4 References

[1]   Kass, M., Witkin, A.,and Terzopoulos D. Snakes: Active Contour Models, Int. J. Computer Vision 1. 321 - 331, 1987

[2]   Terzopoulos D., Witkin A., Kass M. Constraints on Deformable Models: Recovering 3D shape and nonrigid Motion. AI, 36, 91-123, 1988

[3]   Amini, A.A., Tehrani S., and Weymouth T.E. Using dynamic programming for minimising the energy of active contours in the presence of hard constraints, In Second International Conference on Computer Vision p 95-99, Computer Society Press, December 1988

[4]   Karaolani, P., Sullivan, G. D., Baker, K.D., Baines, M. J. A Finite Element Method for Deformable Models, Proceedings of the fifth Alvey Vision Conference, 1989

[5]   Karaolani, P., Sullivan, G. D., Baker , K.D.. Parabolic and Hermite Cubic Elements: A Flexible Technique for Deformable Models,Proceedings of the British Machine Vision Conference , 325-330, Oxford, September 1990

[6]   Strang, G., and Fix, G. J. An Analysis of the Finite Element Method, Prentice-Hall series in automatic computation, N.J. 1973

[7]   Press, W. H., Flannery, B. P., Teukolsky, S. A., Vetterling W. T. Numerical Recipies, Cambridge University Press, 1986

[8]   Golub, G. H. and Van Loan C.F. Matrix Computations, Hopkins Press, Baltimore, 1990

[9]   Zienkiewicz, C.O., and Morgan, K. Finite Elements and approximation, John Wiley & Sons, 1983

[10]   Sullivan, G D, Worrall, A D, Hockney, R W, and Baker K D. Active Contours in Medical Imag Processing using a Networked SIMD Array Processor. British Machine Vision Conference, 1990, pp 395-400.

Automatic Face Location to Enhance Videophone Picture Quality

T.I.P. Trew, R.D. Gallery, and D. Thanassas
Philips Research Laboratories
Redhill, UK

E. Badiqué
Philips Kommunikations Industrie AG
Nürnberg, FRG

Abstract

New video communication and multi-media products open up a range of machine vision applications, in which the potential size of the market can justify a substantial investment in the development of sophisticated algorithms. Face location can be used to enhance the subjective performance of videophones, while still conforming with international video compression standards. This paper gives an overview of the location techniques employed, describes a real-time implementation, and presents the results of the subjective tests which confirmed the improvement in picture quality.

1 Introduction

Although many companies have been active in machine vision research for over 20 years, there are few commercial products based on this technology, other than the ubiquitous bar-code reader. This is changing with the introduction of video communication products, such as videophones, and this paper describes how techniques which solve the real-world machine vision task of face location may be applied to commercial advantage.

For videophones to become accepted as a universal business tool, as the fax machine is today, they must conform rigorously to international standards to ensure compatibility between different manufacturers' equipment. This standardisation has been achieved for videophones operating over the Integrated Services Digital Network (ISDN). There are many CCITT[1] standards which apply to the various video, speech and network access sub-systems of the videophone, but we need only consider H.261, which describes the techniques to be used for video compression.

Since all manufactures must conform to the standard, novel approaches are required to introduce product differentiation, and it is in this area that machine vision techniques are applicable. Although H.261 defines the core of the compression algorithm, so that the transmitted bit-stream can be interpreted by any decoder, additional pre- and post-processing stages are permitted. Furthermore, the standard allows high-level knowledge to be applied to control the parameters of the algorithm, so as to increase

---

[1] Comité Consultatif International Télégraphique et Téléphonique

the fidelity of parts of the picture of interest to the user, at the expense of the remainder. If applied correctly, this will increase the overall subjective image quality.

In the majority of cases, it is the user's face and, in particular, the eyes and mouth, which is of greatest interest. People are particularly sensitive to degradations in these areas and a small improvement in the objective picture fidelity, if it makes it possible to read expressions, can yield a substantial increase in the attractiveness of the product.

This paper first gives an overview of the H.261 video compression standard, since this is necessary to explain how machine vision techniques can be useful. It then describes both a simple face location system, which has been realised and interfaced to an H.261-compliant videophone, and a more sophisticated approach, which can locate individual facial features, but which has not yet been fully realised in real-time. Finally the results of subjective tests in which non-experts assessed the overall picture quality are presented.

2   H.261, the International Videophone Standard

H.261 is an international standard, developed by the CCITT Study Group XV[1] for videophone transmissions over digital networks at low bit rates (multiples of 64kbit/s). Consider that a compression ratio of 300:1 is required for the lowest rate (64kbit/s). Using current coding techniques it is not possible to achieve such a huge reduction without introducing a visible deterioration in the fidelity of the decoded image.

The basis of the H.261 coding algorithm is a hybrid of several well known techniques, and it might be described as a hybrid motion-compensated DPCM/DCT coder, where DPCM is differential pulse coded modulation, and DCT is the discrete cosine transform. Figure 1 shows a block diagram for such a system, in which the "facial area" input should be ignored at this stage. The algorithm, after initialisation, proceeds as follows. The frame store contains the image which was displayed during the previous frame period and the motion estimator, which uses block matching with 16×16 pixel blocks, termed "macroblocks", finds the best match for each block in the current frame. The motion vectors are used to displace the image in the frame store,

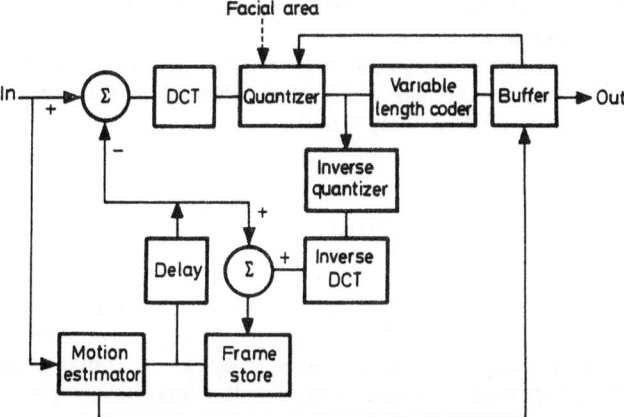

Figure 1.   :   Block diagram of an H.261 coder, showing how the facial area is used to modify its operation.

which is replicated in the decoder, to form the DPCM prediction. The difference between this prediction of the current image and the actual image is calculated by subtracting the two images, to give a motion compensated frame difference. This has exploited the temporal correlation within the image sequence to reduce the amount of data to be transmitted.

The next stage of the algorithm seeks to exploit the intraframe, or spatial, correlation, within the motion compensated frame difference by taking its discrete cosine transform, on an 8×8 pixel block basis. The coefficients of the DCT are quantised (introducing error), and also thresholded to discard the smaller coefficients in any block. The output of this stage is then Huffman coded [2], and fed into a buffer which matches the instantaneous data rate of the encoder to the fixed rate of the transmission channel. At this stage the data need only have error protection coding added before being transmitted. The amount of data within the buffer is monitored, and a signal is fed back to control the step size and threshold of the quantiser, which will determine the resolution and number of the transmitted DCT coefficients.

The subjective quality of the images produced by the above algorithm is dependent upon both the complexity of the image (and how suited this complexity is to the basis functions of the DCT), and also to the extent and type of motion in the image (i.e. block matching can handle 2-D planar motion, but motion involving rotation, or motion parallel to the camera axis, will reduce the correlation of the matching process), resulting in a degradation of the subjective image quality. People using videophones cannot have their movement unduly constrained, and there might be movement in the background of a typical office environment, so the problem of the degradation of picture fidelity due to motion over a significant portion of the image needs to be addressed.

In typical videophone conversations the participants are talking to each other and looking at each others faces, and are not greatly interested in the appearance of the background. Therefore, instead of using a constant quantisation step size for the whole image, the quantisation step used in the facial area can be decreased, so that more bits will be used in this area. The background will of course now receive fewer bits, and hence degrade, but, as it is not the centre of attention, the overall subjective picture quality should improve. The "face area" input, shown dashed in figure 1, leads to a modified H.261 coder, in which a mask indicating the facial region is used to control the quantiser. Figure 2 illustrates the improvement in the decoded picture quality obtained from the modified encoder, in which the step size in the facial area has been halved, compared with the standard encoder.

Having established how the videophone coder will be modified, it is now necessary to devise a machine vision system which is capable of locating the user's face automatically in a normal office environment.

3 Low-level Face Location

Initial experiments concentrated on enhancing scenes in which a single person was present (head-and-shoulder scenes). One can suppose that an especially high quality will be expected when two speaking partners have an eye-to-eye conversation. In these situations, it will be important to be able to have a clear image of the partner's face in order to help in interpreting his expression. The quality of the facial image will be far less critical in situations where several people are in the field of view and the scene activity is high. The first step, prior to the realisation of a facial area recognition

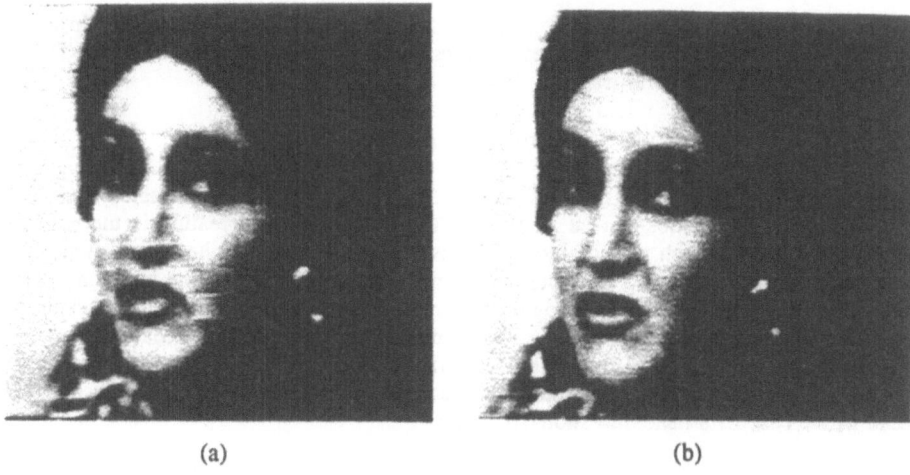

(a)	(b)

Figure 2. : A comparison of the decoded pictures from standard (a) and modified encoders (b).

algorithm which is simple enough for integration within a codec, is the automatic identification of scenes of one person's head and shoulders.

3.1 Recognition of Head-and-shoulder Scenes

The recognition algorithm is based on the detection of the motion within the scene. The different steps are detailed below:

3.1.1 Generation of an Object Silhouette Based on Image Differences

To generate an object silhouette characterising the moving object, the image difference is built over blocks of a given size (8×8 or 16×16). The decision as to whether a block belongs to the moving object or not is then taken based on the temporal activity within the block, characterised by the conjunction of two criteria. The first criterion is the sum of the absolute value of the image difference. The second criterion is the number of points within a block, for which the absolute value of the image difference is larger than a given threshold.

3.1.2 Spatial and Temporal Filtering of the Object Silhouette

Due to a number of factors, the raw silhouette generally cannot be used for analysis without being filtered. Spatial and temporal filters are used in order to improve the smoothness of the silhouette.

As a first step, the silhouette is passed through a non-linear spatial filter. The eight neighbouring blocks of a given moving block are analysed and, according to their status, the moving block is set to "fixed". As a result, most of the isolated "moved" blocks appearing in the background due to camera noise are suppressed and the analysis of the silhouette is simplified.

The spatial filter is followed by temporally smoothing the motion silhouette. In this step the temporal stability of the bitmap is improved by ORing the status bit for a given spatial position over several frames.

3.1.3 Transformation of the Silhouette into a Feature Vector

After filtering the motion silhouette, a 1-D feature vector can be extracted and its topological characteristics can be analysed. The most important feature is the width of the silhouette in each row of blocks. This feature in a typical head-and-shoulder scene will usually have two characteristic regions, corresponding to the head and the shoulders. In order to ensure a recognition performance as size-insensitive as possible, the feature vector is normalised between 0 and 1. Together with the feature vector, an average value of the position of the vertical axis, and also of the top of the moving object, are calculated. These are stored to be used later during the localisation of the head in the case of a head-and-shoulder scene.

3.1.4 Training of the One-dimensional Pattern Associator

Through the transformation described above, the problem was reduced to the analysis of the 1-D vectors. This type of problem can be solved with the help of a pattern associator[4], trained on features contained in a training set.

Let us assume that N typical head-and-shoulder silhouettes are recorded during training and their 1-D feature vectors are obtained. The vectors are described as f_i; $i \in \{1, \dots, N\}$. In order to be able to differentiate between the "true" head-and-shoulder and false silhouettes we also need feature vectors representing "false" (i.e. non-head-and-shoulder) silhouettes. We generate these "false" silhouettes by, for example, inverting the silhouette about its horizontal axis. We add these "false" 1-D feature vectors to the training set and call them f_i; $i \in \{N+1, \dots, 2N\}$. The constraint equation determining a filter H is then:

$$\mathbf{f}_i^T.\mathbf{H} = C_1 = 1; \quad i \in \{1, \dots, N\} \tag{1}$$

$$\mathbf{f}_i^T.\mathbf{H} = C_2 = -1; \quad i \in \{N+1, \dots, 2N\} \tag{2}$$

The filter **H** is determined iteratively with the help of a delta-rule training algorithm. It is written as follows (at the tth iteration):

$$\mathbf{H}_t = (\mathbf{f}_i^T \mathbf{H}_t - C_k)\mathbf{f}_i \quad \begin{matrix} k=1; \quad i \in \{1, \dots, N\} \\ k=2; \quad i \in \{N+1, \dots, 2N\} \end{matrix} \tag{3}$$

For successful training it is necessary that a training set contains 1-D feature vectors which are representative of typical videotelephony situations, with different subjects and various subject-camera distances, etc.. It is also possible to generate a training set from whole sequences, using either simple solutions, such as averaging the different silhouettes, or more sophisticated ones, such as principal component analysis.

3.1.5 Filtering the Feature Vector

Having trained the pattern associator off-line, it is then combined with the current feature vector by a simple dot-product operation, as in equation (1). A head is detected if the resulting value exceeds a threshold.

A number of temporal consistency rules can then be used to smooth the decision and ensure stable head-and-shoulders recognition.

3.2 Localisation and Tracking of the Facial Area

Having established that a head-and-shoulders silhouette is present in the scene, it is necessary to localise the area which is to be enhanced. Since the H.261 standard only allows the quantisation step size to be changed on the boundaries of 16×16 pixel macroblocks, the face is represented as a rectangle containing a whole number of macroblocks. Only limited accuracy is required because of this coarse scale and experiments have shown that only three different rectangle sizes are required to cover the range of head sizes observed in normal videotelephony situations. The rectangle size is a function of the total silhouette size and it is positioned using the vertical axis and top-of-object position determined during the silhouette analysis.

For those frames in which an acceptable face cannot be found, possibly because of complex background activity, the facial area can be tracked from previous frames. A tracking system has therefore been developed which segments the motion vector field to identify areas which are moving coherently. Those segments which correspond to the facial area are identified and the centroid and mean motion vector for this group are calculated. This vector is used to project the centroid into the following frame. The segmentation of the motion vectors is repeated for that frame and a group of segments is grown around the projected centroid, observing spatial and temporal coherence criteria, according to the hypothesis that the motion and position of the head change smoothly in consecutive frames. If these criteria cannot be met then the system takes into account past information about the position and location of the head, as it was last detected in the previous frames. This assumes that the head has not moved enough to generate a coherent vector field, hence its position and motion characteristics can be recovered using past motion history.

This system can be used as a general purpose object tracker provided that the initialisation stage is changed accordingly. Initial simulations, with no code optimisation, run on a 12MIP RISC processor at the rate of 6 frames/sec.

4 Overview of Feature-Based Face Location

Although the face location system described in section 3 operates successfully for most of the time, and has been realised in real-time, the silhouette is very noisy, so that the system fails occasionally, and is coarse, so that the size of the face cannot be determined precisely. Also the positions of facial features, required for other enhancements to the videophone, cannot be obtained.

An alternative, bottom-up approach has been developed in parallel with the previous system. This attempts to identify features in the scene which may be the eyes or the mouth, and then finds triplets of these with the correct geometry to form a complete face. The principles of this approach have already been described[3], but there have been several extensions since that time which have made the technique more robust.

The principle of the approach is that potential faces are located from the current image at regular intervals, determined by the processing power available, and the probability of each of these potential faces being a true face is estimated. The potential faces are tracked through subsequent frames, allowing the probabilities from the individual frames to be smoothed with a non-linear temporal filter. The potential face with the highest smoothed probability is selected. The following sections describe this procedure in more detail.

4.1   Initial Face Location

The technique initially analyses the images to find spatio-temporal features which might be an eye or mouth. The criteria for selecting these points are not very stringent since the scale of the face is unknown initially. Triplets of these points are then considered, using rather strict geometric constraints, to determine which of these groups might form a face. These constraints are set so that a face will only be accepted when the line of sight of the user is close to the camera.

It is now necessary to assign a probability to each of these triplets. This is achieved by assessing the image around the vertices which we postulate may be the eyes. For each triplet, this area is normalised, both geometrically using the size and orientation of the triplet, and for illumination, and is then applied to a perceptron. The perceptron has been previously trained on true eyes and on other features which the low-level analysis has incorrectly identified as potential eyes. Note that it is not possible to recover the full 3-D pose of the face from the triplet geometry obtained from a single camera. The training regime accounts for this ambiguity by repeating each true face several times, to have an equal number of true and false training examples, but adding a random value to each of the geometric normalisation parameters. The probability is assigned to the potential face by using the probability distribution function of the perceptron for the eyes of example faces.

4.2   Tracking and Temporal Filtering

Having instantiated potential faces, as described above, they are tracked through subsequent frames. Tracking allows triplets instantiated in different frames to be associated, so that their probabilities can be smoothed, reducing the number of occasions on which spurious triplets are selected. It also allows triplets to persist, even if feature points are not found at all of the vertices in every frame. These propagated triplets are subjected to the same types of geometric tests described above, but now the criteria are relaxed so that the face will not be discarded, even when the face is turned away from the camera.

Although it would be convenient to use the motion vectors available from the H.261 codec, these vectors are not ideal for this process since they are computed to project the current frame back into the previous frame to satisfy the requirements of the inter-frame coding mode of the H.261 compression standard. In contrast, triplet tracking requires vectors which are estimated looking forwards from the current frame. If the sense of the vectors from the codec are reversed then they can give rise to ambiguities caused by covering and uncovering, and more reliable results can be obtained by using a separate tracking system. Satisfactory results can be obtained using the technique proposed by Seeling [5,6], which also explicitly accounts for lighting changes and occlusion.

4.3   Face Location

For simple sequences, with only a single face in view, the face can be located by selecting an area around the triplet with the highest smoothed probability. For more complex sequences with several people, a higher level mechanism is required to select the individual who should be located. Currently, simple heuristics, such as selecting the person closest to the centre of the screen, are used, but more sophisticated approaches might select the person who is currently speaking.

5 Experimental Hardware and Subjective Tests

The low-level analysis algorithm described in section 3 was implemented in real-time and was extensively tested as a standalone system, as well as being connected to an H.261 coder. The purpose of the real-time experiment was to complement earlier computer simulations and to find out in "real life" situations whether the recognition and improved coding of the facial area in an H.261 codec did lead to an improvement in the overall subjective image quality.

5.1 Hardware Description

The facial area recognition system consists of a videophone module (camera and display), an A/D converter, a block-based motion analysis board and a PC for the analysis and display of the facial region, as shown in figure 3.

The data from the motion analysis board (a 36×44 bitmap corresponding to 8×8 blocks) is sent to the PC through a parallel port. The data is displayed on the PC screen using one pixel per block to guarantee a real-time display. The data is filtered and analysed with a 1-layer neural network and, if a head-and-shoulders scene is recognised, a rectangular box of variable size corresponding to the facial region is superimposed onto the displayed data. The vertices of this rectangle are sent to the codec through an RS232 serial port, causing the codec to code the blocks within that region with the quantisation step size set to a fraction of the value determined by the buffer regulation. In order to avoid buffer overflow, the rest of the scene is coded with a step size which has been correspondingly increased.

In order to facilitate experimentation and optimisation, both the parameters of the analysis program on the PC and the coefficient multiplying the step size in the codec can be changed interactively. The coefficient can be varied between 0.1 and 1.0 in 0.1 steps by hitting a key on a terminal keyboard. The optimal range for this coefficient was found experimentally to be between 0.3 and 0.5.

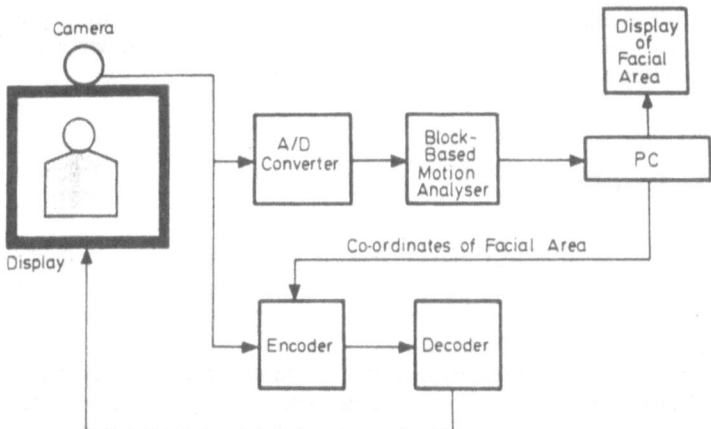

Figure 3. : Block diagram of the sub-systems in a videophone incorporating face location.

5.2 Experimental Results and Subjective Tests

The first results of the real-time test confirmed the indications given by computer simulations. The improved coding of the facial region produced videophone images which tend to have much less blocking in areas of high subjective significance, such as the eyes and mouth. The blocking in such areas is often strongly perceived as being annoying because it inhibits eye contact or lip reading. With the improved coding of the facial area, the image of the speaking partner is reproduced in a way which is more acceptable to the untrained observer.

In order to objectively evaluate the significance of the improvement, more than 30 untrained and unaware observers were asked to judge the difference in image quality between standard coding and adaptive coding. The observers were either asked to sit in front of the camera and to observe their own coded image (self-image mode) or to observe the decoded image of a second person on a monitor (conversation mode). Some of the observers were given a telephone receiver and were asked to simulate a telephone conversation. They could toggle between the two modes, but they were not aware what differences were expected in the two modes, nor which of the modes was currently selected. The observers were then asked whether they could detect any difference between the modes and, if so, to rank the picture quality between them. The experiment was recorded on videotape for later analysis.

The experiment lead to the following observations. One half of the group of observers (18 people) were only asked to observe the scene in self-image mode. The majority of the observers (14 persons) characterised the improvement as "small" to "significant". The other four persons either saw no difference (generally they had not moved much during the test) or a very small improvement without significance. The other half of the group (19 persons) were asked to first observe a self-image and then an image in conversation mode. In this case all the observers made the same pattern of judgements. When asked to evaluate a self-image, they all saw a small to significant improvement. They all described the improvement in the conversation mode images as "significant" to "very significant". This showed that the quality improvement in the facial area is perceived more strongly when the observer is communicating with another person than when he is observing his own self-image. Self-image observers tend to concentrate less on the face and to pay more attention to other parts of the body or to the background. Conversation-image observers tend to concentrate on the eyes and mouth of the speaking partner, thus being more sensitive to quality improvements in these regions. As expected, the quality improvements are most appreciated when they are most needed, that is during a telephone conversation.

6 Conclusions

New video communication and multi-media products open up a range of machine vision applications, in which the potential size of the market makes it worthwhile to make a substantial investment in the development of sophisticated algorithms to obtain an advantage in the market. The US videoconferencing market is expected to be worth $8.3 billion by 1995[7] and any developments which can increase a company's market share are clearly worthwhile. At the same time, the advent of very fast video processing devices, which can be programmed in high level languages, such as the Philips LIFE VLIW device[8], make it possible to realise complex algorithms in real time and at low cost, with short development times.

The subjective tests in this paper demonstrate the potential for machine vision to enhance the performance of an H.261-compatible videophone, and research is continuing to use the capability to locate the user's head to enhance other aspects of the equipment.

References

[1] CCITT Recommendation H.261, "Codec for Audiovisual Services at $n \times 384$kbit/s", *CCITT IXth Plenary Assembly*, 1989.

[2] D.A. Huffman, "A Method for the Construction of Minimum Redundancy Codes", *Proc. IRE*, **40** (10), pp.1098-1101, 1952.

[3] E. Badiqué, "Knowledge-Based Facial Area Recognition and Improved Coding in a CCITT-Compatible Low-Bitrate Video-Codec", *Picture Coding Symposium*, Boston, March, 1990.

[4] J.L. McClelland and D.E. Rumelhart, "Explorations in parallel distributed processing", MIT Press, 1986.

[5] G.C. Seeling, "Tracking 3-D Moving Objects", Imperial College Msc Thesis, Philips Research Laboratories internal report, Redhill, UK, September, 1990.

[6] "Tracking a Moving Object" European Patent EPA 0474307, March, 1992.

[7] M. Nakamoto, "Phones with a View to Profit from Vision", *Financial Times*, London, 28th April, 1992.

[8] G.A. Slavenburg, A.S. Huang and Y.C. Lee, "The LIFE Family of High Performance Single Chip VLIW's", *Hotchips III*, Palo Alto, California, 1991.

Acknowledgements

Sincere thanks are due to J. Hempel and B. Krupa (PKI) for their great help in realising the real-time recognition system. The subjective tests would not have been possible without the help of Mr. Kummerfeld and Mr. Wolf (Daimler Benz Research) who kindly reprogrammed a DB codec for this purpose. Part of this work was realised with the financial support of the European Community RACE-HIVITS project.

Face Recognition by Computer

Ian Craw and Peter Cameron

Department of Mathematical Sciences *

University of Aberdeen AB9 2UB, Scotland

Abstract

We describe a coding scheme to index face images for subsequent retrieval, which seems effective, under some conditions, at coding the faces themselves, rather than particular face images, and uses typically 100 bytes. We report tests searching a pool of 100 faces, using as cue a different image of a face in the pool, taken 10 years later. In two of three tests with different faces, the target face *best* matches the corresponding cue.

Our codes are obtained by texture mapping the face image to a standard shape and then recording both shape and texture. Principal component analysis both reduces the data to be stored, and also improves its effectiveness in describing the face itself, rather than a particular image of the face.

1 Recognising Faces

We are all familiar with the problem of identifying a person cued only with a picture of their face. The difficulty that people have is an indication of the problems that arise when seeking a computer-based solution. The face has to be identified from many different angles, under different lighting conditions, and with different facial expressions. Greater difficulty is usually found if the face is seen in an unusual context, or is a different age from the familiar one.

It is thus not surprising that there have been few successful automatic face recognition devices, despite the commercial potential, particularly for security-based applications. Almost all attempts at computer face recognisers have explored one of two very different strategies to do this.

- One method proceeds via edges to feature recognition, and thus to a full "understanding" of the face as defined by some model, which is then used to match against a database. Each face to be recognised must have a distinct, usually hand-built model, which may just contain relative feature locations, or may contain more detail. An early success following this strategy was by Kanade [12].

- The other approach, via neural nets, goes directly to the desired identification. Specific instances of recognition are provided during the training

*PC is supported by an SERC studentship. Both authors acknowledge the help of other members of the Aberdeen "Faces" group, and particularly that of Andrew Aithison, whose software for placing points aided much of the work described here.

phase, and there is no attempt to identify any abstract intermediate representation held by the "hidden units". Such schemes have successfully recognised single faces [19], and distinguished between a pool of faces cued from a degraded copy of an image in the pool [14], [16]. Training can be a problem when working with a single copy of many different faces if the cue is a new image, but Starkey and Aleksander report some success here (cf the 1992 version of [18]).

Recently, hybrid methods have emerged, which use a preprocessing stage to register the face image and provide more homogeneous input to a recognition component. Examples of this approach include Kirby and Sirovich [13], representing faces using principal components; Cottrell and Fleming [5], attempting net based recognition; Turk and Pentland [22] with recognition based on principal components; and Brunelli and Poggio [4], performing gender classification.

In this paper, we extend this hybrid approach by making the preprocessing stage explicit and, in principle, automatic. We also recognise that more preprocessing, even to the stage when the input image no longer resembles the raw image, brings additional power. Other hybrid methods proposed recently, such as [20] and [9] have a less elaborate coding stage.

Our model of face recognition consists of a number of face images in a *pool*, together with a new image, the *cue*. The interesting case occurs when one or more images of a face (the *targets*) are in the pool, and a different image of that face is used as cue. The aim is to retrieve from the pool those images which best match the cue. This model includes the case of *single person* recognition, in which all the images in the pool are of a single face; however, in what follows, all our tests are done with a single image as target, in a pool of 100 faces. Of course this model says nothing about how the cue and each member of the pool are matched; the major novelty of our work lies in the choice of coding scheme to permit this matching.

2 Coding Faces

Our method of coding a face is a two stage process, reflecting the hybrid nature of the processing. The face is first located in the image, and features found well enough so that a mesh can be drawn as in Fig. 1, with each vertex or *control point* at a known position — the left corner of the mouth, or the centre of the eye etc. Once the control points are located, we distort the face by moving each control point to its position on the average face, and allow the texture to follow. This "full anticaricature" [2] is also shown in Fig 1.

For the pilot studies we describe here, the individual control points are located manually. Making this information available prior to processing means that a lot of knowledge about the shape of each face has been given. In a parallel project, software (FindFace) is being developed, capable of locating these points automatically [7]. At present FindFace finds the control points shown in Fig. 5; these points were needed for a different demonstration, and the same methods yield the control points required here. In a "typical" set of 64 images, the eyes were located accurately in every image, and most features can be found with a high degree of reliability. Even in much more difficult images the overall location is usually correct. Work continues to improve FindFace's robustness and integrate it to provide completely automatic face coding. We regard this capability as an important part of the work described here.

Once the control points have been located, the image is texture mapped to a standard position, as shown in Fig. 1. We refer to the resulting images as *shape-free* faces, because corresponding features occur in corresponding positions in each image, and we have removed the shape information. A shape-free face is no longer a normal square image, but at each point within the standardised shape, we associate, by texture mapping, a grey level.

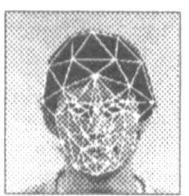  Distort

to

average

shape.

 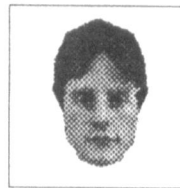

Figure 1: *Control points are located and the mesh mapped to the average shape; the grey levels follow giving a shape-free face or texture vector.*

We now describe the two codes we use subsequently for both matching and retrieval. The first of these, the *shape vector*, consists simply of the locations of each of the control points. The second, which we refer to as the *texture* vector, consists of the vector of grey levels used to texture the corresponding shape-free face. Apart from distortions introduced by imperfect texture mapping, the original face may be reconstructed from this representation. Our proposed coding scheme gives two distinct methods of recognition:-

Texture: since all the shape-free faces are "registered", so that like features occur in the same place in each image, they form an appropriate set of images on which to perform standard template matching, using the cue as template. Although the template may have 30K pixels, this is practicable for small pools. We describe in Section 5 how this difficulty can be overcome; indeed we can not only drastically reduce the size of this code, but improve performance at the same time.

Shape: another way uses the shape vector itself (strictly, relative positions from which global position, scale and orientation effects have been removed), and searches the pool for those faces which best match. Similar faces have face features in the same relative positions; and we have a simple anthropometric system.

For convenience we have described the texture vector as that arising from mapping to the shape of the average face. This is familiar from the work on caricaturing of Benson and Perrett [2], who distort individual faces *away* from the average by an amount proportional to the deviation of the particular face from the average. In fact there is no need in our methodology to use an average; we simply need each face to be distorted to the *same* shape. We are experimenting with old idea of Baron [1], in which proportionately larger areas are devoted to those parts of the face, such as the eyes, which are most important for recognition. A variant is to remove the hair completely from consideration; although hair is important for recognition, it varies a great deal over long periods of time. We present results with such a template in both Table 1 and Table 3.

3 Template Matching

In this section we describe recognition results obtained using just the texture vector as a code on which to perform matching. A *pool* of 100 images was used, all drawn from the Aberdeen face database. This collection of face photographs was made in 1982, in order to test a methodology for mugshot retrieval, and the collection was a simulation of such a mugshot database [17]. Subjects were photographed in a uniform way, ensuring that differences such as background and lighting were effectively eliminated. The images were subsequently digitised in as uniform a manner as possible with location and scale fixed so that each left eye and each right eye occupied a common position. The images were subsequently normalised to the same mean intensity, control points were located on each image by hand, and the corresponding texture vector was stored.

Figure 2: *Modern images of Ian, Ken and Harry, used as cue to search the database.*

In order to test recognition, modern (1991) images were obtained of three of the faces in the original collection. These images (we refer to them as Ian, Ken and Harry) were collected under very different conditions, in a different laboratory, with the lighting in use there. No attempt was made to imitate the conditions that obtained in the earlier collection. In Fig. 2 we show these images, and in Fig. 3, the originals, together with the images of three other faces from the pool to give an idea of the variation between images.

Figure 3: *Six images from the pool of 100, including images of Ian, Ken and Harry taken nearly ten years before those in Fig. 2.*

The cue images were distorted to shape-free form, and the intensity normalised as in the pool. A matching test was then performed using each of the modern images in turn as a cue. The dot product with each image in the pool was calculated, and the magnitude used as (naive) notion of similarity to rank the images in the pool as matches for the cue image. We reproduce in Table 1 the rankings of each of the target images, as matches for the corresponding cue. We report also the rankings obtained in a similar test, in which the portion of the shape-free face used as a template was restricted by excluding the hair.

It is hard to present results in a way which guarantees that bias has been excluded, but the distractors in the pool are believed to be "random". All the tests we report here have been with a fixed pool consisting of faces photographed under very similar conditions in order to avoid difficulties in knowing whether a match occurred for accidental reasons associated with the image, or because the face itself was recognised. At the very least, our results suggest the potential of a shape-free face as a code for recognition. We return to this in Section 5 after presenting results matching on shape alone.

Cue face	Ian-now	Ken-now	Harry-now
Complete Face	1	39	18
Ignoring Hair	1	13	22

Table 1: *Template matching on a pool of size 100. The table gives the ranking of the match by the corresponding "then" image when cued by a "now" image.*

4 Matching on Shape

We defined the shape vector of a face as the vector of (at present) 59 control points used in distorting the face image to the average shape. We describe here recognition results, similar to those above, matching on an invariant form of the shape vector, in which the effects of position and isotropic scale change have been removed. In principle we would expect to remove at least one other rotation parameter, to ensure that the resulting face was level, and perhaps a second, so the face is looking straight ahead, but in practice neither of these proved necessary for our images at present.

One way to remove position and scale parameters is to regard them as the only free parameters in a model of the face using all 59 control points. A fixed standard model is then chosen — typically the average shape again, and the fit between this model and the actual face shape calculated using euclidean distance between corresponding control points. A least squares minimisation of this error then provides "best fit" values of the free parameters, and the resulting "normalised" shape vector is one candidate on which to perform matching.

There are obvious problems with this approach, and we choose to first concentrate on those of the 59 points which are both "significant" and reliably located. Thus we generate a model consisting of only five points, obtained from the average of the 6 left eye points, the average of the 6 right eye points, the average of the 4 points at the end of the nose, the average of the 2 points at the ends of the mouth; and the point in the middle of the chin. We then perform the least squares minimisation described above on this simple "eyes, nose, mouth and chin" model to remove three parameters associated with position and scale. A reduced shape vector was is a point in $\mathbf{R}^7$, and the matches are ranked using euclidean distance in this space. As before, we report how well each target face matched the corresponding cue; the resulting ranks are given in Table 2.

Cue shape	Ian-now	Ken-now	Harry-now
Shape match	5	4	30

Table 2: *Matching faces based on their configuration vector. The table gives the rank, from a pool of 100 faces, of the match between the cue face and the corresponding target.*

We regard these results as preliminary and are currently exploring shape matching methods using explicit invariants, such as the ratio of the distance between the eyes to the mouth width. Such invariants, when combined with psychological ratings, have already been used successfully for recognition on fairly large sets of faces [17]. An important point is that matching is taking place on a very different characteristic from that contained in the texture vector, and that faces confused on one criterion are not confused on another; specifically, no face in our pool matches the cue better than the corresponding target on both shape and texture.

5   Principal Component Coding of Faces

We have presented our methodology as simple template matching, but in practice, the size of a texture vector (perhaps 30Kbytes for a 256×256 image) makes this unrealistically slow for large pools. The problem can be avoided using principal component analysis. As a first step, we choose a fixed set of shape-free faces, or an initial *ensemble*, and approximate all other faces as linear combinations of these. In the work we describe here, a total of 150 shape-free faces were first created. Of these, the three images destined to be targets were put directly in the pool, together with another 97, chosen using a random number generator. The remaining 50 images then became our initial ensemble, with which to represent all our other faces.

Our gain in efficiency comes from coding all shape-free faces in terms of the faces in the ensemble. Rather than code directly, we apply a principal component analysis to the shape-free faces in the ensemble. More precisely, we first obtain the mean image of the ensemble, and then the principal components, the eigenvectors of the covariance matrix, built from the deviations from the mean of each face in the ensemble. This gives a new basis of the subspace spanned by the ensemble; we refer to the basis elements (not themselves faces) as eigenfaces because of the way they are derived. The first six eigenfaces are shown in Fig. 4.

A shape-free face, even one not in the initial ensemble, can be approximated as a linear combination of these eigenfaces. The weights used in this sum, a total of 50 bytes, provide the succinct *eigenface representation* (cf [13], [22],

504

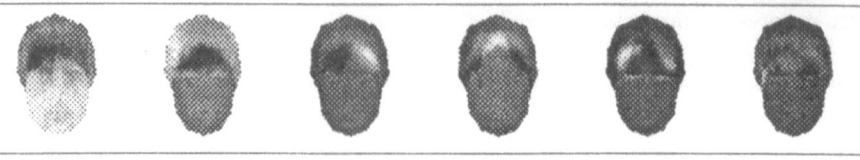

Figure 4: *The first six eigenfaces. Shape-free faces are represented as a linear combination (of typically 20) of these.*

[6]) which we use for matching. Note also that the errors in the approximation process are available. The whole process is illustrated in Fig. 6. We display the corresponding weighted sum of eigenfaces in the third image of Fig. 6. The fourth image then recombines the eigenface representation and the shape vector, to give the version of the original face that our representation effectively stores. A resemblance between the original and the reconstruction makes our ability to match using the reduced codes (see Table 3) more plausible.

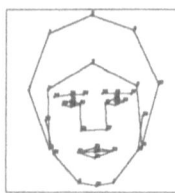    

Figure 5: *The set of 40 control points found by FindFace.*

Figure 6: *The original is first distorted to a shape-free face to give the second image. This is approximated to give the third image; the distortion is then inverted to get the final reconstruction.*

We can both reduce the size of the representation still further, and improve its utility, by using only the most significant eigenfaces. The eigenfaces are initially chosen so that those corresponding to eigenvalues of large magnitude are good at discriminating between ensemble members while the bottom few code similarity. We thus choose to ignore the "unimportant" eigenfaces. There is significant debate on how many eigenvectors to take (eg [11], page 93); rather than keeping enough to capture 95% of the total variance, we first rescale the eigenvalues to have product 1, and discard those eigenfaces with corresponding rescaled eigenvalues less than 1. This will typically select 20 out of 50 eigenfaces, and in the process, capture over 80% of the variance in the ensemble.

We present results which are comparable to those in Table 1, using a pool of 100 faces as described above, and again using as cue each of the images shown in Fig. 2. All the images were first transformed to their shape-free state and then projected onto the subspace spanned by the most significant 20 eigenfaces obtained from the (totally disjoint) ensemble of 50 faces chosen above. The dot product between cue and each image in the pool was calculated, and the magnitude used as (naive) notion of similarity to rank the images in the pool as matches for the corresponding cue image. Table 3 gives these rankings.

Again we also give results when the portion of the image to be first coded, and then matched, is restricted by excluding the hair. To do so, the ensemble of 50 faces used to generate the eigenfaces was also restricted in this way; and in this case, after normalising the eigenvalues, we were left with 21 eigenfaces (rather than 20) of magnitude at least 1, which were used for coding. The results seem significantly better than those in Table 1, although each image is now represented with approximately 20 bytes.

Cue face	Ian-now	Ken-now	Harry-now
Complete Face — top 20 components	1	27	10
Ignoring Hair — top 21 components	1	1	8

Table 3: *Matching shape-free faces in the Reduced Eigenface Representation. The rank, from a pool of 100 faces, of the match between a cue face and the corresponding target is shown.*

For comparison, Turk and Pentland [22], work with a pool containing many images of 16 individuals, build an ensemble with a different image of each individual, and extract 7 eigenfaces. They then classify each face in the pool as one of the sixteen individuals. However the registration, which we regard as vital, is only implicit in their work, and is only done with global position and scaling parameters, rather than with a full anticaricature; indeed size and position are variables they seek to recognise across.

One can view the passage from the full shape-free face, to the reduced representation, in terms of only a few significant eigenfaces: either as a useful approximation to the "true" full template, in which the loss of accuracy is compensated for by the speed with which matching is performed; or as an improvement over crude template matching, in which, by passing to a few codes chosen for their ability to describe variability between faces, we have obtained the ability to generalise from the particular image used as a cue. We believe the evidence above supports the latter view; and that the generalisation occurring is similar to that which can occur in a neural net We note also, again in comparison with net-based systems, that no knowledge of the cue or target is available to the coding system, and no training is needed to extract "suitable" codes.

Finally we note that even our initial description of recognition can be phrased in terms of principal component analysis: using the full encoding, with an ensemble which co-incides with our pool, is equivalent (up to a fixed choice of weights) to the template matching we described in Section 3.

6   Theoretical Discussion and Conclusions

We now place our coding scheme in a theoretical context. Our two stages of coding can be considered as a two stage description of the underlying geometry, (see Craw and Cameron [6]) in which the "face manifold" is first subjected to non-linear perturbations, yielding a linear structure, on which familiar linear operations can be applied [3]. This need for linearity may explain why the codes used for recognition also arise when trying to perform realistic merges (or linear averages) between two faces.

We have explicitly described processing needed to perform recognition; it may be that similar processing occurs within a neural net dedicated to face recognition. The first phase, in which features are identified, and the image is linearised to a shape-free face, is a familiar non-linear warping step, frequently observed and well understood (eg [10], page 235). Since we follow this by principal component analysis, essentially the function performed by linear neural nets [15], the overall functioning of our proposed method may be sufficiently similar to that of a neural net to provide insight into their functioning. Learning the need for the non-linear warp will requires much training, and it is here we obtain a computational advantage: at present by essentially performing this time-consuming operation manually; and in principle by applying FindFace, whose individual modules can be tuned for performance.

We claim to have demonstrated a useful recognition performance even when there is a much more significant difference between target and cue than is usual. Further, we have the possibility of doing this completely automatically. Although our main interest is in the principal component analysis based texture recognition, our experiments with simple template matching suggest that subsequent results are not an artifact, and that the concept of a shape-free face is useful. Our results on shape-based recognition are more tentative, although it is a much more familiar method; it's interest is in providing a relatively independent measure of recognition, to combine with our texture measure yielding a single measure which is more discriminating than either.

We believe our coding scheme has promise, but wish to know more of its limitations. In part these arise from inaccurate control point locations, in part from the approximation inherent in principal component analysis, and in part from the sheer variability of the human face on different occasions; the relative importance of these errors is not yet clear. Template matching methods are notoriously dependent on lighting, and we have done little to explore this. One approach incorporates so much lighting variation in the ensemble that no particular condition is accurately encoded. Control of lighting for the pool images, as used for our tests, provides a way to finesse this difficulty. Another aspect to explore is the choice of initial ensemble; it should be selected to provide good representations of the types of faces being coded, rather than randomly. We noted in Section 4 the possibility of incorporating a limited view-invariance in our recognition process; by using a number of ensembles, each tuned to a specific angle of view, we may provide a much greater degree of view invariance.

References

[1] R. J. Baron. Mechanisms of human facial recognition. *International Journal of Man–Machine Studies*, 15:137–178, 1981.

[2] P. J. Benson and D. I. Perrett. Perception and recognition of photographic quality facial caricatures: Implications for the recognition of natural images. *European Journal of Cognitive Psychology*, 3(1):105–135, 1991.

[3] V. Bruce, A. M. Burton, and I. Craw. Modelling face recognition. *Philosophical Transactions of the Royal Society of London, Series B*, 335:121–128, 1992.

[4] R. Brunelli and T. Poggio. HyberBF networks for gender classification. Preprint, 1991.

[5] G. W. Cottrell and M. Fleming. Face recognition using unsupervised feature extraction. In *Proceedings of the International Neural Net Confernce*, pages 322–325. Dordrecht Kluwer, 1990.

[6] I. Craw and P. Cameron. Parameterising images for recognition and reconstruction. In P. Mowforth, editor, *British Machine Vision Conference 1991*, pages 367–370, London, 1991. Springer Verlag.

[7] I. Craw, D. Tock, and A. Bennett. Finding face features. In G. Sandini, editor, *Proceedings of ECCV-92*, number 588 in Lecture Notes on Computing Science, pages ??–?? Springer-Verlag, 1992.

[8] R. Gallery and T. I. P. Trew. An architecture for face classification. In *Colloquium: Machine Storage and Recognition of Faces. IEE Digest 017*, 1992.

[9] J. Hertz, A. Krogh, and R. G. Palmer. *Introduction to the Theory of Neural Computing*. Computation and Neural Systems series. Santa Fe Institute and Addison Wesley, 1991.

[10] I. T. Jolliffe. *Principal Component Analysis*. Springer-Verlag, New York, 1986.

[11] T. Kanade. *Computer Recognition of Human Faces*, volume 47 of *Interdisciplinary Systems Research*. Birkhäuser, Basel,Stuttgart, 1977.

[12] M. Kirby and L. Sirovich. Application of the karhunen-loève procedure for the characterisation of human faces. *IEEE: Transactions on Pattern Analysis and Machine Intelligence*, 12(1):103–108, 1990.

[13] T. Kohonen, E. Oja, and P. Lehtiö. Storage and processing of information in distributed associative memory systems. In G. Hinton and J. Anderson, editors, *Parallel models of associative memory*, chapter 4. Erlbaum, Hillsdale N.J., 1981.

[14] R. Linsker. From basic network principles to neural architecture: Emergence of orientation columns. *Proceedings of the National Academy of Sciences*, 83:8779–8783, 1986.

[15] R. M. Rickman and J. Stonham. Coding facial images for database retrieval using a self organising neural network. In *Colloquium: Machine Storage and Recognition of Faces. IEE Digest 017*, 1992.

[16] J. W. Shepherd. An interactive computer system for retrieving faces. In H. D. Ellis, M. A. Jeeves, F. Newcombe, and A. Young, editors, *Aspects of Face Processing*, chapter 10, pages 398–409. Martinus Nijhoff, Dordrecht, 1986. NATO ASI Series D: Behavioural and Social Sciences - No. 28.

[17] R. B. Starkey and I. Aleksander. Facial recognition for police purposes using computer graphics and neural networks. In *Colloquium: Electronic Images and Image Processing in Forensic Science. IEE Digest 087*, 1990.

[18] T. Stonham. Practical face recognition and verification with WISARD. In H. Ellis, M. Jeeves, F. Newcome, and A. Young, editors, *Aspects of Face Processing*, pages 426–441. Martinus Nijhoff, Dordrecht, 1986.

[19] K. Sutherland, D. Rensham, and P. B. Denyer. A novel automatic face recogntion algorithm employing vector quantization. In *Colloquium: Machine Storage and Recognition of Faces. IEE Digest 017*, 1992.

[20] M. Turk and A. Pentland. Eigenfaces for recognition. *Journal of Cognitive Neuroscience*, 3(1):71–86, 1991.

A Comparison of Vector Quantization Codebook Generation Algorithms Applied to Automatic Face Recognition.

C. S. Ramsay†, K. Sutherland† D. Renshaw and P.B. Denyer

Integrated Systems Group, Electrical Engineering Department,
University of Edinburgh, Edinburgh, EH9 3JL.

Abstract

Automatic facial recognition is an attractive solution to the problem of computerised personal identification. In order to facilitate a cost effective solution, high levels of data reduction are required when storing the facial information. Vector Quantization has previously been used as a data reduction technique for the encoding of facial images.

This paper identifies the fundamental importance of the vector quantizer codebooks in the performance of the system. Two different algorithms – the Linde-Buzo-Gray algorithm and Kohonen's Self Organising Feature Map – have been used to obtain two sets of facial feature codebooks. For comparison, the system performance has also been analysed using a codebook dedicated to the test population. It has been shown that by using a *good* codebook generation algorithm it is possible to substantially reduce the dimensionality of the vector codebooks, with remarkably little degradation in system performance.

1   Introduction

Automatic facial recognition is now receiving an increasing amount of research interest [1], largely due to its possible applicability to the automatic personal identification task. As such, automatic face recognition is attempting to stake its claim among the other biometric systems available (notably fingerprint, hand geometry, dynamic signature matching and voice pattern recognition).

In order to establish biometric systems as likely tools for personal identification we must instill public confidence in the concept of biometric storage and retrieval. In this area automatic face recognition scores heavily over the other likely biometrics, as the least intrusive and the most *natural* personal identification process. It is also important that any prototype system has a demonstrably high accuracy. To obtain such high accuracy a likely advancement would be to incorporate the use of a number of different biometrics into one identification system, making the final response conditional on all the available information.

However, in practice, the overriding factor in achieving a *marketable* system will be the data reduction obtained, since this controls the speed at which multiple comparisons can be performed and, ultimately, the cost, as data storage

---

†These authors have SERC studentships.

requirement is a prime factor in this area. The available level of data reduction is now of particular importance given the increasing use of smart-card technology and the ever increasing likelihood that we will all eventually carry personal biometric information with us in this way. If a facial image is to be one of many pieces of biometric information stored on our smart-card, then accurate, high data reduction, parametrisation of the face is required.

The authors have previously introduced a Vector Quantization (VQ) based facial recognition technique and presented results for a recognition experiment using 30 individuals [2]. In this paper we would like, firstly, to present further results on a 33% larger data set and, secondly, to investigate the use of VQ codebook generation techniques to reduce further the data space required to store the intrinsics of the face (or the *facial signature*). The codebook generation techniques used here to reduce the overall number of vectors are Kohonen's neural Self-Organising Feature Map and the Linde-Buzo-Gray algorithm. However, firstly, a brief overview of the entire facial recognition algorithm will be presented.

2 System Overview

The intrinsic function of a pattern recognition process is that of parameterisation, *i.e.* the extraction of the fundamental characteristics of the object to be recognised. In [2] the authors presented an algorithm which distilled the face into a compact facial signature, while still maintaining much of the *recognisability* of the initial input face.

In brief, the system identifies seven fundamental features; each eye, the nostrils, the bridge of the nose, the mouth, the chin and the hair. In addition to storing these individual facial parts, the facial signature also incorporates a coarsely sub-sampled view of the entire face. The partitioning of the face used here is shown in Figure 1. These facial features are automatically located using a template matching algorithm based on Fischler and Elschalger's *feature embedding* approach [3]. Manual correction of the location failures was performed.

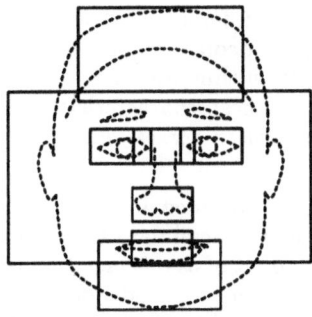

Figure 1: Facial Features Exploited to Differentiate between Individuals.

Having extracted these eight facial parts for storage and discarded the rest of the image data, there still remains the task of data reduction. The signal processing function of VQ has been chosen to provide the requisite reduction. The use of VQ for image coding is widespread [4]. However, recognition analysis using VQ is much more novel.

Applied to faces, VQ performs a function analogous to the police *photofit* system. In turn, each of the eight selected features is compared with a codebook of standard examples (or *vectors*) of only that feature; using the normalised euclidean distance as a metric, the most similar of these standard examples is chosen as the most likely match. Then to store that feature, we need only to keep the *index* of the standard feature and not all the data points which constitute that vector.

It is desirable to train a facial recognition device on a number of examples of each person it is expected to recognise, in order to allow it to construct internal representations of each person. To facilitate this, a system of *feature histograms* has been devised; these reflect the ways in which the subject's face has varied during training. To obtain the histograms we use the VQ technique described above to encode each feature, from each training image, then that *match* is recorded in the histogram. Thus, given 20 possible vectors and ten training images one particular *histogram* could look like this, Figure 2.

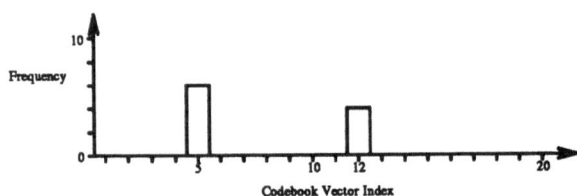

Figure 2: One Feature Histogram for One Person's Training Images.
This feature has been mapped six times to codeword 5 and four times to codeword 12 during training.

In order to differentiate one person from another their respective histograms would have to show significant differences. If we look at the histograms of all eight features for two different people, Figure 3, it can be seen that there is only a low level of variability within training for each person, yet there is a substantial difference between these characteristics between these two people. Low *within person* and high *between person* variabilities are essential for a good recognition system.

By storing the histograms obtained in the manner described above it is argued that we have the requisite facial data to perform recognition. Thus, the feature histograms can be thought of as our facial signatures. The problem with this approach is that, to date, the VQ codebook entries have been extracted from a control image of each member of the test population. In this way the data storage requirement, as determined by the number of vectors, is dependent on the population size. Thus, if we were to enlarge the population there would be a consequent rise in data storage requirements. To combat this, the following section outlines the ways in which the codebook dimensionality can be reduced.

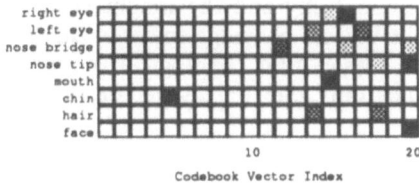 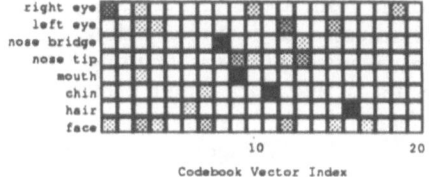

Figure 3: Feature Histograms for Two Population Members.
In this graphic all eight feature histograms are shown, with grey tone signifying the frequency of use of each codebook vector. *Black* represents a well used vector, with lighter tones indicating less use.

3 Codebook Generation

The process of VQ relies on having a good set of vectors in its codebook. For image compression, these vectors are chosen to minimise the overall pixel level error introduced. However, for face recognition, other considerations are also important. For example, distinctiveness may be more significant than pixel error when encoding facial features. Fortunately, there are many different algorithms available which perform codebook generation [5].

As an initial starting point, each feature codebook has been constructed using one sample vector drawn from a control image of each member of the test population, thus reflecting only the variation present within this test population. To obtain a reduction in data storage, the two most common VQ codebook generation techniques have been used to reduce the codebook dimensionality. In this study the test population contained 40 members and thus the initial codebooks (for each of the eight features) contained 40 vectors. A reduction to half this number has been chosen as a sufficiently demanding test of the codebook generation algorithms available. The task of the codebook generation technique is to perform the requisite dimensionality reduction while still containing the system's error rate within acceptable limits.

3.1 Kohonen's Self-Organising Feature Maps

Neural network clustering techniques have been employed in a variety of application areas, such as pattern recognition, optimisation and, notably, VQ codebook design for image compression [6, 7]. Kohonen developed his Self-Organising Feature Maps (KSOFM) in order to model the neural feature maps which are thought to form in the human brain [8]. KSOFMs result in a network where neighbouring output units have similar responses : *topological ordering* has occurred. This ordering can be used to reduce search requirements in VQ applications, though this is not an aspect of the KSOFM which has been exploited here. As a clustering algorithm, KSOFM allows a reduction in the dimensionality of the input vector space to a smaller number of reference vectors.

KSOFM defines a matrix (usually two dimensional) of output units, or neurons, each of which has a weight vector, w_j, associated with it. Input vectors from a training set are presented sequentially and the weight vectors are adjusted as described below. The weight vectors converge towards cluster centres after sufficient training time. The reason that topological ordering occurs is that each input vector adjusts not only one weight vector, but a *neighbourhood* of weight vectors.

The KSOFM algorithm is defined thus :

Step 1. Initialise all weight vectors to random values.

Step 2. Apply new input vector, $x(t)$.

Step 3. Calculate the Euclidean distance from $x(t)$ to all output nodes j

$$d_j = \sum_{i=0}^{N-1} (x_i(t) - w_{ij}(t))^2$$

where N is the dimensionality of the input vector.

Step 4. Select the output node with the smallest distance d_i and label it as the winning unit, j^*.

Step 5. Update weight vectors of all nodes in the matrix according to the equation

$$w_{ij}(t+1) = w_{ij}(t) + \eta(t, \mathcal{D})\alpha(t)(x_i(t) - w_{ij}(t))$$

where $\eta(t, \mathcal{D})$ is the neighbourhood gain function, and $\alpha(t)$ is the adaption gain function.

$\eta(t, \mathcal{D})$ is a function which defines a neighbourhood on the matrix round the winning neuron, decreasing exponentially with distance $\mathcal{D}$ from the winning neuron, and which shrinks over time. $\alpha(t)$ decreases exponentially with time, and $0 \leq \alpha(0) \leq 1$.

Step 6. Repeat **Step 2** to **Step 5** until the entire training set has been presented e times. e is the number of *epochs*, which is set before training starts.

In this application the input vectors, x, come from the original feature codebooks containing 40 vectors, and the matrix was a 5×4 array of neurons producing a codebook of 20 vectors.

3.2 Linde-Buzo-Gray Algorithm

The Linde-Buzo-Gray (LBG) algorithm [9] is the most commonly used codebook design algorithm due to the fact that it was the earliest proposed method and consistently outperforms other methods in a variety of applications [10, 11]. It is an iterative technique which repeatedly moves codewords to cluster centroids in an effort to find a codebook which will display the lowest error when encoding the training data (*i.e.* a modified version of the K-Means clustering algorithm). The basic algorithm is as follows :

Step 1. Initialise the codebook.

Step 2. For each vector, x, in the training set, calculate the Euclidean distance from it to each vector v_j in the codebook.

$$d_j = \sum_{i=0}^{N-1} (x_i - v_{ij})^2 \tag{3}$$

where N is the dimensionality of the vectors.

The minimum distance selects the closest vector, v_j^*, in the codebook. Assign x to the cluster around v_j^*.

Step 3. Replace each codeword with the centroid of the vectors in the training set that have been assigned to it. If any codewords are unused, they are discarded and replaced with new codewords which are more likely to be used in the next iteration. In this work, this has been done by taking the most commonly used codewords and splitting them to create two close copies of the original.

Step 4. If the total error in clustering the training data is still decreasing by a significant amount, return to **Step 2.** Otherwise, stop.

This algorithm should optimise the codebook so that the sum of the distances from each vector of the training set to its nearest codeword is a minimum. It is possible, however, that in certain conditions the algorithm will reach a local minimum rather than a global minimum. This can be seen to be true due to the fact that the final codebook will change if the initial codebook is different.

There are a number of recognised methods for generating an initial codebook, the most common of which is to populate the codebook with vectors chosen randomly from the training set. Another method is to use a codebook generated from other clustering techniques, such as the KSOFM technique described above, as the initial codebook. The method utilised here was to populate the codebook with 20 replicas of the centroid of the entire training set. In this way the codebook is gradually filled with useful codewords, as unused codewords are replaced in **Step 3**.

There are some similarities between the ways in which the two techniques outlined above perform dimensionality reduction. However, the codebooks produced can be significantly different. To illustrate this Figures 4 and 5 show the two 20 vector *face*[†] codebooks generated from the same original input of 40 vectors. It can be clearly seen that both codebooks contain composite faces formed by the merging of several of the input faces together, but that they are quite different from each other.

[†]There are parallel codebooks for each of the other seven features used in the recognition algorithm.

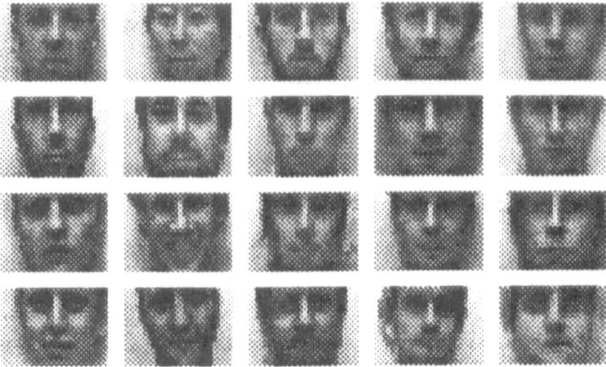

Figure 4: Codebook Generated using LBG.

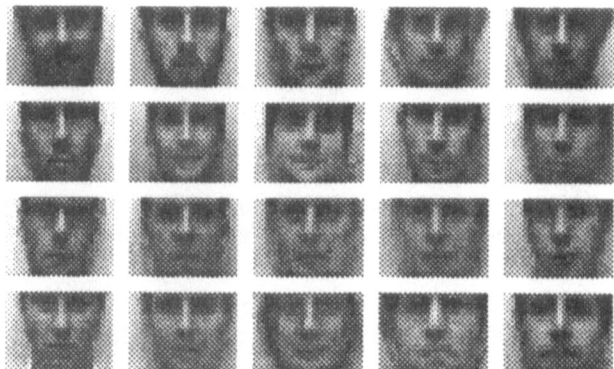

Figure 5: Codebook Generated using KSOFM.

4 Experimental Results

At present the algorithm for facial recognition is implemented as a suite of software programs. The system cannot yet function in real-time and thus the test images used are stored on disk for repetitive analysis. The images used were captured with a video camera under largely controlled conditions. However, the subjects were allowed to vary their expressions during the several days during which images were being captured. In this way, the results obtained for the system will be closer to a real-world implementation than some other, highly controlled, studies.

As mentioned earlier, a test population of 40 males was used, with ten images of each person used for training (*i.e.* the generation of the facial signature) and ten images kept for testing. The trial thus consisted of 400 test presentations. For each presentation, the output ranking of likelihood was recorded. From this data, and from knowing the correct response, it is possible to obtain a first place recognition rate (*ie* the proportion of tests in which the correct signature was selected as the most similar to the test stimulus). By analysing the output ordering of responses an average rank figure can also be obtained.

4.1 Dedicated codebook

To provide a benchmark, and to demonstrate the best possible performance, a set of dedicated codebooks were used. These codebooks were constructed using the same vectors as were used to train the other codebook generation algorithms. The feature histograms were thus 40 vectors wide for each of the eight features. For this experiment the average rank and the cumulative success rates of the first three output positions is given in Table 1.

Average Rank	1.37
1st Place Recognition	88.5%
2nd Place Recognition	92.75%
3rd Place Recognition	95.5%

Table 1: Recognition rates for a 40 member population with 400 test presentations.

4.2 Dimensionality Reduction

Essentially, here, we are performing the same experiment as above, but using codebooks of half the size. Such a significant reduction would be expected to cause a substantial reduction in recognition performance unless, as described in section 3, the algorithms used to derive the new codebook entries reflect the initial characteristics of all 40 vectors. The experiment is thus a parallel comparison between the two different sets of codebook vectors when applied to the recognition function. Table 2 gives the relative performances between the two approaches.

	KSOFM	LBG
Average Rank	1.99	1.46
1st Place Recognition	70.2%	83.3%
2nd Place Recognition	81.5%	90.0%
3rd Place Recognition	88.5%	94.7%

Table 2: Two methods of codebook design.

4.3 Discussion

If we consider the first experiment using a dedicated codebook, the performance results are very encouraging. The first place recognition rate of 88.5% does not yet rival the other biometric systems. However, the results reported here represent one of the very few significant population studies of a practical automatic face recognition system reported to date and, as such, give an important indication of the future viability of automatic face recognition.

Considering the dramatic reduction in codebook dimensionality, the second set of recognition results are remarkably good. The LBG approach performs significantly better than the KSOFM method. Its overall recognition rate is approaching that of the 40 vector system reported in section 4.1. To explain the variation in the performance of these two codebook generation algorithms more detailed consideration of their mechanics is required.

In general the codebook generation algorithms perform reduction by averaging the most similar vectors together, while still maintaining good coverage of the initial vector space. If the most similar vectors are also the most frequently used vectors, then the clustering process may reduce the good spread of vector choice required to maintain good differentiation when encoding the population. To investigate whether this is in fact the case feature histograms have been drawn up for the entire population.

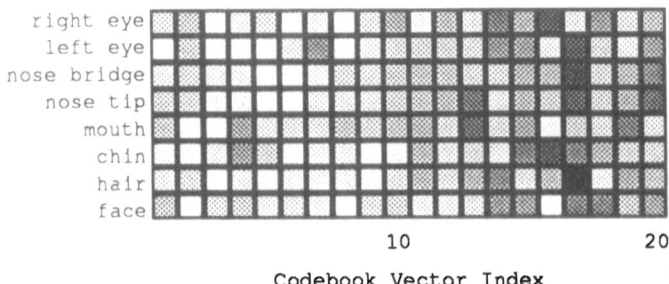

Figure 6: Population Feature Histograms for LBG.

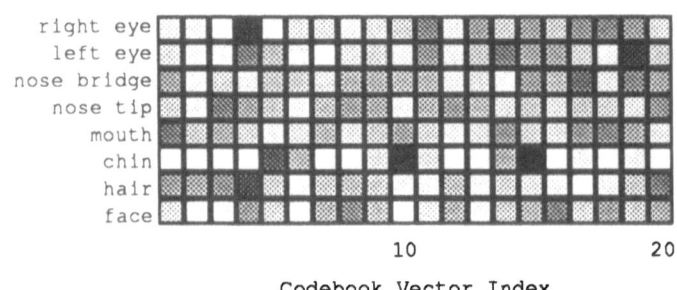

Figure 7: Population Feature Histograms for KSOFM.

Figures 6 and 7 illustrate the feature histograms for the LBG and KSOFM algorithms respectively. The spread of vector choice is much more even for the LBG vector set. We believe this is the underlying factor which explains the relatively poor performance of the KSOFM approach. In effect, KSOFM has maintained *too good* coverage of the entire feature space at the expense of differentiation between the most common feature types.

5 Conclusions and Future Work

The results of the initial research into the use of VQ for automatic facial recognition are encouraging. This approach has been shown to work for facial recognition and undoubtedly has uses in other areas of image pattern recognition.

The dimensionality of the codebooks used by VQ based recognition has been identified as a potential limiting factor, however, this research has shown the important improvements which can be obtained in this area by using different codebook generation algorithms. However, we accept that further experimentation, with different populations, is required to validate the results reported here. It is further recommended that much larger populations are required to adequately test potential facial recognition systems.

References

[1] V Bruce and M Burton. Computer recogniton of faces. In A W Young and H D Ellis, editors, *Handbook of Research of Face Processing*, pages 487–506. North-Holland, 1989.

[2] K Sutherland, D Renshaw, and P B Denyer. A novel automatic face recognition algorithm employing vector quantization. In *Digest of the IEE Colloquium on Facial Recognition and Storage*, London, January 1992.

[3] M A Fischler and R A Elschalger. The representation and matching of pictorial structures. *IEEE Transactions on Computers*, 22:67–92, 1973.

[4] N M Nasrabadi and R A King. Image coding using vector quantization : A review. *IEEE Trans. on Comms.*, COM-36(8):957–971, August 1988.

[5] R M Gray. Vector quantization. *IEEE ASSP Mag.*, 1(2):4–29, April 1984.

[6] N M Nasrabadi and Y Feng. Vector quantization of images based upon the Kohonen self-organising feature maps. In *Proc. Int. Joint Conf. on Neural Networks (IJCNN-88)*, pages 101–108, July 1988.

[7] T-C Lee and A M Peterson. Adaptive vector quantization using a self-development neural network. *IEEE J. on Selec. Areas in Commun.*, 8(8):1458–1471, October 1990.

[8] T Kohonen. Clustering, taxonomy, and topological maps of patterns. In *Proc. 6th Int. Conf. on Pattern Recognition*, pages 114–128. IEEE, October 1982.

[9] J Linde, A Buzo, and R M Gray. An algorithm for vector quantizer design. *IEEE Trans. on Comms.*, COM-28(1):84–95, January 1980.

[10] J D McAuliffe, L E Atlas, and C Rivera. A comparison of the LBG algorithm and Kohonen neural network paradigm for image vector quantization. In *Proc. Int. Conf. on Acoustics, Speech and Signal Processing (ICASSP-90)*, pages 2293–2296. IEEE, April 1990.

[11] W Equitz. Fast algorithms for vector quantization picture coding. In *Proc. Int. Conf. on Acoustics, Speech and Signal Processing (ICASSP-87)*, pages 725–728. IEEE, April 1987.

Blink Rate Monitoring for a Driver Awareness System

David Tock and Ian Craw

Department of Mathematical Sciences *

University of Aberdeen, Scotland

1 Introduction

We describe a working prototype computer vision system for measuring the blink rate of car drivers. It combines an intelligent face recognition system with feature tracking, and statistical image analysis methods to locate and track the head position of a driver in a car, and extract specific measurements from the eyes. This is done using standard hardware (a Sun IPC workstation and frame grabber) without the use of image processing or other specialised hardware. The information obtained is used to give an indication of driver *alertness* which forms one of the inputs to a complete driver monitoring system.

2 Background

Recent studies [10, 8] (and others in Germany, Israel, the UK and USA) have shown that the majority of road accidents take place in good driving conditions, and can not be attributed to bad weather, alcohol/drugs or mechanical failure. The obvious conclusion is that such accidents are caused by driver error. The clustering of accidents around the hours 1am to 6am further indicates a relationship with driver fatigue. As part of the European PROMETHEUS programme, a system is being developed to monitor the driver's status. Although the aim is to obtain this information directly from sensors on the vehicle and its controls, there is no obvious correlation between these sensor inputs and driver status. Currently researchers at Stirling University are developing a neural network system to integrate these sensory inputs, but that still leaves the problem of correlation to driver alertness

Past research has shown a good correlation between a persons alertness and their blink rate, so this has been chosen to provide the necessary reference input with which to train the neural network.

As the system must work in a car, and will be used for extensive testing both on the road and on simulators, a non intrusive, non contact method of measurement was essential. This ruled out the use of special glasses or helmets, or methods which involved the driver keeping their head in a fixed position. The method chosen was to mount a video camera on the dashboard pointing towards the driver and to extract the measurements visually.

---

*This research was supported by a grant from the Ford Motor Company Limited

3 Design Considerations

The system we describe deals with two aspects of the problem independently. Firstly the position of the drivers head, or more specifically, the drivers eyes must be determined; and secondly, the eyes must be measured in some way to identify blinks.

A typical blink lasts approximately 200ms, so a frame rate of at least 5 frames per second must be achieved to avoid blinks occurring entirely between frames[1]. Obviously, a higher frame rate is desirable, and would make detection both easier and more reliable.

One possible approach was simply to develop FindFace, our existing face recognition system [17, 4], to perform the task. This may have been possible with considerable refinement and tuning, but would still fail to address some of the problems associated with working within a vehicle. Alternatively we could have developed systems designed to locate eyes in images, such as Nixon [12], Yuille *et al* [19], Hallinan [6] or Bennett and Craw [2]. All of these systems are designed to work with single images, and would not exploit the benefits of working with an image sequence.

Image sequence analysis is a well documented area of machine vision ([9, 13, 16] give general overviews) which attempts to extract particular information from an image sequence. Some of the more common objectives include;

- to determine the pose (position and attitude) of a known object moving in an otherwise stationary scene (e.g.[7]);

- to determine the shape and structure of an unknown object (or objects) moving in an image sequence (e.g. [14]); and

- to determine the position and movement of the camera within the environment it is viewing (e.g. [3]).

Although our objective differs from these, it shares some common ground with each.

We know approximately what we are looking for in an image, but not well enough to predict its appearance – everyone's face is different. We do not need to use movement to determine the 3D shape of the face as we are only interested in 2D appearance, and we must contend with the the problems of the image changing due to camera movement. Furthermore, the drivers head may remain stationary for long periods of time, or move only very slowly. This makes the *optical flow* approach inappropriate, and the *point correspondence* approach requires as input the very data we are hoping to obtain via the sequence analysis, namely the location of fixed points on the moving object. Despite these problems, there are advantages to be gained from the redundancy of data inherent in such a sequence.

Under most driving conditions camera vibration can be eliminated by mounting it securely to the car's bodywork. We then treat the interior of the vehicle as a stationary scene in which the driver moves. No suitable camera position could be found that gave adequate coverage of the driver without

---

[1] Failing to detect occasional blinks will not upset the results – the important measurement is the change in frequency of blinks over a (relatively) long period of time.

including window regions. The camera can therefore *see* out of the side and back windows. A number of idea were considered, such as polarising the windows and camera lens so as to eliminate most of the outside light, but these ideas were rejected due to the possibility of interference with the driver's visibility. The system must therefore cope with a considerable amount of changing background.

A further complication that the system must contend with is poor contrast and rapidly changing lighting on the drivers face. We have not tackled the problems of operating in night conditions[2]; but even ignoring these problems, because of the environment, the drivers face is usually in the shadow caused by the vehicle. The images obtained have either very low contrast when in the shade, or very high contrast when directly illuminated. These conditions change rapidly when a vehicle is in motion. Obviously, adding visible light sources to the vehicle is inappropriate on the safety grounds. Adding IR or UV illumination may be acceptable.

The camera's field of vision and the lighting characteristics thus limit the effectiveness of using image sequence information for tracking the head. Furthermore, they also presented problems for performing the eye measurements on single images using the FindFace system. Initial experiments with our usual technique for outline location [2] failed to perform satisfactorily, locating onto regions of higher contrast in the background. Similarly the eye detectors, given the poor contrast of the eye region, produced results of unacceptably low accuracy and reliability. We investigated some inter-frame relationships, such as simple differencing between images, but as expected this produced unacceptable levels of noise and spurious response. More significantly, these early efforts revealed a deficiency in our equipment which resulted in corresponding pixels in successive eight bit images varying by up to ± 15. This was believed to be due to an interaction between camera and frame grabber, and although no cure could be found, the work had to proceed (at least initially) with this equipment, despite the atrocious signal/noise level. The variation was not random noise, rather a slow undulation making single images look perfectly normal, with the problem only apparent on sequences.

4 System Overview

The system that developed has two independent subsystems; a fast and simple algorithm which performs the tracking and eye measurement functions based on certain assumptions, and a slower, more reliable *watchdog* system that confirms these assumptions and hence that the fast system is performing correctly.

4.1 Feature Tracking System - Stage I

Due to the problems mentioned above, simple inter-frame differencing produces poor results. We do however know what the image without a driver should look like. The system can be initialised at regular intervals when no driver is present, either automatically by the watchdog system, or manually. Alternatively, a

---

[2] There are plans to investigate both NIR (near infra-red) and UV (ultra violet) illumination, and high sensitivity cameras.

number of images obtained under various conditions can be stored, and the most appropriate selected. These constitute *static* images, and give the system a base on which to build. As the system is only being used under controlled conditions at present, the driver can be ask to re-initialise it manually from time to time; for simplicity this approach has been adopted at present. In contrast we refer to images obtained at run time as *dynamic* images, and these are obtained as required by the system.

Rather than performing a simple difference between two dynamic images, or a dynamic image and a static image, a three way interest operator is applied.

We identify an image m with an array $\{m_{ij}\}$ and as usual refer to m_{ij} as a pixel value. Our initial images have $0 \leq m_{ij} \leq 255$ and $0 \leq i, j \leq 127$, but recall the bottom three or four bits are subject to high levels of noise. We write $\bar{m}$ for the mean of the pixel values of the image m, and m^σ for the corresponding standard deviation. Our system is concerned with images s, p and c, respectively the static, previous and current images. Typical static and dynamic images are shown in figures 1 and 2; due to the noise, these must be preprocessed before the location/tracking stage.

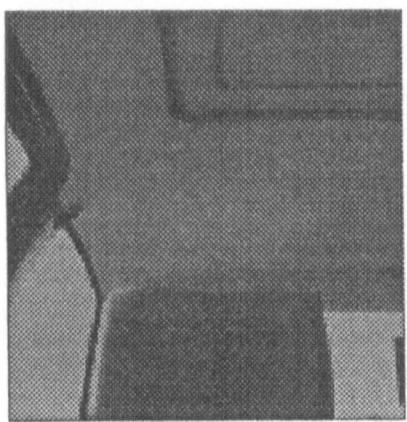

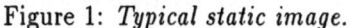

Figure 1: *Typical static image.*   Figure 2: *Typical dynamic image.*

One way to reduce the noise would be to use one of the many region operators, such as Sobel or Prewitt, or simply to blur the image over a small region. Instead we choose to reduce the resolution by pixel averaging. We refer to the reduced resolution images as *meta-images*. Given an image m, we write M_{ij} for the 4×4 image $\{m_{4i+k,4j+l} : 0 \leq k, l \leq 3\}$ and then call $M = \{\bar{M}_{ij}\}$ the meta-image given by M. Thus M_{ij} is the 4×4 image and $\bar{M}_{ij}$ the corresponding pixel of M; by abuse of language, we refer to both M_{ij} and $\bar{M}_{ij}$ as a meta-pixel.

This provides a number of additional benefits over a simple region operator;

- it reduces the size of the image we are working with which helps maintain performance. At this stage we are simply aiming to locate the position of the eyes, and provided the head occupies a reasonable proportion of the image, 32 pixels square is sufficient for this [1];

- we calculate in addition to the mean, the variance for each M_{ij}. This is largely independent of lighting levels, reflecting more the nature of the meta-pixel region; this reflects whether the meta-pixel is a homogeneous region (such as roof lining or seat) or quite variable, such as a window border.

Additionally, we perform an operation similar to the Moravec operator [11] for each meta-pixel to obtain both a direction (one of eight) and *strength* rating, M^θ and M^ι. This gives us a primitive *structure* description of the inside of the car which varies little with lighting.

In order to determine the location of the head, frame differencing is performed on the static meta-image S and current meta-image C using the variance and direction information. Due to lighting changes and the variability of the windows, this is biased with the location information obtained from processing the previous frame. Each meta-pixel is given a score according to the following conditions, each scoring one point if true;

- $|S_{ij}^\sigma - C_{ij}^\sigma| > T_{dev}$ where T_{dev} is the deviation threshold,

- $|\bar{S}_{ij} - \bar{C}_{ij}| > T_{mean}$ where T_{mean} is the mean threshold,

- $S_{ij}^\theta \neq C_{ij}^\theta$ and either $S^\iota > T_{resp}$ or $C^\iota > T_{resp}$ where T_{resp} is the edge response threshold,

- a bonus point if P_{ij} was a potential face pixel.

The thresholds T_{dev} and T_{mean} depend on whether the meta-pixel P_{ij} was determined to be part of the face or not, and were determined empirically. Meta-pixels scoring two or more points are considered potential head pixels, a typical meta-image with meta-pixels scoring two or more marked is shown in figure 3. *Salt and pepper* noise is reduced by a dilation/erosion stage yielding an image such as figure 4.

The next stage attempts to match a head outline to the region marked as potential head. There have been a number of model matching systems described that would be capable of locating an head outline in out image (e.g. Waite and Welsh [18], Taylor and Cooper [15] or Davis and Taylor [5]). In each case, the desired shape can be deformed within statistically determined limits. We adopt the methods used by FindFace, and specifically, the method described by Bennett and Craw [2]. The end result of this system is the approximate location of the head at low resolution in a representation compatible with the quality assurance system discussed below. Bennett and Craw use edge strength to guide the outline, as their system must contend with a lot of extraneous edges. We have a very clear, albeit irregular, outline to match, which allows for a number of simplifications to their method.

Specifically, the outline is defined by 10 points, rather than the 20 originally used. A continuous outline is generated by connecting the points with straight lines. To overcome the lack of gradient information used to guide the original towards the correct edge (the original uses a large region operator to allow approximate matches to be drawn towards the correct position) each of the 10 model points records its orientation. The intervening points that are created

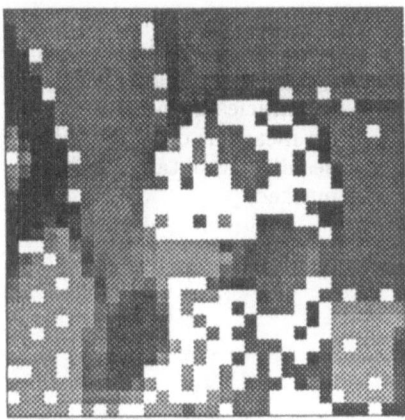

Figure 3: *Meta-image before noise reduction. Meta-pixels scoring more two or more are indicated.*

Figure 4: *Figure 3 after dilation/erosion to remove salt and pepper noise.*

derive their orientation from the two end points of the connecting line. Each edge point then knows, given that it is either on the face or on the background area, in which direction it must move to locate the edge. This approach does not allow for the subtlety of movement of the original, but performs satisfactorily at the resolution being used. The model is only allowed to deform in a symmetrical way by independent scaling; the refinment obtained from the second stage of the original algorithm would be wasted at these low resolution.

Having positioned the head outline model on the meta-image, the meta-pixels within the model area are counted; at least 80% must be correctly classified for the system to believe the presence of a head.

4.2 Feature Tracking System - Stage II

Having identified the head location, part of the statistical model from FindFace can be used to estimate the position of the eyes. The face model used, and the methods used to determine face location, do not allow for lateral rotation of the face, and allow for only a small vertical rotation. This is considered acceptable, as frequent large movements, as observed during driving in heavy traffic for example, would be detected by the QA stage. Furthermore, this degree of movement can generally be taken as an indication that the driver is awake, so the degradation of the system under these circumstances is not critical.

With a predicted location of the eyes, the full resolution (128x128) image can now be used, in particular the eye regions can be examined. The most appealing way of obtaining the required eye measurements is to use one of the eye recognition systems already mentioned [12, 19, 6, 4], thus obtaining a *direct* eye lid separation reading from which to deduce blinks. Two reasons preclude the use of these systems at present; the poor contrast of the image which leads to unreliable results, and the algorithms speed.

A more naive approach is used which simply calculates the grey level histogram for the eye region. When the eye is open, a characteristic bimodal histogram is obtained. When the eyelid is shut, this reduces to a single peak. The profiles differ sufficiently between open and closed eyes (see figures 5 and 6) to detect blinks. The overall profile change is largely independent of lighting levels, the image noise we experience, and slight inaccuracies in location of the eye region. Although the information obtained by no means as detailed as that from the deformable template model, the desired results are obtained. Figure 7 shows figure 2 with the eye regions and outline marked.

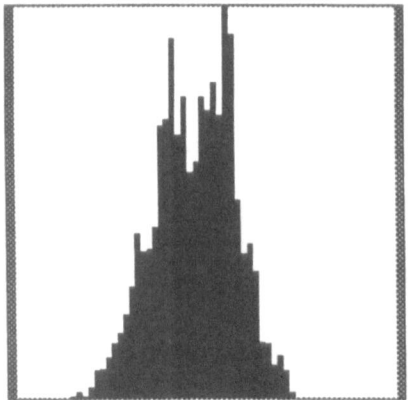 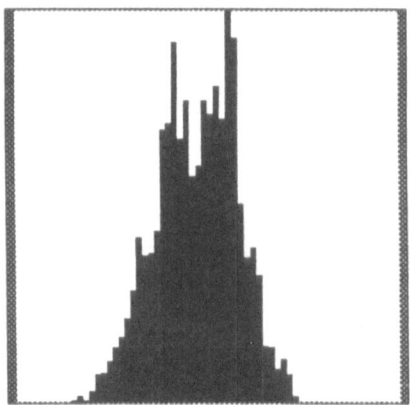

Figure 5: *Histogram of region containing open eye.*

Figure 6: *Histogram of region containing closed eye.*

An alternate method currently being investigated involves template matching with the a template extracted from one of the images with a specific driver in it. Using the QA system described below, the eye regions can be reliably extracted, and simple correlation performed. Although this does not prove satisfactory if a generic template is used, correlation with the drivers own eye region should be much more reliable. Whether this proves more reliable than the statistical method will be determined following testing.

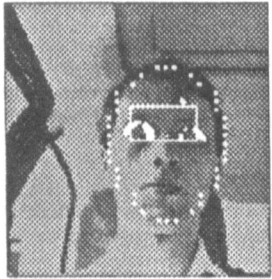 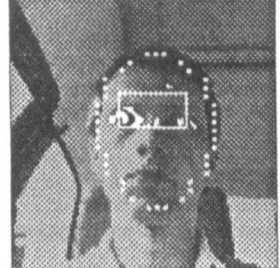 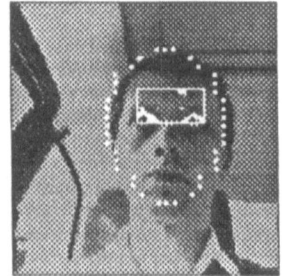

Figure 7: *Series of three processed images showing located head outline and estimated eye region from which the eye histograms are calculated.*

4.3   Quality Assurance System

The quality assurance system is essentially a slimmed down version of FindFace, described in [17], working at a somewhat faster rate (how fast?). The input image sequence is subsampled and processed by the modified FindFace system. As this takes quite a few seconds, it is not possible to use this system directly for obtaining eye measurements. The results of the processed image can be used for a number of purposes.

The location of the eyes as determined by FindFace can be checked with the location obtained by the tracking system for that particular image. This ensures that the tracking system has not been distracted and is still measuring the right region. As the results from FindFace can not be used to initialise the tracking due to the delay between image capture and the availability of results, this is more useful as a measure of reliability rather than as part of a feedback loop.

Feedback can be obtained by using the FindFace results to customise the head model. The model used by the tracking system is a modified version of that used by FindFace; specifically, it contains only the outline and eye locations, and the outline points are used to produce a complete, low resolution, outline, rather than specific points. As mentioned above, sideways rotations of the head are not allowed for, but the tracking model can be adjusted to provide better eye locations by generating a dynamic model from the outline and eye positions obtained from FindFace.

Also as mentioned above, the alternate eye measuring technique relies on the position of the eyes as produced by FindFace to extract the eye template.

5   In Operation

The complete system is written in C, and runs on a Sun IPC workstation. The original system used our elderly Imaging Technology FG100-V frame grabber which required the use of a bus protocol converter (we used a BIT-3 system). This was fed a video signal from a Pulnix TMC-516 remote head camera mounted on the dashboard of a Ford Granada, which housed the equipment in the boot.

In this configuration, and without the QA system running in parallel, approximately six frames per second could be processed, with the head and eye locations displayed on a monitor for confirmation. Ultimately, the monitor output will be sacrificed to gain additional speed.

As new equipment has become available, we have moved to an SBus based system using a DataCell S2200 frame grabber with TMC-6 camera. This is a colour system and offers a number of new ideas to be tried. No attempt has been made yet to optimise the code originally written for FG100-V, the result of which is only a marginal improvement in performance at present. A significant performance increase is expected when the S2200 is treated as a memory mapped device rather than a serial data stream. Early results with the new hardware do however suggest that we no longer have the problems with high levels of noise in the images.

6 Future Developments

Although the QA system has been proven, this still needs further work before being fully integrated into the system; in particular, the conversion of the relevant parts from POP11 into C will be needed to obtain the required performance. The FindFace system is designed to process single images of different individuals. Although some scope is included to adapt dynamically to changing classes of faces, this characteristic must be more fully explored if many images of the same person are to be processed. There are a number of ways the efficiency of the QA system could be improved by using such adaptability.

Although the use of colour must be explored carefully, as moving to NIR or UV illumination would almost certainly render such systems useless, a number of ideas will be considered. The eyes are often of a colour significantly different to the rest of the face. Looking for the change in colour in the eye region may provide blink indication. Cars tend not to have *flesh* coloured interiors, so this could be used to aid location/tracking of the driver.

With better quality data, we will investigate further the different methods of sequence analysis and feature tracking mentioned earlier. Hopefully this would allow a more efficient initial location leading to both a more accurate location of, and more frequent measure of the eye. The ultimate development of this would be to measure the eye lid separation rapidly enough to generate an analogue output. This may allow trends in driver alertness to be determined at an earlier point in time.

References

[1] Talis Bachmann. Identification of spatially quantised tachistoscopic images of faces: How many pixels does it take to carry identity? *European Journal of Cognitive Psychology*, 3(1):87–103, 1991.

[2] Alan Bennett and Ian Craw. Finding image features using deformable templates and detailed prior statistical knowledge. In Peter Mowforth, editor, *British Machine Vision Conference 1991*, pages 233–239, 1991.

[3] D A Castelow and A J Rérolle. A monocular ground plane estimation system. In *Proceedings of the British Machine Vision Conference*, pages 392–395, 1991.

[4] Ian Craw, David Tock, and Alan Bennett. Finding face features. In *Proceedings of the Second Eurpoean Conference on Computer Vision*, 1992. To be published.

[5] D N Davis and C J Taylor. An intelligent segmentation system for lateral skull x-ray images. In *Proceedings of the British Machine Vision Conference*, pages 251–255, 1989.

[6] Peter W Hallinan. Recognizing human eyes. *SPIE - Geometric Methods in Computer Vision*, 1991.

[7] Chris Harris and Carl Stennett. Rapid - a video rate object tracker. In *Proceedings of the British Machine Vision Conference*, pages 73–77, 1990.

[8] Jim Horne. Stay awake, stay alive. *New Scientist*, pages 20–24, 1992. 4th January.

[9] T S Huang, editor. *Image Sequence Analysis*. Springer Series in Information Sciences. Springer-Verlag, 1981.

[10] Merrill Mitler. Catastrophes, sleep and public policy. *Sleep*, 11, 1988.

[11] Hans P Moravec. Towards automatic visual obstacle avoidance. In *5th International Joint Conference on Artificial Intelligence*, 1977.

[12] Mark Nixon. Eye spacing measurement for facial recognition. *Proceedings of SPIE*, August 1985.

[13] A Rosenfeld. *Motion: Analysis of Time-varying Imagery*, pages 173–183. Cambridge University Press, 1983.

[14] T N Tan, G D Sullivan, and K D Baker. Structure from constrained motion using point correspondences. In *Proceedings of the British Machine Vision Conference*, pages 301–309, 1991.

[15] C J Taylor and D H Cooper. Shape verification using belief updating. In *Proceedings of the British Machine Vision Conference*, pages 61–66, 1990.

[16] Graham Thomas. *Image Processing*, chapter 3, pages 40–57. McGraw-Hill, 1991.

[17] David Tock, Ian Craw, and Roly Lishman. A knowledge based system for measuring faces. In *Proceedings of the British Machine Vision Conference*, pages 401–406, 1990.

[18] J B Waite and W J Welsh. An application of active contour models to head boundary location. In *Proceedings of the British Machine Vision Conference*, pages 407–412, 1990.

[19] Alan Yuille, David Cohen, and Peter Hallinan. Feature extraction from faces using deformable templates. In *Proceedings of the IEEE Conference on Computer Vision and Pattern Recognition*, 1989.

Online Calibration of a 4 DOF Stereo Head

N. A. Thacker* and P. Courtney.

AI Vision Research Unit, University of Sheffield

Sheffield, UK

(* now at the Department of Electrical and Electronic Engineering,
University of Sheffield, Sheffield, UK)

Abstract

This paper addresses the problem of recovering accurate 3D geometry
from a 4 degree of freedom stereo robot head. We argue that successful
implementation of stereo vision in a real world application will require
a self tuning system. This paper describes a statistical framework for
the combination of many sources of information for the calibration of a
stereo camera system which would allow continual recalibration during
normal use of the cameras. The calibration is maintained using modules
at three levels: fixed verge, variable verge and pan/tilt/verge calibration. Together these modules provide the means to fuse data obtained at
various head positions into a single coordinate frame.

1 Introduction

Computer vision systems which can deliver an accurate estimate of 3D geometry from stereo are now relatively commonplace in the computer vision
literature. Our own stereo vision system TINA has been demonstrated to have
useful 3D vision capabilities [5]. The vision system, which makes use of edge
based representation and stereo matching, relies upon calibration in both the
determination of epi-polars for the matching process and the calculation of 3D
position from disparity. Implementing these algorithms on a moveable head rig
poses a real problem of multiple parameter calibration.

Calibration has generally been achieved by a procedure whereby the calibration parameters are recovered once from a known stimulus with no concern
for updating this calibration in future by any other means other than total
replacement. In a practical vision system which is to be in continual use, such
one-off calibration methods are inadequate. A moving camera system would
have to 'look' at a calibration stimulus every time it was moved. For a practical stereo vision system, recalibration must be an integrated activity working
with data available during normal use [9]. In addition, we cannot expect there
to be sufficient information at any one time to obtain a completely accurate
calibration of the system, so we require a method of combining data collected
over a period of time into a consistent calibration. Ideally this method would
also allow the integration of information from different image sources and can
be described as "online calibration". The idea is not new, other authors have
suggested mathematical frameworks for self calibrating systems [2]. Here we describe our own practical framework and a three stage calibration system which

maintains the full head calibration (figure 1) and demonstrate its use with hand eye coordination of our head and a robot arm.

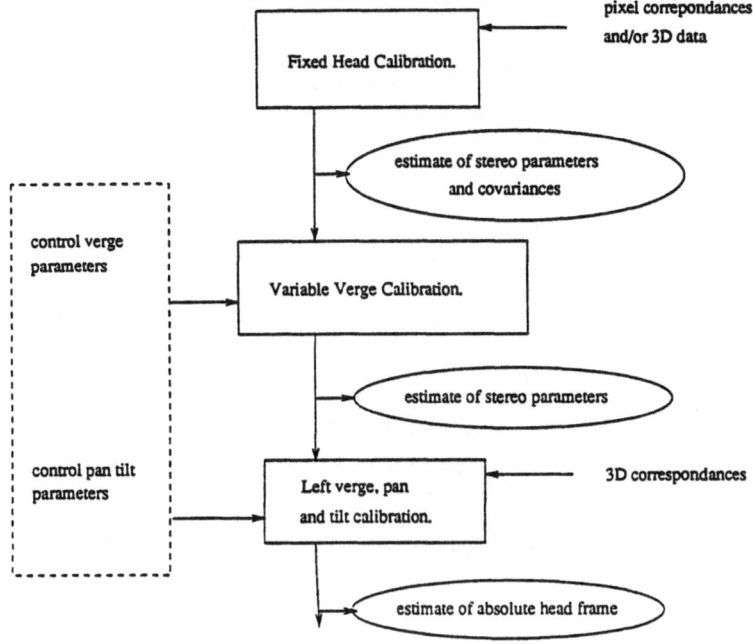

Figure 1. Head Calibration modules.

2 A Unified Mathematical Framework

A method is required for data combination, and this can be achieved via standard statistical methods by minimising a least-squares error measure $(\chi_t{}^2)$:

$$\chi_t{}^2 = (a - a_t)^T C_a{}^{-1}(a - a_t) + \sum_i (y_i - \phi_i(a_t))^T W_i{}^{-1}(y_i - \phi_i(a_t))$$

with respect to the parameters a_t , where χ_t^2 is a summed error criterion comprising a constraint term on the parameters a derived from previous data (which can be called a regularisation term) and a term for the current set of data y_i. C_a is the covariance matrix for the measurement vector a and the t subscripts denotes the iteration. W_i is the data measurement variance. This last term involves the data model ϕ, if this is linear then the model parameters can be estimated using a Kalman filter. If it is approximately linear then it can be linearised locally and solved using the Extended Kalman Filter (EKF). Both of these approaches are common in the computer vision literature [1].

The EKF uses the assumption that if ϕ is approximately linear then the χ^2 can be modeled as a quadratic around the current estimate at the minimum $\chi_0{}^2$:

$$\chi^2(\delta a) = \chi_0{}^2 + \delta a C^{-1} \delta a^T$$

where δa is the difference vector from the current estimate to the chosen point in calibration space. If this is true then the combined estimate of the total χ^2 from all combined data is also valid and should give the same result as if all the data had been minimised simultaneously. If however, the model is very non-linear the EKF may have to be iterated several times and depending on the degree of linearity this process may be unstable. Alternatively, the optimal combined estimate can be obtained directly by finding the parameters that minimise $\chi_t{}^2$. This is the method that we have adopted on the basis of increased robustness. We minimise the function iteratively using the downhill simplex method [4]. This gives the maximum amount of freedom for model parameter change and the inclusion of robust statistical measures. We limit the maximum contribution to the error score from each data point during minimisation, this effectively protects against outliers. Parameter tracking can be achieved by limiting the size of the covariance matrix so that new data takes preference over old [9].

To obtain a covariance matrix we must be minimising a χ^2 variable, this rules out a lot of calibration algorithms as candidates for optimal combination. Generally we cannot combine results unless the method takes correct account of the errors in the measurement system. In a stereo camera system we believe that the errors are mainly due to sampling noise and pixelation, and therefore best modelled in the image plane. The elements of the inverse covariance matrix are defined in terms of the *Hessian* by

$$C_{nm}{}^{-1} = \frac{1}{2} \frac{\partial^2 \chi^2}{\partial e_n \partial e_m}$$

which when close to the minimum of the function can be approximated using the *Jacobian*

$$C_{nm}{}^{-1} = \sum_i \frac{\partial \chi_i}{\partial e_n} \frac{\partial \chi_i}{\partial e_m}$$

We estimate the derivatives using numerical methods for purposes of model parameterisation flexibility. Any method that minimises an error metric in the image plane can be formulated as a χ^2 minimisation and combined within this statistical framework.

2.1 Fixed Stereo Camera Calibration

Obtaining a reliable stereo camera calibration is a particularly difficult task. A full camera model comprises both intrinsic parameters f (internal to the cameras) and extrinsic parameters e
(relative camera transformation specification). These two sets of parameters are often strongly correlated. For example the rotation of the camera and the translation of the centre of image co-ordinates will have virtually identical effects on our error criterion. The same is also the case for translation of the camera and changing the focal length. Correlations between parameters make it impossible to determine an isolated subset of the parameters properly without accurate prior knowledge of the remaining parameters. These correlations produce extended minima in the error surface so that many dissimilar sets of

calibration parameters may be equally valid, making statistical combination difficult.

The EKF and related methods (including ours), which take into account the covariance of the calibration parameters, can overcome these problems. New data is incorporated close to the current estimate where the covariance for the previous set of data is most reliable.

We can identify two sources of calibration data: known 3D measurements (for example the movement of a robot arm or known 3D objects) and epi-polar alignment of matched stereo correspondences. Both of these sources can be used to construct a χ^2 measured in the image plane. For calibration from 3D data it is assumed that data is provided on an accurately measured 3D object and that features on this object have been identified in the image planes of either camera. Such data can be obtained in a working system from the known motion of a robot arm or accurately known rigid 3D objects. A χ^2 is formed as the difference between the observed position of the image features and the predicted position, given the current estimates of the model. As the object is measured in an arbitrary co-ordinate frame the parameters describing the absolute transformation are redundant and only those relevant to the stereo camera system s are required. For calibrating from image correspondences the χ^2 is formulated following the numerical method of Trivedi [7]. The method can accommodate correspondance data either from matched epi-polar tangencies [3] or matched corners [8].

Thus for combination of calibration from these results with data from each source the stereo camera inverse covariance matrix $C_s{}^{-1}$ is needed and the total combination cost function is given by :

$$\chi_t{}^2 = \delta s C_s{}^{-1} \delta s + \chi^2$$

Once this is minimised the stereo camera covariance matrix $C_s{}^{-1}$ must be updated with the inclusion of all new data [10].

2.2 Variable Verge Stereo Camera Calibration

A full parametric model of the vergence camera system would be capable of describing the whole space of possible configurations of the left right verge system. However, a global model is only applicable to a well engineered robot head. Moreover, this method of calibration would have to be developed almost independently of any other solutions for the fixed camera geometry so that we cannot build on previous methods.

Alternatively, we might adopt a look up table solution, this has the advantage that we can use the previous methods for calibrating fixed head configurations to fill the entries of the table. For a head like our own which has a movement resolution of 8 minutes of arc over a range of 60 degrees for two cameras there are 202500 possible configurations of which about 44300 may be regarded as viable stereo vision configurations (figure 2).

This degree of freedom we call an "asymmetric vergence" control paradigm. A simpler moveable head system is where the fixation of an object requires pan tilt and verge angles so that the left and right verge motors have equal and opposite motor position control parameters. This we call a "symmetric vergence" control paradigm. With this method however, the full number of possible configurations is 112. We may safely assume that each configuration of the head would require an independent estimate of the extrinsic stereo camera parameters. Such independent entries for each configuration of the head would need extensive modification each time the system was disturbed.

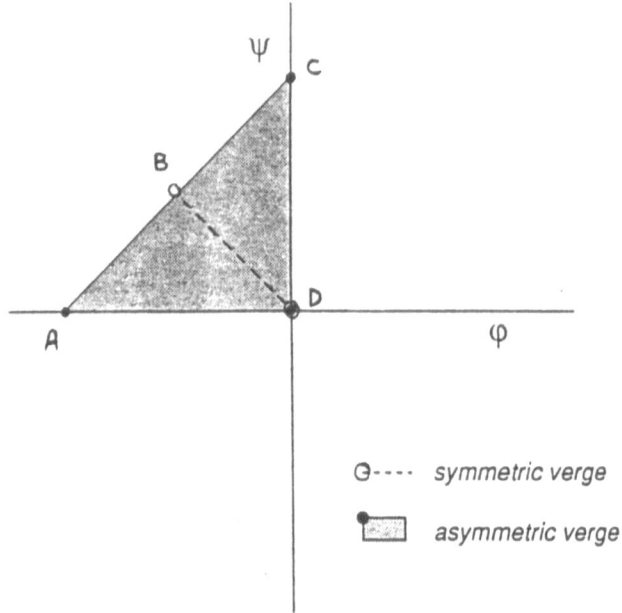

Figure 2. Defining interpolative lines and planes for symmetric and asymmetric calibration interpolation.

What we require is a compromise which combines the speed of training of global methods with the increased generality of the look-up table. The solution we have adopted is one which allows us to specify a local model that can be applied over a large number of possible verge configurations. We make the assumption that between two fixed vergence configurations the stereo camera parameters can be linearly interpolated on the basis of the motor control parameters. In our camera model the relative stereo camera geometry e is stored as a quaternion and translation[9].

$$e = (q, t)^T$$

Linear interpolation would imply that the net effect of a change in verge rotation for both cameras about their verge axes can be approximated by a rotation about one axis and that small rotations about this axis are approximately linear in the left and right verge position control parameters Φ and Ψ. The range over which these assumptions are valid have been tested by simulation and it was found that the approximation will not measurably degrade 3D geometry in our system over a rotation range of 0.2 radians (Figure 3).

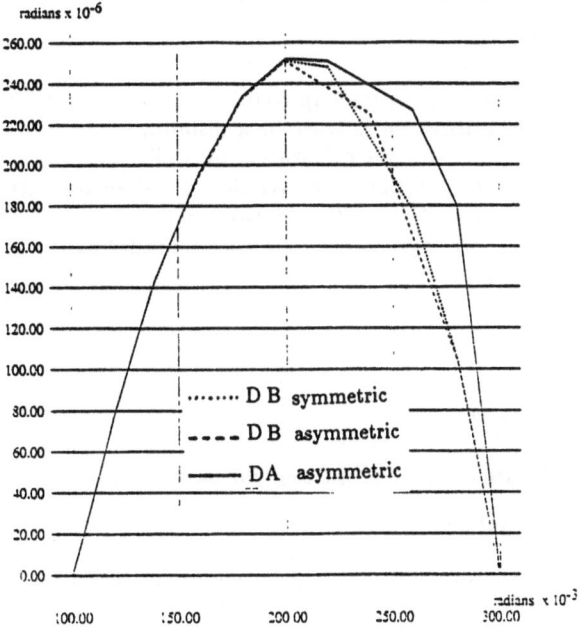

Figure 3. Simulated estimate of the error on interpolated vergence rotation
for symmetric and asymmetric control paradigms.

Rotation errors are approximately quadratic and a maximum at the centre of the interpolated region. Thus, for the symmetric verge paradigm the
whole useful range of the system can be defined with two calibration configurations defining the endpoints of the interpolation line e_1e_2 at Φ_1 and Φ_2. The
interpolated estimate of the relative camera geometry is given by

$$\hat{e} = \alpha_1 e_1 + \alpha_2 e_2$$

with

$$\alpha_1 = (\phi - \phi_1)/(\phi_2 - \phi_1)$$

$$\alpha_2 = (\phi_2 - \phi)/(\phi_2 - \phi_1)$$

Thus we have two stored calibrations as opposed to the 112 entries needed
for a look-up table.

Similarly, for the asymmetric vergence paradigm, a large part of the useful
range of the cameras can be defined using only three calibration configurations
e_1, e_2 and e_3, which define an interpolative plane in configuration space.

$$\hat{e} = \beta_1 e_1 + \beta_2 e_2 + \beta_3 e_3$$

with

$$\beta_2 = \frac{(\Psi_3 - \Psi_1)(\Phi - \Phi_1) - (\Psi - \Psi_1)(\Phi_3 - \Phi_1)}{(1 + (\Psi_2 - \Psi_1)(\Phi_2 - \Phi_1))(\Phi_2 - \Phi_1)(\Psi_3 - \Psi_1)}$$

$$\beta_3 = \frac{(\Phi_2 - \Phi_1)(\Psi - \Psi_1) - (\Phi - \Phi_1)(\Psi_2 - \Psi_1)}{(1 + (\Phi_2 - \Phi_1)(\Psi_2 - \Psi_1))(\Psi_3 - \Psi_1)(\Phi_2 - \Phi_1)}$$

$$\beta_1 = 1 - \beta_2 - \beta_3$$

By comparison, for verge rotations up to 0.2 radians this same region would require 3700 separate calibration entries in a standard look up table. More of the configuration space can be calibrated by defining similar three-point regions and calibrating these separately. We do not advocate extrapolation of the calibration outside the defined interpolation triangle.

The full symmetric and asymetric models g including intrinsic parameters f_l and f_r can be written as

$$g = (f_l, e_1, e_2, f_r)^T$$

or

$$g = (f_l, e_1, e_2, e_3, f_r)^T$$

The current estimate of the calibration can be written as

$$\hat{s} = (f_l, \hat{e}, f_r)^T = g \nabla_g(s)$$

Our fixed camera calibration methods provide an estimate of the intrinsic and extrinsic camera parameters s and a full covariance matrix C_s^{-1}. These are combined into the model g using standard statistical methods as follows

$$g_t = g_{t-1} + C_{gt}(\nabla_g(s))^T C_s^{-1}(s - \hat{s}_{t-1})$$

$$C_{gt}^{-1} = C_{g(t-1)}^{-1} + (\nabla_g(s))^T C_s^{-1} \nabla_g(s)$$

It is vital for the stability of these methods that estimates of parameters s from new data is constrained with estimates of these parameters from previous data $\hat{s}$. This constraint must then be removed from the combined estimate of g to prevent double counting of data [10].

2.3 Calibration of the Pan/Tilt/Verge rotation axes

The above methods provide an interpolative estimate of the relative camera geometry and the intrinsic camera parameters for a restricted range of verge configurations. This can be used to provide estimates in 3D of the location of any observed stereo correspondance in the left camera coordinate system. We now need to be able to relate these coordinate frames for any configuration of the head. In practice this requires the determination of the translations between the pan tilt and left verge rotation axes and rotation scale factors. We call this set of of parameters the head calibration parameters j. In practice only the parameters defining the rotation scale factors and transformation of the left camera into the left verge co-ordinate frame require calibration. The remaining transformation parameters can be determined by direct measurement.

We have implemented two methods for achieving this, the first again uses the robot and minimises error in the back projected position in the left image plane of the robot arm subject to j. The second uses the estimate of the verge calibration and a stereo/temporal corner matcher [8] to provide estimates of temporally matched 3D image points in the left camera coordinate system. A χ^2 is then formed from the summed squared error between the first point and the second point transformed back into the first head configuration in scaled disparity space $(x/Az, y/z, I/\sqrt{2}.z)$ where A is the aspect ratio of the cameras and I the interoccular separation. Disparity space errors provide a scaled approximation to the image plane error but are much easier to compute [6].

3   Results and Conclusions

The data combination method for fixed camera geometry was tested using data from robot motion, matched stereo corner correspondances and a calibration tile. The accuracy of estimated epipolar geometry was found to improve with the inclusion of new data as expected (figure 4). This system will allow the online recalibration of a fixed verge stereo camera system.

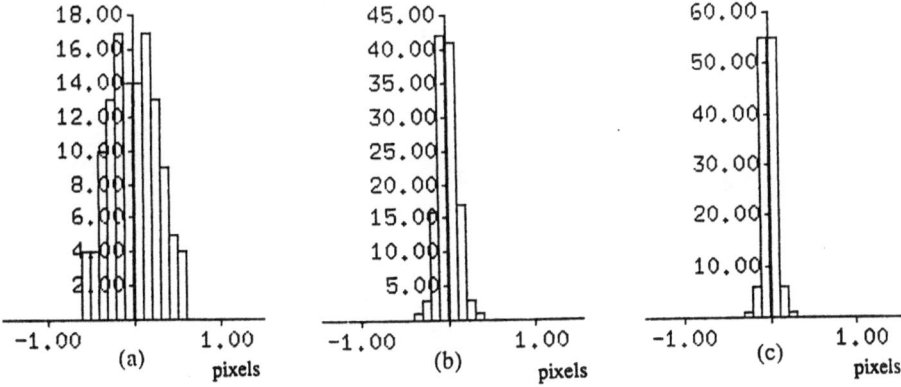

Figure 4. Epi-polar errors on test data after combination of data from
(a) robot position,
(b) matched stereo correpondances and
(c) a calibration tiles.

The variable verge interpolation scheme was initialised with three fixed verge calibrations at the points (0,-30), (30,0) and (30,-30) in verge motor configuration space (one motor count = 2.5 mRadian). This was tested by interpolating the camera geometry at the point (25,-25), the epi-polar accuracy was found to be consistent with a fixed verge calibration at the same point (figure 5). This system supports online recalibration of a variable verge stereo camera system using data from the fixed verge calibration method over a relatively large range of rotations.

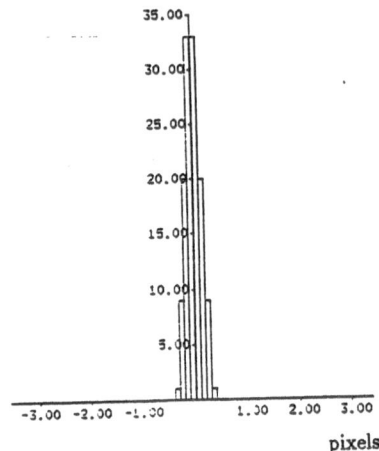

Figure 5. Epi-polar errors for interpolated camera calibration at (25,-25).

The pan/tilt and left verge parameters were obtained by calibrating on the back projected robot motion and matched static 3D points. The resulting full camera model was tested on unseen robot positions (figure 6). Outliers can be seen which are generated by several causes including stereo mis-matching and undershoot of the robot arm. This parameterisation of the system now permits 3D data from different head configurations to be combined into one coordinate frame and computation of head configurations for fixation of 3D world points.

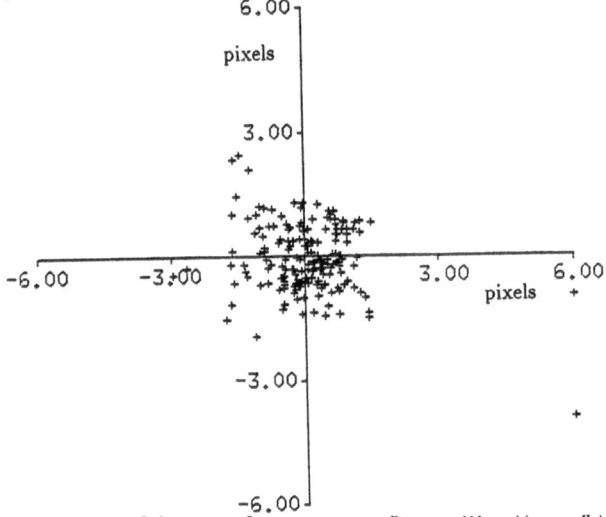

Figure 6. Back projected image plane errors after calibration of the full 4 DOF head system.

These results show that accurate calibration of a 4 DOF stereo head is feasible using a robust statistical framework which will permit online recalibration. Methods in stereo computer vision like those in our own TINA vision system requiring accurate stereo camera calibration can be supported with such a system.

Acknowledgements

We gratefully acknowledge the grant holders Prof. John E.W. Mayhew Dr. Paul Dean and Prof. John Frisby, the support of ESRC/MRC/SERC and the EEC, under the ESPRIT VOILA project, for the funding of this work. We also wish to thank our colleagues at AIVRU especially Li-Dong Cai, John Porrill and Stephen Pollard.

References

[1] Ayache, N., **Artificial Vision for Mobile Robots**, MIT Press. 1991.

[2] Faugeras, O.D. and G. Toscani, The Calibration Problem for Stereo, Proceedings of the CVPR, pp.15-20 1986.

[3] Porrill, J. and S. Pollard, Curve Matching and Stereo Calibration, Image and Vision Computing, pp. 45-50. vol 9 no. 1 feb. 1991.

[4] W.H.Press B.P.Flannery S.A.Teukolsky W.T.Vetterling, Numerical Recipes in C, Cambridge University Press 1988.

[5] Rygol,M., S.B. Pollard, C.R. Brown, A Multiprocessor 3D Vision System, Concurrency vol 3, no 4 1991.

[6] Sparks,E. and M Stephens, Integration of Stereo and Motion, Proceedings of BMVC 90.

[7] Trivedi,H.P., Estimation of Stereo and Motion Parameters using a Variational Principle, Image and Vision Computing, v5, n2 ,pp.181-183 May 1987.

[8] Thacker,N.A., Y.Zheng and R. Blackbourn, Using a Combined Stereo/Temporal Matcher to Determine Ego-motion, Proceedings of BMVC 90.

[9] Thacker,N.A. and J.E.W.Mayhew, Optimal Combination of Stereo Camera Calibration from Arbitrary Stereo Images, Image and vision computing, pp.27-32 vol 9 no 1 Feb., 1991.

[10] Thacker,N.A. and P. Courtney, Online Stereo Camera Calibration, AIVRU internal memo 62.

Visibility Scripts for Active Feature-Based Inspection

E. Trucco, E. Thirion, M. Umasuthan and A.M. Wallace
Department of Computer Science, Heriot-Watt University,
Edinburgh, Scotland

Abstract

We report the first stage of a project aimed at computing *visibility scripts* for active inspection applications, in which a robot-mounted sensor observes a known object from different viewpoints. Visibility scripts describe the optimal sensor position for a given inspection task and may involve different visibility requirements, e.g. achieving optimal visibility of a single object feature or simultaneous visibility of a set of features. We discuss also *stereo visibility*, or the optimal placement of a stereo head within the visibility region of a feature. Stereo visibility is a novel feature in the panorama of comparable systems and may prove nontrivial in some situations. Script generation is based on an approximate visibility space, the *property sphere*. We take into account several constraints imposed by most real systems, for instance the limited workspace of a real sensor or the desired resolution at which a feature must be observed.

1 Introduction

This paper addresses the problem of *optimal sensor placement* for inspection applications. The class of applications considered involves inspection systems which can observe an object from different viewpoints by moving either the sensor or the object. The sensor is required to acquire an optimal image of one or more features, or a sequence of images. Optimality is defined by various factors. noticeably feature visibility and reliability of feature detection. Since the workspace of any robot is constrained in practice, one might have to contend with suboptimal sensor placements. We call the sequence of optimal sensor placements for a given inspection task a *visibility script*.

The first issue in computing visibility scripts is *feature visibility*: from which region of the 3-D space around an object is a feature visible. This problem goes hand-in-hand with the topics of *viewer-centered representations*, of which *aspect graphs* are perhaps the most popular form (see [6] for a recent survey and introduction). Research in the field has considered mostly image models based on line drawings with edges as main features ([7],[14],[8]); a few surface-based [9] and component-based [15] aspect graph algorithms have been reported recently. The main problems are that implementations of exact techniques are few and the algorithms confined to rather limiting shapes: complexities are very high. up to $O(n^9)$ in the number of object features [6]; very few authors consider surfaces. which are interesting features for inspection in practice: and exact aspect graphs can be redundant in practice.

Notice also that aspects are maximally connected set of viewpoints, whereas feature visibility regions can contain holes or be disconnected. In practical applications, therefore, *approximate visibility representations* are adopted instead of exact aspect graphs. Approximate representations ([3],[4],[16],[6]) restrict the set of possible viewpoints to a sphere of large but finite radius, centered around the object. The approximate space is a discrete grid of viewpoints, obtained by computing a *quasi-regular tessellation*, or *geodesic dome*, of the sphere. Raytracing is used to compute visibility from each viewpoint. The main reason for using approximate representations in applications is that they are a well-understood class of methods, applicable to every object shape. The price to be paid is that there is no guarantee that every significant view is captured given the number of viewpoints (the *resolution* of the tessellation). The particular representation we consider in this paper is the *property sphere* ([3],[4]), briefly detailed in Section 3.

Many inspection tasks are *feature-oriented*: one is interested in inspecting object parts which correspond to model features. One would also like to predict *how reliably* a feature will be detected from a given viewpoint. This leads to the definition of *optimal viewpoint* for a feature inspection task ([10],[11],[1],[12]). We have designed a representation which expresses explicitly the visibility region of a feature, associates an optimality coefficient to each viewpoint and makes it possible to access information by feature index. *Constraints* on the extension of the visibility space are imposed by the characteristics of the sensing devices, by the feature detection techniques and by the workspace of the robot on which the sensor is mounted ([1],[2],[11],[13],[12]). Sometimes the sensor adopted is a stereo camera system, for example as part of a triangulation-based range finder. Computing the optimal sensor placement for a stereo head so that visibility is guaranteed from both cameras can prove nontrivial. Although important for active inspection, this problem has not received much attention in the literature of visibility-based sensor placement. We describe a technique for computing *stereo visibility* and finding the optimal placement for a stereo head in Section 7.

2 Definitions

A few key terms and concepts are defined at this point.

Sensors: the techniques described in this paper apply to several types of sensors, including cyclopean cameras, stereo heads and range finders. In the following, the sensor's type and geometry is explicitly mentioned when necessary. We will refer for simplicity to the case of a mobile sensor moving around a fixed object, although the object could be moved instead.

Features: the features considered are *surface patches*, corresponding to planar and curved object faces.

Models: we generate models with the RoboSolid solid modelling package. The examples in this paper use the model of a widget shown in Figure 1, a moderately complex industrial part.

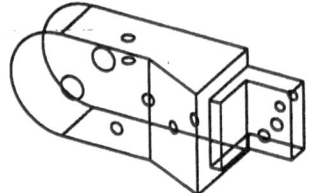

Figure 1: Line-drawing rendering of the widget model.

3 Building the approximate visibility space

The approximate visibility space adopted in this work is the *property sphere*, introduced in [3]. A property sphere is built by subdividing each face of an icosahedron in four equilateral triangles and "pushing out" the new vertices obtained onto the surface of the sphere circumscribed to the icosahedron. By iterating this operation on each new facet, a set of 20 quadtrees can be generated. The resulting sphere tessellation is *quasi-regular* in the sense that it approximates the regularity properties of the platonic polyhedra. The depth of the quadtrees, which is assumed the same for all the faces, is called the *resolution* of the dome. If the resolution of the initial icosahedron is 0, the total number of facets is $20 * 4^{res}$. The main question about geodesic domes is what resolution should be used. Unnecessarily high resolutions result in a wastage of memory; too low resolutions may miss important views. Typical resolutions used in the literature are 2 and 3 (320 and 1280 dome facets respectively).

4 The FIR representation

Inspection tasks require information about the visibility region of a feature and the optimality of the viewpoints inside the region. The representation adopted must make such information explicit and easily accessible. We have designed such a representation, called *FIR* (for Feature Inspection Representation). Computing the FIR solves directly the optimal single-feature inspection problem for *all* features in the model and makes other tasks easy by maintaining explicitly the desired information for all the features. The FIR consists of an array of *feature visibility region descriptors* (henceforth FVRDs). Each FVRD refers to one feature and consists of two components. The first is a *list of viewpoints* from which the feature is visible, which specifies the feature's visibility region under perspective projections. The second is the region's *stability*, which depends on the percentage of the property sphere covered by the region and is given by $\frac{r}{N}$, where r is the number of viewpoints in the region and N the total number of viewpoints in the property sphere. Unstable regions are poor candidates for sensor positioning even if their feature visibility is satisfactory. Each viewpoint descriptor in the FVRD list includes the viewpoint's cartesian and spherical coordinates as well as the following attributes. All attribute values are in the range $[0,1]$.

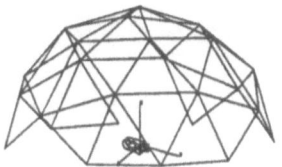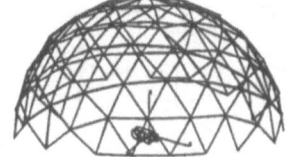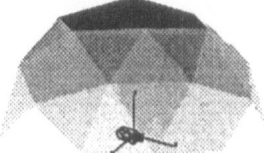

Figure 2: Visibility region for the top plane of the widget at resolution 1, 2 and shaded according to viewpoint optimality (resolution 1, the darker the better). The widget is oriented as in Figure 1.

Visibility: the absolute visibility of the feature in pixels, normalized by the image resolution.

Reliability: the expected reliability with which the feature will be detected from the viewpoint. Computation of this coefficient depends on the characteristics of the sensor and feature detector adopted: for instance, low-curvature cylindrical patches might be confused with planes and assigned low reliability.

Optimality: the global merit of the viewpoint, obtained by combining visibility v and reliability r:

$$o = o(v, r) = k_v v + k_r r$$

where the weights k_v, k_r satisfy $k_v, k_r \in [0, 1]$ and $k_v + k_r = 1$. These weights express the relative importance of visibility and reliability according to the particular task. For instance, if a sequence of images must be acquired to be inspected by an operator (no feature detection involved), a convenient choice is $kr = 0, k_v = 1$.

The essential algorithm for computing a FIR involves generating a geodesic dome and raytracing (perspective projections) from each viewpoint on the dome, from which visibility and reliability are computed for each feature. If too few pixels of a feature are visible from a viewpoint, no viewpoint descriptor is created. Finally, the the algorithm evaluates the stability of each FVRD.

The size of a FIR depends on the object considered and on the resolution of the property sphere. For the widget model in Figure 1 the size is about 20k bytes at resolution 2 (320 viewpoints) and about 80k bytes at resolution 3 (1280 viewpoints). Table 1 in Section 8 gives further examples. The time taken varies with the object, the dome resolution and the image resolution adopted. With 64x64 images, building the FIR for the widget took about 25 mins with 320 viewpoints and 2.5 hours with 1280 viewpoints on a SPARC workstation. With 128x128 images, the time is about 3 hours with 320 viewpoints. Notice that an approximate aspect graph is also computable from the FIR by using a region-growing algorithm on the viewsphere (see for instance [4]).

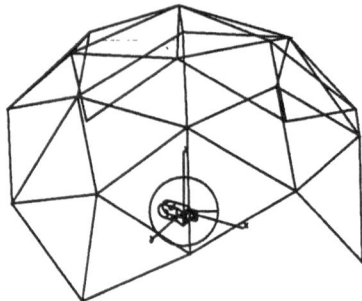

Figure 3: Covisibility region for front cylindrical patch and top L-shaped plane of the widget (referring to Figure 1) at dome resolution 1.

5 Basic visibility scripts: optimal feature visibility

The basic visibility script consists of moving the sensor to the optimal viewpoint for observing a given feature. Our representation has been designed to make this task particularly easy; computing the representation solves directly the basic visibility problem for *all* features. It suffices to pick the best viewpoint in the interesting feature visibility region. Figure 2 shows the visibility region for the side plane of the widget (top plane with hole in Figure 1) as a partial geodesic dome. By associating optimality weights to viewpoints, the FIR supports also the inspection of sets of features from optimal positions. It is sufficient to identify the set of optimal viewpoints for all the features involved. The sensor trajectory can then be planned under appropriate constraints. e.g. that the total distance covered by the sensor is minimum.

6 Optimal covisibility

Knowledge of covisibility is essential whenever several features must be observed simultaneously. The shape of a covisibility region is easily found, thanks to the FIR, by intersecting the visibility regions of all the features involved. The problem reduces to list intersection. Figure 3 shows the covisibility region of the front cylindrical patch of the widget and the side L-shaped plane facing up in Figure 1. The *stability* of a covisibility region is the same as that of a single-feature visibility region. The definition of the region's *optimality* requires more attention. The optimality o_i^{cov} of a viewpoint i belonging to the covisibility region of features $\{f_1, \ldots f_N\}$ is computed as a function of the optimalities o_{ij} of viewpoint i for single-feature visibility of feature j. Any candidate function for o_i^{cov} must meet two requirements. First, the same number of pixels should be visible simultaneously for all features: it is no good to see the whole of feature f_1 and nothing of feature f_2. Second. o_i^{cov} should increase with the number of pixels visible. We assume $o_{ij} \in [0. 1]$ for all i and j. The function satisfying these requirements we adopted is

$$o_i^{cov} = \sqrt{\bar{o}_{ij}^2 - \sigma_o^2}$$

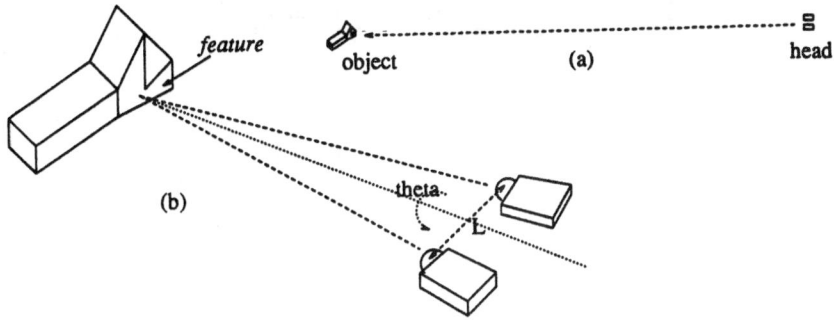

Figure 4: Stereo visibility can require consideration of the sensor geometry (b) or not (a). See text.

where $\bar{o}_{ij}$ and σ_o^2 are respectively the mean and variance of the o_{ij}.

7 Stereo visibility

Stereo visibility can be necessary when the sensor used is a stereo head. Computing stereo visibility requires that a few conditions are satisfied. Firstly and obviously, a given feature must be observable from both cameras: therefore both cameras must lie inside the visibility region of the feature. Secondly, it may be required that the camera-to-camera distance (or *interocular* distance) must be compatible with the distance between two adjacent viewpoints on the property sphere. Thirdly, constraints imposed by the robot used might limit the possible attitudes that the stereo head can assume. All or some of these constraints must be considered according to the task at hand, as discussed below.

In some cases the stereo system can be approximated with one point and we can assume that both cameras observe the same image. This happens usually for tasks requiring global visibility i.e. that the *whole* object is visible from all viewpoints, as for instance when checking for missing subparts. The conditions are that the object-sensor distance is much greater than both the interocular distance and the object size (see Figure 4a). In this case the probability that both cameras are in the same visibility region is high and the problem can be reduced to one of single-camera visibility. This assumption is adopted implicitly in [2].

In some tasks, however, the sensor cannot be approximated by a point. This happens when the probability that the two cameras end up in different visibility regions is not negligible. This can occur when the the interocular distance is comparable with the camera-object distance and therefore with the radius of the viewsphere (Figure 4b), as is the case in close inspection tasks. In such cases positioning the stereo head can be nontrivial.

If objects are known *a priori*, it might be possible to predefine an *optimal inspection direction*. For a planar patch, for instance, this can be the normal to the plane taken through the feature's baricentrum. However, the use of a single direction can be unsatisfactory: for instance, range-based HK curvature estimators might distort cylindrical patches according to the angle formed by the local patch normal and the viewing direction [5]. Moreover, for a stereo head, a single direction does not guarantee visibility from *both* cameras. Our approach is to guarantee stereo visibility

Bytes allocated	Saved	Threshold
84432	62160 (42%)	4
82632	63960 (44%)	6
80784	65808 (45%)	8
76368	70224 (48%)	10

Table 1: Minimum visibility constraint: size of the allocated FIR in bytes, memory saved by visibility thresholding in bytes and as a percentage of the size of the unthresholded representation, threshold enforced in pixels. The widget model was raytraced at a resolution of 64x64 from all viewpoints of a medium-resolution property sphere (320 nodes).

by finding the optimal position of the stereo head inside the visibility region of a feature to be inspected, as described below.

First the radius of the property sphere is determined according to the constraints imposed by the task (see Section 8) and a property sphere generated. If the sensor is to be placed at the minimum distance from the object which guarantees no collision, the radius of the property sphere is taken to be the radius of the minimum sphere enclosing the object. Then the visibility region of the desired feature is computed as described in Section 4. Notice that the region of space from which it is necessary to raytrace can be rather small and only a partial dome is generated at close distance from a feature.

We then try to find the optimal unconstrained position for the stereo head within the visibility region. To do this, the stereo head is approximated with a linear segment of length L equal to the camera-to-camera distance (see Figure 4). The algorithm selects pairs of viewpoints which are distant $L \pm \varepsilon$ from each other, where ε expresses the tolerance introduced by the approximate visibility space and depends on the resolution of the property sphere. We adopted Korn and Dyer's algorithm [4] (complexity $O(f)$, f number of viewpoints in the property sphere) for finding all viewpoints at a fixed distance from a given one. The *combined optimality* of the pairs is then evaluated. Optimalities are combined as described for covisibility (Section 6). The viewpoint pair maximising the combined optimality is the optimal sensor position.

Finally, the solution is checked against workspace constraints, which restrict the possible head rotations around its axis (angle θ in Figure 4). If the optimal unconstrained solution does not satisfy the constraints. the first suboptimal solution which does is chosen. We assume that it is always possible to adjust the cameras' vergence so that they point to the centre of the feature being inspected (as shown in Figure 4). In this case, we can assume that the images actually acquired by the stereo head differ from the images predicted by the FIR only by a rotation.

Notice that only a limited number of candidate viewpoint pairs is usually considered by the algorithm, thanks to the combined effect of the head geometry constraint and the close distance implying small partial domes.

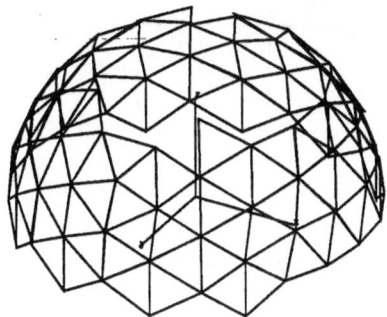

Figure 5: Constrained visibility space for a turntable-based inspection setup.

8 Constraints

In most practical applications, several constraints are imposed on sensor placement. Constraining factors include the sensor's geometry, the robot's workspace, the feature detectors adopted. Constraints lead to a reduction of the number of viewpoints to be considered in any task, but imply additional computation to be enforced. We describe here only the constraints adopted in the present prototype implementation: *workspace, minimum resolution* and *minimum visibility*. More constraints will be added to the system in the future.

Robot workspace. Depending on the characteristics of the robot adopted, certain regions of space will not be accessible to the robot-mounted sensor. The radius and reachable areas on the property sphere must be constrained accordingly, and sensor placement computed within the resulting constrained area. At the moment workspace contraints are expressed through systems of inequalities in spherical coordinates. As an example, Figure 5 shows the area of the property sphere satisfying the workspace constraints imposed by an inspection system whereby the object sits on a turntable and is observed by a camera free to move around the object but in a limited elevation range. The resulting workspace can be easily described in spherical coordinates by the inequalities $\Theta_{min} < \theta < \Theta_{max}$. where Θ_{min} and Θ_{max} depend on the installation and θ is the elevation (spherical coordinates).

Minimum visibility. The minimum visibility constraint imposes that the number of pixels of any feature visible from any viewpoint cannot be less than a threshold specified by the user. The threshold depends on the requirements of the task considered. This constraint is enforced during the construction of the FIR (see Section 4) and inhibits the allocation of viewpoint descriptors for which feature visibility is unacceptably low. The benefit introduced by this simple constraint in terms of memory saved is remarkable. Table 1 shows some figures obtained in our experiments with the widget model.

Minimum resolution. In some tasks it is desirable to ensure that the interesting feature appears in the image at a minimum resolution. for instance for close inspection or because the feature detector would not yield reliable results at coarser resolution. This leads to an upper bound for the distance between the feature and the camera which can be expressed as follows. Consider the geometry shown in Figure 6 and let D be the maximum linear dimension of the interesting feature in millimiters, d the distance between the camera and the feature in millimeters, and

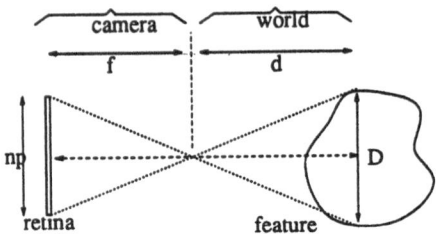

Figure 6: The minimum resolution constraint geometry.

n the feature's size on the retina in pixels. Then the requirement that the feature's linear resolution in pixels on the retina is at least n implies $d \leq k_{cam}\frac{D}{n}$ where $k_{cam} = \frac{f}{p}$ is a constant defined by the camera parameters, f the focal length and p the linear resolution of the retina in millimiters per pixel (supposed homogeneous for rows and columns). Images taken from a property sphere generated around the center of the desired feature, with a radius satisfying the above inequality, will meet the minimum resolution constraint.

9 Conclusions

We have presented some initial developments of a project aimed at generating inspection scripts, or sequences of optimal sensor positions for feature inspection under the constraints imposed by a real setup. The problem is of considerable interest for several active inspection applications. We see the main attractive features of this work in its explicit consideration of stereo visibility spaces and of a set of constraints which, although partial at present, do occur in real installations. Extensions are planned in order to cope with more complex workspaces, to incorporate further constraints, to obtain optimal scripts for sequential inspection of lists of features.

Acknowledgments

Thanks to Bob Fisher. John Hallam and Tim Newman for useful discussions and suggestions. This work was supported by the LAIRD project (Location and Inspection with Range Data), a collaboration between British Aerospace, Bae SEMA Ltd, the National Engineering Laboratory, Heriot Watt University and the Universities of Edinburgh and Surrey. The project is led by British Aerospace and is funded by the SERC/Information Engineering Directorate (GR/F38327:1551).

References

[1] S. Sakane, M. Ishii and M. Kakikura: *Occlusion Avoidance of Visual Sensors Based on a Hand-Eye Action Simulator System: HEAVEN.* Advanced Robotics **2**, pp. 149 – 165, 1987.

[2] S. Sakane and T. Sato: *Automatic Planning of Light Source and Camera Placement for an Active Photometric Stereo System.* Proc. IEEE Int. Conf. on Robotics and Automation, 1991, pp. 1080 –1087.

[3] G. Fekete and L.S. Davis: *Property Spheres: a New Representation for 3-D Object Recognition,* Proc. IEEE Workshop on Computer Vision, Representation and Control, pp. 192 – 201, 1984.

[4] M.R. Korn and C.R. Dyer: *3-D Multiview Object Representations for Model-Based Object Recognition,* Pattern Recognition **20**, pp. 91 – 103, 1987.

[5] E. Trucco: *On Shape-Preserving Boundary Conditions for Diffusion Smoothing.* Proc. IEEE Int. Conf. on Robotics and Automation, 1992, pp. 1690 – 1694.

[6] K. Bowyer and C.R. Dyer: *Aspect Graphs: an Introduction and Survey of Recent Results,* International Journal of Imaging Systems and Technology **2**, 1990, pp. 315 – 328.

[7] Z. Gigus, J. Canny and Seidel: *Efficiently Computing the Aspect Graph of Polyhedral Objects,* Proc. IEEE International Conference on Computer Vision, 1988, pp. 30 – 39.

[8] S. Petitjean, J. Ponce and D. Kriegman: *Computing Exact Aspect Graphs of Curved Objects: Algebraic Surfaces.* Tech. Rep. UIUC-BI-AI-RCV-92-02, University of Illinois at Urbana-Champaign, 1992.

[9] Kaiser, K. Bowyer and Goldgof: *On Exploring the Definition of a Range-Image Aspect Graph,* Proc. 7th Scandinavian Conference on Image Analysis, 1991, pp. 652 – 656.

[10] J. Ben-Arie: *Probabilistic Models of Observed Features and Aspects with Applications to Weighted Aspect Graphs,* Pattern Recognition Letters **11**, 1990, pp. 421 – 427.

[11] K. Ikeuchi and T. Kanade: *Modeling Sensors: Toward Automatic Generation of Object Recognition Program,* Computer Vision, Graphic and Image Processing **48**, 1989, pp. 50 – 79.

[12] H.-S. Kim, R.C. Jain and R.A. Volz: *Object Recognition Using Multiple Views,* Proc. IEEE Conference on Robotics and Automation, 1985, pp. 28 – 33.

[13] C.K. Cowan and P.D. Kovesi: *Automatic Sensor Placement for Vision Task Requirements,* IEEE PAMI **10**, 1988, pp. 407 – 416.

[14] D. Eggert and K. Bowyer: *Perspective Projection Aspect Graphs of Solids of Revolution: an Implementation.* Proceedings 7th Scandinavian Conference on Image Analysis, 1991, pp. 299 –306.

[15] S. Dickinson, A. Pentland and A. Rosenfeld: *From Volumes to Views: An Approach to 3-D Object Recognition,* Proc. IEEE Workshop on Advances in CAD-Based Vision, Hawaii, 1991, pp. 85 – 96.

[16] T. M. Silberberg, L. Davis and D. Harwood: *An Iterative Hough Procedure for Three-Dimensional Object Recognition.* Pattern Recognition **17**, 1984, pp. 621 – 629.

Ground Plane Obstacle Detection under variable Camera Geometry Using a Predictive Stereo Matcher.

Stuart Cornell, John Porrill, John E W Mayhew.
Artificial Intelligence Vision Research Unit,
University of Sheffield, Sheffield, S10 2TN,England

Abstract

A scheme is proposed for ground plane obstacle detection under conditions of variable camera geometry. It uses a predictive stereo matcher implemented in the PILUT architecture described below, in which is encoded the disparity map of the ground plane for the different viewing positions required to scan the work space. The research is the extension of Mallot et al's (1989) scheme for ground plane obstacle detection which begins with an inverse perspective mapping of the left and right images that transforms the image locations of all points arising from the ground plane so that they have zero disparity: simple differencing of the resulting images then permits ready detection of obstacles. The essence of this physiologically-inspired method is to exploit knowledge of the prevailing camera geometry (to find epipolar lines) and the expectation of a ground plane (to predict the locations along epipolars of corresponding left/right image points of features arising from the ground plane).

1 Introduction

The research described here is part of a project investigating adaptive control of a four degree of freedom stereo camera rig [1] mounted on an autonomous vehicle. The research has strived for both psychological and physiological plausibility in both the specification of the particular competences involved, and the adaptive self-tuning methodologies used for their real- time implementation. The component of the work to be described here is ground plane obstacle detection using a predictive stereo matcher to encode the disparity map of the ground plane (see Figure 1). For a vehicle operating in a limited operating enviroment such as a factory floor, simplest form of obstacle is a point in space which is

---

[1] The stereo camera rig used for this work comprises a 3-link kinematic chain, whose degrees of freedom are rotations around the following axes: i) Pan: a vertical axis corresponding to the 'neck'; ii) Tilt: an axis at right angles to the neck; and iii) Verge: each camera ('eye') can rotate independently around an axis at right angles to the tilt axis. The rig has been constructed so that the centres of rotation of the tilt and pan links coincide, and the centres of rotation of left and right verge and the tilt links coincide. It has been a principle of the project not to use measurements of the geometry of the 'head' either in the control of the head or in the development of the predictive stereo matcher to be described here. The length of the tilt link is approximately 12.5 cm for each eye (i.e. the head is about 25 cm wide); the length of the verge link (i.e. approximately how far the centre of rotation is from the focal centre of the camera) is 5 cm. so that tilting the eye also produces a small translation. It is also of note that the right camera has been mounted with a 5 degree heterophoria and about 2.5 degrees of cyclotorsion.

not on the ground plane. By reproducing the mapping of corresponding image points obtained from the stereo camera pair the obstacles can be detected by identifying violations in it. The complexity of this stereo correspondence problem is large as the disparity flow fields vary significantly with camera geometry. Also the disparity vectors have significant components in the X and Y directions presenting a complex non-linear problem of high dimensionality. Solutions to this have been obtained without the use of camera calibration or any dependency upon the dynamics of the camera rig. The approximation of simple local mappings to give more general global results has been extensively used throughout the work in the form of Parametrised Interpolated Look-Up Tables (PILUT). These have been implemented in the form of variably organised Neural Nets. They provide a constantly updateable result in a form which is easily combined into the subsumptive head control architecture of our mobile vehicle.

2 Disparity Fields

To illustrate the problem, Figure 2 shows ground plane disparity maps for different directions of gaze and elevation of the cameras (i.e. different viewing positions). For human vision, these viewing posiitons would roughly correspond to looking at the bottom left and right corners and the central fold of an open book lying on a table at about the normal reading distance. The disparities are the output of local networks serving regions of the motor states of our vehicle when the cameras are directed left, right and straight ahead at points at approximately the same distance away on the ground plane (in fact, about 150 cm; cameras about 75 cm above the ground plane).

The data used to train the nets to deliver (predict) these ground disparities was collected by moving a small light source around on the floor of the laboratory, and, with the head still, tracking the light stereoscopically in realtime using a small ROI window and a centre of gravity process on the images. This procedure offered a simple temporal solution to the stereo correspondence problem which obviated the need for a sophisticated stereo algorithm, with considerable benefits in reducing training time while developing the PILUT. The data sets so obtained were used to train the neural nets described below so as to generate the coordinates of the corresponding point in theone eye's view when given as input the retinal coordinates of the points in other eye. For the variable camera geometry methods the motor positions encoding camera states were also used as input data.

The interpolation achieved by the nets is brought out in the figures as they show a mapping from a grid of retinal locations in the left eye to the corresponding locations in the right eye for points lying on the ground plane.

Figure 3 Shows real data traces (bottom left) of the ground plane and the obstacle, the predicted field for the same position (top), and the difference between the two. This clearly distinguishes the obstacle from the ground plane data points. [2] There are some important points to be made here.

---

[2] The obstacle was a book which stood about 3cm off the ground, and was placed such the head was fixated on position approx 1m in front of the vehicle and slightly to the left.

1. The disparities involved are in general large. The visual angle subtended by the images is almost exactly 30 degrees. Thus the eccentricity of the points lying towards the periphery are not at all excessive, and yet the disparities are often more than a degree in magnitude.

2. There are both vertical and horizontal components to the disparities. The vertical components are often very large, and at some locations, larger than the horizontal disparities.

3. The pattern of disparities is markedly affected by changing the direction of gaze.

It is important to note that the above are quite general points and are not an artefact of using a planar retina.

3 Projective Stereo Mapping

The convenience of using a planar retina such as we have in our camera rig, is that there is a relatively simple, but non-linear, projective relationship (termed here the Projective Stereo Mapping, PSM) between the positions of corresponding retinal points when a planar surface is viewed. This relationship may be represented as the homogeneous projective matrix S, where $x_l.S = x_r$ up to a proportionality.(See figure 1)

The non-linearities arise from the division with the coefficients in the bottom row of the S matrix. The coefficients of the S matrix are a function of the cross products of the retinal coordinates, and can be found by solving a simple linear least squares problem given the coordinates of corresponding points as the input data. Thus a simple linear net can be used to estimate the coefficients but a division must be performed to use them.

The PSM is applicable only to planar retinae however, and the PILUT architecture described below assumes only that the function can be locally approximated by a blending of planar patches and is therefore more general (but, in the case of planar retinae, necessarily sub-optimal).

A PILUT for the stereoscopic ground plane mapping that makes no concession to biological plausibility but is economical both in storage and in training overhead was created as follows:

1. 13 sets of ground plane data were collected as described each for a different head position fixated on the ground.

2. at each position the coefficients of the PSM were found using the simple linear net training regime (See appendix); and

3. a least squares minimisation using Cholesky decomposition was used to find the best fitting quadratic surface for each of the 8 variable coefficients in the S matrix as a function of the head position parameters.

This method has been used extensively not only to prove the principle but also as a source of training and test data for experiments on the different variations of other PILUT architectures. See figure 5 for results.

4 PILUT's

The principal of the PILUT architecture at its most general is to use local linear approximations to multi-dimensional functions. It may be regarded as similar to the tensor-product 3D surface interpolation schemes used in computer graphics (and computer aided design) but in a PILUT the interpolating function is a local hyper-planar patch approximation. An alternative way to regard the architecture is as two levels of neural networks. The first is the indexing or parameterising network: it is coded relatively coarsely and generally has few dimensions, often simply serving to act as a blending function for the local piece-wise approximations carried out by the second level. The latter is is constructed on demand in a particular context and has higher resolution inputs and, in general, more dimensions than the indexing level, possibly including the indexing dimensions at a higher resolution. It is generally appreciated that the phase space trajectories of multi-dimensional systems lie on sub-manifolds which locally may be of a very much lower dimensionality than the system (Potts and Broomhead, 1991). This is because, in general, the physics just does not allow the full combinatorial explosion to occur.

The PILUT architecture is an attempt to provide a similar reduction in dimensionality while at the same time allowing high resolution local approximations to the full dimensional surface. There are 7 dimensions in the problems under consideration here: three for the degrees of freedom of the cameras in the saccade system (head tilt, left and right verge), and four for the x and y retinal coordinates of a target in the left and right images. At first sight it appears there is little potential for a reduction in these dimensions. The insight, however, is to recognise constraints provided by the stereo problem. In this situation the 7 dimensions are immediately reduce to 4 because we are dealing with just head positions which are fixations on the ground plane which make one verge redundant. It is also possible to use, as the indexing level, the coarsely-coded information of the position of only one of the eyes. The sensitivity of stereo to small differences can then be recaptured by using the retinal coordinates again as input to the second-level network that provides a local approximation of the full 7-dimensional hyper plane Figure 4.

This can be reduced to 4 because all the head positions are fixated on the ground plane so two verges are not needed and we can use only the left cameras X and Y because of the small differences between the left and the right images.

The coefficients of the local interpolating hyper plane are stored in a matrix (it is a simple linear net). When the network is accessed through the coarse indexing scheme, a composite matrix is formed by blending together the matrices in a region of the indexing parameter values. Two blending schemes have been explored, both biologically plausible. One method uses radial basis functions (RBFs) to populate the indexing parameter space with a number of centres positioned according to the coarse indexing scheme. Associated with each of the centres is a gaussian weighting function, in addition to a linear approximation to the surface. On indexing the network, a composite local approximation is constructed by adding together the coefficients associated with the individual centres in proportion to the distance they are from the input. During the training phase the errors are propagated back as for simple linear networks, to adjust the coefficients of each matrix associated with each centre in proportion to its contribution to the composite network.

The second architecture we have explored for blending or interpolating across the parameterisation is the CMAC (Albus 1976). The CMAC is generally used for the representation of continuous multi-dimensional scalar functions. We choose to use the CMAC architecture to carry the coefficients of the matrices. The CMAC uses a coarse-coding strategy for the discretisation of the parameter space, and movement in the parameters maybe regarded as equivalent to the discrete differentiation of the function at the resolution of the coarse coding. As for the RBF network, when accessed the CMAC builds a composite matrix by integrating the individual matrices indexed by the different layers of the coarse coding. During training the coefficients of the individual matrices are adjusted using the usual gradient descent methods. Both these architectures have been used separately and in conjunction with each other as the indexing and blending level of the PILUTs, and details of the implementation and training of them may be found elsewhere (Mayhew et al, 1992).

The appeal of the PILUT architecture is that it is in principle simple, readily customised, local and hence stable, easily trained and biologically plausible. Its disadvantages are that it is potentially expensive in memory. It seems to be a simplification of the Hyper Basis Function network representation proposed by Poggio and Girosi (1989, 1990), and a similar idea recently proposed by Lane et al (1991). Physiologically one can regard the LUTs and interpolating nets as a matrix of receptive fields whose configuration or kernel is modulated by eye position information. The latter information may be determined from stereo itself or from the oculomotor system.

5 Results

Figure 5 shows results of the different methods for a fixed head position: looking ahead with symmetrical vergence. The results show that:

1. a simple linear net is unable to capture the disparity mapping;

2. there is little difference between the optimal PSM network and a local PILUT using a 3 by 3 planar tessellation; and

3. the same PILUT can be used to detect small obstacles lying on the floor about a metre and a half away, as they show up as departures from the disparities predicted for the ground plane. It is well known that errors in stereoscopic depth vary as the square of the viewing distance. Hence, though the system can detect obstacles as small as a centimetre high when fairly nearby, the resolution rapidly decreases at greater distances.

Figure 5 shows the change in distribution of errors for non-ground plane points. A clear distinction between the different heights is easy to discern. Note also the that a 4cm obstacle does not cause twice the distribution shift of a 2cm one. This illustrates further the non-linearities involved.

6 Summary

The paper has described part of a project to explore the use of a biologically plausible neural network architectures as part of the system to exploit stereopsis under the variable camera geometry of a four degree of freedom stereo camera rig. A rather simple, but seemingly, adequate neural network architecture for representing high dimensional surface approximations (PILUTs) was evaluated as a method of encoding the projective stereo mapping of the ground plane for different head positions. This has been shown to be succesful as a primitive Ground Plane Obstacle detection device, and we are pursuing an analysis of these results to determine more sophisticated ways of using the predicted mappings.

References

[1] J. Albus, "A new approach to manipulator control: The cerebellar model articulation controller (CMAC)", Trans. ASME - J. Dyn. Syst. Meas. Control, 1975,vol. 97, pp 220-227.

[2] J. Albus,"Data storage in the cerebellar model articulation controllelr (CMAC)", Trans. ASME - J Dyn. Syst. Meas. Control, 1975,vol 97, pp 228-233.

[3] P. Dean, J.E.W. Mayhew, N. Thacker, & P.M. Langdon, "Saccade control in a simulated robot camera-head system: neural net architectures for efficient learning of inverse kinematics.", Biological Cybernetics,1991, 66, 27-36.

[4] F. Girosi, T. Poggio, (1989) "Representation properties of networks: Kolmogorov's theorem is irrelevant", Neural Computation, 1989, vol. 1, no. 4 pp 465-469.

[5] S.H. Lane, M.G. Flax, D.A. Handelman, J.J. Gelfand, "Function approximation using multi-layered neural networks with B-spline receptive field functions.", CSL Report 47,1991, 1-37

[6] H.A. Mallot, E. Schulze, & K. Storjohann, "Neural network strategies for robot navigation.", Proc. n'Euro,In G. Dreyfus & L. Personnaz (Ed.), 1988 ,Paris:

[7] J.E.W. Mayhew, P. Dean,P. Langdon, "Artifical neural networks for the kinematic control of a stereo camera head", (in preparation),1992

[8] J.E.W. Mayhew, H.C. Longuet-Higgins, "A computational model of binocular depth perception.", Nature, 1982, 297 (5865) 376-379.

[9] T. Poggio,F. Girosi, "A theory of networks for approximation and learning.",A.I. MEMO NO. 1140. ,Atifical Inteligence Laboratory, Massachusetts Institute of Technology,1989

[10] T. Poggio, & F. Girosi, "Networks for approximation and learning.", Proceedings of the IEEE, 78(9),1990, 1481-1497.

[11] T. Poggio, & F. Girosi, "Regularization algorithms for leaning that are equivalent to multilayer networks.", Science,1990, 247, 978-982.

[12] M.A.S. Potts, D.S. Broomhead, "Time series prediction with a radial basis function neural network.", Adaptive Signal processing, Simon Haykin (Ed), Proceedings of SPIE 1565,1991, 255-266

[13] N.A. Thacker, J.E.W. Mayhew, "Optimal combination of stereo camera calibration from arbitrary stereo images.", Image and Vision Computing (feb 1991). vol 9 no 1 27-32.

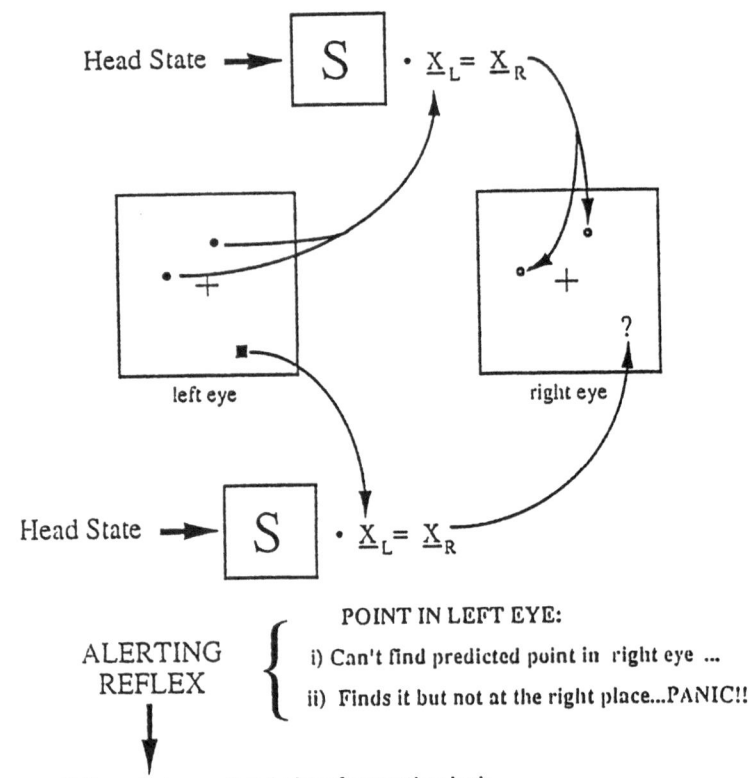

Head State → S · $\underline{X}_L = \underline{X}_R$

left eye

right eye

?

Head State → S · $\underline{X}_L = \underline{X}_R$

ALERTING REFLEX {

POINT IN LEFT EYE:

i) Can't find predicted point in right eye ...

ii) Finds it but not at the right place...PANIC!!

1) Saccade to predicted place for another look
Inverse kinematics for fast eye saccade
Predictive stereo matching at new configuration
VOR to null off vehicle motion

2) Keep an eye on it
2D trajectory modelling for microsaccade tracking
3D trajectory collision evaluation

• Figure 1. Ground plane obstacle detection under variable camera geometry. The scheme uses a predictive stereo matcher which encodes, for each head state. the map from the left eye to the right eye of corresponding points lying on the ground plane. Points that deviate from their predicted coordinates by more than is allowed by the error model are subject to further inspection as potential targets.

Left to Right Disparity Maps for the Whole Retina

Top Centre Head Position

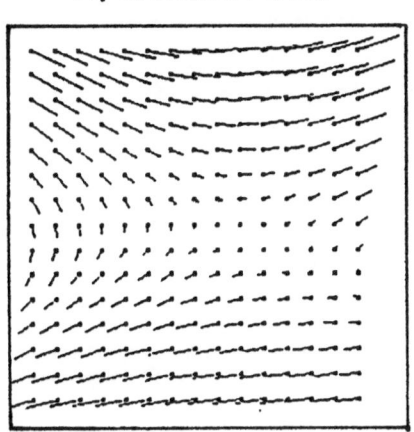

Middle Right Head Position

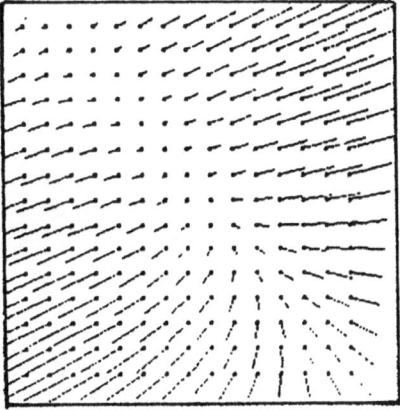

Bottom Left Head Position

- Figure 2. Ground plane disparity maps for different head positions: looking ahead, to the left and to the right. The sampling spacing corresponds to approximately one degree of visual angle (see text for details). Note that changing the eye position radically changes the pattern of disparities, the disparities contain both vertical and horizontal components, and often the vertical components are of the order of a degree (the particular pattern of vertical disparities is determined by the position on the retina, and the camera geometry, and is independent of scene structure to first order).

Effect of a 3cm. Obstacle on Disparity Maps for the Bottom Left Position

Ground Plane Data

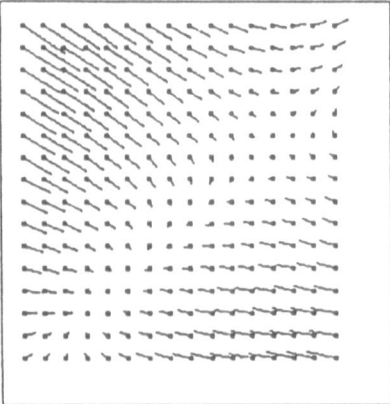

Ground Plane and Obstacle

Differences from predicted map for ground plane

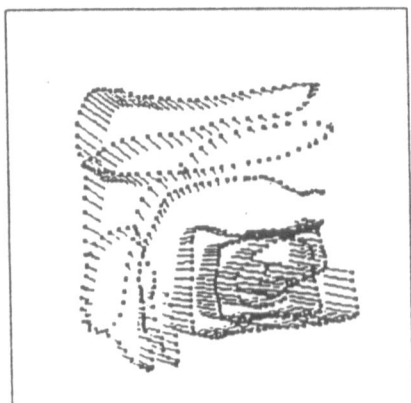
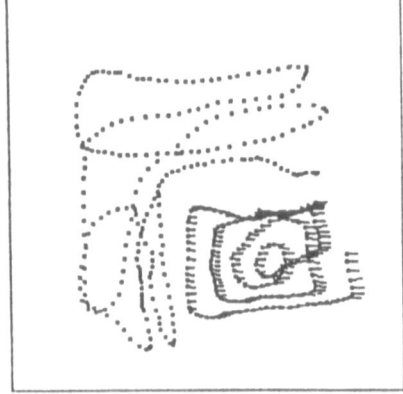

- Figure 3. Disparity Maps for obstacle detection: top: the predicited disparity map for a fixed head position. left: the actual disparity data for ground and obstacle. right: actual - predicted showing disparity error for obstacle points. 3cm obstacle at distance of 1m approx.

PILUT: for stereo correspondence projection

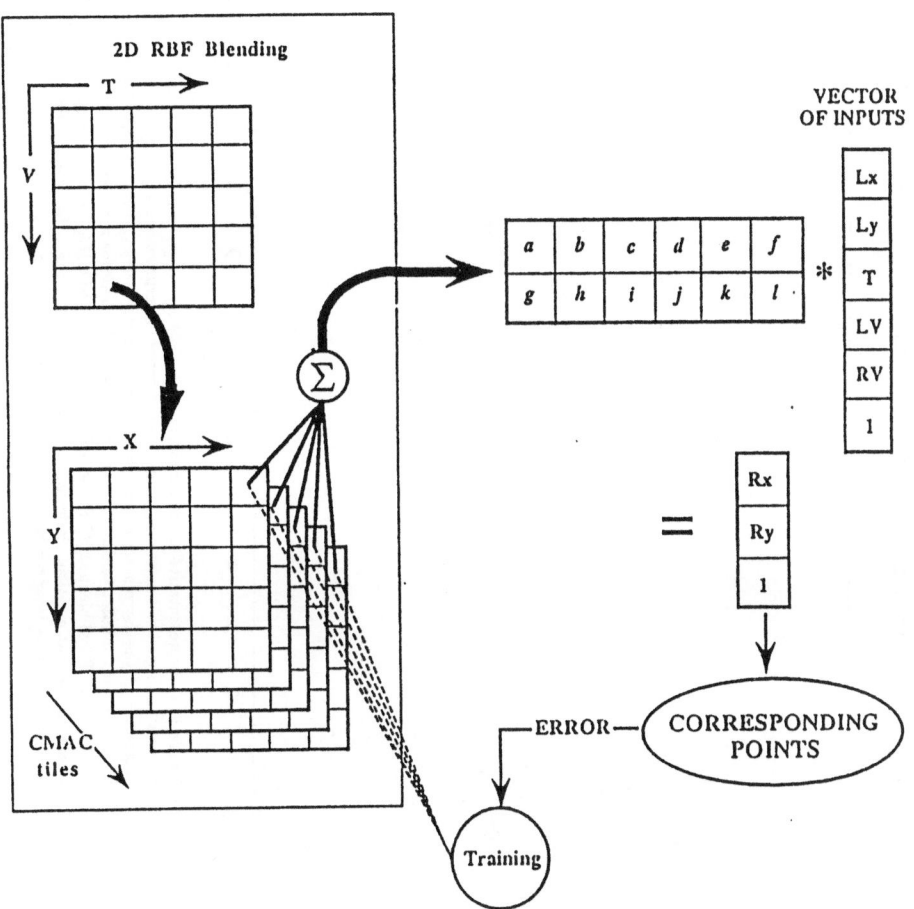

- Figure 4. The PILUT architecture applied to stereopsis under variable camera geometry. The principle is to project the higher dimensional space onto a subspace, possibly a subset of the original dimensions, then to use a coarse coding scheme of the subspace, and full dimensional linear interpolation schemes to encode the hyperplanar approximation of the surface up to the required resolution.

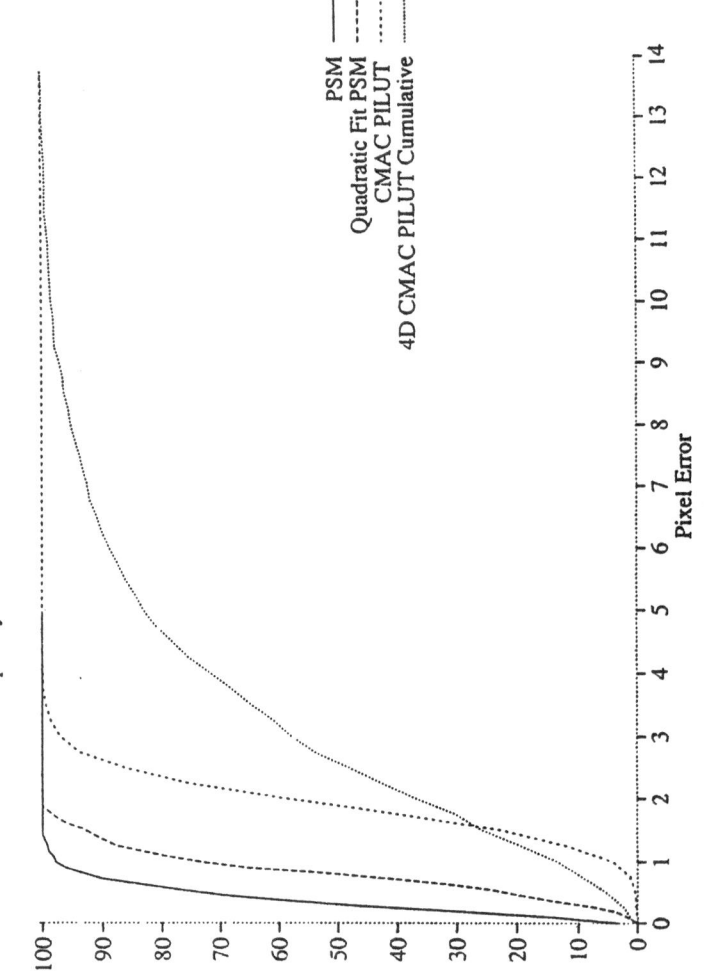

Cumulative Frequency distribuion of retinal errors for Various Methods

PSM
Quadratic Fit PSM
CMAC PILUT
4D CMAC PILUT Cumulative

Pixel Error

Frequency / Cumulative

• Figure 5. a) Representative experimental results evaluating the ground plane stereo predictor at a single head position (we find no difference in performance dependent on the
magnitude of the asymmetry of vergence). a Error distributions shown as normalised
frequency and cumulative distributions in pixels of disparity for a single linear net, a
PSM, a PILUT 3x3 blended RBF, and a PILUT 3x3 blended CMAC all as compared to
a data set for that position. Performance of the linear net is clearly inadequate whereas
the stereo mapping is solved both by the PSM and the PILUTs.

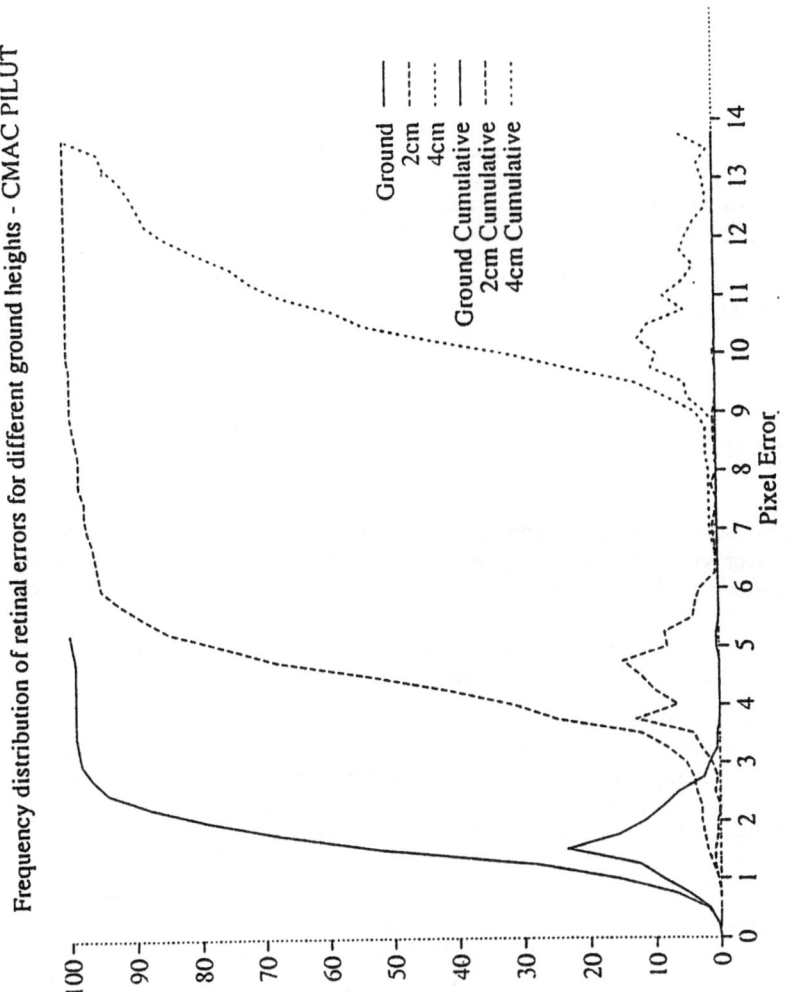

Frequency distribution of retinal errors for different ground heights - CMAC PILUT

• b) Superimposed normalised frequency and cumulative error distributions for the PI-LUTs at a different head position from the data shown in a. The clearly distinguishable distributions correspond to the ground plane, and two obstacles, one 2 cm and the other 4 cm high. A simple statistical decision measure would be sufficient to trigger an alerting reflex in the control architecture.

Non-Wildcard Matching Beats The Interpretation Tree

Robert B. Fisher

Dept. of Artificial Intelligence, University of Edinburgh

5 Forrest Hill, Edinburgh EH1 2QL, Scotland, United Kingdom

Abstract

Probably the best known control algorithm for high-level model matching in computer vision is the *Interpretation Tree* expansion algorithm, popularized and extended by Grimson and Lozano-Perez. This algorithm has been shown to have a high computational complexity, particularly when being applied to matching problems with large numbers of features. This paper introduces a non-wildcard variation on this algorithm that has an improvement of about 4-10 in performance over the standard Interpretation Tree algorithm.

1 Introduction

Probably the most well-known control algorithm for high-level model matching in computer vision is the *Interpretation Tree*(IT) expansion algorithm, as used by Grimson and Lozano-Perez[2]. The IT algorithm searches a tree of model-to-data correspondences, such that each node in the tree represents one correspondence and the path of nodes from the current node back to the root of the tree is a set of simultaneous pairings.

Unfortunately, this algorithm has the potential for combinatorial search explosion. This has prompted researchers to develop techniques for pruning the trees, thus limiting the number of matches considered. The main technique commonly used is based on pruning constraints[2] (which locally reject pairings that are inconsistent, and hence eliminate all of the search that might further extend this inconsistent pairing) and early termination[4] which stops search: (1) at the first hypothesis with a given number of pairings, or (2) at any time that it is impossible to make sufficient pairings with the remaining potential matches. However, even with these effective forms of pruning, the algorithms still can have an exponential complexity, making them unsuitable for use in scenes with many features.

As reported by Grimson[4], the main cause of the exponential complexity is the use of a "wildcard" match feature. This paper discusses and analyses a variation to the standard IT algorithm that explores a different tree without using a wildcard and requires 4-10 times less work.

2 The Standard Interpretation Tree Algorithm

Consider a set $\{\, d_i \,\}$ of D data features and a set $\{\, m_i \,\}$ of M model features. Then, the root of the interpretation tree has no pairings. The first level expands the root node to pair all of the M model features with data feature d_1. The second level in the tree expands each of these nodes to pair all model features with data feature d_2 (multiple pairings are allowed), and so on. The expansion continues for all D data features. At each node at level k in the tree, therefore, there is a hypothesis with k features matched.

If this IT were explored completely, there would be M^D "leaf" nodes at the bottom of the tree (i.e. these many complete interpretations) and

$$\sum_{i=0}^{D} M^i = \frac{M^{D+1} - 1}{M - 1} \doteq M^D$$

nodes in the full tree. If either M or D are of any reasonable size (e.g. larger than 5), then we can expect to have excessively large search trees.

An additional complication is that one usually wishes to include a "wild-card" model feature that will match with any data feature. This is necessary because it may not always be possible to find a model feature that matches the data feature at the current level of the tree (because of fragmentation, bad segmentation, noise, unrelated features, etc.).

One way to reduce the amount of searching is to 'prune whole branches of the tree', by showing that a given pairing or sequence of pairings is inconsistent. Therefore, all descendents from that node in the tree will also be inconsistent and need not be explored. The most common approach uses unary and binary pruning constraints. Unary constraints eliminate model-to-data pairings when some shared property is inconsistent. Binary constraints eliminate hypotheses when a relative property between a pair of model features is inconsistent with the same property between the corresponding pair of data features. For example, Grimson and Lozano-Perez[2] provide a set of *binary* constraints useful for three-dimensional scene analysis, based on *pairwise consistency constraints*, that compare quantities such as relative distance, orientation and direction. Similar constraints can be developed for higher-order consistency (e.g. vector triple products). Of particular importance is the local nature of the consistency tests, based on the assumption that a few simple, fast tests on partially generated hypotheses will eliminate large numbers of globally inconsistent hypotheses.

In the discussion below, the following quantities are used:

- there are M model features in the model.

- on average, $p_v M$ of these are visible in the scene (less than M by occlusion, being on the back side of the object, etc.). In 2D scenes, $p_v \doteq 1$ and, in 3D scenes, $p_v \doteq 0.5$ as about half of the features are back-facing and hence not visible.

- of the visible model features, only p_r of these are recognizable (less than those visible because of segmentation failures, etc.) forming $C = p_r p_v M$ correct observable data features. (If the model chosen for this scene is

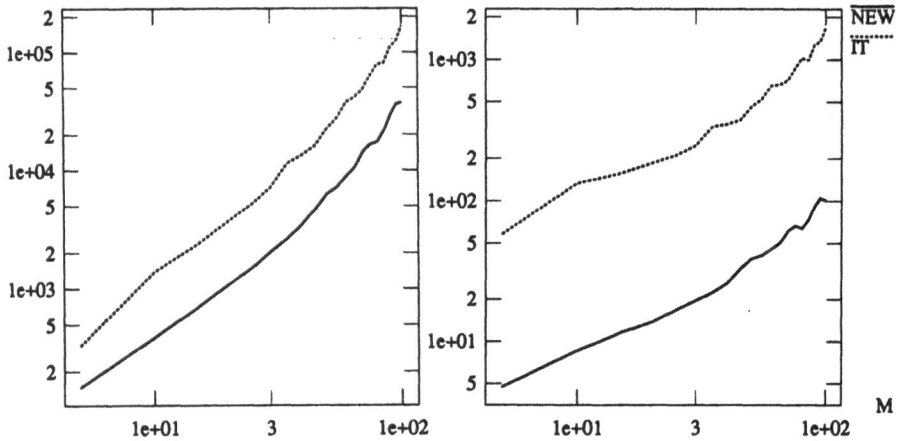

Figure 1: Generated and Accepted Nodes versus Number of Model Features (M) with $S = 20$ $p_r = 0.95$ $p_1 = 0.1$ $p_2 = 0.01$ $p_v = 0.5$ $\tau = 0.5$ (loglog plot)

incorrect, $p_r = 0$.) Which C of the M model features are matchable is not known initially.

- there are also S spurious data features (including noise features and visible model features that are not recognizable); hence altogether there are $D = C + S$ data features.

- the probability that a randomly chosen model feature matches with an incorrect random data feature is p_1 (correct pairings alway match).

- the probability that a random pair of model features is consistent with an incorrect random pair of data features (given that the individual model-to-data pairings are consistent) is p_2.

- an acceptable set of model-to-data pairings must have at least $T = \tau p_v M$ non-wildcard correspondences ($\tau \in [0..1]$). Whenever this many are achieved, then the whole matching process terminates successfully immediately. Any set of matches that can never get T matches (because insufficient potential matches remain) is terminated immediately and the matching process proceeds to considering other matches.

In the discussion that follows, the term *generated* refers to nodes and paths that are created prior to testing the consistency of the node or path, and *accepted* refers to nodes or paths that pass the consistency tests.

Grimson[3] analyzed the combinatorics of the standard algorithm, and showed that, without wildcards, the algorithm tends to accept (Proposition 5, pg 274) a single path with many pairings (i.e. the correct one), and generates (Proposition 6, pg 274) a number of nodes that is quadratic in the number of model features. However, allowing a wildcard means that the algorithm will accept an exponential number of correctly matchable features. One key term is 2^C, arising from the power set of the C matchable features. The complexity occurs because each matchable data feature can be either matched with the correct model feature or the wildcard. Examination of a typical search tree shows that most of the tree consists of paths containing either members of this power set

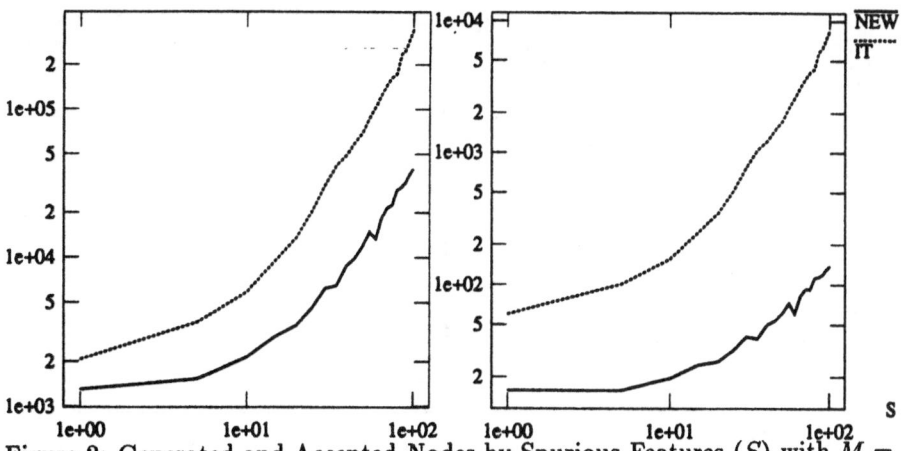

Figure 2: Generated and Accepted Nodes by Spurious Features (S) with $M = 40$ $p_r = 0.95$ $p_1 = 0.1$ $p_2 = 0.01$ $p_v = 0.5$ $\tau = 0.5$ (loglog plot)

or wildcards. Many of these paths can be eliminated by using the *termination threshold* described above. This can only apply when the search is sufficiently advanced, but it does make a significant improvement.

Grimson[3] analyzed the consequences of this termination condition and showed (Corollary 3.2, pg 367) that if:

$$p_2 M D < 2$$

then the expected number of nodes generated is bounded by:

$$\frac{M D^2}{C} < num\_generated < aT\frac{M D^2}{C}$$

where a is a small constant. There might be some problems with the exactness these bounds, but the conclusion that the use of a termination condition improves performance is valid.

Unfortunately, the $p_2 M D < 2$ condition given above does not always hold, in which case the algorithm again seems to be exponential. In fact, in the experiments described below, it only holds for the smallest test cases.

3   The Non-wildcard Matching Algorithm

The vast number of nodes in the standard algorithm arises because of the use of wildcards. An alternative search algorithm explores the same search space, except does not use a wildcard. The essence of the difference is the search process skips over all data pairings that use a wildcard, to consider the next true data-model feature pairing. This results in a flattening of the search tree. The algorithm has two phases:

1. The set $\Omega = \{s_k\} = \{(m_{i(k)}, d_{j(k)})\}, k = 1..N$ of all pairs of features satisfying the unary pairing constraints is formed, such that if s_r is before s_s (i.e. $r < s$), then $j(r) \leq j(s)$.

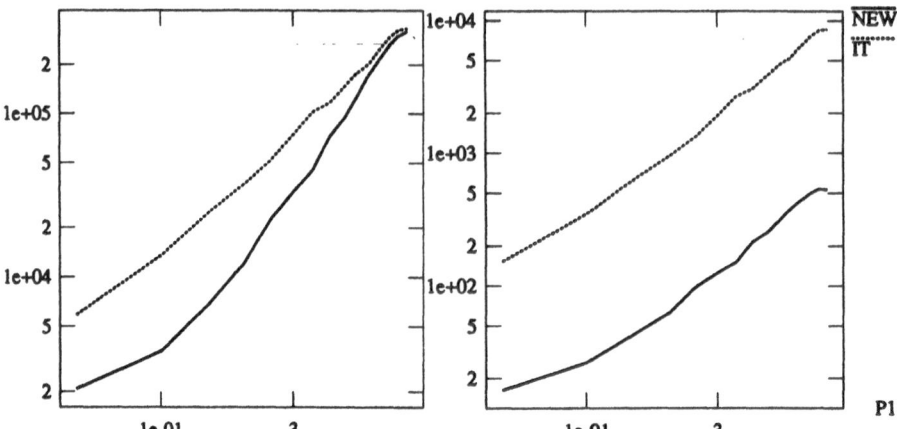

Figure 3: Generated and Accepted Nodes by Unary Match Probability (p_1) with $M = 40$ $S = 20$ $p_r = 0.95$ $p_2 = 0.01$ $p_v = 0.5$ $\tau = 0.5$ (loglog plot)

2. A different search tree is explored, in which each extension of a branch is formed by appending new entries from Ω, subject to the constraints that (1) each data feature appears at most once on a path through the tree and (2) the data features are used in order (with gaps allowed).

Starting from a branch ending with pair s_λ (or nothing at the root of the tree), all pairs $s_{\lambda+1} \ldots s_N$ are possible extensions to the branch. Only extensions that satisfy the normal binary constraints are accepted. Extension stops when the termination number of matches is reached, or on branches where insufficient possibilities remain in the tail of Ω.

For example, if $\Omega = \{s_1, s_2, s_3, s_4\} = \{(m_2, d_1), (m_4, d_2), (m_1, d_2), (m_5, d_4)\}$, the tree:

```
                    X
        s₁        s₂   s₃   s₄
    s₂   s₃   s₄  s₄   s₄
    s₄   s₄
```

is searched depth first following the leftmost branches first (no pruning is shown here to illustrate the shape of the tree). The initial step considers the individual model-data pairings once (i.e. the unary constraints are tested once instead of whenever needed, as in the IT tree). As the second and third levels of the new search tree contain complete matches, the binary constraints eliminate almost all false pairings quickly. The trade off is that the branching factor of the new tree is $sizeof(\Omega)$ instead of M. This search algorithm can produce the same set of hypotheses as the standard IT algorithm, with respect to the data features paired to non-wildcard model features. The order of generation may be different when the termination threshold is used.

4 The Experiments

To demonstrate the effectiveness of the non-wildcard search algorithms, we use the following experimental problem. The approach is designed to allow comparison of methods for which no formal complexity measure has yet been determined, and also to allow comparison of algorithms within the same complexity class. The problem is based on an example described in Grimson[4]. The experiments use simulated data; however, Grimson showed that the model and simulation gave a reasonable characterization of real matching problems. The use of the simulated problems then allows us to compare the algorithm performance on the same data sets of varying sizes.

Based on the problem model given in Section 2, each model-match experiment of the two algorithms will consist of:

1. Initially determining a random selection of C of the D data features to be the solution.

2. For each generated model-to-data pairing, a correspondence that is not part of the solution and does not use a wildcard is accepted if the new correspondence is individually satisfied with probability p_1 and the new correspondence is pairwise satisfied with each previously filled non-wildcard feature with probability p_2. Correspondences that are part of the solution or use the wildcard are accepted.

The experiments with the non-wildcard search tree algorithm are similar. For the experiments described in this paper, we used:

PARAMETER	NOMINAL	RANGE
M	40	5 to 100 by 5
S	20	0 to 100 by 5
p_1	0.1	0.05 to 0.75 by 0.05
p_2	0.01	0.001, 0.002, 0.004, 0.008, 0.01, 0.02, 0.04, 0.06, 0.08, 0.10, 0.12, 0.14, 0.16, 0.18, 0.20
τ	0.5	0.2 to 0.9 by 0.1
p_v	0.5	no variation
p_r	0.95	no variation

In each experiment described in this section, one parameter was varied over the range given above and all others were set to the nominal value. All experiments were run 200 times and the value reported is the mean value. The graphs in Figures 1–5 given show how the number of nodes generated and accepted varied with the parameters for the new and standard IT algorithms.

As we look over the results, which explore a substantial portion of the parameter spaces likely to be encountered in visual matching problems, we can see that the non-wildcard is clearly better than the standard IT algorithm. In search, the non-wildcard algorithm is not bad for most problems, but its performance deteriorates as p_1 increases (this increases the number of possible matches to consider at each stage). For acceptances, the non-wildcard algorithm is also the better, as it does not allow proliferating wildcard hypotheses. Except when p_1 is large, the non-wildcard algorithm did about 4 times less search and 10 times less accepting than the standard algorithm.

One might also consider how the two algorithms perform when there is no instance of the object in the scene. Then, it is unlikely that the early success conditions would occur, and thus almost all of the search space would have to be explored. Figure 6 shows the number of nodes generated and accepted in

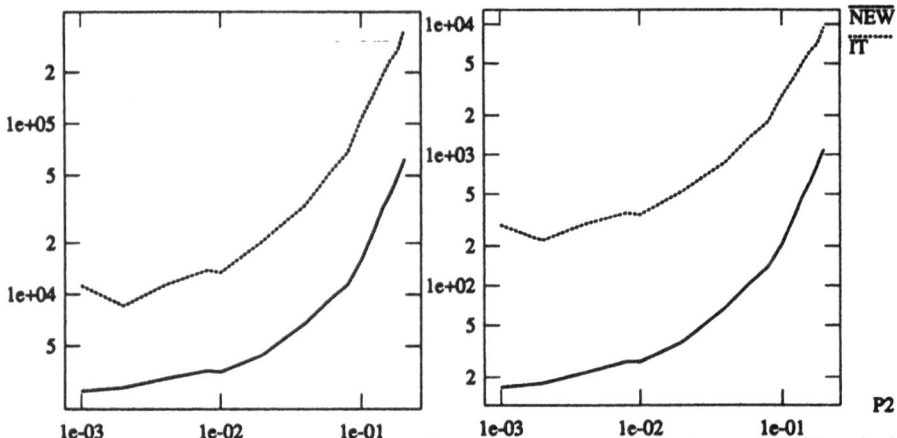

Figure 4: Generated and Accepted Nodes by Binary Match Probability (p_2) with $M = 40$ $S = 20$ $p_r = 0.95$ $p_1 = 0.1$ $p_v = 0.5$ $\tau = 0.5$ (loglog plot)

this case. When there is no true match possible, the non-wildcard is still much better, but in both cases much more work is done (e.g. about 10-30 times more work). Grimson ([3], page 389) shows that the standard algorithm is also much worse when no match is possible.

5 Computational Complexity of the Non-Wildcard Matching Algorithm

Grimson[3] has mainly concentrated on estimating upper and lower bounds for the standard algorithm. As seen in the results from the previous section, the non-wildcard search algorithm looks very promising. Hence, we give here a complexity analysis for that algorithm, except that we state here (without proof) the *mean* performance of the algorithm.

Theorem 1 (Mean Complexity of Non-Wildcard Algorithm) *Given the problem definitions from above, there are expected to be C true pairings and $F = p_1(MD-C)$ false pairings that arise from the initial model to data feature matching. Assume that M and D are very large, so that the effect of matching one feature does not significantly affect the rest of the algorithm. Also assume that no false hypotheses containing 3 or more pairings survive the pruning tests (i.e. $Fp_2 < 1$). Then, the expected amount of search is approximately:*

$$MD + T + \frac{F}{C}(C + F) + p_2\frac{F}{C}(C + F - T)(C + F - T + 1)) \doteq O(M^5)$$

and the expected number of hypotheses accepted is approximately:

$$T + \frac{F}{C} + p_2\frac{F}{C}(C + F - T + 1) \doteq O(M^3)$$

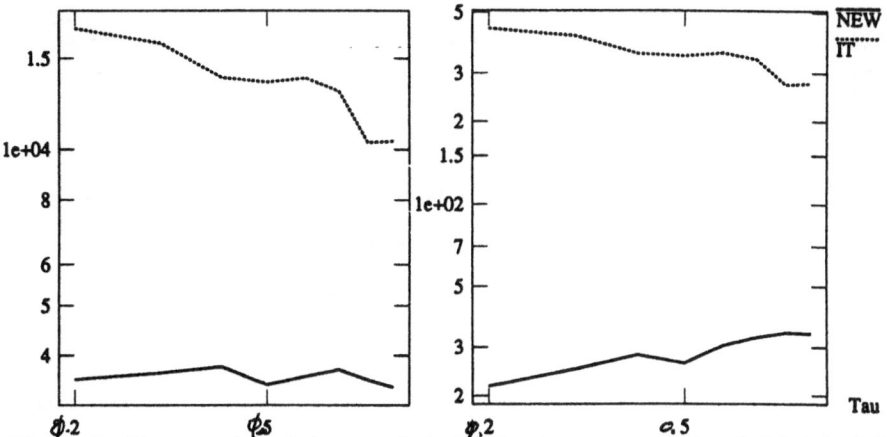

Figure 5: Generated and Accepted Nodes by Acceptance Threshold (τ) with $M = 40$ $S = 20$ $p_r = 0.95$ $p_1 = 0.1$ $p_2 = 0.01$ $p_v = 0.5$ (loglog plot)

6 Discussion and Conclusions

As Grimson observed, most of the complexity of the standard interpretation tree search is a consequence of the use of "wildcards" to overcome missing and erroneous data. However, merely having "good" data does not mean one can avoid the use of the wildcard, because the so-called false features may have arisen from other objects in the scene, or other subcomponents of the object being recognized. Hence, the wildcard is likely to remain a key element of the general interpretation tree search algorithm. If one could assume that there were only a limited amount of scene clutter, then one might limit the use of wildcards to a specific number. However, more than one or two would still allow a considerable number of partially empty hypotheses.

From the experiments, it is obvious that the non-wildcard algorithm produces better performance than the standard IT matching algorithm. For the non-wildcard algorithm, the real work occurs at the first or second step, which effectively requires a comparison between all model and data features. As any model feature might be an explanation for any data feature, it is hard to avoid this complexity, which results in MD initial comparisons and roughly $p_1 MD$ false acceptances, which provides a lower bound on the amount of work required. After that, a reduced search space needs to be considered, but the initial effort is substantial. There does not seem to be much possibility of reducing this amount of effort, unless some additional aspect of the particular problem can be exploited.

Real benefits can be gained by reducing the number of features that need to be considered at a time. If the data features can be partitioned into K subsets, which can be matched independently, and the models features can also be partitioned into L corresponding subcomponents, then the brute-force version of the matching algorithm is reduced from M^D to:

$$KL(\frac{M}{L})^{\frac{D}{K}}$$

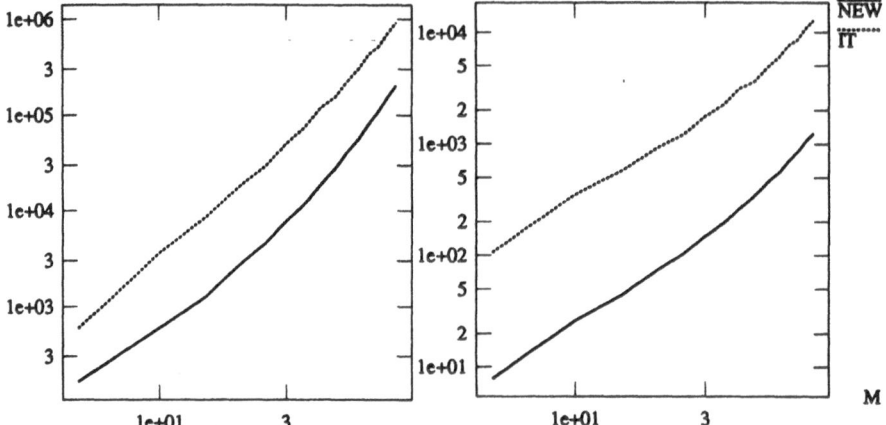

Figure 6: Generated and Accepted Nodes versus Number of Model Features (M) When No Instance of the Model is Present with $S = 20$ $p_r = 0.95$ $p_1 = 0.1$ $p_2 = 0.01$ $p_v = 0.5$ $\tau = 0.5$ (loglog plot)

which is considerably less. This requires perceptual organization [5], such as a region or surface patch grouping (e.g. [1] Chapter 5).

The analysis above also assumed that only one model needed to be considered when matching. If all models must be considered, then the computational complexity will be high, as the results in Section 4 showed. Hence, some form of model invocation method is needed to reduce the number of candidate models (e.g. [1] Chapter 8, [3] Chapter 15).

The net conclusion is that by using the non-wildcard algorithm as an alternative to the standard interpretation tree visual matching algorithm, it is possible to reduce the amount of search by a factor of about 4 and number of partial interpretations accepted by a factor of 10, where the precise amount of improvement depends on the problem parameters. Both factors are important, because, depending on the particular matching algorithm, the savings achieved depend on relative costs of each action (e.g. the pairwise consistency checking costs may high relative to final verification costs).

The relative speed difference of the implemented matching algorithms might overcome this reduction in theoretical search complexity. However, the $M = 100$ case from Figure 1, matching requires 1.27 seconds for the non-wildcard algorithm, as compared to 5.4 seconds for the standard algorithm (on a Sparc-Station 1+). Hence, the speed of the non-wildcard algorithm is also significantly better than the standard algorithm in the implementations compared.

Acknowledgements

This research was funded by SERC (IED grant GR/F/38310). Other facilities provided by University of Edinburgh. This paper benefited greatly from discussions with Dibio Borges, John Hallam, Howard Hughes, Mark Orr, Kristian Simsarian, Manuel Trucco and Mike Uschold.

References

[1] Fisher, R. B., From Surfaces to Objects: Computer Vision and Three Dimensional Scene Analysis, John Wiley and Sons, Chichester, 1989.

[2] Grimson, W. E. L., Lozano-Perez, T., *Model-Based Recognition and Localization from Sparse Range or Tactile Data*, International Journal of Robotics Research, Vol. 3, pp 3-35, 1984.

[3] Grimson, W. E. L., Object Recognition By Computer: The Role of Geometric Constraints, MIT Press, 1990.

[4] Grimson, W. E. L., *The Combinatorics of Heuristic Search Termination for Object Recognition in Cluttered Environments*, Lecture Notes in Computer Science, ECCV-90, Springer-Verlag, pp 552-556, 1990.

[5] Witkin, A. P., Tenenbaum, J. M., *What Is Perceptual Organization For?*, Proceedings 8th IJCAI, pp1023-1026, 1983.

A Non-Wildcard Search Algorithm

```
// Non-wildcard expansion variation on standard algorithm:
//    expand tree by members of valid_pairs (not by data levels),
//    subject to not reusing data features.
searchtree(treesofar, valid_pairs)
{   if empty(valid_pairs) return fail
    trylist = valid_pairs
    do {
        if can never get enough return fail
        extension = head(trylist)
        trylist = tail(trylist)
        if data feature in extension already appears in treesofar
            then skip this extension
        if compatible(extension, treesofar)
        {   if enough matches return success
            if success(searchtree(append(treesofar,extension),
                trylist)), then return success}
    } while non-empty(trylist)
    return fail}

// test for compatibility of new pairing with rest of pairings:
boolean compatible(new_pair, treesofar)
{   // check pairwise with previously filled slots of this hyp
    for each pair in treesofar
        if not compatible2(pair, new_pair) then return false
    return true}
```

Modelling Data Complexity for Model-based Vision

L. Du[1], G D Sullivan and K D Baker

Department of Computer Science, Reading University
Reading, RG6 2AY

Abstract

This paper discusses an issue of wide-ranging importance for computer vision - the systematic consideration of data complexity in assessment of computer vision systems.

We investigate 3D object recognition from 2D features as a typical problem of computer vision. We identify 3 factors which contribute to the complexity of image data: feature truncation, noise and clutter. We propose a modelling scheme for these factors, which allows us both to measure and to simulate each factor.

Using the scheme, a systematic comparison is made between two existing strategies for model-matching as a function of clutter and truncation factors. Post-model-matching object discrimination is then examined as a function of the noise factor. These two examples serve to illustrate the data complexity model, and demonstrate its use for formal assessment of model-based algorithms.

1   Introduction

This work was motivated by the desire to compare two methods for matching 3D models to 2D image features, under conditions of data complexity likely to be met in practice. Models of the factors which confound model-based vision were needed. Similar problems arise in many circumstances in model-matching and the analysis may be used generally.

The model-based approach to 3D recognition from 2D image features has inspired a number of systems and studies, including those by Roberts [13], Brooks [3], Goad [8], Ikeuchi [10], Lowe [11], Sullivan [15], Bodington et al [1], Bray [2], Zhang et al [17], Du et al [4]. These systems have adopted many different strategies for interpreting image features by matching model features to them. It is relatively easy to compare different strategies in terms of their computational costs, but they prove very difficult to compare with respect to their robustness in the presence of errors of the data. A second problem in model-based vision is that of assessing the totality of the evidence for a recognised object. Robust criteria for evidence assessment underpin the decision processes used to detect or reject an object, as well as those used to discriminate between similar objects.

The lack of any accepted basis for testing algorithms is an important obstacle towards any systematic treatment of these issues. In practice, image features suffer from

---

1. Currently a SERC research fellow at the Department of Electrical and Electronic Engineering, University of Surrey, Guildford, GU2 5XH, (email: L.Du@ee.surrey.ac.uk)

3 main errors: (i) truncation, possibly due to occlusion, (ii) clutter, due to features from irrelevant objects and (iii) noise due to digitisation and measurement errors. Such errors act as confounding factors which contribute to the image complexity. We propose a modelling scheme providing measures for the factors and allowing controlled simulation of them. Different model-matching algorithms can thereby be compared objectively with respect to data complexity on a formal statistical basis.

We illustrate the use of the truncation model and clutter model by briefly comparing Lowe's incremental model-matching strategy and the VCA strategy proposed by Du et al [4]. We also illustrate the use of the noise model by an example of object discrimination in model-based vision.

2 Ideal and observed image features

In the domain of 3D recognition from 2D features, ideal image features may be defined as the result of noise-free feature extraction from images containing nothing but the target object. Assuming that the pose of a target object in the scene is known, the ideal image features are in exact agreement with the instantiation of the object model.

Observed image features are the product of an imperfect feature extraction process applied to a complex image. Any difference between the observed and ideal features can be thought of as a form of data complexity. We decompose the data complexity into 3 factors, (i) truncation of ideal features, (ii) noise contamination of ideal features, (iii) clutter caused by extraneous features.

These observations are relevant both to edge and region features, but in the remainder of this paper we consider only the more commonly used edge features. A fourth type of error, due to false extension of an image feature because of accidental alignment with other structures is relatively rare in edge-based methods (although its equivalence for regions, false merging, is more troublesome). False extension of edge features is not considered here.

3 A model of image feature truncation

Truncation of image features is common in feature extraction. It occurs when part of the object is occluded so that only a part of an ideal feature is observed, where the edge detector loses edge points near a junction (as is frequently the case with the Canny detector), when an edge ha uneven contrast, or when an occluding edge is seen against a chequered background.

3.1 Truncation and its measure

Let an ideal feature be represented as $F_I = \overline{(x_1, y_1)\,(x_2, y_2)}$ and its corresponding observed feature by $F_o = \overline{(x'_1, y'_1)\,(x'_2, y'_2)}$. Truncation can be represented as perturbation of the end points along the line of the ideal feature, with only one degree of freedom (see Figure 1).

Let L_I and L_o be the lengths of F_I and F_o respectively. A measure of truncation for each fragmented ideal feature is defined as,

$$k_t = \frac{L_I - L_o}{L_I} \qquad (1)$$

We define a measure of overall truncation of a set of simulated or observed features as the

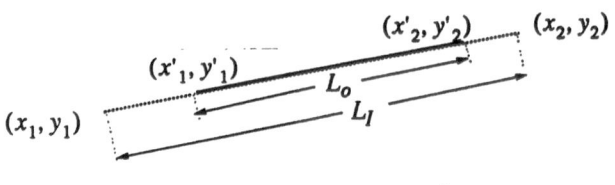

Figure 1   Feature truncation

RMS of all truncations,

$$K_t = \sqrt{\frac{\sum (k_t^i)^2}{n}} \quad (i=1, 2,....n) \tag{2}$$

where n is the total number of features.

3.2   Simulation of truncation

To simulate truncated ideal features, the coordinates of each end point of an ideal feature are randomly shrunk along the feature, with a bound on the maximum of truncation. Figure 2 gives three examples of simulated image feature truncations. The examples are displayed in increasing order for the truncation measure K_t.

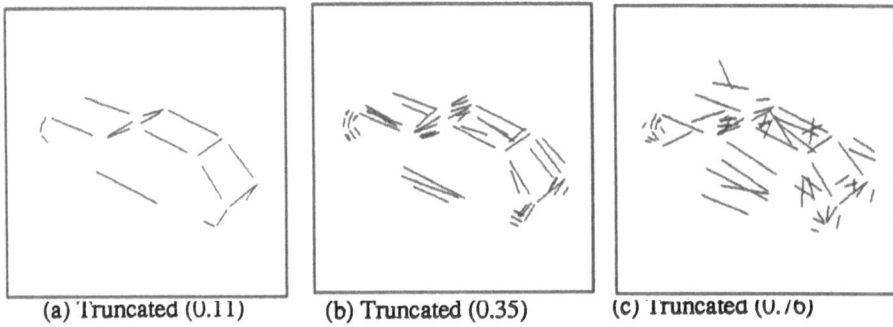

(a) Truncated (0.11)   (b) Truncated (0.35)   (c) Truncated (0.76)

Figure 2   Examples of simulated truncation

4   A model of noise contamination for image features

Noise contamination arises from the perturbation of observed features caused by imperfect feature extraction and from the use of object models of limited accuracy.

4.1   Noise contamination and its measure

Noise contamination of an image feature is described as a random perturbation of the end points of observed feature F_o, with two degrees of freedom (see Figure 3). For a pair of ideal (dotted) and observed (continuous) features we use k_n as the measure of noise, where

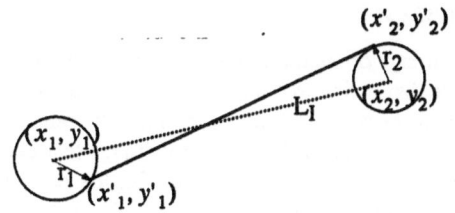

Figure 3   Noise contamination to image features

$$k_n = \frac{\sqrt{r_1^2 + r_2^2}}{L_I} \qquad (3)$$

The overall measure of noise contamination is again defined as the RMS

$$K_n = \sqrt{\frac{\sum_i^n (k_{nc}^i)^2}{n}} \qquad (4)$$

where n is the total number of features.

4.2   Simulation of noise contaminated features

Feature noise contamination can be simulated as described in equation (5), where k is a bound on the perturbation. Examples are shown in Figure 4 for a range of the perturbation measure K_n.

$$x'_1 = x_1 + random\,(sign) \times random\,(k) \times L_I$$

$$(x'_2, y'_1, y'_2, \text{likewise}) \qquad (5)$$

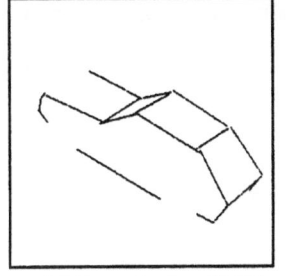 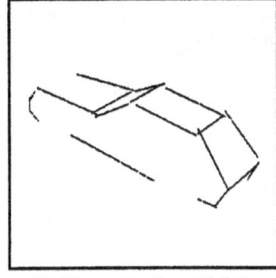 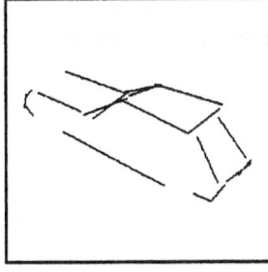

(a) Perturbation (0.02)   (b) Perturbation (0.04)   (c) Perturbation (0.08)

Figure 4   Examples of simulated noise contamination

5   A model of clutter

The third confounding influence is that of extraneous features which arise from other objects in the scene, or from parts of the target object not described by the model.

5.1   Clutter and its measure

Extraneous features of arbitrary orientation, scattered over the image (as used for example by Bray [2]) may be distracting to human perception, but cause little distraction to most machine matching processes. Object recognition usually relies on a hypothesised initial pose, and this enables immediate exclusion of any image features falling outside a tight neighbourhood of the expected feature positions. On the other hand, extraneous features which are located very close to a ideal feature make little difference to either recognition or final pose determination. The main difficulty arises from extraneous features which are close enough to be included in the focus neighbourhood but far enough away to cause substantial error if mis-matched. Extraneous features caused by shadows, patterned backgrounds or adjacent objects are common examples.

As a measure of clutter, we therefore need to take into account only those extraneous features which have the potential to distract model matching and lead to error of pose determination. The following measure of clutter is designed to reflect the potential of an extraneous feature to mislead the model matching. It is a non-monotonic with the geometrical displacement of an extraneous image feature from a corresponding ideal feature.

Consider an ideal feature F_i ($i = 1,..., n$) and clutter features O_{ij} ($j=1,... n_{ei}$) all of which are associated with F_i by being in its focus neighbourhood. We define the measure of clutter to be (b is explained further below)

$$k_c^{ij} = \begin{cases} (g - bg^5) & \text{if } g - bg^5 > 0 \\ 0 & \text{if } g - bg^5 \leq 0 \end{cases} \tag{6}$$

where g is defined as

$$g = \sqrt{(1.15d)^2 + \theta^2} \tag{7}$$

and combines the two geometrical displacements shown in Figure 5 (the factor 1.15 was introduced to ensure equivalent end points shift for numerically equivalent translation and rotation, as for a feature of typical length 50 pixels).

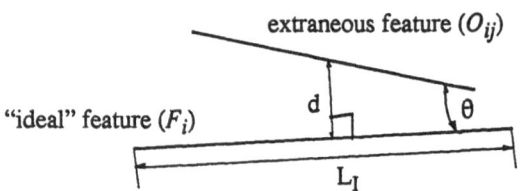

Figure 5   Measure of interference

The measure (6) is essentially an aggregate measure of the displacement subject to clipping. It falls sharply beyond the maximum, reflecting the fact that extraneous features outside the focus neighbourhood have little distracting potential. Parameter b allows its spatial extent to be tuned to suit the size of the focus neighbourhood used in the matching algorithm.

It is often the case that there are multiple extraneous features, all associated with a single ideal feature, in this case we take the maximum of their individual measures, as the measure of their collective potential for distraction

$$k_c^i = MAX\{k_c^{ij}\} \quad (\text{for } j=1,...., n_{ei}) \tag{8}$$

where n_{ei} is the number of extraneous features, each associated with the ith feature.

A measure of the total distraction caused by all extraneous features is defined as the RMS, as before

$$K_c = \sqrt{\frac{\sum\limits_{i}^{n} k_c^{\ i}}{n}}$$
(9)

where n is the total number of visible ideal features.

5.2   The simulation of extraneous features

Observed features are simulated by applying random truncation and noise contamination to instantiated model features (see Figure 6 (a)). Clutter features are obtained by simulating a truncated, noise-perturbed feature (as in section 4), and then further displacing this randomly (in a manner similar to Figure 5) with d and θ drawn from a flat bounded distribution. Note that in simulation experiements the total set of observed features comprise the perturbed ideal features plus the clutter features. The number of associated clutter features n_{ei} is randomly chosen between 1 and 5.

Figure 6 (b), (c) gives 2 examples of a set of truncated features with simulated clutter. The two images show examples of simulated clutter generated using different bounds on d and θ. In the two cases the measure of the clutter (equation 9) was computed using $b{=}0.047$ so that k_c maximises at $g{=}30.5$ pixels, (a typical size of the focus neighbourhood).

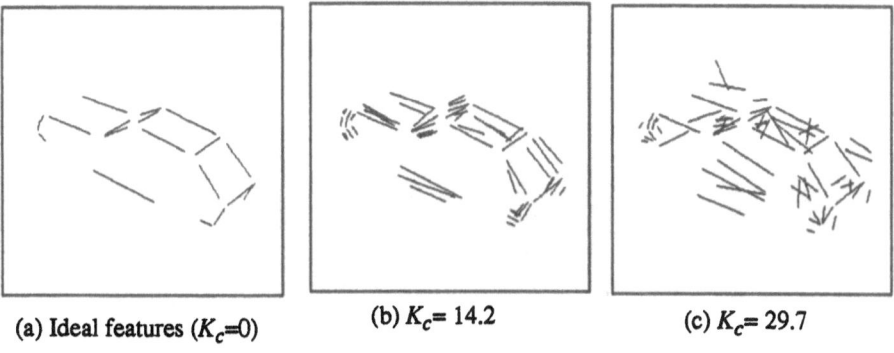

(a) Ideal features ($K_c{=}0$)   (b) $K_c{=}$ 14.2   (c) $K_c{=}$ 29.7

Figure 6   Examples of simulated clutter

6   Using the model of data complexity

This section describes two investigations of model-based vision, one for model-matching algorithms and the other for object discrimination, using the 3 measures of confounding factors.

6.1   Robustness of 2D-3D model matching strategies

We have recently proposed a model matching strategy which relies on viewpoint consistency ascent (VCA) [4]. This makes explicit and precise use of the viewpoint consistency constraint. In previous work it was shown to be superior to Lowe's incremental model-matching strategy [11] (which only weakly exploits viewpoint

576

consistency), for a few example cases. This superiority is explored here formally, using the data complexity model.

Using the truncation model and the clutter model, we conducted two Monte-Carlo experiments, each of 500 trials, to compare the sensitivity of both strategies to data complexity. Both experiments used a cube model to simulate the target object. A measure for the inaccuracy of resultant matches, D_f, is defined as the RMS of the geometrical differences between the matched features and the ideal features, as defined in equation (7).

In the first experiment, each trial used a fixed pattern of clutter and generated randomly perturbed ideal features. The measure of truncation K_t and D_f were then collected after each trial. Figure 7(a) shows the result of 500 trials and Fig. 7(b) shows histograms of inaccuracy.

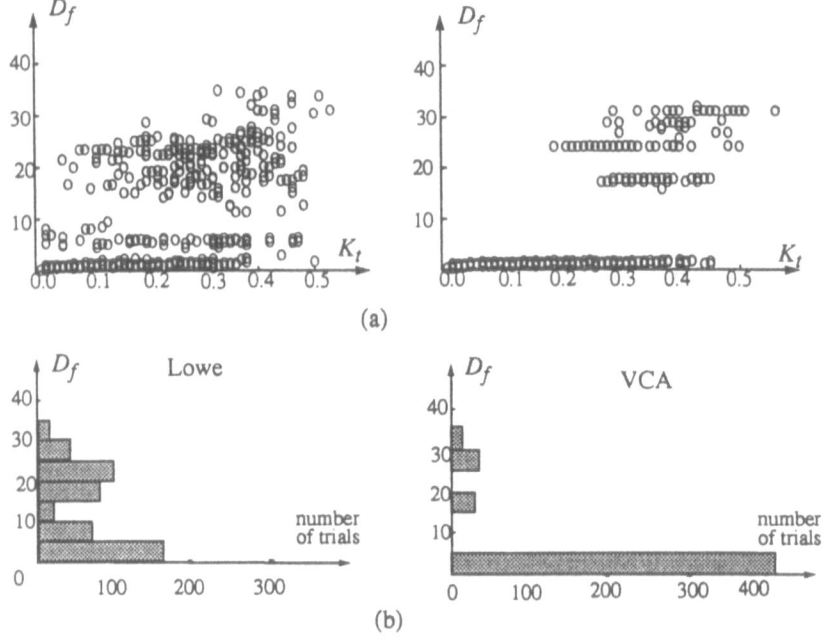

(a)

(b)

Figure 7   Model Matching Error as a function of the truncation measure

The second experiment fixed the pattern of perturbed ideal features, but allowed the measure of clutter K_c to vary and again recorded the output inaccuracy D_f. The result of two sets of 500 trials are plotted in Figure 8(a), and (b) shows the histograms.

This study reveals significant instability in Lowe's method even at very low levels of data complexity. The VCA shows a significant improvement of robustness in model-matching at all levels of data complexity examined. The study demonstrates systematically and quantitatively the benefit of a rigid enforcement of the viewpoint consistency constraint.

6.2   Object Discrimination

In model-based vision, after a set of feature matches is first established, it must be tested to decide whether the matches provide sufficient evidence for recognition or not. The

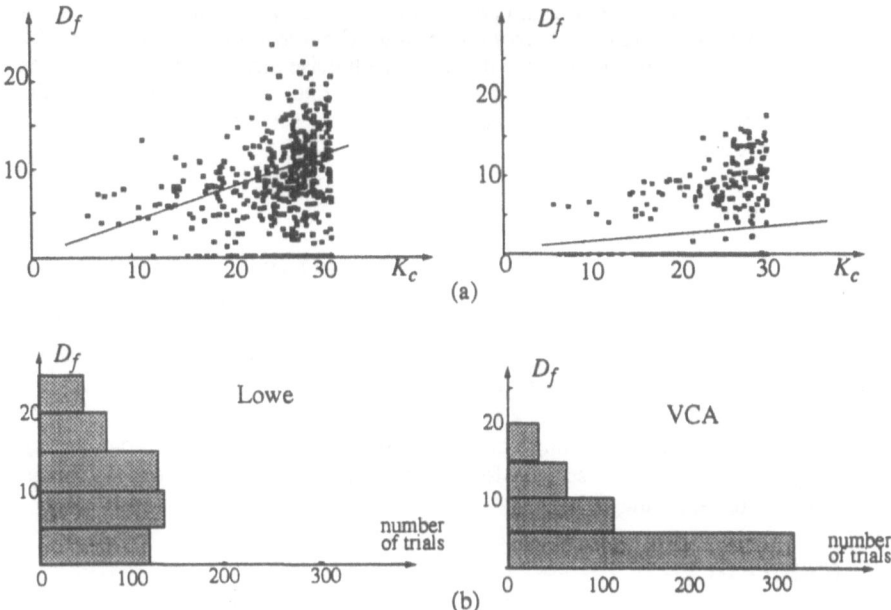

Figure 8   Matching error as a function of the clutter measure

problem is particularly acute if there exist other objects which must be discriminated from the target object.

We have investigated the object discrimination task by using the viewpoint consistency measure (VCE as defined in [4]) and adopting the using signal detection paradigm Here we use this example to show the usefulness of the data complexity model. A companion paper in this volume contains more detailed discussion [5].

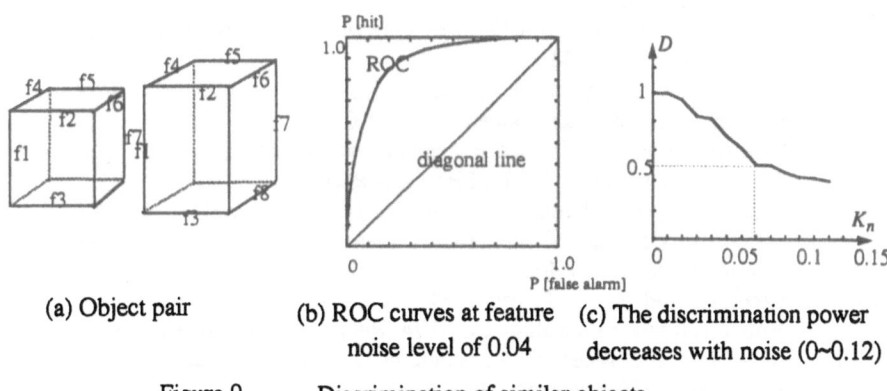

(a) Object pair   (b) ROC curves at feature   (c) The discrimination power
                        noise level of 0.04   decreases with noise (0~0.12)

Figure 9   Discrimination of similar objects

It turns out that feature noise is the main confounding factor for this task. The theory of Signal Detection [9] and the noise model provides a general method to relate the discrimination power of the VCE with the noise level for any given pair of similar objects.

Figure 9 demonstrates the discrimination of an example pair of objects, a cube and a cuboid, (a) shows the models, (b) shows a typical ROC curve at the noise level of 0.04 (c) shows how the discrimination power decreases with noise level. [NB: D is defined as twice the area between the ROC curve and the diagonal line]. The result shows that the viewpoint-consistency-based method is able to discriminate (D=0.5) a cube and a cuboid up to the noise level of 0.06.

7 Conclusion

A modelling scheme has been described for the data errors encountered by algorithms for 3D object recognition based on edge-based features. Use of this model has enabled us systematically to demonstrate that a method that enforce viewpoint consistency rigidly is superior to systems such as Lowe's [11], which enforces it weakly. The data complexity model also made it possible to consider the impact of data noise on object discrimination with VCC-based object recognition algorithms. The studies illustrate the benefits of adopting better methods for testing computer vision, by modelling the confounding factors explicitly. We put this scheme forward as a standard method for model-based vision using edge based image features.

8 References

[1]   Bodington, Sullivan & Baker, Experiments on the use of the ATMS to label features for object recognition, Computer Vision-ECCV'90, Spring-Verlag, 1990.

[2]   Bray, A., Recognising and Tracking Polyhedral Objects, Ph.D Thesis, Sussex University, UK, 1991.

[3]   Brooks, A. B., Model Based Computer Vision, UMI Research Press, 1984.

[4]   Du, L, G D Sullivan and K D Baker, 3D grouping by viewpoint consistency ascent, Image and Vision Computing, Special Issue on BMVC'91, 1992

[5]   Du, L, G D Sullivan and K D Baker, On evidence assessment for model-based vision, Proc. BMVC'92

[6]   Grimson, et al, The combinatorics of object recognition in cluttered environments using constrained search, ICCV'88

[7]   Grimson, et al, Model-based Recognition and Localization from Sparse Range or Tactile Data, Int. J. of Robotics Research.

[8]   Goad, C., Special Purpose Automatic Programming for 3D model-based vision, Proceeding of the ARPA image understanding Workshop, Arlington, Virginia, 1983.

[9]   Green, D and Swets, J., Signal Detection Theory and Psychophysics, Robert E Krieger Publishing Co. 1974 (Reprint of Wiley & Sons 1966)

[10]   Ikeuchi, K and Takeo, K Automatic Generation of Object recognition Programs, IEEE Proceeding, Aug. 1988

[11]   Lowe, D., The viewpoint consistency constraint, International Journal of Computer Vision, 1987

[12]   Lowe, D., Fitting Parameterised 3D Models to Images, IEEE PAMI, No 5, 1991

[13]   Roberts, L Machine Perception of Three-Dimensional Solids, Chapter 9, Optical and Electro-Optical Information Processing, MIT Press, 1965

[14]   Rydz, A., et al, Model based Vision Using a Planar Representation of the Viewsphere, Alvey Vision Conference'88, Manchester, 1988.

[15]   Sullivan, G D, Alvey MMI-007 Vehicle Exemplar: Performance and Limitations, Proc. Alvey Vision Conference'97, Cambridge, 1987, pp39-45

[16]   Worrall, A., et al, Model Based Tracking, BMVC'91, Glasgow, 1991

[17]   Zhang, S, Sullivan, G D, Baker, K D, Relational Model Construction and 3D Object Recognition from Single 2D Monochromatic Image, Proceeding of the British Machine Vision Conference'91, Glasgow, 1991, pp233-239

Practical Aspect-Graph Derivation Incorporating Feature Segmentation Performance

Andrew W. Fitzgibbon

Department of Artificial Intelligence, Edinburgh University
5 Forrest Hill, Edinburgh EH1 2QL

Robert B. Fisher

Department of Artificial Intelligence, Edinburgh University
5 Forrest Hill, Edinburgh EH1 2QL

Abstract

A procedure is described for the automatic derivation of aspect graphs of surface-based geometric models. The object models are made of finite, typed, second order surface patches – allowing the representation of a large number of complex curved objects while retaining ease of recognition. A new representation, the detectability sphere, is developed to encode feature detectability constraints. The detectability metric is directly related to the performance of the imaging system, allowing the generated aspect graph to more truthfully represent the scene's relationship with the vision system.

An algorithm is described which fuses information from several views of the object to produce a small number of characteristic views which cover some desired portion of the viewsphere, and annotates these fundamental views with pose-verification hints. The procedure is compared with previous analytic and approximate solutions to the aspect-graph problem regarding relevance to the vision process, range of applicability, and computational complexity.

1 Introduction

An important problem in model-based 3D computer vision systems is the complexity and number of interpretations possible within a given scene. Even with a limited number of models in the system's modelbase, the many distinct 2D appearances that a complex 3D object may have can give rise to many thousands of possibilities. To categorise these views and their interconnections, Koenderink and van Doorn [13] defined the **aspect graph**, linking 2D object views (nodes) and visual events (arcs). Given a model's aspect graph, the matching process is aided by the constraints placed on the initial generation of hypotheses (invocation) and the ability to perform detailed symbolic verifications once pose has been determined.

Since its introduction, much effort has been expended in analytically deriving exact aspect graphs [3, 14, 15], which yield a complete description of the object's viewsphere. Analytic methods, however, have been limited to simple models and may be expensive if the model contains many features. Gigus and Malik [10] describe an algorithm for nonconvex polyhedra which has a worst case complexity of $O(n^8)$ in the number of vertices. Ponce's algorithm [15] applies to a general class of object but no implementation is reported, and no complexity measure is given.

Moreover, the exact aspect graph is generally too detailed to be useful in model matching with real data (Gigus and Malik suggest up to $O(n^6)$ distinct views). Koenderink and van Doorn recognised that many nodes in the aspect graph will correspond to "unstable" views where an infinitesimal camera motion will change the topological properties of the view, and exact aspect graphs are usually generated without such views, or they are pruned on completion of the graph. In a real application, not only these views, but also many where a non-infinitesimal but suitably small camera movement changes the aspect must be pruned. Ben-Arie [2] introduces the "probability sphere" where the individual and joint probabilities of feature visibility are represented as the areas of regions on the gaussian sphere, and this to some extent considers this situation.

However, the opposite may also happen: poor sensor performance may cause a view which is theoretically unstable to actually cover a significant portion of the viewsphere. For example, the face-on view of a cube under orthographic projection will be classified as unstable by an exact algorithm, whereas with certain range sensors, it will often be the only face visible for up to 15 degrees in any direction.

In summary, exact aspect graphs do not provide the information required for real scene analysis applications.

This report describes a practical approach which combines the geometric object models with empirically-derived sensor models to provide a probability of successful detection (**detectability**) measure for each model component. The significance of this approach is that thereby a viewsphere is derived containing precisely the features that will be viewable when the object is observed. By incorporating the sensor model we can generate fewer, more significant views which are directly usable in the matching process. Because sensor performance is described using experimentally measured performance graphs and rules, the technique is applicable to a very general class of sensors, and can encode not only the physical sensor, but also the effects of the image processing stages (smoothing, segmentation) that generally precede the symbolic matching.

2 Context

This work was done in the context of the IMAGINE II 3D vision system, described in [7]. This system uses a three-stage approach to the model-matching task:

Invocation pairs scene features with model subparts in an $m \times d$ network [6]. Hypotheses generated at this stage consist of pairings between model visibility groups and scene perceptual groupings. Using coarse visibility groupings at this stage allows constraints such as "Surfaces A and B will never be seen together" to be cheaply encoded, simply because they never appear in the same visibility set. A priori knowledge such as "The object is sitting on its bottom" and hints like "B might be visible but don't depend on it" can also be included, improving the accuracy and speed of recognition.

Interpretation tree pruning applies geometric constraints similar to those described in [11] to sets of hypotheses, further reducing the number of hypotheses. As this is a combinatorially explosive process, visibility information is used to reduce the numbers of features considered.

Geometric reasoning creates a position estimate for each assembly hypothesis, using a Kalman filter-based stochastic fitting process to minimise the effects of noise in the data [4, 1].

2.1 Object Models

The models considered for analysis are assemblies of non-infinite 2nd order surfaces, described using the Suggestive Modelling System ([5], [9]). An **assembly** is a set of surfaces and reference-frame transformations:

$$\mathcal{A} = \left\{ (\mathcal{S}_i, {}^A T_{\mathcal{S}_i}) \right\}_{i=1}^m$$

In this notation, the transformation ${}^A T_{\mathcal{S}}$ transforms points in the surface's reference frame into the assembly's reference frame. The surface primitive $\mathcal{S}$ is on one side of the infinite quadric surface parameterized by a function $s(u, v)$. The surface's finite extent is represented by a **parameter-space mask** — a subset of (u, v) space. Because the model stores the transformation from surface to assembly coordinates, the surfaces are represented in canonical positions. In practice we are additionally limited to those surfaces which can be reliably extracted from the sensor data. Currently this means restriction to Plane, Cylinder, Ellipsoid, Cone, and Elliptical and Hyperbolic Paraboloids.

3 Overview of the Method

An outline of the procedure for generating the visibility information for the matcher is as follows. Several sample viewpoints are chosen by tessellating the gaussian sphere into tesserae (typically around 800), the choice of tessellation being designed to maximize the coverage of the viewsphere while maintaining tessera connectivity and solid angle. Each tessera defines a camera position, from which the geometric model is raycast, generating visibility statistics for each subcomponent in each view. The amount of information which must be included in the visibility statistics is determined by the form of the detectability rules used in the next stage.

The visibility statistics are then analyzed on a per-feature basis, using empirically derived detectability rules to generate **detectability spheres** for each subcomponent, which encode the likelihood that the subcomponent will be correctly sensed and segmented from each sampled viewpoint.

Combining the per-feature detectability spheres and applying an overall desired reliability criterion yields a set of "reliably detectable" surfaces at each viewpoint. Merging views from which the same subcomponents are so classified divides the viewsphere into regions, labelled by the list of features which are detectable from within the region.

Collecting the viewsphere regions gives a list of visibility groups which is pruned based on stability and likelihood criteria to give a smaller list of "viewgroups" which is added to the SMS model.

3.1 Choice of Tessellation

We now consider the problem of choosing viewpoints from which to take sample views. Because the sensor used (a laser range finder) produces orthographic images, only the direction of the viewing vector need be considered, reducing the problem to finding an appropriate tessellation of the gaussian sphere. The naive approach, dividing the normal elevation (ϕ) and azimuth (θ) evenly, produces a rectangular array of normals which is easy to deal with in a computer. This, however, produces a non-uniform tessellation, which means that near to the poles of the sphere, each direction represents a smaller proportion of azimuth than an equivalent at the equator.

A number of techniques to avert this effect have been suggested in the literature. The most commonly seen solution is to take a solid such as an icosahedron and recursively subdivide its faces until a desired resolution has been achieved [12]. This approach, while pleasing, still fails to generate a perfectly uniform tessellation. Alternative approaches include various random techniques which try to ensure that the solid angle subtended by each tessera remains constant.

The solution adopted in this work takes a similar, but simpler, view. First, what are the difficulties we will have with a naive regular tessellation? The connectivity of regions will still give us the same characteristic views, but they will be 'larger' near the poles (occupying more tesserae), and work will have been wasted in generating views at more than the required resolution. We can correct for both the size (in solid angle) and the amount of wasted work simply by reducing by a factor of $\sin(\phi)$ the number of subdivisions of azimuth at each elevation ϕ.

3.2 Ray Casting the Finite Surfaces

To determine the visibility of objects in a single view, the depth of each surface at each point in the scene is required. These depths are calculated using a technique similar to graphical ray tracing using an image plane divided into $X \times Y$ pixels. The line of sight passing through each pixel is intersected with each subcomponent of the model, producing a list of intersection reports of the form (i, z), saying that surface S_i intersected at depth z. The subcomponent which reports an intersection at the nearest point along the line of sight will be visible (the **front surface**), all others are occluded by the front surface.

3.3 Determining Detection Reliability for One Surface

Analysis of the information produced by the raycasting algorithm uses the concept of a detectability sphere to represent feature visibility over the range of possible camera positions. The detectability sphere for a single surface S_i is the distribution

$$P\{S_i\}(\hat{\mathbf{v}})$$

which defines, for each viewing direction, the probability that S_i will be correctly sensed, segmented and classified. This probability is on a binary event: "Will the symbolic data description produced by the early vision processing be good enough to allow us to modelmatch?" To answer this question, we first note that successful object regognition is a function of several factors:

Occlusion alters the size of a surface and its 2D appearance. Three dimensional shape is not affected but significant occlusion can hamper the segmentation process, impacting the reliability of shape parameter estimation. Cylinder radii, for example, are estimated poorly as less of the cylinder is seen.

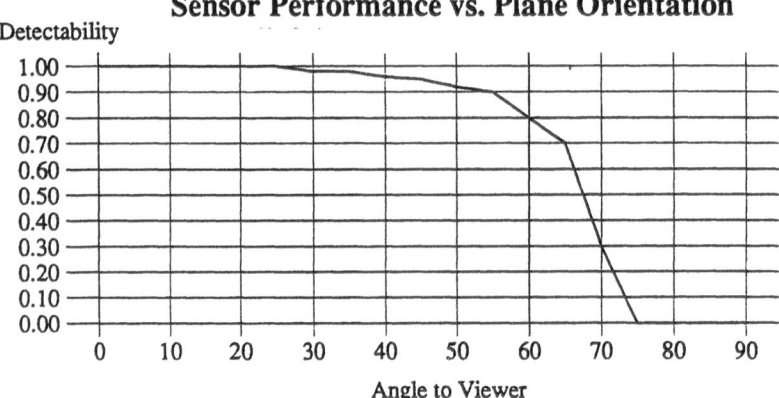

Figure 1: Graph relating detectability to the angle between the line of sight and a planar surface normal. The material used was anodized aluminium, the data acquired using a stereo laser striper in a $1m^3$ workspace. Note the sudden catastrophic failure at about $65°$.

The **quality of sensor data** depends on surface finish, shape and position with respect to the sensor. Figure 1 illustrates the variation of planar surface detectability over a range of orientations. For cylindrical patches a similar dependency is seen, but on the orientation of the cylinder axis.

The choice of **algorithm parameters** will affect system reliability at all stages of processing. For example, the HK curvature-based range data segmentation algorithm described in Trucco [16] has performance curves relating cylinder radius R and the H_0 and N_c segmentation parameters to the number of region pixels correctly classified. Hence, given R, H_0 and N_c for a particular cylinder, we can predict the percentage of image pixels that will be correctly classified, and thus the likelihood that the surface patch itself will be detected. At a higher level, if a simple search-tree pruning constraint uses a χ^2 test to compare model and data areas, say, then the choice of confidence interval is a parameter which will alter the detectability of surfaces whose area measurements are deviate significantly from the true values.

In addition, some general criteria can be applied — very small patches (as measured by raw pixel area) will not segment, and patches with a high compactness number ($perimeter^2/area$, again measured in raw pixel values) will also prove difficult. These criteria apply to all surface types and play an important role in removing from consideration small, rarely detectable subparts which will not aid recognition.

3.4 Detectability Rules

To represent these constraints in our algorithm, "detectability rules" are encoded for each surface type which relate the surface's pose, the parameters of the segmentation algorithm, and the segmentation performance data to give a probability that the feature will be correctly classified. The rules are represented as functions of the

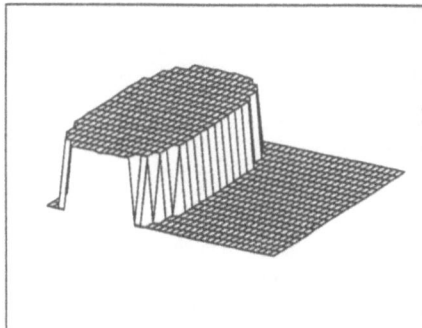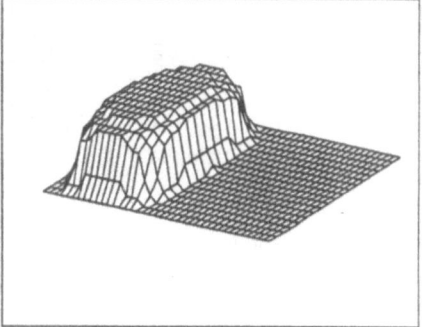

Figure 2: Detectability sphere (left) for a plane based on ratio of visible area to model area. The detectability sphere has been flattened onto a rectangular grid, indexed by camera azimuth and elevation, with height above the plane representing likelihood of correct segmentation. Azimuth varies between $-\pi$ and π. Elevation is between 0 and π. The figure on the right is the detectability sphere based on probability of correct classification.

single-view, per-surface, visibility statistics:

• The total number of rays which intersected the forward-facing portion of the finite surface.

• The number of rays along which the surface was the nearest to the camera. This defines the actual surface visibility, taking account of occlusion.

• The visible surface area. This is calculated at each image pixel, by correcting the pixel area for foreshortening by the dot product of the surface normal and line of sight.

3.4.1 Representing Self-Occlusion for Matching

The per-surface data indicates how much a particular surface has been occluded by all other surfaces in the view. For pose verification, however, it is useful to know in greater detail which surfaces have occluded others. This information is supplied in an **occlusion matrix** (Table 1), where the entry at cell (i, j) is the number of rays along which surface S_i was the front surface, and occluded a forward-facing portion of surface S_j.

Occlusion information is useful in a quick model verification process once pose estimation is complete [7, 5]. The verification process can use the pose to index information about feature visibility (nearly fully visible, partially obscured, tangential) and surface ordering ("Surface A partially obscures surface B and hence expect a depth discontinuity boundary between the two.").

4 Merging Information from Multiple Views

The process of deriving the characteristic views may be separated into two sub-processes. First, each model surface S is taken individually and its detectability sphere, $P\{S\}(\hat{v})$ is generated. The individual surface probabilities are then combined to produce group probabilities, giving codetectability likelihoods for groups

		Front Surface				
		1	2	3	4	...
Occluded Surface	1	–	7			
	2	9	–	300		
	3			–		
	4				–	
	⋮					

Table 1: Example occlusion matrix for a single view. The 300 figure is the number of rays cast where both surface 3 and surface 2 were intersected, but surface 3 occluded surface 2. The 7 and 9 illustrate quantisation noise caused by surfaces 1 and 2 sharing a boundary. Also, the diagonal is empty as a second-order surface cannot self-occlude (Consider the gradients in the case where swo intersections are reported.)

of subcomponents.

The second stage is to process the plausible groupings in order to filter out pathological or unstable views, before augmenting the SMS model with the new visibility information.

4.1 Joint Detectability for Each View

Applying the detection rules to each model surface in a particular view produces a set of detection probabilities $P\{S_i\}$. These are combined to obtain the likelihood of codetectability for a set of surfaces $\{S_{i_j}\}_{j=1}^n$. In general, we want to find the largest set whose joint probability exceeds some reliability threshold. We would expect, then, to have to calculate joint probabilities for all sets of surfaces in the view. However, because the probability that a surface will segment correctly depends only on the viewing direction, and not on the detectability of the other surfaces, the joint probability calculation simply amounts to taking the minimum of the individual probabilities[1]:

$$P\{S_1, ..., S_n\} = \min_{i=1}^{n} P\{S_i\}$$

This in turn means that the joint probability of a group of surfaces will exceed the reliability threshold iff each individual probability does so. Thus, the surfaces can be individually thresholded and the successful surfaces grouped into a "reliably detectable" set (called a **viewgroup** in SMS). A connected-components algorithm then divides the viewsphere into regions from which groups of subcomponents are reliably detectable.

4.2 Pruning Insufficiently Stable Viewgroups

The list of viewgroups produced by the reliability thresholding will still contain undesirable views which are too unstable or contain too few surfaces. Unstable views are identified as those which are represented by very "thin" regions on the flattened viewsphere. This corresponds to a discretisation of the usual concept of viewpoint instability; essentially we are now defining an unstable group as one

---

[1]This is similar to Ben-Arie's [2] use of set intersection on the gaussian sphere in the exact case.

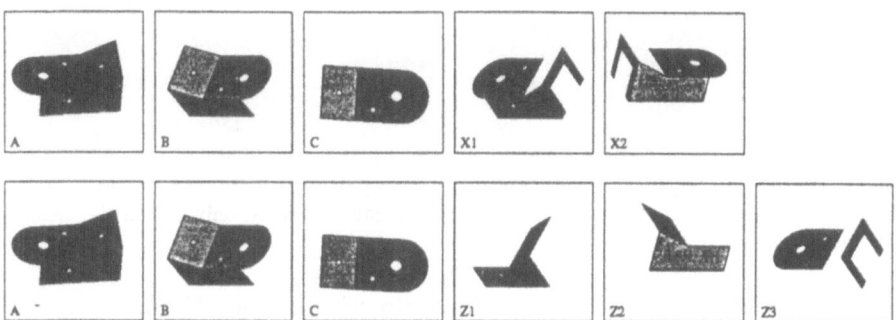

Figure 3: The B.Ae. Widget. This part is taken as an example of a moderately complex object for industrial inspection tasks. The upper sequence shows the original hand-selected viewgroups, with only the visible subcomponents drawn. The lower sequence shows the groups selected by the algorithm. Also, only the views from above are shown, as the widget is assumed to be constrained in this case to lie on its base.

from which a small, but no longer infinitesimal, camera movement will change the object's aspect. Thin regions are pruned by finding their bounding box at several orientations. Regions whose minimum width over all orientations is less than $\sqrt{2}$ are regarded as pathological views and are rejected.

After this pruning, the viewgroups are converted to SMS format and added to the supplied geometric model.

5 Results

When evaluating the system, our first consideration is to see how well its predicted viewgroups agree with the views previously chosen by hand. The part considered is shown in Figure 3. The views shown are the views expected when the part is observed from above, its base being flat on a workbench.

Comparing the automatic and hand generated versions, views A, B and C correspond, but views D_1 and D_2 in the hand-edited model have been split into three by the program. This is explained by the diagnostic that D_1 and D_2, when evaluated by the program, proved to occupy a very small portion of the viewsphere. Instead, the three groups Z_1, Z_2 and Z_3 have been created, explaining significantly more of the sampled views. These three views are all the possible two-surface subgroups of the erroneous three-surface groups.

The algorithm was also run on the partially symmetric object shown in Figure 4, again producing a sensible list of views. However, because the object is symmetric, the program has produced two copies of each group, where two model surfaces have the same shape, but are separately described.

6 Conclusions

The presented algorithm shows a number of advantages over exact techniques. Using second order surface patches as the basic primitive allows the incorporation of many

Figure 4: Renault part. Due to the object's rotational symmetry, only half of the generated views are shown.

more types of models, and leads to a viewsphere with a reasonably small number of important views. In contrast, exact edge based systems produce tremendously large viewspheres, which are difficult to index, and which may consist of many pathologically unlikely views. For example the B.Ae. part, with 45 modelled edges and 22 surfaces, would be expected to generate 8×10^9 views, as opposed to the 13 generated by our algorithm.

Using actual performance rules empirically derived from the segmentation software allows the viewgroups to more accurately represent what will be detected from the scene. For example, under orthographic projection, the top view (view 'C') in figure 3 should be deemed unstable, as a small camera movement will reveal the other surfaces. The segmentation performance, however, is such that the top view covers quite a large section of the viewsphere before the sides will pass the reliability threshold.

The ability to decide the threshold on detectability gives an easy tradeoff between reliability and speed in the model matcher – setting a high threshold means more distinct viewgroups and longer matching times. The modelmatcher with which the system is currently used is now being placed under knowledge-based control, which means that this is a useful property.

7 Further Work

The described algorithm works well for 'flat' (single-assembly) models with a small number of surfaces. When the number of surfaces becomes large, or when articulated parts are considered, the viewsphere quickly becomes very complex. This complexity is significantly reduced by using hierarchical object models. Fisher [8] shows that for a sample articulated part, the number of viewgroups is reduced to $O(n)$ in the number of surfaces by converting a flat assembly of surfaces to a hierarchical assembly of assemblies. The program could be extended to automatically deduce subcomponent hierarchies which would reduce the complexity of the viewsphere.

Symmetric objects cause problems in that too many viewgroups are produced when the object has separately modelled, but similar, subcomponents. Currently, matcher performance is improved by hand-editing the model file to remove 'duplicate' groups. The program could possibly detect such symmetries and remove them automatically.

References

[1] N. Ayache and O.D. Faugeras. Maintaining representations of the environment of a mobile robot. In *Robotics Research 4*, pages 337 – 350. MIT Press, USA,

1988.

[2] J. Ben-Arie. Probabilistic models of observed features and aspects with applications to weighted aspect graphs. *Pattern Recognition Letters*, 11(6):421 – 427, June 1990.

[3] D. Eggert and K. Bowyer. Perspective projection aspect graphs of solids of revolution: an implementation. In *Proceedings, 7th Scandinavian Conference on Image Analysis*, pages 299 – 306, 1991.

[4] R.B. Fisher and M.J.L. Orr. Geometric reasoning in a parallel network. *International Journal of Robotics Research*, 10(2):103–122, 1991.

[5] Robert B. Fisher. SMS: A suggestive modelling system for computer vision. *Image and Vision Computing*, 5(2):98–104, May 1986.

[6] Robert B. Fisher. Model invocation for three dimensional scene understanding. In *Proceedings, 10th International Joint Conference on A.I.*, pages 805–807, 1987.

[7] Robert B. Fisher. *From Surfaces to Objects: Computer Vision and Three Dimensional Scene Analysis*. John Wiley, UK, 1989.

[8] Robert B. Fisher. Reducing viewsphere complexity. In *Proceedings, European Conference on Artificial Intelligence*, pages 274–276, 1990.

[9] Andrew W. Fitzgibbon. *The Suggestive Modelling System: Reference Manual for Version 2*. University of Edinburgh, 1990.

[10] Z. Gigus and J. Malik. Computing the aspect graph for line drawings of polyhedral objects. *IEEE T-PAMI*, 12:113 – 122, 1990.

[11] W. Eric L. Grimson and Tomas Lozano Pérez. Model-based recognition and localization from sparse range or tactile data. *Image and Vision Computing*, 3(3), 1984.

[12] K. Ikeuchi. Recognition of 3d objects using the extended gaussian image. In *Proceedings 7th IJCAI*, pages 595–600, 1981.

[13] J.J. Koenderink and A.H. van Doorn. The internal representation of solid shape with reference to vision. *Biological Cybernetics*, 32:211 – 216, 1979.

[14] D.J. Kriegman and J. Ponce. Computing exact aspect graphs for curved objects: solids of revolution. *International Journal of Computer Vision*, 5(2):119 – 135, November 1990.

[15] J. Ponce and D.J. Kriegman. Computing exact aspect graphs for curved objects: parametric patches. In *Proceedings, 8th National Conference on Artificial Intelligence (AAAI)*, pages 1074 – 1079, 1990.

[16] E. Trucco and R. B. Fisher. Shape- and discontinuity-preserving segmentation of range images using curvature sign estimates. In *Proceedings, IEEE International Conference on Image Processing*, 1992.

Recognising Polyhedral Objects from a Single Perspective View

K. C. Wong and J. Kittler

Dept. of Electronic and Electrical Engineering,

University of Surrey,

Guildford, Surrey GU2 5XH, United Kingdom

Abstract

This paper considers the problem of recognising 3D polyhedral objects from a single perspective image. A hypothesis-verification paradigm based on a local shape representation is presented. In the framework, 2D vertices interpreted as the projection of a trihedral vertex which is a 3D spatial vertex with three line emanating from the tip are employed as seed features for model invocation and hypothesis generation. To simplify the perspective analysis, Kanatani [7] has proposed an intuitive and elegant technique. Using the technique, we derive a fourth-degree polynomial for interpreting a trihedral vertex. The contribution of our solution is that there are no restrictions on angles between the vertex edges. To reduce the number of hypotheses generated from scene-model vertex assignments, and recover the complete object pose, we propose a composite feature, namely vertex-CS feature by combining a trihedral vertex and a V-junction which share a common edge. The geometric constraint of this composite feature is derived. A matching strategy used in the recognition system is discussed. The feasibility of the proposed method is illustrated on real data.

1   Introduction

In general, there are several distinct phases in model-based matching of rigid objects. Two off-line stages are model generation and model analysis. The former is required for constructing a CAD-like database of models. The latter is exploited to identify and organise model features into structures for matching and for developing strategies for the execution of the matching task. The two main run-time stages are hypothesis generation and verification. The first of these involves extracting interesting 2D geometric features from an image and then generating possible poses of scene objects using the geometric cues of the plausible model-scene correspondences. The subsequent object verification process is thus provided with tight constraints on where to search for confirmatory evidence of model existence. The model verification process performs a detailed check of the description of the projection of 3D features and 2D image data, confirming the feature presence and accounting for features which are not observed. Most of the existing recognition systems which use the above approach to accomplish the image interpretation task, rely on cues derived from the geometric relationships between model-scene correspondences [8], [3], [2].

In this paper, we describe a model-based polyhedral object recognition system for identifying the scene-model correspondences and estimating the poses of the scene object from a single perspective image. A hypothesis-verification paradigm based on local shape properties is presented. In the framework, trihedral vertices and their composite with V-junctions are employed as key features for model invocation and hypothesis generation. There are several reasons for choosing this feature : the number of vertices extracted from a scene is generally manageable; they are robust in the presence of moderate noise; they are qualitative invariants over a wide range of view points [5]; they can constraint the transformation between the model and camera frames. Although our approach is inspired by Kanatani [7] the proposed method advances the state of the art in at least three important respects. Firstly, we demonstrate that the pose of a scene object can be recovered using a very intuitive formulation, by analytically solving a quartic equation derived from the geometric constraint of a model-scene vertex pair. Moreover, our method is not restricted to objects, with 3D vertices involving at least two right angles. Secondly, we have derived the geometric relationship of a composite feature formed by a vertex and a V-junction to reduced the search space and recover the translation of the scene object accurately. Thirdly, the process of estimating the pose of a scene object is broken down into two stages whereby in the first stage no quantitative information is required about edge length. Many false hypotheses can be pruned out at this first stage without the need for full edge visibility. This greatly simplifies the problem of verification.

Many researchers have attempted to solve the problem of determining the pose of a spatial 3D vertex from its orthographic [5], [6] or perspective [1], [4], [7] projection. Among these approaches, we find that the formulation and framework proposed by Kanatani [7] are most elegant and intuitive. However, the analytical solution derived by him can only deal with a corner with a minimum of two right angles out of three.

2 Solving vertex edge orientations

To determine the relative pose of a scene object with respect a camera frame, we first concentrate on solving the edge orientations of a spatial vertex measured in the camera frame. The formulation and analysis of 3D geometric primitive features being

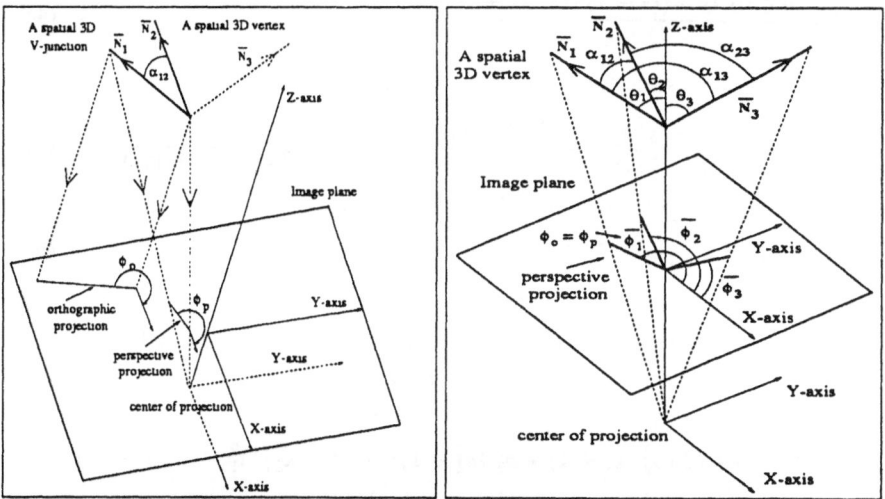

Figure 1: (a) A scene vertex being viewed in general position. (b) A configuration of a vertex transformed to a canonical position

viewed under perspective projection in general position is very complicated. Hence, the geometric meaning of the equations derived from these formulations is in general implicit and non-intuitive. To resolve the perspective geometry problem, Kanatani [7] proposed a technique to move scene features to be analysed into an image origin, namely the canonical position where the analysis of the scene feature can be greatly simplified. In particular, the difficulty of understanding the perspective projection of a spatial V-junction can be reduced significantly as the perspective effect is reduced to orthographic in the canonical position. For example in Fig. 1(a), the image angles of the edges of a spatial V-junction observed under perspective (ϕ_p) and orthographic (ϕ_o) projection are different in general position. However, the distinction between these two disappears ($\phi_p = \phi_o$) in the canonical position where the optical axis of the camera intersect the tip of the V-junction (see Fig. 1(b)). Therefore, the orientation of the edge $\bar{N}_1 = \Re^t N_1$ of a V-junction can be simply expressed as $\bar{N}_1 = \sin\theta_1 \cos\bar{\phi}_1 \, \tilde{i} + \sin\theta_1 \sin\bar{\phi}_1 \, \tilde{j} + \cos\theta_1 \tilde{k}$, where θ_1 is the angle between the edge $\bar{N}_1$ and an optical axis of a camera and $\bar{\phi}_1$ is an orientation of the edge under *perspective or orthographic projection*. $\Re$ is a *standard transformation* which maps an image point to the origin $(0,0)$ (see [7] for more details).

Having expressed the edge orientations of a V-junction, the 3D true angle α_{12} between the edges $\bar{N}_1$ and $\bar{N}_2$ shown in Fig. 1(b) can easily be written as $\bar{N}_1 \cdot \bar{N}_2 = \cos\alpha_{12}$. Similarly, a system of trigonometric equations can be derived in the same manner for a spatial vertex shown in Fig. 1(b). Replacing the $\sin\theta_i$ and $\cos\theta_i$ in the resultant equations with $\frac{2\,t_i}{1+t_i^2}$ and $\frac{1-t_i^2}{1+t_i^2}$ respectively, where $t_i = \tan\frac{\theta_i}{2}$, for $(i = 1,2,3)$, we find

$$(k_1 - 1)\, t_1^2\, t_2^2 + (k_1 + 1)\, (t_1^2 + t_2^2) - 4\, q_1\, t_1\, t_2 + (k_1 - 1) = 0 \tag{1}$$

$$(k_2 - 1)\, t_2^2\, t_3^2 + (k_2 + 1)\, (t_2^2 + t_3^2) - 4\, q_2\, t_2\, t_3 + (k_2 - 1) = 0 \tag{2}$$

$$(k_3 - 1)\, t_1^2\, t_3^2 + (k_3 + 1)\, (t_1^2 + t_3^2) - 4\, q_3\, t_1\, t_3 + (k_3 - 1) = 0 \tag{3}$$

where $k_i = \cos\alpha_{i,(i \bmod 3)+1}$ and $q_i = \cos(\phi_i - \phi_{(i \bmod 3)+1})$. From Eq.(3), we find that

$$t_3^2 = \frac{4\, q_3\, t_1\, t_3 - (k_3 + 1)\, t_1^2 - k_3 + 1}{(k_3 - 1)\, t_1^2 + k_3 + 1} \tag{4}$$

Substituting this expression for t_3^2 into Eq.(2) gives,

$$t_3 = \frac{(k_2 - k_3)\, t_1^2\, t_2^2 + (k_3 + k_2)\, t_1^2 - (k_3 + k_2)\, t_2^2 + k_3 - k_2}{2\, (q_3\, ((k_2 - 1)\, t_2^2 + (k_2 + 1))\, t_1 + q_2\, ((1 - k_3)\, t_1^2 - (k_3 + 1))\, t_2)} \tag{5}$$

Substituting back the expression of t_3 into Eq.(3) yields,

$$\delta_4\, t_2^4 + \delta_3\, t_2^3 + \delta_2\, t_2^2 + \delta_1\, t_2 + \delta_0 = 0 \tag{6}$$

where,

$$\delta_4 = (k_2 - k_3)^2\, t_1^4 + 2\, ((k_3^2 - k_2^2) + 2\, q_3^2\, (k_2^2 - 1))\, t_1^2 + (k_2 + k_3)^2$$

$$\delta_3 = 8\, q_2\, q_3\, ((1 - k_2\, k_3)\, t_1^2 - (1 + k_2\, k_3))\, t_1$$

$$\delta_2 = 2\, (2\, q_2^2\, (k_3^2 - 1) + (k_2^2 - k_3^2))\, (t_1^4 + 1)$$
$$\qquad + 4\, (2\, (q_2^2\, (k_3^2 + 1) + q_3^2\, (k_2^2 + 1)) - (k_3^2 + k_2^2))\, t_1^2$$

$$\delta_1 = 8\, q_2\, q_3\, ((1 - k_2\, k_3) - (1 + k_2\, k_3)\, t_1^2)\, t_1$$

$$\delta_0 = (k_2 + k_3)^2\, t_1^4 + 2\, ((k_3^2 - k_2^2) + 2\, q_3^2\, (k_2^2 - 1))\, t_1^2 + (k_2 - k_3)^2$$

Equation (1) can be rewritten,

$$\rho_2 \, t_2^2 + \rho_1 \, t_2 + \rho_0 = 0 \tag{7}$$

where, $\rho_2 = (k_1 - 1) \, t_1^2 + (k_1 + 1)$; $\rho_1 = -4 \, q_1 \, t_1$; $\rho_0 = (k_1 + 1) \, t_1^2 + (k_1 - 1)$; and consequently from eq. (7), we get $t_2^2 = -\frac{\rho_1 \, t_2 + \rho_0}{\rho_2}$. Substituting this expression for t_2^2 into Eq.(6) yields,

$$t_2 = \frac{(\rho_1^2 - \rho_0 \, \rho_2) \, \rho_0 \, \delta_4 - \rho_0 \, \rho_1 \, \rho_2 \, \delta_3 + \rho_0 \, \rho_2^2 \, \delta_2 - \rho_2^3 \, \delta_0}{(2 \, \rho_0 \, \rho_2 - \rho_1^2) \, \rho_1 \, \delta_4 + (\rho_1^2 - \rho_0 \, \rho_2) \, \rho_2 \, \delta_3 - \rho_1 \, \rho_2^2 \, \delta_2 + \rho_2^3 \, \delta_1} \tag{8}$$

Substituting this expression for t_2 into Eq.(7) we obtain,

$$\begin{aligned}
&(((\rho_1^2 - 2 \, \rho_0 \, \rho_2)^2 - 2 \, \rho_0^2 \, \rho_2^2) \, \delta_0 + (3 \, \rho_0 \, \rho_2 - \rho_1^2) \, \rho_0 \, \rho_1 \, \delta_1 \\
&+ (\rho_1^2 - 2 \, \rho_0 \, \rho_2) \, \rho_0^2 \, \delta_2 - \rho_0^3 \, \rho_1 \, \delta_3) \, \delta_4 + ((3 \, \rho_0 \, \rho_2 - \rho_1^2) \, \rho_2 \, \rho_1 \, \delta_0 \\
&+ (\rho_1^2 - 2 \, \rho_0 \, \rho_2) \, \rho_2 \, \rho_0 \, \delta_1 - \rho_0^2 \, \rho_1 \, \rho_2 \, \delta_2) \, \delta_3 \\
&+ ((\rho_1^2 - 2 \, \rho_0 \, \rho_2) \, \rho_2^2 \, \delta_0 - \rho_0 \, \rho_1 \, \rho_2^2 \, \delta_1) \, \delta_2 - \rho_1 \, \rho_2^3 \, \delta_0 \, \delta_1 \\
&+ \rho_2^4 \, \delta_0^2 + \rho_0 \, \rho_2^3 \, \delta_1^2 + \rho_0^2 \, \rho_2^2 \, \delta_2^2 + \rho_0^3 \, \rho_2 \, \delta_3^2 + \rho_0^4 \, \delta_4^2 = 0
\end{aligned} \tag{9}$$

Substituting the known constants $\delta_0, \delta_1, \delta_2, \delta_3, \delta_4, \rho_0, \rho_1$ and ρ_2 into Eq.(9), we obtain

a polynomial equation of degree 16 with no odd terms in one unknown t_1. Replacing t_1^2 with the trigonometrical $\frac{1-\cos(\theta_1)}{1+\cos(\theta_1)}$ yields the following fourth-degree polynomial in one unknown $\cos^2(\theta_1)$:

$$A_4 \, \cos^8(\theta_1) + A_3 \, \cos^6(\theta_1) + A_2 \, \cos^4(\theta_1) + A_1 \, \cos^2(\theta_1) + A_0 = 0 \tag{10}$$

This quartic equation can be solved analytically or using iterative numerical method. t_2 and t_3 can be obtained by substituting t_1 into Eq.(1) and Eq.(3) respectively. All the solutions of t_1, t_2 and t_3 are then verified using the Eq.(2). The angles between the trihedral vertex edges can be easily determined from the *t-formula*, $\theta_i = 2 \, \tan^{-1}(t_i)$, where $0 < \theta_i < \pi$. Many hypotheses can be pruned away during the recovery of the edge orientations of a vertex leaving very few hypotheses to be verified. Having determined the vertex edge orientations, we will describe the pose determination problem in next section.

3   Pose Estimation

Formally, the problem of pose estimation may be defined as follows : *Given a set of N three dimensional vectors with respect to an inherent object model coordinate system, and the 2D perspective projection of the corresponding N vectors of a scene object with respect to a camera frame, estimate the relative rotation R_{MC} and the translation vector T_{MC} to define the relationship between the object model and camera frame.* To solve this problem, we first concentrate on solving the relative rotation transform consisting of three degrees of freedom. The translation vector can easily be recovered once the former is determined. Consider the perspective geometry of a camera model depicted in Fig. 2(a), the image plane is assumed to be in front of the center of projection so as to acquire an upright scene image. The focal length, foc is the normal distance from the center of projection to the image plane. Based on the above configuration, the position of the scene vertex P_s can be expressed in a camera frame centered at the origin E as $P_s = R_{MC} \, P_w + T_{MC}$, where R_{MC} is the relative orientation between the model and camera frame and T_{MC} is a translation vector. In the next subsections, we will describe the methods for computing these parameters.

3.1 Relative Rotation

To determine the relative rotation, we decompose the rotation transform into model-to-vertex R_{MV} and camera-to-vertex R_{CV} transforms. Consider the edges of a trihedral vertex E_i, $i = 1, 2, 3$, described with respect to an object model coordinate system. An orthogonal vertex-based frame can be constructed by using the Gram-Schmit orthogonalization process, $M'_i = E_i - \sum_{j=1}^{i-1} \frac{E_i \cdot M'_j}{||M'_j||^2} M'_j$, and then normalised to obtain unit vectors $M_i = \frac{M'_i}{||M'_i||}$. The transformation R_{MV} of vertices P_w with respect to an object model coordinate system to vertices P_v with respect to a vertex-based coordinate system (see Fig. 2(a)) can be expressed as $P_v = R_{MV} P_w$ where $R_{MV} = \begin{pmatrix} m_{1x} & m_{1y} & m_{1z} \\ m_{2x} & m_{2y} & m_{2z} \\ m_{3x} & m_{3y} & m_{3z} \end{pmatrix}$. Having determined the edge orientations of a vertex in the canonical position, the orientations of the edges in the scene can be recovered by $N_i = \Re \ \tilde{N}_i$, $i = 1, 2, 3$, which are described with respect to the camera coordinate system. An orthogonal vertex-based frame is constructed using the corresponding recovered edges by taking $C'_i = N_i - \sum_{j=1}^{i-1} \frac{N_i \cdot C'_j}{||C'_j||^2} C'_j$, and then normalised to obtain unit vectors $C_i = \frac{C'_i}{||C'_i||}$. The transformation R_{CV} of vertices P'_s with respect to a camera coordinate system, which is centered at the origin of the world coordinate system O, to vertices P_v with respect to a vertex-based coordinate system can be expressed as $P_v = R_{CV} P'_s$ where $R_{CV} = \begin{pmatrix} c_{1x} & c_{1y} & c_{1z} \\ c_{2x} & c_{2y} & c_{2z} \\ c_{3x} & c_{3y} & c_{3z} \end{pmatrix}$. Having determined R_{MV} and R_{CV}, the rotation transform which maps an object model P_w to the scene feature point P'_s with respect to the camera coordinate system, which is centered at the origin O of the model frame can be written as $P'_s = R_{MC} P_w$ where $R_{MC} = R^t_{CV} \times R_{MV}$.

3.2 Translation

To determine the translation from an object model to a scene, one of the three line segments of an image vertex must be the projection of the full edge of a spatial 3D vertex. We will defer a detailed description of this issue to Section 4. Here we shall assume that one of the line segments of length l_i is the true projection of an edge of length L_i of a spatial vertex in canonical position. The orientation of the edge orientation θ_i can be evaluated from the solution derived in Section 2. The positional vector D_i of the tip of the vertex in canonical position can be easily expressed as $D_i = 0 \ \tilde{i} + 0 \ \tilde{j} + L_i \ (\frac{foc}{l_i} \sin \theta_i - \cos \theta_i) \tilde{k}$, where foc is the focal length of the camera. Let the positional vector of the tip of the corresponding model vertex be D_m. The translation from the model to the scene can then be computed using, $T_{MC} = \Re \ D_i - R_{MC} \ D_m$ Having determined the complete pose of the scene object, the results can be employed for predicting the description of the the 2D scene features in the verification phase of the recognition system. In the next section, we will introduce a feature primitive which can be reliably used for computing the translation vector. Furthermore, the geometric constraint of the composite feature will be derived.

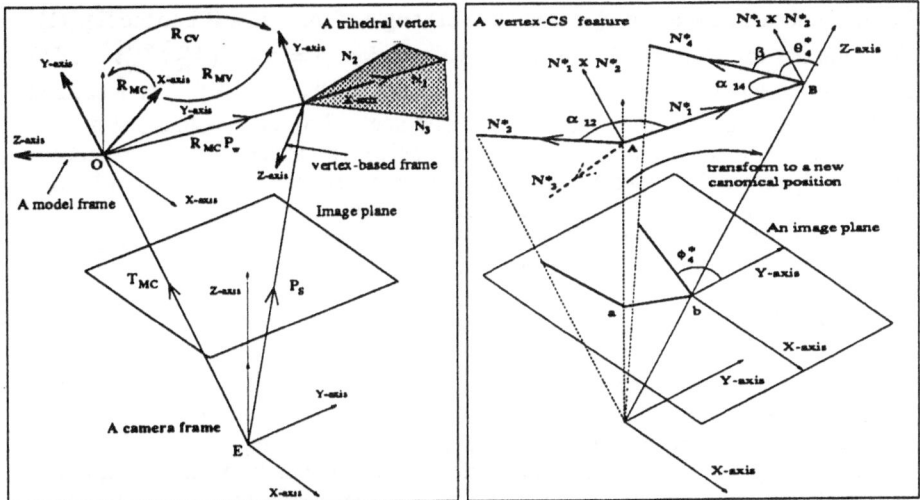

Figure 2: (a) A simplified pin-hole camera. (b) A composite vertex-CS feature

4 Composite Vertex-CS feature

To determine the translation from an object model to a scene, one of the three line segments of an image vertex must be the projection of the full edge of a spatial 3D vertex. In order to select at least one plausible line segment of a vertex extracted from a real image, we introduce a primitive, namely vertex-CS feature, by combining a vertex and a V-junction which share a common line segment as shown in Fig. 2(b). This feature description can be reliably extracted from image data. The common line **ab** of the vertex-CS feature is taken as the true projected 3D edge **AB** as the V-junctions **a** and **b** are most simply interpreted as the projections of the two spatial vertices **A** and **B** respectively. The geometric constraint imposed by the vertex-CS feature can be employed to discard the inconsistent hypotheses generated from the model-scene vertex pairs.

Now, we shall derive the geometric constraint of the composite vertex-CS feature. The edge orientation $\bar{N}_1, \bar{N}_2$ and $\bar{N}_3$ of the vertex at the canonical position **a** can be determined by solving the equations described in Section 2. All the edge orientations N_i^* are represented as unit vectors. To analyse the V-junction **B** of the composite feature, the camera is transformed to a new canonical position by pointing the optical axis to the vertex **B**. In another words, transforming the V-junction **b** to the origin of the image plane. Let $\Re_{ab}$ be the *standard transform* which maps the image point **b** to **a**. The edge orientations N_1^*, N_2^* and N_3^* under the new canonical position **B** can then be written as $(\Re_{ab}^t \ \bar{N}_1), (\Re_{ab}^t \ \bar{N}_2)$ and $(\Re_{ab}^t \ \bar{N}_2)$ respectively, where t denotes a transpose. The angle α_{14} of the vertex can be expressed as a dot product $N_1^* \cdot N_4^* = \cos(\pi - \alpha_{14})$. After some manipulation, the angle θ_4^* between the edge N_4^* and the optical axis can be expressed as :

$$\theta_4^* = \cos^{-1} \frac{\cos(\pi - \alpha_{14})}{\sqrt{\mu^2 + (n_{1z}^*)^2}} + \tan^{-1} \frac{\mu}{n_{1z}^*} \tag{11}$$

Where $\mu = n_{1x}^* \cos \phi_4^* + n_{1y}^* \sin \phi_4^*$. The edge angle θ_4^* can be determined by substituting the projected edge angle ϕ_4^* and the true 3D angle of the vertex **B** into the Eq.(11). The solutions can then be substituted into the equation, $\beta = \cos^{-1}(\frac{(N_1^* \times N_2^*)}{|N_1^* \times N_2^*|} \cdot N_4^*)$, which is the angle between the normal $N_1^* \times N_2^*$ and the edge N_4^*. The measured angle β should correspond to the pre-computed angle β of the hypothesised model. The model-scene vertex pair hypotheses which agree with angle β will be considered in the verification. It is worth noting that the geometric constraint imposed by the composite feature does not require quantitative information about the edge length. In the next section, we will discuss the matching strategies used for recognising polyhedral objects from a single perspective image.

5 Matching Strategy

Several modules are integrated into our system to accomplish the task of polyhedral object recognition. They can be briefly described as follows. Vertices extracted from the image satisfy some predefined criteria such as junction region size, the length of the radiating line segments and the angles between them. In order to control the combinatorial explosion associated with unconstrained association of model-scene vertex pair assignments, high quality vertices with small region size, relatively long segments and reasonable angle sizes between them are extracted from the given scene first. All 3 possible combinatorial assignments of corresponding edges between a model and image vertex are considered. Scene vertices which match the geometric configuration of at least one vertex stored in the model base will be considered in the subsequent process. The model with at least one vertex satisfying the geometric constraints will be registered as a consistent interpretation.

The remaining vertex candidates will be employed to provide a tight constraint on where to search for V-junctions which share one of three line segments of the feasible vertices. When searching for plausible V-junctions to form feasible composite features, scene vertices were processed in the descending order of their line length. If no plausible composite features are found, we can then relax the threshold on the junction region size of the V-junctions. In the very worst case, we may treat those T-junctions when one of the vertex line segments is either the cap or the bar, as required V-junction. The pose of hypothesised object models is then estimated using the geometric relationships derived from the model and scene composite features. A simple visibility test is performed on the residual hypotheses. A trihedral vertex which is interpreted as a scene vertex should contain at least two visible surfaces otherwise the associated hypothesised model can be removed from the candidate list for further consideration.

The 2D description of each backprojected model is compared with the features extracted from the image. First, we count the number of 2D junctions of the hypothesised objects that overlap a junction extracted from the scene image. Two junctions are said to be overlapping if they are within a proximity threshold value and their angles and orientation match to an allowable tolerance. After comparing every projected junction of the hypothesised objects with the 2D junctions extracted from the scene, the hypotheses with the greatest number of matched junctions will be invocaked. The aim of this stage is to select the hypothesis with the greatest number of features overlapping with the scene data. To achieve this, the nearest scene line from each projected 2D model line length is identified. If it is within an allowable threshold, then the line length is divided by the corresponding projected line length. These computed quotients are then summed up and divided by the number of visible projected edges of the hypothesised model yielding the confidence measure for each hypothesis.

6 Experimental Results

Real images were employed to test the reliability, robustness and computational efficiency of the polyhedral recognition system described in this paper. The model and camera frames employed in all the experiments are designated as right-handed coordinate system. There are two pyramid models and a roof model used in this experiment (see Fig. 3(a)). The images were taken with a standard CCD camera. Fig. 3(b) and (c) show the grey-level image and lines extracted by Hough process, respectively. In this example, 15 vertices identified from the scene are shown in Fig. 3(c) marked **Tn**. Some of these vertices, were generated by extraneous lines due to effects such as shadowing. All the vertices of each model were compared exhaustively with each vertex extracted from the test scene. The number of admissible solutions generated from matching the pyramid model #1, #2 and the roof model #3 against all the scene vertices are 243, 256 and 313, respectively. In some cases, there were no feasible solutions found when establishing the geometrical relationships between individual vertices of the pyramid models and the scene vertices. For hypothesised candidates which do satisfy the geometrical constraint of a model-scene vertex pair, the grouping process generates feasible composite features around the line segments of the scene vertices. In this example, 59 vertex-CS composite features were extracted from the scene. These composite features were checked using the geometric constraint.

Hypotheses remaining after applying the vertex-CS geometric constraint and simple visibility test for the model #1, #2 and #3 were reduced by about 55.6%, 57.8% and 66.8%, respectively. Using the information of the vertex-CS composite feature, the poses of all the admissible hypotheses were computed and their 2D predictions were compared with the geometric primitives such as line segments and V-junctions extracted from the scene shown in Fig. 3(c). Both the correct and wrong candidates were generated and their confidence measures or the close correspondence between backprojected model lines and lines extracted from a given scene image were computed. Some of the incorrect hypotheses generated from the pyramid and roof model-scene vertex assignments are shown in Fig. 3(d) and (e) respectively.

In this instance, the pyramid model #2 and the roof model #3 matched the scene object #2 and #5 correctly. The number of matched junctions in each case are 5 (out of 6) and 6 (out of 7) respectively. The confidence measures of both cases are are 90.2% and 86.8%. Unfortunately, in the case of computing the confidence measures for the hypotheses generated from the matching of model pyramid #1 against scene vertices, the most plausible candidate among the admissible solutions generated by matching against scene vertex T_2 was 5.6% lower than the best hypothesis (81.7%) generated from matching model #1 against the scene vertex T_8. Many experiments were performed using the test scene containing the two pyramid models. The two pyramid models always matched to the scene pyramid with better quality 2D features. This was due to the fact that the 2D descriptions of the two pyramid models under perspective projection were very similar. Futhermore, the 2D description of the scene object #1 was *degraded significantly relative* to the scene object #2. In this case, the correct hypothesis can only be found if the distance from the camera to the *table top* is known a priori. The computed furthest distance for the five hypothesised pyramid models for object #1 at the top of the list differed from the edge of the table by a factor of two. We acknowledge that in general the distance to the table from the the camera may not be known in advance. However, the correct model for the scene pyramid object #1 could only be identified by making this extra assumption. Fig. 3(f) shows the superimposed models onto the scene objects using the computed transformation.

Next, the proposed method was explored on a cluttered scene shown in Fig. 4(a).

The target objects are the pyramid and the roof model. They are labelled with **S1** and **S2** respectively (see Fig. 4(a)). One of the two visible trihedral vertices of the roof model was occluded by a "computer mouse". There are 8 vertices extracted from this scene (see Fig. 4(b)). The numbers of hypotheses generated by matching pyramid #2 and roof model #3 vertices against the scene vertices were 84 and 119 respectively. Some of the incorrect hypotheses generated from the roof model-scene vertex assignments are shown in Fig. 4(c). After applying vertex-CS constraints, the number of hypotheses for each case were reduced by 42.9% and 68.1%, respectively. The correct models for the scene objects were identified. The confidence measures for each case were 72.6% and 66.1%. The superimposed models onto the scene objects using the computed transformation are shown in Fig. 4(d).

7 Conclusion

In this paper, we have presented a paradigm based on local shape properties for identifying the scene-model correspondences and estimating the poses of the scene object from a single perspective image. In the framework, vertices and composite vertex-CS feature are employed as seed features for model invocation and hypothesis generation. We have derived an analytical quartic equation for describing geometric relationships of a model-scene vertex pairs, with not restriction on the angles between vertex edges. We have introduced a vertex-CS feature of which the effectiveness and the geometric constraint are presented. Using the seed features, Many false hypotheses can be pruned away without concerns about full edge visibility which greatly simplifies the problem of computational intensive verification process. The experimental results reported confirm the feasibility of the proposed paradigm.

Clearly, the robustness of the method depends entirely on the extraction of the vertices and composite features. Some features may however be missing due either to occlusion or inadequate low-level processing. To cope with these problems, the low confidence or poor quality features can be enhanced by modifying the thresholds on proximity and orientation checks. However, if the tolerance is too large, the number of features extracted from a scene may cause the model-scene correspondences to grow exponentially. This is one of the important issues which can be solved by developing an adaptive control mechanism for providing an interactive environment between the matching phase and low-level or feature grouping process.

Acknowledgement

This work was carried out as part of the **ESPRIT** Basic Research Action Project BR3038, "Vision as Process". KCW is supported by an (UK) **ORS** Award.

References

[1] Stephen T. Barnard, "Interpreting Perspective Images", *Artificial Intelligence*, **21**, pp 435-462, 1983.

[2] M. Dhome, M. Richetin, J. T. Lapresté and G. Rives, "Determination of Attitude of 3-D Objects from a Single Perspective View", *IEEE Trans. Pattern Anal. and Machine Intell.*, Vol. 11, No. 12, Dec 1989.

[3] R. Horaud, "New Methods for Matching 3-D Objects with Single Perspective Views", *IEEE Trans. on Pattern Anal. and Machine Intell.*, PAMI-9, No. 3, pp 401-412, May 1987.

[4] R. Horaud, B. Conio, O. Leboulleux and B. Lacolle, "An Analytic Solution for the Perspective 4-Point Problem, *CVGIP 47*, pp 33-44, 1989.

[5] T. Kanade, "Recovery of the three-dimensional shape of an object from a single view" *Artificial Intelligence*, 17, pp 409-460, 1981.

[6] K. Kanatani, "The Constraints on Images of Rectangular Polyhedra", *IEEE Trans. on Pattern Anal. and Machine Intell.*, PAMI-8, No. 4, pp 456-463, 1986.

[7] K. Kanatani, "Constraints on Length and Angle", *CVGIP 41*, pp 28-42, 1988.

[8] D. G. Lowe, "Three-Dimensional Object Recognition from Single-Two Dimensional Images", *Artificial Intelligence*, 31, pp 355-395, 1987.

[9] T. Shakunaga and H. Kaneko, "Perspective Angle Transform : Principle of Shape from Angles" Int. Journal of Compute Vision, 3, pp 239-254, 1989.

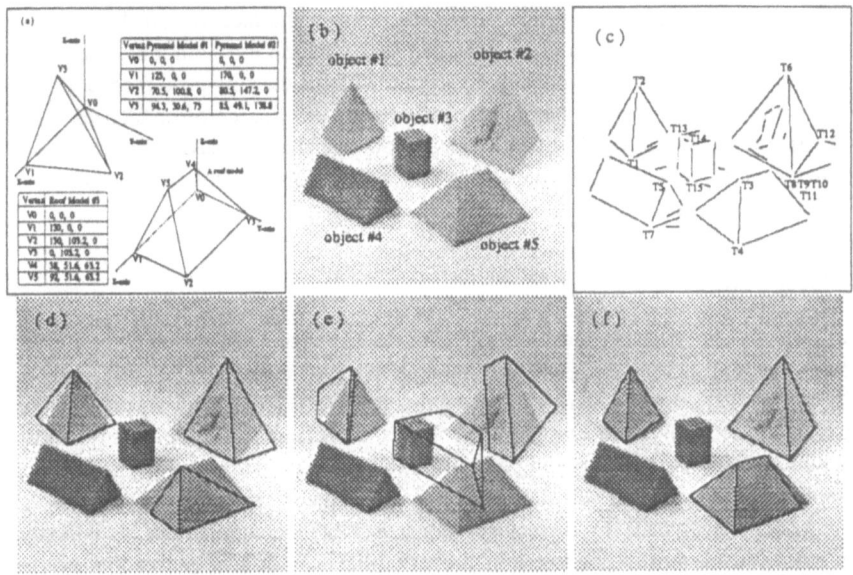

Figure 3: (a), (b), (c), (d), (e) and (f) *see text*

Figure 4: (a), (b), (c) and (d) *see text*

Linear Algorithms for Object Pose Estimation[†]

T. N. Tan, G. D. Sullivan and K. D. Baker

Intelligent Systems Group
Department of Computer Science
University of Reading, ENGLAND

Abstract

This paper concerns the estimation of object pose in scenes where objects are located on the ground plane which has known orientation and position w.r.t. the camera. Novel algorithms are described, based on the concept of interpretation planes and that of pencils of planes. The methods are linear, computationally simple, and give unique and closed-form solutions, thus eliminating many of the problems associated with the existing pose recovery algorithms. They require a minimum of two 2D-3D line correspondences. Experimental results are included which show that the proposed algorithms are robust to noise, and capable of accurate pose recovery using real images of outdoor scenes.

1 Introduction

Object pose recovery from monocular light intensity images is a major objective for computer vision. Most previous work in the area has used either 2D-3D point [1-4] or 2D-3D line [5-12] correspondences. Although some success has been reported in the literature (e.g., [5-6]), there are many problems associated with the existing pose estimation algorithms [16], most importantly: high or unknown sensitivity to sensory data noise, non-uniqueness of the solution, requirement of good initial guesses, unguaranteed convergence to correct solutions, and high computational complexity. These problems are mostly due to the inherent non-linearity of pose estimation and the fact that the existing pose recovery algorithms are (rightly) concerned with the general case of six degrees of freedom, and treat special cases as having marginal interest.

Nevertheless, special cases abound in the real-world. Their occurrence, if properly ascertained, can help to avoid the above problems, and dramatically simplify pose estimation. The particular special case considered here arises in traffic scenes where objects (e.g., cars) are located on a ground surface which is, at least locally, substantially planar. Other potential applications include model-based vision for industrial parts on conveyor belts or other robot working surfaces.

We approximate the ground surface by the X-Y plane of a world coordinate system (WCS) whose Z-axis points towards the sky. The pose of an object in this WCS is uniquely determined by three independent pose parameters (assuming the X-Y plane of

---

†. This work was carried out as part of the ESPRIT project P2152 (VIEWS).

the object-centred coordinate system coincides with that of the world coordinate system): the rotation angle θ about the (vertical) Z-axis, and the two translations T_x and T_y on the ground plane. The other three parameters are all zero:

$$\phi, \quad \psi, \quad T_z = 0 \qquad (1)$$

where ϕ and ψ are the rotation angles around the X and Y axes, and T_z is the translation along the vertical axis. Equation (1) is called the *ground plane constraint* (GPC).

It is shown in this paper that by expressing object pose in the world coordinate system and by incorporating the GPC into the formulation of pose constraint equations, the pose estimation problem can be linearized. As a consequence, simple and robust pose recovery algorithms can be developed, and the problems associated with the existing algorithms listed at the beginning of this section are avoided. Moreover, closed-form solutions can be obtained. This choice of the WCS simplifies the analysis since it directly eliminates 3 of the 6 pose parameters; any other coordinate system could be used but would introduce additional parameters with their corresponding constraint equations computable from the GPC. The potential importance of the GPC in simplifying pose recovery has been discussed in several articles (e.g., [1, 11, 17]), but either no use has been made of the constraint [11], or it has not been used fully to avoid the problems listed at the beginning of this section [1, 17]. It should be pointed out that this paper concerns the use of the GPC in model-based object pose recovery. The application of the GPC in 3D structure and motion estimation is described in a companion paper [18].

In the following sections, we first describe the imaging model used in this paper, then discuss pose estimation under the GPC using 2D-3D line correspondences, and finally summarize the experimental results obtained.

2 Coordinate systems and imaging model

We assume a pinhole camera model with perspective projection and no lens distortion. This is sketched in Fig.1. It can be shown that under this imaging model and the GPC,

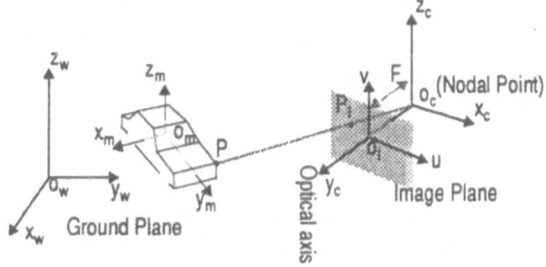

Figure 1. *Coordinate systems and the imaging model.*

the image coordinates (u, v) and the model or object coordinates (x_m, y_m, z_m) of point P are related to each other by [16]:

$$
\begin{bmatrix} su \\ sF \\ sv \\ s \end{bmatrix}^T = \begin{bmatrix} x_m \\ y_m \\ z_m \\ 1 \end{bmatrix}^T \begin{bmatrix} \cos\theta & \sin\theta & 0 & 0 \\ -\sin\theta & \cos\theta & 0 & 0 \\ 0 & 0 & 1 & 0 \\ T_x & T_y & 0 & 1 \end{bmatrix} \begin{bmatrix} m_{11} & m_{12} & m_{13} & m_{14} \\ m_{21} & m_{22} & m_{23} & m_{24} \\ m_{31} & m_{32} & m_{33} & m_{34} \\ m_{41} & m_{42} & m_{43} & m_{44} \end{bmatrix} \tag{2}
$$

where θ, T_x and T_y are the three pose parameters of the object in the WCS, F the camera focal length, s a non-zero scale, and m_{ij}, $i, j = 1, 2, 3, 4$, are the elements of the 4x4 homogeneous perspective transformation matrix which maps 3D world coordinates into 2D image coordinates, and which is assumed to be known in the subsequent discussions.

3 Pose recovery algorithms

A 3D model line L_i can be described either by two points $\vec{P}_{i1}$ and $\vec{P}_{i2}$:

$$
L_i: \begin{cases} \vec{P}_{i1} = (x_{i1m}, y_{i1m}, z_{i1m}) \\ \vec{P}_{i2} = (x_{i2m}, y_{i2m}, z_{i2m}) \end{cases} \tag{3}
$$

where subscript m indicates model coordinates, or by a unity directional vector $\vec{N}_i$ and a point $\vec{P}_i$:

$$
L_i: \begin{cases} \vec{N}_i = (\alpha_i \ \beta_i \ \gamma_i)^T \\ \vec{P}_i = (x_{im}, y_{im}, z_{im}) \end{cases} \tag{4}
$$

or by two intersecting planes Π_{i1} and Π_{i2}:

$$
L_i: \begin{cases} \Pi_{i1}: & x_m + b_{i1}y_m + c_{i1}z_m = 0 \\ \Pi_{i2}: & x_m + b_{i2}y_m + c_{i2}z_m + d_{i2} = 0 \end{cases} \tag{5}
$$

Equation (5) implies that Π_{i1} and Π_{i2} are assumed not to be parallel to the X-axis, and Π_{i1} passes through the origin of the model (object-centred) coordinate system (MCS). Let the corresponding image line l_i of L_i be specified by

$$
a_i u + b_i v + c_i = 0 \tag{6}
$$

where a_i, b_i and c_i are known constants. Substituting u and v defined in (2) into (6) and writing the resultant equation in terms of model coordinates (x_m, y_m, z_m) yields [16]

$$
\delta_i x_m + \varepsilon_i y_m + \zeta_i z_m + \eta_i = 0 \tag{7}
$$

where

$$\delta_i = (\dot{r}_1 \bullet \hat{n}_i)\cos\theta + (\dot{r}_2 \bullet \hat{n}_i)\sin\theta$$

$$\varepsilon_i = (\dot{r}_2 \bullet \hat{n}_i)\cos\theta - (\dot{r}_1 \bullet \hat{n}_i)\sin\theta$$

$$\zeta_i = \dot{r}_3 \bullet \hat{n}_i$$

$$\eta_i = (\dot{r}_1 \bullet \bar{n}_i)T_x + (\dot{r}_2 \bullet \bar{n}_i)T_y + \dot{r}_4 \bullet \hat{n}_i \qquad (8)$$

$$\bar{n}_i = (a_i \ b_i \ c_i)^T$$

$$\dot{r}_k = (m_{k1} \ m_{k3} \ m_{k4}); \quad k = 1, 2, 3, 4$$

Equation (7) defines the plane Π_i in the MCS on which L_i must lie. We call this plane the *interpretation plane* of L_i. The normal vector of Π_i is $\vec{N}_{\Pi i} = (\delta_i \ \varepsilon_i \ \zeta_i)^T$.

In terms of the relationship between L_i and its interpretation plane Π_i, we have the following simple geometrical observations:

Observation I: If L_i is described by (3), then $\vec{P}_{i1}$ and $\vec{P}_{i2}$ must lie on Π_i.

Observation II: If L_i is described by (4), then $\vec{N}_i$ must be normal to $\vec{N}_{\Pi i}$, and $\vec{P}_i$ must be on Π_i.

Observation III: If L_i is described by (5), then Π_{i1}, Π_{i2} and Π_i must all be members of the pencil of planes containing L_i.

These three observations suggest three different algorithms for obtaining very similar pose constraint equations (see Equations (10), (12) and (16)). The algorithms differ only in the numerical stability of the calculation for deriving the coefficients of the equations.

From Observation I and for N 2D-3D line correspondences, $L_i \leftrightarrow l_i,\ i = 1, 2, ..., N$, we have

$$\begin{cases} \delta_i x_{i1m} + \varepsilon_i y_{i1m} + \zeta_i z_{i1m} + \eta_i = 0 \\ \\ \delta_i x_{i2m} + \varepsilon_i y_{i2m} + \zeta_i z_{i2m} + \eta_i = 0 \end{cases}, \quad \forall i \in \{1, 2, ..., N\} \qquad (9)$$

which, by recalling the variables defined in (8) and by isolating the pose parameters, can be rewritten as

$$\begin{cases} A_{i1}\cos\theta + B_{i1}\sin\theta + C_{i1}T_x + D_{i1}T_y = E_{i1} \\ \\ A_{i2}\cos\theta + B_{i2}\sin\theta + C_{i2}T_x + D_{i2}T_y = E_{i2} \end{cases}, \quad \forall i \in \{1, 2, ..., N\} \qquad (10)$$

where $A_{ij}, B_{ij}, C_{ij}, D_{ij}$ and E_{ij} are terms computable from known constants [16].

From Observation II and for N 2D-3D line correspondences, $L_i \leftrightarrow l_i,\ i = 1, 2, ..., N$, we can write

$$\begin{cases} \alpha_i \delta_i + \beta_i \varepsilon_i + \gamma_i \zeta_i = 0 \\ \\ \delta_i x_{im} + \varepsilon_i y_{im} + \zeta_i z_{im} + \eta_i = 0 \end{cases}, \quad \forall i \in \{1, 2, ..., N\} \qquad (11)$$

which, by using the variables defined in (8) and by isolating the pose parameters, can be equivalently written as

$$\begin{cases} F_i\cos\theta + G_i\sin\theta = H_i \\ \\ A_i\cos\theta + B_i\sin\theta + C_iT_x + D_iT_y = E_i \end{cases} , \quad \forall i \in \{1, 2, ..., N\} \tag{12}$$

where $A_i, B_i, C_i, D_i, E_i, F_i, G_i$ and H_i are terms computable from known variables [16].

To show how Observation III can be used to derive pose constraints, we represent planes Π_{i1}, Π_{i2} and Π_i by three 4x1 column vectors $\vec{\Pi}_{i1}$, $\vec{\Pi}_{i2}$ and $\vec{\Pi}_i$:

$$\Pi_{i1}: \quad \vec{\Pi}_{i1} = (1 \quad b_{i1} \quad c_{i1} \quad 0)^T$$

$$\Pi_{i2}: \quad \vec{\Pi}_{i2} = (1 \quad b_{i2} \quad c_{i2} \quad d_{i2})^T \tag{13}$$

$$\Pi_i: \quad \vec{\Pi}_i = (\delta_i \quad \varepsilon_i \quad \zeta_i \quad \eta_i)^T$$

then Observation III implies that there exist real λ_1, λ_2 and λ_3 (not all equal to zero) such that [14]

$$\lambda_1\vec{\Pi}_{i1} + \lambda_2\vec{\Pi}_{i2} + \lambda_3\vec{\Pi}_i = 0 \tag{14}$$

from which the following two pose constraint equations can be derived (for details, see [16]):

$$\bar{F}_i\cos\theta + \bar{G}_i\sin\theta = \bar{H}_i$$
$$\bar{A}_i\cos\theta + \bar{B}_i\sin\theta + \bar{C}_iT_x + \bar{D}_iT_y = \bar{E}_i \tag{15}$$

where $\bar{A}_i, \bar{B}_i, \bar{C}_i, \bar{D}_i, \bar{E}_i, \bar{F}_i, \bar{G}_i$ and $\bar{H}_i$ are terms computable from known constants such as the coefficients of the two intersecting planes [16]. Thus for N line correspondences, $L_i \leftrightarrow l_i$, $i = 1, 2, ..., N$, there are $2N$ equations of the form of (15):

$$\begin{cases} \bar{F}_i\cos\theta + \bar{G}_i\sin\theta = \bar{H}_i \\ \\ \bar{A}_i\cos\theta + \bar{B}_i\sin\theta + \bar{C}_iT_x + \bar{D}_iT_y = \bar{E}_i \end{cases} , \quad \forall i \in \{1, 2, ..., N\} \tag{16}$$

It is clear from (10), (12) and (16) that, no matter what line representation method is used, we need $2N \geq 3$ or a minimum of two 2D-3D line correspondences to ensure that the number of equations is no less than the number of unknowns. This contrasts with the minimum of three lines in the general case of six degrees of freedom [6-8, 10-11].

3.1 The linear solution technique

If $\cos\theta$ and $\sin\theta$ are regarded as two independent unknowns, then each of (10), (12) and (16) becomes a set of $2N$ linear equations in four unknowns: $\cos\theta$, $\sin\theta$, T_x and T_y, and can be solved by the standard linear least squares technique [16]. We then compute the rotation angle by $\theta = \tan^{-1}(\sin\theta/\cos\theta)$. The correct quadrant of θ can be determined from the senses of $\cos\theta$ and $\sin\theta$.

3.2 The iterative solution technique

If $\cos\theta$ and $\sin\theta$ are not treated as two independent unknowns, then each of (10), (12) and (16) is a set of $2N$ non-linear equations in the three pose parameters θ, T_x and T_y. If an approximate value for θ is given, we can linearize $\cos\theta$ and $\sin\theta$, and thus transform each of (10), (12) and (16) into a set of $2N$ linear equations which can then be solved using an iterative linear least squares technique [16].

Several remarks can be made before concluding this section. The concept of pencil of planes has been used in structure from motion algorithms (e.g., [15]), but has not appeared in previous pose recovery algorithms to the best of our knowledge. (12) and (16) show that the recovery of the rotation angle and the translational parameters can be decoupled since the first N equations of (12) and (16) only involve the rotation angle.

4 Summary of experimental results

The algorithms described in the above section have been tested under a variety of conditions. For convenience, we call the linear solution technique associated with the two points representation, the directional vector and point representation, and the intersecting planes representation of 3D lines, Algorithms A, B and C respectively. They are applied to solve (10), (12) and (16) respectively. Due to space limitation, full and detailed experimental results are not included here but can be found in [16].

4.1 Robustness against image measurement errors

The robustness of the algorithms against image measurement errors has been investigated using synthetic data. Lines were randomly generated from within a cuboid of dimension 8x4x2 m^3 (=length*width*height) located on the ground plane with a depth of about 20 meters from the camera. Images were of size 512x512 pixels. To generate a noisy 2D-3D line correspondence, each 2D noise-free line segment was first translated in its normal direction by Δd pixels, and then rotated about its middle point by $\Delta\alpha$ degrees, where Δd was uniformly distributed over $[-\Delta D, +\Delta D]$ (in pixels), and $\Delta\alpha$ uniformly distributed over $[-\Delta A, +\Delta A]$ (in degrees). The combination $(\Delta D, \Delta A)$ determines the noise level of the synthetic image data. Monte Carlo simulations were conducted at various noise levels with a fixed number (=10) of lines, and the absolute errors in the three pose parameters were recorded. As an example, Figure 2 shows the absolute error curves of the pose parameters vs. translational (Fig.2(a)) and directional (Fig.2(b)) errors of 2D image lines. (Tests carried out under a wider range of conditions are reported in [16]). Comprehensive Monte Carlo simulation results show that

- The performances of Algorithms A and B are very similar.
- Pose estimation by Algorithm C is much less accurate than that of Algorithms A and B. This may be due to the assumptions made in (5). If the randomly generated 3D lines during Monte Carlo simulation do not satisfy these assumptions, then the resultant representation of the two intersecting planes becomes very unstable, and may cause large errors in the recovered pose. Such *accidental* large errors can easily be avoided by adopting more stable representations for the intersecting planes.

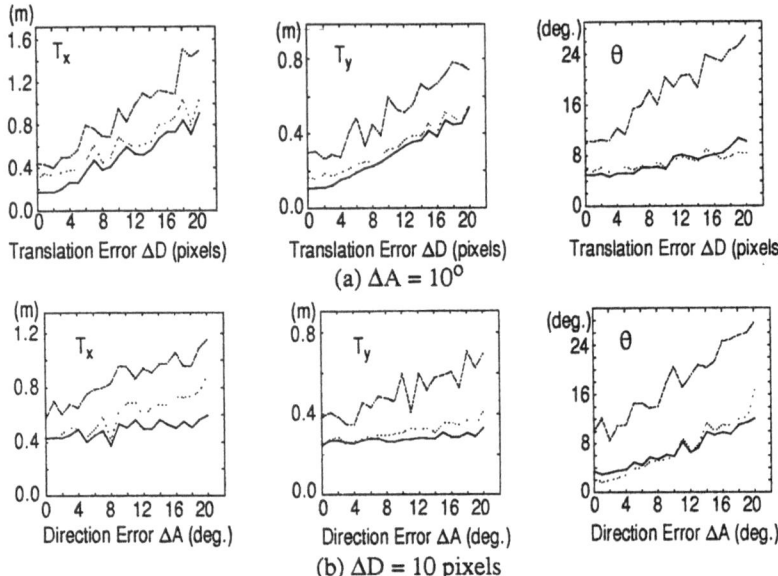

Figure 2. *Absolute error curves of pose parameters in Algorithms A (solid), B (light), and C (dark).*

- As a whole, the proposed algorithms are robust to image measurement errors. Even at unrealistically high noise levels $(\Delta D = 20$ pixels, $\Delta A = 20^{o})$, the mean absolute errors of Algorithms A and B are below 1.20 meters in T_x, 0.60 meter in T_y, and 22° in θ.

As an experiment, the iterative solution technique described in Section 3.2 was also applied to solve (10), using the rotation angle given by Algorithm A as starting conditions. Monte Carlo simulation results [16] show that under severe noise conditions, the iteration actually increased the error. This illustrates the effectiveness of the (simpler) linear solution technique. It also provides a clear case of Aggarwal's taunt that "Often it is better to keep [a] good initial guess and forget about the [non-linear constraint] equations!" [13].

4.2 Effectiveness of using more lines in controlling noise

Monte Carlo simulations were also conducted to investigate the effectiveness of using more line correspondences in combating noise. The noise level was fixed at $\Delta D = 5$ pixels and $\Delta A = 5^{o}$, and the number of lines was increased from the minimum of 2 to 40. The results are summarized in Fig.3. As expected, the accuracy of the estimated pose parameters is consistently improved by using more lines. The improvement is most dramatic when the number of line correspondences is increased from the minimum of 2 to 4 or 5, and beyond 8 there is little improvement. Therefore 5 to 8 line correspondences may be regarded as adequate for robust pose estimation using these algorithms.

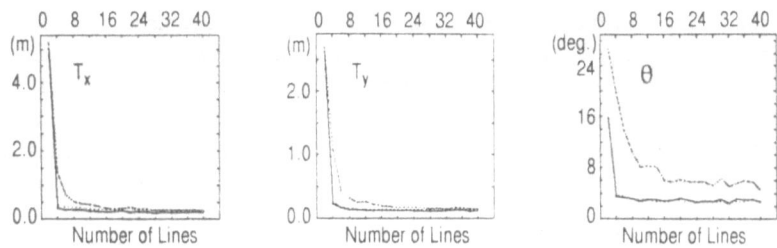

Figure 3. *Effectiveness of using more lines in improving the noise-robustness of Algorithms A (solid), B (light), and C (dark).*

4.3 Performance under real outdoor images

All three algorithms have also been applied to locate a saloon car in 9 frames taken at 0.48 Hz (i.e. every 12th video frame) of an outdoor image sequence. Because of space limitation, only results related to Algorithm B are included. Three out of the nine frames are shown in the left column of Fig.4, where the white car is the object to be located. The

Figure 4. *Pose estimation in real outdoor images (see text for detailed captions).*

gradient images inset in the left column of Fig.4 are the outputs of the Canny operator applied to the small regions surrounding the car. Thresholding and a simple curve segmentation operation were then applied to the gradient images to obtain a set of straight line segments. A subset of these segments were retained which correspond to contour segments of the car, and are inset in the right column of Fig.4. The number of the retained line segments is between 5 and 10. No efforts were made to "tune" the various parameters (such as the scale of the Canny operator) involved in the iconic to symbolic data transformation. The correspondences between the selected 2D lines and the 3D saloon model lines were established manually. These correspondences were used as the input to the pose recovery algorithms. The saloon model was then instantiated at the recovered poses, and superimposed on the original intensity images as shown in the right column of Fig.4. The accurate matching between the model and the image shown in Fig.4 indicates the high accuracy of the recovered poses. The performance of the proposed algorithms can be further appreciated from Fig.5 which shows the recovered X

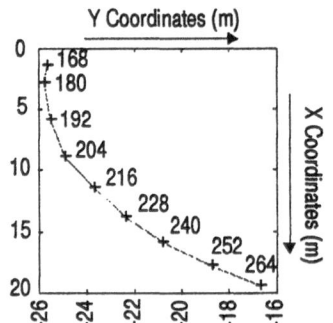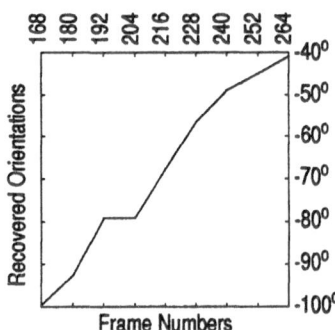

Figure 5. *Recovered car path on the ground plane. The numbers along the curve are the frame numbers, and the symbol "+" marks the recovered poses.*

(i.e, T_x) and Y (i.e., T_y) coordinates of the car on the ground plane for all 9 frames. The recovered orientations of the car are also shown in Fig.5. The resultant path is very smooth and is physically plausible.

5 Conclusions

This work has been concerned with object pose estimation under the ground plane constraint. A number of novel algorithms have been presented which make use of the constraint. The algorithms are linear, computationally simple, and give unique and closed-form solutions, thus eliminating many of the problems associated with the existing pose recovery algorithms. They require a minimum of two 2D-3D line correspondences, and are highly robust with 5 or more correspondences. The algorithms are also applicable to point correspondences.

Experimental results show that the proposed algorithms are robust to noise, and capable of accurate pose recovery in real images of outdoor scenes. The algorithms provide practical and efficient methods for pose recovery which can be applied in a wide range of industrial applications in which the objects move on a known plane.

References

[1]   T. M. Silberberg, D. A. Harwood and L. S. Davis, Object Recognition Using Oriented Model Points, CVGIP, vol.35, 1986, pp.47-71.

[2]   R. M. Haralick, et. al., Pose Estimation from Corresponding Point Data, IEEE Trans. Systems, Man and Cybern., vol.19, 1989, pp.1426-1446.

[3]   S. Ullman and R. Basri, Recognition by Linear Combinations of Models, PAMI,

vol.13, 1991, pp.992-1006.

[4] S. T. Barnard, Interpreting Perspective Images, Artif. Intell., vol.21, 1983, pp.435-462.

[5] D. G. Lowe, Three-Dimensional Object Recognition from Single Two-Dimensional Images, Artif. Intell., vol.31, 1987, pp.355-395.

[6] D. G. Lowe, Fitting Parameterized Three-Dimensional Models to Images, PAMI, vol.13, 1991, pp.441-450.

[7] M. Dhome, et. al., Determination of the Attitude of 3-D Objects from a Single Perspective View, PAMI, vol.11, 1989, pp.1265-1278.

[8] S. Linnainmaa, D. Harwood and L. S. Davis, Pose Determination of a Three-Dimensional Object Using Triangle Pairs, PAMI, vol.10, 1988, pp.634-647.

[9] Y. C. Liu, T. S. Huang and O. D. Faugeras, Determination of Camera Location from 2-D to 3-D Line and Point Correspondences, PAMI, vol.12, 1990, pp.28-37.

[10] A. D. Worrall, K. D. Baker and G. D. Sullivan, Model Based Perspective Inversion, Image and Vision Comput., vol.7, 1989, pp.17-23.

[11] R. Horaud, New Methods for Matching 3-D Objects with Single Perspective Views, PAMI, vol.9, 1987, pp.401-412.

[12] C. Goad, Special Purpose Automatic Programming for 3D Model-Based Vision, in From Pixels to Predicates, A. Pentland, Ed., Norwood, NJ: Ablex, 1986, pp.371-391.

[13] J. K. Aggarwal and A. Mitiche, Structure and Motion from Images: Fact and Fiction, 3rd IEEE Workshop on Vision: Representation and Control, October 1985, pp.127-128.

[14] R. Hartshorne, Foundations of Projective Geometry, Benjamin, 1967.

[15] O. D. Faugeras, F. Lustman and G. Toscani, Motion and Structure from Motion from Point and Line Matches, Proc. of 1st Int. Conf. Comput. Vision, London, 1987, pp.25-34.

[16] T. N. Tan, Locating Objects on the Ground Plane Using 2D-3D Line Correspondences, ESPRIT II Project (P2152) Research Report, RU-03-WP.T3137-TNT-01, January 1992.

[17] J. L. Mundy and A. J. Heller, The Evolution and Testing of A Model-Based Object Recognition System, Proc. of ICCV90, December 4-7, 1990, Osaka, Japan, pp.268-282.

[18] T. N. Tan, K. D. Baker and G. D. Sullivan, 3D Structure and Motion Estimation from 2D Image Sequences, Proc. of British Machine Vision Conference 1992, Springer-Verlag, 1992.

Author Index

Author Index